QB843.W6
INT

WITHDRAWN
FROM STOCK
QMUL LIBRARY

OM Library

D1187812

DATE DUE FOR

WOLF–RAYET STARS AND INTERRELATIONS WITH OTHER MASSIVE STARS IN GALAXIES

INTERNATIONAL ASTRONOMICAL UNION

UNION ASTRONOMIQUE INTERNATIONALE

WOLF–RAYET STARS AND INTERRELATIONS WITH OTHER MASSIVE STARS IN GALAXIES

PROCEEDINGS OF THE 143RD SYMPOSIUM OF THE
INTERNATIONAL ASTRONOMICAL UNION,
HELD IN SANUR, BALI, INDONESIA, JUNE 18–22, 1990

EDITED BY

KAREL A. VAN DER HUCHT

SRON Space Research Laboratory, Utrecht, The Netherlands

and

BAMBANG HIDAYAT

Observatorium Bosscha, ITB, Lembang, West-Java, Indonesia

KLUWER ACADEMIC PUBLISHERS

DORDRECHT / BOSTON / LONDON

Library of Congress Cataloging-in-Publication Data

International Astronomical Union. Symposium (143rd : 1990 : Sanur,
  Indonesia)
   Wolf-Rayet stars and interrelations with other massive stars in
  galaxies : proceedings of the 143rd Symposium of the International
  Astronomical Union held in Sanur, Bali, Indonesia, June 18-22, 1990
  / edited by Karel A. van der Hucht and Bambang Hidayat.
      p.   cm.
   Includes index.
   ISBN 0-7923-1086-1 (alk. paper)
   1. Wolf-Rayet stars--Congresses. 2. Supergiant stars--Congresses.
  I. Hucht, Karel A. van der. II. Hidayat, Bambang.
  III. International Astronomical Union. IV. Title.
  QB843.W6I57 1990
  523.8'8--dc20   90-26480
ISBN 0-7923-1086-1 (HB)

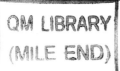
QM LIBRARY
(MILE END)

*Published on behalf of
the International Astronomical Union
by
Kluwer Academic Publishers, P.O. Box 17, 3300 AA Dordrecht, The Netherlands.*

*Kluwer Academic Publishers incorporates
the publishing programmes of
D. Reidel, Martinus Nijhoff, Dr W. Junk and MTP Press.*

*Sold and distributed in the U.S.A. and Canada
by Kluwer Academic Publishers,
101 Philip Drive, Norwell, MA 02061, U.S.A.*

*In all other countries, sold and distributed
by Kluwer Academic Publishers Group,
P.O. Box 322, 3300 AH Dordrecht, The Netherlands.*

Printed on acid-free paper

*All Rights Reserved
© 1991 International Astronomical Union*

*No part of the material protected by this copyright notice may be reproduced or utilized
in any form or by any means, electronic or mechanical including photocopying,
recording or by any information storage and retrieval system, without written permission
from the publisher.*

Printed in the Netherlands

TABLE OF CONTENTS

SESSION III. INTRINSIC VARIABILITY – *Chair: Allan J. Willis*

SESSION V. MASS LOSS – *Chair: Lindsey F. Smith*

Session V Poster Summaries

SESSION VIII. INVENTORY AND DISTRIBUTION – *Chair: JiPi De Greve*

Session VIII Poster Summaries

SESSION IX. SUMMARY – *Chair: Karel A. van der Hucht*

CLOSING ADDRESS

OBJECT INDEX

PREFACE

The first ideas for the symposium were generated in Brussels in the summer of 1986, during exquisite lunches in between HST proposal consortium meetings. At the time it was expected that soon after the previous IAU symposium (No. 116) devoted to luminous hot massive stars, a bonanza of new exciting observational material would become available, together with significant advances on the theoretical front. Also it was felt that Wolf-Rayet stars should feature predominantly, because that had not been the case since IAU Symposium No. 99 in 1981.

Tradition requires that IAU symposia on hot massive stars take place in high luminosity beach resorts, and after Buenos Aires, Qualicum Beach, Cozumel and Porto Heli, Bali sounded like a reasonable place. Therefore we were only too pleased with the invitation of the Indonesian astronomical community to host the symposium in Sanur (Denpasar).

The aim of the symposium was to bring together both observers and theoreticians active in the field of Wolf-Rayet stars and related objects, to present and discuss their recent results, in order to expose to what extent consensus exists as to the physical and chemical properties of Wolf-Rayet stars, their evolutionary status and their interrelations with other massive stars in galaxies.

A candidate Scientific Organizing Committee was invited in 1987, and following approval by the IAU Executive Committee a year later, quora of the SOC met in 1988 during the IAU General Assembly in Baltimore and during IAU Colloquium No. 113 in Val Morin, and again in 1989 during the First Boulder-Munich Workshop. Those gatherings and extensive email communication subsequently established the programme for the symposium. The basic structure was: Atmospheres, Intrinsic Variability, Binaries, Mass Loss, Enrichment, and Inventory and Distribution; to be preceeded by an overview of the relevant literature of the past decade, and to be concluded by a summary of the symposium.

In this framework 20 invited reviews, 38 invited contributions, and 76 poster papers were presented at the symposium, entertaining 118 participants from 18 countries, including some of the finest brains in the solar system. Thanks to the cooperation of the authors, most manuscripts arrived in time and in good shape, thus alleviating the task of the editors. The discussions were recorded in the usual way. Decipherization of discussion forms and transcription of tapes as well of some of the manuscripts have skillfully been performed by Ms. Karin Meertse in Utrecht. The object index has been compiled scrupulously and accurately by Ms. Diah Yudiawati Anggraeni in Lembang.

The Local Organizing Committee made an early start in finding not only a very efficient and pleasant venue, but also in gathering generous support from numerous national and international sources. The symposium was supported and funded by the Directorate General of Higher Education of the Department of Education and Culture of the Republic of Indonesia. The LOC operated under the wings of the Institut Teknologi Bandung (ITB) in Bandung (West Java) and of the Universitas Udayana (UNUD) in Denpasar (Bali). A record of appreciation is more than appropriate here. We thank the Indonesian Institute of Sciences (LIPI), the adhering body to the IAU, for securing the permit to hold the meeting in Indonesia; the Indonesian National Committee for UNESCO (KNI-UNESCO) for its invaluable support which facilitated the organization; and the National Planning Board (Bappenas), whose moral as well as financial support made preparation and execution possible.

The Office of the President (Binagraha) provided means for the preparation as well as for the finishing touch of the symposium, for which we are very grateful. We are happy to recall the most welcome 'intervention' of the Department of Tourism, Post and Telecommunication, in particular Ir. Soedjono Kramadibrata, and of PT Hotel Indonesia, which made it possible for us to make use of Hotel Bali Beach in Sanur, as the venue for the symposium. We thank them very much for the many facilities which were generously provided to us under the direction of Mr. Alfons D. Viera and the charming and efficient guidance of Ms. Made Pudjayanti, Mr. Nyoman Megeg and Mrs. Suyasih Mudita.

It would be a sin of omission if we did not record here the trust and thrust which have been provided by the International Astronomical Union. Its financial contributions have enabled quite a number of colleagues to attend the symposium. Many astronomers also received support from the International Center for Theoretical Physics (Trieste, Italy) and from the European Space Agency, which also provided the poster announcing the symposium. We owe our indebtedness to those highly respected organizations.

The seed money which catapulted the LOC into motion was provided already in 1988 by the Leids Kerkhoven-Bosscha Fonds (the Netherlands). We greatly appreciate her support. Due regards are directed also to the British Council, the Australian Embassy's Cultural Office, and the Deutsches Akademisches Austausch Dienst, all in Jakarta. Those offices have, upon request of the SOC, helped many astronomers with travel support.

We are greatly indebted to many friends and sympathizers in Indonesia, whose contributions, in one way or another, have made it possible to realize the symposium. The official carrier to the symposium *GARUDA Indonesia* provided a very generous reduction in airfares; the newspaper Kompas helped to support some of our colleagues from abroad, as well as to meet some local expenses. The ITB Alumni Association has provided support for Indonesian astronomers to attend this meeting. Bir Bintang – literally meaning Stellar Beer – provided support for one foreign astronomer.

All participants who enjoyed the special evening events may want to know that those were made possible by generous support from many Indonesian corporations. We would like to record our appreciation to: ASTRA International Corporation for organizing the cultural Evening at the Puri Anyer Kerambitan; the cigarette company PT Jarum Kudus (Central Java) and the PT Perkebunan XIII (West Java) for their contributions in organizing the conference dinner. PT Perkebunan XIII is a tea plantation south of Bandung, whose influence on the development of astronomy in Indonesia has been felt ever since the 1920-ies. Directed by the late Karel A.R. Bosscha, the tea plantation sponsored the founding of the observatory in Lembang – whence the name Bosscha Observatory.

Our special thanks go to the cigarette factory PT Gudang Garam (East Java), which continued to extend its support for international astronomy meetings in Indonesia, as it had done already in two occasions in the past. Financial and other facilities have been kindly provided by PT Bukaka Teknik Utama, PT Unilever Indonesia, PT Tambang Timah Indonesia, PT Electrindo Nusantara, PT Perkebunan XXI-XXII, Kodak-Interdelta and Pudak Scientific. PT Dwitunggal Jaya Sakti contributed a slide projector. We thank them all very much.

The Organizing Committee wishes also to express her thanks to individuals whose encouragement has always been felt: Prof.Ir. Wiranto Arismunandar, Prof.Dr. C. de Jager, Dr.Ir. Arifin Wardiman, Ki Probosutedjo, Dr. Ariono Abdulkadir, Ir. Fadel Mohammed, Mr. Riyanto Gozali, Col. Soelarso, Mr. Ansari, and Miss Maria Natawira of Interlink. In addition, interest and support from the Governement of the Bali province and the Department of Foreign Service are gratefully acknowledged.

Last but not least, the Organizing Committee was in her tasks extensively supported by an able army of assistants from Universitas Udayana and Institut Teknologi Bandung, and by her very effective Administrative Officer Ms. Diah Yudiawati Anggraeni, without whom this symposium would not have been the same.

Karel A. van der Hucht & Bambang Hidayat

THE ORGANIZING COMMITTEE

SCIENTIFIC

D.A. Allen (Australia)
W. Hagen Bauer (U.S.A.)
J.P. Cassinelli (U.S.A.)
A.M. Cherepashchuk (U.S.S.R.)
J.-P. De Greve (Belgium)
C.D. Garmany (U.S.A.)
B. Hidayat (Indonesia)
K.A. van der Hucht (The Netherlands, chairman)

M.J. Jerzykiewicz (Poland)
F. Matteucci (Italia)
V.S. Niemela (Argentina)
M.R. Rosa (Germany)
M. Rosado (Mexico)
B.S. Shylaja (India)
N.R. Walborn (U.S.A.)
A.J. Willis (U.K.)

LOCAL

B. Hidayat (ITB, chairman)
W. Sutantyo (ITB, secretary)
S.D. Wiramihardja (ITB, treasurer)
A.A. Ngurah Made Agung (UNUD)

B. Gultom (LAPAN)
F. Ruskanda (LIPI)
D. Sukartadiredja (Planetarium, DKI)

Administrative Officer:

Diah Yudiawati Anggraeni (ITB)

The symposium was sponsored by IAU Commission 35
and co-sponsored by IAU Commissions 25, 27, 28, 29, 36, 37 and 44.

The Local Organizing Committee operated under the auspices of
the National Committee for UNESCO,
the Ministry of Education and Culture,
the Office of the President (Binagraha),
the Ministry of Tourism, Post and Telecommunication,
the National Planning Board (Bappenas), and
the Leids Kerkhoven-Bosscha Fonds.

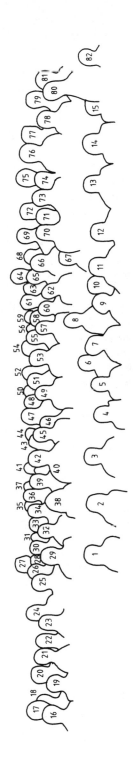

1. Wiramihardja
2. mrs. Felli
3. Felli
4. Underhill
5. Shylaja
6. Heydari-Malayeri
7. Humphreys
8. Lamontagne
9. mrs. van der Hucht
10. Bart van der Hucht
11. van der Hucht
12. van Genderen
13. Walborn
14. Dawanas
15. Kogure
16. Admiranto
17. Lequeux
18. Linda Smith

19. Raharto
20. Conti
21. Crowther
22. Maeder
23. Willis
24. Deacon
25. Spurzem
27. Schmutz
28. Nugis
29. Baratta
30. Turolla
31. The
32. Nobili
33. Annuk
35. de Groot
38. Matteucci
39. Lewis
40. Kingsburgh

42. Kunth
45. Breysacher
46. Taylor
47. Hillier
48. Georgiev
49. Sreenivasan
51. Pollock
52. De Greve
53. Shara
54. Moffat
55. Drissen
57. Robert
58. Esteban
59. Prantzos
60. Niemela
61. Armandroff
62. St-Louis
63. Doherty

64. Vanbeveren
65. Vilchez
66. Schulte-Ladbeck
67. Sukartadiređja
68. Langer
69. Stickland
70. Cohen
71. Chu
72. Debray
73. Barlow
74. Zhang
75. Eenens
76. Richter
77. Dopita
78. Lozinskaya
79. Cherepashchuk
80. Hidayat
81. Yungelson
82. Sutantyo

LIST OF PARTICIPANTS

A.G. Admiranto, Lembaga Penerbangan dan Antariksa Nasional, Bandung, Indonesia

A.A. Ngurah Made Agung, Program Studi Fisika, UNUD, Denpasar, Bali, Indonesia

Y. Andrillat, Lab. d'Astronomie, Université de Montpellier II, Montpellier, France

K. Annuk, W. Struve Tartu Astrophysical Observatory, Toravere, Estonia, U.S.S.R.

T.E. Armandroff, Kitt Peak National Observatory, Tucson, AZ, U.S.A.

K. Artawan, Program Studi Fisika, Universitas Udayana, Denpasar, Bali, Indonesia

G.B. Baratta, Osservatorio Astronomico di Roma, Roma, Italia

M.J. Barlow, Department of Physics & Astronomy, UCL, London, U.K.

J. Breysacher, European Southern Observatory, Garching-bei-München, B.R. Deutschland

H. Budiningarti, Program Studi Fisika, Universitas Udayana, Denpasar, Bali, Indonesia

K. Cananzi, Observatoire de Marseille, Marseille, France

J.P. Cassinelli, Washburn Observatory, University of Wisconsin, Madison, MD, U.S.A.

A.M. Cherepashchuk, Sternberg Astronomical Institute, Moscow State University, U.S.S.R.

Y.-H. Chu, Department of Astronomy, University of Illinois, Urbana, IL, U.S.A.

M. Cohen, Radio Astronomy Laboratory, University of California, Berkeley, CA, U.S.A.

P.S. Conti, JILA, University of Colorado, Boulder, CO, U.S.A.

P. Crowther, Department of Physics & Astronomy, UCL, London, U.K.

D.N. Dawanas, Jurusan Astronomi, ITB, Bandung, Jawa Barat, Indonesia

J.R. Deacon, Department of Physics & Astronomy, UCL, London, U.K.

B. Debray, Astrophysics Division SSD, ESTeC, Noordwijk, Nederland

J.P. De Greve, Astrofysisch Instituut, VUB, Brussel, België

L.R. Doherty, Washburn Observatory, University of Wisconsin, Madison, U.S.A.

M.A. Dopita, Mt. Stromlo & Siding Spring Observatories, Private Bag, ACT, Australia

L. Drissen, Département de Physique, Université de Montréal, Montréal, Canada

P.R.J. Eenens, Royal Observatory, Edinburgh, Scotland, U.K.

C. Esteban Lopez, Instituto de Astrofisica de Canarias, La Laguna, Tenerife, Espana

M. Felli, Osservatorio Astrofisico di Arcetri, Firenze, Italia

A.V. Filippenko, Department of Astronomy, University of California, Berkely, CA, U.S.A.

J.G. Gandhiadi, Program Studi Fisika, Universitas Udayana, Denpasar, Bali, Indonesia

A.M. van Genderen, Sterrewacht, Rijksuniversiteit Leiden, Nederland

L.N. Georgiev, Department of Astronomy, Sofia University, Sofia, Bulgaria

P. Giannone, Instituto Astronomico, Università "La Sapienza", Roma, Italia

M.J.H. de Groot, Armagh Observatory, Armagh, Northern Ireland, U.K.

B. Gultom, LAPAN Aerospace Research Center, Bandung, Jawa Barat, Indonesia

W.-R. Hamann, Inst. f. Theor. Physik und Sternwarte der Univ., Kiel, B.R. Deutschland

M. Heydari-Malayeri, European Southern Observatory, La Silla, Chile

B. Hidayat, Observatorium Bosscha, ITB, Lembang, Jawa Barat, Indonesia

D.J. Hillier, JILA, University of Colorado, Boulder, CO, U.S.A.

I.D. Howarth, Department of Physics & Astronomy, UCL, London, U.K.

K.A. van der Hucht, SRON Space Research Utrecht, Utrecht, Nederland

R.M. Humphreys, Department of Astronomy, Univ. of Minnesota, Minneapolis, MN, U.S.A.

M.A. Jura, Department of Astronomy, University of California, Los Angeles, CA, U.S.A.

R. Kingsburgh, Department of Physics & Astronomy, UCL, London, U.K.

Kirwiyanto, Program Studi Fisika, Universitas Udayana, Denpasar, Bali, Indonesia

G. Koenigsberger, Instituto de Astronomia, UNAM, Mexico City, Mexico

T. Kogure, Astronomy Department, Kyoto University, Sakyoku, Kyoto, Japan

L.V. Kuhi, Provost and Vice President, University of Minnesota, Minneapolis, MN, U.S.A.

D. Kunth, Institut d'Astrophysique, Paris, France

R. Lamontagne, Département de Physique, Université de Montréal, Montréal, Canada

N. Langer, Universitäts-Sternwarte Göttingen, B.R. Deutschland

C. Leitherer, Space Telescope Science Institute, Baltimore, MD, U.S.A.

J. Lequeux, Observatoire de Paris, Section de Meudon, Meudon, France

K.-C. Leung, Dept. of Physics & Astronomy, University of Nebraska, Lincoln, NE, U.S.A.

D. Lewis, Département de Physique, Université de Montréal, Montréal, Canada

C.W.H. de Loore, Astrofysisch Instituut, VUB, Brussel, België

M.-C. Lortet, Observatoire de Paris, Section de Meudon, DAEC, Meudon, France

T.A. Lozinskaya, Sternberg Astronomical Institute, Moscow State University, U.S.S.R.

A. Maeder, Observatoire de Genève, Sauverny, Switzerland

I. Bagus Sujana Manuaba, Program Studi Fisika, UNUD, Denpasar, Bali, Indonesia

D. de Martino, ESA IUE Observatory, Viilafranca del Castillo, Spain

P.L. Massey, Kitt Peak National Observatory, Tucson, AZ, U.S.A.

F. Matteucci, Istituto di Astrofisica Spaziale CNR, Frascati, Italia

C.F. McCain, Observatorium Bosscha, ITB, Lembang, Jawa Barat, Indonesia

A.F.J. Moffat, Département de Physique, Université de Montréal, Montréal, Canada

T. Montmerle, Service d'Astrophysique, CEN-Saclay, Gif-sur-Yvette, France

I.G.L. Muliarta, Jurusan Astronomi, ITB, Bandung, Jawa Barat, Indonesia

C.L. Neese, Kitt Peak National Observatory, Tucson, AZ, U.S.A.

V.S. Niemela, Instituto de Astronomia y Física del Espacio, Buenos Aires, Argentina

L. Nobili, Departimento di Fisica "Galileo Galilei", Università Degli Studi di Padova, Italia

K. Nomoto, Department of Astronomy, University of Tokyo, Tokyo, Japan

T. Nugis, W. Struve Tartu Astrophysical Observatory, Toravere, Estonia, U.S.S.R.

C. Nurwendaya, B.P. Planetarium DKI Jakarta, Jakarta, Indonesia

S.P. Owocki, Bartol Research Institute, University of Delaware, Newark, DE, U.S.A.

M.W. Pakull, Observatoire de Besancon, Besancon, France

K. Pandiarta, Program Studi Fisika, Universitas Udayana, Denpasar, Bali, Indonesia

L.P. Paramita, Jurusan Astronomi, ITB, Bandung, Jawa Barat, Indonesia

A.M.T. Pollock, Computer & Scientific Co. Ltd, Sheffield, U.K.

N. Prantzos, Institut d'Astrophysique, Paris, France

R.K. Prinja, Department of Physics & Astronomy, UCL, Gowerstreet, London, U.K.

M. Raharto, Observatorium Bosscha, ITB, Lembang, Jawa Barat, Indonesia

O.-R. Richter, Space Telescope Science Institute, Baltimore, MD, U.S.A.

C. Robert, Département de Physique, Université de Montréal, Montréal, Canada

A. Rosana, Observatorium Bosscha, ITB, Lembang, Jawa Barat, Indonesia

M.S.A. Sastroamidjojo, Solar Physics Lab., Univ. Gadjah Mada, Yogyakarta, Indonesia

H. Schild, Department of Physics & Astronomy, UCL, London, U.K.

W. Schmutz, JILA, University of Colorado, Boulder, CO, U.S.A.

R.E. Schulte-Ladbeck, Space Astronomy Lab., Univ. of Wisconsin, Madison, WI, U.S.A.

E. Sedlmayr, Inst. f. Astron. & Astroph. der Techn. Univ., Berlin, B.R. Deutschland

M. Shara, Space Telescope Science Institute, Baltimore, MD, U.S.A.
B.S. Shylaja, Physical Research Laboratory, Navrangpura, Ahmedabad, India
S. Siregar, Jurusan Astronomi, ITB, Bandung, Jawa Barat, Indonesia
L.F. Smith, Mt. Stromlo & Siding Spring Observatories, Canberra, ACT, Australia
L.J. Smith, Department of Physics & Astronomy, UCL, London, U.K.
O. Soemantri, Observatorium Bosscha, ITB, Lembang, Jawa Barat, Indonesia
R. Spurzem, Institut für Astronomie und Astrophysik, Würzburgh, B.R. Deutschland
S.R. Sreenivasan, Dept. of Physics & Astronomy, The University of Calgary, AB, Canada
I.R. Stevens, NASA Goddard Space Fligth Center, Greenbelt, MD, U.S.A.
D.J. Stickland, Space & Astrophysics Division, RAL, Chilton, U.K.
N. St-Louis, Department of Physics & Astronomy, UCL, London, U.K.
Y.B. Sugiarto, Program Studi Fisika, Universitas Udayana, Denpasar, Bali, Indonesia
D. Sukartadiredja, B.P. Planetarium DKI Jakarta, Jakarta , Indonesia
W. Sutantyo, Jurusan Astronomi, ITB, Bandung, Jawa Barat, Indonesia
M.-J. Taylor, Space Astronomy Laboratory, University of Wisconsin, Madison, WI, U.S.A.
L.-S. The, Department of Physics and Astronomy, Clemson University, Clemson, SC, U.S.A.
R. Turolla, Departimento di Fisica "Galileo Galilei", Università di Padova, Italia
A.B. Underhill, Dept. of Geophys. & Astron., Univ. of BC, Vancouver, Canada
D. Vanbeveren, Fysisch Instituut, VUB, Brussel, België
J.M. Vilchez, Instituto de Astrofisica de Canarias, La Laguna, Tenerife, Espana
S. Wahyono, Program Studi Fisika, Universitas Udayana, Denpasar, Bali, Indonesia
N.R. Walborn, Space Telescope Science Institute, Baltimore, MD, U.S.A.
U. Wessolowski, Inst. f. Theor. Physik und Sternwarte der Univ., Kiel, B.R. Deutschland
N. Widana, Program Studi Fisika, Universitas Udayana, Denpasar, Bali, Indonesia
A.J. Willis, Department of Physics & Astronomy, UCL, London, U.K.
A.S. Wilson, Astronomy Program, University of Maryland, College Park, MD, U.S.A.
Windaryoto, Program Studi Fisika, Universitas Udayana, Denpasar, Bali, Indonesia
S.D. Wiramihardja, Observatorium Bosscha, ITB, Lembang, Jawa Barat, Indonesia
Diah Yudiawati Anggraeni, Observatorium Bosscha, ITB, Lembang, West Java, Indonesia
L.R. Yungelson, Astronomical Council, USSR Academy of Sciences, Moscow, U.S.S.R.

WELCOMING ADDRESSES and AFTER-DINNER SPEECH

Opening ceremony

SAMBUTAN KETUA LOCAL ORGANIZING COMMITTEE

Bapak Gubernur, Prof. Oka, Yth.,
Bapak Direktur Jenderal Pendidikan Tinggi, Yth.,
Hadirin Yth.,

Perkenankanlah kami melaporkan bahwa kami bersyukur dan bergembira, pertemuan yang telah direncanakan semenjak 2 tahun lalu, pada hari ini dapat terlaksana. Semenjak General Assembly International Astronomical Union di Baltimore pada bulan Agustus 1988 menyetujui tema dan susunan anggota Scientific Organizing Commitee, komunikasi antara anggota SOC yang tersebar di 13 negara tampak menjadi lebih intensif untuk memilih tema dan subtema. Pemilihan subtema itu sangat penting karena pertemuan ini berharap dapat:

1. memberikan hasil pengamatan dan interpretasi teoritis tentang kelahiran dan evolusi bintang Wolf-Rayet dan bintang masif, dari fasilitas pengamatan landas bumi; landas layang maupun dari satelit.

2. menentukan masalah strategis dan terpadu untuk mengetahui sumbangan bintang terhadap evolusi galaksi; serta evolusi bintang berat.

Peserta pertemuan ini berjumlah 118 astronom dari 18 negara, akan membahas 20 buah invited reviews, 36 buah invited contributions. Sayang sekali karena keterbatasan waktu memaksa kami mempergunakan metode penyajian - yang tidak kalah penting - "poster session" sebanyak 73 karya ilmiah. Sebagian besar sumbangan ilmiah poster in merupakan hasil pengamatan dan pengembangan teori dalam tempo 12 bulan terakhir ini.

Dalam merealisasikan pertemuan ini kami telah memperoleh banyak dukungan moril dari Bapak Mendikbud, Menparpostel, dan bantuan materil dari berbagai pihak.

Pada bulan Oktober 1988 kami menghadap Bapak Gubernur Bali, Prof. Oka, dan Bapak Rektor Universitas Udayana, Prof. Adnyana. Dari pertemuan awal itu kami merasa memperoleh dorongan untuk maju mempersiapkan pertemuan ini di Denpasar. Kami hendak menyampaikan banyak terima kasih kepada Pemda yang telah memberikan dukungan moril selama ini.

Secara operasional Local Organizing Committee, yang terdiri dari 7 orang dari 5 instansi bekerja dibawah naungan 2 buah universitas: Udayana dan ITB dan Direktur Jendral Pendidikan Tinggi. Kehangatan naungan sayap mereka membuat LOC lebih bergairah untuk membuat pertemuan ini berharga bagi ilmu pengetahuan.

Dari buku "Biru" dapat dilihat bahwa pertemuan ini dapat terselenggara berkat sentuhan tangan banyak pihak. Tanpa itu kami, LOC dan SOC, hanya seperti kertas saja tanpa daya. Kami mohon maaf. Karena keterbatasan waktu pagi ini kami tidak dapat menyebut satu persatu nama pendukung pertemuan ini, namun rekaman tertulis dalam "buku program" - yang juga merupakan prakata "Proceedings" kami nanti - adalah pengejawantahan rasa terima kasih kepada semua pihak yang telah ikut menyelenggarakan pertemuan ini. Namun begitu, kami mohon ijin untuk menyebut kerjasama KNI-Unesco, Departemen Luar Negeri, LIPI dan Bappenas serta Garuda Indonesia Airlines, yang benar-benar telah merupakan rantai vital bagi terselenggaranya pertemuan ini.

Akhirnya, kami hendak menyampaikan banyak terima kasih kepada Bapak Alfons D. Viera, G.M. Hotel Bali Beach, beserta staf yang telah banyak memberikan kemudahan bagi kami. Dalam pada itu sudah sepantasnya kalau kami mohon maaf kepada semua pihak atas segala kekurangan L.O.C. dalam penyelenggaraan ini.

Hadirin yth., ijinkan saya sekarang menyampaikan pesan dalam bahasa Ingris, untuk kepentingan rekan yang tidak paham bahasa Indonesia dan untuk memberikan rasa international pada pertemuan ini.

K. A. van der Hucht and B. Hidayat (eds.),
Wolf-Rayet Stars and Interrelations with Other Massive Stars in Galaxies, 3–4.
© 1991 *IAU. Printed in the Netherlands.*

Distinguished guests, colleagues and participants,

I have just reported to the Minister of Education and Culture what we are going to talk about, discuss and, hopefully, solve during the next five days here on the island of gods, goddesses and demons. It is not my intention to repeat what I have said about our business, but, instead, I would like to extend to you all a few words on behalf of the Local Organizing Committee and all our Indonesian friends.

Dr. van der Hucht, the Chairman of SOC, who chose the date of this meeting may have a 6th sense. The calendar indicates that today is June 18th, 1990. This is all right, we all know it. But the Balinese have many different systems of time designation. According to their calendrical system today is the 25th week of Caka 1912, or Sasih Sadha Ngunya Karo, or the 28th day of Mintuna Rasi, etc., which all coincides with the best time to sow seeds. Everybody, man, woman and hermaphrodites, are promised success if they plant on this day. I happened to discover the significance of this day only two weeks ago when I consulted an old-time Bali calendar. In this frame of mind I wish you all success in your deliberations. We will plant good seeds, which we hope will bear good useful fruit in the decade of the 90's and beyond.

We would like to wish all participants, first of all, enjoy your scientific topics and our meeting venue, and naturally, your short stay here in Bali, Indonesia. If I stress the word "meeting" in the first place, I am really not exaggerating, because the natural surroundings of this hotel are a really great attraction and distraction which might lure you all to forget the prime business of coming to Bali.

Now please forgive our shortcomings and omissions. You know our word-processors and type writers have occasional lapses when typing English or foreign names. So there may be some spelling mistakes. But, anyway please ask if there is something which is not clear. Our desk and office, respectively on the first and third floor (rm 304) are open to you all time.

We are really delighted to welcome you all here. Bali is well-known for its cosmical and cosmogonical perception, which is interwoven with day-to-day affairs. This is reflected in the smiling, contended faces of the Balinese people and by their constant observation of their religious duties which blesses their lives. The rolling green of the countrysides, towered by the puras for offerings, are manifestations of their harmonious life with the nature, which, I hope, you may have a chance to experience.

Saudara - I am fond of this word as it brings a sense of a brotherhood -, Selamat Datang di Indonesia dan Selamat Bekerja.

Bambang Hidayat

ADDRESS BY THE CHAIRMAN OF THE SCIENTIFIC COMMITTEE

Yang sangat saya hormati:
Bapak Gubernur Kepala Daerah Tingkat Satu Bali, Prof. Dr. Ida Bagus Oka,
Bapak Direktur Jenderal Pendidikan Tinggi, Departemen Pendidikan dan Kebudayaan,
Prof. Dr. Sukadji Ranuwihardjo,
Bapak Rectores Magnifici Institut Teknologi Bandung dan Universitas Udayana, masing-
masing Wiranto Arismunandar dan I.G.N.P. Adnyana, dan hadirin yang mulia,

Your Excellency, Ladies and Gentlemen,

It gives me great pleasure to participate in this inaugural ceremony of the 143rd Symposium
of the International Astronomical Union and to speak on behalf of the Scientific Organizing
Committee, welcoming all of you to the symposium.

Today we are especially honoured to have this symposium opened by Prof.Dr. Sukadji
Ranuwihardjo, Director General of Higher Education, and in the presence of the Governor
of Bali, Prof.Dr. Ida Bagus Oka, which demonstrates the thorough understanding of the
Indonesian Government to the importance of astronomy and of this symposium. It is a
privilege to have this symposium in your country.

Astronomical research of the present century has had many contributions from Indonesia, both from the results that have originated from the Observatorium Bosscha in
Lembang, as well as from the solar eclipse stations located in this country within the path
of total eclipses. Most of us are aware of the work carried out in the past that is associated
with important personalities in astronomical history. The names of Pannekoek, Wallen-
quist, de Sitter and van Albada bring to our minds the efforts of the past. In the trail
blazed by these men of distinction come the efforts of the present. Twelve astronomers and
over 70 students in Lembang and Bandung, with the vigour of confidence and enthusiasm
that comes from accomplishment. Located on the equator, Indonesia is privileged to have
access to nearly the whole sky, which offers so much promise. And with the continued
support that astronomy in this country has from the authorities, your accomplishments
of the future are bound to be even greater. Among your accomplishments of the past
are the organization of the 2nd Asian Pacific Regional IAU Meeting in 1981, and of IAU
Colloquium No. 80 on "Double Stars, Physical Properties and Generic Relations" in 1983,
both hosted in Bandung.

In many ways, this symposium symbolizes the spirit of Astronomy, where friendly
cooperation overrides the inhibiting influences of national boundaries. Seated here are 118
astronomers from 18 different countries, covering a range of latitude and longitude. They
come from institutions having diverse facilities. Their principle common motive is to get
together with their counterparts from different places, to learn, to discuss, and to plan new
approaches that will take us further in the question of understanding the appearance and
evolution of massive stars. Here is astronomy, the forerunner of all sciences, at its best.

Symposia of this kind have played a very important role in the activities of the International Astronomical Union. Hot massive stars were discussed in Argentina (1971, IAU
Symposium No. 49), Canada (1978, IAU Symposium No. 83), Mexico (1981, IAU Symposium No. 99) and Greece (1985, IAU Symposium No. 116). Those symposia, together with
additional colloquia and workshops on the subject have been the principle fora that have
witnessed the announcements of discovery and contributed to the stimulus for the future.
This symposium will follow that tradition.

123 years ago, the French astronomers Wolf and Rayet discovered the spectacular appearance of the emission-line-dominated spectra of three stars in the constellation Cygnus.
Ever since, those stars, and the numerous other ones which have been discovered later, have

5

K. A. van der Hucht and B. Hidayat (eds.),
Wolf-Rayet Stars and Interrelations with Other Massive Stars in Galaxies, 5–6.
© 1991 *IAU. Printed in the Netherlands.*

been called Wolf-Rayet (WR) stars. It took astronomers and physicists 65 years (Edlén) to figure out the identification of the emission line spectra, and another 10 years (Gamow) to realize the meaning of the two spectral sequences of WR stars, the nitrogen and the carbon sequences.

Because it takes time for light to travel, looking at stars means looking back in time. In that respect astronomers could be considered historians. When we look at the Sun, we see the light that left the Sun 8 minutes ago. The light from the nearest WR star, however, the star γ in the constellation Vela, left that star more than 1400 years ago, that is, at the time when Indonesia was a Hindu-Javanese empire, governed by raja's. The light from the most remote known WR star in our Galaxy travelled to us during as many as 40.000 years. The light of the WR stars in the Large Magellanic Cloud, our neighbouring galaxy, travelled about 170.000 years, which means, that light left the LMC at the time that alongside the Solo River in Central Java, in the village of Trinil, the Homo Trinilensis was living, also called the Phitecantropus Erectus, an early variety of the Homo Sapiens.

Homo Sapiens nowadays knows that WR stars represent a conspicuous phase in the evolution of hot massive stars and that they have many extreme properties. They are among the most luminous stars: with an energy output typically 100.000 times that of the Sun, and very hot: with surface temperatures of about 5 to 20 times that of the Sun. They have very strong stellar winds wherein their characteristic emission line spectra originate, and which make them lose matter to their environments at rates of one solar mass in only 100.000 to 10.000 years. Their surface chemical composition looks highly anomalous, dominated by helium, nitrogen and carbon rather than hydrogen. In the past decades it has been recognized that mass loss due to stellar wind and mass transfer in double stars significantly influences the evolution of massive stars. WR stars release so much matter by their strong stellar winds or during bulk-ejection of shells of matter, that some of them show extended nebulae around them, which give us additional handles to study the energy house keeping and evolution of their central WR stars. WR matter can even be traced on Earth. Observations of cosmic rays and of meteorites have revealed isotopic ratios of elements which can only have formed in the interior of massive stars like WR stars. This could imply that the interstellar cloud in which our solar-system once formed, contained traces of matter of nearby pre-solar WR stars.

Ladies and Gentlemen, we are indeed grateful to the Indonesian authorities, to the Rectores of the Institut Teknologi Bandung and of the Universitas Udayana in Denpasar, and to our Indonesian colleagues, for organizing this symposium here. I am confident that we will have a very useful, constructive and delightful week of lectures, discussions and deliberations. On behalf of the Scientific Organizing Committee, I express our gratitute to our hosts and extend my good wishes for a highly successful symposium.

Terima kasih: Thank you very much.

Karel A. van der Hucht

ADDRESS BY THE MINISTER OF EDUCATION
AND CULTURE, R.I.

REPUBLIC OF INDONESIA
MINISTER OF EDUCATION AND CULTURE

Permit me first of all to bid a "Selamat Datang" to the participants of this congress and I hope that your stay in Bali will be sufficient to enjoy the tranquility and pleasures that can be found on the isle of the gods. Hopefully during your stay you will be introduced to the plethora of cosmic phenomena that is described in Bali's unique culture, including its perceptions on celestial objects, especially the stars.

Before continuing, I must be the first to admit that as a layman with very little knowledge of matters concerning astronomy, I should in fact not have been so bold a to accept to talk before this forum. I am certain that I will not be able to contribute any new ideas to this congress. Nevertheless, it is because I am a layman that I may be able to touch upon matters based on naive and innocent observations rather than based on a scientific framework. As one who enjoys the rapport of academic intercoures, I must admit that at times I am fascinated by naive observations which sometimes are accompanied by interesting interpretations on various phenomena and happenings, especially those relating to the cosmos.

Since prehistoric times man and nature have been so cognate that various natural phenomena have been given certain meanings which have been woven into his existence as part of the totality of life. No matter how naive that assumption is, it would be difficult to deny that mere perception and interpretation is in fact the element that gives comfort and stability to man. Man's preparedness to become one with nature - *i.e.* by giving special meaning to the various natural phenomena - makes him more sensitive to the many signs that it augurs. *Sympatheia* was a distinctive feature of man's comprehension of the world around him before he began to slowly distance himself and treat it as an opposing pole; the world slowly changed and became different from himself, and for that reason he had to dominate and control it to his advantage. The unity between man and his world slowly became a dichotomy with the poles further growing apart. In his development, man has come to realise himself as being the center of all things. Hominocentrism or anthropocentrism was a stage which in the end radically changed man's attitude towards the world around him, and finally towards the cosmos. Indeed, the various cosmic occurrences and phenomena - although still interesting - no longer arouse the same feelings as before. This is probably what differentiates the scientist from the layman in observing and comprehending the many occurrences and phenomena. Both may concur in trying to "transcend the real into the possible", nevertheless the meaning given to what is concealed behind reality may differ. Scientists may regard this as an unknown X factor, whereas the layman would very easily consider this as a mystery.

Indeed, so many cosmic occurrences and phenomena are regarded by the layman as a

7

K. A. van der Hucht and B. Hidayat (eds.),
Wolf-Rayet Stars and Interrelations with Other Massive Stars in Galaxies, 7–8.
© 1991 *IAU. Printed in the Netherlands.*

sign or symptom which holds within it something mysterious, and for that reason is always viewed as *mysterium fascinosum*. This is valid whenever we talk about the moon and stars as a source of light in the obscurity of the night. In the darkness, the moon and stars are not merely a source of light, but rather a guide and a source of inspiration. The moon and stars are distinctive manifestations of the night; nevertheless, whereas there is only one moon, the stars are many, scattered across the sky. However, rather than being merely scattered, there is an order in its formation, brightness, distance, etc. It is therefore not surprising if people are fascinated by these natural phenomena which are most interesting to study.

In all cultures, stars have always held the symbolic meaning of excellence or transcendence. There is no single example which I know of that has disparaged the star as a symbol - except, of course, by those who would seek to disparage those who use that star as a symbol. The stars are considered superior because they are beyond the reach of man. The stars seem so distant, and yet their presence is clearly felt. They will always have a special appeal. This symbolic meaning has continued until today, when man is continuing forward in wanting to understand them by scientific research through astronomy.

Although astronomy does not have an image of eminence such as the other ancient disciplines, nevertheless people will always be attracted to it given the opportunity to observe the constellations. It is for this reason that astronomers should endeavour to create an interest and awareness amongst members of the lay community towards this science. With the rapid development of science and technology in the field of outer space today, knowledge of astronomy could also support the growing interest in outer space in general. I believe that astronomers will receive a positive response if they are able to present their knowledge in an understandable manner to the genereal public. Together with developments in space science and technology as well as the success of many space missions, astronomy can become an important source of information in future explorations. My great hope is that this forum - apart from discussing various scientific matters - will also find a mode in which to inform the general public on the achievements of modern astronomy. Indeed, the increased interest in astronomy will in turn attract the interest of the young to choose this subject as a field for research and study. Today, interest in astronomy is problably limited in comparison to other disciplines. Nevertheless, I believe that this is a result of lack of information concerning the study of astronomy. Therefore I believe that it is one of the responsibilities of this congress to enhance the interest of astronomy, especially among the young.

I hope that this congress will proceed smoothly and that the participants will also be able to enjoy the star-lit nights on the isle of the gods.

Prof.Dr. Fuad Hassan

WELCOMING REMARKS OF THE RECTOR OF INSTITUTE OF TECHNOLOGY, BANDUNG

Distinguished members of the International Astronomical Union from all over the world,
Distinguished scientists and researchers in astronomy, astrophysics, geophysics, physics, and other sciences, present in this important meeting,
Distinguished guests,
Ladies and gentlemen,

It is indeed an honour and a pleasure for me to welcome you all to this International Astronomical Union Symposium Number 143 on Wolf-Rayet Stars and Interrelations with other Massive Stars in Galaxies, which is held in Denpasar, Bali, from June 18 to 22, 1990.

Welcome to Indonesia and welcome to the island of Bali. I am happy that this gathering is held in Indonesia.

I wish to express my thanks to the IAU and the Government of Indonesia and other sponsoring organizations for the opportunity, the trust, and assistance given to Indonesia, including the Institute of Technology Bandung (ITB) and the Udayana University, to host this international symposium.

It is indeed appropriate and fortunate that this scientific meeting is held at this particular time, because today an increasing emphasis is given to strengthening and developing mathematics and natural sciences in Indonesia. For science and technology to contribute meaningfully to the development of Indonesia, such an effort to developing the basic sciences is mandatory.

Ladies and gentlemen,

Astronomy, as a part of science that tries to unlock the mysteries of the origins of our stellar system and the galaxies, I know, can and will give answers to questions relating to the existence of our earth and life on earth. It is my sincere believe that ideas from deliberations in this meeting will lift the spirit and awaken inspirations to find answers to scientific questions about our stars and galaxies and to make conjectures and predictions of the future of our universe and the future of science and mankind. Above all, I hope cooperation and mutual understanding will be renewed among all the participants in this meeting so that new and existing scientific data can be shared and transformed into information for the benefit of man.

On behalf of the academic and research community of the Institute of Technology Bandung and joining the Indonesian community of scientists and the host committee, I wish you a succesful symposium that will contribute new information for the sciences represented here.

May I also wish you a pleasant stay in Bali.

Thank you.

Prof.Ir. Wiranto Arismunandar

9

K. A. van der Hucht and B. Hidayat (eds.),
Wolf-Rayet Stars and Interrelations with Other Massive Stars in Galaxies, 9.
© 1991 *IAU. Printed in the Netherlands.*

ADDRESS BY THE DIRECTOR GENERAL OF HIGHER EDUCATION
MINISTRY OF EDUCATION AND CULTURE, REPUBLIC OF INDONESIA

Distinguished guests, and participants,

First of all I would like to second the Chairman of the Local Organizing Committee to welcome you in this very particular place here on the Island of Bali. Last night you already had the first experience of the richness of Balinese arts and dances, at the courtesy of our Governor of Bali. Of course you would want to know more about Bali and our cultural features of other regions as well. Our coat of arms mentions "Bhineka Tunggal Ika", meaning Unity in Diversity. This reflects the richness of our socio-economic and cultural heritage. With more than thirteen thousand islands that comprise the modern state of Indonesia, located at the crossroads of the Pacific Rim community, Indonesia indeed offers an endless variety of cultural life.

It is with great pleasure and anticipation that I am here to open the International Union Symposium No. 143 on Wolf-Rayet Stars and Interrelations with other Massive Stars in Galaxies. From what I have been informed, these types of stars are playing a very important role in modifying their galactic environments. In their short lifetime, of course on galactic time-scale, these stars have shown in their vigorous and efficient ways of revealing, that, at least according to some theories, nuclear reactions have actually taken place inside the stars. The products are brought out to the upper atmosphere, showing "bizarre" spectral features that attract the astronomers' interest. The relative youthness of the stars is important in many different fields of astronomy. They can, for example, be used to trace the spiral features of our Galaxy.

Indonesians live practically under the shadow of the center of the Milky Way, which we call *Bimasakti*, named after the ancient hero from the great Hindu epic *Mahabharata*. We are enchanted by the glamorous radiation from that part of the sky: the Milky Way band that stretches from South to North in our summer sky, which creates spectacular views since times immemorial. Many of us do not know what it is. It is gratifying, therefore, to learn that you in modern time, equipped with modern techniques and theoretical knowledge, are here to unravel one of the mysteries of the *Bimasakti*.

Here in Bali, where Hindu religion is predominant, life and death are viewed as an unending process. I wish all of you will be inspired by the wisdom of the Balinese, and will also find the studying of the life and death processes of stars in the galaxies a rewarding undertaking.

It is at this point that I would like to express our gratitudes, on behalf of the Government of Indonesia, that the International Astronomical Union has entrusted to one of our components of higher education to organize this important gathering. I hope that the Bali meeting will contribute substantially to the existing body of knowledge of astronomy. I would also like to express my sincerest thank to many organizations, here and abroad, that have made this meeting possible.

Finally, I wish you all a pleasant stay and fruitful meeting here on the Island of Paradise. Allow me to declare the beginning of the Symposium on Wolf-Rayet Stars and Interrelations with other Massive Stars in Galaxies.
Thank you.

Prof.Dr. Sukadji Ranuwihardjo

K. A. van der Hucht and B. Hidayat (eds.),
Wolf-Rayet Stars and Interrelations with Other Massive Stars in Galaxies, 10.
© 1991 *IAU. Printed in the Netherlands.*

Wolf-Rayet Stars in the Old Days

Leonard V. Kuhi
University of Minnesota
213 Morrill Hall
100 Church St. S.E.
Minneapolis, Minnesota 55455

This morning, as part of the welcoming ceremonies opening this conference, we were reminded that Indonesia consists of over 13,000 islands and that the coat of arms carries the phrase: "Unity in Diversity." We might take the same theme for the study of Wolf-Rayet Stars. There has certainly been no shortage of diversity over the 120 years since their discovery and we all keep hoping for unity; perhaps IAU Symposium No. 143 will provide it! The conference was opened this morning with three strokes of the ceremonial gong to chase away all the bad spirits. So far that symbolic action has been most successful but perhaps more likely due to the kind hospitality of our hosts and the special warmth of Bali. We owe them all a vote of thanks!

What does the title of this talk: "Wolf-Rayet Stars in the Old Days" actually mean? My first reaction was that it reflects a sign of old age when one is invited to be the after dinner speaker. After all, only "old" astronomers are afforded that pleasure if memory serves me. Perhaps the title means that the study of Wolf-Rayet stars is a mature subject and that there are actually old days to talk about. Or perhaps it goes back to my old work on Wolf-Rayet stars in the 1960's but that would be vanity. More likely it goes back to two conferences held in 1968 and 1971. The first was held at Boulder and was simply entitled "Wolf-Rayet Stars"; the second was the IAU Symposium No. 49 held in Buenos Aires on "Wolf-Rayet and High Temperature Stars." The Boulder symposium rekindled interest in Wolf-Rayet stars with several major presentations: Lindsey Smith (1969) on classification, luminosities, binary nature and galactic distribution; L. V. Kuhi (1969) on continuous energy distributions, spectral scans, line profiles and infrared observations; A. B. Underhill (1969) on spectroscopic diagnostics and atmospheric models; and R. N. Thomas (1969) on the physical structure of Wolf-Rayet stars.

The Buenos Aires conference in 1971 reflected a two year period of intense activity following up on many of the problems discussed at Boulder. R. N. Thomas (1971) started off the symposium with an overview of the general problems of extended atmospheres and non-

11

K. A. van der Hucht and B. Hidayat (eds.),
Wolf-Rayet Stars and Interrelations with Other Massive Stars in Galaxies, 11–15.
© 1991 *IAU. Printed in the Netherlands.*

classical stellar atmospheric models. Major theoretical presentations were made by D. Van Blerkom (1971) on the theory of Wolf-Rayet spectra and by B. Paczynski (1971) on evolutionary aspects of Wolf-Rayet stars. L. F. Smith (1971) reviewed the classification and galactic distribution of Wolf-Rayet stars, the interpretation of the WN sequence and of the WC stars. L. V. Kuhi (1971) discussed Wolf-Rayet binaries and used the observed changes in C III and IV line profiles to determine the atmospheric stratification and run of temperature with radius. Much new information was provided at that meeting which basically set the stage for the ensuing research of the last twenty years. It is gratifying today to see how much of that original work has withstood the test of time and the onslaught of countless young Ph.D.'s all eager to publish and not perish.

But we really shouldn't get too smug in our reflections about the recent past; after all twenty years can hardly be "the old days." Also, as Peter Conti remarked earlier: "The old astronomers twenty years ago looked so much older than the old astronomers do today." Therefore, I will take as the crucial timescale to "the old days" as 50 years -- my lifetime to astrophysical accuracy. It is only then that we come to the real pioneers of the field, namely those astronomers who published before 1940.

Van der Hucht in his opening remarks has already referred to the discovery of Wolf-Rayet stars by Wolf and Rayet (1867) who had conducted a visual survey of the spectra of stars in Cygnus. They noted that three stars, HD 191765, 192103 and 192641, showed bright emission bands and colors which were "all yellow, orange yellow and greenish yellow." More such stars were discovered by Respighi (1872), Pickering (1881) and Copeland (1884). The earliest wavelength lists of the emission lines were published by Campbell (1894). Wright (1918) produced further wavelength lists and provided some identifications; he also noted that the nuclei of planetary nebulae were often Wolf-Rayet stars. A major contribution was made by Plaskett (1924) who investigated all northern Wolf-Rayet stars brighter than 9th mag. He presented spectra, wavelength lists, further identifications and suggested a classification scheme. Perrine (1920) and Payne (1926, 1927) (she was referred to as Miss Payne) extended the investigations to the southern sky and also provided additional wavelengths of emission lines.

However, the classic paper was published by Beals in 1929 entitled "On the Nature of Wolf-Rayet Emission" based on a study of Wolf-Rayet stars down to 11th mag. He described the spectra as "atomic emission in the form of broad bands superposed on a comparatively faint continuous background" with half-widths ~10 to 100 A and a very large ratio of line to continuum intensity. He concluded that the broadening was due to the Doppler effect and reflected the large velocity of emitting gases in the line of sight. He ruled out other broadening agents such as Stark effect, Zeeman effect or high pressure being responsible for the large line widths. Instead he put forth the expansion hypothesis which is still with us today. He compared the Wolf-Rayet spectra with those of novae and noted that the great concentration of energy in the emission lines suggested that the origin of the emission must be the same. Also the violet-displaced absorption features on many emission lines were due to the outward motion of absorbing gases, especially He I. He proposed an expanding envelope or nebula fed by the continuous ejection of material from the central star. A constant velocity of expansion for the envelope produced a flat-topped profile and the model qualitatively accounted for all of the observed features. As for the agency driving the ejection he suggested selected radiation pressure following Milne's (1926) work in connection with the solar chromosphere where ejection velocities of the order of a few 100 km/sec were required. Johnson (1926) had shown the importance of radiation pressure and the fact that the ratio of radiation pressure to gravity could have widely different values for different ions, e.g., He II, C III, Si IV, O III and N III had very high ratios. Hence, the intensity of emission lines also reflected large ratios of radiation pressure to gravity. Beals also estimated the mass loss $\Delta M/M \sim 4 \times 10^{-9}$ per year assuming $V_{ejection} = 2000$ km/sec., $\rho = 3 \times 10^{-17}$ gm/cm^3, $R = 10^8$ km and $M = 30$ M$_\odot$. The key factors were a high surface temperature and a low surface gravity. It was also "logical to assume that Wolf-Rayet stars are very massive."

Beals and Plaskett (1935, 1938) under the auspices of Commission 29 of the IAU provided the basic classification scheme for Wolf-Rayet spectra that we use today: WC stars with lines of C and O and WN stars with lines of N with both types having lines of He I and He II. The scheme was based on the relative intensities of emission lines from two stages of ionization as well as the widths of the lines in the WC sequence. Beals (1940) also used the Zanstra mechanism to estimate temperatures for Wolf-Rayet stars ranging from 59,000 to 110,000°K. This assumed an optically thin atmosphere with no collisions.

The first theoretical attempts to explain the Wolf-Rayet stars came in 1933 when Gerasimovic discussed the contours of the emission lines and in 1934 when Chandrasekhar published a paper entitled "On the Hypothesis of the Radial Ejection of High-Speed Atoms for the Wolf-Rayet Stars and the Novae." Chandrasekhar assumed an extensive envelope with radial ejection with $V_{ejection} \sim 500$ to 3000 km/sec. He also stated that with a dynamical theory of ejection specifying $\rho(r)$ and $v(r)$ one could calculate the emission per unit volume and hence the line profile. He considered two possible theories: 1) the atoms are repelled from the surface by some kind of force (=fg) which falls off like gravity -- this would include Milne's radiation pressure mechanism; or 2) the atom receives a large initial outward velocity and is then decelerated by gravity. The velocity could be imparted by an explosion or perhaps via a mechanism suggested by Plaskett (1924), i.e., atoms near the boundary receive a large outward momentum on being ionized with the emission line being produced by subsequent recombination. Chandrasekhar calculated the expected emission line profiles from these two theories including occultation effects since the star is not of negligible dimensions with respect to the envelope. He concluded that the repulsive force model produced profiles in better agreement with those observed.

Finally, another fruitful area of Wolf-Rayet study was opened up with Wilson's 1940 paper on the spectroscopic binary HD 193576 (V444 Cygni). The spectral types were WN 5 and B1 with masses of 9.74 and 24.8 M⊙, respectively. He calculated the absolute magnitude from the interstellar absorption lines of CA II H and K: $M_V = -1.7$ and a distance of 740 parsecs. The Wolf-Rayet star was 4 mag. brighter than the absolute magnitude estimated from its mass and the standard mass-luminosity relationship. He also noted the variations in the line profile of He II 4686 as a function of phase. This work led to a series of papers on Wolf-Rayet binaries and the general conclusions that $R_* \sim$ 2-3Ro, $R_{env} \sim$ 10R_* and $T_{surface} \sim$ 50,000 to 100,000°K.

The rest is history! But these early astronomers and physicists were the real pioneers who reached conclusions which were not far off the mark. They made brilliant deductions with a minimum of analytic equipment or even the necessary physics. We might very well ask what could they have done with the benefit of modern day computers, detectors and technology? Who's to say? This symposium in Bali is really a tribute to these pioneers who set the field on its modern course. Van der Hucht, in his talk today, said that we have had 120 years of papers on Wolf-Rayet Stars and that was a "large number of papers under the bridge." That is true, but the first few were probably the most important and we should not forget that.

References

Beals, C. S. (1929) M.N. 90, 202.
Beals, C. S. (1930) Pub. D. A. O. 4, 271.
Beals, C. S. (1940) JRASC 34, 169.
Beals, C. S. and Plaskett , J. S. (1935) Trans. I.A.U. 5, 184.
Campbell, W. W. (1894) Astr. Ap 13, 448.

Chandrasekhar, S. (1934) M.N. 94, 522.

Copeland, R. (1884) Copernicus 3, 206.

Gerasimovic, B. P. (1933) Z.f. Ap. 7, 335.

Johnson, M. C. (1926) M.N. 86, 300.

Kuhi, L. V. (1969) in K. B. Gebbie and R. N. Thomas (eds.), Wolf-Rayet Stars, NBS Pub. #307, p. 101.

Kuhi, L. V. (1971) in M. K. V. Bappu and J. Sahade (eds.), Wolf-Rayet and High Temperature Stars, I.A.U. Symposium #49 Reidel Publ., Dordrecht, Holland, p. 205.

Milne, E. A. (1926), M.N. 86, 459.

Paczynski, B. (1971) in M. K. V. Bappu and J. Sahade (eds.), Wolf-Rayet and High Temperature Stars, I.A.U. Symposium #49, Reidl Publ., Dordrecht, Holland, p. 143.

Payne, C. H. (1926) H.B. #834, 836.

Payne, C. H. (1927) H.B. #842, 843.

Payne, C. H. (1933) Z.Ap. 7, 1, 143.

Perrine, C. D. (1920) Ap. J. 52, 39; MN 81, 142.

Pickering, W. H. (1881) Nature 23, 604.

Plaskett, J. S. (1924) Pub. D. A. O. 2, #16.

Respighi, M. (1872) Comptes Rendues 74, 516.

Smith, L. F. (1969) in K. B. Gebbie and R. N. Thomas (eds.), Wolf-Rayet Stars, NBS Pub. #307, p. 21.

Smith, L. F. (1971) in M. K. V. Bappu and J. Sahade (eds.), Wolf-Rayet and High Temperature Stars, I.A.U. Symposium #49, Reidl Publ., Dordrecht, Holland, p. 15.

Thomas, R. N. (1969) in K. B. Gebbie and R. N. Thomas (eds.), Wolf-Rayet Stars, NBS Pub. #307, p. 237

Thomas, R. N. (1971) in K. B. Gebbie and R. N. Thomas (eds.), Wolf-Rayet and High Temperature Stars, NBS Pub. #307, p. 3.

Underhill, A. B. (1969) in K. B. Gebbie and R. N. Thomas (eds.), Wolf-Rayet Stars, NBS Pub. #307, p. 181.

Van Blerkom, D. (1971) in K. B. Gebbie and R. N. Thomas (eds.), Wolf-Rayet and High Temperature Stars, NBS Pub. #307, p. 165.

Wilson, O. C. (1940) Ap. J. 91, 379.

Wolf, C. J. E., and Rayet, G. (1867) Comptes Rendues 65, 292.

Wright, W. H. (1918) Lick Pub. 13, 224.

Len Kuhi

SESSION I. INTRODUCTION – *Chair: Bambang Hidayat*

"*... But quite otherwise was the case of γ Argûs when on April 24th I first viewed its spectrum in the open field of the prismatic eye-piece. Its intensely bright lines in the blue, and the gorgeous group of three bright lines in the yellow and orange, render its spectrum incomparably the most brilliant and striking in the whole heavens . ...*"

Ralph D. Copeland, 1884, *Copernicus* **3**, 193, 'Experiments in the Andes'.

A DECADE OF WOLF-RAYET LITERATURE

Karel A. van der Hucht
SRON Space Research Utrecht
Sorbonnelaan 2, 3584 CA Utrecht
The Netherlands

ABSTRACT. To set the stage for the symposium, a brief overview is given of the main thrust of the about 1250 papers on Wolf-Rayet star research published in the past decade, broken down into inventory, basic parameters and galactic distribution, atmospheres, binaries, intrinsic variability, mass loss, enrichment and evolution.

1. Introduction

This year we celebrate the 123rd anniversary of the discovery by Wolf and Rayet (1876) of the stars bearing their names. Before, among those stars of which astronomers started to study their brightness with the help of a prism, only one case was known, *i.e.* γ Cas, of which the spectrum showed constantly bright lines. In their discovery paper Wolf and Rayet reported to *l' Académie* the existence of similar lines in three stars in the constellation of Cygnus, now known as WR134, WR135 and WR137, which distinguished themselves from neighbouring stars by their bright lines and yellow colours: pure yellow, orange yellow and greenish yellow, respectively. So far for the classification.

Since then about 2500 papers on Wolf-Rayet (WR) stars have been published, meaning a lot of papers under the bridge, but also many still outstanding. Those 2500 papers put together occupy less then three metres on a book shelf. Each of us makes a living by adding a few millimeters per year, some of us more than others. During the coming days we will add some three centimeters of literature to the subject, which is a fine prospect, because they will constitute a new milestone in the history of massive star research.

Figure 1 on the number of WR papers *vs.* time shows that since the early 60-s the production rate has increased sharply, and that half of the total number of 2500 WR papers has been published in past decade (van der Hucht 1990). Peaks in the distribution can readily be identified with conference years, or more often, the year after. In 1989 authors were apparently holding their breath for this symposium, which will give 1991 a kick-off with 134 papers.

In the past ten years a fair number of reviews and proceedings of conferences related to WR stars and other hot massive stars have appeared, and without claiming completeness we list here some of them:

1981, van der Hucht, Conti, Lundström & Stenholm, 'The Sixth Catalogue of Galactic Wolf-Rayet stars, their Past and Present'; Appendix: 'A bibliography on galactic Wolf-Rayet literature, 1867-1980'.

K. A. van der Hucht and B. Hidayat (eds.),
Wolf-Rayet Stars and Interrelations with Other Massive Stars in Galaxies, 19–36.
© 1991 *IAU. Printed in the Netherlands.*

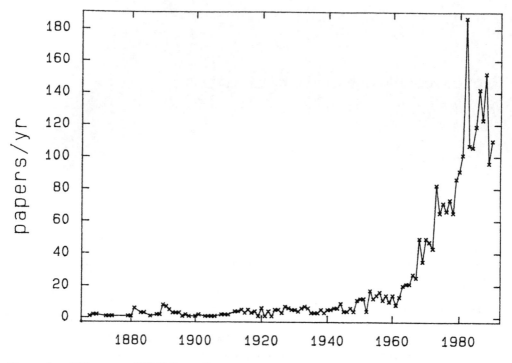

Figure 1.   123 years of Wolf-Rayet literature.

1981, D'Odorico *et al.* (eds.), Proc. ESO Workshop, 'The Most Massive Stars'.
1981, Chiosi & Stalio (eds.), Proc. IAU Coll. No. 59, 'Effects of Mass Loss on Stellar Evolution'.
1982, de Loore & Willis (eds.), Proc. IAU Symp. No. 99, 'Wolf-Rayet Stars: Observations, Physics, Evolution'.
1983, Lortet & Pitault (eds.), Proc. Paris Workshop, 'Wolf-Rayet Stars: Progenitors of Supernovae?'
1985, Massey, 'WR stars in Nearby Galaxies'.
1986, de Loore *et al.* (eds.), Proc. IAU Symp. No. 116, 'Luminous Stars and Associations in Galaxies'.
1986, Chiosi & Maeder, 'The Evolution of Massive Stars with Mass Loss'.
1987, Abbott & Conti, 'Wolf-Rayet Stars'.
1987, Willis & Garmany, 'Wolf-Rayet Stars'. (*IUE*)
1987, Lamers & de Loore (eds.), Proc. *Cornelis de Jager* Workshop, 'Instabilities in Luminous Early Type Stars'.
1988, Conti & Underhill (eds.), 'O Stars and Wolf-Rayet Stars'.
1988, Nugis & Pustyl'nik (eds.), Proc. El'va All-Union Conference, 'Wolf-Rayet Stars and Related Objects'.
1989, Davidson *et al.* (eds.), Proc. IAU Coll. No. 113, 'Physics of Luminous Blue Variables'.
1990, Garmany (ed.), Proc. Boulder-Munich Workshop, 'Intrinsic Properties of Hot Luminous Stars'.
1990, Willson & Stalio (eds.), Proc. Ames Workshop, 'Angular Momentum and Mass Loss for Hot Stars'.

Table 1. Wolf-Rayet literature 1980 – 1989: number of papers per topic

	1980	1981	1982	1983	1984	1985	1986	1987	1988	1989
basic parameters	14	9	16	14	13	11	13	12	9	6
atmospheres	17	11	61	21	16	28	20	13	26	21
variability	1	2	4	2	4	6	16	26	16	12
binaries	18	15	35	14	18	14	16	9	27	12
mass loss	5	5	14	4	2	9	15	12	13	7
shells, nebulae	10	17	17	14	14	11	12	8	16	9
extragalactic	-	6	11	9	7	8	10	9	7	3
evolution	15	26	24	24	29	29	34	29	26	25
interstellar	7	6	1	2	3	1	1	-	2	-
review or book	4	4	4	3	-	2	4	5	10	1
total	91	101	187	107	106	119	142	123	152	96

The distribution of papers over the various topics as listed in Table 1 has generally been stable, with always the largest fraction devoted to evolutionary studies. But there are certain features. *E.g.*, 1987 shows a sudden peak in variability studies (a topic of increasing interest in the 80-s) at the expense of studies of (alleged) binaries, though the next year binaries are up again. Extragalactic studies seem to be reaching, for the time being, technical limits at the end of the 80-s.

To set the stage for the symposium, we briefly overview the *status quo* of the various topics.

2. Inventory

In the Galaxy the WR census by van der Hucht *et al.* (1988) has been complemented with the discovery of one new WR star, listed by Thé (1964) as planetary nebula Th3-28, but classified as a new WN2.5-3 star without a nebula by Acker & Stenholm (1990); with the discovery by Cohen *et al.* (1991 and these proceedings) of IRAS 17380-3031 as a new WC 9 star; and with the discovery of 13 new WR stars (11 WN and 2 WC) by Shara *et al.* (these proceedings). This brings the number of known galactic WR stars to 172, among which there are 91 WN, 6 WN/WC (Conti & Massey 1989), 73 WC and 2 WO (Barlow & Hummer 1982) stars.

For the LMC, Lortet (these proceedings) refers to 27 new WR stars since the census by Breysacher (1981). Excluding the 12 Of/WN stars she lists, this brings the number of LMC WR stars up to 115. Of particular importance among the new LMC WR stars is one WC9-type (Heydari-Malayeri *et al.* 1990 and these proceedings), a subtype which was hitherto unknown in the LMC.

In the SMC no new WR stars have been found since the census by Azzopardi & Breysacher (1979), keeping the number at 8, among which there are 7 WN and 1 WO stars.

A new topic of the past decade form the numerous discoveries of extragalactic WR stars, in Table 2 broken down into the Local Group and beyond. Only in Local Group galaxies WR stars have been studied individually, as reviewed, *e.g.*, by Massey (1985). In more remote galaxies, primarily irregular dwarf galaxies and blue compact emission line galaxies,

Table 2. Wolf-Rayet stars in galaxies

Local Group galaxies	d (kpc)	type	N(WR)	subtype (WC/WN)	References
Milky Way		Sb/c	172	0.77	14,31,62
LMC	52	Irr I	115	0.26	6,29,43,53
SMC	63	Irr I	8	0.14	5
NGC 6822	470	Irr	2	WN	2,7,16,49,51,58,64,67
M31 = NGC 224	670	Sb	46	2.8	3,16,32,47,49,51,52,61,64
— NGC 206			11		47,52
M33 = NGC 598	730	Scd	96	0.7	7,8,16,17,28,42,45,48,49,59,64,66,68
— NGC 588					15,58
— NGC 592					15,58
— NGC 595			9		15,25,58
— NGC 604			50	WN7	13,15,21,22,25,30,54,57
— IC 132				WNE	9,23
IC 1613	740	Irr I	1	WN	2,7,16,18,24,42,49,64
beyond the Local Group	d (Mpc)	type	N(WR)	subtype (WC/WN)	References
C1148-203		BC	10000	WNE	11,12
ESO 148-IG02	200	pec	many	WNL	35
Haro 2		BC			44
He2-10 = ESO 495-G21	1.5	Irr II	3300	2.2	1,22,33,34,38,54,69
IC 4662		Irr			59
IRAS 01003-2238	470	FIR	100000	WNL	4
ISZ 59	25	BC			38
M83 = NGC 5236	3.2	Sbc	180	WN,WC	59,60
M101 = NGC 5457	3.8	Scd			
— NGC 5447					22,58
— NGC 5455					22,58
— NGC 5461				WN6-7	22,55,58
— NGC 5471					41,44,58,63
— Ho 40				WN7	22
— Searle 12					22
Mkn 33	24	BC			38
Mkn 59		BC			44
Mkn 309	160	BC	50000	1.5	38,54
Mkn 724	18	BC	120	WN4-5	40
Mkn 750	11	BC			38
Mkn 1236	26	BC	900		40
NGC 55	2.3	Sd-Irr			59
NGC 300	1.4	Sc			19,22,47,59
— #2				WC5	22
— #5					22
— #7				WN	22
— #13					22
— #24				WC	19
— #53c				WN	19
— #77				WN	19
— #137D				WC	19
NGC 1741 = Mkn 1089	58	pec	19500	WN	40
NGC 3049 = Mkn 710	20	SBbp	15700	WN	40,44
NGC 3125 = Tol 3	17	Sa	300-10000	WN	12,38,39,44,54,58

Table 2 (cont'd)

beyond the Local Group	d (Mpc)	type	N(WR)	subtype (WC/WN)	References
NGC 4038		Sd-Irr			59
NGC 4214		Irr			44
NGC 4216		Sb			59
NGC 4385 = Mi 499 = Mkn 52	50	SBbc	1300/4200	WN/WC8	11,27
NGC 4670		Irr			44
NGC 4861	9	Irr			20
NGC 5068		Sc			59
NGC 5128 = Cen A	4.4	SO			50,58,59
NGC 5253		E-SO	10000	WNE	11,12,59
NGC 5398: Tol 89	20	Sd	180	WN7-8	26,38
NGC 5430 = Mkn 799	60	SBb	30000		36,37
NGC 6764	480	Sey	3000	2.0	38,54
Pox 4	51	BC			38,39
Pox 120	89	BC			38
Pox 139	30	BC			38
Pox 186	15	BC			38
Tol 2	15	BC			38
Tol 9	52	BC		WN,WC	40
Tol 35 = T1324-276	24	BC		WNE	11,38
IZw18	10	BC			41,58
IIZw33	39	BC			38
IIZw40		BC			39
IIZw62	56	BC			38
IIZw187	169	BC			38
IIIZw107S	79	BC			38
VIIZw19	69	BC			38
VIIZw156NE = Mkn 8	49	BC			38

References: 1. Allen *et al.* 1976; 2. Armandroff & Massey 1985; 3. Armandroff *et al.* 1986; 4. Armus *et al.* 1988; 5. Azzopardi & Breysacher 1979; 6. Azzopardi & Breysacher 1985; 7. Azzopardi *et al.* 1988; 8. Bohannan *et al.* 1985; 9. Boksenberg *et al.* 1977; 10. Breysacher 1981; 11. Campbell & Smith 1986; 12. Campbell *et al.* 1986; 13. Clayton 1988; 14. Cohen *et al.* 1991 and these proceedings; 15. Conti & Massey 1981; 16. Conti 1988; 17. Corso 1975; 18. Davidson & Kinman 1982; 19. Deharveng *et al.* 1988; 20. Dinerstein & Shields 1986; 21. D'Odorico & Rosa 1981; 22. D'Odorico *et al.* 1983; 23. D'Odorico & Benvenuti 1983; 24. D'Odorico & Rosa 1982; 25. Drissen *et al.* 1989a; 26. Durret *et al.* 1985; 27. Durret & Tarrab 1988; 28. Freedman 1985; 29. Heydari-Malayeri *et al.* 1990; 30. Hippelein & Fried 1984; 31. van der Hucht *et al.* 1988; 32. Hutchings *et al.* 1987; 33. Hutsemekers & Surdej 1984; 34. Johansson 1987; 35. Johansson & Bergvall 1988; 36. Keel 1982; 37. Keel 1987; 38. Kunth & Joubert 1985; 39. Kunth & Sargent 1981; 40. Kunth & Schild 1986; 41. Lequeux *et al.* 1981; 42. Lequeux *et al.* 1987; 43. Lortet, these proceedings; 44. Mas-Hesse & Kunth, these proceedings; 45. Massey & Conti 1983; 46. Massey 1985; 47. Massey *et al.* 1986; 48. Massey *et al.* 1987b; 49. Massey *et al.* 1987a; 50. Möllenhoff 1981; 51. Moffat & Shara 1983; 52. Moffat & Shara 1987; 53. Morgan & Good 1990; 54. Osterbrock & Cohen 1982; 55. Rayo *et al.* 1982; 56. Rosa & D'Odorico 1982; 57. Rosa & Solf 1984; 58. Rosa *et al.* 1984; 59. Rosa & D'Odorico 1986; 60. Rosa & Richter 1988; 61. Shara & Moffat 1986; 62. Shara *et al.*, these proceedings; 63. Skillman 1985; 64. Smith 1988; 65. Terlevich & Melnick 1985; 66. Wampler 1982; 67. Westerlund *et al.* 1983; 68. Wray & Corso 1972. 69. Zinnecker, 1987.

only integral WR features are found, though sometimes seen grouped into individual (giant) *HII* regions. The inferred numbers of WR stars are based on, given the distance of a galaxy, the amount of energy in its *HeII* λ4686 emission line compared to that of a single (galactic) WR star. Properties of extragalactic WR features will be discussed in section 9.

3. Basic parameters and galactic distribution

Intrinsic parameters of Wolf-Rayet stars, like intrinsic colours, colour excesses and absolute visual magnitudes, have been derived and discussed by Hidayat *et al.* (1982; 1984), Lundström & Stenholm (1984a), Massey (1984), van der Hucht *et al.* (1988), Torres-Dodgen & Massey (1988) and Vacca & Torres-Dodgen (1990). New methods, based on emission line strengths and ratios, have been developed by Conti & Massey (1989), Conti & Morris (1990) and Smith *et al.* (1990) to determine absolute visual magnitudes of WN stars, interstellar reddening of WN stars and distances of WC stars, respectively.

Hidayat *et al.* (1982, 1984) and van der Hucht *et al.* (1988) used available filter photometry of Smith (1968a) and Lundström and Stenholm (1979, 1984a), which has some emission line contamination in particular for WC stars. They derived colour excesses, intrinsic colours and absolute visual magnitudes for the Galaxy only, adopting distances and colour excesses found in studies of WR stars in galactic open clusters and associations by Lundström & Stenholm (1984a) as the basis of their calibration. In spite of the limited number of WR stars in galactic clusters and associations (42, distributed over 18 WR subtypes), van der Hucht *et al.* (1988) warn against *a priori* mixing of galactic and LMC basic parameters. Their fear is confirmed by Hamann (these proceedings), who finds from quantitative spectroscopy that LMC WN stars have on the average lower luminosities than their galactic counterparts.

Massey (1984) and Torres-Dodgen & Massey (1988) were able to derive synthetic line-free photometry from spectrophotometry. In addition, Vacca & Torres-Dodgen (1990) derived colour excesses by nulling the 2200Å feature in low resolution *IUE* spectra of galactic and LMC WR stars. The latter averaged the parameters found per subclass in the Galaxy (based on nulling for the colour excesses and on cluster and association distances for the absolute visual magnitudes) and in the LMC (based on nulling and on the LMC distance).

Studies of the galactic distribution of WR stars as a diagnostic for the determination of their initial masses and (subtype) evolution have been attempted by Maeder *et al.* (1980), Hidayat *et al.* (1982), Meylan & Maeder (1983) and Hidayat *et al.* (1984) on the basis of the galactic data of van der Hucht *et al.* (1981); by Conti *et al.* (1983) based on the LMC data of Prévot-Burnichon (1981); by van der Hucht *et al.* (1988) using galactic data only; and by Conti & Vacca (1990) mixing galactic and LMC data. In all those studies the *overall* galactic WR distributions found agree remarkably well among each other, in spite of the different data sets and methods used. Notably the concentration of WR stars in galactic spiral arms and their absence toward the galactic anti-center (Orion spur) as known since the work by Roberts (1962), and the concentration of late-WC stars toward the galactic center as known since the work by Smith (1968b), are confirmed by the recent studies.

Conti & Vacca (1990), using WR absolute visual magnitudes determined by Vacca & Torres (1990) by averaging galactic data (from cluster and association distances) and LMC data, *re-calculated* the distances of the galactic WR stars first used as members of open clusters and associations, which constitute about 50% of the number of WR stars within 2.5 *kpc* from the Sun. This inconsistency causes the main discrepancy between their WR galactic distribution and the one found by van der Hucht *et al.* (1988), the latter showing better agreement in subtype ratios *vs.* galactocentric distance with evolutionary calculations of Maeder (1990b and these proceedings). Comparison between the two observational studies also indicates that distances of WR *field* stars are presently determinable with an accuracy of only about 50%.

4. Atmospheres

Spectrophotometric atlases of WR stars have become available from the ultraviolet to the infrared. Ultraviolet (*IUE*; no single observatory ever had such a profound effect on hot star research) low and high resolution atlases have been published by Nussbaumer *et al.* (1982) and Willis *et al.* (1986), respectively, atlases of low resolution *IUE* spectra combined with optical spectra have been given by Smith and Willis (1983) for LMC WR stars, and by Garmany *et al.* (1984) for galactic WR stars. Optical spectral atlases have been presented by Smith & Kuhi (1981), by Sivertsen (1981), by Lundström & Stenholm (1984b,c), by Jeffers & Weller (1985), by Torres & Conti (1984) for WC9 stars, by Torres & Massey (1987) for WC stars, and by Lundström & Stenholm (1989). Near-infrared spectral atlases have been published by Vreux *et al.* (1983), Vreux *et al.* (1989), Conti *et al.* (1990) and Vreux *et al.* (1990).

Infrared photometric fluxes originating in WR stellar winds or, for the late-WC types, in their circumstellar dust shells are discussed in sections 7 and 8, respectively. *Free-free* radio and normal X-ray fluxes are related to stellar winds in section 7, while non-thermal radio and strong X-ray fluxes are related to colliding winds in binaries in section 5.

The development of model atmospheres for WR stars has taken great strides in recent years, as demonstrated in reviews by Hamann *et al.* (1990), Hillier (1990) and Schmutz (1990), in turn reviewed by Kudritzki & Hummer (1990). Multilevel non-LTE codes have been developed for hot stratified expanding pure helium model atmospheres by Hamann & Schmutz (1987), Hamann *et al.* (1988), Schmutz *et al.* (1989) and Hillier (1987a,b, 1988, 1989), the latter two papers also including nitrogen and carbon. Those models, using the Sobolev approximation and treating line and continuum formation simultaneously, are able to calculate continuum energy distributions, line-strengths and line-profiles for WN and WC stars, which match observations to a satisfactory degree of accuracy. They prove that the WN and WC sequences arise from different surface abundances of *He* and *CNO*, thus confirming the suggestion of Gamow (1943) that WN and WC stars display at their surfaces products of different phases of thermo-nuclear processing.

Typical number abundance ratios derived from those detailed model atmospheres in comparison with observations are $N/He \approx 4 \times 10^{-3}$ and $C/N \approx 0.07$ for the WN5 star WR6 (Hillier 1988) and $C/He > 0.1$ for the WC5 star WR111 (Hillier 1989). Smith & Hummer (1988) and Smith & Maeder (1990) find a continuous increase in the $(C+O)/He$ number ratio for the sequence WCL $\rightarrow$ WCE $\rightarrow$ WO, ranging from 0.03 to > 1.

The effective temperatures found by comparison between pure helium model atmospheres and observations range from 30 to 35 kK for WNL and WC stars, and from 35 to 90 kK for WNE stars (Schmutz *et al.* 1989), while higher temperatures for WC stars are found when carbon is introduced (Hillier 1989). Such temperatures have been confirmed independently for WR stars with ring nebulae (Rosa & Mathis 1990). Corresponding bolometric corrections range from -3 *mag* for model atmospheres with some hydrogen content to -5 *mag* for model atmospheres without hydrogen. This corresponds reasonably well with average B.C.

values of -4.2 *mag* derived by van der Hucht *et al.* (1988) and of -4.5 *mag* derived by Smith & Maeder (1989) from studies of WR binaries in open clusters and associations, inferring their luminosities from their observed masses and adopted theoretical M-L relations for evolved massive stars, *e.g.*, by Maeder & Meynet (1987). These results lend additional confidence to the reliability of the model atmospheres and the interior models used, to the WR cluster/association memberships adopted, as well as to the reality that almost all WR stars are hydrogen deficient.

At variance with the results mentioned above, which demonstrate that WR stars are evolved massive stars with larger-than-cosmic abundances of helium and nitrogen in the atmospheres of WN stars and of helium, carbon and oxygen in the atmospheres of WC stars, are the conclusions of Underhill (1990 and references therein). She favours cosmic abundances and a pre-MS evolutionary status for WR stars, and argues that the observed differences in line strength of nitrogen and carbon lines in WN and WC stars are caused by different wind densities and electron temperatures and not by differences in surface abundances. Her one-representative-point models have been refuted, *e.g.*, by Nugis (these proceedings) who by applying her models found discrepancies of an order of magnitude between calculated and observed *CIII* and *CIV* line ratios. Arguments proving the evolved evolutionary status of WR stars, like: *(i)* the lack of hydrogen in most WR stars as found in their Pickering decrement (Smith 1973); *(ii)* the WR atmospheric abundance determinations mentioned above and the high nitrogen content of WR ring nebulae (*e.g.*, Rosa & Mathis 1990); *(iii)* the abundance predictions by interior models of, *e.g*, Maeder & Meynet (1987) and Maeder (1990b), which also match the observed WR/O star number ratios in the Galaxy (van der Hucht *et al.* 1988); and *(iv)* the fact that some WR binaries have evolved companions; have been reiterated by Lamers *et al.* (1990). Even more compelling arguments in favour of the evolved evolutionary status of WR stars are *(v)* the continuous carbon dust formation around WC stars *only* (Williams *et al.* 1987); and *(vi)* a fractional abundance $n_C \approx 0.06$ derived from the X-ray spectrum of the WC7 binary WR140 (Williams *et al.* 1990a), further discussed in section 5.

5. Binaries

After the numerous studies of WR binaries in the early 80-s by Massey (1981) and co-workers, the main breakthrough has been the determination of orbital inclinations by Moffat and co-workers from linear polarization modulation measurements, as summarized by St.-Louis *et al.* (1988), Moffat (1988) and Schulte-Ladbeck & van der Hucht (1989). Resulting masses listed by those authors range from 6 to 40 $M_\odot$ for WN stars and from 5 to 20 $M_\odot$ for WC stars in binaries. Other new studies of established binaries (*i.e.*, those with radial velocity solutions) have been summarized, *e.g.*, by Smith & Maeder (1989) for the Galaxy and by Moffat *et al.* (1990) for the LMC and SMC.

WR stars which may be accompanied by a compact companion left after a supernova explosion of the primary, *cf.* the evolutionary scheme of van den Heuvel (1976), are listed by Hidayat *et al.* (1984). Their number has dwindled somewhat since, where the interest has drifted from compact companions to intrinsic variability (*cf.* Vreux 1985). Among the dozen or so left, the two most likely candidates are still WR6 (WN5 + neutron star?, P = 3.76 d) (Firmani *et al.* 1980; Drissen *et al.* 1989b; van der Hucht *et al.* 1990) and WR148 (WN7 + black hole?, P = 4.32 d) (Drissen *et al.* 1986).

The binary frequency in the solar neighborhood was found to be 37% (van der Hucht *et al.* 1988), and this number is still increasing. A fascinating but time consuming new way of detecting binaries has been discovered by Williams *et al.* (1985, 1990a,b, these proceedings) in the case of long period (P $\approx$ 10 *yr*) eccentric WC binaries with periodic dust formation (section 8). Because of their relatively large component separations, these binaries can also be discovered as strong X-ray and non-thermal radio sources, which emissions are believed to be associated with their colliding winds (Williams *et al.* 1990a). From their X-ray luminosity and non-thermal radio characteristics, also possible WN binaries with large separations may be discovered, for which WR25 (WN7+a) (Pollock 1987 and these proceedings) and WR147 (WN7) (Caillaut *et al.* 1985; Moran *et al.* 1989) are likely candidates.

Studies of colliding winds have also been performed for the WR binary V444 Cyg (WR139, WN5+O6, P = 4.21 *d*) by Shore & Brown (1988), explaining the variability of its UV P-Cygni profiles in terms of shock dominated wind-wind interactions, and by Luo *et al.* (1990) and Usov (1990), addressing its X-ray luminosity modulation.

New short-period binaries (P < 1 *d*) may be discovered from high time-resolution photometric observations as demonstrated, *e.g.*, by van Genderen *et al.* (1990 and these proceedings), although also intrinsic variability may play a role here (section 6).

6. Intrinsic variability

Interior models of WR stars indicate that the fundamental radial mode could be excited due to nuclear-burning instabilities (Maeder 1985; Cox & Cahn 1988) with periods of about 1 to 2 *hr*. Non-radial modes, which have longer periods, have found to be generally stable. Vreux (1985, 1987) attacked the credibility of published periods of emission-line profile variations of alleged WR+c systems, and suggested the observed variability to be due to single-star non-radial pulsations. In order to find radial or non-radial pulsations in WR stars, observations have to be of a quality which allows searches for periods of less than one day and for multiple periodicities.

From intensive photometry encouraging new results have become available. *E.g.*, Taylor *et al.* (1988) and Taylor (1990) find a P = 1.26 *d* in γ^2 Vel (WC8+O9I); Gosset & Vreux (1990) find P_A = 6.25 *d* and P_B = 2.5 *d* in WR40 (WN8); Gosset *et al.* (1990) find P_A = 9.09 *d*, P_B = 15.15 *d*, and P_C = 5.46 *d* in WR16 (WN8); and van Genderen *et al.* (1990 and these proceedings) find tentatively P = 0.28 *d* for WR46 (WN3p), P = 1.06 *d* for WR50 (WC6+a), P = 2.44 *d* for WR55 (WN7), and P = 1.94 *d* for WR123 (WN8); but for each of these stars duplicity as the cause of the observed variability can not be excluded. Also Balona *et al.* (1989) present a number of cases.

From intensive spectroscopic observations of line-profile variations, *e.g.*, Vreux *et al.* (1985) find P_A = 0.45 *d* and P_B = 0.31 *d* in WR136 (WN6+c?); for the same object St.-Louis *et al.* (1989) find P $\approx$ 1 *d* from *IUE* high resolution monitoring; and Willis *et al.* (1989) find P $\approx$ 1 *d* for WR6 (WC5+c?), also with *IUE*.

Variability indicative of blob ejection in the winds of WR stars was suggested first from photometry by, *e.g.*, van Genderen *et al.* (1987) and has subsequently been confirmed in high-time- and high-spectral resolution spectroscopy by Moffat *et al.* (1988), McCandliss (1988) and Robert & Moffat (1990). Linear polarization variability, possibly also related to blob ejection, has been reported by Moffat (1988, and references therein) and Schulte-Ladbeck & van der Hucht (1989), and appears to be largest in the late-WR subclasses.

7. Mass Loss

Wind characteristics and mass loss rates of WR stars have been elaborated on by Willis (1981), updated by Willis & Garmany (1987 and references therein). Mass loss rates can be determined for many WR stars from IR (Barlow *et al.* 1981) and radio (Abbott *et al.* 1986) *free-free* measurements for the electron densities, from UV P-Cygni profiles for the wind terminal velocities (Willis & Garmany 1987), and from their distances, applying the theory of spherically symmetric mass loss by Wright & Barlow (1975).

In the past years such determinations have been affected by the following realizations: *(i)* a high carbon abundance and a low ionization leads to a higher mass loss rate, especially for late-type WC stars (van der Hucht *et al.* 1986); *(ii)* an outward decreasing ionization degree leads to a higher mass loss rate (Schmutz & Hamann 1986; Hillier 1987a); *(iii)* improved distances (*e.g.*, by van der Hucht *et al.* 1988) are of importance because $\dot{M} \propto d^{3/2}$; and *(iv)* $CIV\lambda1550$ edge velocities yield overestimates of terminal velocities (Williams & Eenens 1989; Prinja *et al.* 1990), leading to lower mass loss rates. Taking those effects into account, Prinja *et al.* (1990) derive mass loss rates for WR stars in the range $\dot{M} = [2 - 10] \times 10^{-5} \ M_\odot \ yr^{-1}$.

A new method to derive the electron density in the winds of WR binaries from polarization modulation measurements has been developed by St.-Louis *et al.* (1988), confirming the electron densities derived from radio *free-free* measurements. Since the same observations yield also orbital inclinations (section 5), together they allow a consistent observational determination of a relation between mass and mass loss. An overall dependence $\dot{M} \propto M^{1-2}$ seems compatible with most of the data (St.-Louis *et al.* 1988).

An independent method to find mass loss rates of WR binaries from observed period changes has been developed and applied to V444 Cyg (WN5+O6) by Khaliullin (1974) and has been improved by Khaliullin *et al.* (1984). The latter determined that $\dot{M} = 1.0 \times 10^{-5} \ M_\odot \ yr^{-1}$, whereas Prinja *et al.* (1990) found for the same star $\dot{M} = 2.4 \times 10^{-5} \ M_\odot \ yr^{-1}$ by applying the Wright & Barlow method. From her own data set of V444 Cyg, Underhill *et al.* (1990) find a 60% smaller period change and a 50% smaller mass loss rate than found by Khaliullin *et al.* (1984). Since this method has been applied up to now for just one WR binary, its results can not be generalized to be representative for all WR stars.

WR stars have the largest constant mass loss rates of all stellar classes. A problem not solved in the 80-s is that of the WR wind momentum, conveniently expressed in the ratio $\eta = (\dot{M}v_\infty)/(L_*/c)$. Although the problem has come down from $\eta = 5 - 50$ (Barlow *et al.* 1981) to $\eta = 1 - 30$ (Willis, these proceedings), it is not clear which mechanism, apart from radiation pressure, drives the WR stellar winds. A possible way out has been offered by Poe *et al.* (1989), involving effects of rotation, differences in equatorial and polar outflows and open magnetic fields, but needs further development and observational evidence. Another solution, involving a pure radiation driven wind, is suggested by Bandiera & Turolla (1990), who require a $\dot{M} \propto M^{2.3}$ dependence.

8. Enrichment

Enrichment by WR stars of the interstellar medium in energy and mass has been discussed by Abbott (1982) and was up-dated by van der Hucht *et al.* (1986). During its lifetime a WR star emits an amount of energy comparable to that released during a supernova event.

Although, within 3 *kpc* from the Sun, they comprise less than 5% of the total number of early-type stars having stellar winds (B- and A-type supergiants, O-type and WR stars), WR stars provide more than half of the mass and wind energy input into the interstellar medium.

Infrared photometric studies of WR stars since Allen *et al.* (1972) have demonstrated that many late-type WC stars have dust characteristics on top of their *free-free* energy distributions. Energy distributions combining groundbased IR photometry and *IRAS*-LRS spectra of late-type WC stars are presented by van der Hucht *et al.* (1985). Williams *et al.* (1987) show that 85% of the WC9 stars and 50% of the WC8 stars have heated ($T_d \approx 1300K$) circumstellar amorphous carbon dust (soot), which is being formed continuously at typical radii of $\sim 350\ R_*$ for WC9 stars and $\sim 800\ R_*$ for WC8 stars, *i.e.*, within their stellar winds. They argue that a critical wind density $\rho \approx 8 \times 10^{-18}\ gcm^{-3}$, corresponding to $n \approx 6 \times 10^5\ cm^{-3}$, needs to be exceeded for dust to form at distances from a central WC star where its radiation field is sufficiently diluted to allow dust formation. This explains why stars earlier than WC8 generally do not show dust formation: their wind densities at radii where their radiation fields would generally allow dust formation are below the critical wind density given above. Exceptions are provided by some long period binaries with eccentric orbits, like the WC7 binaries WR137 (Williams *et al.* 1985) and WR140 (Williams *et al.* 1990a), the possible WC8 binary WR48a (Williams *et al.* and these proceedings) and the possible WC4 binary WR19 (Williams *et al.* 1990b and these proceedings). In those cases local density increments during periastron passage are believed to give rise to periodic *c.q.* episodic dust formation. In WC9 stars with very pronounced dust characteristics, the 7.7 *μ*m PAH emission feature, attributed to CC stretching modes in carbonaceous materials, has been detected (Cohen *et al.* 1989).

The formation of amorphous carbon in the winds of late-type WC stars is a clear indication of the larger-than-cosmic carbon abundance of WC stars, and can directly be compared to cases of carbon dust formation in other evolved objects like novae (Sedlmayr & Gass, these proceedings).

Some 10% of the known galactic and LMC WR stars are surrounded by shell nebulae, more often called ring nebulae, (semi-)spherical *HII* regions with radii of the order of parsecs, as a manifestation of past mass release of their central stars, be it steady wind or episodic events. Reviews have been given by Chu *et al.* (1983), Rosado (1986), Dufour (1989) and Smith (1990). WR ring nebulae closely resemble the symmetric *HII* regions seen around some Luminous Blue Variables (LBV), *e.g.*, AG Car (Dufour 1989), and those seen around some Of stars (Lozinskaya 1982; Rosado 1986), in one case (NGC 6164/5) being bipolar (Leitherer & Chavarria 1987; Dufour *et al.* 1988). It is well possible that at least the ejecta-type WR ring nebulae were created during a pre-WR (*e.g.*, LBV) phase.

From their morphology, WR ring nebulae have been classified W-type (stellar wind blown bubble), E-type (stellar ejecta), R_s-type (shell structured *HII* region), and R_a-type (amorphous *HII* region). Most W-type nebulae are associated with WNE stars; most R_a-type nebulae surround WNL stars; most E-type nebulae surround WN8 stars; and most R_s type nebulae surround WC stars (Chu *et al.* 1983).

Matter in WR ring nebulae can be a combination of (pre-WR) stellar wind *c.q.* ejecta and swept-up interstellar matter. Since Kwitter (1981, 1984), as confirmed again by Rosa & Mathis (1990), it is known that WR ring nebulae are chemically enhanced in helium and nitrogen, revealing the larger-than-cosmic abundances of the (pre-)WR wind and ejecta. Observed ionization ratios in the nebulae yield effective temperatures for the central stars (Rosa & Mathis 1990) which are in good agreement with those found by comparing WR spectra with model atmospheres (Schmutz *et al.* 1989). *IRAS* imagery shows that the dust in WR ring nebulae has a bi-modal composition, *i.e.*, consisting mainly of large cool particles (Van Buren & McCray 1988), but mixed with less cool small particles (Cassinelli *et al.*, these proceedings). Recent ring nebula studies concentrate on their dynamics and energetics, *e.g.*, Chu (1988) and Smith *et al.* (1988).

Because of their production of heavy elements, WR stars may to a certain degree be reponsible for anomalous *C*, *Ne* and *Mg* isotope ratios found in cosmic rays and certain meteorites (Maeder 1983). The importance of a possible WR contribution to the amount of ^{26}Al near the galactic center, derived from the intensity of the observed 1.81 MeV γ-ray emission observed in that direction, is still a matter of debate, *cf.* Blake & Dearborn (1989) and Signore & Dupraz (1990).

9. Evolution

Since the review of Chiosi & Maeder (1986) on the evolution of single massive stars and the productions of WR stars, arguing for the evolutionary sequences

O → Of → BSG and LBV → WR → SN   for $M_i > 60 \ M_\odot$ and
O → BSG → RSG → WR → SN   for $25 \ M_\odot < M_i < 60 \ M_\odot$

(the supernovae most likely being of Type Ib, *cf.* Ensman & Woosley 1988), and since the comparative study of de Loore (1988), new major evolutionary studies have appeared, most recently those of Langer (1989a,b, 1990) and Maeder (1990a,b). The authors differ in that the former applies mass loss and semi-convection (the effect of molecular weight gradients on convection) as mixing process to bring nuclear burning products to the surface, while the latter relies on mass loss and convective overshooting (the Schwarzschild criterion and an increase of the convective core size). Both authors agree on the theoretical necessity of a $\dot{M} \propto M^\alpha$ dependence, with $\alpha = 2.5$ for WNE and WC stars and $\alpha = 0$ for WNL stars.

Maeder (1990a,b) calculates grids of evolutionary models for various values of the metallicity Z, notably those found in the Galaxy and the Magellanic Clouds, and scales $\dot{M} \propto Z^{0.5}$. He finds, depending on the value of Z, WR lifetimes of the order of 5×10^5 yr and initial masses M_i as low as 20 $M_\odot$, to be compared with the observed minimum value of 25 $M_\odot$ found by van der Hucht *et al.* (1988). WR/O, WC/WR and WC/WN number ratios following from Maeder's models compare remarkably well with the ratios observed in the Galaxy by van der Hucht *et al.* (1988), in the LMC (Azzopardi & Breysacher 1985), and in M31 and M33 (Smith 1988). The comparison also indicates that in regions with lower Z an increasing fraction of the observed WR stars has to originate from another 'channel' of WR formation, *e.g.*, from binary evolution. The observed WN/WC ratio in the LMC requires $\dot{M}$-rates which are about 50 % smaller than those in the solar neighborhood.

Maeder's model data confirm the overall general evolutionary sequence WNL → WNE → WCL → WCE → WO. Explicitly for binaries, studies by Moffat et al. (1990 and references therein) also point toward a continuous WCL → WCE/WO subtype evolution. Of the allowed subtype-evolution trails, based on the observed WR galactic distribution (van der Hucht *et al.* 1988), *i.e.*:

• at galactocentric radius $R < 8.5 \ kpc$: WNL → WCL

- at $R > 6.5\ kpc$: WNL $\rightarrow$ WCE $\rightarrow$ WO
- and in general: WNE $\rightarrow$ no WC stars,

the first two trails agree very well with models with high M_i and low Z values, respectively. The third trail corresponds only to lower M_i at low Z, but may be somewhat relaxed since WNE stars and WCE stars do share some common galactocentric distances (Maeder 1990b). In spite of the different physics, the results of the models by Langer agree generally very well with those Maeder, both predicting similar surface abundances and, *e.g.*, also the ~ 3 % fraction of WN/WC 'transition' stars classified by Conti & Massey (1989).

Evolutionary models of massive close binaries including mass transfer and convective core overshooting, linking WR binaries with O-type binaries and X-ray binaries, have been calculated by, *e.g.*, Doom & De Greve (1983), De Greve & Doom (1988), and De Greve *et al.* (1988), Although direct observational evidence of mass transfer is lacking, observed mass ratios and the non-evolved luminosity class of O-type components in some WR binaries point to some degree of rejuvenation by mass transfer. Even periodic mass transfer seems required, to explain the existence of the very eccentric ($e = 0.84$) WC7+O4 binary WR140 (De Greve, these proceedings). Duplicity constituting somehow an additional incentive for WR formation, be it by mass transfer or by tidal mixing, is also needed to explain the observed number ratios in regions with low Z (Maeder 1990b). In a comparative study, Schulte-Ladbeck (1989) finds that present WR binary observations can better be reconciled with non-conservative models without convective core overshooting.

The indications for numerous WR stars in some irregular dwarf galaxies and blue compact emission line galaxies (see Table 2), as inferred from the intensity of their $HeII\lambda4686$ emission, provide convincing evidence for massive starburst events or intermittent short bursts. The latter, coined 'lazy' galaxies by Kunth & Joubert (1985), show WR stars in sites of higher metallicity and low effective temperature ($T < 35\ kK$). Kunth & Schild (1986) find further evidence for a correlation between the $\lambda4686$ luminosity and the metallicity of the nebular gas, in addition to a correlation between the $\lambda4686$ luminosity and the luminosity of the parent galaxy. Closer to Earth, a relationship between the WR star numbers and subclass distributions with metallicity in Local Group galaxies has been demonstrated by Smith (1988), making extragalactic WR research a powerful instrument for verification of the realism of evolutionary models.

Concluding remarks

All the above will be superseded this week by your own excellent reviews and contributions. It is with good confidence that I express the hope that this symposium will make proper use of the bonanza of observational and theoretical achievements of the past, and that we, on this inspiring island, together will come to new improvements and refinements of our understanding of those arresting physics laboratories, called Wolf-Rayet stars.

References

Abbott, D.C. 1982, *Astrophys. J.* **259**, 282.
Abbott, D.C., Bieging, J.H., Churchwell, E., Torres, A.V. 1986, *Astrophys. J.* **303**, 239.
Abbott, D.C., Conti, P.S. 1987, *Ann. Rev. Astron. Astrophys.* **25**, 113.

Acker, A., Stenholm B. 1990, *Astron. Astrophys. Suppl.* **86**, 219.

Allen, D.A., Swings, J.P., Harvey, P.M. 1972, *Astron. Astrophys.* **20**, 333.

Allen, D.A., Wright, A.E., Goss, W.M. 1976, *Monthly Notices Roy. Astron Soc.* **177**, 91.

Armandroff, T.E., Massey, P. 1985, *Astrophys. J.* **291**, 685.

Armandroff, T.E., Massey, P., Conti, P.S. 1986, in: C. de Loore, A.J. Willis, P. Laskarides (eds.): Luminous Stars and Associations in Galaxies, *Proc. IAU Coll. No. 116* (Dordrecht: Reidel), p. 239.

Armus, L., Heckman, T.M., Miley, G.K. 1988 *Astrophys. J.* (Letters) **326**, L45.

Azzopardi, M., Breysacher, J. 1979, *Astron. Astrophys.* **75**, 120.

Azzopardi, M., Breysacher, J. 1985, *Astron. Astrophys.* **149**, 213.

Azzopardi, M., Lequeux, J., Maeder, A. 1988, *Astron. Astrophys.* **189**, 34.

Balona, L.A., Egan, J., Marang, F. 1989, *Monthly Notices Roy. Astron. Soc.* **240**, 103.

Bandierra, R. Turolla, R. 1990, *Astron. Astrophys.* **231**, 85.

Barlow, M.J., Smith, L.J., Willis, A.J. 1981, *Monthly Notices Roy. Astron. Soc.* **196**, 101.

Barlow, M.J., Hummer, D.G. 1982, in: C. de Loore, A.J. Willis (eds.), Wolf-Rayet Stars: Observations, Physics, Evolution, *Proc. IAU Symp. No. 99* (Dordrecht: Reidel), p. 387.

Blake, J.B., Dearborn, D.S.P. 1989, *Astrophys. J.* (Letters) **338**, L17.

Bohannan, B., Conti, P.S., Massey,P. 1985, *Astron. J.* **90**, 600.

Boksenberg, Willis, A.J., Searle, L. 1977, *Monthly Notices Roy. Astron. Soc.* **180**, 15P.

Breysacher, J. 1981, *Astron. Astrophys. Suppl.* **43**, 209.

Caillault, J.-P., Chanan, G.A., Hefland, D.J., Patterson, J., Nousek, J.A., Takalo, L.O., Bathun, G.D., Becker, R.H. 1985, *Nature* **313**, 376.

Campbell, A.W., Smith, L.J. 1986, in: C. de Loore, A.J. Willis, P. Laskarides (eds.): Luminous Stars and Associations in Galaxies, *Proc. IAU Symp. No. 116* (Dordrecht: Reidel), p. 499.

Campbell, A.W., Terlevich, R., Melnick, J. 1986, *Monthly Notices Roy. Astron. Soc.* **223**, 811,

Chiosi, C., Stalio, R. (eds.) 1981, Effects of Mass Loss on Stellar Evolution, *Proc. IAU Coll. No. 59* (Dordrecht: Reidel).

Chiosi, C., Maeder, A. 1986, *Ann. Rev. Astron. Astrophys.* **24**, 329.

Chu, Y.-H., Treffers, R.R., Kwitter, K.B. 1983, *Astrophys. J. Suppl.* **53**, 937.

Chu, Y.-H. 1988, *Publ. Astron. Soc. Pacific* **100**, 986.

Clayton, C.A. 1988, *Monthly Notices Roy. Astron. Soc.* **231**, 191.

Cohen, M., Tielens, A.G.G.M., Bregman, J.D. 1989, *Astrophys. J.* (Letters) **344**, L13.

Cohen, M., van der Hucht, K.A., Williams, P.M., Thé, P.S. 1991, *Astrophys. J.* (Letters), in press.

Conti, P.S., Massey, P. 1981, *Astrophys. J.* **249**, 471.

Conti, P.S., Garmany, C.D., de Loore, C., Vanbeveren, D. 1983, *Astrophys. J.* **274**, 302.

Conti, P.S. 1988, in: V.M. Blanco, M.M. Phillips (eds.), Progress and Opportunities in Southern Hemisphere Optical Astronomy, *Proc. CTIO 25th Anniversary Symposium, A.S.P. Conf. Series Vol. 1*, p. 100.

Conti, P.S, Underhill A.B. (eds.) 1988, O Stars and Wolf-Rayet Stars, *NASA SP-497*.

Conti, P.S., Massey, P. 1989, *Astrophys. J.* **337**, 251.

Conti, P.S., Morris, P.W. 1990, *Astron. J.* **99**, 898.

Conti, P.S., Vacca, W.D. 1990, *Astron. J.* **100**, 431.

Conti, P.S., Massey, P., Vreux, J.-M. 1990, *Astrophys. J.* **354**, 359.

Corso, G.J. 1975, Thesis, Northwestern University.

Cox, A.N., Cahn, J.H. 1988, *Astrophys. J.* **326**, 804.

Davidson, K., Kinman, T.D. 1982, *Publ. Astron. Soc. Pacific* **94**, 634.

Davidson, K., Moffat, A.F.J, Lamers, H. (eds.) 1989, Physics of Luminous Blue Variables, *Proc. IAU Coll. No. 113* (Dordrecht: Kluwer).

De Greve, J.-P., Doom, C. 1988, *Astron. Astrophys.* **200**, 79.

De Greve, J.-P., Hellings, P., van den Heuvel, E.P.J. 1988, *Astron. Astrophys.* **189**, 74.

Deharveng, L., Caplan, J., Lequeux, J., Azzopardi, M., Breysacher, J., Tarenghi, M., Westerlund, B. 1988, *Astron. Astrophys. Suppl.* **73**, 407.

Dinerstein, H.L., Shields, G.A. 1986, *Astrophys. J.* **311**, 45.

D'Odorico, S., Baade, D., Kjär, K. (eds.) 1981, The Most Massive Stars, *Proc. ESO Workshop* (Garching: ESO).

D'Odorico, S., Rosa, M. 1981, *Astrophys. J.* **248**, 1015.

D'Odorico, S., Rosa, M. 1982, *Astron. Astrophys.* **105**, 410.

D'Odorico, S., Benvenuti, P. 1983, *Monthly Notices Roy. Astron. Soc.* **203**, 157.

D'Odorico, S., Rosa, M., Wampler, J. 1983, *Astron. Astrophys. Suppl.* **53**, 97.

Doom, C., De Greve, J.-P. 1983, *Astron. Astrophys.* **120**, 97.

Drissen, L., Lamontagne, R., Moffat, A.F.J., Bastien, P., Seguin, M. 1986, *Astrophys. J.* **304**, 188.

Drissen, L., Moffat, A.F.J., Shara, M.M. 1989a, in: K. Davidson *et al.* (eds.), Physics of Luminous Blue Variables, *Proc. IAU Coll. No. 113* (Dordrecht: Kluwer), p. 308.

Drissen, L., Robert, C., Lamontagne, R., Moffat, A.F.J., St-Louis, N., van Weeren, N., van Genderen, A.M. 1989b, *Astrophys. J.* **343**, 426.

Dufour, R.J., Parker, R.A.R., Henize, K.G. 1988, *Astrophys. J.* **327**, 859.

Dufour, R.G. 1989, in: S. Torres-Peimbert, J. Fierro (eds.), Star Forming Regions and Ionized Gas, *Proc. 2nd Mexico-Texas Conf. on Astrophysics, Rev. Mexicana Astron. Astrof.* **18**, 87.

Durret, F., Bergeron, J., Boksenberg, A. 1985, *Astron. Astrophys.* **143**, 347.

Durret, F., Tarrab, I. 1988, *Astron. Astrophys.* **205**, 9.

Ensman, L.M., Woosley, S.E. 1988, *Astrophys. J.* **333**, 754.

Firmani, C., Koenigsberger, G., Bisiacchi, G.F., Moffat, A. F.J., Isserstedt, J. 1980, *Astrophys. J.* **239**, 607.

Freedman, W.L. 1985, *Astron. J.* **90**, 2499.

Gamow, G. 1943, *Astrophys. J.* **98**, 500.

Garmany, C.D., Massey, P., Conti, P.S. 1984, *Astrophys. J.* **278**, 233.

Garmany, C.D (ed.) 1990, Intrinsic Properties of Hot Luminous Stars, *Proc. Boulder-Munich Workshop, A.S.P. Conf. Series Vol. 7.*

van Genderen, A.M., van der Hucht, K.A., Steemers, W.J.G. 1987, *Astron. Astrophys.* **185**, 131.

van Genderen, A.M., van der Hucht, K.A., Larsen, I. 1990, *Astron. Astrophys.* **229**, 123.

Gosset, E., Vreux, J.-M. 1990, *Astron. Astrophys.* **231**, 100.

Gosset, E., Vreux, J.-M., Manfroid, J., Remy, M., Sterken, C. 1990, *Astron. Astrophys. Suppl.* **84**, 377.

Hamann, W.-R., Schmutz, W. 1987, *Astron. Astrophys.* **174**, 173.

Hamann, W.-R., Schmutz, W., Wessolowski, U. 1988, *Astron. Astrophys.* **194**, 190.

Hamann, W.-R., Wessolowski, U., Schwarz, E., Dünnebeil, G., Schmutz, W. 1990, in: C.D. Garmany (ed.), Intrinsic Properties of Hot Luminous Stars, *Proc. Boulder-Munich Workshop, A.S.P. Conf. Series Vol. 7*, p. 259.

van den Heuvel, E.P.J. 1976, in: P. Eggleton, S. Mitton, J. Whelan (eds.), Structure and Evolution of Close Binary Systems, *Proc. IAU Symp. No. 73* (Dordrecht: Reidel), p. 35.

Heydari-Malayeri, M, Melnick, J., Van Drom, E. 1990, *Astron. Astrophys.* (Letters) **236**, L21.

Hidayat, B., Supelli, K., van der Hucht, K.A. 1982, in: C. de Loore, A.J. Willis (eds.), Wolf-Rayet Stars: Observations, Physics, Evolution, *Proc. IAU Symp. No. 99* (Dordrecht: Reidel), p. 27.

Hidayat, B., Admiranto, A.G., van der Hucht, K.A. 1984, in: B. Hidayat, Z. Kopal, J. Rahe (eds.), Double Stars: Physical Properties and Generic Relations, *Proc. IAU Coll. No. 80, Astrophys. Space Sci.* **99**, 175.

Hillier, D.J. 1987a, *Astrophys. J. Suppl.* **63**, 947.

Hillier, D.J. 1987b, *Astrophys. J. Suppl.* **63**, 965.

Hillier, D.J. 1988, *Astrophys. J.* **327**, 822.

Hillier, D.J. 1989, *Astrophys. J.* **347**, 392.

Hillier, D.J. 1990, in: C.D. Garmany (ed.), Intrinsic Properties of Hot Luminous Stars, *Proc. Boulder-Munich Workshop, A.S.P. Conf. Series Vol. 7*, p. 340.

Hippelein, H., Fried, J.W. 1984, *Astron. Astrophys.* **141**, 49.

van der Hucht, K.A., Conti, P.S., Lunström, I., Stenholm, B. 1981, *Space Sci. Rev.* **28**, 227.

van der Hucht, K.A., Jurriens, T.A., Olnon, F.M., Thé, P.S., Wesselius, P.R., Williams, P.M. 1985, in: W. Boland, H. van Woerden (eds.), Birth and Evolution of Massive Stars and Stellar Groups, *Proc. Symp. in honour of Adriaan Blaauw* (Dordrecht: Reidel), p. 167.

van der Hucht, K.A., Cassinelli, J.P., Williams, P.M. 1986, *Astron. Astrophys.* **168**, 111; **175**, 356.

van der Hucht, K.A., Hidayat, B., Admiranto, A.G., Supelli, K.R., Doom, C. 1988, *Astron. Astrophys.* **199**, 217.

van der Hucht, K.A. 1990, A Bibliography of Wolf-Rayet Literature 1980 - 1990, preprint.

van der Hucht, K.A., van Genderen, A.M., Bakker, P.R. 1990, *Astron. Astrophys.* **228**, 108.

Hutchings, J.B., Massey, P., Bianchi, L. 1987, *Astrophys J.* (Letters) **322**, L79.

Hutsemekers, D., Surdej, J. 1984, *Astron. Astrophys.* **133**, 209.
Jeffers, S., Weller, W.G. 1985, *Astron. Astrophys. Suppl.* **61**, 173.
Johansson, L. 1987, *Astron. Astrophys.* **182**, 179.
Johansson, L., Bergvall, N. 1988, *Astron. Astrophys.* **192**, 81.
Keel, W.C. 1982, *Publ. Astron. Soc. Pacific* **94**, 765.
Keel, W.C. 1987, *Astron. Astrophys.* **172**, 43.
Khaliullin, Kh.F. 1974, *Astron. Zh.* **51**, 395 (= *Sov. Astron.* **18**, 229).
Khaliullin, Kh.F., Khaliullina, A.I., Cherepashchuk, A.M. 1984, *Pis'ma Astron. Zh.* **10**, 600 (= *Sov. Astron. Letters* **10**, 250).
Kudritzki, R.P., Hummer, D.G. 1990, *Ann. Rev. Astron. Astrophys.* **28**, 303.
Kunth, D., Sargent, W.L.W. 1981, *Astron. Astrophys.* (Letters) **101**, L5.
Kunth, D., Joubert, M. 1985, *Astron. Astrophys.* **142**, 411.
Kunth, D., Schild, H. 1986, *Astron. Astrophys.* **169**, 71.
Kwitter, K.B. 1981, *Astrophys. J.* **245**, 154.
Kwitter, K.B. 1984, *Astrophys. J.* **287**, 840.
Lamers, H., de Loore, C. (eds.) 1987, Instabilities in Luminous Early Type Stars, *Proc. Workshop in honour of Cornelis de Jager* (Dordrecht: Reidel).
Lamers, H., Maeder, A., Schmutz, W., Cassinelli, J.P. 1990, in: L.A. Willson, R. Stalio (eds.), Angular Momentum and Mass Loss for Hot Stars, *NATO ASI Series Vol. C316* (Dordrecht: Kluwer), p. 349.
Langer, N. 1989a, *Astron. Astrophys.* **210**, 93.
Langer, N. 1989b, *Astron. Astrophys.* **220**, 135.
Langer, N. 1990, in: C.D. Garmany (ed.), Intrinsic Properties of Hot Luminous Stars, *Proc. Boulder-Munich Workshop, A.S.P. Conf. Series Vol. 7*, p. 328.
Leitherer, C., Chavarria, K. 1987, *Astron. Astrophys.* **175**, 208.
Lequeux, J., Mancherat-Joubert, M., Deharveng, J.M., Kunth, D. 1981, *Astron. Astrophys.* **103**, 305.
Lequeux, J., Meyssonnier, N., Azzopardi, M. 1987, *Astron. Astrophys. Suppl.* **67**, 169.
de Loore, C., A.J. Willis (eds.) 1982, Wolf-Rayet Stars: Observations, Physics, Evolution, *Proc. IAU Symp. No. 99* (Dordrecht: Reidel).
de Loore, C.W.H, Willis, A.J., Laskarides, P. (eds.) 1986, Luminous Stars and Associations in Galaxies, *Proc. IAU Symp. No. 116* (Dordrecht: Reidel).
de Loore, C. 1988, *Astron. Astrophys.* **203**, 71.
Lortet, M.C., Pitault, A. (eds.) 1983, Wolf-Rayet Stars: Progenitors of Supernovae? (Paris: l'Observatoire).
Lozinskaya, T.A. 1982, *Astrophys. Space Sci.* **87**, 313.
Lundström, I., Stenholm, B. 1979, *Astron. Astrophys. Suppl.* **35**, 303.
Lundström, I., Stenholm, B. 1984a, *Astron. Astrophys. Suppl.* **58**, 163.
Lundström, I., Stenholm, B. 1984b, *Astron. Astrophys. Suppl.* **56**, 43.
Lundström, I., Stenholm, B. 1984c, *Astron. Astrophys.* **138**, 311.
Lundström, I., Stenholm, B. 1989, *Astron. Astrophys.* **218**, 199.
Luo, D., McCray, R., Mac Low, M.-M. 1990 *Astrophys. J.* **362**, 267.
Maeder, A., Lequeux, J., Azzopardi, M. 1980, *Astron. Astrophys.* (Letters) **90**, L17.
Maeder, A. 1983, *Astron. Astrophys.* **120**, 130.
Maeder, A. 1985, *Astron. Astrophys.* **147**, 300.
Maeder, A., Meynet, G. 1987, *Astron. Astrophys.* **182**, 243.
Maeder, A. 1990a, *Astron. Astrophys. Suppl.* **84**, 139.
Maeder, A. 1990b, *Astron. Astrophys.*, in press.
Massey, P. 1981, *Astrophys. J.* **246**, 153.
Massey, P., Conti, P.S. 1983, *Astrophys. J.* **273**, 576.
Massey, P. 1984, *Astrophys. J.* **281**, 789.
Massey, P. 1985, *Publ. Astron. Soc. Pacific* **97**, 5.
Massey, P., Armandroff, T.E., Conti, P.S. 1986, *Astron. J.* **92**, 1303.
Massey, P., Conti, P.S., Armandroff, T.E. 1987a, *Astron. J.* **94**, 1538.
Massey, P., Conti, P.S., Moffat, A.F.J., Shara, M.M. 1987b, *Publ. Astron. Soc. Pacific* **99**, 816.
McCandliss, S.R. 1988, Thesis, Univ. of Colorado.
Meylan, G., Maeder, A. 1983, *Astron. Astrophys.*, **124**, 84.

Moffat, A.F.J., Shara, M.M. 1983, *Astrophys. J.*, **273**, 544.

Moffat, A.F.J., Shara, M.M. 1987, *Astrophys. J.* **320**, 266.

Moffat, A.F.J. 1988, in: G.V. Coyne, A.M. Magalhães, A.F.J. Moffat, R.E. Schulte-Ladbeck, S. Tapia, D.T., Wickramasinghe (eds.), Polarized Radiation of Circumstellar Origin (Vatican: Vatican Observatory), p. 607.

Moffat, A.F.J., Drissen, L., Lamontagne, R., Robert, C. 1988, *Astrophys. J.* **334**, 1038.

Moffat, A.F.J., Niemela, V.S., Marraco, H.G. 1990, *Astrophys. J.* **348**, 232.

Möllenhoff, C. 1981, *Astron. Astrophys.* **99**, 341.

Moran,J.P., Davis, R.J., Bode, M.F., Taylor, A.R., Spencer, R.E., Argue, A.N., Irwin, M.J., Shanklin, J.D. 1989, *Nature* **340**, 449.

Morgan, D.H., Good, A.R. 1990, *Monthly Notices Roy. Astron. Soc.* **243**, 459.

Nugis, T., Pustyl'nik I. (eds.), 1988, Wolf-Rayet Stars and Related Objects, *Proc. All-Union Conf., El'va, USSR, 1986, Tartu Astrofüüs. Obs. Teated* No. 89.

Nussbaumer, H., Schmutz, W., Smith, L.J., Willis, A.J. 1982, *Astron. Astrophys. Suppl.* **47**, 257.

Osterbrock, D.E., Cohen, R.D. 1982, *Astrophys. J.* **261**, 64.

Poe, C.H., Friend, D.B., Cassinelli, J.P. 1989, *Astrophys. J.* **337**, 888.

Pollock, A.M.T. 1987, *Astrophys. J.* **320**, 283.

Prévot-Burnichon, M.L., Prévot, L., Rebeirot, E., Rousseau, J., Martin, N. 1981, *Astron. Astrophys.* **103**, 83.

Prinja, R.K., Barlow, M.J., Howarth, I.D. 1990, *Astrophys. J.* **361**, 607.

Rayo, J.F., Peimbert, M., Torres-Peimbert, S. 1982, *Astrophys. J.* **255**, 1.

Robert, C., Moffat, A.F.J. 1990, in: C.D. Garmany (ed.), Intrinsic Properties of Hot Luminous Stars, *Proc. Boulder-Munich Workshop, A.S.P. Conf. Series Vol. 7*, p. 271,

Roberts, M.S. 1962, *Astron. J.* **67**, 79.

Rosa, M., D'Odorico, S. 1982, in: C. de Loore, A.J. Willis (eds.), 'Wolf-Rayet Stars: Observations, Physics, Evolution', *Proc. IAU Symp. No. 99* (Dordrecht: Reidel), p. 555.

Rosa, M., Solf, J. 1984, *Astron. Astrophys.* **130**, 29.

Rosa, M., Joubert, M., Benvenuti, P. 1984, *Astron. Astrophys. Suppl.* **57**, 361.

Rosa, M., D'Odorico, S. 1986, in: C.W.H. de Loore, A.J. Willis, P. Laskarides (eds.), Luminous Stars and Associations in Galaxies, *Proc. IAU Symp. No. 116* (Dordrecht: Reidel), p. 355.

Rosa, M.R., Mathis, J.S. 1990, in: C.D. Garmany (ed.), Intrinsic Properties of Hot Luminous Stars, *Proc. Boulder-Munich Workshop, A.S.P. Conf. Series Vol. 7*, p. 135.

Rosa, M., Richter, O.-G. 1988, *Astron. Astrophys.* **192**, 57.

Rosado, M. 1986, *Astron. Astrophys.* **160**, 211.

Schmutz, W., Hamann, W.-R. 1986, *Astron. Astrophys.* (Letters) **166**, L11.

Schmutz, W., Hamann, W.-R., Wessolowski, U. 1989, *Astron. Astrophys.* **210**, 236.

Schmutz, W. 1990, in: C.D. Garmany (ed.), Intrinsic Properties of Hot Luminous Stars, *Proc. Boulder-Munich Workshop, A.S.P. Conf. Series Vol. 7*, p. 117.

Schulte-Ladbeck, R.E. 1989, *Astron. J.* **97**, 1471.

Schulte-Ladbeck, R.E., van der Hucht, K.A. 1989, *Astrophys. J.* **337**, 872.

Shara, M.M., Moffat, A.F.J. 1986, in: C.W.H. de Loore, A.J. Willis, P. Laskarides (eds.), Luminous Stars and Associations in Galaxies, *Proc. IAU Symp. No. 116* (Dordrecht: Reidel), p. 231.

Shore, S.N., Brown, D.N. 1988, *Astrophys. J.* **334**, 1021.

Signore, M., Dupraz, C. 1990, *Astron. Astrophys.* (Letters) **234**, L15 .

Sivertsen, S. 1981, *Astron. Astrophys. Suppl.* **43**, 221.

Skillmann, E.D. 1985, *Astrophys. J.* **290**, 449.

Smith, L.F. 1968a, *Monthly Notices Roy. Astron. Soc.* **140**, 409.

Smith, L.F. 1968b, *Monthly Notices Roy. Astron. Soc.* **141**, 317.

Smith, L.F., 1973, in: M.K.V. Bappu, J. Sahade (eds.), Wolf-Rayet and High-Temperature Stars, *Proc. IAU Symp. No. 49* (Dordrecht: Reidel), p. 15.

Smith, L.F., Kuhi, L.V. 1981, An Atlas of Wolf-Rayet Line Profiles, *JILA Report* No. 117 (Boulder: Univ. of Colorado).

Smith, L.F. 1988, *Astrophys. J.* **327**, 128.

Smith, L.F., Hummer, D.G. 1988, *Monthly Notices Roy. Astron. Soc.* **230**, 511.

Smith, L.F., Maeder, A. 1989, *Astron. Astrophys.* **211**, 71.

Smith, L.F., Shara, M.M., Moffat, A.F.J. 1990, *Astrophys. J.* **358**, 229.

Smith, L.F., Maeder, A. 1990, *Astron. Astrophys.*, in press.

Smith, L.J., Willis, A.J. 1983, *Astron. Astrophys. Suppl.* **54**, 229.

Smith, L.J., Pettini, M., Dyson, J.E., Hartquist, T.W. 1988, *Monthly Notices Roy. Astron. Soc.* **234**, 625.

Smith, L.J. 1990, in: *Proc. XIth Eur. Regional Astron. Meeting IAU* (Cambridge: Univ. Press), in press.

St-Louis, N., Moffat, A.F.J., Drissen, L., Bastien, P., Robert, C. 1988, *Astrophys. J.* **330**, 286.

St-Louis, N., Smith, L.J., Stevens, I.R., Willis, A.J., Garmany, C.D., Conti, P.S. 1989, *Astron. Astrophys.* **226**, 249.

Taylor, M.-J., Chen, K.-Y. McNeill, J.D., Merrill, J.E., Oliver, J.P., Wood, F.B. 1988, *Publ. Astron. Soc. Pacific* **100**, 1544.

Taylor, M.-J. 1990, *Astron. J.* **100**, 1264.

Terlevich, R., Melnick, J. 1985, *Monthly Notices Roy. Astron. Soc.* **213**, 841.

Thé, P.S. 1964, *Contr. Bosscha Obs.* No. 26.

Torres, A.V., Conti, P.S. 1984, *Astrophys. J.* **280**, 181.

Torres, A.V., Conti, P.S., Massey, P. 1986, *Astrophys. J.* **300**, 379.

Torres, A.V., Massey, P. 1987, *Astrophys. J. Suppl.* **65**, 459.

Torres-Dodgen, A.V., Massey, P. 1988, *Astron. J.* **96**, 1076.

Underhill, A.B. 1990, in: L.A. Willson, R. Stalio (eds.), Angular Momentum and Mass Loss for Hot Stars, *NATO ASI Series* Vol. *C316* (Dordrecht: Kluwer), p. 353.

Underhill, A.B., Grieve, G.R., Louth, H. 1990,

Usov, V.V. 1990, *Astroph. Space Sci.* **167**, 297. *Publ. Astron. Soc. Pacific* **102**, 749.

Vacca, W.D., Torres-Dodgen, A.V. 1990, *Astrophys. J. Suppl.*, in press.

Van Buren, D., McCray, R. 1988, *Astrophys. J.* (Letters) **329**, L93.

Vreux, J.M., Dennefeld, M., Andrillat, Y. 1983, *Astron. Astrophys. Suppl.* **54**, 437.

Vreux, J.-M. 1985, *Publ. Astron. Soc. Pacific* **97**, 274.

Vreux, J.-M., Andrillat, Y., Gosset, E. 1985, *Astron. Astrophys.* **149**, 337.

Vreux, J.-M. 1987, in: H. Lamers, C. de Loore (eds), Instabilities in Luminous Early Type Stars, *Proc. Workshop in honour of Cornelis de Jager* (Dordrecht: Reidel), p. 81.

Vreux, J.-M., Dennefeld, M., Andrillat, Y., Rochowicz, K. 1989, *Astron. Astrophys. Suppl.* **81**, 353.

Vreux, J.-M., Andrillat, Y., Biemont, E. 1990, *Astron. Astrophys.* **238**, 207.

Wampler, E.J. 1982, *Astron. Astrophys.* **114**, 165.

Westerlund, B.E., Azzopardi, M., Breysacher, J., Lequeux, J. 1983, *Astron. Astrophys.* **123**, 159.

Williams, P.M., Longmore, A.J., van der Hucht, K.A., Talevera, A., Wamsteker, W.M., Abbott, D.C., Telesco, C.M. 1985, *Monthly Notices Roy. Astron. Soc.* **215**, 23P.

Williams, P.M., van der Hucht, K.A., Thé, P.S. 1987, *Astron. Astrophys.* **182**, 91.

Williams, P.M., Eenens, P.R.J. 1989, *Monthly Notices Roy. Astron. Soc.* **240**, 445.

Williams P.M., van der Hucht, K.A., Pollock, A.M.T., Florkowski, D.R., van der Woerd, H., Wamsteker, W.M. 1990a, *Monthly Notices Roy. Astron. Soc.* **243**, 662.

Williams P.M., van der Hucht, K.A., Thé, P.S., Bouchet, P. 1990b, *Monthly Notices Roy. Astron. Soc.* **247**, 18P.

Willis, A.J. 1981, in: C. Chiosi, R. Stalio (eds.), Effects of Mass Loss on Stellar Evolution, *Proc. IAU Coll. No. 59* (Dordrecht: Reidel), p. 27.

Willis, A.J., van der Hucht, K.A., Conti, P.S., Garmany, C.D. 1986, *Astron. Astrophys. Suppl.* **63**, 417.

Willis, A.J., Garmany, C.D. 1987, in: Y. Kondo (ed.), Exploring the Universe with the *IUE* satellite (Dordrecht: Reidel), p. 157.

Willis, A.J., Howarth, I.D., Smith, L.J., Garmany, C.D., Conti, P.S. 1989, *Astron. Astrophys. Suppl.* **77**, 269.

Willson, L.A., Stalio, R. (eds) 1990, Angular Momentum and Mass Loss for Hot Stars, *NATO ASI Series* Vol. *C316* (Dordrecht: Kluwer).

Wolf, C.J.E., Rayet, G. 1867, *Comptes Rendus* **65**, 292.

Wray, J.D., Corso, G.J. 1972, *Astrophys. J.* **172**, 577.

Wright, A.E., Barlow, M.J. 1975, *Monthly Notices Roy. Astron. Soc.* **170**, 41.

Zinnecker, H. 1987, in: T.X. Thuan, T. Montmerle, J. Tran Thanh Van (eds.), Starbursts and Galactic Evolution, *Proc XXIInd Rencontres de Moriond* (Gif-sur-Yvettes: Editions Frontières), p. 165.

SESSION II. ATMOSPHERES – *Chair: André Maeder*

André Maeder chairing, Tiit Nugis

OBSERVATIONS VERSUS ATMOSPHERIC MODELS OF WR STARS

WERNER SCHMUTZ
Joint Institute for Laboratory Astrophysics
University of Colorado and National Institute of Standards and Technology
Boulder, CO 80309-0440, USA

ABSTRACT. Current knowledge of Wolf-Rayet continua is reviewed and observed continuum energy distributions are compared with model predictions. Good agreement is found. For WN stars a large range in intrinsic color is found for each subclass and the so far unexplained large range in the observed reddening-free continuum indices can be understood on the basis of atmospheric models. The strength of the continuum discontinuities determines the value of the index. Because of severe blending by the emission lines in WC spectra it can not be decided whether the intrinsic colors of WC stars of a given spectral type vary over a range of values or whether all stars of a given spectral type are similar. The reliability of model calculations for Wolf-Rayet stars is further tested by comparing observed and predicted correlations of helium line strengths. Satisfactory agreement is found for the Kiel/JILA models but not for the Bhatia-Underhill models. Methods used to determine the reddening due to interstellar extinction are reviewed and their accuracies estimated.

1. Introduction

This paper aims to compare observations of Wolf-Rayet spectra with theoretical predictions. The observations are taken from the literature and since most of the recently published papers that are of relevance to Wolf-Rayet atmospheres are cited here, this paper is also an observational review. The results of atmospheric calculations are taken mostly from the model grids published by Schmutz, Hamann, and Wessolowski (1989) and Schmutz et al. (1990) (Kiel models); but atmospheric models calculated with the code of Hillier (1987a) (JILA models) could have been used as well since both codes—although developed independently—yield the same results. The theory of Wolf-Rayet atmospheres is reviewed by Hillier (this symposium) who discusses model assumptions and calculation techniques and the reader is referred to his paper for more information about the model atmospheres. Here, results from models with atmospheres that consist only of pure helium are used. We may term these models "first generation" or "standard" models, in contrast to the more elaborate calculations that now also include metals (see e.g. Hamann or Hillier, this symposium).

Theory and observations can be compared in many different ways. Here we focus on three aspects: 1) What is known about the continua of Wolf-Rayet stars; 2) How the predicted model energy distributions compare to the observations; and 3) How the predicted helium line strengths compare with those observed. These comparisons allow one to judge the reliability of the model calculations.

39

K. A. van der Hucht and B. Hidayat (eds.),
Wolf-Rayet Stars and Interrelations with Other Massive Stars in Galaxies, 39–52.
© 1991 *IAU. Printed in the Netherlands.*

2. Non-Variability

Every Wolf-Rayet star monitored with high precision has shown some variability. Before starting with comparisons between model calculations and observations it is useful to think about what the observed variabilities in line profiles and continuum fluxes imply for an analysis. The crucial question is how large are these variations. For most single Wolf-Rayet stars the observed line profile variabilities are only minor perturbations on a well defined mean profile. This is true for lines in the UV (see e.g., St-Louis *et al.* 1989) as well as in the optical (e.g. McCandliss 1988). The deviations from the mean profile are generally not larger than the deviations of the synthetic profiles from the observed (mean) profile. Thus, it is obvious that model atmospheres are not complex enough to reproduce such profile variations and that the inferred stellar parameters are not affected by the line profile variations.

Short time-scale (hours) photometric variations of single Wolf-Rayet stars are usually below 0.01 mag (e.g. Monderen *et al.* 1988). Some stars have larger amplitudes, (e.g. HD 104994, van Genderen, van der Hucht, and Larsen 1990), but even in these cases the amplitudes are only of the order of 0.02 mag or less. To recalibrate old photometric measurements, Schmutz and Vacca (1990) compared the photometric fluxes of northern Wolf-Rayet stars determined by Westerlund (1966), with those measured by Massey (1984). They found that the differences in the $b - v$ colors of most (73%) of the Wolf-Rayet stars have a standard deviation of only 0.012 mag and that the differences in the v magnitudes have $\sigma = 0.038$ mag. This agreement over a time-span of about 18 years implies that: (1) the two authors observed very carefully (in fact the agreement is even slightly better than the authors claimed their data would be) and (2) most Wolf-Rayet stars are stable within ≈ 0.02 mag not only over short time-scales but also over longer time-scales (20 years).

It is a common belief that the observed variations should be interpreted as (small?) deviations from a homogeneous spherically symmetric atmosphere rather than reflecting variations of the stellar parameters. But even if the variations reflect changes of the stellar parameters they are so small that they can be ignored. If we assume that the bolometric luminosity remains constant, an amplitude of 0.03 mag in v implies only a variation of the effective temperature of about 1%. This is certainly much less than the uncertainty of the results from spectroscopic analysis obtained with currently available model atmospheres.

The result that Wolf-Rayet stars are basically stable is not a trivial one. Luminous Blue Variable stars are generally thought to be in a pre-Wolf-Rayet evolutionary stage. Obviously their stellar parameters are variable, yet their mass-loss rates—which could be regarded as a signature for instability—are similar to that of Wolf-Rayet stars.

3. X-Ray Emission

In principle, X rays are part of the stellar continuum energy distribution. However, there is not yet a quantitative theory that predicts X rays from single hot stars. Therefore, there is nothing to compare with the observations. What is known about X-ray emission from Wolf-Rayet stars has been summarized by Pollock (1987). However, in order to calculate a X-ray luminosity for a given star he had to adopt a distance, a quantity which is often only poorly known. For a subset of stars with measured X-ray fluxes the bolometric corrections are known (Schmutz, Hamann, and Wessolowski 1989). For these stars we can calculate the distance independent-ratio L_x/L_{bol}. In Figure 1 this ratio is given for 14 stars (excluding two non-detections). In several cases the original luminosity values given by the two references had to be adjusted for different assumptions of the stellar distance. From Figure 1 it is obvious that the L_x/L_{bol} ratio for most Wolf-Rayet stars is of the order of (a few) 10^{-7},

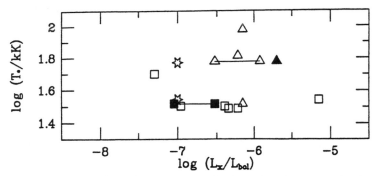

Figure 1: Ratio of X-ray luminosity (as given by Pollock 1987) to bolometric luminosity (as calculated by Schmutz, Hamann, and Wessolowski 1989 except for WR111 where L_{bol} is from Hillier 1989) for Wolf-Rayet stars common to both sources. Triangles denote WN E stars, squares WN L stars, and asterisks WC stars. Filled symbols denote binaries. The lines connect upper and lower limits of X-ray variables.

similar to that found for O stars (Long and White 1980; Cassinelli *et al.* 1981). One of the strong sources ($L_x/L_{bol} > 10^{-6}$) is a binary (V444 Cyg), and one is known to be variable and to have other peculiar properties as well (WR6). Only the extremely large X-ray flux from HD 93162 (WR25) is puzzling. For most of the Wolf-Rayet stars the conclusion of White and Long (1986), that the L_x/L_{bol} ratio for Wolf-Rayet stars is similar to that for OB stars, is confirmed.

4. Continuum Colors

The reason we are interested in knowing the intrinsic continuum energy distribution of a star is that the determination of the reddening due to interstellar extinction depends directly on it. For Wolf-Rayet stars it is not unrealistic to end up with an error of 0.15 mag in reddening if due care is not given to its evaluation (see below); this translates into an uncertainty of 50% in the luminosity determination. Even if this is not the largest source of uncertainty—usually the distance is—it is clear that one would like to avoid such a large error, if possible. There are several papers that have attempted to find a relation between spectral type and absolute magnitude and intrinsic colors for Wolf-Rayet stars (e.g. Torres-Dodgen and Massey 1988; Vacca and Torres-Dodgen 1990). These studies revealed some correlation but they also found a large scatter around the mean value for a given spectral type. The intrinsic scatter of the absolute magnitude-spectral type calibration seems to be as large as 1 mag. This implies an uncertainty of 60% if such a calibration is used for distance determinations. The intrinsic scatter of the $(b - v)_0$-spectral type calibration is of the order of 0.15 mag. Although we can add nothing to previous results regarding the absolute magnitudes, we are able to understand the origin of the scatter of the observed colors on the basis of theoretical models.

4.1 CONTINUUM COLORS OF WC STARS

Vacca and Torres-Dodgen (1990) found that the intrinsic $(b - v)_0$ colors of WC4 stars in the LMC span the range of 0.0 to −0.5 mag. In view of this result it appears that the suggestion put forward by Smith, Shara, and Moffat (1990), that all WC4 stars are similar

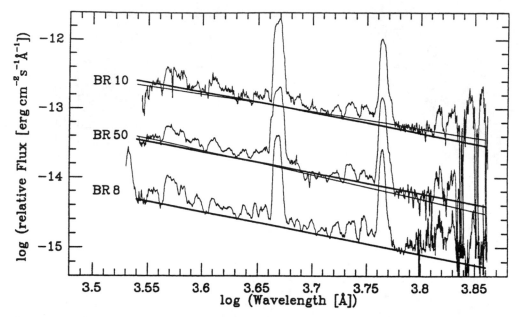

Figure 2: Dereddened relative fluxes of BR8, BR10, and BR50. The observations are from Torres and Massey (1987) and the E_{B-V} values from Vacca and Torres-Dodgen (1990). The lines drawn indicate the estimated continua with slopes corresponding to $(b-v)_0 = -0.2$ (thick lines) and -0.1 (thin line, BR10) and -0.3 (thin line, BR50).

and they all have an intrinsic color of $(b-v)_0 = -0.3$, cannot be correct. However, because of the extreme blending of the numerous emission lines throughout the UV and the optical wavelength regions in WC stars—especially in WC4 stars which have the broadest line profiles—it is very difficult to define a continuum level. Thus, it could be that the large range in $(b-v)_0$ found by Vacca and Torres-Dodgen (1990) is due to incorrect determination of the continuum level. We can demonstrate that this is possible with the aid of results from atmosphere calculations. By examining the reddening-free indices Δ and η, as defined by Smith (1968) and Massey (1984), we have a means of estimating the errors in the continuum definition. From the work of Hillier (1989) and unpublished calculations we expect WC4 stars to have a power law continuum energy distribution in the UV/optical wavelength region with only small ($\lesssim 0.03$ mag) continuum discontinuities. Therefore, atmospheric models predict the line-free reddening-free indices to be close to 0.0 mag for all early type WC stars. This theoretical prediction is not confirmed by the observations. Massey (1984) found the line-free reddening-free indices of early WC stars to deviate by up to 0.3 mag from 0.0. We believe that these deviations are dominated by errors in defining the continuum level. The reddening-free indices are obtained from color differences, hence we may adopt an uncertainty of about 0.2 mag in the $b-v$ colors. If we also allow for about 0.1 mag error in reddening (this relatively large error is adopted because reddening determinations from the 2200 Å feature are also affected by the line blending problem) then uncertainties of up to 0.3 mag in determining the intrinsic colors seem to be possible. Thus, the suggestion of Smith, Shara and Moffat (1990) that all LMC WC4 have similar colors cannot be refuted.

We may investigate their suggestion further by plotting in Figure 2 the dereddened fluxes of three WC4 stars, including the two with the most extreme $(b-v)_0$ colors, as determined by Vacca and Torres-Dodgen (-0.0 and -0.5 mag), and one for which an intermediate color (-0.2 mag) was determined. As mentioned above, according to the theoretical predictions WCE continua should follow power laws. Accordingly, in Figure 2 the continuum levels are approximated by straight lines in the log-log diagrams. The thick lines have slopes corresponding to $(b - v)_0 = -0.2$. From this figure it is obvious that defining a continuum level for some WC stars is so difficult that the possibility that all three stars have the same intrinsic color of -0.2 cannot be excluded. However, a slope of -0.3, as suggested by Smith, Shara, and Moffat (1990), is not consistent with the observed continuum of BR8 which has the best defined slope of the three stars shown in Figure 2. An intrinsic color of -0.2 would also agree with the results of the only published model for a WC star (Hillier 1989 and personal communication). The possibility of a similar slope for all three stars is strengthened by the intrinsic slope in the UV. For all LMC WC4 stars Vacca and Torres-Dodgen (1990) determined similar values of α, where α is the slope in the UV $\log \lambda$-$\log F_\lambda$ diagram. Nevertheless, the stars may still be different. It might be that continua with slopes of -0.1 and -0.3 fit two of the stars in Figure 2 slightly better than -0.2. Additional evidence that the three stars are not identical is provided by their absolute magnitudes. The three stars BR8, BR10, and BR50 have $M_v = -3.4$, -4.6, and -4.1 mag, respectively. The differences in absolute magnitude are significant since the reddening corrections are not likely to be in error more than by -0.1 mag, and therefore the absolute magnitudes are good within 0.3 mag.

To summarize, for LMC WC4 stars we cannot exclude the possibility that their intrinsic colors are similar. However, if their colors are identical then $(b-v)_0$ is more likely to be -0.2 rather than -0.3. There is evidence that there are differences between the WC4 stars—but this evidence is not conclusive enough to be considered a proof. For WC subtypes other than WC4 the same difficulty in defining a continuum level makes it unclear whether the range in for the $(b - v)_0$ values found by Vacca and Torres-Dodgen (1990) for each subtype is real or not.

4.2 CONTINUUM COLORS OF WN STARS

As in the case of WC stars, Vacca and Torres-Dodgen (1990) determined a large range of intrinsic $(b - v)_0$ colors within each WN spectral type. However, in contrast to the situation for WC stars, we are much more confident that it is possible to locate correctly the continuum levels in the spectra of WN stars. Therefore, there is no reason to question the reality of an intrinsic scatter in $(b - v)_0$ for each WN subtype. From the theoretical point of view this is a natural result since the predicted $(b - v)_0$, as well as equivalent widths of helium lines, is a function of two parameters, temperature and wind density (Schmutz 1988; see also Fig 3 of Hamann and Schmutz 1987), whereas the spectral type is a one-dimensional classification. Obviously, for WN stars there are different combinations of these two parameters realized in nature. Actually, it is more difficult to understand why it is apparently sufficient to use a one-dimensional description for WC stars.

If we trust the line-free colors measured from the absolutely calibrated spectra of WN stars by Massey (1984) and Torres-Dodgen and Massey (1988) then there is another observational fact to explain: Why do the reddening-free line-free indices of WN stars show a large range of values? Note that a range in $(b - v)_0$ colors, which we claim to be real, is not sufficient to explain this observation. If the continuum followed a power law then the reddening-free indices would still be close to zero. The answer comes naturally from the theoretical calculations. If the true opacity in an atmosphere is dominated by helium, as is the case for most WN stars, then for certain model parameter combinations the He II n=4 bound-free continuum introduces a considerable discontinuity at 3645 Å. The reddening-free index Δ is essentially a measure of this continuum jump. Thus, the observed range

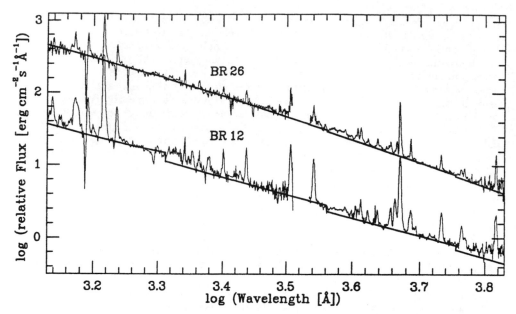

Figure 3: Dereddened relative fluxes of BR12 ($E_{B-V} = 0.05$ mag) and BR26 ($E_{B-V} = 0.04$ mag) compared with theoretical flux distributions predicted for these stars. The optical observations are from Torres-Dodgen and Massey (1988) and the combined *IUE* spectra from Vacca and Torres-Dodgen (1990).

of values of this parameter implies that there are WN stars with no significant continuum jumps and there are other WN stars with discontinuities of up to 0.3 mag in emission.

Atmospheric models predict a close correlation between $(b-v)_0$ and D_{3645}, the continuum discontinuity at 3645 Å. If the model calculations yield a flat continuum (small $(b-v)_0$) then the jump at 3645 Å is found to be in emission. This correlation also implies that the range in $(u-b)_0$ is much smaller than that in $(b-v)_0$:

$$\text{WN stars:} \quad \begin{aligned} (u-b)_0 &= -0.22 \ldots -0.27 \, \text{mag} \\ (b-v)_0 &= -0.05 \ldots -0.35 \, \text{mag} \end{aligned}$$

The small range in $(u-b)_0$ and the relation between $(b-v)_0$ and D_{3645} allow us to formulate an expression to calculate the interstellar reddening from observed $u-b$ and $b-v$ colors alone (Vacca and Schmutz, this symposium).

5. Continuum Fits

In section 4 we used theoretical continuum energy distributions to predict intrinsic colors, but we have not yet shown that the predictions actually fit the observations. Two examples of observed and theoretical continuum energy distributions are shown in Figure 3. The examples shown are among the best at hand. Other comparisons are usually of similar quality, except that for some either the UV or the optical observations are not available, or in other cases the UV and optical fluxes do not agree very well. In some cases the

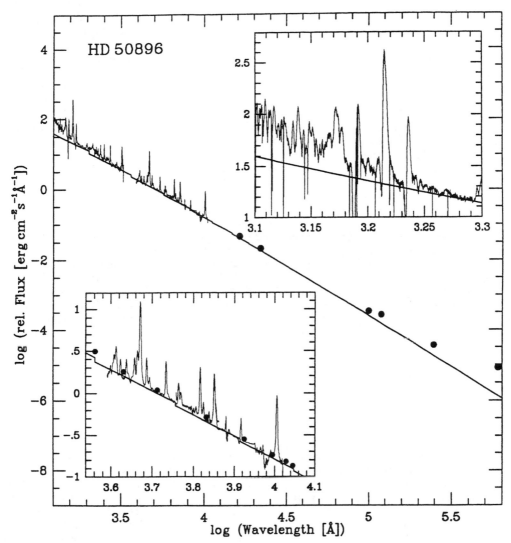

Figure 4: Observed spectrum of HD 50896 (WR6) dereddened by 0.07 mag. The UV spectrum is a low-resolution *IUE* observation. The *H* and *K* photometric values are from Allen, Swings, and Harvey (1972), and the 10 μm flux is obtained from Barlow, Smith, and Willis (1981). The *IRAS* observations at 12, 25, and 60 μm (van der Hucht *et al.* 1985) do not agree with either an extrapolation of the observations or with the predicted spectrum. Probably the flux measured through the large aperture of *IRAS* is contaminated by (dust?) emission from the ring nebula S308. *Inset top*: Enlargement of the *IUE-SWP* wavelength region. The high-resolution spectrum published by Howarth and Phillips (1986) is the average from 52 *IUE* spectra. *Inset bottom*: Enlargement of the optical/near-IR wavelength region. The optical spectrum is an observation by the author and the near-IR observation is taken from Vreux, Dennefeld, and Andrillat (1983). The dots mark photometric *ubv* fluxes as measured by Smith (1968) and near-IR values from Cohen, Barlow, and Kuhi (1975).

models fit most of the observed wavelength range but deviate over small spectral regions (e.g., around 2200 Å). In Figure 3 we show fits to two LMC WN stars. Both stars are only slightly reddened and therefore the adopted reddening law has practically no influence on the observed continuum energy distribution. The stellar parameters for the models of the two stars are from Koesterke et al. (1990) and the corresponding continuum energy distributions have been taken from the flux-grid published by Schmutz et al. (1990). The fits shown illustrate one continuum in which there are essentially no discontinuities and one that exhibits strong emission jumps. Note that if there is a discontinuity at 3645 Å there are also discontinuities expected at all other observable He II edges. The models predict a correlation not only between $(b - v)_0$ and D_{3645}, but also between these quantities and the strengths of the He II lines. Therefore, if the discontinuity is strongly in emission the emission lines of the corresponding series will be strong, will merge longward of the continuum edges, and will form a pseudo-continuum. This will give the impression that the discontinuity is at a longer wavelength than theoretically expected.

The predicted continua are clearly excellent fits to the observations with the exception of deviations at each end of the displayed wavelength region. The disagreement in the UV will be discussed below; the discrepancy in the red wavelength range is due to the deteriorating quality of the observation. The latter can be demonstrated to be true because the theoretical spectrum fits additional observations at even longer wavelengths (JHK fluxes).

For the lightly reddened Galactic star HD 50896 (WR6) a comparison over a large wavelength range—3.7 dex in wavelength—is shown in Figure 4. For this comparison of observation and theory a model continuum of Hillier (personal communication; see also Hillier 1987b) was used. This figure demonstrates again how well theory fits the observation. From such fits the reddening due to interstellar extinction can be determined to an accuracy of 0.01 mag, provided the UV extinction law is known and there are no systematic errors in the predicted spectrum. As an illustration that theoretical spectra also fit the red wavelength region, and in particular the discontinuity at 5695 Å, the lower inset in Figure 4 gives an enlargement of the optical/near-IR wavelength region. The theoretical spectrum shown in this figure is taken from the work of Hamann, Schmutz, and Wessolowski (1988).

6. UV Pseudo-Continuum Formed by Fe Lines

For all stars that display strong emission lines the theoretical continua do not fit the observations for wavelengths shorter than about 1500 Å, the predicted continua being always too low. Examples of this inconsistency can be seen in Figures 3 and 4. This is not a result of insufficient resolution of the UV observations. The upper inset of Figure 4 shows that the discrepancy remains even if high-resolution observations are used for the comparison. The reason for the disagreement has been found independently by two groups (Koenigsberger and Auer 1985; Nugis and Sapar 1985): In the wavelength region shortward of 1500 Å numerous Fe v and Fe vi lines are present and form a pseudo-continuum. In Figure 5 a comparison is shown between the normalized spectrum of HD 50896 and the combined Fe v/Fe vi laboratory spectrum (Ekberg 1975a, b). The basic shape of the emission spectrum is very similar to that of the iron intensity distribution. This figure demonstrates that it is quite plausible that iron emission is the source of the pseudo-continuum. Most of the remaining (strong) emission lines have well-known identifications (Willis et al. 1986).

7. Comparison Between Predicted and Observed Helium Line Strength Ratios

The model that fits a given observation is determined by the strength of one He I and one

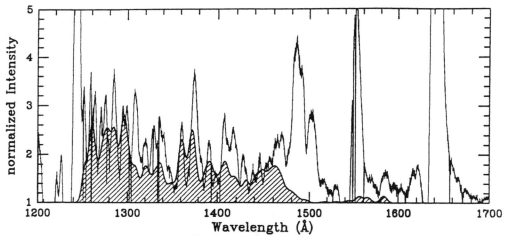

Figure 5: Pseudo-continuum in the spectrum of HD 50896. The plotted spectrum is the observed dereddened flux divided by the predicted continuum energy distribution (see Fig. 4). The shaded area is the emission spectrum of Fe V and Fe VI as observed in the laboratory. The Fe V/Fe VI spectrum broadened to 1000 km s^{-1} was obtained from Koenigsberger (1988).

He II line, e.g. He I λ5876 and He II λ4686 (see Schmutz, Hamann, and Wessolowski 1989). However, there are usually more He I and He II lines observed than just the minimum of two lines needed for the analysis. Conti and Massey (1989) demonstrated that there is a well defined correlation between the observed line strengths of He II transitions. It is possible therefore, to test the reliability of the models by comparing the observed and predicted correlations.

In Figure 6a the theoretical and observed correlations between the equivalent widths of He II λ1640 and He II λ4686 are compared. The theoretical correlation is tighter than the observed one. This can be understood as the result of errors in measuring the equivalent widths. However, the observed values appear to differ systematically from the predicted ones for Galactic WN E stars with strong lines. Presumably, the iron pseudo-continuum is stronger in the Galactic stars than in the LMC stars due to the higher metal abundance in the Galaxy and, therefore, the He II λ1640 equivalent widths in Galactic stars may be underestimated.

In Figure 6b the theoretical and observed correlations between He II λ5411 and He II λ4686 are compared. Both correlations are about equally well defined. This indicates that the observational scatter in Figure 6a is predominantly due to errors in the He II λ1640 equivalent width measurements. In Figure 6b a slight discrepency between the observation and theory is perceptible. An example of the disagreement between the observed and predicted line ratios is found in the case of HD 50896. The profile fits published by Hamann, Schmutz, and Wessolowski (1988) show good agreement for the He II $\lambda\lambda$1640, 4686 lines, a result which implies that the ratio between these two lines is predicted correctly. However the theoretical profile for He II λ5411 is too narrow, and thus, the theoretical equivalent width is too small. A plausible explanation for the disagreement would be a contribution to the He II λ5411 equivalent width by another line. However, there is no good candidate for such a line. Although Hillier (1987b) fits the observed width of He II λ5411, his predicted He II line ratio is incorrect as well. He calculates a He II λ1640 line which is too broad, and thus finds an equivalent width which is too large.

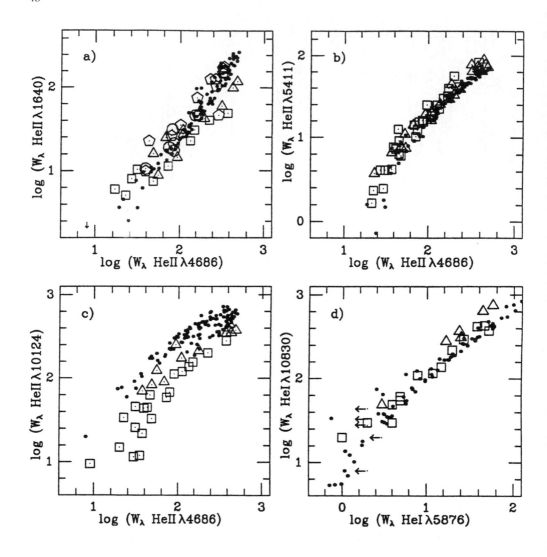

Figure 6: Comparison between observed and theoretical correlations of helium line equivalent widths; *a*) upper left panel: He II λ1640 versus He II λ4686; *b*) upper right panel: He II λ5411 versus He II λ4686; *c*) lower left panel: He II λ10124 versus He II λ4686; *d*) lower right panel: He I λ10830 versus He I λ5876. The theoretical equivalent widths are marked with small filled dots. The observed values are marked with open symbols: squares—Galactic WN L stars; circles—LMC WN L stars; triangles—Galactic WN E stars; pentagons—LMC WN E stars. The arrow in panel (*a*) marks a value where the theoretical equivalent width of He II λ1640 is in absorption; the arrows in panel (*d*) mark upper limits for the observed equivalent widths of He I λ5876. The references for the observed equivalent widths are: Smith and Willis (1983); Conti and Massey (1989); Conti and Morris (1990); Vreux, Dennefeld, and Andrillat (1983); Vreux *et al.* (1989); Conti, Massey, and Vreux (1990); Vreux, Andrillat, and Biemont (1990).

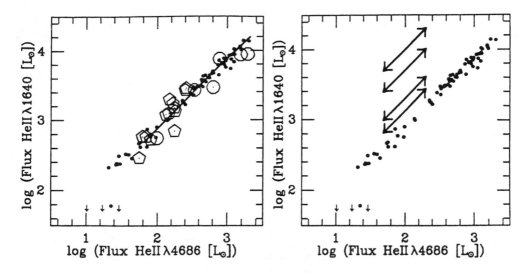

Figure 7: Correlation of He II λ1640 and He II λ4686 line fluxes: a) Comparison of observed and theoretical correlations; b) Comparison between the correlation predicted by the Kiel models (dots) with the flux ratios predicted by Bhatia and Underhill (1986) (arrows).

In Figure 6c the theoretical and observed correlations between He II λ10124 and He II λ4686 are compared. It is obvious that the two correlations do not agree. Since He II λ4686 is involved in the two comparisons where agreement is found, the problem must lie with the He II λ10124 line. The comparisons in Figures 6a and b demonstrated that the populations of He II n=2, 3, 4, and 7 are calculated correctly in the models. The He II λ10124 transition is between levels n=4 and 5. Theoretically, it is difficult to understand how the model population of He II n=5 could be wrong while other He II level-populations are correct. Yet, there is no obvious reason why the observations should be in error. To make the disagreement even more mysterious, the detailed analysis of HD 50896 by Hamann, Schmutz, and Wessolowski (1988) yields the correct intensity for this line. The entry in Figure 6c corresponding to this star lies in the upper right corner at about the only place where observational and theoretical values overlap.

In the case of He I lines data are available only for He I λ10830 and He I λ5876. The comparison between the theoretical and observed correlations of these lines is shown in Figure 6d. Good agreement is found except for strong-line stars where theory predicts He I λ5876 too strong. The reason for this discrepency can be traced to the fact that the blending of He I λ5876 with the lines of the He II n–5 series, which form a pseudo-continuum at the wavelength of the He I line (see lower inset Fig. 4), is not taken into account in the model calculations. Calculations simulating the effect of the blends by artificially including the He II n=5 continuum at the wavelength of the He I line yield a smaller He I λ5876 equivalent width. The theoretical He I line ratio then agrees with the observed one for WN stars with strong lines as well.

Conti and Morris (1990) also measured line fluxes. In Figure 7a the correlation between the observed dereddened line fluxes is compared with the theoretical correlation. The solid line is the mean correlation adopted by Conti and Morris (1990) and corresponds to an average line flux ratio of about 7.6. However, theory predicts a range of line flux ratios. This range translates into an intrinsic uncertainty of 0.07 mag in E_{B-V} if the average line flux ratio is used to derive the color excess due to interstellar reddening.

In Figure 7b the theoretical line flux correlation between He II $\lambda 1640$ and He II $\lambda 4686$ is compared with some of the line-flux ratios predicted by the models of Bhatia and Underhill (1986); the smallest ratio predicted is $(F(1640)/F(4686) = 13.4)$. Obviously, Bhatia and Underhill (1986) predict ratios different from those calculated with the model atmospheres used elsewhere in this paper. The reason for the difference is that, given the physical conditions assumed by Bhatia and Underhill (1986) in the line emitting region, collisional rates play an important role in the equations for statistical equilibrium of the level populations. This contrasts with the inherent assumptions of the Kiel/JILA models, in which the rate equations are dominated by radiative processes. Since neither the large ratios nor the wide range of He II line flux ratios calculated by Bhatia and Underhill (1986) agree with the observations we conclude that their model does not describe correctly the physical conditions in Wolf-Rayet atmospheres.

8. Conclusion

In Section 5 it was found that the predicted continuum energy distributions fit the observations and in Section 7 it was demonstrated that the observed correlations between several helium lines are reproduced well by the "standard" model calculations. This agreement provides strong evidence for the reliability of the model calculations. Therefore, we may use the model results to judge the accuracy of different methods of determining reddening. We conclude by presenting a summary of the various methods, their intrinsic errors, and the expected uncertainty in their application to actual observations (Table 1).

Table 1. Reddening determinations for WN stars.

method	intrinsic uncertainty (not Gaussian)	σ in practice (Gaussian)
E_{B-V} from fit to model continuum	0.	± 0.02 but: UV reddening law
$E_{B-V} = \mathcal{F}(u-b, b-v)$ (Vacca and Schmutz, this symposium)	± 0.02	± 0.07 but: only for helium dominated atmospheres and $T_* \gtrsim 35$ kK
E_{B-V} from 2200 Å feature (e.g. Vacca and Torres-Dodgen 1990)	0.	$\pm 0.05 \ldots 0.1$ but: UV reddening law
E_{B-V} from $\frac{F(1640)}{F(4686)}$ (Conti and Morris 1990)	± 0.07	± 0.1 but: UV reddening law
E_{B-V} from $< (b-v)_0 >_{\text{Sp. T.}}$	± 0.15	± 0.15

Acknowledgements. I thank my colleagues Drs. John Hillier, Pete Storey, and Bill Vacca for many discussions and valuable suggestions. Thanks also go to Bill Vacca and Lorraine Volsky for editorial assistance. I acknowledge Gloria Koenigsberger for providing the Fe V and Fe VI data, Ian Howarth for sending the *IUE* high-resolution spectrum of HD 50896, and the Astronomical Data Center at the Goddard Space Flight Center for making available the spectra published by Torres and Massey (1987) and Torres-Dodgen and Massey (1988). This work was supported by the Swiss National Foundation and by the National Science Foundation (U.S.) through Grant AST88-02937.

References

Allen, D. A., Swings, J. P., and Harvey, P. M. 1972, *Astr. Ap.*, **20**, 333.

Barlow, M. J., Smith, L. J., and Willis, A. J. 1981, *M.N.R.A.S.*, **196**, 101.

Bhatia, A. K., and Underhill, A. B. 1986, *Ap. J. Suppl.*, **60**, 323.

Cassinelli, J. P., *et al.* 1981, *Ap. J.*, **250**, 677.

Conti, P. S., and Massey, P. 1989, *Ap. J.*, **337**, 251.

Conti, P. S., Massey, P., and Vreux, J. M. 1990, *Ap. J.*, **354**, 359.

Conti, P. S., and Morris, P. W. 1990, *A. J.*, **99**, 898.

Cohen, M., Barlow, M. J., and Kuhi, L. V. 1975, *Astr. Ap.*, **40**, 291.

Ekberg, O. 1975*a*, *Phys. Scripta*, **11**, 23.

———. 1975*b*, *Phys. Scripta*, **12**, 42.

Hamann, W.-R., and Schmutz, W. 1987, *Astr. Ap.*, **174**, 173.

Hamann, W.-R., Schmutz, W., and Wessolowski, U. 1988, *Astr. Ap.*, **194**, 190.

Hillier, D. J. 1987a, *Ap. J. Suppl.*, **63**, 947.

———. 1987b, *Ap. J. Suppl.*, **63**, 965.

———. 1989, *Ap. J.*, **347**, 392.

Howarth, I. D., and Phillips, A. P. 1986, *M.N.R.A.S.*, **222**, 809.

Koenigsberger, G. 1988 *Rev. Mex. Astr. Astrof.*, **16**, 75.

Koenigsberger, G., and Auer, L. H. 1985 *Ap. J.*, **297**, 255.

Koesterke, L., Hamann, W.-R., Schmutz, W., and Wessolowski, U. 1990, in preparation.

Long, K. S., and White, R. L. 1980, *Ap. J. (Letters)*, **239**, L65.

Massey, P. 1984, *Ap. J.*, **281**, 789.

McCandliss, S. R. 1988, Ph. D. Thesis, University of Colorado.

Monderen, P., De Loore, C. W. H., van der Hucht, K. A., and van Genderen, A. M. 1988, *Astr. Ap.*, **195**, 179.

Nugis, T., and Sapar, A. 1985, *Sov. Astr. Let.*, **11**, 188.

Pollock, A. M. T. 1987, *Ap. J.*, **320**, 283.

Schmutz, W. 1988, in *Lecture Notes in Physics* Vol. **305**, *IAU Colloquium 108*, p. 133.

Schmutz, W., Hamann, W.-R., and Wessolowski, U. 1989, *Astr. Ap.*, **210**, 236.

Schmutz, W., and Vacca, W. D. 1990, *Astr. Ap. Suppl.*, in preparation.

Schmutz, W., Vogel, M., Hamann, W.-R., and Wessolowski, U. 1990, *Astr. Ap.*, in preparation (model fluxes are available from the author).

Smith, L. J., and Willis, A. J. 1983, *Astr. Ap. Suppl.*, **54**, 229.

Smith, L. F. 1968, *M.N.R.A.S.*, **140**, 409.

Smith, L. F., Shara, M. M., and Moffat, A. F. J. 1990, *Ap. J.*, **348**, 471.

St-Louis, N., *et al.* 1989, *Astr. Ap.*, **226**, 249.

Torres, A. V., and Massey, P. 1987, *Ap. J. Suppl.*, **65**, 459.

Torres-Dodgen, A. V., and Massey, P. 1988, *A. J.*, **96**, 1076.

Vacca, W. D., and Torres-Dodgen, A. V. 1990, *Ap. J. Suppl.*, **73**, in press.

van der Hucht, K. A., *et al.* 1985, in *Birth and Evolution of Massive Stars and Stellar Groups*, ed. W. Boland and H. van Woerden (Dordrecht: Reidel), p. 167.

van Genderen, A. M., van der Hucht, K. A., and Larsen, I. 1990, *Astr. Ap.*, **229**, 123.

Vreux, J. M., Andrillat, Y., and Biemont, E. 1990, *Astr. Ap.*, preprint.

Vreux, J. M., Dennefeld, M., and Andrillat, Y. 1983, *Astr. Ap. Suppl.*, **54**, 437.

Vreux, J. M., Dennefeld, M., Andrillat, Y., and Rochowicz, K. 1989, *Astr. Ap. Suppl.*, **81**, 353.

Westerlund, B. E. 1966, *Ap. J.*, **145**, 724.

White, R. L., and Long, K. S. 1986, *Ap. J.*, **310**, 832.

Willis, A. J., van der Hucht, K. A., Conti, P. S., and Garmany, C. D. 1986, *Astr. Ap. Suppl.*, **63**, 417.

DISCUSSION

Koenigsberger: Do you have any means of relating the physical position within the wind of the gas producing the spectrum you model with the radius of the underlying stellar core?
Schmutz: Yes, we do, most of John Hillier's papers include graphs that show the radial distribution of the line emissions.

Underhill: It is strange that your calculations indicate that collisional transitions may be neglected in your model atmospheres because most of your line emission comes from volume elements where the density is greater than $10^9 cm^{-3}$. From debugging the Bhatia and Underhill programs I know that the collisional rates are comparable to the radiative rates for many transitions when $N_e > 10^9 cm^{-3}$ in our one-point models. You should always carry all terms in the equations of statistical equilibrium and let the computer find which are negligible. Bhatia and Underhill included 5 different kinds of transition. This is the most complete study of statistical equilibrium made for modelling WR atmospheres.
Schmutz: I did not say collisions can be neglected. Collisional rates are of course included in our model calculations and they play a role to recover LTE at large optical depth. However, in the line-formation regions it turns out that for the physical conditions we are modelling the formation of the $HeII$ lines is dominated by recombination and photo-ionization. The special physical conditions you assume are needed in order that collisions become important. This is exactly the reason why you and we predict different ratios for the $HeII$ lines. Thus, we have a tool to discriminate between the models. I am afraid this test is not in favour of the ideas you are putting forward.

Owocki: (1) Your "nonvariability" refers only to overall luminosity at very long time scales. There are numerous spectral observations indicating few % variability on dynamical (~ 1 day) time scales. (2) X-rays are normally thought to arise from variable wind situations (*i.e.* shocks). What do you mean by "normal" X-ray?
Schmutz: (1) I am well aware that there are real variations in profiles and magnitude. The point that I wanted to make is that most vary only so little that it does not affect the results of spectroscopic analyses. (2) By "normal" I mean the X-ray level of "most" of the stars.

Schulte-Ladbeck: I have a question about the continuum energy distribution. Some WR stars may have flattened atmospheres as evidenced by polarimetry (maybe due to rotation)! Your models, I assume, use spherical symmetry. Do you have any idea how the continua would change for a flattened model?
Schmutz: Our models do assume spherical symmetry and we do not (yet) know what the influences of a flattened atmosphere would be. However, for some stars polarimetric measurements indicate basically zero polarisation, implying that at least some WR atmospheres are spherically symmetric, *i.e.* our models are valid for these stars.

Pollock: I would advise you against taking too literally the normality of the X-ray emission from WR stars because most of the single WR stars, those that are not binaries or non-thermal radio sources, are pretty weak in X-rays and *e.g.* the point for WR136 (that was about the only one from a single star that appeared on your graph) is in fact a non-detection and I would not put it there if I were you.
Schmutz: I just wanted to say, there is a belief that WR stars are not normal, *i.e.* below the limit of normal O stars, and actually they are not. Most of them are in the $L_x/L_{bol} = 10^{-7}$ region, with large scatter.

EVIDENCE THAT WOLF-RAYET STARS ARE PRE-MAINSEQUENCE OBJECTS

ANNE B. UNDERHILL
Department of Geophysics and Astronomy
University of British Columbia
Vancouver, B. C., V6T 1W5, Canada

ABSTRACT. The evidence is reviewed that Population I Wolf-Rayet stars have solar abundances, that they are surrounded by remnant disks formed from their natal clouds, and that their rate of mass loss is moderate. These properties are consistent with Wolf-Rayet stars being young objects recently arrived on the main-sequence rather than the evolved, peeled-down remnants of massive stars.

1. Introduction

Many people believe that Population I Wolf-Rayet stars are highly evolved massive stars which show anomalous abundances on their surfaces as a result of mass loss at a rate of the order of 5×10^{-5} $M_\odot$ yr^{-1}. I shall present evidence (1) that they have solar abundances, (2) that they are surrounded by remnant disks formed from their natal cloud, and (3) that their rate of mass loss is uncertain, but possibly of the order of or smaller than 10^{-6} $M_\odot$ yr^{-1}. These properties together with the fact that Wolf-Rayet stars are associated with O and early B stars in regions which radiate interstellar CO lines points toward them being young objects with ages no greater than about 5×10^6 yrs.

The predominant emission-line spectra of Wolf-Rayet stars are suggestive of the spectra of Herbig Ae/Be stars and T Tauri stars. The major difference is that in the line-emitting regions (LERs) of Wolf-Rayet stars the electron temperature (T_e) is of the order of 10^5 K whereas in the case of the less massive pre-mainsequence stars it is of the order of or less than 10^4 K. To generate a high T_e in the LER of a Wolf-Rayet star one may postulate that mechanical energy is transformed to heat by MHD effects which may occur in a low-beta plasma. Whenever larger than normal magnetic fields are occluded as a massive star ($M_* \geq 10M_\odot$) is formed, a high excitation, emission line spectrum may be generated as a result of the deposit of non-radiative energy and momentum in the LER. Such spectra are the criteria by which Wolf-Rayet stars are recognized. Analysis of these spectra reveals the physical state of the LER, not of the underlying photosphere. Additional information and postulates must be made to relate the underlying stars to the theory of evolution of massive stars. These postulates and information are different for Population I than for Population II Wolf-Rayet stars.

Study of binary Wolf-Rayet stars and of those in clusters shows that typically Population I Wolf-Rayet stars have masses of the order of $10 - 15M_\odot$ and that their luminosities are typical of hydrogen-burning stars of these masses. Their effective temperatures estimated from integrated fluxes (Underhill 1983) and from the radiation temperatures (T_*) which generate their spectra (Bhatia and Underhill 1986, 1988) are appropriate for their

K. A. van der Hucht and B. Hidayat (eds.),
Wolf-Rayet Stars and Interrelations with Other Massive Stars in Galaxies, 53–58.
© 1991 IAU. Printed in the Netherlands.

masses, i.e. of the order of 25,000–30,000 K.

The Population I Wolf-Rayet stars have 6-cm fluxes of the order of or less than a mJy, and all show significant infrared excess fluxes. These properties suggest that the Population I Wolf-Rayet stars are buried in much circumstellar plasma. Many show significant polarisation, see, for example, St. Louis et al. (1987) and Drissen et al. (1987). This indicates that a Wolf-Rayet star is buried in a cloud of electrons which is not spherically symmetric. Binary Wolf-Rayet stars tend to show polarisation changes which correlate with the orbital period, see, for instance, Drissen et al. (1986), St. Louis et al. (1988), as well as Schulte-Ladbeck and van der Hucht (1989). Some Wolf-Rayet stars show small random changes in polarisation.

2. The Analysis of the Emission Lines of Wolf-Rayet Spectra

The first challenge is to find what range of parameters is significant for creating Wolf-Rayet type spectra. I use the one-representative-point theory of Castor and van Blerkom (1970) for a wide range of parameters to predict the relative energies radiated in lines in the visible spectral range for comparison with observations. This theory sets up the equations of statistical equilibrium for a model atom and provides expressions for the needed radiation field in the case where a velocity gradient exists in the LER. All emission lines are assumed to be formed by photons which escape from the same body of plasma.

It is essential to carry terms in the equations of statistical equilibrium for all types of radiative and collision transitions which occur in a model atom as a result of the presence of the atom in the ambient radiation field of an LER. Bhatia and Underhill (1986, 1988) tell how to evaluate photoionization from and radiative recombination rates to all levels of the model atoms, rates for collisional ionization and three-body recombination, also photo and collision excitation rates via line transitions and their reverse rates, and how to estimate the effect of dielectronic recombination on the degree of ionization of each atom/ion sensitive to dielectronic recombination.

The "Of" emission lines are strong in Wolf-Rayet spectra. It is now known that these lines are sensitive to the selected cascade routes which may follow dielectronic recombination. It has been impossible so far to carry such details in the equations of statistical equilibrium for simple model Wolf-Rayet LERs. Because the relative intensities of "Of" emission lines to other lines have been empirically selected as criteria for the various WC and WN subtypes, it is impossible at this time to predict patterns of line emission to match what is observed for the various subtypes. However, it has been possible for Bhatia and Underhill to establish the major properties of WC and WN LERs from their studies of the statistical equilibrium of H, He, C, N, and O model atoms under a wide range of conditions and for several compositions, see BU86, BU88, and BU89.

The results are as follows:

QUANTITY	WC LER	WN LER
Radiation temperature, T_*:	25,000 K	25,000 K
Electron temperature, T_e:	$5 \times 10^4 \leq T_e \leq 10^5$ K	$10^5 \leq T_e \leq 2 \times 10^5$ K
Electron density, N_e:	$10^{10} - 10^{11}$ cm^{-3}	$10^9 - 10^{10}$ cm^{-3}
Composition:	solar	solar

These parameters allow one to predict successfully the observed relative intensities in many of the key lines of Wolf-Rayet spectra. See the papers by Bhatia and Underhill for details. The energy of the electrons in the LER, thus the electron temperature, is the dominant factor which causes WC spectra to be different from WN spectra. Density is a second important parameter.

3. Discussion

Five properties of atmosphere models appropriate for Wolf-Rayet stars can be inferred by studying the results of the one-point analyses carried out by Bhatia and Underhill.

3.1 SOLAR ABUNDANCES

Previous conclusions that anomalous abundances are present on the surfaces of Wolf-Rayet stars are a result of the use of incomplete equations of statistical equilibrium or of the assumption of LTE and of failure to explore a wide enough range of T_*, T_e, and N_e such that the true pattern of predicted line intensities as functions of the parameters can be seen. Bhatia and Underhill (1986, 1988, 1989) have discussed the strengths and limitations of their studies. All in all, good agreement with key line ratios used to classify Wolf-Rayet spectra is obtained. The few discrepancies can be understood in terms of the neglect by Bhatia and Underhill of the details of dielectronic recombination, and the inadequate treatment of detailed balance in a few resonance lines. If it was all right to neglect photoionization from and radiative recombination to excited levels, the results of BU88 and BU89 would have confirmed those obtained in the NLTE studies by others. These results were not confirmed. Local thermodynamic equilibrium is not a viable hypothesis for line formation in Wolf-Rayet LERs.

3.2 SIZE AND SHAPE OF THE LER

BU88 find that the emissivity in He II $\lambda 5411$ of typical Wolf-Rayet LERs is such that the LER has a volume of $10^{43} - 10^{44}$ cm^3. Because a contiguous sphere of this volume at the typical N_e in an LER is nearly opaque in electron scattering, (which would imply larger apparent optical continuum radii for Wolf-Rayet stars than are estimated, see Underhill 1983), BU88 suggest the LER may be a thin ring-like disk at a radius of about 10^{15} cm. Evidence that line formation in a rotating, thin ring occurs is given by the observed profile of He I $\lambda 5876$ in HD 191765, WN6. This line has an unchanging double-peaked shape typical of line formation in a ring-like disk where the lines are broadened by macroturbulence, see Underhill et al. (1990).

3.3 T_e IN AN LER

A high value of the order of 10^5 K requires the deposit of non-radiative energy in the LER. BU88 suggest that turbulent mechanical energy in the disk is transformed to heat by MHD effects in the presence of a small magnetic field of interstellar origin. A field of 4–8 gauss is sufficient to create a low-beta plasma which could be heated by MHD effects.

3.4 LOW RATE OF MASS LOSS

If the observed 6-cm fluxes of Wolf-Rayet stars are transformed to rates of mass loss assuming thermal bremsstrahlung in a spherical wind, there is a shortfall of radiative momentum to drive the wind by a factor of 20–50. Underhill (1984) has suggested that part of the 6-cm flux could be due to gyroresonance magnetic bremsstrahlung. Some of the 6-cm flux could also be generated in a disk-driven wind, cf. Pudritz and Norman (1986). Consequently the true M *from the star* is probably much less than $5 \times 10^{-5} M_\odot$ yr^{-1}. The fact that *sharp* subpeaks are seen on the profile of He II $\lambda 5411$ in the spectrum of HD 191765 (Moffat et al. 1988; Underhill et al. 1990) indicates that this star is not surrounded by a dense sphere of electrons. If it were, the subpeaks would be greatly broadened by electron scattering. The rate of mass loss from a Wolf-Rayet star alone may be $\leq 10^{-6} M_\odot$ yr^{-1}.

3.5 T_{eff} OF WOLF-RAYET STARS

The calculations of BU86, BU88, and BU89 clearly indicate that the radiation temperature (T_*) of the continuous spectrum shortward of the Lyman limit is about 25,000 K. Model atmospheres for massive stars of solar composition indicate that this condition is met for $T_{eff} = 25,000$–30,000 K. Such effective temperatures are appropriate for stars having the masses of Population I Wolf-Rayet stars. Solutions of the radial-velocity orbit of the eclipsing Wolf-Rayet binary CQ Cephei (Underhill, Gilroy, and Hill 1990) show that an acceptable solution can be obtained for stars having $M_{WR} = 13.6 M_\odot$ and $M_{comp} = 16.0 M_\odot$. Both stars are in the hydrogen-burning stage. If the Wolf-Rayet star was an evolved object such as postulated by Maeder and Meynet (1987, 1988) and by Langer (1989), its surface abundances would be anomalous and its effective temperature would be about 1.4×10^5 K. The effective temperature of the companion star would be about 30,000 K. Such effective temperatures would not produce nearly equally deep primary and secondary light minima as are observed from far ultraviolet to visible wavelengths (Stickland et al. 1984). The high effective temperatures of the evolved model stars of Maeder and Meynet (1988) and of Langer (1989) make these models unsuitable for representing the Wolf-Rayet star of CQ Cephei and, for that matter, of V444 Cygni. The known properties of the light curves cannot be reproduced at 1300 Å and at 5500 Å when one model star is taken from the sets of highly evolved, peeled down models of Maeder and Meynet or of Langer, and the other from sets of hydrogen-burning models.

4. Conclusions

The model for Wolf-Rayet stars considered by many, see Section 1, to be appropriate is supported neither by reliable analysis of the spectra of Wolf-Rayet stars nor by the properties of eclipsing spectroscopic binaries containing Wolf-Rayet stars.

The model put forward by Underhill, see Section 1, is responsive to the results of spectrum analysis and to the results from eclipsing spectroscopic binaries containing Wolf-Rayet stars. It is a viable proposition to consider Wolf-Rayet stars to be massive pre-mainsequence stars still surrounded by the remnant of a natal disk. Small magnetic fields appear to be present in the disk with the result that MHD effects create the high electron temperatures which Wolf-Rayet spectra imply.

The location of Wolf-Rayet stars in a galaxy indicates those regions where massive stars have formed in the presence of larger than normal interstellar magnetic fields. If these magnetic fields were not present, the newly formed stars would show spectral types in the range B1–O8. Wolf-Rayet spectra imply high T_e in the LER and moderate T_*. This state of affairs is a result of the deposit of non-radiative energy in the LER surrounding the star. It is not a normal result of stellar evolution.

5. References

Bhatia, A. K., and Underhill, A. B. 1986, *Ap. J. Suppl.*, **60**, 323. (BU86)

Bhatia, A. K., and Underhill, A. B. 1988, *Ap. J. Suppl.*, **67**, 187. (BU88)

Bhatia, A. K., and Underhill, A. B. 1989, *Ap. J.*, submitted. (BU89)

Castor, J. I., and van Blerkom, D. 1970, *Ap. J.*, **161**, 485.

Drissen, L., Moffat, A. F. J., Bastien, P., Lamontagne, R., and Tapia, S. 1986, *Ap. J.*, **306**, 215.

Drissen, L., St. Louis, N., Moffat, A. F. J., and Bastien, P. 1987, *Ap. J.*, **322**, 888.

Langer, N. 1989, *Astr. Ap.*, **210**, 93.

Maeder, A., and Meynet, G. 1987, *Astr. Ap.*, **182**, 243.

Maeder, A., and Meynet, G. 1988, *Astr. Ap. Suppl. Ser.*, **76**, 411.

Moffat, A. F. J., Drissen, L., Lamontagne, R., and Robert, C. 1988, *Ap. J.*, **334**, 1038.

Pudritz, R. E., and Norman, C. A. 1986, *Ap. J.*, **301**, 571.

Schulte-Ladbeck, R. E., and van der Hucht, K. A. 1989, *Ap. J.*, **337**, 872.

Stickland, D. J., Bromage, G. E., Budding, E., Burton, W. M., Howarth, I. D., Jameson, R., Sherrington, M. R., and Willis, A. J. 1984, *Astr. Ap.*, **134**, 45.

St. Louis, N., Drissen, L., Moffat, A. F. J., and Bastien, P. 1987, *Ap. J.*, **322**, 870.

St. Louis, N., Moffat, A. F. J., Drissen, L., Bastien, P., and Robert, C. 1988, *Ap. J.*, **330**, 286.

Underhill, A. B. 1983, *Ap. J.*, **266**, 718.

Underhill, A. B. 1984, *Ap. J.*, **276**, 583.

Underhill, A. B., Gilroy, K. K., and Hill, G. M. 1990, *Ap. J.*, **351**, 651.

Underhill, A. B., Gilroy, K. K., Hill, G. M., and Dinshaw, N. 1990, *Ap. J.*, **351**, 666.

DISCUSSION

Niemela: There are also disks around evolved stars, this could be the case for WR stars.
Underhill: I postulate that WR stars are young stars just approaching the ZAMS. The disks around cataclysmic variables have a different origin. Those around WR stars are the remnant of the natal cloud; those around evolved stars are from a companion.

Montmerle: Could you give more details on the analogy you make between WR stars and T Tauri stars? There is no evidence for disks around T Tauri stars from their emission-line spectra; thus evidence comes - indirectly - mostly from their IR excess. Their emission lines are thought to come from a boundary layer between the star and the accretion disk, not from the disk itself.
Underhill: The T Tauri lines have similar shapes as WR lines, mostly a broad rather flat-topped emission feature. Such lines can be formed in rotating turbulent disks, see the numerical experiment of Underhill and Nemec (1989). Also the energy in T Tauri emission lines is $0.1 - 0.3L$ as in WR stars, and the electron temperature in the line-emitting region is higher than in the photosphere in both types of stars. Some investigators suggest the line emission of T Tauri stars comes from a wide ring-like disk. This is what I think is possible for WR stars. The disk is a remnant of the birth cloud.

Maeder: If WR stars would be pre-MS stars, their number ratio to O stars should be about 0.01 rather than 0.1. Of course, I know your argument that WR stars originate in a relatively low mass range, but then one should not only find them in very young clusters.
Underhill: I believe the number of WR stars is no more than 0.01 of all the O stars in the Galaxy. My point is that WR stars are found only where there is massive star formation $(M > 10M_\odot)$, but also weak interstellar magnetic fields must be present to create the necessary weak magnetic field which in turn permits mechanical energy to be transformed to heat energy. This results in the high electron temperature needed to generate a WR-type spectrum.

Dopita: How do you explain the relative paucity of WR stars in low-metallicity systems which are seen to be undergoing a higher specific star formation than, for example, our galaxy?
Underhill: According to my scenario WR stars form only in those molecular clouds where larger than normal interstellar magnetic fields are present.
Dopita: The magnetic field in the LMC appears to be exactly the same as in the solar region.
Underhill: The magnetic field in the LMC surely is not the same everywhere.

Cassinelli: The difficulty that I have with the pre-main sequence model is that WR stars do not have the properties of known pre-main sequence massive stars. The ultracompact HII ($UCHII$) regions that have been studied by Churchwell and Wood are clearly young O stars deep in their natal molecular cloud. These objects are heavily extincted, unobservable at optical wavelengths and have infrared excesses from surrounding dust cocoons. In contrast WR stars often have small $E(b - v)$ ($\lesssim 0.1$), and have a moderate IR excess that can be explained by free-free emission from the wind.
Underhill: A Wolf-Rayet spectrum of emission lines and f-f continuum from a disk-like line-emitting region can be seen only after the dust and other material has dissipated. According to my scenario, WR stars are practically on the ZAMS; their lifetimes as emission-line stars may be as little as $10^4 - 10^5$ years. Many are heavily reddened, possibly in part by the remnant of a cocoon.

THEORY OF WOLF-RAYET ATMOSPHERES

D. John Hillier
Joint Institute for Laboratory Astrophysics
University of Colorado and National Institute of Standards and Technology
Boulder, Colorado 80309-0440
USA

ABSTRACT.   Theoretical modeling of line and continuum formation in Wolf-Rayet (W-R) atmospheres is reviewed. We examine the basic premises on which the models are built, and critically compare theoretical models with observation. Discrepancies between theory and observation are analyzed to determine their implication for our understanding of W-R atmospheres. Line formation, continuum formation, and the ionization structure of W-R envelopes are examined in detail. Abundance determinations are briefly reviewed.

1. INTRODUCTION

The idea that W-R stars consist of a hot core surrounded by a dense stellar wind has been in existence for many years (e.g. Beals 1944, Rublev 1964) but only recently have techniques been developed to allow this theory to be examined in detail. With the assumptions of spherical geometry and homogeneity, it is now possible to self consistently solve the transfer problem in moving atmospheres, allowing us to make a detailed quantitative comparison of theoretical models with observation.

Below we examine recent theoretical modeling of W-R atmospheres. In §2 we describe the *Standard Model*, and examine the basic premises on which it is built. Pure helium models of W-R stars are introduced in §3. This lays the foundation for more detailed discussion of continuum formation (§4), and line formation (§5). The ionization structure is examined in §6, with particular regard to comparison of theory with observation. In §7 we present a critical discussion regarding the current status of the *Standard Model* — we describe its successes and its failures. The current status of abundances in W-R stars is briefly reviewed in §8, while in §9 we critically examine the accuracy of the stellar parameters.

2. STANDARD MODEL

The models to be examined are primarily those of Hillier (1983,1987a,b,1988,1989) and the Kiel group — Hamann, Schmutz, and Wessolowski — in Germany (Hamann 1985, Schmutz and Hamann 1986, Hamann and Schmutz 1987, Wessolowski,

59

K. A. van der Hucht and B. Hidayat (eds.),
Wolf-Rayet Stars and Interrelations with Other Massive Stars in Galaxies, 59–73.
© *1991 IAU. Printed in the Netherlands.*

Schmutz, and Hamann 1988, Hamann, Schmutz, and Wessolwoski 1988, and Schmutz, Hamann, and Wessolowski 1989). In these models the observed W-R emission line spectrum arises in a dense stellar wind which is photoionized by a hot core. The *Standard Model* assumes

(i) spherical geometry,
(ii) a monotonic velocity law,
(iii) homogeneity,
(iv) and time independence (stationary).

In the models of Hillier (1983,1987a,b), radiative equilibrium was assumed, although the temperature in the outer regions of the stellar wind could be adjusted to allow for cooling by metal lines, or mechanical energy deposition. The models by the Kiel group assume a gray temperature distribution (slightly modified at large radii) (*e.g.*, Wessolowski, Schmutz, and Hamann 1988) which is in rough agreement with the radiative equilibrium temperature found in more sophisticated calculations. Both groups can now compute full radiative equilibrium models which include cooling due to N, C and O (*e.g.*, Hillier 1989). In the models to date, line blanketing has been neglected.

The *Standard Model* is probably the simplest model that can be offered to explain the spectral appearance of W-R stars. Its assumptions are well defined and importantly, the radiation transfer equation and the statistical equilibrium equations are solved self consistently. These models provide important insights into line and continuum formation processes in W-R atmospheres — insights which can only be gained from detailed transfer calculations. Only by generating such models can we hope to understand W-R atmospheres: They are a prerequisite to all future modeling.

3. THE He I/He II MODELS OF WOLF-RAYET STARS

The first detailed model of a W-R star was that of Hillier (1983, 1987a,b), who modeled the WN5 star HD 50896. He found that the continuous energy distribution, the He I spectrum, and the He II spectrum could be matched by a model with $R_* = 2.5R_\odot$, $\dot{M} = 5.0 \times 10^{-5} M_\odot$ yr^{-1} and an effective temperature of $55000\,K(L = 5 \times 10^4 L_\odot)$ to $65000\,K(L = 10^5 L_\odot)$. The distance was a free parameter (the above model corresponds to 1.2 kpc) and could not be derived spectroscopically — models with different luminosities, mass loss rates and core radii yielded similar spectra (see below).

The relative strengths of the He I and He II lines are primarily determined by the ionization structure and it is the ratio of these two strengths that allows us to determine the effective temperature of the underlying star. Stars that show appreciable emission in He I $\lambda5876$ in their spectra will have He$^+$ as the dominant ionization state of helium in the radio region (Hillier 1983,1987a, Schmutz and Hamann 1986). The presence of He I thus provides an upper limit to the effective temperature of the underlying star (Schmutz and Hamann 1986).

To match the He I line strengths in HD 50896 the low luminosity model was preferred, however with the inclusion of metal cooling (Hillier 1989) the high luminosity model is now preferred. Cooling due to lines of N and C decreases the electron temperature in the He I formation region from the 30000K adopted by Hillier (1987b) to approximately 15000K (Hillier 1989) which increases the strength of the He I lines. (The He I lines are formed by recombination, and thus their strength scales roughly as $1/T$.)

Hamann, Schmutz, and Wessolowski (1988) have also modeled the spectrum of HD 50896 in detail. They considered three models corresponding to different adopted distances between 0.7 and 2 kpc — all models have a similar effective temperature of approximately 60000 K. Their model B is similar to that of Hillier, and they reach essentially the same conclusions regarding the mass loss rate, luminosity, and core radius. Based on analysis of the interstellar NaD line, a distance of > 1.6 kpc, corresponding to Hamann, Schmutz, and Wessolowski's model A, is to be preferred (Schmutz and Howarth, these proceedings).

Hamann and Schmutz (1987) computed a systematic grid of He II spectra for W-R stars, illustrating the variation of line profiles and line strengths with various parameters, $e.g.$, $\dot{M}$, T_{eff}, R_*. A similar analysis for He I was performed by Wessolowski, Schmutz, and Hamann (1988).

From their detailed numerical calculations Schmutz, Hamann, and Wessolowski (1989) found that approximately the same line equivalent widths could be produced in different models provided that $\dot{M}$, V_∞, and R_* were related. They used a transformed radius R_t to relate these parameters:

$$R_t = R_* \left(\frac{V_\infty}{2500 \mathrm{km\ s^{-1}}} \right)^{2/3} \left(\frac{10^{-4} M_\odot \ \mathrm{yr^{-1}}}{\dot{M}} \right)^{2/3} .$$

Properties of individual stars could thus be read from a single grid of R_t versus T_{eff}. To derive the stellar parameters of WN stars (T_{eff}, $\dot{M}$, R_*) four scaler parameters are required — the absolute flux in one continuum band, the equivalent width of one He II line, the equivalent width of one He I line, and the terminal velocity (as indicated by the He I and He II line profiles) of the stellar wind.

The inclusion of N and C in WN stellar models has only a small effect on the derived parameters of WN stars (Hillier 1989, Hamann and Wessolowski 1990) — the inclusion of these species primarily affects the energy balance and hence the equilibrium wind temperature.

On the other hand, inclusion of C in WC star models has a significant effect. This is not surprising as the C/He abundance (by number) in WC stars is greater than 0.1 (Nugis 1982b, Torres 1988, Smith and Hummer 1988, Hillier 1989). Schmutz, Hamann, and Wessolowski (1989) derived an effective temperature of 35000K for HD 165763 (WC5). With the inclusion of carbon, Hillier (1989) found that a model with an effective temperature of 60000 K gave the best agreement with observations.

4. CONTINUUM FORMATION

Hillier (1983,1987a) has made a detailed investigation into formation of the continuum in HD 50896. These models emphasize the insensitivity of the continuum to effective temperature (T_{eff}), and the importance of other processes in determining the observed continuum shape. We cannot derive effective temperatures from the continuous energy distribution — a comparison of the model energy distribution with that observed can only rule out models, it cannot validate them. The insensitivity of the continuum to the effective temperature for O stars has been discussed by Abbott and Hummer (1985).

We define T_{eff} relative to the stellar radius where the velocity has become subsonic. This may be considerably different from the classic definition in which the radius is chosen at Rosseland optical depth 2/3. This radius can occur well out in the stellar wind, and the temperature so defined in no way reflects the quality of the radiation field — the T_{eff} defined at R_* is much better. Note that T_{eff} is merely a parameterization — it does not mean that the emitted spectrum can be approximated by a blackbody of this temperature. W-R spectra clearly show excesses in the visual and IR, and most do not emit any flux shortward of the He II Lyman limit.

At low mass loss rates the visual flux is proportional to R_*^2 and T_{eff} (as on Rayleigh-Jeans tail of blackbody), but as the mass loss rate increases the visual magnitude becomes increasingly influenced by the stellar wind (see Hillier 1987a). For the hot W-R stars (e.g. WN2-WN6, WC4-WC7) the bulk of the flux is emitted shortward of 912Å — in the $10^5 L_\odot$ model of HD 50896, 75% of the stellar flux is emitted below the Lyman limit, and hence is unobservable. Schmutz, Hamann, and Wessolowski (1989) provide a detailed grid of models illustrating the variation of absolute visual magnitude as a function of $\dot{M}$, R_*, and T_{eff}.

The agreement with observation (see Schmutz, this symposium) is excellent, especially when allowance is made for the neglect of line blanketing and observational error (particularly the correction due to reddening). The radio and IR fluxes are also predicted by these models (e.g., Hillier 1987b) — there is excellent agreement with the IR fluxes ($\lambda \leq 10\mu m$) while the radio fluxes may be systematically over-estimated by up to a factor of 2 (Hamann, Schmutz and Wessolowski 1988, Hillier 1989). For HD 50896 (with d=1.2 kpc) Hillier (1987b) derived a mass loss rate of $5.0 \times 10^{-5} M_\odot$ yr^{-1}, whereas from the radio data of Hogg (1989) a mass loss rate of $3.5 \times 10^{-5} M_\odot$ yr^{-1} is derived. The reason for the discrepancies is unknown — better agreement is obtained if we adopt the larger terminal velocities derived from UV P Cygni profiles rather than those obtained from fitting He I profiles. This problem needs to be re-examined using detailed spectral analysis with the more advanced model calculations now available.

There is a suggestion in some recent observations by Hogg (1989) of some peculiar spectral indices in the radio region (considering only those stars believed to have thermal winds) but as these results are only 2σ, higher signal-to-noise observations are required. It is very important that high signal-to-noise IR and radio flux measurements, such as the 1100 μm IR flux for the WC8 star γ Vel recently measured by Williams et al. (1990), be obtained in order to provide checks on the modeling. Such measurements can place important constraints on the ionization structure, the velocity law, and inhomogeneities in the stellar wind. Likewise, important constraints can be found from radio observations which resolve the stellar wind (e.g., Hogg 1985).

5. LINE FORMATION, AND THE STRATIFICATION OF THE LINE EMISSION WITHIN THE EXTENDED W-R ATMOSPHERE

Although we are discussing line and continuum formation separately I wish to stress that these processes are not independent, and must be modeled simultaneously to match observations.

The *Standard Model* is capable of matching both the line strengths and emission line profiles seen in W-R spectra. Model fits of line profiles with observation are given by Hillier (1987a,1988) and Hamann, Schmutz and Wessolowski (1988), while a comprehensive comparison of predicted line equivalent widths is given by Hillier (1987b,1989). Both the Gaussian He II profiles, and the flat topped He I profiles are reproduced by the *Standard Model*.

Neither Hillier nor the Kiel group has attempted to fine tune the models to systematically fit the observed profiles — even if such a procedure could be done it is unwarranted because of limitations in the models (*e.g.*, the neglect of line blanketing), because of line variability, because of line blending and observational error, and because fine tuning would not affect the results of the analyses. In §7 we examine those differences between theoretical and observed profiles which we believe to be of significance for our understanding of W-R stellar winds.

Using the Sobolev approximation, it is possible to illustrate the regions from which various emission lines arise (*e.g.*, Hillier 1987b,1988,1989). This is very useful since it allows one to determine the local properties affecting the strengths of particular emission lines. In a given series (*i.e.*, n-4 Pickering series), the optical depth decreases with upper principal quantum number, and consequently these lines are formed deeper within the stellar envelope (see Hillier, Jones, and Hyland 1983).

The principal processes leading to W-R emission lines are:

(i) radiative recombination,
(ii) collisional excitation,
(iii) continuum fluorescence and
(iv) dielectronic recombination.

These processes, together with examples of lines formed by each mechanism, are discussed in detail by Hillier (1990a). By understanding line formation processes we are in a better position to determine those lines best suited for abundance determinations, and those lines best suited to provide diagnostics of the atmospheric structure.

The variable ionization structure (§6) has important implications for line formation in the stellar wind. Lines from different ions generally arise in different regions of the stellar wind, and hence their strengths are influenced by different physical conditions. In early W-R stars the He I recombination lines arise at larger radii, and at lower electron temperatures than He II recombination lines. As the recombination rates have an approximate inverse dependence on T_e, the He I emission lines are enhanced relative to He II. In addition, in early W-R stars the He I lines tend to be optically thin, whereas the strengths of the He II lines are influenced by optical depth effects. Similarly, in HD 165763 (WC5), Hillier (1989) found that the C II and C III recombination lines originate at larger radii than the C IV recombination lines.

6. IONIZATION STRUCTURE

In the models of Schmutz and Hamann (1986), Hamann and Schmutz (1987), Schmutz, Hamann, and Wessolowski (1989) and Hillier (1987a,b,1988,1989) the ionization of the envelope decreases outward. This is a firm prediction of the models, and provides a simple explanation for the flat-topped profiles of He I and C III, the observed $10\,\mu$m $-$ 6 cm spectral index, and the observed line strengths.

The models predict a correlation of ionization potential with line width (Hillier 1988,1989), as observed (Kuhi 1973). This is in agreement with the standard picture that the ionization in a W-R envelope decreases outward (Beals 1944, Kuhi 1968). Note this is not an indication of the wind temperature — rather the outward decreasing ionization state of the envelope arises from dilution of the continuum radiation and the important role of photoionizations from excited states.

Ionizations from excited states are of critical importance in determining the ionization of the envelope — arguments by Antokhin, Kholtygin, and Cherepaschuk (1988) that the observed He I and He II line strengths require all W-R stars to have similar effective temperatures are invalid. This is convincingly demonstrated above, and in the paper by Schmutz, Hamann, and Wessolowski (1989) where a spectral analysis of 30 W-R stars is presented. Likewise the observed ionization of other species in W-R spectra does not, at this time, provide convincing evidence that an alternative model of the emission line region is required.

Close to the surface of a WN5 or WC5 star, He^{2+}, C^{4+}, N^{5+} and O^{6+} are the dominant ionization stages. As we move further out in the envelope the ionization systematically decreases so that in a WN5 star, He^{+}, C^{3+}, N^{3+}, and O^{3+} (or O^{2+}) eventually become the dominant ionization stages. In HD 165763 (WC5), He^{+} and C^{2+} are likely to be the dominant ionization stages in the radio region. The large optical depth in the He II Lyman continuum (i.e., $\lambda < 228$Å), the large optical depths in the resonance lines coupled with the extreme UV continuous radiation field, and ionizations from excited states play a fundamental role in determining the ionization structure of the envelope. The predicted ionization structure is illustrated by Hillier (1988) for WN stars, and by Hillier (1989) for WC stars.

The inclusion of metals in the H/He W-R models of Hillier, and the Kiel group is extremely significant as it provides an important consistency check of the models. Discrepancies in model N/C/O line strengths (as a function of ionization state) will indicate deficiencies of the stellar model.

The present calculations are unable to simultaneously match all stages of ionization — in particular in the WC model of HD 165763 the O VI doublet at $\lambda\lambda 3811, 3834$ is too weak, indicating that a higher ionization model is required. A model with twice the luminosity matches better the O VI lines (but is still too weak) but then underestimates the strength of the He I lines by a factor of 2. Similarly, a model that fits the He I/He II ratio in HD 50896 underestimates the strength of the N V $\lambda\lambda 4604, 4620$ doublet (Hillier 1988). The inclusion of line blanketing may resolve these discrepancies.

7. STANDARD MODEL — SUCCESSES AND FAILURES

The *Standard Model* is very elementary, and it has been realized from the outset that this model is only an approximation to the true situation in W-R stars. Line profile variability, or the production of X-rays, cannot, for example, be addressed with the *Standard Model*. The model will stand or fall, or undergo modification based on its agreement with observation. Below we summarize the successes and failures of the *Standard Model* and discuss their implications for our understanding of W-R atmospheres.

7.1 Successes of *Standard Model*

(i) The *Standard Model* provides a good match to the observed continuum of W-R stars (from 6cm to 1500Å), to the strengths of the He I and He II lines, and to their general profile shapes. The ionization stratification predicted in the models is consistent with observation, and in conjunction with optical depth effects provides an explanation for the different profiles illustrated by different ionization species.

(ii) The ionization structure for metals is in rough agreement with observation. The discrepancies for individual lines are generally at the factor of 2 level (a few lines differ by up to a factor of 5) (Hillier 1988,1989; Wessolowski, Hamann, and Schmutz — these proceedings). In the WC model of Hillier (1989) the C II, C III, C IV, He I and He II spectra were generally reproduced to better than a factor of 2.

7.2 Problems With *Standard Model*

(i) Incorrect metal line strengths: In the WC model of Hillier (1989), several lines of carbon were not produced. For example, the strength of C IV $\lambda5805$ was underestimated by a factor of 5. The reason for this discrepancy is unknown. Likewise, some C III line strengths did not agree with observation. For some C III lines (*e.g.*, C III $\lambda6740$) this is due to inadequate atomic data. New atomic data from the opacity project, and specifically new C III atomic data computed by Peter Storey (private communication) should help rectify this problem.

Likewise, in WN models it is impossible at the present time to fit all N lines simultaneously (Hamann, Wessolowski, and Schmutz — these proceedings).

At present there are three basic deficiencies of calculations with the *Standard Model*. We neglect line blanketing, and line overlap, and accurate atomic data are only now becoming available. As a consequence it is difficult to decide whether these discrepancies provide useful insights into the structure of W-R winds — more detailed and advanced calculations are required.

(ii) P Cygni absorption on the He I and He II lines: For the strong lined WN stars, the models tend to produce a flat topped He I $\lambda5876$ line profile (as observed), but with a P Cygni absorption much deeper than observed. This is significant, and may indicate a real problem with the *Standard Model*. The P Cygni profile arises from the large column density along the line of sight to the stellar core — the velocity gradient along this line of sight is small. At present two possible solutions present themselves: we need to decrease the column density, or alternatively, allow for a fluctuating velocity field (macro-turbulence). Evidence for fluctuations in the velocity field come from UV lines — the P Cygni absorption is seen to vary in some stars (*e.g.*, St-Louis *et al.* 1989). As neutral helium is always an impurity species, the presence of density inhomogeneities is unlikely to solve this problem.

A similar problem applies to the He II lines, principally He II λ1640. The *Standard Model* yields a black absorption on the blue side — the observed P Cygni profiles are broader, and are generally not black.

(iii) P Cygni absorption on C III λ1909: As for He I, and He II, the *Standard Model* produces a deep absorption of the C III λ1909 intercombination line which is much stronger than observed (Hillier 1989). Absorption is seen but it is weaker. This may be related to the problem with the P Cygni absorption dip on the He I and He II lines, but could also be a problem with the ionization structure. In the WC model of Hillier (1989) C^{2+} becomes the dominant ionization state of carbon in the outer regions of the stellar wind and thus density inhomogeneities might help solve this problem (as column density is less).

(iv) Electron scattering wings: The strength of the electron scattering wings provides a consistency check on the mass loss rates derived from line profile fitting. If the mass loss rates are higher than 10^{-5} $M_\odot$ yr^{-1} these wings must exist (Auer and Van Blerkom 1972, Hillier 1984). Because of the expanding atmosphere, and because the scattering is coherent in the frame of the electron, we preferentially see a red wing on W-R emission lines. In P Cygni stars, which have considerably lower terminal velocities, the electron scattering profile is much more symmetric (see Hillier, McGregor, and Hyland 1988 for an example).

In WN stars two lines provide the best evidence for the wings — He II λ1640, and He II λ5411. Alternate explanations for the wings have been provided but at the present time these are unsatisfactory. Suggestions that the wing on the He II line at λ5411 line is due to OV, C IV, N III, or O III are not satisfactory as lines expected to be stronger are not seen (*e.g.*, Bappu 1973). A peculiar excitation mechanism may operate, but none has been offered.

The electron scattering wings produced in the models of Hillier (1987b, 1988, 1989) tend to be stronger than observed — possibly by a factor of 2. If the wing on λ5411 is a blend, the discrepancy is even larger. Such discrepancies may be due to density inhomogeneities in the stellar wind — the emission processes scale as the density squared whereas the electron scattering optical depth scales as the density. Electron scattering wings of reduced strength can be obtained in this way (Hillier 1990b).

8. ABUNDANCES

From the studies of Willis and Wilson (1978), Nugis (1975,1982a,b), Smith and Willis (1982,1983), and others, and more recent work by Torres (1988), Smith and Hummer (1988), de Freitas Pacheco and Machado (1988), and Hillier (1988,1989) there can be little doubt that the WN and WC sequences have different elemental abundances. As the status of abundances in W-R stars has recently been critically discussed by Hillier (1990a) I will review the situation only briefly.

WN stars have enhanced N and He (relative to solar abundances), and are deficient in O, C, and H. Extensive mass loss has revealed, at the surface, matter processed by the CNO nuclear burning cycle. For WNE stars Smith and Willis (1982) found C/N ratios of 0.01 to 0.09 by number. For HD 50896, Hillier found from detailed modeling a C/N ratio of 0.07 (within a factor of 3). These values are in excellent agreement with evolutionary models which give a range from 0.02 to 0.05 (e.g., Maeder 1983, 1987, Prantzos et al. 1986). The N/He ratio is more difficult to determine because of the different sensitivity of the N and He lines to the wind electron temperature. More detailed calculations are required.

WC stars have enhanced C, O, and He, and are deficient in H, and probably N. In this case, the products of He nuclear burning have been revealed at the surface. C/He ratios of > 0.1 are now favored (Nugis 1982b, Torres 1988, Smith and Hummer 1988, Hillier 1989). Earlier determinations by Smith and Willis (1982,1983) that the C/He ratio was 0.01 to 0.04 underestimate the C/He abundance. Smith and Willis used the strength of C III λ2296 to constrain the electron temperature but this line is now known to be significantly influenced by dielectronic recombination (Storey 1981, Hillier 1989). Smith and Willis overestimated the wind electron temperature, and hence underestimated the C/He abundance. In the WC model of Hillier (1989) there is no disagreement between the strong UV collisionally excited lines, and the optical/IR recombination lines.

The above abundance determinations are entirely consistent with the observed properties of W-R stars. At least four observational arguments lend strong support to the interpretation that W-R stars are evolved objects in which mass loss has revealed nuclear processed material at the surface, that many W-R stars (but not all) contain no H, and that the WN and WC sequences are a direct result of different surface compositions.

Massey and Conti (1980) found that the WN8 star HD177230 (which shows emission due to N II, N III, N IV, N V and He I) is severely hydrogen deficient. This star has an excitation similar to other WN8 stars that clearly show the presence of hydrogen. The only non-contrived means of explaining this observation is on the basis of different H abundances.

Bhatia and Underhill (1986,1988) argue that the difference in strengths of N emission lines in WN and WC stars is primarily due to different wind electron temperatures, and is not an abundance indicator. This is contradicted by observation. In HD 165763 emission due to O VI λ3811, 3834 is present (Bappu 1973) but in HD 50896 (WN5) such emission is not detectable. If HD 50896 has a higher excitation than HD 165763 (as argued by Bhatia and Underhill on the basis of the strong N IV, N V emission line spectrum), why are the higher excitation O VI emission lines not seen (Hillier 1990)?

The WC5 star HD 165763 (WC5) and the WN5 star HD 50896 are characterized by very different C and N spectra (as expected by their spectral types). However, the Fe V and Fe VI emission spectrum at UV wavelengths (Fig. 1) is very similar. As Fe V has an ionization potential very similar to that of N IV, while that of Fe VI is virtually identical to N V, it is difficult to see how excitation effects can be invoked to explain the observed spectral differences between these two stars. The importance of Fe IV, Fe V, and Fe VI in W-R stars has previously been emphasized by Nugis and Sapar (1985a,b), Eaton, Cherepaschuk, and Khaliullin (1985a,b) and

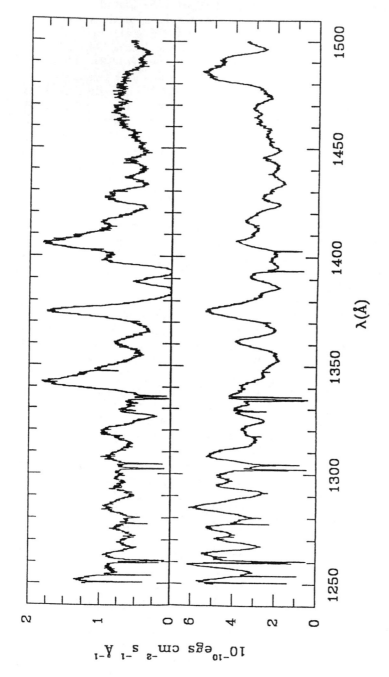

Fig. 1.—High resolution spectra of the WC5 star HD 165763 (top) and the WN5 star HD 50896. The spectra are strikingly similar in this spectral region — most of the emission features must be due to Fe V and Fe VI. In HD 165763 the emission at 1342Å is probably due to O IV λ1338.6 − 1343.5, while the complex near 1400 Å is due to Si IV and O IV]. In the spectrum of HD 50896, N IV] λ1486, is readily apparent. The spectra were kindly supplied by Ian Howarth (Howarth and Phillips 1986, and Howarth, private communication).

Koenigsberger and Auer (1985).

Finally, we note that N lines are seen in the spectra of some WC and WO central stars of planetary nebulae (Heap 1982, Barlow and Hummer 1982). This suggests that the WC sequence is not a result of excitation — rather the absence of N in population I WC spectra indicates a low N/C abundance ratio.

9. ACCURACY OF STELLAR PARAMETERS

The accuracy of current stellar parameter determinations is very difficult to ascertain. There are several reasons for this:

(i) The bulk of the flux in the early W-R stars is unobservable, and hence model calculations must be used to constrain the luminosity and effective temperature.

(ii) The mass loss mechanism is still not understood. What drives the wind, and why can the observed wind momenta exceed the single scattering limit by a factor of 40?

(iii) It is difficult to determine the significance of discrepancies between theory and observation.

(iv) The models to date neglect line blanketing which must play an important role in determining the atmospheric structure.

Firstly, model mass loss rates are comparable to radio mass loss to better than 50% , although it appears that the model mass loss rates may be upper estimates. The principal uncertainty with the mass loss rates is the importance of clumping — hopefully analysis of electron scattering wings and other density sensitive diagnostics will allow strong constraints to be placed on the importance of this effect.

With regard to bolometric corrections, Schmutz (1990) argues that they are incorrect by at most 1 magnitude. He bases his conclusion on a comparison of different models made with different assumptions, and on the basis of preliminary line blanketing calculations. Bolometric magnitudes and effective temperatures are likely to be more accurate for the cooler W-R stars, and those with relatively weak (*i.e.*, low density) winds (*e.g.*, WN 7 stars).

Support for the luminosities and effective temperatures derived from the *Standard Model* comes from analyses of nebulae. Smith and Clegg (1990), Rosa and Mathis (1990), and Vogel (1990) find T_{eff}'s compatible with that found by the Kiel group.

Of fundamental concern for the radii and effective temperatures we derive is the difficulty in relating these parameters to those adopted in evolutionary calculations. Until we understand the wind dynamics this cannot be achieved. It is the strong lined WN stars, and the WC stars that are most affected by the atmospheric extension.

This problem is highlighted by some recent calculations made by Hillier to address observational and theoretical constraints on the velocity law in W-R stars. A model with $L = 10^5 L_{\odot}$, $R_* = 2.5 R_{\odot}$ (*i.e.*, $T_{eff} = 65,000 K$), $\dot{M} = 5 \times 10^{-5} M_{\odot} \, \mathrm{yr}^{-1}$, $V_{\infty} = 1700 \, \mathrm{km \, s^{-1}}$, and $\beta = 1$ (exponent in standard velocity law) was found to have a continuum, and line profiles (for selected lines of He I, He II, C IV, N IV, and N V) observationally indistinguishable (see Fig. 2) from a model of the same luminosity and mass loss rate, but with $R_* = 1 R_{\odot}$

(*i.e.*, $T_{eff} = 100,000K$), $V_\infty = 1800\,\mathrm{km\,s^{-1}}$, and a slow velocity law characterized by $\beta = 3$. Clearly, we need a dynamic model to help constrain the velocity
law, and to allow reliable determinations of effective temperatures for comparison
with evolutionary calculations.

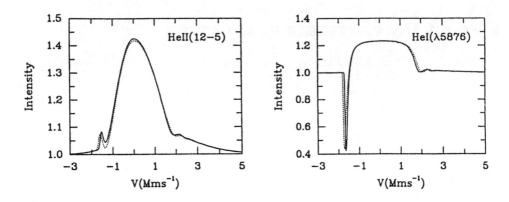

Fig. 2.— Example line profiles computed for two models characterized by
different velocity laws and core radii. The solid line has $R_* = 2.5R_\odot$, $V_\infty = 1700\,\mathrm{km\,s^{-1}}$, and $\beta = 1$, while the broken line is for a model with $R_* = 1.0R_\odot$,
$V_\infty = 1800\,\mathrm{km\,s^{-1}}$, and $\beta = 3$. The profiles are observationally indistinguishable.
Similar comments apply to other lines such as N V $\lambda1240$, N V $\lambda4609$, N IV $\lambda1486$,
and C IV $\lambda1549$.

10. CONCLUSION

Since the last IAU Symposium on W-R stars our knowledge and understanding of
these stars has dramatically increased. Sophisticated models can now be used to
provide essential diagnostic information regarding W-R stellar parameters, and the
structure of W-R stellar winds. Improvements to the codes are in progress, and it
is hoped that these, combined with high quality observations over a wide spectral
range, will allow further significant advances.

The *Standard Model* used to represent the extended atmosphere of a W-R star is
an excellent first approximation. The ionization stratification, and the importance
of ionizations from excited states will also appear in more complicated models.
Inferences concerning the structure of W-R stellar atmospheres which are based on
models that do not allow for these effects must be treated with caution.

ACKNOWLEDGEMENTS

The author wishes to thank Drs. Peter Storey, David Hummer, and Werner Schmutz for invaluable discussions. Thanks to Lorraine Volsky for editorial assistance. This work was supported by National Science Foundation grant AST88-02937 through the University of Colorado. The computations were done on the JILA VAX 6440.

REFERENCES

Abbott, D. C., and Hummer, D. G. 1985, *Ap. J.*, **294**, 286

Antokhin, I. I., Kholtygin, A. F, and Cherepaschuk, A. M. 1988, *Astr. Zh.*, **65**, 558 (*Sov. Astr.*, **32**, 285)

Auer, L. H., and Van Blerkom, D. 1972 *Ap. J.*, **178**, 175

Bappu, M. K. V. 1973, in IAU *Symposium 49, Wolf-Rayet and High Temperature Stars*, ed. M. K. Bappu and J. Sahade (Boston: Reidel), p. 59

Barlow, M. J., and Hummer, D. G. 1982, in *IAU Symposium 99, Wolf-Rayet Stars: Observations, Physics, Evolution*, ed. C. W. H. de Loore, and A. J. Willis (Dordrecht: Reidel), p. 387

Beals, C. S. 1944, *M.N.R.A.S.*, **104**, 205

Bhatia, A. K., and Underhill, A. B. 1986, *Ap. J. Suppl.*, **60**, 323

Bhatia, A. K., and Underhill, A. B. 1988, *Ap. J. Suppl.*, **67**, 187

Eaton, J. A., Cherepaschuk, A. M., and Khaliullin, Kh. F. 1985a, *Ap. J.*, **296**, 222

Eaton, J. A., Cherepaschuk, A. M., and Khaliullin, Kh. F. 1985b, *Ap. J.*, **297**, 266(*erratum* 1988, **334**, 1076)

de Freitas Pacheco, J. A., and Machado, M. A. 1988, *A. J.*, **96**, 365

Hamann, W.-R., 1985, *Astr. Ap.*, **145**, 443

Hamann, W.-R., and Schmutz, W. 1987, *Astr. Ap.*, **174**, 173

Hamann, W.-R., Schmutz, W., and Wessolowski U. 1988, *Astr. Ap.*, **194**, 190

Hamann, W.-R., and Wessolowski, U. 1990 *Astr. Ap.*, **227**, 171

Heap, S. R. 1982, in *IAU Symposium 99, Wolf-Rayet Stars: Observations, Physics, Evolution*, ed. C. W. H. de Loore, and A. J. Willis (Dordrecht: Reidel), p. 423

Hillier, D. J. 1983, Ph. D. thesis, Australian National University, Canberra

Hillier, D. J. 1984, *Ap. J.*, **280**, 744

Hillier, D. J. 1987a, *Ap. J. Suppl.*, **63**, 947

Hillier, D. J. 1987b, *Ap. J. Suppl.*, **63**, 965

Hillier, D. J. 1988, *Ap. J.*, **327**, 822

Hillier, D. J. 1989, *Ap. J.*, **347**, 392

Hillier, D. J. 1990a, in *Properties of Hot Luminous Stars*, ed. C. D. Garmany (Astron. Soc. Pacific Conf. Series, Vol. 7), p. 340

Hillier, D. J. 1990b, *in preparation*

Hillier, D. J., Jones, T. J., and Hyland, A. R. 1983 *Ap. J.*, **271**, 221

Hillier, D. J., McGregor, P. J., and Hyland, A. R. 1988, *Mass Outflows from Stars and Galactic Nuclei*, ed. L. Bianchi and R. Gilmozzi, (Dordrecht: Kluger), p. 215

72

Hogg, D. E. 1985, in *Radio Stars*, ed. R. M. Hjellming and D. M. Gibson (Dordrecht: Reidel), p. 117

Hogg, D. E. 1989, *A. J.*, **98**, 282

Howarth, I. D., and Phillips, A. P. 1986, *M.N.R.A.S.*, **222**, 809

Koenigsberger, G., and Auer, L. H. 1985, *Ap. J.*, **297**, 255

Kuhi, L. V. 1968, in *Wolf-Rayet Stars*, ed. K. B. Gebbie and R. N. Thomas (NBS SP-307), p. 101

Kuhi, L. V. 1973, in IAU *Symposium 49, Wolf-Rayet and High Temperature Stars*, ed. M. K. Bappu and J. Sahade (Boston: Reidel), p. 205

Maeder, A. 1983, *Astr. Ap.*, **120**, 113

Maeder, A. 1987, *Astr. Ap.*, **173**, 247

Massey, P., and Conti, P. S. 1980, *Ap. J.*, **242**, 638

Nugis, T., 1975, in *IAU Symposium 67, Variable Stars and Stellar Evolution*, ed. V. E. Sherwood and L. Plaut (Dordrecht: Reidel), p. 291

Nugis, T., 1982a, in *IAU Symposium 99, Wolf-Rayet Stars: Observations, Physics, Evolution*, ed. C. W. H. de Loore and A. J. Willis (Dordrecht: Reidel), p. 127

Nugis, T., 1982b, *IAU Symposium 99, Wolf-Rayet Stars: Observations, Physics, Evolution*, ed. C. W. H. de Loore, and A. J. Willis (Dordrecht: Reidel), p. 131

Nugis, T., and Sapar, A. 1985a, *Pis'ma Astron. Zh.* **11**, 455 (*Sov. Astron. Lett.* **11**, 188)

Nugis, T., and Sapar, A. 1985b, *Tartu Astrofüüs. Obs. Teated* Nr. 80

Prantzos, N., Doom, C., Arnould, M., and de Loore, C. 1986, *Ap. J.*, **304**, 695

Rosa, M. R., and Mathis, J. S. 1990, in *Properties of Hot Luminous Stars*, ed. C. D. Garmany (Astron. Soc. Pacific Conf. Series, Vol. 7), p. 135

Rublev, S. V., 1964, *Astr. Zh.*, **41**, 63(*Sov. Astr.*, **8**, 45)

Schmutz, W. 1990, in *Properties of Hot Luminous Stars*, ed. C. D. Garmany (Astron. Soc. Pacific Conf. Series, Vol. 7), p. 117

Schmutz, W., and Hamann, W.-R. 1986, *Astr. Ap.*, **166**, L11

Schmutz, W., Hamann, W.-R., and Wessolowski, U. 1989, *Astr. Ap.*, **210**, 236

Smith, L. F, and Clegg, R. E. S. 1990, in *Properties of Hot Luminous Stars*, ed. C. D. Garmany (Astron. Soc. Pacific Conf. Series, Vol. 7), p. 132

Smith, L. F., and Hummer D. G. 1988, *M.N.R.A.S.*, **230**, 511

Smith, L. J., and Willis A. J. 1982, *M.N.R.A.S.*, **201**, 451

Smith, L. J., and Willis A. J. 1983, *Astr. Ap. Suppl.*, **54**, 229

St-Louis, N., Smith, L. J., Stevens, I. R., Willis, A. J., Garmany, C. D., and Conti, P. S. 1989, *Astr. Ap.*, **226**, 249

Storey, P. J. 1981, *M.N.R.A.S.*, **195**, 27P

Torres, A. V. 1988, *Ap. J.*, **325**, 759

Vogel, W. 1990, in *Properties of Hot Luminous Stars*, ed. C. D. Garmany (Astron. Soc. Pacific Conf. Series, Vol. 7), p. 129

Wessolowski, U., Schmutz, W., and Hamann, W.-R. 1988, *Astr. Ap.*, **194**, 160

Williams, P. M., van der Hucht, K. A., Sandell, G., and Thé, P. S. 1990, *M.N.R.A.S.*, **244**, 101

Willis, A. J., and Wilson, R. 1978, *M.N.R.A.S.*, **182**, 559

DISCUSSION

Pollock: With regard to abundance measurements, I would encourage you to have a look at the X-ray measurements of HD 193793 of Williams *et al.* (1990, MNRAS, <u>243</u>, 662) where an X-ray source suffers increased absorption as it moves deeper into the wind of the WC7 star. This gives an essentially ionization independent abundance measurement of $N_c \sim 0.06$.

Hillier: It would be of use to compare the above abundance with that derived from a detailed spectroscopic analysis. While the abundance you derive is ionization independent, it will contain other model dependencies.

Underhill: You remarked that the large discrepancies with the absorption components predicted by you are "no serious limitation"! I disagree entirely. They are a significant test of your model's physical state; they show that the models fail. The fact that $\dot{M}v_\infty c/L$ is of the order of 40-50 in many of the Kiel models and your models is a disaster. You imply that there is a strong force, unknown to physics, present to drive the wind. I refuse to believe in presently unknown long-range forces of physics.

Hillier: We are well aware of the discrepancies of the model with observation, however the "Standard Model" describes the spectrum too well for it to be seriously in error. One purpose of the "Standard Model" is to identify such discrepancies so that we can further our understanding of WR stellar winds. The problem with the absorptions is principally one of degree - strong P Cygni absorption is seen in the $HeI\lambda5876$ line in HD 192163. The observed variability of the P Cygni profiles, both in the UV and optical, clearly show that the intrinsic instability of radiation winds is having a significant influence on the line profiles. We are currently addressing the implications for our analyses. As regards the momentum problem, it must be emphasized that the single scattering limit is not an absolute limit to the momentum that can be delivered to the stellar wind (see Cassinelli, this symposium). In WR stars multiple scattering will be very important in determining the wind dynamics. We also need to consider line overlap, the diffuse radiation field and the large optical depth of the stellar wind at the sonic point. Whether these effects can lead to the large momentum depositions required is still an open question. We also need to consider the errors in the determinations of M and L. It is possible that the single scattering limit is exceeded by only a factor of 10, rather than the factor of 40 currently found for some objects.

Langer: Many WR stars have wind densities as high that the optical depth (continuum) close to the stellar surface must be high and you cannot observationally determine the velocity law in this part of the wind. How does this uncertainty affect your results, especially results on R_*?

Hillier: This question was already answered in my talk.

John Hillier

CHEMICAL COMPOSITION IN THE ENVELOPES
OF DIFFERENT WR SUBTYPES

T. NUGIS
W.Struve Astrophysical Observatory of Tartu
202444 Tõravere
Estonia, USSR

ABSTRACT. Valuable model–independent information about the chemical composition of WR stars can be obtained by using lines of different ions arising at transitions between highly excited energy levels. We determine the abundances of He, H, C, N, O using theoretical line intensities obtained by solving statistical equilibrium equations for level populations for different subtype WR envelope conditions.

The chemical composition of WR stars strongly differs from the mean cosmic composition. In the case of WN stars nuclear processed CNO products of H-burning are revealed in the envelopes and He-burning products in the WC envelopes.

The abundance of oxygen in WC stars differs from the predictions of recent evolutionary calculations which account for the new $^{12}C(\alpha, \gamma)\,^{16}O$ reaction rate.

1. Model–dependence

In recent years new observational evidence has been obtained which point to the complicated nature of the envelopes (winds) of WR stars (clouds, deceleration and coronal zones). This, together with the unexplained nature of their intensive mass loss, does not allow the development of physically self–consistent atmospheric models for this type of star. Therefore, it is very important to find out in every case whether the results depend on concrete models and physical parameters. For the determination of the chemical composition of WR stars it is possible to use for some most abundant elements the subordinate lines which arise from transitions between highly excited energy levels (hydrogenic transitions). Our special study shows that the ratios of theoretical intensities of these lines depend weakly on a concrete model of the envelope and physical conditions. Therefore, these lines are very suitable for abundance determinations.

Our present determinations of chemical composition are based on solutions of the statistical equilibrium equations of level populations for the ions of H, He, C, N and O. All important collisional and radiative processes of population and depopulation of levels are considered. Radiation transfer effects in line frequencies were accounted by the Sobolev escape probability method as formulated by Castor

75

K. A. van der Hucht and B. Hidayat (eds.),
Wolf-Rayet Stars and Interrelations with Other Massive Stars in Galaxies, 75–80.
© 1991 *IAU. Printed in the Netherlands.*

(1970) for spherically symmetric expanding media. The partial nontransparency of WR envelopes in the continua of the lowest series of abundant ions is treated approximately as described by Nugis (1990). Usually reliable results for theoretical line intensities can be obtained by solving statistical equilibrium equations either for full transparency of the envelope in continua or by using "on-the-spot" approximation. Before the presentation of our results for the chemical compositions of WR stars, we want to prove by concrete examples that quite reliable abundances for dominant elements can be determined even in the case of quite inadequate models for the real envelopes of WR stars and even by using very simple methods of calculation of theoretical line intensities. This becomes evident when analyzing the logic of the nearly normal (solar) chemical composition for WR stars proposed by Bhatia and Underhill (1986, 1988). We repeated the calculations of level populations of some ions for their WR envelope models (HeII, HI, CIV, CIII) and obtained for the lines, used by them nearly the same theoretical intensities. They used for the abundance study strong lines arising from transitions between low energy levels (CIV $\lambda5806$, CIII $\lambda5696$, NIV $\lambda3482$ and so on). Models of Bhatia and Underhill result in strong overpopulation of low levels of the studied ions. We included in statistical equilibrium calculations a larger number of levels for C, N ions and found that for the Bhatia and Underhill models ($T_* \approx 25000$ K, $T_e \approx 100000$ K, $N_e \approx 10^{10}$cm^{-3}, $dv/dr \approx 10^{-5}$) the ratios of intensities of subordinate lines arising from transitions between low energy states and of highly excited states are in conflict with the observed data (Figs. 1, 2).

The HeII line intensity run of the Pickering series, as predicted by the calculations of Bhatia and Underhill (1986), is also in serious conflict with the observed run. In the case of every WR star the lower members of the Pickering series have intensities much smaller than the LTE prediction for an optically thin case (the mean observed coefficients $b_k\beta_{ik}$ are lower than unity (b_k is the Menzel departure coefficient and β_{ik} is the escape probability coefficient for line quanta)) (Figs. 3, 4). Bhatia and Underhill (1986) models predict $b_k\beta_{ik} > 1$ for lower members of the Pickering series. From our investigation it can be concluded that the models of Bhatia and Underhill (1986, 1988) are in serious conflict with the observed line spectra, but the line intensity ratios of subordinate lines arising at transitions between high energy levels are quite close to the ratios predicted by our WR wind models. Very wrong models can be used for obtaining reliable results, only "selected lines" must be used. For subordinate lines arising at transitions between high energy levels even very simple calculation schemes (the recombination theory) predict quite close results to those obtained by much more realistic envelope models and appropriate statistical equilibrium calculations for level populations.

2. Chemical composition

For determinations of abundance ratios of most dominant ions we used the theoretical line intensities found from the solutions of the statistical equilibrium equations for level populations for wind models of different WR subtypes. We used the mean parameters of WR stars and their winds as derived by Nugis (1989). For concrete estimates the envelope was divided into different effective subzones having different T_e values as derived from energy balance equations for free electrons. The mean

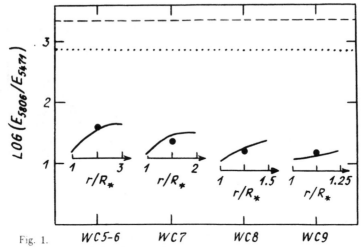

Fig. 1.

● – mean ratios of the observed CIV line energies (the data are taken from Torres, 1985).

– – – theoretical prediction of the model envelope of Bhatia and Underhill (1988) ($T_* = 25000$ K, $T_e = 100000$ K, $R_* = 10$ $R_\odot$, $r = 3$ R_*, $dv/dr = 10^{-5}$, normal C abundance).

··· – 10^3 times higher carbon abundance with the other parameters being same as for the previous (– –) model.

—— – the predictions of our wind models at some different r values.

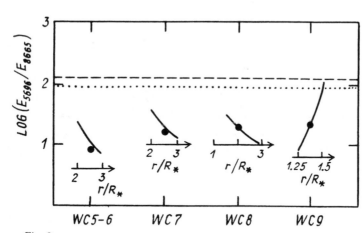

Fig. 2.

● – mean ratios of the observed CIII line energies (the data are taken from Torres, 1985 and Conti et al., 1989).

– – – theoretical prediction of the model envelope of Bhatia and Underhill (1988) ($T_* = 25000$ K, $T_e = 100000$ K, $R_* = 10$ $R_\odot$, $r = 3$ R_*, $dv/dr = 10^{-5}$, normal C abundance).

··· – 10^3 times higher carbon abundance with the other parameters being same as for the previous (– –) model.

—— – the predictions of our wind models at some different r values.

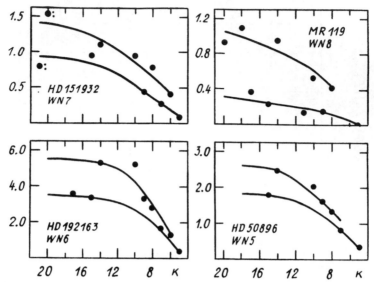

Fig. 3. Observed energies of the Pickering series lines divided by theoretical intensities corresponding to an optically thin medium having LTE populations. The difference between odd and even members of the Pickering series gives us information about the presence of hydrogen.

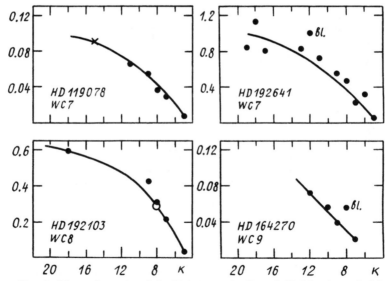

Fig. 4. Observed energies of the Pickering series lines divided by theoretical intensities corresponding to an optically thin medium having LTE populations. The difference between odd and even members of the Pickering series gives us information about the presence of hydrogen.

o — lines of HeII K-3 series;

x — lines of HeII K-5 series;

bl. — blended line.

value of the expansion velocity was adopted to be $v_{max}/2$ for early subtypes and outer regions $(r \geq 1.5R_*)$ of the envelopes of later subtypes (WC8-9, WN7-8). In the inner regions of late subtype stars we adopted $v = v_{max}/3$. The velocity gradient was taken to correspond to the law $v \propto r$.

From our present calculations and also from our other recent studies we infer the presence of a coronal zone in the inner part of an envelope. This coronal zone ought to be opaque at wavelengths $\lambda < 228\mathring{A}$. The radiation temperature in that region must be higher than for $\lambda > 228\mathring{A}$ spectral region. The results of our chemical composition estimations are given in Table 1. For final determinations we used the higher members of the Pickering and Balmer series, and the lines: HeI (4471.5, 5875.6), CIV (5471), CIII (8665, 4070), CII (4267), NV (4945), NIV (2646, 6219), NIII (4379), OVI (3811, 3834), OV (2783, 5600), OIV (1343, 3063+3071, 3412), OIII (3265, 3760).

The oxygen abundance derived for 6 WC stars differs from the prediction of the latest calculations of WR evolution (Prantzos et al., 1986; Langer and El Eid, 1986; Maeder, 1987). Interestingly enough, the abundances of He, C, O for WC stars derived by us are in accord with earlier evolutionary calculations where a lower rate of reaction $^{12}C(\alpha,\gamma)$ ^{16}O was used. The discrepancy with evolutionary calculations was found also in the case of Ne abundance in the envelope of WC8 component of binary γ^2 Vel (Barlow et al., 1988).

TABLE 1. Chemical composition of WR stars. Mean values of the ratios of most abundant elements are presented. The numbers of the stars used for finding the mean values are given in brackets (in the case of a single estimate the HD number of that star is given).

Sp	N(He)/N(H)		N(N)/N(He)		N(C)/N(He)	N(O)/N(He)	
WN3	5–10	(9974)	0.003–0.01	(9974)			
WN4	3–10	(187282)	0.003–0.01	(187282)			
WN5	3–6	(2)	0.003–0.01	(2)	≤ 0.004 (2)		
WN6	1.5–6	(2)	0.003–0.01	(2)	≤ 0.001 (2)		
WN7	1.5–3	(151932)	0.003–0.01	(151932)			
WN8	0.7–1.5	(2)	0.003–0.01	(2)			
WC5–6	≥ 10	(2)			0.3–0.7 (3)	0.05	(3)
WC7	≥ 10	(156385)			0.2–0.4 (3)		
WC8	≥ 10	(192103)			0.1–0.3 (2)	0.02	(2)
WC9	≥ 10	(164270)			0.1–0.2 (9)	0.02	(164270)

We found that the carbon abundance is increasing from later WC classes to earlier classes. This tendency was found also by Smith and Hummer (1988) from IR carbon lines by using the recombination theory for predicting line intensities. Torres (1988) has not found such a trend for WC stars, but in the error limits (factor $\approx$ 2–3) all the results are in accord. In the case of WN classes some chemical evolution from later classes to earlier classes seems to be present as well.

The chemical composition of WR stars strongly differs from the mean cosmic composition. In the case of WN stars nuclear processed CNO products of H–burning are revealed in the envelopes and He–burning products in WC envelopes. The transition from WN type to WC type probably does not take place for most WN subclasses.

References

Barlow, M.J., Roche, P.E. and Aitken, D.K., 1988. *M.N.R.A.S.*, **232**, 821.
Bhatia, A.K. and Underhill, A.B., 1988. *Ap.J. Suppl.* **67**, 187.
Bhatia, A.K. and Underhill, A.B., 1986. *Ap.J. Suppl.* **60**, 323.
Castor, J.I., 1970. *M.N.R.A.S.* **149**, 111.
Conti, P.C., Massey, P. and Vreux, J.-M., 1989. *Preprint.*
Langer, N. and El Eid, M.F., 1986. *Astron. Astrophys.* **167**, 265.
Maeder, A., 1987, *Astron. Astrophys.* **173**, 247.
Maeder, A., 1983, *Astron. Astrophys.* **120**, 113.
Nugis, T., 1990. *Astrofiz.* **32**, 85.
Nugis, T., 1989. *Tartu Astrofüüs. Obs. Teated* No 94, 3.
Prantzos, N., Doom, C., Arnould, M. and de Loore, C, 1986. *Ap.J.* **304**, 695.
Smith, L.F. and Hummer, D.G., 1988. *M.N.R.A.S.* **230**, 511.
Torres, A.V., 1988. *Ap.J.* **325**, 759.
Torres, A.V., 1985. *PhD thesis, JILA*, University of Colorado.

DISCUSSION

Smith, Lindsey: I am glad to see you get increasing C/He from WC9 to WC4. This confirms the result by myself and Hummer.
Nugis: Yes, the results of your and Hummer's study are in accord with my estimates.

Langer: Can you exclude a vanishing hydrogen abundance in early WN stars?
Nugis: Hydrogen to helium ratio is not more than 1/3 for the studied WNE stars. For the earliest WN types, in the limits of observational errors hydrogen may be absent at all.

ANALYSES OF WOLF-RAYET SPECTRA

W.-R. HAMANN
Institut für Theoretische Physik und Sternwarte der Universität,
Olshausenstraße 40, D-2300 Kiel, Federal Republic of Germany

1. Introduction

Wolf-Rayet stars represent an important stage in the evolution of massive stars, but are only poorly understood so far. For a better knowledge one might determine their effective temperatures, luminosities and atmospheric compositions. But the emission-line dominated Wolf-Rayet spectra were not accessible to a quantitative analysis for a long time, because "standard" model calculations for static, plane-parallel stellar atmospheres are not adequate to that type of stars.

In a systematic approach to that problem we developed elaborate computer codes for non-LTE radiation transfer in expanding atmospheres (Hamann and Schmutz 1987; Wessolowski *et al.* 1988). Basic features of these models are briefly summarized in the following section. With this tool we then analyzed the helium line spectra of WR stars in our Galaxy and in the LMC (Sect. 3). A special study concerns the WR+0 eclipsing binary V444 Cygni (Sect. 4). In the next step we started to determine the hydrogen abundances in WN atmospheres (Sect. 5). Recently we proceeded to the next generation of models which include further elements, namely nitrogen (for WN stars) and carbon (WC stars), accounting for very complex model atoms and for non-grey radiative equilibrium (Sect. 6).

2. The Model Atmospheres

Prerequisite for the intended analyses of WR spectra is the adequate calculation of the non-LTE spectrum formation. We assume that the WR atmospheres are expanding, spherically symmetric, homogeneous and stationary. (Of course, the validity of these simplifying assumptions can only be justified *a posteriori* from the achieved agreement with the observation.) The equation of continuity then reads

$$\dot{M} = 4 \pi r^2 \rho(r) v(r) \quad , \tag{1}$$

where the mass-loss rate $\dot{M}$ is one of the essential free parameters of the models. The supersonic part of the velocity field is pre-specified by a law of the form

$$v(r) = v_\infty (1 - r_0/r) \quad . \tag{2}$$

The "terminal velocity" v_∞ is an adjustable parameter. At small velocities, a smooth transition towards static layers is ensured by suitable assumptions (for the definition of r_0 and other details see Hamann and Schmutz 1987). The "stellar radius" R_*, corresponding approximately to the location of the sonic point, is a further basic parameter which specifies a WR model.

The temperature structure is derived from the assumption of radiative equilibrium, until recently (cf.

81

K. A. van der Hucht and B. Hidayat (eds.),
Wolf-Rayet Stars and Interrelations with Other Massive Stars in Galaxies, 81–86.
© 1991 *IAU. Printed in the Netherlands.*

Sect. 6) evaluated in grey LTE approximation with the additional assumption that the temperature may never drop below 10 kK. A temperature parameter T. is defined from the luminosity and the stellar radius R. by

$$L = 4 \pi \sigma R_*^2 T_*^4 \tag{3}$$

and may be considered as an "effective temperature", related to the radius R. of the hydrostatic stellar "core".

Hence a specific WR model is characterized by the four parameters R., T., $\dot{M}$, v_∞, and by the chemical composition. Note that no self-consistent hydrodynamics are included in our semi-empiric models. Actually, not even the physical nature of the accelerating forces is known at present.

The radiation transfer in the spherically expanding atmosphere is treated in our code with a co-moving frame formalism. The simultaneous solution of the equations of statistical equilibrium, which may represent very complex model atoms, is achieved by "iteration with approximate lambda operators" (cf. Hamann 1985, 1986, 1987). This new technique allows for the first time to treat such formidable non-LTE problems.

3. Analyses with Pure-Helium Models

In a first stage we worked with models which only account for pure helium. For a number of galactic WN stars we performed detailed fits of the helium line profiles and of the continuous energy distribution. A typical example was presented by Hamann *et al.* (1988), concerning the WN5 star HD50896 alias WR6. Meanwhile we have performed eight "fine analyses" of that kind, and we always arrived at the following conclusions. The models can reproduce well the general features of the observed helium lines. The theoretical continua are always in good agreement with observation, when accounting for the interstellar reddening. Hence our models are basically adequate to describe real WR atmospheres. However, some discrepancies in details of the line profiles might indicate that the models are not yet perfect.

Encouraged by these considerations, we established an extensive grid of models which now allows to determine the parameters of various WR stars with little effort. (For a "coarse analysis" it is sufficient to account for equivalent widths only, instead of a detailed line-profile fitting.) 30 galactic WR stars, covering the different subtypes, have been analyzed in this way (Schmutz *et al.* 1989). Fig. 1 (open symbols) gives the positions of the WN stars in the HR diagram. We find that the stellar temperatures and luminosities are smaller than most authors had expected previously. It turned out that there is no simple correlation between subtype (e.g. WN5, WN6 ...) and T., i.e. a one-dimensional classification is obviously insufficient. For instance, some of the WNE-A stars (i.e. early-type WN with weak lines) are actually as "cool" as WNL ("late") types, namely ≈ 35 kK.

From the theoretical models we learned that models with the same ratio $R_*/\dot{M}^{2/3}$ show very similar spectra. Consequently, the unique determination of stellar radius and mass-loss rate requires the knowledge of one distance-dependent quantity, e.g. the absolute visual magnitude M_v. Fortunately the distances of many WR stars are known from cluster or association memberships, but for individual stars this assignment may be in error.

For WR stars in the Large Magellanic Cloud the absolute magnitudes are more reliable, because their uniform distance is known and the interstellar reddening is generally small. Moreover, it is interesting to compare their parameters with the galactic members. Therefore we extended our analysis of WN stars to the LMC (Koesterke *et al.*, in preparation). Spectra were obtained with the ESO 3.6m telescope and EFOSC. The analysis of the helium lines is performed in the same way and with the same grid of models as used for the galactic stars. Thus the comparison should be unbiased from systematic errors of the models. The - still somewhat preliminary - results for the LMC stars are also indicated in

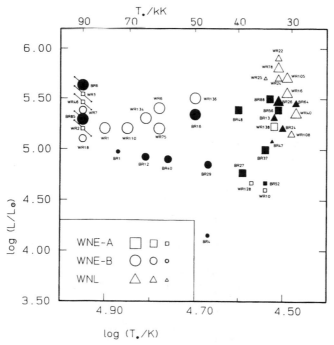

Fig. 1. Positions of WN stars in the HR diagram, as obtained by analyses with pure-helium models. The individual stars are identified by their catalogue number (van der Hucht et al., 1981; Breysacher, 1981). LMC stars (filled symbols) tend to smaller luminosities than galactic WN stars (open symbols). Different subtypes are distinguished by different symbols, as explained in the inlet. The different sizes of the symbols indicate the mass loss rate, depending on whether log $\{\dot{M}/(M_\odot/yr)\}$ is larger than -4.2 (large symbols), smaller than -4.8 (small symbols) or in between (medium size).

the HR diagram (Fig. 1, filled symbols). The comparison with the galactic WN stars reveals that, on the average, the LMC stars lie at lower luminosities. Strikingly different is the position of the early-type WN stars with strong lines. Among our sample we find five such WNE-B stars in the LMC with luminosities below 10^5 $L_\odot$ which have no galactic counterparts. But, of course, we must be cautious with statistical conclusions because our program stars are selected arbitrarily.

4. The Eclipsing Binary V444 Cygni (WN5 + O6?)

The binary system V444 Cygni provides a crucial test of our WR models. We synthesize the light curve, accounting for the mutual eclipse of the extended WR atmosphere and the O star. The former is given by our models, i.e. we neglect any disturbance of its spherical symmetry by the companion. The basic result is that we can reproduce the observed light curve of V444 Cygni with our model, but the solution is not unique (Schwarz et al., in preparation; Hamann et al., these proceedings; Hamann et al., 1990).

We can also satisfactorily fit the helium lines in the observed composite spectrum. The stellar parameters of such a spectral analysis depends, however, on the brightness ratio between the binary components which is a free parameter. We have not yet succeeded in fitting both, the composite spectrum *and* the light curve, with *exactly* the same set of parameters, but this point must still be further investigated.

5. Hydrogen Abundances in WN Stars

The most comprehensive study about hydrogen abundances in WN stars was performed by Conti *et al.* (1983). Their results confirmed that WN atmospheres consist predominantly of helium, while hydrogen is more or less depleted or even absent within the detection limit. A histogram (Fig. 12 in Conti *et al.* 1983) illustrates that there is no clear correlation between hydrogen detection and spectral subtype: stars with hydrogen and stars without detectable hydrogen are found among both, the WNE ("early") as well as the WNL ("late").

This picture drastically changes when the hydrogen detection is not correlated with the subtype, but instead with the stellar parameters as obtained by our spectral analyses. Fig. 2 shows a clear separation between the two groups: the "cool" stars with $T_* \approx 35$ kK exhibit hydrogen, while the "hot" stars ($T_* > 60$ kK) do not. WR136 is intermediate in temperature and in hydrogen abundance. The only exception, WR3, is a peculiar object in many aspects. Note that four of the stars classified as WNE-A (WR10, WR128, BR52, BR88) are actually "cool", and just these stars show hydrogen.

Conti *et al.* (1983) derived the hydrogen abundances from a "semiquantitative" study of the Balmer-Pickering decrement. We now repeat this analysis with our detailed model calculations in order to obtain more precise results (Hamann *et al.*, these proceedings; Hamann *et al.*, 1990). Up to now we have completed four fine analyses of the described kind. In one case (WR22) the hydrogen mass fraction reaches 40%, while for the other two "cool" stars ($T_* < 35$ kK) the hydrogen mass fraction is of the order of $\beta_H = 20\%$. WR136, which is considerably hotter ($T_* \approx 50$ kK), has only 6% hydrogen, and thus appears as a link to the "hot" WN stars in which hydrogen is not detectable (cf. Fig. 2).

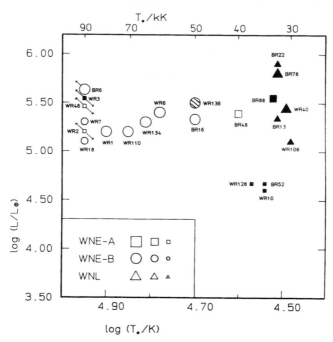

Fig. 2. Hydrogen in WN stars. The individual stars are identified as in Fig. 1. Open symbols denote stars without hydrogen, filled symbols stars with hydrogen detected (according to Conti et al. 1983). WR136 got a hatched symbol because Conti et al. could not detect any hydrogen, while we found a very small abundance ($\beta_H = 6\%$). The stellar parameters (L, T$_*$) are from our spectral analyses (Schmutz et al. 1989, partly improved from "fine analyses"; Koesterke et al., in preparation). Different subtypes are distinguished by different symbols, as explained in the inlet. The different size of the symbols indicates the mass-loss rate: $\log\{\dot{M}/(M_\odot/yr)\} > -4.2$ (large symbols), < -4.8 (small symbols), or in between (medium size).

6. WR Models with Nitrogen and Carbon

The non-LTE models for WR atmospheres reach their next stage of sophistication when not only helium (and possibly hydrogen), but also the most important "metals" are taken into account. Hillier (1988) was the first who introduced nitrogen into his calculations, and meanwhile he proceeded to carbon in order to synthesize WC spectra (Hillier, 1989).

Our recent models now also include nitrogen, represented by a model atom with 90 levels and about 1500 lines, 350 of them treated explicitely. Low-temperature dielectronic recombination via hundreds of transitions is allowed. Together with the nitrogen we introduced an improved temperature structure which now fully accounts for radiative equilibrium, instead of the grey LTE approximation applied so far. Test calculations with these improved models (Hamann and Wessolowski, 1989), setting the nitrogen abundance to 1.5% by mass, show no significant reaction of the helium spectrum. Thus the previous results obtained from models with pure helium and "grey" temperature structure remain valid, as far as WN subtypes are concerned.

As our next step we try to reproduce in detail the nitrogen line spectrum of one WN star (WR 136; Wessolowski *et al.*, these proceedings). From preliminary results we conclude that some general agreement can be achieved, while remaining discrepancies can be attributed to problems with the very complex nitrogen model atom.

Calculations for WC stars which account for a complex carbon model atom are also available now in our group (Leuenhagen *et al.*, in preparation). A first analysis of WR111 confirms the results of Hillier (1989), especially concerning the carbon mass fraction of about one half. As the carbon opacities control the radiative transfer, the analysis of WC spectra with pure-helium models (Schmutz *et al.*, 1989) could only be preliminary. The improved models with carbon now yield, e.g., for WR111 a stellar temperature of about 60 kK, instead of 35 kK obtained previously. Therefore we will refrain from discussing the parameters of WC stars until we have analyzed more of them with the adequate models.

7. Conclusions: the Evolutionary Status of Wolf-Rayet Stars

The spectral analyses indicate that most WR stars are considerably cooler and less luminous than predicted by the standard evolutionary calculations, which generate WR stars as a post-red-supergiant stage (cf. Fig. 3). The hydrogen abundances in WN stars provide a further check of the evolutionary szenario. The most luminous WNL star in our sample (WR22) has about 40% hydrogen, indicating that this star has either not yet evolved to the red, or - as it is known as a spectroscopic binary - was formed by mass transfer. For the two other "cool" WN stars (T. ≈ 35 kK), irrespective whether classified as WNL or WNE-B, we find typical hydrogen mass fractions of about 20%, just as predicted for the phase of rapid blueward post-RSG evolution. The observed disappearance of hydrogen towards higher stellar temperatures also agrees qualitatively with the theoretical tracks.

The agreement with the evolutionary calculations ends, however, when the stellar parameters are considered. The observed luminosities are smaller than predicted by more than a factor of ten. Very small luminosities are determined for a couple of WNE-B stars in the LMC, possibly indicating that under low-metallicity conditions even less massive stars can reach the WN phase, compared to our Galaxy. Recall that the ratio between the number of WN stars and WC stars also strongly differs between the Galaxy and the LMC.

Summarizing, we state that we can perform now a quantitative analysis of Wolf-Rayet spectra. For a large number of stars we already obtained stellar parameters and atmospheric abundances, and further results are to be expected soon. This empirical basis strongly constrains any evolutionary explanations. Hopefully it will help us to a better understanding of the evolution of massive stars.

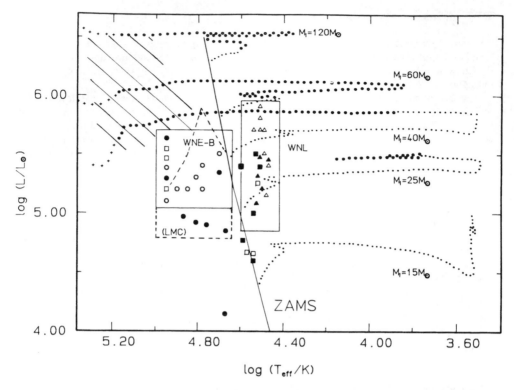

Fig. 3. Loci of WR stars in the HR diagram, as obtained from spectral analyses. Different symbols denote the subgroups WNL (Δ), WNE-A (□), and WNE-B (O), while open or full symbols indicate whether the stars belong to our Galaxy or the LMC, respectively. The overall location of the WNL and the WNE-B subgroups is additionally emphasized by large boxes. The box for the WNE-B stars in the LMC (broken line) extends to smaller luminosities than for our galactic sample. Superimposed are evolutionary tracks (from Maeder and Meynet, 1987), with indication of the predicted surface abundances: ··· "normal", ••• WN-type, + + + WC-type. Only in the hatched region the evolution proceeds slowly enough so that WR stars are likely to be observed.

References

Breysacher, J.: 1981, Astr. Ap. Suppl. Ser. **43**, 203

Conti, P.S., Leep, E.M., Perry, D.N.: 1983, Ap. J. **268**, 228

Hamann, W.-R.: 1985, Astr. Ap. **148**, 364

Hamann, W.-R.: 1986, Astr. Ap. **160**, 347

Hamann, W.-R.: 1987, in "Numerical Radiative Transfer", ed. W. Kalkofen, Cambridge University Press, p. 35

Hamann, W.-R., Schmutz, W.: 1987, Astr. Ap. **174**, 173

Hamann, W.-R., Schmutz, W., Wessolowski, U.: 1988, Astr. Ap. **194**, 190

Hamann, W.-R., Wessolowski, U.: 1990, Astr. Ap. **227**, 171

Hamann, W.-R., Wessolowski, U., Schwarz, E., Dünnebeil, G., Schmutz, W.: 1990, in "Properties of Hot Luminous Stars: Boulder-Munich Workshop", ed. C. D. Garmany, The A.S.P. Conference Series, Vol. 7, p. 259

Hillier, D.J.: 1988, Ap. J. **327**, 822

Hillier, D.J.: 1989, Ap. J. **347**, 392

van der Hucht, K.A., Conti, P.S., Lundström, I., Stenholm, B.: 1981, Space Sci. Rev. **28**, 227

Maeder, A., Meynet, G.: 1987, Astr. Ap. **182**, 243

Schmutz, W., Hamann, W.-R., Wessolowski, U.: 1989, Astr. Ap. **210**, 236

Wessolowski, U., Schmutz, W., Hamann, W.-R.: 1988, Astr. Ap. **194**, 160

VLBI OBSERVATIONS OF WR STARS

M. FELLI, M.MASSI
Osservatorio Astrofisico di Arcetri, Florence, Italy

ABSTRACT. We present the first VLBI radio observations of two WR stars (WR140 and WR146) and of two early type stars (HD167971 and θ^1 ORI A). The derived size and brightness temperature suggest that in all cases the radio emission originates from two distinct components: a steady weak emission due to the ionized wind of the primary and a smaller diameter brighter non-thermal component, probably associated to a small region within the ionized envelope of the star.

1. Introduction

The radio emission from WR and early type stars can either be thermal, i.e. it originates from the free-free emission of the envelope created by the mass loss and ionized by the stellar uv photons, or non-thermal. In the last case the emission mechanism is not yet fully understood and several alternatives have been proposed. Sources exhibiting both types of emission have also been found.

The radio data up to now came almost entirely from single dish or aperture synthesis observations, which only in few cases were able to resolve the thermal wind structure. In fact, for a typical mass loss of a WR star of $4\ 10^{-5}$ $M_\odot$ y^{-1}, a wind terminal velocity of 10^3 Km s^{-1} and a distance of 1 kpc the expected size at 6 cm is ~ 215 mas, which can barely be resolved with the VLA in the A configuration. Only in these sources the low brightness temperature found ($< 10^4$ K) are consistent with a thermal interpretation.

For the unresolved sources the distinction between the two types of emission comes from the radio spectral index (a spectral index of 0.6 is expected for a thermal wind expanding at constant velocity, Panagia and Felli, 1975), or from the variability of the radio emission. However, neither of the two methods is known to be unambiguous. For instance the "bona fide" thermal wind of P Cygni is known to vary in the radio (Felli et al, 1985). Alternatively, if the mass loss derived from radio data is much greater than that derived from other diagnostics at optical and infrared wavelengths, the radio emission is suspected to be non-thermal.

Only VLBI observations are capable to reach the resolutions (~ 10 mas with the EVN and 1-2 mas with transatlantic baselines at 6 cm) necessary to resolve the non-thermal components and to measure the source size and brightness temperature. In fact:

$$T_b/K = 4.9*10^7 S_{6cm}(\text{mJy})(\lambda/6\text{cm})^2 / (\theta_{6cm}\ (\text{mas}))^2.$$

K. A. van der Hucht and B. Hidayat (eds.),
Wolf-Rayet Stars and Interrelations with Other Massive Stars in Galaxies, 87–93.
© 1991 *IAU. Printed in the Netherlands.*

For a 6 cm flux density of a few mJy (the present VLBI detection threshold) and a source size of a few mas the brightness temperature is $\sim 10^6$-10^7 K.

However, due to the weakness of the radio emission, VLBI observations of WR stars and early type stars are still in the pioneering stage and, at present, mapping is very difficult. Only for a small number of stars a rather crude analysis of the fringe visibility curve is feasible.

We present VLBI observations at 6 cm with the European VLBI Network (EVN) of W146, as well as observations of WR140 (HD193793) which include also transatlantic baselines. This WR has also been monitored with the VLA and Westerbork in the last years and it is now in the rising part of its periodic radio emission light-curve.

These results are compared with recent EVN observations of HD167971 and θ^1 ORI A, the last one observed also with transatlantic baselines.

All the above stars were suspected to be non-thermal radio emitters on the basis of the previously mentioned criteria and were selected from extended VLA surveys (Becker and White, 1985; Abbott et al, 1986; Churchwell at al, 1987; Bieging, Abbott and Churchwell, 1989) as the most favorable candidates for VLBI observations.

2. VLBI and VLA results of individual stars

2.1 WR146

WR146 is a WC4 star in the Cyg OB2 cluster (distance 2.0 kpc). Its radio emission was recently detected by Becker, Helfand and White (private communication) during a VLA survey of the galactic plane. It has a clear non-thermal spectral index ($\alpha = -1$) and has remained bright on the few occasions that it was observed. On November 13 1989 the VLA 6 cm flux density was 32.4 mJy.

VLBI observations were carried on September 9 1989 with the five EVN antennas (Effelsberg, Jodrell, Westerbork, Medicina and Onsala, hereon E,J,W,M and O), with four integrations (13 min each) with Mark III tapes, mode A (i.e. total bandwith of 56 MHz for maximum sensitivity). The source is clearly detected with a high S/N (between 18 and 10) on all the four integrations with the shortest and most sensitive baseline (E-W). The mean flux density is 7 ± 0.5 mJy, where the error represents the scatter around the mean.

A pure thermal wind explanation (already unlikely from the spectral index) can be ruled out from a comparison of VLA and EVN results. In fact, if all the 32.4 mJy observed at the VLA came from a wind, the derived mass loss would be $1.2 \ 10^{-4} \ M_\odot \ y^{-1}$, much greater than the mean value found for WR ($4 \ 10^{-5} \ M_\odot \ y^{-1}$). The implied angular size of such a wind would be ~ 105 mas and the expected flux density of a gaussian source of this size at the E-W baseline would be at least one order of magnitude lower than observed.

The most plausible hypothesis seems to be that of a wind plus a small diameter non-thermal component. For a mass loss of WR146 equal to the mean value of a WR the expected flux density of the thermal wind would be ~ 5.5 mJy and the size ~ 43 mas. The non-thermal flux density would then be ~ 27 mJy. This value reduces to 7 mJy on the E-W baseline for a source size ≤ 30 mas. The brightness temperature would be $\geq 1.5 \ 10^6$ K, clearly indicating the non-thermal nature of this component. The proposed two component configuration, which should be regarded only as indicative, has the advantage that the spectral index of the entire source is dominated by the non-thermal component, in agreement with the observations.

2.2 WR140

WR140 (HD193793) is a binary system composed by a WC7 and an O4-5, at a distance of 1.3 kpc. The orbital period is 7.94 years, with periastron of 2.35 AU (1.8 mas) and apastron of 27.0 AU (20.8 mas). Its near IR light-curve has a well defined periodicity with peak at periastron and period equal to the orbital one. For a complete description of the radio, X-ray and IR properties we refer to Williams et al (1990).

The radio emission is also known to vary with the same period, but shifted in phase. The peak occurs at $\phi \sim 0.8$. The emission appears to be quiescent for most of the orbital period, at a level of 1.5 mJy at 6 cm, and has a spectral index compatible with the canonical 0.6 value for a thermal wind. A thermal wind of 1.5 mJy at 6 cm with a terminal velocity of 3000 Km is produced by a mass loss of $5 \cdot 10^{-5}$ $M_\odot$ y^{-1} and has an angular size of ~ 90 mas. Clearly, the entire orbit of the O4-5 companion is within the the wind of the WR. At $\phi = 0.3$ the emission begins slowly to increase and reaches a maximum as high as 25 mJy at 6 cm at $\phi = 0.8$. The spectral index is non-thermal at the peak. The next maximum should occur in 1991.

On December 1989 the 6 and 2 cm flux densities were 13.5 and 19.2 mJy, respectively (White and Becker, private communication). If we subtract from the observed fluxes the steady thermal wind component, the spectral index of the variable one is 0.23, clearly changing to non-thermal values, although remaining positive. However, we remind that if the non-thermal component is produced within the WR wind, due to the wind optical depth the observed spectral index may differ noticeably from the intrinsic one. For instance, if the optical depth between the observer and the source due to the wind is $\tau_{wind}(6 \text{ cm}) = 1$, the intrinsic 2 cm - 6 cm spectral index will be increased by 0.8.

To study the relative contribution of the two components before the light-curve maximum we performed on March 7 1990 a MarkIII mode A 56 MHz bandwidth VLBI experiment with the EVN plus the VLA in the phased array mode and the NRAO 140' (Green Bank). Since the observations were reduced only few weeks ago, the results are very preliminary. Software problems at J did not allow a proper recording of the signal. No fringes were detected with O due to the low S/N. The NRAO 140' recorded both polarization with 28 MHz bandwidth each, leaving only one polarization for the correlation. At the EVN stations, where the high density was available, 16 scans were obtained, each of 13 minutes. Only one scan was available with transatlantic baselines. Fringes with high S/N were detected on all scans on the E-W baseline and on 10 scans on the E-M baseline, the lost ones being due to instrumental failure. By using the rate and delay values of these two baselines, the M-W baseline was refringed using a smaller search window: from -1 to -11 mHz for the rate and from 5 to 50 nsec for SBD, narrower than the default values (-200 to 200 mHz and -2000 to 2000 nsc, respectively).

For no intercontinental baseline there was a detection in the first fringe output based on the default values for delay and rate. The scans were refringed with a narrower search window derived from the calibrator's delay and rate. The results of all the baselines with the NRAO 140' (including that with the VLA) have been rejected as completely unreliable due to the high closure rate and delay for the available triangles. On the contrary, for the same arguments the results for the baselines with the VLA are acceptable. Since we have only one scan with the VLA it is hard to assess the reliability of these results.

To convert the observed fringe visibility into flux density we need to evaluate the b-factors. This is of no problem for the EVN baselines, but becomes difficult for the intercontinental baselines since the calibrator (3C418) is resolved. For the baselines with the VLA we have assumed a b-factor of 1.5, equal to the mean value on the EVN baselines.

The final observed flux densities on the EVN baselines are 17 ± 1 mJy for E-W, 9 ± 3 mJy for E-M, 14 ± 2 mJy for M-W, with a mean value of 13.5 mJy. On the trasatlantic baselines the flux densities are 2.0 ± 0.5 mJy for E-VLA, 4 ± 1 for W-VLA and 4 ± 1 for M-VLA. For comparison, the VLA 6 cm flux density on March 3 1990 was 15 mJy (White and Becker, private communication) and the Westerbork flux density on May 3 1990 was 16 ± 1 mJy (de Bruyn, private communication).

Using 13.5 mJy as the indicative flux density of the variable component the derived angular size is 1.2 mas and the implied brightness temperature $4.9 \ 10^8$ K.

Even though the number of visibilities that we have is large, the gap between the EVN and the transatlantic baselines is so large that a two gaussian components fit to the visibilities (i.e. a ~ 40 mas wind with 1.5 mJy and a smaller diameter source) gives a confidence factor undistinguishable from that of a single component. The existence of a 1.5 mJy wind (although formally equal to the difference between VLA and mean EVN fluxes) is within the noise of the VLBI data and its presence has to rely more securely on the previous history of the radio emission and on the spectral index of the steady component.

The separation of the two stars at the time of the VLBI observations (corresponding to a phase $\phi = 0.62$) is 26.1 AU, or 20 mas. Given the parameters of the orbit, the projected separation is ~ 17 mas, greater than the source size.

We will tentatively associate the non-thermal component with the region of interaction between the wind of the O star and that of the WR. In fact, the absorption of the WR wind favours positions more distant than periastron. Even for moderate mass losses the WR wind may be completely opaque to the non-thermal component near periastron passage, regardless of its intrinsic intensity. This optical depth effect can give an explanation to the disappearance of the non-thermal component after $\phi = 0.8$ and the phase offset of the radio light-curve (Williams et al, 1990). In all the above it is assumed that the region of non-thermal emission is much closer to the O star than to the WR, as expected form the lower intensity and velocity of the mass loss of the O star.

2.3 HD167971

HD167971 is thought to be a triple system with an O7.5 If primary. It was selected from the survey of Bieging et al (1989) for being classified as non-thermal on the basis of the spectral index ($\alpha = -0.7$) and weak variability. The VLA flux density at 6 cm is 14 mJy.

The EVN observations of September 9 1989 could not have worst luck since E had problems with the field system and M was down for repair. We report the result here for the benefit of documentation and in the hope that they may trigger further VLBI observations.

Only for the remaining most sensitive baseline, i.e J-W, we have a S/N ~ 5.5 and fringe and delay rate consistent with that of the calibrator. The derived flux density is 10 ± 2 mJy. Comparison with the VLA flux implies a source size of 6 mas and a brightness temperature of 2 10^7 K. The VLBI experiment confirms the non-thermal nature of the radio emission, but more observations with transatlantic baselines are needed to get a better description of this source.

2.4 θ^1 ORI A

θ^1 ORI A is one of the four Trapezium stars in the Orion Nebula. It has a spectral type B0.5 and hence it is much less luminous and emits a lower flux of ionizing photon than the brightest member θ^1 ORI C, which is the main responsible for the ionization of the nebula.

Its radio emission was detected during VLA studies of the Orion Nebula (Gary, Moran and Reid, 1985; Churchwell et al, 1897) and re-observed in several other occasions. Its flux density can vary between 2 mJy and 73 mJy at 2 cm. The spectral index, when measured, is ~ 0.0. For a complete description of the radio properties of the star we refer to Felli, Massi and Churchwell (1989).

The star is known to be a close binary system with a period of 65.43 days. The maximum and minimum separation between the primary and the secondary (presumed to be a TTauri) are 0.4 and 2.73 mas.

6 cm EVN observations on September 28 1987 were able to detect the star and to establish that all the VLA flux density (14.8 mJy at that epoch) was contained in the VLBI structures, with an upper limit of 4 mas to the source size and a lower limit of 4 10^7 K to the brigthness temperature.

We present here new VLBI observations with transatlantic baselines aimed to resolve the source structure. The experiment was performed on April 13 1989 and included the EVN antennas and the VLA. The VLA flux density was, at that epoch, 18 ± 1 mJy. Similarly to our previous EVN run, the VLA flux density was equal to the visibility on the EVN baseline. The only transatlantic baseline for which we have a meaningful detection is E-VLA, which gives a flux density of 2.2 ± 0.6 mJy. Also in this case we had to search in a smaller window of ± 5 mHz in fringe rate and of ± 50 nsec in SB delay.

A gaussian fit to the visibilities gives a source size of 1.3 mas and a brightness temperature of 6 10^8 K. Any steady extended thermal component, if present, must have a flux density ≤ 1 mJy.

With our VLA data base on Orion and with observations found in the literature we tried to see if there is any periodicity in the variable emission with a period similar to that of the binary. The data are rather scattered, partly for instrumental reasons since they come from VLA observations with different configurations and/or frequencies and, consequently, with different resolutions. However, they indicate, on the average, an increase of about a factor 3 with respect to the mean value (12 mJy) at $\phi = 0.85$, not dissimilar to what found for WR140.

On the last VLBI run the binary phase was 0.52, and the separation of the two components was maximum. The fact that the radio source size is less than this separation suggests that the non-thermal source may be associated with one of the two stars. Following arguments similar to those used for WR140, we believe this is associated with the secondary and reaches maximum slightly before periastron.

In fact, early type stars are often surrounded by ionized envelopes produced by mass loss, even though less intense than for WR. For a B0.5 the expected mean mass loss is

$\sim 10^{-8}$ M$_\odot$ y^{-1}. The expected flux density from this thermal source would be ~ 0.002 mJy, clearly below detectability. However, the source size of this wind would be 0.9 mas, greater than the minimum separation of the two stars. Optical depth effects of this wind could be important when the secondary passes inside the primary wind and, similarly to WR140, make it more likely that the non-thermal radio emission is closer to the secondary. A model radio light-curve obtained taking into account the optical depth effects of the B0.5 wind and assuming an emissivity of the non-thermal component proportional to r^{-2} gives a satisfactory fit of the observations.

Conclusions

We have presented VLBI observations of WR and early type stars which establish the size and surface brightness of the non-thermal variable component. This is usually much smaller that the maximum binary separation.

The observed radio light-curves indicate a periodicity equal to the orbital period, suggesting a possible association between the non-thermal component and the secondary. Optical depth effects of the primary wind may strongly influence the observed intensity and spectral index and be the main responsible for the shape of the light-curve.

An other WR which shows two separate thermal and non-thermal components is WR147 (Moran et al, 1989), a binary system with much greater separation (1100 AU, the two components are clearly separated) and a much longer period (10^3 y, no information on the radio variability or periodicity is available).

The common binary nature of all these systems seems to suggest that θ^1 ORI A, WR140 and WR147 are scaled up version of the same phenomenon, i.e. they are composed by a steady component due to the thermal wind and a variable non-thermal one due to the interaction between the mass loss of the primary and the weaker wind of the secondary.

More sensitive VLBI observations are needed to set the position of the non-thermal component with respect to the faint elusive extended thermal wind.

References

Abbott,D.C., Bieging,J.H., Churchwell,E., 1986, Ap.J., 303, 236.

Becker, R.H., White, R.L., 1985, Ap.J., 297, 649.

Bieging,J.H., Abbott,D.C., Churchwell,E., 1989, Ap.J., 340, 518.

Churchwell,E., Felli,M., Wood,D.O.S., Massi,M., Ap. J., 321, 516.

Felli, M., Stanga,R., Oliva,E., Panagia, N.. 1985, Astron. Astrophys 151, 27.

Felli,M., Massi,M., Churchwell,E., 1989, Astron. Astrophys., 217, 179.

Garay,G., Moran, J.M., Reid,M.J., 1985, in Radio Stars, R.M. Hjellming and D.M. Gibson eds, 131.

Moran,J.P., Davis,R.J., Bode,M.F., Taylor,A.R., Spencer,R.E., Argue,A.N., Irwin,M.J., Shanklin,J.D., 1989, Nature, 340, 449.

Panagia, N., Felli, M., 1975, Astron. Astrophys., 39, 1.

Williams,P.M., van der Hucht,K.A., Pollock,A.M.T., Florkowski,D.R., van der Woerd,H., Wamsteker,W.M., Mon.Not.R.Astr.Soc., 1990, 243, 662.

DISCUSSION

Owocki: All the objects you mentioned were binaries. Is it not possible to also use this technique to interpret proporties of radio emission from single stars?

Felli: The radio stars included in the VLBI program were selected only for having the largest flux density and being unresolved with the VLA. The fact that they are binary systems is a pure coincidence. VLBI techniques can be used for any radio star (single or binary) provided that there are radio components with sufficient flux density at resolutions of a few mJy.

Cassinelli: Can you use your brightness temperature to estimate the magnetic field associated region, where the radio flux is coming from?

Felli: We have not yet started to model which is the source of emission, so, I can not give you an answer.

Montmerle: MERLIN has resolved AS 431 into two components, one thermal and the other non-thermal. Do you see any indication of this in your stars?

Felli: The main goal of these VLBI observations was to establish if there were small diameter non-thermal components. We tried a map of WR140 with the limited visibilities available. The first map clearly indicates the presence of two distinct components. However the UV coverage and the sensitivity are too small to allow a proper mapping of the extended 1.5 mJy thermal wind. New VLBI observations (by us and the van der Hucht group) with better UV coverage are planned for WR140.

Anne Underhill, Balinese dancer

HELIUM IN WOLF-RAYET SPECTRA[*]

Y. ANDRILLAT[1] and J.-M. VREUX[2]
1. Laboratoire d'Astronomie, Université Montpellier II Sciences et
   Technique du Languedoc, Place Eugène Bataillon,
   F-34095 Montpellier, Cedex 5, France.
   CNRS : U.R.A. 1281, Observatoire du Pic du Midi, France.
2. Institut d'Astrophysique, 5, avenue de Cointe,
   B-4200 - Cointe-ougrée, Belgium.

The observation of Helium lines in the near-infrared proved to be a powerfull tool to investigate the Wolf-Rayet stars and more particularly the WN.

Combining new and previous observations of the equivalent widths of HeII λ 10124 and HeI λ 10830, it is possible to empirically divide the presently sampled WNs population within six groups. These are well connected with the "temperature parameter" of the models of Schmutz et al. (1989) and, also, with the classical WN subtypes although the definition of the latters is mainly based on the appearance of NIII, NIV and NV lines (and only very marginally for the latest ones on the aspect of a He I line).

When H is present, the observation of the near infrared lines of the 6-n series of He II, some of them being blended by Paschen emission lines, allows a determination of the H^+/He^{++} ratio. A combination of all the data presently available (optical and near-infrared) indicates that, with one exception (to be confirmed), all the WN8-7 stars have a sizable H content with a tendency towards a stronger H signature in the WN8s relative to the WN7s.

He I lines of the 3 p - nd and 3d - nf series can be strong in the spectra of some WN stars, for example in WR 123. Nevertheless this is not the case in the spectrum of WR 158 and we suggest that the emission we observe at λ 8446 is due to OI. WR 158 being a H rich WN, this OI line could be produced by a fluorescence mechanism induced by Lyβ as it is observed in Be, planetary nebulae, novae, ...

References
Schmutz, W., Hamann, W.-R., Wessolowski, V., 1989: Astron. Astrophys., 210, 236.
Vreux, J.-M., Dennefeld, M., Andrillat, Y., Rochowicz, K., 1989: Astron. Astrophys. Suppl. Ser., 1, 353.
Vreux, J.-M., Andrillat, Y., Biémont, E., 1990: Astron. Astrophys., in press.
(*) All the observations reported here have been collected at the Haute Provence Observatory, CNRS, France.

K. A. van der Hucht and B. Hidayat (eds.),
Wolf-Rayet Stars and Interrelations with Other Massive Stars in Galaxies, 95.
© 1991 IAU. Printed in the Netherlands.

SOME EMISSION LINES IN WCE STARS HAVE NEARLY CONSTANT LINE STRENGTHS

KENNETH R. BROWNSBERGER and PETER S. CONTI
Joint Institute for Laboratory Astrophysics
University of Colorado, Boulder, CO 80309-0440, USA

We present empirical measurements of line strengths (Equivalent Widths-EW) of WCE stars (WC4-WC7). Our database presently includes approximately 50 emission lines stretching from the UV to the NIR. We have data for approximately 40 stars, but the spectral range is not complete for each star. At present, there are approximately 10-20 EW measurements for each line in the database. The lines used in the classification of the WC subtypes, along with a few others, show a dramatic dispersion in strength (as expected) within the WCE subtypes. Many other lines of carbon and oxygen ions indicate only a moderate range among the subtypes. We have found that there are several lines that show almost no dispersion in strength from WC4-WC7. These nearly constant lines can be used to give empirical estimates of the brightness ratio of the W-R to companion star in WCE + OB systems, and can also be used to estimate interstellar reddening. The similarity of line strengths among several C III and C IV emission features in WCE stars suggests that the ionization level in the wind is not fundamentally different among WC4-WC7 subtypes. Another parameter (e.g. C/He?) may have a substantial influence on those lines used for the WCE classification.

Since we have not yet reduced the spectroscopic data for WCL stars, we are uncertain as to whether these same lines are nearly constant in strength for them. We can assert that there are <u>no</u> lines in WN types that have nearly constant line strengths.

K. A. van der Hucht and B. Hidayat (eds.),
Wolf-Rayet Stars and Interrelations with Other Massive Stars in Galaxies, 96.
© 1991 *IAU. Printed in the Netherlands.*

THE H/HE RATIO OF WN STARS

Paul A. Crowther, Linda J. Smith and Allan J. Willis
Department of Physics and Astronomy
University College London
Gower St, London WC1E 6BT
England

ABSTRACT. The observed He II Pickering decrement is modelled using the escape probability method developed by Castor & Van Blerkom (1970) to solve simultaneously the equations of radiative and statistical equilibrium for detailed model hydrogen and helium atoms. All important radiative and collisional processes are incorporated in this nLTE model. We confirm values of $0.0 \leq H/He \leq 0.5$ for WNE stars and $0.0 \leq H/He \leq 3.0$ for WNL stars with considerable spread in each subtype.

1. Model Atom and Grids

The model atom used contains 41 levels of He^0 (every S, P, and D state with $n \leq 8$), 30 levels of He^+, the He^{2+} ion, plus 16 levels of H^0 and the H^+ ion. Several model atoms were tested, including one with the same doubly-excited states incorporated by Bhatia and Underhill (1986, 1988). However, test calculations show that level populations are not significantly affected by these levels.

Since the detailed structure of WR stars remains uncertain, model results for a very large parameter space have been covered: $10^9 \leq n_e \leq 10^{12} cm^{-3}$, $5 \times 10^4 \leq T_e \leq 10^6 K$, $0.1 \leq H/He \leq 10.0$. The level populations were iterated until either 1 part in 10^4 accuracy was achieved or 30 iterations performed. Those runs for which convergence was not achieved are expected to be accurate to within several percent. It is noted that when $T_c = T_e$, and the geometric dilution factor is set to unity, LTE populations are returned as expected.

Since the even quantum numbers of the He II Pickering ($n \rightarrow 4$) series coincide with the H I Balmer lines ($n \rightarrow 2$), the H/He ratio is derived from the excess of the even (He II + H I) over the odd (He II) lines in the series. Pickering decrements ($\log \Delta F_\lambda$) were found from the mean of the logarithmic difference between the odd and even He II optically thin lines ($n = 9$–17).

2. Results

Model values of $\log \Delta F_\lambda$ were compared to those observed in Galactic and LMC WN stars, from Conti, Perry and Leep (1983). We find that for WNE stars $0.0 \leq H/He \leq 0.5$, and for WNL stars $0.0 \leq H/He \leq 3.0$. Comparison of observed Pickering decrements with model grids are consistent over a wide range of n_e, T_e, R_c, T_c, and H/He, and confirm the hydrogen depletion of WN stars, and their chemically evolved nature.

Special thanks are given to Pete Storey for his help with the atomic rate calculations.

97

K. A. van der Hucht and B. Hidayat (eds.),
Wolf-Rayet Stars and Interrelations with Other Massive Stars in Galaxies, 97.
© 1991 *IAU. Printed in the Netherlands.*

OPTICAL CONTINUUM FORMATION IN ROTATING-WIND MODELS OF WR STARS

L. R. DOHERTY
University of Wisconsin-Madison, USA

Wolf-Rayet stars could be rapid rotators. We must ask if a rotationally flattened wind can lead to an observed continuous spectrum that depends on the angle of inclination. Schmid-Burgk (1982) showed that in the radio region the spectral shape is not changed. Here we consider the optical region, where both thermal emission and electron scattering are important. This study uses a simple, idealized model of a rotating wind in which $v(r)$ is from Friend and Abbott (1986), but depends on polar angle in the manner of Poe et al. (1989). Equatorial densities are five times polar values at the same radial distance. The wind consists only of fully ionized He surrounding a spherical core and has a radial optical depth of 2.0 in the equator in the limit of pure electron opacity, essentially at wavelengths less than 2000A. Radiative transfer in the wind is fully treated by a second-order moment method based on the equation of transfer in general spherical coordinates (Doherty 1989). A constant flux at the core boundary is assumed. This flux corresponds to a $T = 4 \times 10^4 K$ blackbody and the wind temperature is 2×10^4. The shape of the emergent spectrum in the region computed (2500-7500A) changes little with inclination and resembles the spectrum of a spherical wind with intermediate density. Thus a rotating wind may masquerade as spherical but, in the present model at least, vary in flux by nearly a factor of two from pole to equator. This can affect the apparent magnitude, wind polarization, and possibly the line spectrum and wind acceleration.

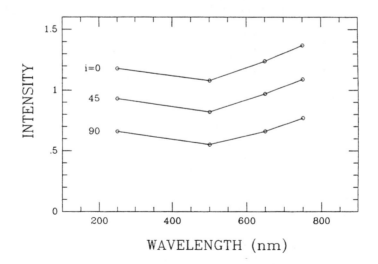

Spectrum of rotating wind model for 3 inclinations of the polar axis to the line of sight. Units of intensity are $\pi R^2 F_c$, the intensity which would be observed for a spherical, pure-scattering wind.

K. A. van der Hucht and B. Hidayat (eds.),
Wolf-Rayet Stars and Interrelations with Other Massive Stars in Galaxies, 98.
© 1991 IAU. Printed in the Netherlands.

IDENTIFICATION AND ANALYSIS
OF INFRARED LINES OF WC STARS

P.R.J. EENENS[1], P.M. WILLIAMS[2] and R. WADE[3]

[1] *Department of Astronomy, University of Edinburgh, Blackford Hill,*
Edinburgh EH9 3HJ, United Kingdom
[2] *Royal Observatory, Blackford Hill, Edinburgh EH9 3HJ, United Kingdom*
[3] *Joint Astronomy Centre, 665 Komohana St, Hilo, HI 96720, USA*

ABSTRACT. We present $1 - 3.4 \, \mu m$ spectra of six Wolf-Rayet stars: WR 146 (WC4), WR 111 (WC5), WR 86 (WC7), WR 140 (WC7+O4), WR 135 (WC8) and WR 88 (WC9). Examination of the relative strengths of the emission lines has enabled us to make over 20 new line identifications. Recombination analyses of the carbon and helium lines yield C/He abundance ratios much in excess of the solar value and correlated with spectral subtype.

The spectra were observed at resolutions of 300 – 600 with the cooled grating spectrometer (CGS2) on the United Kingdom Infrared Telescope (UKIRT). The spectra of C III, C IV, He I and He II are well represented. About 20 lines are identified for the first time, most conspicuously the C IV 4p-4s doublet at $1.435 \, \mu m$, observed previously but not identified. We also identified fine structure lines of C III and (more tentatively) of oxygen (O II, O III, O V); and satellite lines of the C III 7-6 and C IV 10-9, 9-8 and 8-7 arrays. Satellite and fine stucture lines are specially useful to help identify the contributors to the strong hydrogenic transition arrays. Even a few helium lines had not been reported before: He I 5p-3d (at $1.299 \, \mu m$), He II 17-7 (at $1.344 \, \mu m$) and He II 16-7 (at $1.381 \, \mu m$). The 2p-2s He I lines show P Cygni profiles, the absorption component of the triplet ($1.083 \, \mu m$) being stronger in earlier subtypes and the singlet ($2.058 \, \mu m$) in later subtypes. The line ratios correlate with spectral subtype but the spectrum of WR 146 seems anomalous: the C IV lines indicate WC7 or later while the He II lines indicate WC4.

We followed the recombination method of Hummer, Barlow and Storey (*IAU 99*, p.277, 1982) to derive relative abundances for ions of He and C. The analysis shows C/He abundances much in excess of the solar value, as well as a decrease of the C/He ratio with WC subtype, thus qualitatively supporting the predictions of evolutionary models that the later WC subtypes evolve into the earlier.

K. A. van der Hucht and B. Hidayat (eds.),
Wolf-Rayet Stars and Interrelations with Other Massive Stars in Galaxies, 99.
© 1991 *IAU. Printed in the Netherlands.*

THE HYDROGEN ABUNDANCES IN WN STARS

W.-R. HAMANN[1], G. DÜNNEBEIL[1], U. WESSOLOWSKI[1], W. SCHMUTZ[2]
[1] *Institut für Theoretische Physik und Sternwarte der Universität,*
Olshausenstraße 40, D-2300 Kiel, Federal Republic of Germany
[2] *Joint Institute for Laboratory Astrophysics, University of Colorado*
Boulder, Colorado 80309-0440, USA

Hydrogen abundances in WN stars have been derived by Conti *et al.* (1983) from a "semiquantitative" study of the Balmer-Pickering decrement. No clear correlation between hydrogen detection and spectral subtype could be established: stars with hydrogen and stars without detectable hydrogen are found among both, the WNE ("early") as well as the WNL ("late").

This picture drastically changes when the hydrogen detection is not correlated with the subtype, but instead with the stellar parameters as obtained by a detailed spectral analyses with our elaborate models (Schmutz *et al.*, 1989). Now we find a clear separation between two groups: the "cool" stars with $T_* \approx$ 35 kK exhibit hydrogen, while the "hot" stars ($T_* > 60$ kK) do not. WR136 is intermediate in temperature and in hydrogen abundance. Some of those stars classified as "early" subtypes are actually "cool", and just these stars are found to show hydrogen.

We now apply our model calculations for a precise analysis of hydrogen abundances. For that purpose we extended our model code for a fully consistent treatment of line blends. We perform a "fine analysis" of the helium spectrum, giving special weight to the fit of those He II Pickering lines which are not blended by hydrogen. Subsequently, the theoretical profiles of those He II lines which coincide with hydrogen Balmer lines are compared with the observation. The best fit yields the stellar parameters (R_*, T_*, $\dot{M}$, v_∞) and the hydrogen mass fraction, β_H (Table 1).

The hydrogen mass fractions obtained for WR40 and WR128 ($\beta_H \approx 20\%$) agree well with theoretical predictions for a post-RSG evolution. WR136 with its smaller value ($\beta_H = 6\%$) appears as a link to the "hot" WN stars in which hydrogen is not detectable. WR22 is either in pre-RSG stage or represents the result of binary evolution.

Table 1. Fine Analyses Including Hydrogen

	WR 40	WR 128	WR 22	WR 136
Subtype	WN8	WN4	WN7abs	WN6
T_*/kK	31.2	37	31	51
R_*/$R_\odot$	18.	5.3	30	6.0
$\log\{\dot{M}/(M_\odot \, yr^{-1})\}$	-4.05	-5.0	-4.32	-3.85
v_∞/(km/s)	1000	2000	1500	1700
$\log L/L_\odot$	5.4	4.7	5.9	5.5
β_H [%]	16 ± 5	20 ± 10	40 ± 5	6 ± 3

References
Conti, P.S., Leep, E.M., Perry, D.N.: 1983, Ap. J. **268**, 228
Schmutz, W., Hamann, W.-R., Wessolowski, U.: 1989, Astr. Ap. **210**, 236

K. A. van der Hucht and B. Hidayat (eds.),
Wolf-Rayet Stars and Interrelations with Other Massive Stars in Galaxies, 100.
© 1991 *IAU. Printed in the Netherlands.*

THE PROPERTIES OF POPULATION I WO STARS

R.L. Kingsburgh and M.J. Barlow
Department of Physics & Astronomy
University College London
Gower St., London. WC1E 6BT.

ABSTRACT: High and low resolution UV and optical spectra of the four Population I WO stars originally classified by Barlow and Hummer (1982), Sanduleak 1, 2, 4 and 5, have been analyzed. Reddenings, terminal velocities and the relative abundances of He^{2+}, C^{4+} and O^{6+} have been determined. The results are presented in Table 1.

The WO stars show strong OVI 3811,34 Å, CIV+HeII 4658+86 Å and CIV 5801,12 Å emission. The oxygen lines are stronger in these stars than in the WC class and the WO stars are believed to be the next evolutionary stage after WC stars.

P Cyg line profiles in the UV spectra of Sand 1 and 2 yielded wind expansion velocities. In the case of Sand 1, the terminal velocity (v_∞) was obtained from v_{black} of the saturated CIV 1548,50 Å profile in a high resolution IUE spectrum. Sand 1 was used to check which optical lines were most appropriate for determining v_∞. Half the FWZI's of HeII 1640 Å and of CIV 5801,12 Å (corrected for instrumental profile and doublet separation) were found to give the best agreement with the CIV 1548 Å v_{black}, so these were used to derive v_∞ for the other WO stars.

The abundance ratios C^{4+}/He^{2+} and O^{6+}/He^{2+} were derived from recombination lines which were assumed to be optically thin at T=50000K and $\log(n_e)$=11, using the method described by Barlow and Hummer (Proc. IAU Symp. 99, p. 387, 1982).

Table 1: WO Properties	Sand 1	Sand 2	Sand 4	Sand 5
Other names	Sk 188, AB 8	Brey 93, FD 73	WR 102	WR 142, ST 3
Spectral Type	WO4+O7	WO4	WO1	WO2
EW(OIV 3400) (Å)	68±1	299±3	–	–
EW(OVI 3434)	–	–	74±8	60±20
EW(OIV 3811,34)	64±1	336±5	1740±30	990±30
EW(CIV+HeII 4658,86)	90±5	531±5	150±10	380±10
EW(OVI 5290)	10±1	45±3	71±3	62±8
EW(OV 5590)	25±1	110±10	30±5	25±5
EW(CIV 5801,12)	200±30	2450±40	150±5	320±10
E(B–V)	0.05	0.19	1.65	2.04
V	13.52	16.35	14.56	13.37
M_V	−5.4	−2.6	(−2.8)	−2.8
D (*kpc*)	57.5	46.8	(2.9)	0.9
v_∞ (*km s^{-1}*)	4200	4500	4600	5500
n(C^{4+})/n(He^{2+})	0.51	0.38	0.66	0.20
n(O^{6+})/n(He^{2+})	0.08	0.03	0.10	0.03

K. A. van der Hucht and B. Hidayat (eds.),
Wolf-Rayet Stars and Interrelations with Other Massive Stars in Galaxies, 101.
© 1991 IAU. Printed in the Netherlands.

X-RAYS FROM HOT STARS AND THE ROLE OF RELATIVISTIC ELECTRONS

A.M.T.Pollock
Computer & Scientific Co. Ltd., Sheffield S11 7EY, England

This paper challenges the usual view that X-ray emission from hot stars is direct thermal radiation from shocks in the wind. Instead, detailed calculations show that the relativistic electrons in the wind required to explain the non thermal radio flux of some hot stars can also account for the X-rays from all single hot stars via inverse Compton scattering of photospheric radiation. First, there are two essential general points to understand; (1) the heavily absorbed accreting neutron-star X-ray sources orbiting OB companions in binary systems like 4U1700-37 and Vela X-1 show **without doubt** that the instrinsic X-rays from hot stars **cannot** originate in a corona or any other structure confined close to the photosphere; and (2) the observed 4:1 ratio of mean WN:WC X-ray luminosities, much less than the value 50:1 expected for a point source, strongly implies a distributed X-ray source. The non thermal radio components observed by Abbott and others show that relativistic electrons are present far out in the wind of some and perhaps all hot stars. The single most important reason for supposing that these same electrons also account for the ubiquitous X-rays is that the radiation density is so high that Compton losses cause very rapid cooling. It is complicated to work out in any detail the observational consequences of relativistic electrons in a stellar wind because (1) electrons cool via ionisation, adiabatic, synchrotron, inverse-Compton and Bremsstrahlung energy losses of which only the final three give potentially observable radiation at radio, X-ray and γ-ray frequencies respectively; and (2) only electrons in the outer parts of the wind are visible because of free-free radio absorption and photoelectric X-ray absorption. The relative importance of the different energy-dependent cooling processes changes as electrons are advected out with the wind. Near the star, ionisation losses cut off the supply of observable low-energy electrons until Compton losses take over at electron γs of 70-80 in the W-R stars and 20-30 in the OB stars . Further out adiabatic losses are the principal low-energy cooling mechanism before giving way to Compton losses at an electron break energy that increases linearly with radius. Synchrotron and Bremsstrahlung losses are always small. The solution of the transport equation for some supposed electron injection law tempered by these energy-loss mechanisms gives the radial evolution of the electron spectrum that, in turn, allows local radiative emissivities to be calculated. These then combine with the local optical depths to determine how much radiation escapes to the observer. Most of the 20 or so stellar parameters involved in the calculations are reasonably well known. The exceptions are the surface magnitude, B_*, and geometry of the stellar magnetic field and how the electrons are injected. Solving the 1-D electron transport equation with a Lax-Wendroff upwind differencing scheme before numerical integration of the emissivity formulae leads to the following main conclusions and predictions: (1) a small population of relativistic electrons can reproduce simultaneously the observed non thermal radio fluxes and spectra and the X-ray luminosities for $B_* \sim 10 - 100G$; (2) the majority of stars have weaker fields so that the observed X-rays are accompanied by a non thermal radio component too weak to be seen above the thermal flux; (3) X-rays come from a mixture of optically thin and optically thick emission so that the spectra do not show heavy photoelectric absorption; (4) the X-ray spectra show no lines; (5) the W-R stars are soon likely to be detected as γ-ray sources; (6) if electrons are only injected in the inner wind acceleration region orders of magnitude more X-rays and γ-rays than observed are produced. Where do the electrons come from ? From the same acceleration process operating in the solar-wind terrestrial magnetosphere interaction, in the sun, in SNRs, in radio galaxy jets, wherever a magnetised plasma is on the move ?

102

K. A. van der Hucht and B. Hidayat (eds.),
Wolf-Rayet Stars and Interrelations with Other Massive Stars in Galaxies, 102.
© 1991 IAU. Printed in the Netherlands.

DIFFUSE C IV EMISSION AROUND THE WO STAR ST3

V.F. Polcaro(1,2), F.Giovannelli (1), R.K. Manchanda(3),
A. Pollock (4), L. Norci(5,6), C. Rossi(6)

1) Istituto di Astrofisica Spaziale, CNR, Frascati, Italy

2) Dip. Aerospaziale, Univ. "La Sapienza", Rome, Italy

3) Tata Institute of Fundamental Research, Bombay, India

4) Space Science Departement of ESA, Noordwijck, The Netherlands

5) Max-Plank-Institut fur Extraterrestrische Physik, Garching, FRG

6) Istituto di Astronomia, Univ. "La Sapienza", Rome, Italy

A strong diffuse C IV (5801-12 A) emission has been discovered in Berkeley 87 cluster near to the WR (WO) star ST3 (Barlow and Hummer, 1982) using the Loiano 152 cm telescope equipped with a Boller and Chivens 26767 spectrograph and a CCD RCA camera. The diffuse emission seems to interest all the core of the cluster Berkeley 87. In fact it is present around many cluster members sited near to the center.

This young open cluster is embedded in the ON2 molecular clouds complex on the boundary of the Cygnus X region (Turner and Forbes, 1982). A weak hard X-ray source, positionally coincident with the cluster, was identified during a slew by EXOSAT satellite (Warwick et al., 1988). We can thus suppose that the envelope of ST3 is strongly interacting with the cluster interstellar medium.

The C IV diffuse line should originate from the shock- ionized material at the interacting boundary between the wind (our measurements give a 5200 Km/s velocity from the Doppler broadening of the He II 6561 line) and the molecular clouds present in the cluster. The flux in the diffuse line can be evaluated to be at least the 45star at the same wavelength. Given that the C IV is the only carbon ion present in the star spectrum, the classification of ST 3 as WC star can be uncertain.

References

Barlow M.J. and Hummer, D.C., 1982: Proc of the IAU Symposium on "Wolf-Rayet Stars: Observation, Physics, Evolution", C. de Loore and A. Willis eds, p.387.

Turner, D.G. AND Forbes, D., 1982: Publ. Astr. Soc. Pac., 94, 789.

Warwick, R. S., Norton, A. J., Turner, M. J. L., Watson, M. G., Willingale, R., 1988: Mon. Not. R. astr. Soc. 232, 551.

K. A. van der Hucht and B. Hidayat (eds.),
Wolf-Rayet Stars and Interrelations with Other Massive Stars in Galaxies, 103–104.
© 1991 *IAU. Printed in the Netherlands.*

Fig. 1: a) Low dispersion spectrum of ST 3 between 4000-7100 A , taken at the Loiano Telescope on August 31, 1989 (detector CCD RCA coated, exposure time 1000 s).

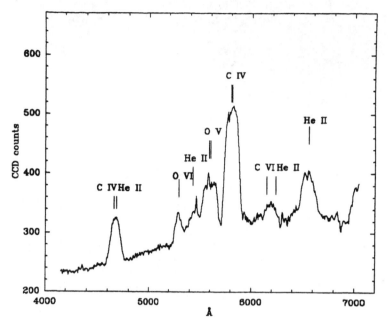

Fig. 2: The sky spectrum around (a) ST3 and (b) the standard star BD+28 4211. Local sky features and the diffuse emission are marked.

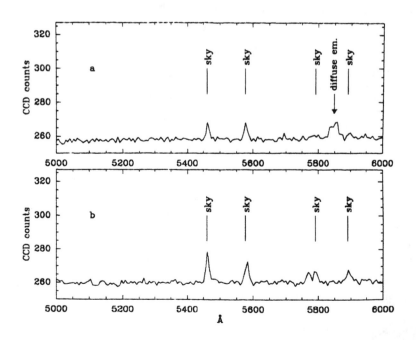

MORE *EINSTEIN* VIEWS OF WOLF-RAYET STARS

A.M.T.Pollock
Computer & Scientific Co. Ltd., Sheffield S11 7EY, England

1. The X-ray spectrum of HD 93162 WR25 WN7

HD 93162 is a star of special interest because of its unexplained and unusually high X-ray luminosity that is roughly 100 times greater than typical W-R values and rivalled only by HD 193793, although in contrast to WR140 it is not variable and despite the concentrated efforts of optical radial velocity observers shows no evidence of binarity. In addition to several *Einstein* X-ray IPC and HRI observations there was also one with the high-resolution SSS non-imaging 0.4-4. keV spectrometer on 1979 July 17. In the 6′-diameter aperture of this instrument the star's spectrum is diluted by a contribution from the Carina Nebula's diffuse emission, that from inspection of the IPC image should be roughly equal in intensity, and by an instrumental background of again roughly the same intensity. The SSS data were extracted and analysed via remote access over the network to the EXOSAT database system. The instrumental-background-subtracted spectrum was modelled with two components : (1) the diffuse component assumed, following Chlebowski et al's η Carinae work, to be a Raymond-Smith thermal plasma of fixed temperature $T = 7 \times 10^6$K and X-ray absorbing column density $N_X = 2 \times 10^{21}$cm^{-2} but unknown intensity ; and (2) the stellar component, assumed to be a Raymond-Smith thermal plasma of unknown temperature and intensity with absortion constrained to be at least the interstellar value. The two best-fit components are of almost equal intensity and give a good fit to the data. For the W-R star, the following parameters apply :

$$\text{HD 93162} \qquad L_X(0.4 - 4.0\text{keV}) = 4.2 \times 10^{33}\text{ergs s}^{-1} \qquad T = 27 \times 10^6\text{K}$$

This is a much higher temperature than those of the OB supergiants like δ Orionis but is something it shares with HD 193793, where the high luminosity is almost certainly caused by colliding winds in a binary system. Perhaps this is a further piece of evidence that HD 93162 is a WR+O binary system seen at low inclination.

2. An X-ray image of NGC 6321 including HD 152270 WR79 WC7+O5

HD 152270 lies in the core of the NGC 6321 cluster in a crowded field with 4 other hot stars brighter than 7^m with spectral types between O6 and B0.5, all potential X-ray sources. The nominal angular resolution of the *Einstein* IPC instrument, with which the cluster was observed on 1980 March 13, was $\sim 1'$ so that the images of all five stars overlap. The only sensible approach is to model the image by fitting the intensity of the background and all five sources simultaneously. A likelihood method was used with the relative source positions fixed at values determined by the known coordinates. The results were as follows :

		V	$L_X(0.2 - 4.\text{keV})$ 10^{32}ergs s^{-1}		
HD 152248	O8	6.11	27.	±	5.
HD 152249	O9Ib	6.47	9.	±	4.
HD 152270	WC7+O5	6.59	10.	±	3.
HD 152234	B0.5Ia	5.45		<	5.
HD 152233	O6	6.59	16.	±	4.

These are more accurate values than previously published: WR79 is twice as bright as previously published putting it among the brightest of the WC stars as common among the WR+O binaries.

105

K. A. van der Hucht and B. Hidayat (eds.),
Wolf-Rayet Stars and Interrelations with Other Massive Stars in Galaxies, 105.
© 1991 *IAU. Printed in the Netherlands.*

THE NITROGEN SPECTRA OF WN STARS:
THE WN6 "STANDARD" STAR HD 192163 (WR136)

U. WESSOLOWSKI[1], W.-R. HAMANN[1], W. SCHMUTZ[2]
[1] Institut für Theoretische Physik und Sternwarte der Universität
Olshausenstraße 40, D-2300 Kiel 1, FRG
[2] Joint Institute for Laboratory Astrophysics
University of Colorado, Boulder, CO 80309-0440, USA

Hitherto our quantitative analyses of WR spectra [2][5] have been based on pure-helium models [1][6]. Now we further improved our non-LTE calculations by including a complex model atom of nitrogen (90 energy levels, 351 line transitions; with low-temperature dielectronic recombination) into our model atmospheres in order to synthesize adequately the spectra of WN subtypes (Wessolowski et al., in preparation). Together with the nitrogen (the most important "metal" in WN atmospheres), we introduced an improved temperature structure into our model calculations [3], now accounting for non-grey radiative equilibrium instead of the grey approximation applied so far. Moreover we took into account the line overlap of the considered elements (here: helium, nitrogen) and also their blanketing effects on the continuous radiation field.

Theoretical line profiles of nitrogen are compared with the observed spectrum of HD 192163 (alias WR136), a well-known WN6 "standard" star (Table 1). Most of the equivalent widths of the observed nitrogen lines can be reproduced within a factor of 2 to 3, but only by two slightly different models (Model 1: $T_* = 50kK$ and $R_* = 6.0R_\odot$, Model 2: $T_* = 60kK$ and $R_* = 5.5R_\odot$; both models with $log[\dot{M}/(M_\odot yr^{-1})] = -3.85$, $v_\infty = 1700\,km/s$ and a nitrogen abundance $\beta_N = 1.5\%$ by mass).

TABLE 1. Equivalent widths [Å] of nitrogen lines for HD 192163 (WN6)

nitrogen lines	N III $\lambda4640$	N IV				N V		
		$\lambda1486$	$\lambda1718$	$\lambda3480$	$\lambda4058$	$\lambda1240$	$\lambda4610$	$\lambda4944$
Observed	79	29	34	60	38	10	10	3
Model 1	30	58	10	12	28	3	1	2
Model 2	13	81	15	41	43	11	1	3

Altogether these results confirm the tendencies of Hillier's cool wind model for the WN5 star HD 50896 [4]. Remaining problems may be attributed to the very complex model atom and minor uncertainties in the stellar parameters and the temperature structure.

References:
[1] Hamann, W.-R., Schmutz, W.: 1987, Astron. Astrophys. **174**, 173
[2] Hamann, W.-R., Schmutz, W., Wessolowski, U.: 1988, Astron. Astrophys. **194**, 190
[3] Hamann, W.-R., Wessolowski, U.: 1990, Astron. Astrophys. **227**, 171
[4] Hillier, D.J.: 1988, Astrophys. J. **327**, 822
[5] Schmutz, W., Hamann, W.-R., Wessolowski, U.: 1989, Astron. Astrophys. **210**, 236
[6] Wessolowski, U., Schmutz, W., Hamann, W.-R.: 1988, Astron. Astrophys. **194**, 160

K. A. van der Hucht and B. Hidayat (eds.),
Wolf-Rayet Stars and Interrelations with Other Massive Stars in Galaxies, 106.
© 1991 IAU. Printed in the Netherlands.

SESSION III. INTRINSIC VARIABILITY – *Chair: Allan J. Willis*

Allan Willis chairing

Baratta, Leung, Underhill, Walborn, Koenigsberger, Humphreys, Chu

INTRINSIC VARIABILITY OF WOLF-RAYET STARS
FROM AN OBSERVATIONAL POINT OF VIEW

ANTHONY F. J. MOFFAT and CARMELLE ROBERT
Département de physique, Université de Montréal, Montréal and Observatoire du mont
Mégantic, Canada

ABSTRACT. Evidence is mounting that the dominant random component of variability in single
WR stars can be explained by one common phenomenon: stochastic formation, propagation and
decay of density enhancements in the winds.

1. WR Variability in General

Although it has often been noted that WR stars are basically rather stable at least on long time-scales (e.g. Schmutz 1991), they nevertheless do show varying degrees of variability particularly on short time-scales, in flux and polarization of continuum light and line emission. Only one review on the subject of variability of WR stars has appeared previously (Vreux 1987). The short interval from Vreux's to the present review suggests that the topic of variability is gaining importance in its own right.

In Vreux's (1987) review, the emphasis was on binary- versus pulsation-generated variations. In this review we will discuss mainly intrinsic variability as it relates to whatever observational wind phenomena may prevail. Intrinsic variability due to supernova explosions of WR stars is not considered here. We briefly contrast intrinsic with extrinsic variability, the latter being limited to periodic effects in binaries.

The principal signature of wind variability is the recently discovered presence of systematically propagating emission bumps superimposed on many WR emission lines. These are interpreted as the consequence of outward propagating blobs or waves (Moffat *et al.* 1988) driven by any or a combination of the following:

- intrinsic wind instabilities (Owocki 1990),
- rotation, possibly with a magnetic field (Underhill 1983; Nerney and Suess 1987; Poe *et al.* 1989),
- pulsation, either radial (Maeder 1985) or non-radial (Vreux 1985).

It is hoped that a deep understanding of the variability phenomenon in WR stars will eventually lead to constraints on the stellar parameters themselves which, because of the dense winds, have proved quite evasive.

2. Recent Observations

2.1 Photometry

By far the most photometric work has been done in the optical, where the highest precision is still attainable. Future work may be directed more to the UV (e.g. with the HRS in HST to probe the hot inner parts of the winds) or the IR (to probe the exterior, cooler parts of the wind, e.g. where dust may be forming in some WR stars). Here, we limit the discussion to the optical continuum, although most filter photometry is polluted to varying degrees by the inevitable presence of emission lines, especially in WC stars.

K. A. van der Hucht and B. Hidayat (eds.),
Wolf-Rayet Stars and Interrelations with Other Massive Stars in Galaxies, 109–123.
© 1991 *IAU. Printed in the Netherlands.*

2.1.1 Binary WR Stars

Lamontagne *et al.* (1991) have shown that essentially all **binary** WR stars show periodic, phase-dependent light variations, as long as the period is not too long or the orbital inclination and the WR mass-loss rate too small. Apart from the few rare cases of real eclipses of the stars (signature: two light minima per orbital cycle separated by 0.5 in phase for circular orbits), most WR binaries show a single V-shaped dip in their light curves, whose apex occurs when the O star is located behind the WR star, i.e. O-star light is most attenuated by free-electron scattering opacity of the intervening WR wind. The amplitude of the dip varies mainly with $1/a$, $1/v_\infty$, i and M_{WR}. The successful modelling of the single-dip light curves by Lamontagne *et al.* (1991) shows that at least our gross understanding of the basic structure of WR winds (e.g. spherical symmetry, density-radius law) cannot be too far off base. In fact it gives us some confidence when we come to discuss intrinsic perturbations in the winds.

A special, probably unique case is the WN5 star WR 6 = HD 50896. Over any two-week period this star reveals a coherent 3.766 day periodicity in light (e.g. Drissen *et al.* 1989) but the shape of the light curve changes remarkably on a time scale of months. Is this a precessing binary with a low-mass, compact companion or a single rotating star with some kind of slowly changing wind asymmetry?

2.1.2 Single WR Stars

While there may be some disputed cases, we take to mean "single" those stars for which no duplicity has been explicitly and unambiguously reported either from spectral morphology or periodic behaviour. On the basis of more extensive surveys for photometric variability of WR stars of different spectral subtype (Moffat and Shara 1986; Lamontagne and Moffat 1987; cf. Fig. 1 for sample light curves), one finds that (1) the time scale of the mostly random photometric variations is typically of the order of a day, in contrast to their possible progenitor luminous blue variable (LBV) stage, where the time scale is typically 10 times longer (cf. P Cyg: de Groot 1990), (2) the amplitude of variability tends to be greater for WN than WC stars and increases rapidly towards later subtypes (cf. Fig. 2; e.g. ~ 0.1 mag for WN8, somewhat less for WC9), (3) the variations tend to be independent (or nearly so) of wavelength for true continuum observations.

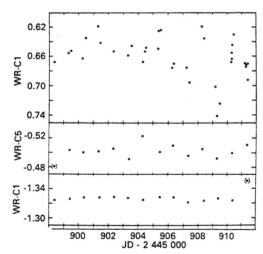

FIGURE 1. Sample B-band photometry of three single WR stars that vary randomly with remarkably different amplitudes: *top* – WR 123 (WN8), $\sigma = 0.030$ mag; *centre* – WR 134 (WN6), $\sigma = 0.008$ mag; *bottom* – WR 135 (WC8), $\sigma = 0.004$ mag (from Moffat and Shara 1986).

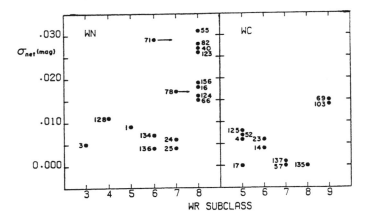

FIGURE 2. Photometric variability of WR stars not known to possess O-type companions, as a function of subclass (Lamontagne and Moffat 1987). The peculiar star WR 6 has been omitted.

Several notable exceptions have emerged from these trends recently. Van Genderen *et al.*'s (1990) intensive photometric monitoring of several WR stars over several nights indicates three stars (WR 46, WR 50 and WR 86) of types WN3p, WC6 + abs and WC7 that vary by 0.03-0.1 mag during intervals of several hours. While WR 50 may be a spectroscopic binary and WR 86 is an equal-magnitude, very close visual binary and thus ambiguous, WR 46 is quite exceptional although its spectrum is peculiar. More intense follow-up of these and other stars is needed, although older studies (e.g. Moffat and Haupt 1974: 7 stars of different subtype; Lamontagne and Moffat 1986: 1 W0 star) found changes below 0.01 mag in ~ 3 hours, more in line with the above global trends. It remains to be seen how typical this large, short-term variability is.

On longer time scales, the WC9 star WR 103 = HD 164270 has twice shown a curious ~ 1 mag dip lasting several weeks, one in 1909, the other in 1980 (Massey *et al.* 1984). Little is known how typical this is (e.g. only WCL stars?), although the WC8 + O binary CV Ser once showed a similar dip (Hjellming and Hiltner 1963) at an orbital phase that does not coincide with the passage of the O star behind the WR star.

In particular, six relatively bright WR stars have enjoyed rather intense photometric monitoring by many different investigators (cf. WR literature compiled for 1980-1990 by van der Hucht 1990): WR 134 = HD 191765 (WN6), WR 136 = HD 192163 (WN6), WR 78 = HD 151932 (WN7), WR 16 = HD 86161 (WN8), WR 40 = HD 96548 (WN8) and WR 103 = HD 164270 (WC9). At first, many of these stars were thought to be low-amplitude, periodic variables and were proposed to have compact companions (cf. Moffat 1982). However, now that much more data have been collected, this is much less certain, with different periods being claimed and periodic signals, if real, generally buried in a much higher level of random noise (cf. Vreux 1985).

The best example is WR 40, which shows large variations and has thus been observed very frequently. In particular, Gosset *et al.* (1989) and Gosset and Vreux (1990) summarize all previous as well as their own attempts to extract periodic sine waves from all data for this star up to the time of publication. In Gosset *et al.* (1989) two global, simultaneous periods are claimed: 2.5d and 6.25d, each with semi-amplitudes of ~ 0.01 mag, compared to the total range of photometric variation of ~ 0.1 mag. This 20% in amplitude translates into ~ 4% power and shows the difficulty of extracting useful information even from long series of data. Figure 3 depicts the plot of all data up to the time of the Gosset *et al.* (1989) publication, phased with the 6.25d period: despite formal statistical tests which strongly support the reality of this period, the figure does little to inspire confidence! Indeed, Gosset and Vreux (1990) revise the 1989 best value of 6.25d to 7d.

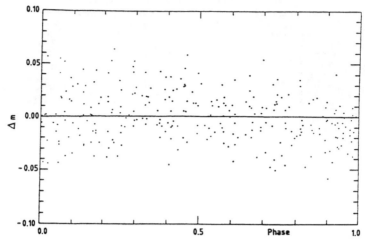

FIGURE 3. A large bank of photometric data for HD 96548 = WR 40 folded in phase with a period of 6.250d (Gosset *et al.* 1989).

2.2 Polarimetry

Although the first serious attempts to look for polarimetric variability in WR stars date to the 1940's (e.g. Hiltner 1950), only recently has there been a great surge in activity. The reason is precision: variations are very small and difficult to detect (< 1%, often < 0.1%) and therefore demand more time per data point than in photometry, something which researchers are less inclined to pursue at first.

2.2.1 Binary WR Stars

The same free electrons in the WR wind that cause a dip in the light curve of WR + O stars when the O star is behind, can scatter O-star light into the line of site. This scattered light will be polarized, depending on the scattering angle; for an ensemble of electrons, the degree of polarization depends on the vectorial sum of all the individual scatterings (cf. Brown *et al.* 1978; Rudy and Kemp 1978). Extensive observations in polarization of WR + O binaries elegantly confirm this simple notion (cf. especially St.-Louis *et al.* 1987; Drissen *et al.* 1987; Robert *et al.* 1989): a double sine-wave in P, θ (or $Q = P\cos2\theta$, $U = P\sin2\theta$) per orbital cycle is normally seen in close, circular-orbit binaries, with largest amplitude for $i = 0°$! Note that of all binaries observed, WR + O stars generally show the greatest amplitudes by virtue of their strong ionized winds. Not only are the binary polarization variations independent of wavelength (cf. Luna 1982; Piirola and Linnaluoto 1988), there is also no component of circular polarization (Robert and Moffat 1989), as expected for electron scattering. Furthermore, the amplitude of polarization modulation yields reliable estimates of $\dot{M}_{\mathrm{WR}}$ (cf. St.-Louis *et al.* 1988), while derivation of the orbital inclination from the polarization modulation can be used with spectroscopic values of $M\sin^3 i$ to calculate the actual stellar masses. The success of this technique on a broad scale again inspires confidence in our understanding of WR winds.

2.2.2 Single WR Stars

As in continuum photometry, there is a range in the level of polarimetric variability for single WR stars, such that WN stars tend to vary more than WC of similar subclass stars and the late-type, cooler subtypes are more variable (cf. Figs. 4 and 5). In a few cases, there is no detected polarization variability at the instrumental level ($\sim 0.015\%$ in P for the best observations). Whether this means that there is also zero net intrinsic polarization of the WR star is not clear until one vectorially subtracts off the interstellar (IS) polarization. This turns out to be unreliable in most cases where one tries to determine the interstellar polarization from field stars along the line of site close to the WR star: the usual chaotic nature of the ISM is the culprit. However, in one case studied by St.-Louis *et al.* (1987), WR 90 = HD 156385 (WC7), the observed polarization vector of the WR star is identical with the IS polarization at the 0.1% level. This leads to an upper limit of the flattening of the WR wind in this case to ~ 1% (Moffat 1988). Hence, we have at least one case in which it can be asserted that the WR wind is likely to be spherically symmetric (unless one has the fortuitously improbable case of a flattened wind seen pole-on).

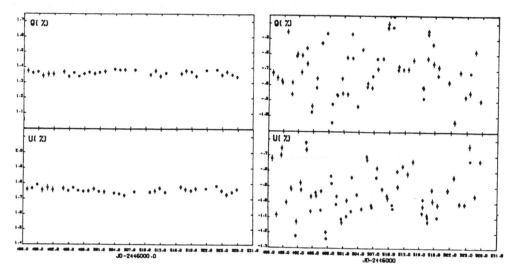

FIGURE 4. Polarization parameters Q and U versus time (separation between ticks is 0.1% and 3 days) for two extremes: *left* – WR 90 (WC7) with $\sigma_P = 0.016\%$, i.e. essentially instrumental (St.-Louis *et al.* 1987); *right* – WR 40 (WN8) with $\sigma_P = 0.155\%$ (Drissen *et al.* 1987). Note that the scales are identical in both pairs.

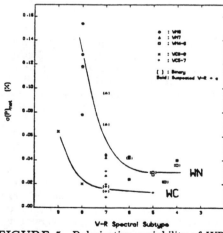

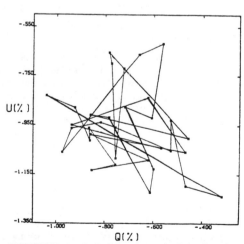

FIGURE 5. Polarization variability of WR stars as a function of subclass (Robert *et al.* 1989).

FIGURE 6. Polarimetric variations for the strongly varying WN8 star WR 40 in the Q-U plane (Drissen *et al.* 1987). Note the lack of a preferred plane.

Another way to subtract off the IS polarization is to decompose the observed continuum polarization at different wavelengths into two distinct components, one wavelength dependent (IS), the other wavelength independent (WR wind scattering). This method has not yet been tried in a systematic way using truly continuum polarization.

For those stars that vary significantly, the variations are clearly dominated by random noise in time (cf. Robert *et al.* 1989). The time scale, as in photometry, is typically about a day, i.e. a factor 10 shorter than seen in polarization variations of potential progenitor LBV's such as P Cyg (Hayes 1985). Furthermore, the general lack of a preferred polarization plane (e.g. a flattened distribution of time-dependent points in the Q-U plane; cf. Fig. 6), favours the idea of spatially random events. Finally, the wavelength-independent nature of the variations (Robert and Moffat 1989; Marchenko *et al.* in preparation) implies that electron scattering is responsible.

Taken together, the photometric and polarimetric variability of binary or single WR stars strongly points to the same phenomenon in each. In binaries one appears to understand that phenomenon quite well. In single stars we will first have to look in detail at the spectral variations before concluding what causes their light and polarization variations.

2.2.3 Flux Variability at Other Wavelengths

Attempts have been made to qualify the nature of the flux variability of WR stars in the radio, IR and X-ray regions. Most of these techniques suffer from a lack of sufficient signal-to-noise (S/N).

2.2.3.1 Radio

Hogg (1989) has discussed radio variability in 5 WR stars for which there exist several repeated VLA flux measurements. On a time scale of months to years only one star shows even a hint of significant variability. Compared to the S/N of typically ~ 300 that is routinely obtained in optical data, the S/N here is only ~ 10-20, so it is probably not surprising that variations have not been clearly found in the radio. On the other hand, clear radio variations of over a factor ten have been detected in the unusual, long-period binary WR 140 (Williams *et al.* 1990).

2.2.3.2 Infrared

The only IR monitoring of any significance has been carried out by Williams *et al.* (1987). So far, they have found three WR stars (all WCL) to show large IR eruption-like light curves lasting months. In one case (WR 140), there is a clear correlation with periastron passage in a very long period binary. In the two other stars, binary wind interaction is also suspected. More systematic monitoring at higher S/N is called for.

2.2.3.3 X-rays

Pollock (1987) has found that X-ray fluxes from WR stars tend to be higher in close binaries (probably like O stars: Chlebowski 1989). Especially variable in X-rays (Pollock 1989; Williams *et al.* 1990) is WR 140, as in the IR. Periastron passage enhances the wind collisions between the WR and the O companion such that X-rays emerge later when the optical depth is diminished. Next in variable level is WR 6, which may also be an elliptical orbit binary like WR 140, but of much shorter period and with a neutron star companion (Firmani *et al.* 1980). As noted above however the true nature of this unique system has yet to be revealed. Apart from these two stars, very little significant variability in X-rays has been detected.

2.3 Spectral Variability

While photometry and polarimetry yield spatially unresolved, global, scalar or vectorial sums, respectively, of the light output, spectroscopy offers the advantage of at least partial spatial resolution via the Doppler effect. For example, in a homogeneous, radially expanding wind, monochromatic line radiation will arise in rings concentric with the line joining the star's centre and the observer, the radii of which vary with depth according to the $v(r)$ law of the wind. Any inhomogeneous clumps of wind material propagating radially at velocity $v(r)$ and angle θ relative to the line of site, should then reveal themselves instantaneously in the form of a narrow emission feature at a specific wavelength $\lambda = \lambda_0 + (v(r)\cos\theta)/c$, where λ_0 is the wavelength of the unperturbed line centre. Such a feature could appear in emission from anywhere in the wind, or in absorption if seen close to the line of site to the WR star (P Cyg profile). The study of pure emission lines thus offers the advantage of probing the **global** wind structure, as opposed to the localized column towards the star that one sees in the absorption edge of a P Cyg profile.

Time series of spectra of the brighter WR stars have been carried out with increased frequency recently in both the optical and the UV. The time resolution is typically ∼ 15-60 min. Using IUE at high spectral dispersion, the UCL and Colorado groups have published detailed UV variability studies of three WR stars so far: WR 40 (Smith *et al.* 1985), WR 6 (Willis *et al.* 1989) and WR 136 (St.-Louis *et al.* 1989). The last two stars have the most intense data runs. With S/N $\lesssim 50$ per 0.1 Å spectral element, these studies are sensitive mainly to the larger variations seen typically in the P Cyg absorption edges, particularly of the strong UV resonance lines. Unlike in O stars (cf. Prinja *et al.* 1990), no narrow absorption-line components (NAC) are seen in these stars. This is surprising, in view of the scaled-up nature of the WR winds compared to O-star winds. However, claims have been made for the presence of NAC's in UV and optical spectra of some WR stars (Koenigsberger 1990). In any case one does tend to see in single WR stars broad UV absorption dips that possibly propagate from intermediate to high negative velocity. The S/N appears to be inadequate to get a definitive handle on this behaviour: the HRS on the HST may be a welcome instrument for scrutiny of this UV phenomenon.

In the optical, an array of WR stars of different subclass is under investigation (Robert in preparation; cf. also Robert *et al.* 1991). So far, only WR 134 (and to a lesser extent WR 136) has been published in any detail (Moffat *et al.* 1988; McCandliss 1988; Underhill *et al.* 1990). All these studies show that (a) the Pickering HeII emission lines behave almost identically, (b) HeII 4686 shows some differences cf. (a), and (c) the NIV 4058 line shows quite different variability. Figures 7 and 8 show an example from Moffat *et al.* (1988); note how the difference spectra allow one best to distinguish different subpeaks by removing the constant background wind emission profile (although with a price: negative artifacts). More details of this technique are given elsewhere (e.g. Robert *et al.* 1991). Suffice it to say here that the overall trend appears to be emerging from the study of different stars that emission subpeaks are accelerating: blue-shifted subpeaks get bluer with time (also applies to absorption dips), while red ones get redder. The intensity of the subpeaks grows and wanes on a time scale of the order of 10 hours.

2.4. Correlation Among Different Modes of Observation

Clearly it would be desirable to monitor some stars intensively and continuously for several days or weeks in as many different simultaneous modes as possible. So far, only a few attempts have been made. Robert and Moffat (1989) used a photo-polarimeter to monitor several stars simultaneously in continuum light, linear and circular polarization. Although truly simultaneous data are rather limited in quantity, it is already reasonably clear that light and linear polarization show little if any correlation, e.g. for WR 40. More recent observations by Drissen *et al.* (in preparation) are being studied to look at this problem in more detail.

Robert *et al.* (1991) are analysing simultaneous photometry (10 nights) and echelle spectroscopy (3 nights) of three WR stars of different subclasses. Again, no clear correlations are evident, although the time scales of the variations are similar, e.g. during one night when the P Cyg absorption components of He I 5876 and He II 5411 in WR 40 increased, the continuum light flux also increased, while other nights of photometric change were not accompanied by an obvious change in absorption edge strength. Clearly this is only the beginning...

2.5 The Peculiar WN8 Subclass

The WN8 subclass is outstanding in its high level of variability compared to all other WR subclasses. Other characteristics make WN8 stars unusual as well (cf. Moffat 1989):

- their spectra reveal narrow, often strong P Cyg profiles even in the optical,
- they avoid star clusters and often show other signs of runaway status,
- they appear to be devoid of O-type binary companions (e.g. in a sample of 9 WN8/9 stars monitored for spectroscopic orbits, **none** showed any orbit that could be attributed to a normal O-type companion, compared to the O stars found to orbit 16 (57%) out of a sample of 28 WN6/7 stars studied for orbital motion).

The mean absolute magnitudes of WN8 stars in the LMC, $< M_v >= -5.6 \pm 0.3$, are only marginally fainter than those of WN6/7 stars, $< M_v >= -6.1 \pm 0.2$ (Moffat 1989), so these two

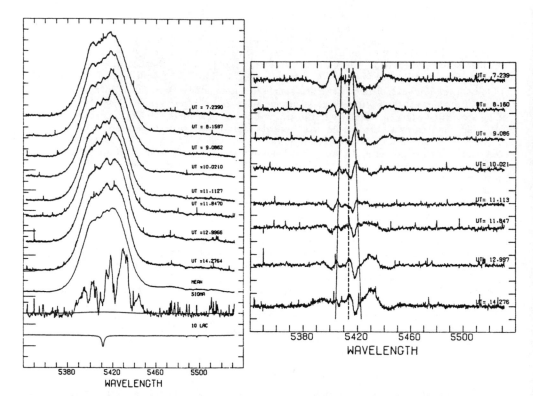

FIGURE 7. Time series of high resolution and high S/N spectra of He II 5411 in WR 134 (Moffat *et al.* 1988).

FIGURE 8. Difference from the mean of the spectra in Fig. 7 (Moffat *et al.* 1988). Two sample blobs are traced using solid straight lines as a guide, compared to the line centre (dashed line).

groups appear to have similar masses now as they must have had during their progenitor stage. The remarkable differences between WN8 and WN6/7 stars must then be caused by some other unknown factor, such as their mode of origin (e.g. WN8 stars could be ejected single stars from young clusters or they could be second stage binary WR + c stars, accelerated by a supernova explosion of the original primary, while WN6/7 stars have not suffered either process). These differences must be kept in mind when attempts are made to generalize the variability pattern of the different subtypes.

WC9 stars are also quite variable, but at about half the level of WN8 stars on the average (cf. Fig. 2). Furthermore, there exists at least one known binary among the WC9 stars: WR 70 (WC9 + B0I; Golombek 1983). Thus, even though WC9 stars share some properties with the WN8 stars, such as narrow emission lines, often P Cyg, the similarity stops there. Clearly, WC9 stars deserve further systematic attention.

3. Interpretation

3.1 Source of Variability

As noted in § 1, there are three plausible physical sources of variability in WR stars: pulsation, rotation and intrinsic wind instability. We discuss each of these in turn.

3.1.1 Pulsation

Table 1 gives a comparison of the expected behaviour of WR stars based on theory or as observed in other related (e.g. OB-type) stars, compared to what is actually observed, based on an assessment of the latest data. The general divergence of the two columns suggests that pulsations (radial or non-radial) are not likely to be responsible for explaining the observed variability. This conclusion is supported by other considerations: Noels and Scuflaire (1986) find that g-mode NRP can only be generated in some WR stars of type WNL for a negligeably short interval ($\lesssim$ 5000 yrs); Matthews and Beech (1987) argue against the reality of the NRP periods claimed by Vreux (1985); and Cox and Cahn's (1988) WR models indicate that g-mode NRP are unlikely in any WR star, while fundamental mode RP are possible in low H/He models.

TABLE 1
Comparison of Pulsation Theory with Observed Variability in WR Stars

Property	Pulsation Theory, or [Observed in Other Stars]	Observed
Periods	15-60 min: RP (Maeder 1985) hours: NRP (Vreux 1985)	$\sim$ a day (time scale, mostly non-periodic)
Most Var. Subtype	WNE, WCE: RP (Maeder 1985) (no H)	WNL
Spectral Subpeaks	[b $\rightarrow$ r only: NRP + rotation]	blue $\rightarrow$ blue+ red $\rightarrow$ red+
NAC's	Start at wind base (?)	$(0.5 - 1.0)v_\infty$ in OB star winds
Variability (pol/light)	[$\sim$ 0.01%/0.1 mag for NRP in β Cep (Watson 1983)]	0.5%/0.1 mag

Note: RP = radial pulsation, NRP = non-radial pulsation

3.1.2 Rotation

Most variability of single WR stars is dominated by random processes. Thus it is difficult to believe that any regular pattern associated with rotation can play a significant role in accounting for the observed variations. If there are real periodicities buried in the noise, they may be related to rotation, but they cannot be an important factor. Indeed, it is the rapid, radially expanding winds which dominate WR spectra. A possible exception is the star WR 6, with a 3.766d period which is normally coherent over at least 2 weeks. As noted above, it is uncertain whether this is due to rotation of a single star or the orbit of a low-mass, binary companion.

3.1.3 Intrinsic Wind Instabilities

The random presence of discrete emission subpeaks and P Cyg absorption dips always propagating to higher velocity makes a strong case for stochastic ejection of inhomogeneities or "blobs" of wind material. Note that ejected shells would not give rise to the same effect in the emission lines. We strongly suspect (but cannot prove yet) that the same phenomenon gives rise to the random photometric and polarimetric variations seen in many single WR stars.

   We illustrate this blob interpretation for one of the most frequently observed stars, WR 134 (WN6). Figures 7 and 8 have already shown the spectroscopic observations during a typical interval of 7 hours. In Figure 9 we show the radial velocities and strengths of the most obvious subpeaks in these same data, as a function of **time** (the straight lines are only guides to match up the same subpeaks from one spectrum to another). In Figure 10 we present expected radial velocity trajectories of localized density enhancements, assuming: (a) simultaneous radial ejection somewhere into a cone of angle θ relative to the direction towards the observer, who is at $\theta = 0$; the projected velocity will then be $v = v_w\cos\theta$, where v_w is the wind velocity directed radially from the star; (b) a velocity law of the form $v_w(r) = v_\infty(1 - R_*/r)^\beta$, with $\beta = 1$. R_* is the radius of the star, where

$v_w = 0$. The velocity law can be converted to depend on time instead of radius from the star, r, by a trivial manipulation.

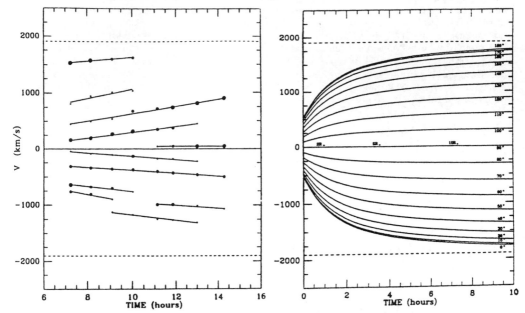

FIGURE 9. Propagation of projected subpeak velocities identified in Fig. 8, with time. Arbitrary straight lines join (assumed) common subpeaks whose strength is proportional to dot size. The terminal velocity from Prinja *et al.* (1990) is indicated by horizontal dashed lines.

FIGURE 10. Propagation model ($R_* = 7R_\odot$ – St.-Louis *et al.* 1988; $\beta = 1$) to be compared with Fig. 9. Angle of ejection varies from $\theta = 0°$ (towards the observer) to $\theta = 180°$ (away from the observer). Sample true subpeak radii (2, 5 and $10R_\odot$) are indicated on the $\theta = 90°$ line; they apply to all lines of different θ.

Comparison of Figures 9 and 10 must first allow for the trivial fact that the ejection times of individual blobs are random. It is then remarkable that we see no strongly curved trajectories in Figure 9, unlike in Figure 10, especially near the start of the ejection near the star. This may be due to the fact that, like the NAC's in OB-star winds, the growth of blobs does not become significant until beyond $v_w \simeq 0.5v_\infty$, i.e. $r \cong 2R_*$. (In any case the winds of all WR stars except some WNL are generally opaque inside this radius, even in the continuum.) Another interesting factor is that the observed trajectories (Fig. 9) appear to slope differently from the $\beta = 1$ curves at the same angle θ indicate in Figure 10. If the blobs are propagating at the same velocity as the general wind (the similar shape of the dispersion and intensity line profiles in Figure 7 would seem to indicate that this is indeed the case), this can only be understood if in fact β is larger than unity (cf. Robert and Moffat 1990), at least in the intermediate region of the wind that is probed here. Independent evidence for larger β values (i.e. softer winds) beyond the inner part of the wind has been noted previously by Koenigsberger (1991) in the UV spectra of WR stars and Fullerton (1990) in the optical spectra of Of stars. If this turns out to be incorrect, the only alternative is that the (overdense) blobs must be trailing the general wind and thus must be optically thick, in order to have suffered less radiative acceleration. If this effect were extreme, then it would be difficult to account for the fact in Figure 9 that one sees blobs with similar projected velocity distribution as the background wind and none is seen unambiguously in absorption (e.g. mini P Cyg profiles for blobs seen in projection towards the central star).

We give below a summary of the most important properties of the blobs, deduced from inspection of several WR stars of different subclass:

- the ejection process (spatial and temporal) is random,
- they generally provide a few % of the total wind emission,
- slow winds appear less stable (but cool WR have a few large blobs while hot WR have many small blobs at any given time),
- low ionization lines are more variable (since they tend to form further out in the wind, this may mean that blobs grow with time),
- propagation time is typically ~ 10 hours,
- blobs might be useful as potential wind tracers e.g. HeII (Pickering) forms blobs from $R = 2$-$10R_*$ in WR 134,
- stochastic blobs probably account for random variations seen also in light and polarization, due mainly to uncorrelated, wavelength-independent electron-scattering off blobs as opposed to line emission from blobs,
- blobs are probably driven by (random) wind instabilities,
- many details were predicted before they were detected (Antokhin *et al.* 1988).

4. Concluding Remarks

Can random radial ejection, growth and decay of blobs explain everything (i.e. light curves, polarization variations, spectral variations) seen in single WR stars? Undoubtedly our model is still quite primitive, with many details yet unexplored (e.g. blob sizes, masses, mass frequency...). Qualitatively, the answer to this question is probably affirmative, but a final answer must await a more quantitative study (Robert in preparation).

Even if the emission from the ejected blobs wanes with time, it is not clear that the blobs themselves necessarily dissipate completely. It is conceivable that some (e.g. the largest occasional "super" blobs) survive for a relatively long time, becoming spatially resolved from the central star. Indeed, some WR ring nebulae show knots and filaments which suggest that this might indeed be the case, presuming of course that the clumps were not already there as part of the ISM or previous ejection episodes. The best example may be the ejection-type nebula RCW 58 around the WN8 star WR 40 (Chu 1982) — a familiar star already in this paper — that shows a high level of variability! This nebula exhibits remarkable filaments pointing towards the central star, even close to it (Fig. 11). This cannot be therefore merely a projection effect. The filaments also reveal large abundance excess variations in He and N with position (Rosa and Mathis 1990). Smith *et al.* (1988) find an expansion velocity of ~ 87 km s^{-1} at $r \simeq 2$ kpc from the central star and claim that the clumps originated at most 3×10^4 years ago from a red supergiant progenitor. (But in view of the high luminosity and therefore mass of WN8 stars, and the poor correlation of WR stars with red supergiants — Maeder *et al.* 1980 — it seems more likely that the progenitor was an LBV). This would explain the relatively low expansion velocity compared to the observed WR wind terminal speed of 975 km s^{-1} (Prinja *et al.* 1990). However, one still has to observe the expansion velocities of the filaments very close to the central star, to check for higher velocities more directly from the WR star. Another ring nebula in which inhomogeneous, high-speed, N-rich stellar ejecta are claimed, is RCW 104 around the WN6 star WR 75 = HD 147419 (Goudis *et al.* 1988). Even the SNR Cas A shows fast-moving, N and H-rich knots outside the main optical/radio shell of the SNR (rich in O and S from the explosion itself). These are claimed to be fragments of a WN8 progenitor (Fesen *et al.* 1987).

Future work on the variability of WR stars will undoubtedly profit greatly from the technique of time-resolved, high S/N, high spectral resolution spectro-polarimetry, i.e. obtaining all four wavelength-dependent Stokes' parameters as a function of time. In particular, the degree of depolarization in the **spectra** of individual blobs being expelled at different angles, should allow one to narrow down the geometry. However, to acquire the necessary high quality data ($\lesssim 0.5$ Å resolution, S/N $\gtrsim 3000$ in polarimetry (!), and time resolution $\lesssim 30$ min) will necessitate the use of 8-10 metre class telescopes or larger, even for the brighter WR stars of 7-8 magnitude.

A. F. J. M. acknowledges financial support from the Conseil de Recherches en Sciences et Génie

120

du Canada and C. R. thanks the Formation de Chercheurs et Aide à la recherche du Québec for assistance in the form of a graduate bursary.

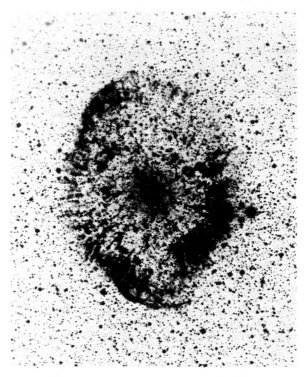

FIGURE 11. Photograph in H_α light of the nebula RCW 58 around the WN8 star HD 96548 = WR 40 (cf. Smith *et al.* 1988).

REFERENCES

Antokhin, I. I., Kholtygn, A. F., and Cherepashchuk, A. M. 1988, *Astron. Zh.*, **65**, 558; *Sov. Astron.*, **32**, 285.
Brown, J. C., McLean, I. S., and Emslie, A. G. 1978, *Astron. Astr.*, **68**, 415.
Chlebowski, T. 1989, *Ap. J.*, **342**, 1091.
Chu, Y.-H. 1982, *Ap. J.*, **254**, 578.
Cox, A. N. and Cahn, J. H. 1988, *Ap. J.*, **326**, 804.
de Groot, M. 1990, in A. S. P. Conference Series, *Properties of Hot Luminous Stars*, ed. C. D. Garmany (San Francisco: BookCrafters), p. 163.
Drissen, L., St.-Louis, N., Moffat, A. F. J., and Bastien, P. 1987, *Ap. J.*, **322**, 888.
Drissen, L., Robert, C., Lamontagne, R., Moffat, A. F. J., St.-Louis, N., van Weeren, N., and van Genderen, A. M. 1989, *Ap. J.*, **343**, 426.
Fesen, R. A., Becker, R. H., and Blair, W. P. 1987, *Ap. J.*, **313**, 378.
Firmani, C., Koenigsberger, G., Bisiacchi, G. F., Moffat, A. F. J., and Isserstedt, J. 1980, *Ap. J.*, **239**, 607.
Fullerton, A. W. 1990, Ph. D. thesis, U. of Toronto.
Golombek, D. A. 1983, Thesis, U. of Buenos Aires.
Gosset, E., Vreux, J. M., Manfroid, J., Sterken, C., Walker, E. H., and Haefner, R. 1989, *M. N. R. A. S.*, **238**, 97.
Gosset, E. and Vreux, J. M. 1990, *Astron. Astr.*, **231**, 100.

Goudis, C. D., Meaburn, J., and Whitehead, M. J. 1988, *Astron. Astr.*, **191**, 341.
Hayes, D. P. 1985, *Ap. J.*, **289**, 726.
Hiltner, W. A. 1950, *Ap. J.*, **112**, 477.
Hjellming, P. M. and Hiltner, W. A. 1963, *Ap. J.*, **137**, 1080.
Hogg, D. E. 1989, *Astron. J.*, **98**, 282.
Koenigsberger, G. 1990, preprint.
Koenigsberger, G. 1991, these proceedings.
Lamontagne, R. and Moffat, A. F. J. 1986, *Astron. Astr.*, **162**, 114.
Lamontagne, R., and Moffat, A. F. J. 1987, *Astron. J.*, **94**, 1008.
Lamontagne, R., Robert, C., Grandchamps, A., Lapierre, N., Moffat, A. F. J., Drissen, L., and
    Shara, M. M. 1991, these proceedings.
Luna, H. G. 1982, *P. A. S. P.*, **94**, 695.
Maeder, A. 1985, *Astron. Astr.*, **147**, 300.
Maeder, A., Lequeux, J., and Azzopardi, M. 1980, *Astron. Astr.*, **90**, L17.
Massey, P., Lundström, I., and Stenholm, B. 1984, *P. A. S. P.*, **96**, 618.
Matthews, J. M. and Beech, M. 1987, *Ap. J.*, **313**, L25.
McCandliss, S. R. 1988, Ph. D. thesis, U. of Colorado.
Moffat, A. F. J. 1982, in Proc. IAU Symp. No. 99, *Wolf-Rayet Stars: Observations, Physics,
    Evolution*, eds. C. de Loore and A. J. Willis (Dordrecht: Reidel), p. 263.
Moffat, A. F. J. 1988, in *Polarized Radiation of Circumstellar Origin*, eds. G. V. Coyne *et al.*
    (Vatican: Vatican Observatory), p. 607.
Moffat, A. F. J. 1989, *Ap. J.*, **347**, 373.
Moffat, A. F. J. and Haupt, W. 1974, *Astron. Astr.*, **32**, 435.
Moffat, A. F. J. and Shara, M. M. 1986, *Astron. J.*, **92**, 952.
Moffat, A. F. J., Drissen, L., Lamontagne, R., and Robert, C. 1988, *Ap. J.*, **334**, 1038.
Nerney, S. and Suess, S. T. 1987, *Ap. J.*, **321**, 355.
Noels, A. and Scuflaire, R. 1986, *Astron. Astr.*, **161**, 125.
Owocki, S. P. 1990, in *Reviews of Modern Astronomy* (Berlin: Springer), **3**, in press.
Piirola, V. and Linnaluoto, S. 1988, in *Polarized Radiation of Circumstellar Origin*, eds. G. V.
    Coyne *et al.* (Vatican: Vatican Observatory), p. 655.
Poe, C. H., Friend, D. B., and Cassinelli, J. P. 1989, *Ap. J.*, **337**, 888.
Pollock, A. M. T. 1987, *Ap. J.*, **320**, 283.
Pollock, A. M. T. 1989, *Ap. J.*, **347**, 409.
Prinja, R. K., Barlow, M. J., and Howarth, I. D. 1990, *Ap. J.*, **361**.
Robert, C. and Moffat, A. F. J. 1989, *Ap. J.*, **343**, 902.
Robert, C. and Moffat, A. F. J. 1990, in A. S. P. Conference Series, *Properties of Hot Luminous
    Stars*, ed. C. D. Garmany (San Francisco: BookCrafters), p. 271.
Robert, C., Moffat, A. F. J., Bastien, P., Drissen, L., and St.-Louis, N. 1989, *Ap. J.*, **347**, 1034.
Robert, C., Moffat, A. F. J., and Seggewiss, W. 1991, these proceedings.
Rosa, M. R. and Mathis, J. S. 1990, in A. S. P. Conference Series, *Properties of Hot Luminous Stars*,
    ed. C. D. Garmany (San Francisco: BookCrafters), p. 135.
Rudy, R. J. and Kemp, J. C. 1978, *Ap. J.*, **221**, 200.
Schmutz, W. 1991, these proceedings.
Smith, L. J., Lloyd, C., and Walker, E. N. 1985, *Astron. Astr.*, **146**, 307.
Smith, L. J., Pettini, M., and Dyson, J. E. 1988, *M. N. R. A. S.*, **234**, 625.
St.-Louis, N., Drissen, L., Moffat, A. F. J., and Bastien, P. 1987, *Ap. J.*, **322**, 870.
St.-Louis, N., Moffat, A. F. J., Drissen, L., Bastien, P., and Robert, C. 1988, *Ap. J.*, **330**, 286.
St.-Louis, N., Smith, L. S., Stevens, I. R., Willis, A. S., Garmany, C. D., and Conti, P. S. 1989,
    Astron. Astr., **226**, 249.
Underhill, A. B. 1983, *Ap. J.*, **268**, L127.
Underhill, A. B., Gilroy, K. K., Hill, G. M., and Dinshaw, N. 1990, *Ap. J*, **351**, 666.
van der Hucht, K. A. 1990, preprint.
van Genderen, A. M., van der Hucht, K. A., and Larsen, I. 1990, *Astron. Astr.*, **229**, 123.
Vreux, J. M. 1985, *P. A. S. P.*, **97**, 274.
Vreux, J. M. 1987, in Proc. Workshop in honour of C. de Jager, *Instabilities in Luminous Early
    Type Stars*, eds. H. Lamers and C. de Loore (Dordrecht: Reidel), p. 81.
Watson, R. D. 1983, *Ap. Sp. Sc.*, **92**, 293.
Williams, P. M., van der Hucht, K. A., and Thé, P. S. 1987, *Q. J. R. A. S.*, **28**, 248.
Williams, P. M., van der Hucht, K. A., Pollock, A. M. T., Florkowski, D. R., van der Woerd, H.,
    and Wamsteker, W. M. 1990, *M. N. R. A. S.*, **243**, 662.
Willis, A. J., Howarth, I. D., Smith, L. J., Garmany, C. D., and Conti, P. S. 1989, *Astron. Astr.
    Suppl.*, **77**, 269.

DISCUSSION

Conti: (1) Any idea how flattened HD 191765 might be? (2) What fraction of WR stars might have non-spherical winds?
Moffat: (1) Judging from the continuum polarization compared to the maximum depolarized line polarization of HD 191765 of Schmidt (1988), the intrinsic polarization is about 1%. This is like Be stars. (2) According to Schmidt, only 2 out of some dozen or so WR stars show significant depolarization in the lower ionization lines.

Cherepashchuk: Did you try to determine the velocity law from your fine structure spectroscopic observations?
Moffat: Yes. The motion is accelerated but for the detailed interpretation more observations are needed.

Cassinelli: I question your approach in concluding that "rotation is unimportant" based on polarization observations. You isolated intrinsic polarization from interstellar based on variability alone. But as you mentioned only very briefly during your summary, one can also find the intrinsic polarization by observing the change in polarization *vs.* wavelength across a strong emission line. Schmidt and also the Wisconsin group (Schulte-Ladbeck, Taylor, Bjorkman et al.) find from spectropolarimetry that WR polarization properties are often like Be stars, with magnitudes of about 1%. Based on the information we have thus far, I certainly do not think we should conclude that rotation is not important.
Moffat: Only a minority of WR stars show such a depolarization across their emission lines. Indeed, some WR stars (for which the interstellar polarization can be reliably estimated from field stars) show close to zero net-intrinsic polarization.

Koenigsberger: I do not see how you can rule out the existence of radial pulsations in WNE stars. These winds are so dense that the periodic oscillations can not be reflected in the observations.
Moffat: As Owocki points out (see next review talk), any perturbation near the wind base becomes significantly amplified later when it propagates outward with the wind.

Underhill: Recently, Harmanec has interpreted the moving absorption dips in lines of OB stars as due to a rotating spokelike structure rather than as due to NRP. It seems possible that the sets of changing subpeaks (blobs) on emission lines could be interpreted as concentrations of radiating plasma in a set of ever changing rotating filaments in the central hole of a possible disk. It seems significant that the subpeaks are chiefly seen in the velocity range from $-1/2v_\infty$ to $+1/2v_\infty$. If they were freely moving blobs, why do we not see any between $1/2 - 1v_\infty$?
Moffat: But the emission peaks on pure emission-line WR profiles show clear trends of outward acceleration (blue ones get bluer, red get redder). This is not expected from rotation filaments. The subpeaks are seen with similar distribution as the background wind in velocity space, *i.e.* we do see them between 0.5 and $1.0v_\infty$ but fewer in number than between $\pm 0.5v_\infty$.

de Groot: Quite apart from questions of interpretation, if one is looking for interrelations with other massive stars, let me point out that very much the same variations you reported for the latest WN stars are also present in P Cyg: similar brightness variations with somewhat larger amplitude again, similar polarimetric varations as you showed, and similar variations in the optical spectrum.
Moffat: [Due to some limitations, a viewgraph of P Cygni light variations by de Groot (1990) was not shown during the talk. This was rectified after this question.]

Sreenivasan: If the star is rotating and if there is evidence for a flattened structure, such a structure would be rotating very slowly (due to angular momentum loss from the star as well as the size of the flattened structure's radius). You are also not observing a homogeneous structure around a star and one would not see any strong correlation with rotation. Further, if non-radial pulsation is present one usually finds many non-radial modes simultaneously excited in evolved stars and again it would be hard to see specific periodicity. So, you would in fact see what you described, although the converse is not necessarily valid.

Moffat: If rotation of the central core does play a role in WR intrinsic variations, there must be some inhomeogeneity associated with it (*e.g.* magnetic loops, spots, non-radial pulsation). These should eventually propagate outward to the visible part of the wind where some trace of periodicity should be observable. There is no compelling case in which one observes this to happen except for HD 50896, which may be a binary in any case.

Maeder: One cannot rule out pulsations in WR stars. The optically thick wind is unable to respond to the short periods of the interior. Only if one could see deep enough in the winds, one could infirm or confirm the existence of pulsations.

Moffat: But as Owocki points out, a small perturbation deep in the wind should become significantly amplified as it propagates outward.

Owocki: (1) I did not understand your distinction between optically thick *vs.* thin blobs moving slower than *vs.* as fast as wind. In simulations we see very dense structures moving at near v_{wind}. (2) In OB NAC, both the repetition and acceleration times seem to be related to the *vsini*, implying rotation is playing some role.

Moffat: (1) If blobs are truely density enhancements, being optically thick means that they will not "see" the whole radiation field and will be less accelerated than thinner parts of the (ambient) wind. (2) Perhaps!

Tony Moffat

SPORADIC VARIATIONS IN WOLF-RAYET STARS

B.S. SHYLAJA
Indian Institute of Astrophysics
Bangalore 560034
India

ABSTRACT. Many Wolf-Rayet stars display variations which are observable either spectroscopically or photometrically or both. These are not exclusively associated with binary systems. A study is made of the sporadic events in the context of extended atmospheric structure.

1. INTRODUCTION

Intensive observations of Wolf-Rayet stars in the optical, IR and UV have shown that generally all WR stars are variable. The types of variabilities differ from system to system. For example the variability may be observable in fluxes, radial velocity amplitudes, emission line intensities etc. (For a review, see Moffat, 1990). Multiple periodicities (established in some cases) complicate the analysis.

A detailed investigation of each system reveals that there are sporadic variations in many systems. Such variations might have been observed in any region of the spectrum as an increase in flux or as a change in the line profile. An attempt is made to collect the details of such sporadic variations and understand it in terms of the atmospheric structure.

2.1 Spectrum Variability

The well studied binary CQ Cep has interesting sporadic events to be reported. The behaviour of the HeI lines has been changing at some epochs. The line at $\lambda 4471\text{Å}$ showed line splitting in 1943 (Hiltner, 1943) and in 1952 (Bappu and Viswanadhan, 1977). In 1978 similar line splitting became observable for the HeI $\lambda 3888\text{A}$ line (Leung, Moffat and Seggewiss, 1983). In 1950, a sudden change in the strength of HeII line at $\lambda 4686$ was observed (Hiltner, 1950). Spectrophotometric observations show that in 1982 there was probably a similar increase in this line strength (Shylaja, 1986a). It may be noted in all these cases that a corresponding increase in other lines are not seen.

The spectrometric study of HD 50896 (Shylaja, 1986b) showed that only on one occasion there was an increase of flux for HeII line at $\lambda 4860$. Similar spectral variations have been reported - a decrease of flux of N IV line at $\lambda 3482$ and an increase in flux for HeII line at $\lambda 4686$ (Van der Hucht et al., 1990).

The spectrophotometric study of HD 76536 (Shylaja, 1990) showed a decrease in flux for carbon lines at $\lambda 4650$, at one epoch only.

125

K. A. van der Hucht and B. Hidayat (eds.),
Wolf-Rayet Stars and Interrelations with Other Massive Stars in Galaxies, 125–127.
© 1991 IAU. Printed in the Netherlands.

The study on HD 193077 and HD 192103 are reported to show episodic variations (Marshenko, 1988).

2.2 Light Variability

A brightening of the order of 0.03 mag. was reported for CQ Cep (Hiltner, 1950). Later a variation upto 0.1 mag. was observed in a total duration of 10 months (Kartasheva, 1976). Many more systems are being studied for light variability and therefore there is likely to be an increase in such observations.

It may be logical to include here the only one recorded 'eclipse' of WR 113 ≡ CV Ser, since this also may be regarded as a sporadic event in the absence of repeatability. The same argument holds good for WR 22 ≡ HD 92740 (Balona et al., 1989) as well. The absence of periodic variations in HD 152270 reported in the same work also may be included as sporadic nature.

2.3 IR Variability

Williams et al. (1989) have studied most of the stars of the WC subgroup and found that many are variables. Out of the six variables GL 2179 ≡ WR 118 has only one record of variability. Their conclusion is that 'variability in IR from circumstellar dust shells and especially the inner edges is rare', from which they arrive at a constant rate of dust formation.

3. DISCUSSION

It is evident from the recorded episodes of sporadic variations, that these variations are preferential to later subgroups in both WN and WC.

Lamontagne and Moffat (1988) have shown that in general the variability is maximum in the WN 7 subgroup. The WN 7 subgroup has also a larger mass loss rate (Doom, 1988).

From a study of the flux variations of CQ Cep (Shylaja, 1986a) it was shown that the eclipse effects are noticeable for only the line of N V at $\lambda 4603$. Further, on comparing with similar results of V 444 Cyg it was shown that the extent of atmospheres is larger in case of CQ Cep.

This leads us to an important clue that the sporadic variation has its origin in the atmospheric strcuture itself. The stratification prevalent in the atmosphere puts the N V and N IV lines closer to the photosphere and low excitation lines line HeI in the outer regions of the atmosphere. Thus any sporadic phenomena will become immediately apparent in the HeI line structure. This argument is strengthened by looking at the RV curves and flux variations of HeI lines which present a scatter unlike other lines in the same binary system.

The IR sporadic variations can be explained on the basis of nonconstant dust formation rates, although not favoured.

The changes in the mass loss rate associated with a sudden

change in particle density can explain the sporadic episodes in
several cases (for example, HD 152270, Balona et al. 1989 and HD
96548, Gosset and Vreux, 1990). The presence of a companion can
lead to such situations when there is a sudden change in mass transfer.

Recently Moffat et al. (1988) have explained small narrow emission bumps moving across an emission profile in terms of rapid blob
ejection. Such a theory can be extended to explain the sporadic
variations.

Any further work on the interpretation of sporadic events,
therefore, needs more data input. An emphasis on the study of HeI
lines and a regular monitoring of the systems to 'catch' the event,
would facilitate the exclusion of chaotic behaviour as one of the
causes.

At the same time it is very important to report such sporadic
variations instead of excluding it for purposes of publication.

References

Balona, L.A., Egan, J. and Marang, F., 1989, MNRAS, 240, 103.
Bappu, M.K.V. and Viswanadham, N., 1977, KOB, A, 2, 89.
Doom, C., 1988, Astron. Astrophys. 192, 170.
Gosset, E. and Vreux, J.M., 1990, Astron. Astrophys. 231, 100.
Hiltner, W.A., 1943, Ap. J., 99, 273.
Hiltner, W.A., 1950, Ap. J., 112, 477.
Kartasheva, T.A., 1976, Sov. Astron. Lett., 2, 197.
Lamontagne, R. and Moffat, A.F.J., 1987, Astron. J. 94, 1008.
Leung, K.C., Moffat, A.F.J. and Seggewiss, W., 1983. Ap. J., 265, 961.
Marshenko, S.V., 1988, Kinematics Phys. Cel. Bodies.
Moffat, A.F.J., 1990, these proceedings.
Moffat, A.F.J., Drissen, L., Lamontagne, R., Roberts, C., 1988,
    Ap. J., 334, 1038.
Shylaja, B.S., 1986a, J. Astrophys. Astron. 7, 171.
Shylaja, B.S., 1986b, J. Astrophys. Astron, 7, 305.
Shylaja, B.S., 1990, Astrophys. Sp. Sci. 164, 63.
Van der Hucht, K.A., van Genderen, A.M. and Bakker, P.R., 1990, Astron.
    Astrophys. 228, 108.
Williams, P.M., Van der Hucht, K.A. and The, P.S., 1987, Astron.
    Astrophys. 182, 91.

Drissen, Shylaja, Lewis, The, Pollock

(INTRINSIC) VARIATIONS OF WOLF-RAYET STARS

A.M. van Genderen[1], M.A.W. Verheijen[1], E. van Kampen[1],
F.H.A. Robijn[1], R. van der Heiden[1], B.P.M. van Esch[1], H.
Greidanus[1], R.S. le Poole[1], R.A. Reijns[1], K.A. van der
Hucht[2], H.E. Schwarz[3], C.W.H. de Loore[4], E. Kuulkers[5], L.
Spijkstra[5]

[1]Leiden Observatory, Postbus 9513
2300 RA Leiden, The Netherlands

[2]SRON Space Research Utrecht, Sorbonnelaan 2
3584 CA Utrecht, The Netherlands

[3]ESO, Casilla 19001
Santiago 19, Chile

[4]Astrophysical Institute, Pleinlaan 2
1050 Brussels, Belgium

[5]Astronomical Institute, Roetersstraat 15
1018 WB Amsterdam, The Netherlands

ABSTRACT

Twenty - two Wolf-Rayet stars (12 of type WN and 10 of type WC)
were observed in the years 1986-1990 with the VBLUW photometer of
Walraven. Eight (WC and WN) objects appeared to be constant. Five
of the in total 14 variable objects will be discussed in the
present paper: WR46 (WN3 pec), WR50 (WC6+a), WR55 (WN8), WR86
(WC7) and WR123 (WN8).
    Simultaneous spectroscopy is made for WR46 and WR50. These
results will be also discussed in short.

1. Introduction

This paper, based on the oral contribution at the IAU Symposium
No. 143, has a title which probably suggests more than can be
justified by the authors. Consequently the reader might see his
expectations for an enlighting description on "intrinsic
variations" (thus directly caused by the photosphere) blighted.

K. A. van der Hucht and B. Hidayat (eds.),
Wolf-Rayet Stars and Interrelations with Other Massive Stars in Galaxies, 129–145.
© 1991 IAU. Printed in the Netherlands.

Nevertheless the various morphological types of light curves are most interesting and still not quite understood.

Although duplicity among WR stars appears to be an important source of variability (Vreux, 1987; Moffat and Shara, 1986) and many WR stars turn out to be constant, we have not given up the hope that real intrinsic variations, thus directly reflecting photospheric phenomena, once might be established for some of them. A crucial condition is that the suspected specimen (if any) should be observed intensively by various techniques.

If random or large scale turbulence in the envelope causes strong emission line variations (see for example the review paper of Vreux 1987), this could turn up as small light variations if medium broad band photometry is applied . However, they are not the "first order" type of variations we are looking for. They rather could be called variations of the "second order".

We shall discuss here in short five peculiar specimens of variable WR stars. A more detailed analysis will be given in forthcoming papers.

2. Observations and reductions

The observations of the objects listed in Table 1 of which some will be discussed here were made with the 90-cm Dutch telescope at the ESO, La Silla, Chile, by various observers during the interval 1986-1990. The telescope is equipped with the VBLUW simultaneous photometer of Walraven. The last detailed description of this system is given by Lub and Pel (1977).

Each object was alternately measured relative to a nearby comparison star a number of times per night, or during a few hours in a row, or during complete nights. The diaphragm aperture was 16". Typical integration times per measurement varied between 1 and 5 minutes. During the time that an object was monitored, the time resolution varied between 5 and 15 minutes. Calibrations and corrections for differential extinction were applied with the aid of standard stars measured throughout each night.

Twice the average standard deviation (2σ) per data point is indicated in the figures representing light and colour curves (in log intensity scale). All these curves are relative to the comparison star.

For some objects also simultaneous spectroscopy was performed by monitoring them for several hours with the B & C spectrograph of the ESO 1.52 m telescope equipped with a CCD detector. The resolution $R = \frac{\lambda}{\lambda} \sim 450$, spectral range: 4080-7150 A. Part of the calibration and reduction is still underway, part of it will be discussed in the present paper.

3. The objects under investigation

Table 1 lists the 22 WR stars observed by our group during the

Table 1. The WN and WC type Wolf-Rayet stars, indicated by their WR numbers (van der Hucht et al. 1981) investigated by our group with multi-colour photometry. Underlined objects are variable. Objects marked with an asterisk were partly simultaneously monitored with spectroscopy.

WN							WC				
3	4	5	6	7	8	9	5	6	7	8	9
46*[a,b]		6[c,d]	110	22[a,b]	16[d,e]	108	52 23	86*[a,b]		53	103*[f]
97*				24[d]	40[d]		111 50*[b]	90[d]			121
				25[d]	123			93[a]			
				55							

a. Monderen et al. (1988)
b. van Genderen et al. (1990)
c. van der Hucht et al. (1990)
d. van Genderen et al. (1987)
e. van Genderen et al. (1989)
f. van Genderen and van der Hucht (1986)

last four years and still only partly published. Most of the objects were monitored during a number of hours per night, sometimes within a time span of one month, others in a few months and scattered in two or more seasons.

The sample consists of 12 WN type and 10 WC type stars. Nine specimen of the first group and five of the second group are variable. It must be noticed that this is no reliable statistic, since many stars were observed because they were known variables already. Others were simply put on the program just to investigate whether they are variable.

It is of interest to mention that eight objects turned out to be constant. That is to say that the noise is usually smaller than 0^m005 (differentially with respect to a nearby comparison star).

A glance at Table 1 shows that there is obviously no correlation between variability and spectral type. Obviously constancy is a normal habit among an important part of the WR stars. This does not exclude the possibility that the envelope masks any variation of the photosphere.

4. The light- and colour curves of a selected sample

4.1. WR46 = HD104994 (WN3 PEC)

WR46 ($V_J \sim 10.9$) was observed relative to HD108355 (B8IV, $V_J \sim 6.0$) in February and March 1989 by two of us (E.K. and E.v.K, respectively).

132

Typical light and colour curves for one night are shown in Fig. 1.

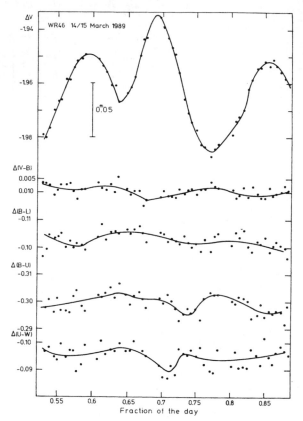

Fig. 1   Typical light and colour curves of WR46 (WN 3 pec) (Observations by E.v.K.)

The brightness and colours relative to the comparison star are plotted as a function of H.J.D.

The time scale of a single wave is ~ 3.3 h, the same as for the observations in 1986 (Monderen et al., 1988) and in 1988 (van Genderen et al., 1990). The light curve often shows per night one deep and one shallow minimum. The maxima do not differ always in height and sometimes bumps are present on one of the branches.

The period search program of Sterken (1977) was used in the range 0^d1 and 0^d4 and with steps of 0^d0001 and 0^d001. For both sets (February and March) the best candidate period is $0^d1412 \pm 0^d0002$, resulting in a one peaked light curve with a large scatter due to the varying shape of each cycle. However the phase diagram for a period twice this length: $P = 0^d2824$ is shown in Fig. 2

(March 1989 observations only), because of the significant difference between the two maxima of the colour curves V-B and V-W (we omit the U-W curve). It could mean that this period is more significant than the previous one. The scatter is of the same order.

If the February 1989 data set is combined with those of March, the scatter in Fig. 2 slightly increases, but apparently the period is still valid. Whether it is stable over an interval of years cannot be checked due to the large time gap between the observations of 1986, 1988 and the present set (1989).

The light curve resembles that of a binary consisting of a deformed bright component (the WR star) and a smaller one. The average distance between the extrema is ~ 0.25 of the period, which is in favour of the binary hypothesis. However, the variation in the times of the extrema varies sometimes by 15% or more. If the binary hypothesis is correct, this variation must be caused by some other effect superimposed on the binary modulation. An analysis of the height variation of the extrema seems to indicate a more or less systematic trend with a periodicity of ~ 0^d30, which is slightly longer than the binary period. Whether this periodicity is real cannot be said due to the relative short time interval in which the March data are collected (eight nights). (The February data set is a much smaller data set and therefore not used). One could think of luminous clouds in a disk around the system rotating slightly slower than the binary, which cause the extra modulation, or we are dealing with a (randomly) varying shape of the continuum emitting region. The peculiar fact that the colours are bluer in the light minima, instead of redder as it should be in the case of normal elongated stars, might be caused by a better transparency of the envelope of the WR star at the place facing the companion and at the opposite side. Due to the lower gravity at the points of a normal distorted star, the temperature should be lower there.

A period of 0^d28 or 6.8 h is very short for a massive binary. With Kepler's law and adopting total masses of 5, 10, 15, 20 and 40 $M_\odot$ we find for the distances between the components 3.1, 3.9, 4.4, 4.9 and 6.2 $R_\odot$, respectively. This is not impossible since the radius of WN type stars can be as small as 0.5-5 $R_\odot$ (Cherepashchuk et al., 1984; Hillier, 1987; Hamann et al., 1988; Langer, 1989).Then V_{orbit} amounts to 520, 650, 700 , 800 and 1000 kms^{-1}, respectively.

The total mass of the system is estimated as follows:
M_v (WN3) = -2.8, BC = -4.2 ± 1.2 (van der Hucht et al., 1988), we find M_{bol} = -7.0 ± 1.2. Then with the aid of the M-L relation of Maeder and Meynet (1987): M = 3.2 (+4.1, - 1.8) $M_\odot$. Consequently, the total mass of the system should lie somewhere in the range 1.4-7.3 $M_\odot$.

The simultaneously obtained spectroscopic observations in three nights in March 1989 by one of us (H.E.S.) are partly analyzed and allow some quantitative discussion. Each night about

40 spectra were obtained with integration times of 3 min.

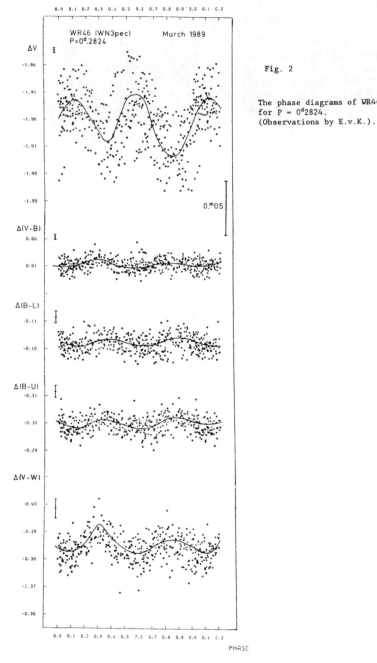

Fig. 2

The phase diagrams of WR46 (WN 3 pec)
for P = 0^d2824.
(Observations by E.v.K.).

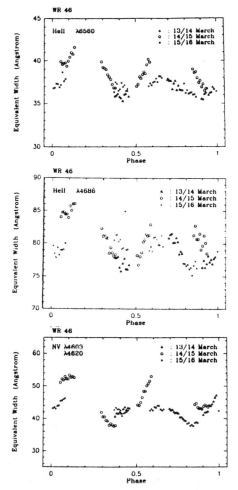

Fig. 3 The variation of the equivalent width for a few
prominent lines of WR46 as a function of the phase.

Figure 3 shows for a few prominent lines the variation of the equivalent width in a phase diagram. The total range of the variation per line amount to 10-25%, (the brightness varies by 10% only). The variations are in phase with the light curve, but they do not vary each night in the same way. Obviously the emission line emitting region shows a <u>stronger distortion</u> than the region which emits the continuum light (the pseudo-photosphere). This has the important consequence that the companion presumably revolves within the envelope, while the emission line emitting region closely surrounds both Roche lobes. Besides both regions are apparently subject to geometrical changes from cycle to cycle in view of the large scatter in the phase diagrams (Figs. 2 and 3). Therefore one may assume that turbulence at a large scale takes place in the pseudo-photosphere as well as in the outer envelope.

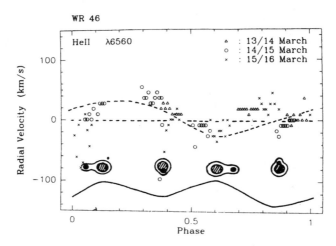

Fig. 4   The radial velocity curve of WR46 for the
HeII λ6560A line as a function of the phase. A
schematic light curve and the geometry of the system
(pseudo-photosphere, outer envelope and companion)
are sketched at the bottom.

Figure 4 shows the radial velocity curve of the HeII 6560 line in a phase diagram. Also here the scatter is large. The dotted curve should therefore be considered with care. The light curve and the geometry of the system (pseudo-photosphere, outer envelope and companion) are sketched at the bottom.

The dotted curve roughly sketched through the data points, indicates $V_1 \sin i \sim 40$ kms^{-1}.

Then with $P=0^d2824$, $a_1 \sin i = 1.6 \ 10^5$km and the mass function $f(M) = 0.002$. With $M_1 + M_2 = 5 \ M_\odot$, $V_{orbit} = V_1 + V_2 = 520$ kms^{-1} (see above) and with the aid of the formula $V_1 = M_2/(M_1+M_2) \ V_{orbit}$, we find $M_1 \leq 4.6 \ M_\odot$ and $M_2 \sin i = 0.4 \ M_\odot$.

For $M_1 + M_2 = 10 \ M_\odot$ these latter values amount to 9.4 and 0.6 $M_\odot$, respectively.

Consequently, if the total mass estimation given above is of the correct order, the mass of the primary lies presumably between 4 and 9 $M_\odot$, its radius between 2 and 3 $R_\odot$, the mass of the secondary between 0.4 and 0.6 $M_\odot$. The latter could then be a white dwarf. However, if we are dealing with a neutron star ($\geq 1.4 \ M_\odot$) then the total mass of the system should lie within the range of 30-40 $M_\odot$.

According to our binary model the moment with the <u>shallow</u> minimum and the <u>bluest</u> colour corresponds with the companion in front! Based on an analysis of the ratio line flux/continuum flux in V and B band-passes as a function of the phase, a correction of the V-B colour index was possible. The effect of the line flux was subtracted from the photometric flux, leaving a pure continuum variation. It appeared that the amplitude of the colour curve V-B doubled up to 0^m02.

4.2 WR50 (WC6a)

WR50 ($V_J \sim 11.9$) was observed relative to the same comparison star and in the same nights as WR46 (sect. 4.1). The schematic light and colour curves for the nine nights in March 1989 are shown together in Fig. 5.
Due to the low signal to noise ratio in the W band pass (λ_{eff} = 3235 A), no U-W curve can be given. In February 1989 the variations showed a smaller range.
One set of typical light and colour curves obtained in 1988 (van Genderen et al., 1990) is shown also for comparison purposes (dotted curve). In 1989 colours are reddest at minimum light. The B-L curve (for B λ_{eff} = 4300 A, for L λ_{eff} = 3840 A) does not vary much. Due to the fact that each night showed more or less the same trend: bright in the beginning and faint at the end of the night, the period or quasi-period may be close to 1^d. By shifting the descending branches on each other we obtain P ~ 1^d06.
The behaviour of especially the colour V-B is often opposite to that from 1988 (van Genderen et al., 1990): then the star is sometimes blue during minimum light. It is possible that those variations concern another phase of the light curve, if we are dealing with an eclipsing phenomenon, see further.
The largest drop in brightness and reddening of the colours occurs in the night of 19/20 March: the drop amounts to 0^m2 in V, 0^m5 in B and L and even 0^m9 in U!
The simultaneously obtained spectroscopic observations in three nights of March 1989 by one of us (H.E.S.), are partly analyzed and allow some quantitative discussion. Each night about 40 spectra were obtained with integration times of 3 minutes. See for a comparison of a typical WC spectrum and the position of the V and B band-passes Fig. 7 in van Genderen et al. (1990).
Figure 6 shows for one night the typical trend of the ratio line flux/continuum flux. It appears that the ratio <u>increases</u> for the V and B band passes by 5-10%, while the amount of light drops (see Fig. 5). Thus the contribution to the descending branch of the light curve in V and B is larger for the continuum light. Applying these ratios and the photometric fluxes in the V and B band-passes, it appears that the line fluxes stop declining midway the descending branch.
It is possible that we are dealing with an eclipsing phenomenon. The peculiar and variable characteristics of the

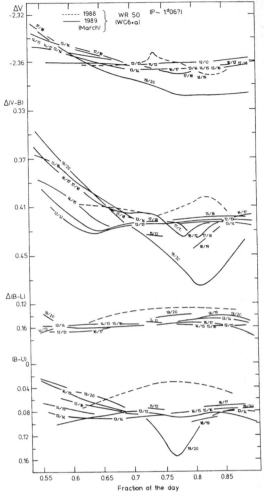

Fig. 5

A schematic representation of the light and colour curves of WR50 (WC6+a). Note that the B-L and B-U scales are compressed by a factor of two, compared to V and V-B. (Observations by E.v.K.)

light curve may be the result of an intricate eclipse of (pseudo)-photospheric light and a luminous disk/envelope with variable geometry. Such an eclipse could cause peculiar center-to-limb effects and the peculiar progressive increase of the light amplitudes to shorter wavelengths. The type of the companion is still unknown.

4.3. WR55 = HD117688 (WN7)

WR55 ($V_J \sim 10.9$) was observed relative to HD116875 (B8V, $V_J \sim 7.8$) during nine nights in April and May 1989 by two of us (F.H.A.R.

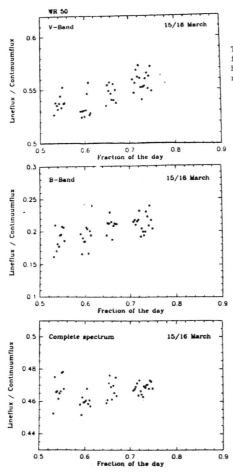

Fig. 6

The trend in one night of the ratio line flux/continuum flux of WR50 for the V and B bandpasses and for the whole spectral range.

and R.v.d.H., respectively). The period search program was applied between 1^d and 3^d (the range in which a possible period should lie) with steps of 0^d02. There are several candidate periods such like 1^d22 and 2^d56, but the light curves are not convincing at all. Twice the first mentioned one: $P = 2^d44$ resulted in a peculiar binary type light curve as shown in Fig. 7. The colour V-B is generally slightly redder in the light maxima than in the light minima.

A binary period of this length is not impossible, adopting a radius of 10 $R_\odot$ for the WR star and a smaller radius for the component. The adopted total mass lies in the range of 20-50 $M_\odot$.

Rotation modulation of a deformed WR star and a smaller massive companion is thus possible, but then one has to assume a

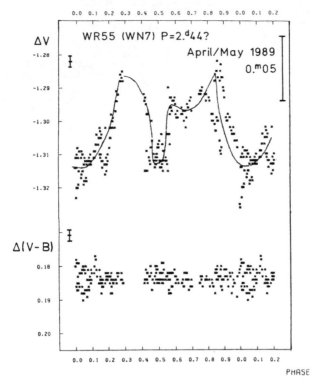

Fig. 7

The phase diagrams of
WR55 (WN7) for the
tentative period
P = 2^{d}44.
(Observations by F.H.A.R.
and R.v.d.H.).

quite peculiar light distribution across the circumference of the
WR star.

More observations are planned to increase the number of
observations. Until then we have to consider the period as highly
uncertain.

4.4 WR86 = HD156327 (WC7), visual companion B0V, sep ≈ 2"

WR86 (V_J ~ 9.3) was observed relative to HD158528 (A5, V_J ~ 8.4)
during six nights in July 1989 by two of us (L. S. and H.v.W.G.).
The spectral type of the companion is determined by Smith (1968),
implying that the magnitude is ~ 1^{m}5 fainter than the WR star (van
der Hucht et al., 1988).
The star was also monitored during one night in 1986 showing an
oscillation of ~ 3h and an amplitude of ~ 0^{m}02 in V (Monderen et
al., 1988). The colour curves are bluer in the light maxima. The
same type of variation is exhibited by the new observations. The
result is shown in the phase diagrams for V, B, L, U and W of
Fig. 8 with P = 0^{d}1385 ± 0^{d}0002. The scatter is partly intrinsic

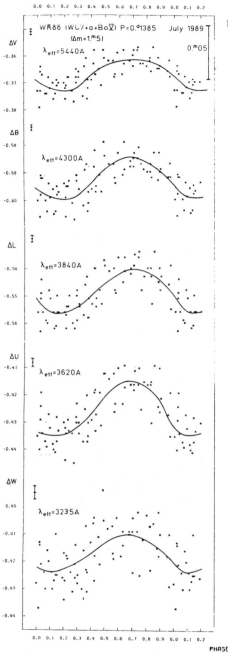

Fig. 8

The phase diagrams of WR86
(WC7 including the visual
companion B0V) for the period
$P=0^d1385$.
(Observations by L.S. and H.G.).

due to the fact that the cycles differ significantly from each other. The maximum amplitude in V amounts to ~ 0ᵐ05. The colour curves (not shown) are bluer during maximum light except for the U-W curve, which is redder. The reason is that the light amplitude in W is smaller than that of the U band pass.

It is possible that the B type companion is the source of this variability, which could then be a ß Cep star. The length of the period, the shape and variability of the light curves and the size of the amplitude (~ 0ᵐ25 if a correction is applied for the effect of the light of the WR star) agree with that supposition (see the review paper of Lesh and Aizenman (1978)).

However, the fact that the progression of the size of the light amplitude to shorter wavelengths is interrupted by the amplitude in the W band pass (see for example BW Vul in Lesh and Aizenman (1978) and τ'Lup observed by Cuypers (1987), disagrees with the ß Cep theory.

New photometry is planned to investigate the variability further. Simultaneous spectroscopy in a few nights is made, but not yet available for a discussion.

4.5 WR123 = HD177230 (WN8)

WR123 (V_J ~ 11.3) was observed relative to HD174916 (A0, V_J ~ 7.5) during 17 nights in September and October 1989 by two of us (R.S.1.P. and R.A.R). The star turned out to be variable with an amplitude of ~ 0ᵐ15.

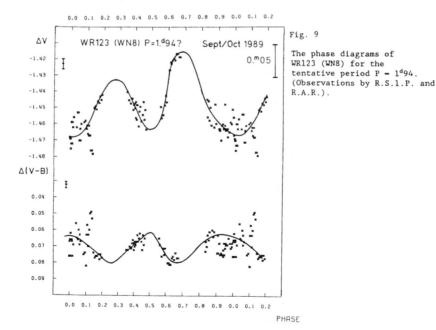

Fig. 9

The phase diagrams of WR123 (WN8) for the tentative period P - 1ᵈ94. (Observations by R.S.1.P. and R.A.R.).

A search of a period was made between 0^d5 and 2^d5 with steps of 0^d01. The best but very tentative result is $P = 1^d94$ of which
The peculiar light curve (with two unequal maxima) looks likethe phase diagrams for V and V-B are shown in Fig. 9. that of a binary with an elongated bright companion (the WR stars) similar to WR46 and WR55. Indeed a RV curve revealed by Lamontagne et al. (1983) proofs that a companion is present. They claim that it might be a neutron star. However their tentative period of 1^d76 and those of Moffat and Shara (1986): 1^d76 or 2^d37 do not fit the present material. Whether the period is correct or not, the light minima correspond with the bluer colour in V-B (not so outspoken in the other colour indices), similar to WR46 and WR55, which also have doubly peaked light curves (Figs. 1, 2 and 7).
New observations are planned to investigate the object further, until then the period should be considered with reserve.

5. Conclusions

Three of the five variable WR stars discussed here show light curves which presumably reflect a rotation modulation of an elongated star (WR46, WR55 and WR123). The presence of a smaller companion for each of them is suggested. For WR123 (WN7) this seems to be the case indeed in view of its variable radial velocity (Lamontagne et al., 1983).
The period of revolution for WR46 (WN3pec) implicates that it is the shortest period known among the WR binaries. The tentative radial velocity curve of the HeII 6560A line indicates a mass function $f(M) = 0.002$. The orbital velocity is at least 500 kms^{-1} and the mass and radius for the WR component likely lie in the range of 4-9 $M_\odot$ and of 2-3 $R_\odot$, respectively. The companion, presumably a white dwarf, could have a mass in the range 0.4-0.6 $M_\odot$. However, if it is a neutron star (~1.4 $M_\odot$) then the total mass of the system should lie in the range 30-40 $M_\odot$. The variation of the light curve may be caused by secondary effects such like a fast varying geometry of the regions from which we receive the continuum and the line emission. The equivalent widths of the most prominent lines show a phase dependent variation of 10-25%, indicating that the line emitting region (the envelope) is more distorted than the continuum emitting region (the pseudo-photosphere). Possibly this envelope surrounds both Roche lobes.
Among the two other (WC type) objects, WR50 (WC7) shows a light variation with a time scale of presumably ~$1^{d\cdot}06$ and a light amplitude which increases strongly to the short wavelengths: ~ 0^m2 in V, ~ 0^m9 in U. A peculiar type of eclipse is possible.
Finally WR86 (WC7) shows a light curve which looks like that of a ß Cep star. If this interpretation is correct the observed variation must be that of the visual B type companion.
We can conclude that the present material does not give us a reason to believe that any of the variations are caused by the

photosphere of the WR star itself. Duplicity and other sources of a "higher order" (Sect. 1) are the main cause.

References

Cherepashchuk, A.M., Eaton, J.A., Khaliullin, K.H.F.: 1984, Astrophys. J. 281, 774

Cuypers, J.: 1987, Astron. Astrophys. Suppl. 69, 445

van Genderen, A.M., van der Hucht, K.A.: 1986, Astron. Astrophys. 162, 109

van Genderen, A.M., van der Hucht, K.A., Steemers, W.J.G.: 1987, Astron. Astrophys. 185, 131

van Genderen, A.M., van der Hucht, K.A., Bakker, P.R.: 1989, Astron. Astrophys. 224, 125

van Genderen, A.M., van der Hucht, K.A., Larsen, I.: 1990, Astron. Astrophys. 229, 123

Hillier, D.J.: 1987, Astrophys. J. Suppl. 63, 947

Hamann, W.R., Schmutz, W., Wessolowski, U.: 1988, Astron. Astrophys. 194, 190

van der Hucht, K.A., Conti, P.S., Lundström, I., Stenholm, B.: 1981, Space Science Rev. 28, 227

van der Hucht, K.A., Hidayat, B., Admiranto, A.G., Supelli, K.R.: 1988, Astron. Astrophys. 199, 217

van der Hucht, K.A., van Genderen, A.M., Bakker, P.R.: 1990, Astron. Astrophys. 228, 108

Lamontagne, R., Moffat, A.F.J., Seggewiss, W.: 1983, Astrophys. J. 269, 96

Langer, N.: 1989, Astron. Astrophys. 210, 93

Lesh, J.R., Aizenman, M.L.: 1978, Ann. Rev. Astron. Astrophys. 16, 215

Lub, J., Pel, J.W.: 1977, Astron. Astrophys. 54, 137

Maeder, A., Meynet, G.: 1987, Astron. Astrophys. 182, 243

Moffat, A.F.J., Shara, M.M.: 1986, Publ. Astron. Soc. Pacific 92, 952

Monderen, P., de Loore, C.W.H., van der Hucht, K.A., van Genderen, A.M.: 1988, Astron. Astrophys. 195, 179

Smith, L.F.: 1968, Monthly. Not. Roy. Astron. Soc. 138, 109

Sterken, C.: 1977, Astron. Astrophys. 57, 361

Vreux, J.M.: 1987, Proc. Workshop C. de Jager: Instabilities in Luminous Early Type Stars, eds. H.J.G.L.M. Lamers and C.W.H. de Loore, Reidel, Dordrecht, p. 81

DISCUSSION

Niemela: How accurately is the distance of WR46 known? With such a short period, could it be one of those "naked planetaries"?

van Genderen: The distance is not known very well. It is not a member of a cluster or associations. Van der Hucht *et al.* (1988) estimate a distance of 3·4 *kpc*, for an adopted average absolute visual magnitude $M_v(WN3) = -2.8$.

Schulte-Ladbeck: You mentioned that the line emitting region in WR46 is more distorted than the continuum region. Could you please specify or describe this in more detail?

van Genderen: The amount of light is proportional with the visible surface. The light curve shows a distortion of 10 % (amplitude is $0\overset{m}{.}1$), that is to say, if we face the points of the elongated star (surface $\sim \pi a^2$), we receive 10% less light than if the star is seen from aside (surface $\sim \pi ab$). Consequently the ratio between the two is $\sim b/a$. Since the amplitude of the curves representing the variation of W_λ ranges from 10 - 25%, the distortion b/a also ranges up to 25%. Therefore the envelope from which we receive the emission lines must be more distorted than the pseudophotosphere from which we receive the continuum light. Therefore we suggest that the envelope also surrounds the companion.

Cassinelli (question directed to Maeder): Your conclusion that pulsation is not occurring in WR stars is a very important one in regard to possible mechanisms for driving mass loss. I recall from Maeder *et al.* (1983) that interior theorists argue that WR stars lose sufficient mass by a wind to remain quasi-homogeneous. I can think of no way for the deep interior to affect the mass loss *except* by way of pulsation. Your conclusion therefore indicates that one cannot use the quasi-homogeneity argument to derive mass loss rates.

van Genderen: My first remark "Do variations of WR photospheres exist?" gives the wrong impression that I do not believe in them. If they exist, it proves that the envelopes are so optically thick, that they cannot be detected at all.

Maeder: If you do not maintain quasi-homogeneity by high mass loss you will have the WR star much too luminous, as shown by Hamann. The model would be too luminous. So in order to have WR models at the location found by Hamann *et al.*, you need to lose a lot of mass. And this is a way to assure homogeneity and of course if there are additional mixing mechanisms this would also favour the homogeneity of the star.

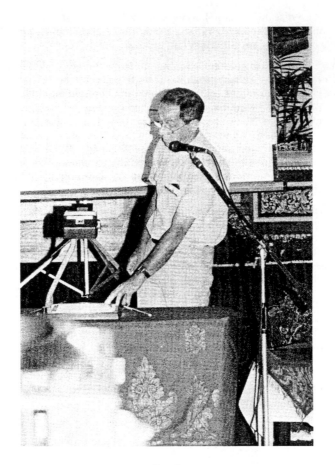

van Genderen

PHYSICAL PARAMETERS OF INHOMOGENEITIES IN WOLF-RAYET WINDS

C. ROBERT[1], A. F. J. MOFFAT[1], and W. SEGGEWISS[2]
[1]Université de Montréal, Département de physique, Montréal, and Observatoire du mont Mégantic, Canada
[2]Observatorium Hoher List, Universitäts-Sternwarte Bonn, F. R. Germany

ABSTRACT. Time-resolved spectroscopy has revealed small, systematically varying features superimposed on the broad emission lines of WR stars. We believe that these structures are due to inhomogeneities of emitting wind material propagating with the general wind. Here we suggest a common, but still unknown, origin for the intrinsic variations observed in WR stars.

INTRODUCTION

Photometric and polarimetric observations of Wolf-Rayet (WR) stars indicate that these objects can show intrinsic variations with time scales of hours to days. Late-type stars of both sequences, WN8 and WC9, are the most variable (e.g. Lamontagne and Moffat 1987; Robert et al. 1989). No clear periodicity can by unambiguously discerned. Therefore, we favor an explanation for these variations as due to stochastically fluctuating inhomogeneities in the winds.

In the spectra, variations in the shapes of the line profiles are also observed. Here we present new spectra of three WR stars of different spectral type: HD 96548 = WR40 (WN8), HD 164270 = WR103 (WC9) and HD 165763 = WR111 (WC5). Along with the WN6 stars HD 191765 = WR134 and HD 192163 = WR136 (McCandliss 1988; Moffat et al. 1988), we describe and compare, in a qualitative way for now, the variations in the lines.

OBSERVATIONS AND ANALYSIS

The new spectra presented here were collected from 1989 May 31 to June 2 (UT) with the echelle spectrograph CASPEC and a CCD detector at the ESO 3.6 m telescope, Chile. High signal-to-noise (300-500) and high resolution (0.07 Å/pix) spectra were obtained in the range 5160 to 6190 Å.

Special care has been taken to eliminate the ripple appearing when merging together the different spectral orders obtained with the echelle. A routine involving Fourier analysis has been developed for this purpose.

Line variability was then assessed by comparing the standard deviation (rms difference from the mean for each night) in the line with the standard deviation in the continuum. A variable line is taken to be one for which $\sigma_{line} > 3\sigma_{continuum}$, in excess of the Poisson errors.

RESULTS AND DISCUSSION

1) The variability

In the WC5 star HD 165763, the C IV 5805 and especially the C III 5696 lines are noticibly variable (see Fig. 1). Small variations are also seen in the He II 5411, C IV 5471 and O III 5592 lines.

Almost all the lines of the WC9 star HD 164270 are strongly variable (see Fig. 2). The HeII 5411 and O III 5592 lines show a P Cyg profile, where changes occur in both the emission and the absorption parts (He I 5876 also has a P Cyg profile but which is strongly blended).

Many lines between 3950 and 6610 Å of the WN6 star HD 191765 were also found to be variable (McCandliss 1988). For the other WN6 star, HD 192163, the He II 5411 line shows less activity during the night of 1986 July 26 (UT) than the same line in HD 191765 (Moffat et al. 1988).

The WN8 star HD 96548 shows strong P Cyg lines of He II 5411 and He I 5876 which are considerably variable both in the emission and the absorption parts of the profile (Fig. 3).

A closer look at the variable lines reveals the presence of small subpeaks or bumps which move with time (e.g. C III 5696 line of HD 165763, Fig. 1). These bumps have a short life-time of the

K. A. van der Hucht and B. Hidayat (eds.),
Wolf-Rayet Stars and Interrelations with Other Massive Stars in Galaxies, 147–153.
© 1991 *IAU. Printed in the Netherlands.*

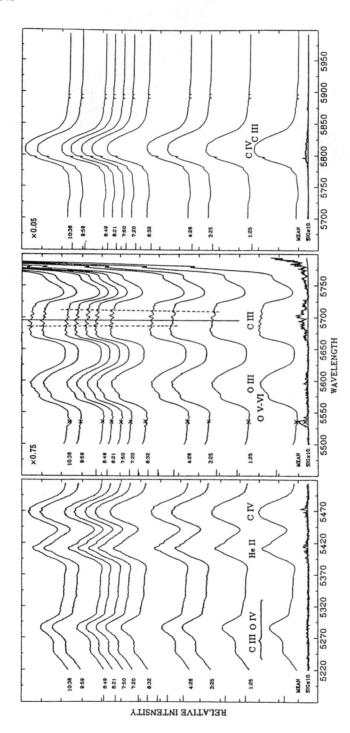

FIGURE 1. Spectra collected for HD 165763 (WC5) on 1989 June 2 (UT). Universal time is given for each observation and increases from bottom to top. The mean spectrum is the average of all spectra collected during the night. The sigma curves (multiplied by 10 relative to the spectra) are the bin-to-bin standard deviations from all spectra for the night (the noisier sigma curve) and the expected Poisson noise (almost a straight line). Line identifications are based on Torres and Massey (1987) and the atlas of Smith and Kuhi (1990). Numbers given in the upper left corner of the middle and right panels are the intensity scale factors relative to the first panel. The vertical line in the second panel indicates the central wavelength. Dashed lines show the outward acceleration of two sample blobs.

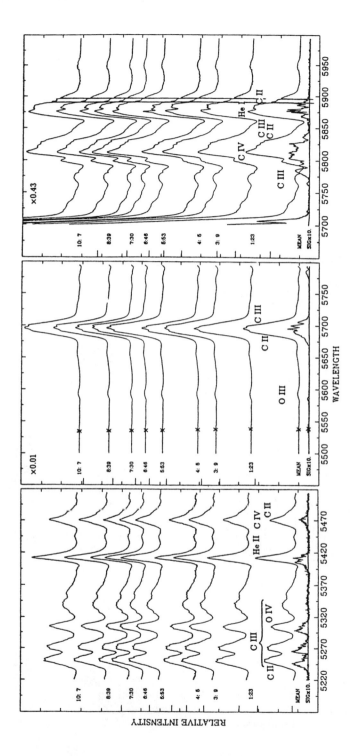

FIGURE 2. Spectra collected for HD 164270 (WC9) on 1989 May 31 (UT). Refer to Figure 1 for details.

150

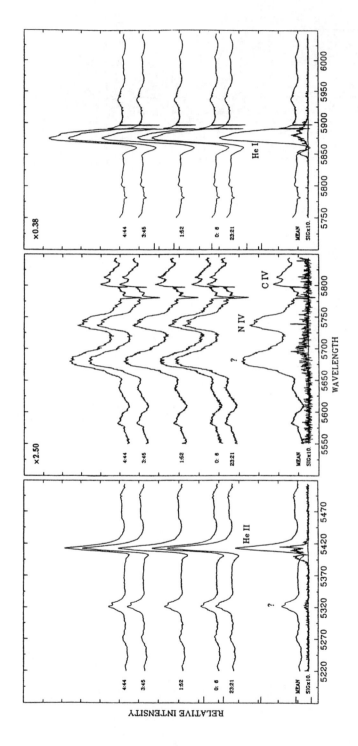

FIGURE 3. Spectra collected for HD 96548 (WN8) on 1989 June 2 (UT). Refer to Figure 1 for details. Line identifications are based on the atlas of Smith and Kuhi (1981).

order of 10 hr for most stars observed to date. One can clearly see that bumps that start red-shifted relative to the line center become even more red-shifted with time, while the blue-shifted bumps become more blue-shifted (see other examples in Robert *et al.* 1988, Robert and Moffat 1989). Figure 4 shows the O III 5592 P Cyg profiles of HD 164270, where one can see moving bumps in emission and dips in absorption (or yet more emission bumps?), both of which are of similar amplitude. These structures are interpreted as a consequence of wind density enhancements or "blobs" moving outwards along with the general wind.

For the three newly observed stars, the same general pattern of bumps is seen in the variable lines at the same time. An example of this behavior for the WC9 star HD 164270 is given in Figure 5. (Further work needs to be done to check for possible time delays between the lines.) In HD 191765, all the Pickering He II emission lines tend to vary in the same way, while He II 4686 shows some differences and N IV 4058 varies quite differently (McCandliss 1988).

2) Quantifying the inhomogeneities

Table I gives the average height, maximum height and the average FWHM of the bumps in the most variable lines for each star. The last column indicates the typical number of inhomogeneities identified at any given time. This parameter is more easily deduced (especially for the narrower lines) by subtracting first the overall average spectrum.

TABLE I

HD	Sp	Variable lines	<Height> (line flux units)	Max. Height	<FWHM> (Å)	No. of "blobs"
96548	WN8	He II 5411 (P Cyg)	5%	13%	2	2-3
		He I 5876 (P Cyg)	6%	14%	3	2-3
164270	WC9	C III 5250 (blended)	11%	22%	2	3
		C III 5272	13%	26%	2	3
		C III 5305 (blended)	14%	36%	2	3
		He II 5411 (P Cyg)	8%	23%	2	3-4
		C IV 5471 (blended)	11%	23%	-	-
		O III 5592 (P Cyg)	15%	40%	2	3
		C III 5696	10%	14%	2	4-5
		C IV 5805 (blended)	-	-	-	-
		He I 5876 (blended)	-	-	-	-
165763	WC5	C III 5696	7%	15%	4	8-9
		C IV 5805	2%	4%	4	8-9
191765	WN6	He II 5411 (Moffat *et al.* 1988)	7%	14%	4	4-5

Early-type stars appear to show a greater number of discrete bumps in their spectral lines at a given time compared to late-type stars. (Here we make no distinction between the different sizes of the blobs. In future work, the mass function of the blobs will be considered.) The inhomogeneities in early-type stars also tend to display a larger spread in velocity (i.e. <FWHM> of the blobs is larger). These same structures are also likely to be responsible for the variations seen in photometry and polarimetry. In the case of early-type, hot WR stars, the combined effect of a larger number of discrete blobs should lead to a lower *net* amplitude in the photometric or polarimetric variability as is indeed seen for these stars. For the late-type, cooler WR stars, the bumps tend to be less frequent, narrower (and more intense in late WC), suggesting a higher over-density of wind material in their blobs. This could imply stronger asymmetry of the electron distribution around late-type WR stars, which will result in a larger amplitude of the polarimetric variations, as seen. Denser but fewer simultaneous knots in the winds of late-type stars may also explain the greater photometric variations observed in these stars.

CONCLUSION

The overall evidence so far suggests a common (but unknown) origin for all the non-binary related, intrinsic variations observed in WR stars. Future work will attempt to quantify this statement in a more rigorous way.

152

REFERENCES

Lamontagne, R., and Moffat, A. F. J. 1987, *Astron. J.*, **94**, 1008.
McCandliss, S. R. 1988, Ph. D. thesis, University of Colorado.
Moffat, A. F. J., Drissen, L., Lamontagne, R., and Robert, C. 1988, *Ap. J.*, **334**, 1038.
Robert, C., and Moffat, A. F. J. 1990, in *Properties of Hot Luminous Stars*, ed. Garmany (San Fransisco: Astronomical Society of the Pacific), p. 271.
Robert, C., Moffat, A. F. J., Bastien, P., Drissen, L., and St.-Louis, N. 1989, *Ap. J.*, **347**, 1034.
Robert, C., Drissen, L., and Moffat, A. F. J. 1988, in *Physics of Luminous Blue Variables*, ed. Davidson *et al.* (Dordrecht: Kluwer), p. 299.
Smith, L. F., and Kuhi, L. V. 1990, preprint.
Smith, L. F., and Kuhi, L. V. 1981, *J. I. L. A. Report*, No. 117.
Torres, A. V., and Massey, P. 1987, *Ap. J. Suppl.*, **65**, 483.

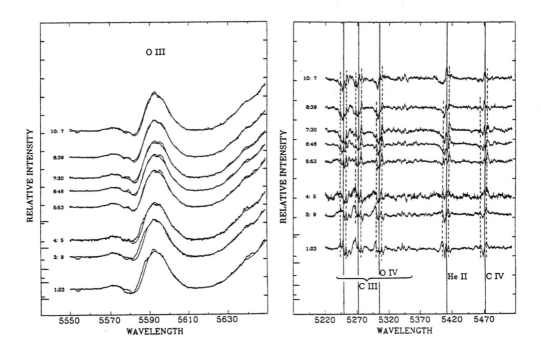

FIGURE 4. P Cyg profiles of the O III 5592 line of HD 164270 on 1989 May 31 (UT). Universal time is given for each observation. The smooth curve superposed on the profiles is the overall mean spectrum for the 3 nights. Bumps and dips are seen on top of the emission and the absorption parts of the profile.

FIGURE 5. Difference from the mean (3 nights) of HD 164270 obtained on 1989 May 31 (UT). Universal time is given for each observation. The vertical solid lines indicate the mean central wavelength for five emission lines shown in Figure 2 (first panel). Dashed lines show the radially outwards acceleration of two sample blobs (one red and one blue relative to the line centers). These two blobs (and others not indicated here) are seen in all lines at the same time in this star.

DISCUSSION

Owocki: I found it very surprising that you see simultaneous absorption dips and emission bumps. Do these emission bumps only occur on the blue, or also on the red side of emission features?

Robert: When we see an absorption dip in the absorption part of a P Cyg profile, there is always a blue emission bump at the same time. The emission part of the profile usually shows many blue and red structures at the same time but it is not clear that there is a correlation (at the same speed on each side) between the blue en red bumps (as expected if there was shell ejection).

Cherepashchuk: Could you compare the amplitudes of fine spectral variability for HeI and $HeII$ lines?

Robert: For WR40, the amplitude seems to be stronger for HeI lines. There are some maximum heights, there are some structures that are higher but maybe it is too soon to say that there are strong differences between the two lines.

Prinja: In our studies of the O stars, one of the main things that enabled us to rule out something like a blob model was the very slow velocity laws that were implied by the motions of the single features. Do you have any estimates for the masses of the radially propagating blobs, and - if time-averaged - how this compares with the mass-loss rate of the WR stars?

Robert: No work has been done yet on the mass of the material in the blobs. But we can see already that the range of masses we expect to find will be large. Also, it seems that the number of large blobs is smaller compared to the number of small blobs.

Koenigsberger: Is there any correlation between the frequency of occurrence of the emission spikes and the intensity of the line for a given star?

Robert: The S/N can be significantly small in weak lines compared to the strongest lines. But, as in the case of WR111, even the weakest lines show many bumps that are easy to count. This is a delicate question, and in the future we will need to strictly define the way to identify each bump in order to make good statistics.

de Groot: Supposing these blobs are real, once you have lined them up you can determine acceleration. Did you do this? And are the accelerations variable too?

Robert: There is an example of this type of study in the Proceedings of the Boulder-Munich Workshop (Robert and Moffat, 1990). There we compare the blob velocities with the general wind velocity law ($v_\infty(1 - R_*/r)^\beta$). This revealed that maybe $\beta > 1$. Further work will be done on this topic in the near future.

Koenigsberger, Willis chairing

THEORY OF INTRINSIC VARIABILITY IN HOT-STAR WINDS

Stanley P. Owocki
Bartol Research Institute
University of Delaware
Newark, DE 19716 USA

ABSTRACT. The winds of the hot, luminous, O, B, and WR stars are driven by the line-scattering of the star's continuum radiation flux. Several kinds of observational evidence indicate that such winds are highly structured and variable, and it seems likely that a root cause of this variability is the known strong instability of the line-driving mechanism. Initial dynamical models of the nonlinear evolution of this instability confirm that the wind indeed becomes highly structured, with large amplitude ($\sim$500 – 1000 km/s) shocks that separate high-speed rarefied flow from lower speed, dense shells. Remarkably, such variability can often have an *intrinsic* character, persisting even in the absence of explicit perturbation, and it now appears that this is a direct consequence of a *degeneracy* in the steady-state solutions for such models. However, recent work indicates that including scattering effects, which have so far been ignored in these *pure-absorption* models, might reduce or even break this steady-state solution degeneracy; through the "line-drag" effect, scattering can also reduce the strength of the instability, possibly rendering it an *advective* character for which wind variability now requires explicit perturbation from below. This review will examine the consequences of these ideas for understanding the likely nature of wind variability among the various kinds of early-type stars.

1. Introduction

Although this conference focuses on the properties of Wolf-Rayet stars and their winds, this review on the theory of intrinsic wind variability will deal primarily with the related O and B stars. The reason for this is that for the massive Wolf-Rayet winds there are still no completely satisfactory models to explain even the mean, time-averaged properties, e.g., the large mass loss rate, and so there is little solid basis on which to build a much more complicated picture of instability and variability. Nonetheless, it is generally thought that the basic driving mechanism for Wolf-Rayet winds is similar to that of OB stars, namely scattering of the star's strong radiation field in spectral lines of heavy minor ions in the wind (Lucy and Solomon 1970; Castor, Abbott, and Klein 1975, hereafter CAK). For these OB stars, steady-state wind theories have reached a quite high level of sophistication (Abbott 1980; Pauldrach, Puls, and Kudritzki 1986) and their predictions of gross wind properties like the time-averaged mass loss rate or terminal flow speed are in quite good agreement with values inferred by observation. (See, e.g., recent reviews by Abbott 1988, Kudritzki and Hummer 1990, and Owocki 1990.)

Despite this general success of steady-state models, there are numerous observational phenomena that seem to imply that such line-driven winds are not at all smooth, time-steady flows, but rather are both spatially and temporally highly variable. These include: soft X-ray emission (Harnden *et al.* 1979); nonthermal radio emission (Abbott, Bieging, and Churchwell 1984); line profile variations in UV spectra of OB stars ("narrow absorption

155

K. A. van der Hucht and B. Hidayat (eds.),
Wolf-Rayet Stars and Interrelations with Other Massive Stars in Galaxies, 155–165.
© 1991 *IAU. Printed in the Netherlands.*

components"; Lamers and Morton 1976; Prinja and Howarth 1988) or in optical spectra of WR stars (Robert and Moffat 1990); and blackness of saturated UV profiles from OB stars (Lucy 1982a).

It now seems likely that this observationally inferred variability is an *intrinsic* property of the wind for which a root cause is the inherent instability of the line-driving mechanism. This means that the wind variability does not require external driving by, say, large-amplitude pulsation of the underlying star. Of course, to the extent that such stars might in fact exhibit pulsations (Baade 1988), these would have a definite impact on any wind variability (Willson and Hill 1979; Castor 1987), but such a circumstance will not be a focus of this review. Rather, the emphasis here will be on the physics of line-driven mass loss, its fundamental instability, and how this can by itself give rise to intrinsic wind variability that persists even without explicit excitation from the underlying star.

2. Linear Theory of the Line-Driven Instability

The basic physics of this line-driven instability is quite simple (Lucy and Solomon 1970; see review by Rybicki 1987): a small-scale increase in radial flow speed Doppler-shifts the local line-frequency out of the absorption shadow of underlying material, leading to an increased radiative force which then tends to further increase the flow speed. However, even in the linear case of small amplitude perturbations, there are two important modifications to this simple picture. First, the perturbation must also be on a small enough scale so as to remain optically thin (MacGregor, Hartmann, and Raymond 1979; Carlberg 1980), which allows one to ignore any change in the absorption shadow itself. As shown by Owocki and Rybicki (1984), the limiting scale is roughly the Sobolev length $L \equiv v_{th}/(dv/dr)$, over which the mean wind velocity v increases by an ion thermal speed v_{th}. Perturbations with a scale larger than this can be analyzed using the same Sobolev (1960) approximation methods that are used to model the mean outflow (Abbott 1980). The line-force is then proportional to the velocity gradient, and since this implies that the perturbed force is out of phase with the perturbed velocity, no net work can be done on such large-scale perturbations, which means they are stable.

The second modification to the above simple picture has to do with its neglect of the diffuse radiation that results from the nearly pure-scattering character of most driving lines. In a frame comoving with the mean flow, this diffuse radiation has a near fore-aft symmetry, implying that its associated net mean force can be neglected in favor of the much stronger direct absorption term. However, Lucy (1984) first pointed out that, as seen by a small-scale velocity perturbation δv, this diffuse field is stronger by a factor $\delta v/v_{th}$ against the direction of the perturbed velocity. As a result, there is a net *drag* force that is on the same order as the perturbed direct force that gives rise to the instability. Very near the star this can in fact completely *eliminate* the instability, but away from the wind base, the combined effects of spherical expansion and the shrinking of the solid angle subtended by the stellar core sharply reduce the relative importance of this line-drag effect. At $r = 1.5R_*$, for example, the net growth rate is about half that obtained from a pure-absorption analysis (Owocki and Rybicki 1984, 1985).

This still implies a strong instability, because the basic growth rate associated with the direct-absorption term is quite large, roughly $\Omega \approx g_{rad}/2v_{th}$, i.e., about half the mean radiative acceleration rate through a thermal speed. Since $g_{rad} \approx v(dv/dr)$, this can alternatively be written as $\Omega \approx v/2L$, i.e., about half the flow rate through a Sobolev length. If we compare this to a typical expansion rate v/H, where $H = v/(dv/dr)$, we see that the growth rate is a factor $H/2L = v/2v_{th}$ higher. In these highly supersonic winds, this implies on the order of $50 - 100$ e-folds of cumulative growth for a typical wind perturbation! In practice, of course, this means that any perturbation with a small but finite amplitude would quickly become nonlinear in the wind. The remaining sections review various attempts to determine the nature of the resulting nonlinear wind structure.

3. Heuristic Models of a Wind with Embedded Shocks

Given the supersonic nature of the wind outflow and the large instability growth rate, it seems inevitable that, whatever its detailed form, the nonlinear wind structure arising from this instability will include shocks. Several of the attempts to study the consequences of this instability have thus centered on determining the properties of these assumed shocks. (See review by Castor 1987).

The first models along these lines (e.g., Lucy 1982b) assumed the wind contains a quasi-periodic train of *forward* shocks. Relatively fast flow exposed to a nearly unshadowed stellar flux is strongly driven, and so is pushed against relatively slow flow, which is shadowed by the fast material and only weakly driven. A *forward* shock thus forms that sweeps up and accelerates this slower material, thereby adding it to the faster flow. A crucial point is that the fast material represents *post-shock* flow, with its associated *high density* and (at least initially) *high temperature*. To maintain the structure, this material must quickly cool and reform the driving ions that line-absorb the radiative momentum.

The question of the ionization and energy balance behind the shock was analyzed by Krolik and Raymond (1985) for a single, forward shock that propagates outward through the wind. For massive winds, they inferred that the cooling would be sufficiently rapid to allow the driving ions to reform and so maintain the structure. For low-density winds, however, they argue that the shock amplitude would be limited by the requirement that the cooling time be less than the flow time.

MacFarlane and Cassinelli (1989) studied the evolution of wind shocks with a phenomenological numerical-hydrodynamics model in which an initially smooth, slow wind is disrupted by a sudden increase in the driving force, thereby accelerating the wind to a much higher speed. For parameters chosen to give initial and final state terminal speeds of $500\,\mathrm{km/s}$ and $2500\,\mathrm{km/s}$, collision between the fast and slow flow forms a forward/reverse pair of shocks of velocity amplitude $\sim 1000\,\mathrm{km/s}$ each. This turns out to be just what's needed to reproduce the observed X-ray properties of their model star, τ Sco.

4. Radiation-Hydrodynamics Simulation of Non-Linear Structure

Owocki, Castor, and Rybicki (1988; hereafter OCR) have recently developed a numerical, radiation-hydrodynamics code aimed at directly simulating the dynamical evolution of the line-driven wind instability, and thereby determining the likely nature of the resulting nonlinear wind structure. Because the instability occurs for perturbations with a length scale near and below the Sobolev length, OCR had to develop a method for computation of the line force which did not use the Sobolev approximation, but which still avoided the inordinate computational expense of solving the full line-transfer problem at each time step. The crucial simplification they adopted was to ignore the diffuse, scattered radiation, and to assume that the flow is driven by an ensemble of nonoverlapping, *pure-absorption* lines. In order to focus on the flow dynamics, they also did not treat a detailed energy balance, but simply assumed that radiative processes would be rapid enough to keep the flow nearly isothermal.

The wind variability simulated by OCR was not exactly intrinsic, but rather was the result of a small-amplitude (1%), periodic perturbation at the wind base. The strong dependence of driving force on flow speed quickly amplifies this initially small perturbation, giving rise to velocity variations $\Delta v \lesssim 1000\,\mathrm{km/s}$ by $r \approx 2R_*$. An important finding was that these high velocity regions were actually extremely *rarefied*. This turns out to be a very robust result, stemming directly from the linear instability property (Owocki and Rybicki 1984) that wave modes with opposite phase in density and velocity are much more unstable. Although advected away from the star by the supersonic flow, these unstable waves actually propagate *inward* relative to the fluid, and so as they steepen they naturally evolve into *reverse*, not forward, shocks. These reverse shocks act to *decelerate* high-speed, rarefied flow as it impacts slower material that has been compressed into dense shells. This is quite different from the Lucy (1982b) model, in which *forward* shocks were assumed to abruptly *accelerate* ambient wind material as it is rammed by a dense, strongly driven flow. In fact, the very low density of this high-speed flow implies that, unlike Lucy's model, only a very small fraction ($\sim 10^{-3}$) of the wind material ever undergoes a shock near the maximum amplitude $\Delta v \approx 1000 \, \mathrm{km/s}$. This type of structure is very favorable for explaining several observational features that are difficult to understand in terms of forward shock models (OCR; Owocki 1990), e.g., the hard X-ray tail (Cassinelli and Swank 1983), and the non-sharp, variable blue edges of saturated UV resonance lines (Henrichs, Kaper, and Zwarthoed 1988).

5. Intrinsic Nature of Wind Variability

The wind structure computed by OCR arose from amplification of an explicit perturbation introduced at the wind base, but subsequent work (Owocki, Poe, and Castor 1990; Poe, Owocki, and Castor 1990, hereafter POC) has shown that such wind models can also often exhibit an *intrinsic* variability that persists even in the absence of such explicit perturbations. The incidence of this intrinsic variability was found empirically to depend on the assumed value of v_{th}/a, the ratio of the ion thermal speed to sound speed. For the usual case of driving by CNO ions, an appropriate value of this ratio is $v_{th}/a \approx 0.3$. However, in order to study the effect of variability arising from explicit perturbations without the complication of a background model that was itself intrinsically variable, OCR found it necessary to assume an artificially high value $v_{th}/a \gtrsim 1/2$. The situation is graphically illustrated in Figs. 1, which show the spatial and temporal variations of velocity and mass loss rate in unperturbed wind models that differ only in the assumed v_{th}/a. In all cases the flow is initially disrupted because the assumed CAK/Sobolev initial condition is not an appropriate steady state for either non-Sobolev model. The model with $v_{th}/a = 0.5$, however, quickly relaxes to a somewhat steeper (POC) steady solution, whereas the models with slightly smaller v_{th}/a never settle down.

POC showed that the qualitative differences among the variability properties of these cases reflect a difference in the nature of the corresponding steady solutions. Fig. 2 illustrates the steady-state solution topology near the critical (sonic) point. Unlike the usual saddle- or X-type solution that applies, e.g., to the solar wind (Fig. 2a) (Parker 1963) the solution topology in this case is of the *nodal* type, with not one but *two* positive slope critical solutions (Fig. 2b). Note that along the steeper slope there is only one distinct solution passing though the critical point, whereas along the shallower slope there is a *degenerate family* of solutions that pass through this point. POC showed that, for reasonable boundary conditions, the distinct, steeper solution applies when $v_{th}/a \gtrsim 1/2$, whereas the degenerate, shallower solution applies when $v_{th}/a \lesssim 1/2$. (These values apply to the simple case of a point star, as was assumed by OCR; subsequent calculations now properly treat a finite stellar disk, and in this case steady, steeper solutions exist at the somewhat lower ratios $v_{th}/a \gtrsim 3/8$.) Apparently, the existence of a well-defined steady solution in the former case is sufficient to enable a time-dependent model without explicit perturbations to relax to this steady state. On the other hand, the lack of a well-defined steady solution for the latter case leads in the time-dependent model to an

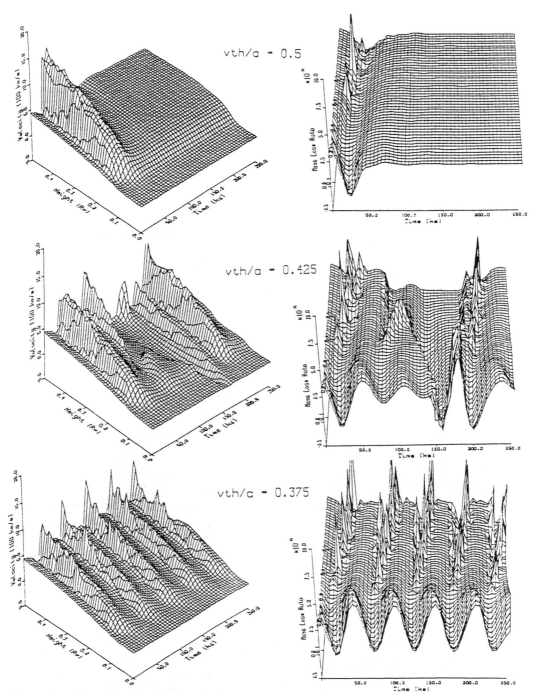

Figure 1:   3-D perspective plots of the height and time variations of velocity (left) and mass loss rate (right) in models with v_{th}/a of 0.5 (top), 0.425 (middle), and 0.375 (bottom). (For clarity, the mass loss rate is plotted from a perspective looking outward along the height axis.)

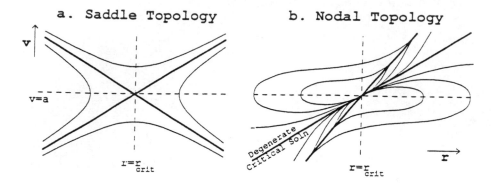

Figure 2: The velocity near the sonic radius for solution topologies of (a) the usual saddle type that applies to the solar wind and (b) a nodal type that applies to absorption-line-driven winds. Note in the nodal case how a large number of solutions converge as they approach the sonic point along the shallower of the two positive critical solutions (heavy solid lines). The implied solution degeneracy leads to the intrinsic variability seen in the lower panels of Fig. 1.

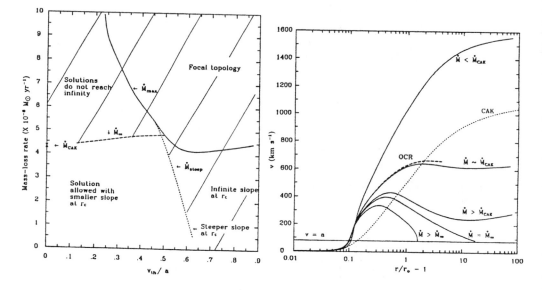

Figure 3: The nature of wind solutions in the parameter plane $\dot{M}$ vs. v_{th}/a. In the hatched region solutions are excluded for various reasons, but flow solutions are possible in the entire non-hatched region.

Figure 4: Velocity v vs. radius r for steeper slope solutions with various $\dot{M}$ and v_{th}/a. Note that models with $\dot{M} > \dot{M}_\infty$ do not extend to a large radius with a finite speed.

intrinsic variability in which the flow, roughly speaking, continuously varies among this degenerate family of possible steady solutions.

This kind of solution degeneracy appears to be endemic to line-driving. It was only not apparent in the original CAK model because this model used the Sobolev approximation, even in the subsonic flow, to eliminate all but one "critical" solution. In a more self-consistent model in which one also assumes a vanishing sound speed $a \to 0$ in the usual Sobolev limit of vanishing thermal speed $v_{th} \to 0$, a similar solution degeneracy exists in that the CAK mass loss rate $\dot{M}_{CAK}$ then only defines an *upper limit* of the mass loss, not a unique value (POC, Owocki 1990). POC showed that in pure-absorption models in which both v_{th} and a are nonzero, transonic solutions likewise exist up to a maximum rate $\dot{M}_{max}$. As shown in Fig. 3, for small thermal speeds ($v_{th}/a \lesssim 0.5$) this can actually exceed $\dot{M}_{CAK}$, but in this case the CAK rate still roughly corresponds to the maximum rate, $\dot{M}_\infty$, that the line force can drive all the way to infinity. Note also that the mass loss rate for the distinct, steeper slope solution, $\dot{M}_{steep}$, can likewise exceed $\dot{M}_\infty$. Fig. 4 then shows that for such steeper solutions with $\dot{M}_{steep} > \dot{M}_\infty$, the velocity reaches a maximum supersonic value and then declines, returning to the sonic speed at a finite "stagnation" radius. Since the steeper solution defines an upper velocity limit for the degenerate family of shallow slope solutions, this means that all solutions with $\dot{M} > \dot{M}_\infty$ must likewise stagnate at a finite radius.

These rather peculiar properties of the steady wind solutions actually provide a useful basis for understanding the intrinsic variability seen in the corresponding time-dependent models. For $v_{th}/a \approx 0.5$, $\dot{M}_{max} \approx \dot{M}_{steep} \approx \dot{M}_\infty$, meaning that the distinct, steeper solution through the sonic point has just the mass flux the line-force can drive all the way to infinity; as shown in Figs. 1a and 1b, this steeper slope solution thus provides a suitable asymptotic steady-state for the time-dependent model.

Figs. 1c and 1d show, however, that for a somewhat smaller value $v_{th}/a = 0.425$, the flow remains highly variable. The initial base oscillation in mass flux now settles onto a value which the line-force can only lift to a finite height $\approx 0.3 R_*$. As matter accumulates at this stagnation point, a shock forms which slows the underlying outflow. By time $t \approx 100$ this turns it into an *accretion shock*, and until $t \approx 150$, when all the accumulated material falls back onto the star, this completely disrupts the wind. The recovery from this accretion is quite irregular, but ultimately the wind will be reestablished. Soon, however, the mass flux again settles onto a value that is too high to reach infinity, and so another disruption interval of accretion begins. Both the accretion and reestablishment phases of this cycle require on the order of several free-fall times from the stagnation point, and so the entire cycle can take a day or more, depending on the stellar gravity and the height of the stagnation (Section 6).

This stagnation height depends sensitively on v_{th}/a. In models with $v_{th}/a \approx 0.45$, for example, material actually reaches a height of more than a stellar radius before falling back on the star. On the other hand, for the model with $v_{th}/a = 0.375$, this stagnation height is only slightly above the sonic point itself. In this particular case, the accretion and reestablishment phases turn out to be nicely synchronized and so give rise to the smooth oscillatory behavior seen in Figs. 1e and 1f.

By focusing on the associated cyclical variation in the mass flux (Fig. 1f), we can formulate the following simple picture of the dynamics of this oscillation: First, inward propagation of a small portion of unstable waves that form near the sonic point causes the mass flux in the subsonic flow to temporarily overshoot the continuously sustainable value. As material near the sonic point becomes shadowed from the driving radiation by the enhanced density of the underlying subsonic flow, it decelerates, which then causes the mass flux to undershoot, even becoming briefly inward (the accretion). Once this wave of diminished mass flux reaches the subsonic flow, the overlying material is now less shadowed and so accelerates, forming another unstable wave to enhance the mass flux and repeat the cycle.

6. Characteristic Time Scales

The time scale for this relatively simple *relaxation oscillation* is found empirically to scale with the acoustic cutoff period $4\pi a/g_*$. This can be understood physically by noting this is also roughly the sound travel time between the unstable flow just below the sonic point, and the stable subsonic base region below the "thermal" point (i.e., where $v < v_{th}$). In the present models, which for reasons of computational expense (see OCR) assume a sound speed $a = 80$ km/s that is about a factor of 3 too high, this yields a relaxation period of $47,000$ s ≈ 13 hours. Analogous models with a sound speed $a = 25$ km/s more appropriate to the assumed stellar temperature $(50,000$ K) exhibit shorter periods $16,000$ s ≈ 4 hours. This is about a factor ten longer than the corresponding growth times for small scale instabilities.

However, this is still significantly shorter than the observed repetition time of discrete absorption features (typically $1/2 - 2$ days; Henrichs, Kaper, and Zwarthoed 1988). Despite this disagreement in scale, the time-variations of synthesized absorption profiles do show moving narrow absorption features quite reminiscent of such discrete absorption components (Prinja and Howarth 1988). These arise from outward accelerating dense "shells" whose formation is repeatedly triggered by the quasi-periodic disruptions. For other selected models in which the stagnation height is quite high above the star, the overall time for the semi-regular cycles of accretion/recovery can become correspondingly longer, resulting in profile variations with time-scales approaching that of observed discrete components.

Interestingly, the time scale for the simple relaxation oscillation is roughly comparable to that of optical line-profile variability that has often been described in terms of "non-radial pulsations" (Baade 1988). This might not be just coincidence since, as noted above, this intrinsic wind variability actually does extend down to the subsonic wind base, where these optical lines are formed. Furthermore, the accretion that occurs when the wind stagnates should surely have a substantial impact on the underlying photosphere, perhaps even inducing variations of sufficient magnitude to appear as "pulsations" (Fullerton 1990). Of course, at this point, such a "tail wags dog" scenario of a wind inducing photospheric variability is quite speculative, since these highly simplified, 1-D models ignore several potentially important effects, e.g., stellar rotation.

7. Role of Scattering

The lines that drive hot star winds are actually more nearly pure-scattering than pure-absorption, and including the dynamical effect of this diffuse radiation may have important effects on the intrinsic variability. In the supersonic regions of a smooth flow, the near fore-aft symmetry of the diffuse radiation field means that the associated net force is typically small; but in the region near and below the sonic point, the radiation escape probability is rapidly increasing, leading to a diffuse field asymmetry. Since the escape probability scales with the velocity gradient, the net asymmetry depends on the second derivative of velocity, and so the effect of the diffuse field can be reasonably modeled as a *radiative viscosity*. Owocki and Zank (1990) show that such a radiative viscous force can alter or even "break" the solution degeneracy found in pure-absorption models, and this might explain why comoving frame models that include scattering appear to have a well-defined solution (Pauldrach, Puls, and Kudritzki 1986).

Through the Lucy (1984) line-drag effect (Section 2), scattering might also stabilize the wind base enough to suppress variations that govern the intrinsic variability. The wind instability would then have an *advective* rather than *absolute* character, (Bers 1983; Owocki and Rybicki 1986; OPC) for which variability would persist only with some explicit driving from the underlying star, and would only become large amplitude in the supersonic wind. Recently, Owocki and Rybicki (1990) have extended their previous linear stability analyses to the case, relevant to Wolf-Rayet stars, in which the continuum remains optically thick into the supersonic flow. Assuming that the continuum radiation thus attains an angular distribution appropriate to the diffusion approximation, they find that a flow driven by pure-scattering lines is then stabilized by this line-drag effect. They also find, however, that a relatively small amount of photon destruction can cause such a flow region to remain unstable. Determining the linear stability properties of the lower regions of Wolf-Rayet winds will thus require knowing the effective mean destruction probabilities for the lines that drive the wind.

For OB stars, the linear stability properties are relatively well understood, but how scattering effects might influence the nonlinear properties like the intrinsic variability is still not clear. Unfortunately, straightforward calculation of the diffuse field transfer at each time step would be prohibitively expensive, with a timing increase of more than a factor 100 over the present pure-absorption calculation. It will thus be essential to develop less costly approaches. One simple technique currently being tested is based on escape probability methods; this can be implemented with only modest (factor 2) timing increase, and, although quite crude, nonetheless appears capable of incorporating several important effects, like the radiative viscosity and line-drag terms mentioned above. Other methods, such as accelerated Lambda iteration (e.g., Olson, Auer, and Buchler 1986), are also being explored, as is the possibility of some kind of hybrid between the two.

8. Future Work

In addition to clarifying the role of scattering in regulating the instability, there is obviously a great deal of work still to be done to include the many other potentially important effects neglected so far, e.g., detailed energy and ionization balance, shock X-ray emission, 2-D or 3-D structure, and stellar rotation. Inclusion of a realistic energy balance in the dynamical wind models is needed to determine the temperature structure and thereby the shock X-ray emission. Likewise, inclusion of ionization balance is needed to determine whether ionization from shock heating and/or X-rays can effect the line-driving and hence the dynamics of the wind and instability. Consideration of rotation and other 2-D (and ultimately 3-D?) effects is needed to determine the likely lateral scales of the dense clumps. For example, can rotation string out such structures in longitude in a manner similar to corotating interaction regions in the solar wind (Mullan 1984, 1985), and thus make it possible for a given dense clump to cover a large enough fraction of the stellar disk to produce the observed discrete absorption components? Finally, perhaps one of the most urgent needs is to develop methods to compare more closely predictions from the theoretical simulations with available observational diagnostics, and thereby guide and test the further theoretical development in this complex but fascinating problem in radiation hydrodynamics.

This work was supported in part by NSF grant AST 88-14580 and NASA grant NAGW-1487. Many of the computations were carried out using an allocation of supercomputer time from the San Diego Supercomputer Center. I thank G. Cooper and A. Fullerton for many helpful discussions, questions, and comments.

References

Abbott, D. C. 1980, *Ap. J.*, **242**, 1183.

Abbott, D. C. 1988, in *Solar Wind VI* , ed. V. J. Pizzo, T. E. Holzer, and D. G. Sime (Boulder: NCAR/TN–306), p.149.

Abbott, D. C., Bieging, J. H., and Churchwell, E. 1984, *Ap. J.*, **280**, 671.

Baade, D. 1988, in *O Stars and Wolf-Rayet Stars* , ed. P. S. Conti and A. B. Underhill (NASA SP–497), p.137.

Bers, A. 1983, in *Handbook of Plasma Physics, Vol. 1: Basic Plasma Physics I*, ed. A. A. Galeev and R. N. Sudan (Amsterdam: North-Holland), p. 451.

Carlberg, R. G. 1980, *Ap. J.*, **241**, 1131.

Cassinelli, J. P., and Swank, J. H. 1983, *Ap. J.*, **271**, 681.

Castor, J. I. 1987, in *Instabilities in Luminous Early Type Stars*, ed. H. J. G. L. M. Lamers and C. W. H. de Loore, (Dordrecht: Reidel), p. 159.

Castor, J. I., Abbott, D. C., and Klein, R. I. 1975, *Ap. J.*, **195**, 157. (CAK)

Fullerton, A. W. 1990, Ph.D. Thesis, University of Toronto.

Harnden, F. R. *et al.* 1979, *Ap. J. (Letters)*, **234**, L51.

Henrichs, H. F., Kaper, L., and Zwarthoed, G. A. A. 1988, in *A Decade of UV Astronomy with the IUE Satellite, Vol. 2*, ed. by E. J. Rolfe (Paris: ESA), p.145.

Krolik, J. H., and Raymond, J. C. 1985, *Ap. J.*, **298**, 660.

Kudritzki, R. P., and Hummer, D. G. 1990, *Ann. Rev. Astr. Ap.*, **28**, in press.

Lamers, H. J. G. L. M., and Morton, D. C. 1976, *Ap. J. Suppl.*, **32**, 715.

Lucy, L. B. 1982a, *Ap. J.*, **255**, 278.

Lucy, L. B. 1982b, *Ap. J.*, **255**, 286.

Lucy, L. B. 1984, *Ap. J.*, **284**, 351.

Lucy, L. B., and Solomon, P. M. 1970, *Ap. J.*, **159**, 879.

MacFarlane, J. J., and Cassinelli, J. P. 1989, *Ap. J.*, **347**, 1090.

MacGregor, K. B., Hartmann, L., and Raymond, J. C. 1979, *Ap. J.*, **231**, 514.

Mullan, D. J. 1984, *Ap. J.*, **283**, 303.

Mullan, D. J. 1985, *Astr. Ap.*, **165**, 157.

Olson G. L., Auer, L. H., and Buchler, J. R., *J. Q. R. S. T.*, **35**, 431.

Owocki, S. P. 1990, *Reviews of Modern Astronomy*, **3**, (Berlin: Springer), in press.

Owocki, S. P., and Rybicki, G. B. 1984, *Ap. J.*, **284**, 337.

Owocki, S. P., and Rybicki, G. B. 1985, *Ap. J.*, **299**, 265.

Owocki, S. P., and Rybicki, G. B. 1986, *Ap. J.*, **309**, 127.

Owocki, S. P., and Rybicki, G. B. 1990, *Ap. J.*, submitted.

Owocki, S. P., Castor, J. I., and Rybicki, G. B. 1988, *Ap. J.*, **335**, 914. (OCR)

Owocki, S. P., Poe, C. H., and Castor, J. I. 1990, in *Properties of Hot Luminous Stars*, C. D. Garmany, ed. (San Francisco: ASP), p.283.

Owocki, S. P., and Zank, G. P. 1990, *Ap. J.*, submitted.

Parker, E. N. 1963, *Interplanetary Dynamical Processes*, (New York: Interscience).

Pauldrach, A., Puls, J., and Kudritzki, R. P. 1986, *Astr. Ap.*, **164**, 86.

Poe, C. H., Owocki, S. P., and Castor, J. I. 1990, *Ap. J.*, **355**, in press. (POC)

Prinja, R. K., and Howarth, I. D. 1988, *M. N. R. A. S.*, **233**, 123.

Robert, C., and Moffat, A. F. J. 1990, in *Properties of Hot Luminous Stars*, C. D. Garmany, ed. (San Francisco: ASP), p.271.

Rybicki, G. B. 1987, in *Instabilities in Luminous Early Type Stars*, ed. H. J. G. L. M. Lamers and C. W. H. de Loore, (Dordrecht: Reidel), p. 175.

Sobolev, V. V. 1960, *Moving Envelopes of Stars*, (Cambridge: Harvard University Press).

Willson, L. A., and Hill, S. J. 1979, *Ap. J.*, **228**, 854.

DISCUSSION

Montmerle: To what extent are the instabilities you find dependent on the fact that you are doing 1-D calculations? Rotation for instance could remove some of the degeneracies?
Owocki: I have not yet investigated this question in detail, but I believe going to 2-D and 3-D will mostly add new instabilities, *e.g.* Rayleigh-Taylor, which will tend to break my "shells" into "clumps". I think it likely that rotation plays a big role in extending the lateral scale of those clumps, helping to make them occult a large enough fraction of the stellar disk to be observable as narrow components. The ultimate structure may be a hybrid of Mullan's CIR picture and this instability model.

Langer: Thinking about a pulsating WR star with a high density (optically thick) wind, would the wind instabilities wipe out the signature of pulsation or would it even amplify it, *i.e.* would you *observe* any variations with the period of the pulsation?
Owocki: On the contrary. I think the effect of any underlying pulsation would ultimately be amplified by the wind, and thus be quite observable.

Spurzem: Could you please give a comment on the method used to numerically solve the hydrodynamic equations, *e.g.* which type of artificial viscosity was utilized?
Owocki: We use van Leer's method of piecewise linear, monotonized advection on a staggered spatial mesh. Quadratic viscosity is used on strong compression to smooth shocks. We tried and abandoned use of linear viscosity on expansion to limit steepness of strong rarefactions. We now limit this by cutting off a few of the very strongest lines from our power-law number distribution of lines vs. opacity. Details can be found in Owocki, Castor and Rybicki (1988).

Sreenivasan: I do not wish to make life any more difficult but I must point out that a realistic energy conservation equation (including rotational effects) must be included to understand really what happens in this case.
Owocki: I agree, and I have a graduate student whose thesis will be to do just that. I should point out, however, that the isothermal assumption is not too unrealistic in the case of dense winds, since then the radiative cooling length behind the shocks can be estimated to be quite small compared to other length scales.

Maeder: What is according to your models the deep physical reason for the difference between the winds (and the mass loss rates) of WR stars and O stars?
Owocki: Because WR winds are optically thick at the sonic point, the used "core-halo" approximations of OB winds breaks down, and any radiative driving must be by the diffuse, scattered radiation field. This already means that the wind dynamics and structure must be entirely different than in the OB star case. As for the "deep" reason, I believe it ultimately stems from the lower H-abundance of WR stars. An atmosphere dominated by He has lower sound speed a, and this increases the ratio v_{th}/a leading to a breakdown of the usual Sobolev approximation near the sonic point. This means the usual direct driving of the core-halo picture fails, and so the diffuse driving must take over.

Stan Owocki

Owocki explaining it
to Hillier once more

THEORETICAL RESULTS ON WR INSTABILITIES

André Maeder
Gérard Schaller
Geneva Observatory
CH-1290 Sauverny, Switzerland

ABSTRACT. The instabilities of WR stars are discussed in the framework of both the quasi–adiabatic and non–adiabatic methods. We show that a very careful treatment of stellar opacities is necessary. Both methods show that WR stars, with little or no H left, are vibrationally unstable in the fundamental radial mode due to the ϵ–mechanism, confirming earlier results by Maeder (1985). Thus, from the point of view of stability, WR stars are fundametally different from O stars. Our new analysis also supports our previous view that WR stars evolve, keeping at the edge of vibrational instability.

1 INTRODUCTION

What is the main physical reason for the differences between an O star and a WR star? Clearly, surface abundances are a key point, but we also show here that with respect to stellar instabilities, WR star models are in a completely different situation than O stars: WR stars have a deep seated instability. Whether this instability is the main source of the WR phenomenon is not known at present. However, this instability is a major physical property of WR models and it is likely to determine the evolution of WR stars keeping at the edge of vibrational instability. We also show that the nuclear evolution of stars at the edge of vibrational instability implies very high mass loss rates $\dot{M}$ as currently observed for WR stars, and also implies a $\dot{M}$ vs. M relation.

2 NUCLEAR ENERGIZING AND RADIATIVE DAMPING IN RADIAL PULSATIONS: QUASI–ADIABATIC APPROACH

Numerous analyses of vibrational stability in helium stars have been performed (e.g. Boury and Ledoux, 1965; Noels–Grötsch, 1967; Simon and Stothers, 1969, 1970; Stothers and Simon, 1970; Noels and Masereel, 1982; Noels and Magain, 1984; Maeder, 1985). These works clearly show that the results about stability or instability of the radial mode in He–stars depend critically on the relative importance of two terms:

1. The nuclear driving
2. The radiative damping

The nuclear driving occurs in the central region: compression implies heating, which in

K. A. van der Hucht and B. Hidayat (eds.),
Wolf-Rayet Stars and Interrelations with Other Massive Stars in Galaxies, 167–173.
© 1991 *IAU. Printed in the Netherlands.*

turn implies an increase of ϵ, the rate of nuclear energy generation, which means mechanical expansion until temperature and ϵ go down and the restoring force of gravity again produces a compression. The radiative damping is the loss of pulsational energy radiated by the temperature excess of the waves. Radiative damping usually dominates in the external stellar layers. If the nuclear driving overcomes the radiative damping, the star is unstable. The fundamental mode of radial pulsation is the most unstable mode in this case.

This instability mechanism is known as the Eddington ϵ-mechanism. For an He-star, Noels and Masereel (1982) found a critical mass of 16 $M_\odot$: above this mass limit, the ϵ-mechanism produces radial pulsation instability. For more realistic WR models, Maeder (1985) found an instability limit of about 12 $M_\odot$. On the zero-age sequence of Pop. I stars, the critical mass is around 100 $M_\odot$.

The relative importance of the nuclear driving and radiative damping depends essentially on the relative size of the pulsation amplitudes ξ_c in the center to the amplitudes ξ_s at the stellar surface. A relatively large ξ_c favours the pumping of nuclear energy in central regions, while a small ξ_c produces little driving. In turn, the amplitude ratio ξ_c/ξ_s essentially depends on the density contrast $\rho_c/\bar{\rho}$ in the star. Here ρ_c is the central density and $\bar{\rho}$ the average density. One has the following connexion:

$$\text{if} \quad \frac{\rho_c}{\rho} \searrow \quad \Longrightarrow \quad \frac{\xi_c}{\xi_s} \nearrow$$

Thus a low density means a relatively high central amplitude, which favours pulsational instability. Usually the ξ-values have been taken from linear adiabatic calculations and this is why this approach is called the quasi-adiabatic approximation; let us also note that in the linear approximation, ξ_s is normalised to 1.

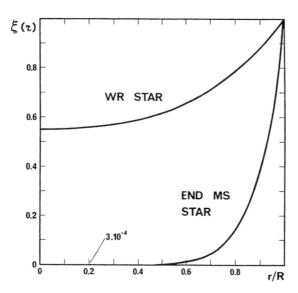

Figure 1. Comparison of the pulsation amplitudes from the center to the surface in a WR star model and a star at the end of the main sequence.

Figure 1 compares the pulsation amplitudes (cf. Maeder, 1985) from the center to the surface in a WR star model (WC star with 33.7 M$_\odot$, resulting from an initial 120 M$_\odot$ model) with the pulsation amplitudes for an initial 120 M$_\odot$ model at the end of the Main Sequence (actual mass 98.7 M$_\odot$). In the WR model, the relatively large amplitude in the center strongly favours the nuclear driving and makes the WR model unstable. On the contrary, the MS star (even for an initial 120 M$_\odot$ model) has very small central amplitudes; thus nuclear driving is negligible there compared to radiative damping in the external layers.

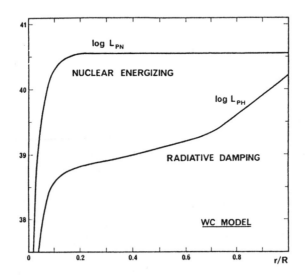

Figure 2. Comparison of the integrated values (from the center to the surface) of the nuclear energizing and radiative damping in a WC star model.

Figure 2 shows the integral of L$_{PN}$, the rate of gain of pulsation energy from nuclear sources, integrated from the center to the point considered. The integral L$_{PH}$ is also shown, where L$_{PH}$ is the rate of loss of pulsation energy by heat leakage. From Fig. 2 we conclude that in a WR star the contribution of the nuclear energizing is very large in central regions and is not overcome by radiation damping in the outer layers. WR stars are likely to be the only case where this situation occurs.

The analyses made in the quasi–adiabatic approximation (cf. Maeder, 1985) have shown the following results:

- The variability of the Hubble–Sandage variables is not due to vibrational instability, since the density contrast of supergiants is too high.

- In the WR star models considered, the nuclear energizing of pulsations largely overcomes the radiative damping due to heat leakage and the stars are radially vibrationally unstable in the fundamental mode.

- For models with observed mass loss rates and standard treatment of convection, entry into the unstable regime occurs when (H/He)=0.3 (in numbers) at the stellar surface.

- The pulsation periods of the inner hydrostatic star (without the wind) are in the range of 15–60 minutes.

- Any mixing process such as overshooting or turbulent diffusion would enhance the vibrational instability of WR stars.

- Vibrational instability characterizes stellar models of WR stars that occupy a well–defined mass–luminosity relation, showing a large overluminosity with respect to the main sequence.

3 THE CASE FOR NONRADIAL PULSATIONS

Let us examine here the theoretical status about the nonradial pulsations of WR stars. The driving of nonradial oscillations by central nuclear energizing is in a very unfavourable situation, because the amplitudes $\delta T/T, \delta\rho/\rho$ tend to be zero at the stellar center and no efficient pumping of energy can occur there (cf. Simon 1957). This explains why Kirbiyik et al. (1984), in an investigation of nonradial oscillations for WR models, found these stars to be stable for the low harmonics ℓ; they noticed, however, the appearance of instabilities for high degrees of harmonics ($\ell = 15$). Noels and Scuflaire (1986) interestingly showed that, while there is no driving of nonradial oscillations in WR stars to be expected from He–cores, the H–burning shells may produce some efficient driving of nonradial pulsations. The periods found are of the order of a few hours. There is, however, a limitation: the H–shell is, if any, only present for a very short time in WR stars. In the case studied by Noels and Scuflaire it lasts only 6000 years. Indeed, most WR models (cf. Maeder, 1981) even do not exhibit an H–burning shell. This is obvious if we remember that WNE as well as WC stars no longer have hydrogen at their surface; thus only a fraction of the WNL stars could have an efficient driving from the H–shell. Studying various WR models, Cox and Cahn (1988) found no unstable g–type nonradial models. The nonradial modes have large amplitudes outside the convective core where there would be an H–burning shell. But even there Cox and Cahn (1988) find the WR models to be stable, with respect to nonradial oscillations, because of the large radiative damping in the outer layers. Thus, on theoretical grounds, nonradial instabilities due to the ϵ–mechanism in WR stars seem to be very unlikely.

4 THE ϵ AND κ–MECHANISMS IN WR STARS

The κ–mechanism, which is responsible for cepheid pulsations, could also play a destabilizing role in WR stars. The study of this problem has been undertaken in the framework of the linear non–adiabatic method for the case of radial oscillations. The pulsation calculations are based on evolutionary models taking into account the entire stellar structure from the very center to the surface allowing us to model an ϵ driving mechanism as well as a κ–mechanism (cf. Cox and Cahn, 1988). An effective opacity as described in Stellingwerf (1975) is taken for the finite–difference scheme of the radiative transfer equation.

The effects of stellar opacities on the radial pulsations were firstly explored. Our standard opacity tables (called Iben XVIII et XIX mixtures) led to no destabilizing contribution by the κ–mechanism and predicted, with that method, a critical mass limit of 43 M$_\odot$. We have also established a set of new opacity tables computed with the Los Alamos Opacity

Programme for abundances in He, C, N, O, Ne etc. well adapted to WR stars. The new tables produce a small increase of the opacity coefficient in the outer layers (κ increases from 0.24 to 0.28), mainly due to partially ionized oxygen.

Using the new opacity tables we found radial vibrational instability at lower masses. For example, we found the critical mass at 32 $M_\odot$ instead of 43 $M_\odot$ with standard tables for an initial model of 120 $M_\odot$ at the WC stage. Figure 4 shows the corresponding changes in the cumulative work integral (here the integration is performed from the surface to the center). In Fig. 4 we clearly see the positive driving due to the κ–mechanism in the outer layers (continuous curve), while with standard opacities no such driving could be observed. The central T and ρ are also slightly increased and thus the nuclear energizing is also larger in this model. It appears that a small change in the values of the opacities, in our case a few percent, plays a determinant role in the stability results. These critical mass values are given only to show the sensitiveness of the stability to a change in the $\dot{M}$–rate or opacity, but the right value is certainly lower, maybe as low as 10 $M_\odot$. This is especially true if we account for the fact that opacity coefficients are expected to be increased by future developments in the field.

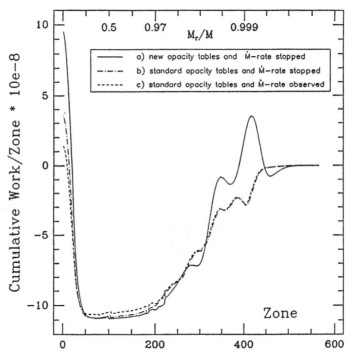

Figure 3. Cumulative work integral normalized by the kinetical energy of the pulsation vs. the zone number labeled from the center to the surface for the fundamental mode in a 56 $M_\odot$ model of an initial 120 $M_\odot$ star.

We have also explored the effects of changes in the mass loss rates $\dot{M}$ on WR stability. If the $\dot{M}$–rate is increased by a high factor the instability declines because central T and ρ are slightly reduced and therefore also the nuclear energizing. On the contrary, reduction of $\dot{M}$–rate leads to an enhancement of the instability. In Fig. 3 we notice the differences between the dashed and the dashed–dotted curves near the stellar center. This numerical result well supports the view (cf. Maeder 1985) that WR stars evolve, keeping to the edge of vibrational instability. It was also shown analytically that if WR stars evolve quasi–homogeneously and keep to the edge of vibrational instability with $M\bar{\mu}^2 \simeq$ const, the pulsations would be able to sustain the observed mass flux. Indeed, for a star evolving at the edge of vibrational instability, we get

$$\dot{M} = \frac{1}{2}\,\bar{\mu}\,M\dot{Y}$$

and by using the relation between the luminosity L and the change $\dot{Y}$ of helium content we obtain

$$\dot{M} \simeq -\frac{1}{2}\,\bar{\mu}\,\frac{L}{E_{3\alpha}q_c}$$

which corresponds numerically to $\dot{M} = -\left(2\cdot10^{-5} - 10^{-4}\right)\,M_{\odot}\mathrm{y}^{-1}$. This demonstrates the ability of a star evolving at the edge of vibrational instability to produce the observed mass loss rates. Moreover, since there is a mass–luminosity relation for WR stars (cf. Maeder, 1983), it turns out that there is also a $\dot{M}$ vs M^{α} relation for WR stars, with $\alpha \simeq 2$ for WR stars with masses $M \leq 10\ M_{\odot}$.

Finally, let us emphasize that these results on WR instabilities are based for now on WR models, which do not account for the optically thick wind. The wind is likely to affect the visibility of the inner pulsation, the boundary conditions, the stability limit as well as the pulsation periods. The study of these effects will demand complete stellar models of both the stellar interior and of the non–static and non–stationary envelope. Years of work may still be needed until we reach that stage of development in pulsation analysis.

5   REFERENCES

Boury A., Ledoux P.: 1965, Ann. Astrophys. **28**, 353

Cox A.N., Cahn J.H.: 1988, Astrophys. J. **326**, 804

Kirbiyik H., Bertelli G., Chiosi C.: 1984, in 25th Liège Colloquium, Ed. A. Noels, M. Gabriel, p. 126

Maeder A.: 1981, Astron. Astrophys. **99**, 97

Maeder A.: 1983, Astron. Astrophys. **120**, 113

Maeder A.: 1985, Astron. Astrophys. **147**, 300

Noels–Grötsch A.: 1967, Ann. Astrophys. **30**, 349

Noels A., Magain E.: 1984, Astron. Astrophys. **139**, 341

Noels A., Masereel C.: 1982, Astron. Astrophys. **105**, 293

Noels A., Scuflaire R.: 1986, Astron. Astrophys. **161**, 125

Simon R.: 1957, Bull. Acad. Roy. Belgique, Cl. Sci. **43**, 610

Simon N.R., Stothers R.: 1969, Astrophys. J. **155**, 247

Simon N.R., Stothers R.: 1970, Astron. Astrophys. **6**, 183

Stellingwerf R.F.: 1975, Astrophys. J. **195**, 441

Stothers R., Simon N.R.: 1970, Astrophys. J. **160**, 1019

DISCUSSION

Owocki: What driving mechanism lifts the mass to infinity in your scenario, in which you assume $\dot{M} \approx 10^{-5} - 10^{-4} M_\odot yr^{-1}$, in order to stay close to your instability line?

Maeder: My statement was that if a WR star keeps at the edge of vibrational instability during its evolution, its mass has to decrease with a mass loss rate comparable to the currently observed ones. In this case of vibrational instability and purbational motions are likely to turn, as shown by Appenzeller in the early '70-s, into outward bulk motions in the very outer layers.

Sreenivasan: Most people who do pulsation calculations now believe that opacities have to be increased by a certain factor. But no explicit calculations of additional sources of opacity are generally available. Did you simply scale the opacities upwards in an *ad hoc* manner? Have you looked for stability of non-radial pulsations in view of the κ-mechanisms in your outer layers? If you made more realistic calculations you would also see non-radial modes to be overstable.

Maeder: On one hand, we have calculated opacity tables appropriate to the WC composi- tions. On the other hand, we are also exploring the effect of enhanced opacities. Non-radial modes driven by the ϵ-mechanism are unlikely as shown by Ledoux.

Cassinelli: I would like to follow up on a question I had after van Genderen's talk yesterday. Observations show no pulsation effects. Yet it seems that it is only by pulsation that the interior of the star can directly influence the atmosphere and cause the star to remain at the vibrational instability limit. Should we not see some effect of the pulsations required?

Maeder: This is a point I am also wondering about. Certainly, the optically thick wind may affect the visibility of pulsations, but it also reflects that the outer boundary conditions are not the usual ones in stellar pulsations. On the whole, I wonder whether the very high wind of WR stars is not just an atmospheric response to internal pulsations.

Underhill: With the eclipsing variable V444 Cyg you can obtain an independent estimate of $\dot{M}$ from $\dot{P}$, see Underhill *et al.* (1990). We find $\dot{M} < 6 \times 10^{-6} M_\odot yr^{-1}$. This makes it doubtful that $\dot{M}$ is of the order of $5 \times 10^{-5} M_\odot yr^{-1}$ for WR stars. [But see van der Hucht, p. 28, and Willis, p. 279-280 (eds.)]

Maeder: I think your comment is more relevant to the problem of the current mass loss determinations, on which I am relying.

Maeder explaining it to Langer once more

MODELLING THE WIND ECLIPSES IN WR+O BINARIES: THE QUALITATIVE PICTURE

Gloria Koenigsberger

Instituto de Astronomía, UNAM

Lawrence H. Auer

Los Alamos National Laboratory, Los Alamos, NM

ABSTRACT. The phase dependent profile variations due to wind eclipses in WR + O binary systems are shown to be a means of establishing the WR wind structure. The results of two model calculations are presented which indicate that there are important qualitative differences between the profile variatons which will be observed in slowly –and rapidly– accelerating winds.

INTRODUCTION

Recent developments in the treatment of radiative transfer in the winds of WR stars now provide a means of diagnosis of the intrinsic properties of these stars (Schmutz and Hammann, and Wessolowski 1989; Hillier, 1988). However, one indeterminate quantity in these analyses which introduces uncertainties in the results is the distance scale within the wind; that is, the location of the gas forming the spectra that are being modelled cannot be related to the center of the star (Hummer, 1990). Binary systems in which the orbital parameters are well determined provide such a "yardstick" and the line profile variations which occur as a result of the wind eclipse of the companion O-star by the WR wind can be used to derive information on the structure of the WR wind. With a broad spectral coverage, it is possible in principle to derive the velocity and ionization structure (c.f. Koenigsberger 1990).

WIND ECLIPSES AND PROFILE VARIATIONS

As first suggested by Munch (1950), when the O-star companion to a WR star is behind some portion of the WR wind, its continuum radiation is absorbed by this wind at line wavelengths, enhancing the P Cyg-type absorptions present in the spectrum. In addition, however,as modelled by Khaliullin and Cherepashchuk (1976) the strength of the emission component is reduced due to absorption by gas traveling with a velocity component

175

K. A. van der Hucht and B. Hidayat (eds.),
Wolf-Rayet Stars and Interrelations with Other Massive Stars in Galaxies, 175–178.
© 1991 *IAU. Printed in the Netherlands.*

towards the O-star. These wind eclipses give rise to periodic profile variations which are observed to be stronger in the UV that in the optical spectra (Koenigsberger and Auer 1985 and refs. therein).

Given the orbital parameters of the binary system, one can calculate the impact para-meter of the ray along the line-of sight to the O-star as a function of the orbital phase. If one assumes that the phase dependent variations result exclusively from the wind eclipse mechanism, then it is possible to associate the observed profile variations with changes in the wind structure along each ray as a function of impact parameter. This allows a contraint to be placed on the physical position within the wind of the absorbing/emitting gas, and on the manner in which the velocity and density structures change with distance from the WR core. Thus, the change in the P Cygni structure of the line, from one orbital phase to another, provides information on the change in the structure of the wind from one impact parameter to another.

In order to be able to interpret the observations it is necessary to model the radiative transfer. In the simplest approximation, this just requires ray tracing, and a code which interactively permits this analysis is now available. The program treats the full velocity dependence of the radiation transfer along the line of sight to the O-star. It does not assume the transsonic or "Sobolev" limit, and can be used to estimate empirically the velocity and the density structure of the WR wind. The details of the calculations and a grid of models will be presented elsewhere.

One of the basic questions which may be addressed with this method is the velocity structure within the wind.

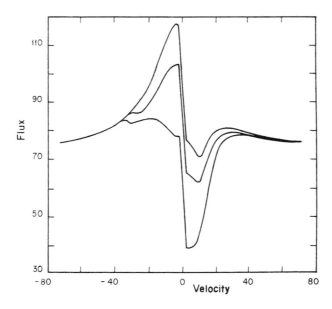

Figure 1: Results of a model calculation showing the phase dependent profile variations which result from wind eclipse for the case of a linear velocity law. The three curves correspond to impact parameters (from top to down) $p/r_* = 9.9$, 5.9 and 3.1. In this reference frame positive velocities correspond to gas moving towards the observer.

In Figure 1 we illustrate the predicted profile variations for the case of a linear velocity law, $v(r) = v(r/r_*)$ out to $20r_*$. This is a "slow" velocity law, in which terminal speeds are achieved far from the WR core, and the density falls off as r^{-3}. In Figure 2 we illustrate the predicted variations for a velocity law of the form $v(r) = v_\infty(1 - r/r_*)^\beta$, in this case with $\beta = 0.5$. This is a very fast velocity law, since terminal speeds are achieved within less that a couple of stellar radii. In both calculations we have taken the opacity $\chi(r) = F(r)/v(r)r^{-2}$ with $F(r) = $ constant throughout, an O-star to WR-star continuum luminosity ratio of 1, an O-star radius of 3 r_*, and an orbital separation of 10 r_* .

Disregarding the intrinsic shape of the profiles, but concentrating only on the nature of the phase-dependent variations, there are two major differences which emerge from a comparison of Figures 1 and 2:

1. There is a "narrow", high velocity absorption spike which becomes stronger as a function of decreasing impact parameter in the case of the $\beta = 0.5$ velocity law which is completely absent in the case of a linear velocity law. This spike becomes less prominent as β becomes larger , but is present even at $\beta = 3$, and results from the large column density of gas traveling at constant velocity and which is projected onto the O-star.

2. The profile changes are much more prominent in the case of a linear velocity law and occur at larger impact parameters than for the $\beta = 0.5$ law; i.e., at $p = 3.09r_*$ the emission component in Fig. 1 has been completely overwhelmed by the absorption. This is not the case in Fig. 2 because of the (Sobolev) optical depth falls off more rapidly due to the larger velocity gradients along the line of sight to the O-star.

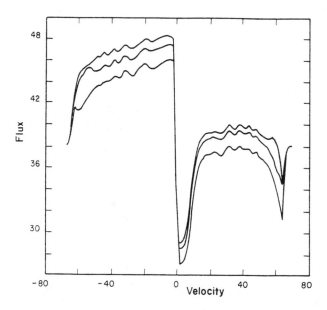

Figure 2: same as Fig. 1 but for a $\beta = 0.5$ velocity law.

CONCLUSIONS

Given the above results, it would appear to be relatively simple to descriminate between a fast and a slow velocity law based on a comparison of observed profile changes with model calculations . In practice, however, the picture is not that simple. In the above examples we have not taken into account the presence of emission lines arising in a wind associated with the O-star, nor the effects introduced by a possible wind-wind collision. In addition, local fluctuations, such as those produced by inhomogeneities in the WR wind, will introduce uncertainties in the interpretation. These problems, however, are soluble or may be avoided by chosing a binary system with a late-type Main Sequence companion and by averaging the observations over several orbital cycles. Thus, a quantitative application of the use of wind eclipses may provide the tool required for determining the wind structure in those WR stars where it is applicable.

REFERENCES

Hillier, J.D. 1988, *Ap. J.*, **327**, 822.

Hummer, D. 1990, private communication.

Khaliullin, Kh. and Cherepashchuk, A.M. 1976, *Soviet Astron.*, **20**, 186.

Koenigsberger, G. 1990, *Astro. Astrof.*, in press.

Koenigsberger, G. and Auer, L.H. 1985, *Ap. J.*, **297**, 255.

Munch, G. 1950, *Ap. J.*, **112**, 266.

Schmutz, W. and Hamman, W.-R., Wessolowski, V. 1989, *Astron. Astrof.*, **210**. 236.

DISCUSSION

Owocki: What are the relative $\dot{M}$-values of the O and WR stars? This will indicate the scale of the interaction zone - if $\dot{M}_{WR} \gg \dot{M}_O$ then this interaction will be confined to very near the O star.

Koenigsberger: The ratio of total wind momentum in V444 Cyg is 30 (Pollock, Blondin and Stevens, this symposium).

Leitherer: You find a relatively shallow velocity law with accerelation up to $\sim 5R_*$. On the other hand, we heard yesterday from Hamann and Hillier that they find optimum fits to optical recombination lines and UV resonance lines with a very steep CAK-type law. Do you have an explanation for this difference?

Koenigsberger: My analysis does not rule out a rapidly accelerating zone near the WR core. At this stage all I can say is that significant acceleration must still be occurring at a distance of several stellar radii.

Shylaja: There is a general problem in fixing the continuum of WR energy distribution. How was this avoided?

Koenigsberger: I always try to use ratios of spectra: the spectrum of the system taken at quadrature is used in the denominator and that at other phases in the numerator.

THE UV VARIABILITY OF THE WR STAR HD 50896 – A ONE DAY RECURRENCE TIMESCALE

N. ST-LOUIS, A.J. WILLIS, L.J. SMITH
Department of Physics and Astronomy, University College London
Gower Street, London WC1E 6BT, U.K.

C.D. GARMANY, P.S. CONTI
Joint Institute for Laboratory Astrophysics, University of Colorado
Boulder CO 80309, U.S.A.

1. Introduction

The WN 5 star HD 50896 has been the subject of many variability studies in the optical. A period of 3.766 days is frequently associated with this variability although it is by no means clear whether it reflects, for example, binary motions or rotation.

In previous work (*cf.* Willis *et al.* 1989) we have found substantial variations in the P Cygni absorption components of N V λ1240, C IV λ1550, He II λ1640 and N IV λ1718. No convincing periodicity was found and the variations were interpreted as reflecting physical changes in the velocity, density or ionization structures of the WR wind. The dataset was, however, insufficient in time resolution to accurately determine the timescale of the changes as well as any possible recurrence timescale. We have therefore obtained a further extensive time sequence of IUE spectra with the aim of gaining more insight into the nature of the changes. The preliminary results of the analysis of these new data are presented here.

2. Observations

In a joint NASA-Vilspa project, we have obtained a total of 130 HIRES SWP IUE spectra of HD 50896 over a period of 6 consecutive days in December 1988. The spectra were uniformly extracted from the PHOT images provided by the ground stations using the IUEDR software package available on the UK Starlink network of VAX computers. Subsequent measurements and analysis were performed using the Starlink DIPSO software package.

3. Results

Inspection of the spectra reveals variations in the N V λ1240, C IV λ1550, He II λ1640 and N IV λ1718 P Cygni profiles. We have quantified the changes by measuring the equivalent width of the absorption components for the last three ions (N V and C IV show a similar behaviour). The variation of the equivalent widths as a function of Julian Date is presented in Figure 1. For all three lines at least three events are clearly visible, each lasting for $\sim$ 1 day, which is of the order of the flow time through the wind. The first and third events are similar for all lines while, for the second, the onset of the He II decrease seems to

179

K. A. van der Hucht and B. Hidayat (eds.),
Wolf-Rayet Stars and Interrelations with Other Massive Stars in Galaxies, 179–180.
© 1991 *IAU. Printed in the Netherlands.*

180

be delayed by ~ 6 hours. We have subjected all these measurements to a power spectrum periodicity search between $\nu_{min}=1/(\Delta t)=0.19\ d^{-1}$ where Δt is the time interval over which the data were obtained and the Nyquist frequency $\nu_{max}=\Delta t/2(N-1)=12.44\ d^{-1}$ where N is the number of data points. The periodogram for the NIV, CIV and HeII equivalent width variations show very clear peaks at 0.98, 0.97 and 1.03 days respectively.

One important characteristic of the variations is the velocity range over which they occur. This is typically between -1600 and -2400 km s^{-1} for HeII ; -1800 and -2800 km s^{-1} for NIV ; and -2200 and -3150 km s^{-1} for CIV. This seems to reflect the line forming mechanisms of each transition and indicates that large regions of the wind are being affected at one particular time. Furthermore, most of these variations occur at velocities exceeding the bulk outflow terminal velocity of the wind (~ -1900 km s^{-1}). This is reminiscent of the radiatively driven blob model of Lucy and White (1980) or the propagating shock models of Lucy (1983) and Owocki *et al.* (1988).

We conclude that this extensive new dataset for HD 50896 presents the best evidence so far for intrinsic stellar wind variability, affecting large regions of the wind, and having a recurrence timescale of the order of one day.

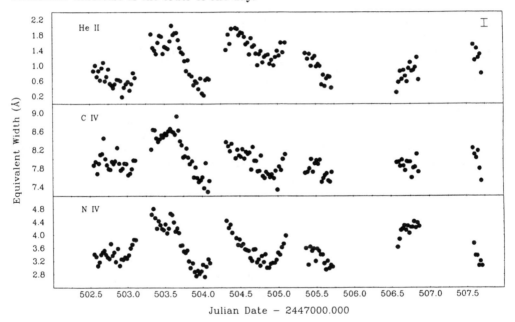

Fig. 1 : *Equivalent width of absorption components as a function of Julian Date*

References

Lucy, L.B. 1983, *Ap. J.*, **274**, 372.

Lucy, L.B., & White, R.L. 1982, *Ap. J.*, **241**, 300.

Owocki, S.P., Castor, J., & Rybicki, G.B. 1988, *Ap. J.*, **335**, 914.

Willis, A.J., Howarth, I.D., Smith, L.J., Garmany, C.D., & Conti, P.S. 1989, *Astr. Ap. Suppl.*, **77**, 269.

EZ (EASY?) CANIS MAJORIS

C. ROBERT[1], A. F. J. MOFFAT[1], L. DRISSEN[2], V. S. NIEMELA[3], P. BARRETT[4], W. SEGGEWISS[5], and R. LAMONTAGNE[1]
[1]Université de Montréal, Montréal, Canada; [2]Space Telescope Science Institute, Baltimore, U.S.A.; [3]IAFE, Buenos Aires, Argentina; [4]SAAO, Cape, South Africa; [5]Observatorium Hoher List, Universitäts-Sternwarte Bonn, F. R. Germany

EZ CMa (HD 50896, WN5) is an enigmatic object. New photometry and polarimetry of EZ CMa are presented in the figure. Again the 3.77 day period is found but, as observed at previous epochs (e.g. Drissen et al. 1989, Ap. J., **343**, 426), the shapes of the curves change. The new photometry can also be interpreted in terms of a shorter period, of 1.254 days. A period of about one day is also claimed in other sets of photometric data (e.g. van der Hucht et al., 1990, A. A., **228**, 108) and in the IUE spectra of St.-Louis et al. (1990, this symposium). However, despite the complex nature of the light curve, the 3.77 day period is strongly supported by the polarimetry, which shows no evidence for the shorter period.

If EZ CMa is a binary WR + c system (Firmani et al. 1980, Ap. J., **239**, 607), then the short-term phase-dependent variations could be an indication of interaction between the WR wind and the 3.77 day orbiting companion. Long term changes may be related to the precession of an accreting disk around the companion; however, no long periodicity is obvious yet. The possibility of a single rotating or pulsating star must also be considered.

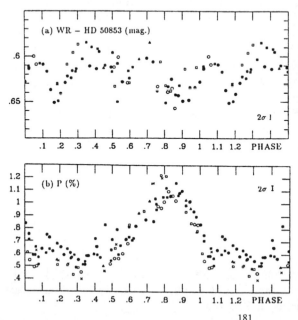

FIGURE. (a) Light curve (WR − HD 50853) and (b) polarization curve (P, the degre of polarization) of EZ CMa in 1990. The phase has been calculated from the Lamontagne et al. (1986, Astron. J., **91**, 925) ephemeris. Measurements obtained at the San Juan Observatory (Argentina) with VATPOL and a 30 Å wide filter centered on 4700 Å (continuum) are plotted for the observing runs Jan. 13 to Feb. 1 (∗) and Mar. 19 to 27 (×). Filled symbols are data from the University of Toronto Southern Observatory (Chile) collected with MINIPOL and a G filter from Feb. 5 to 17 (●), Feb. 18 to Mar. 1 (■) and Mar. 17 to 22 (▲). Open symbols refer to the data collected with the South African Astronomical Observatory photo-polarimeter from Feb. 19 to 27 (○) and Mar. 20 to Apr. 1 (□), both with a Johnson V filter. Each group of data has been shifted in ordinate to give the same mean value.

181

K. A. van der Hucht and B. Hidayat (eds.),
Wolf-Rayet Stars and Interrelations with Other Massive Stars in Galaxies, 181.
© 1991 *IAU. Printed in the Netherlands.*

Emission Line Variations in γ^2 Velorum

MaryJane Taylor
Space Astronomy Laboratory, University of Wisconsin
1150 University Avenue, Madison, WI 53706, U.S.A.

In this paper, we present results of an intensive search for periodic variations in the strength of the HeII emission line of γ^2 Vel relative to the nearby continuum. The data were obtained in 1986 and 1988 using an automated optical system located at the Amundsen-Scott South Pole Station. Ten of the data sets have a time resolution between 2 and 3 minutes and consist of 2.3 - 20.4 hours of continuous photometry (Taylor, *AJ*, in press). Three of the data sets have a time resolution of about 1 minute and span between 1.2 and 6.5 hours.

Power spectrum analyses of the lower resolution data suggest a period of about 1.26 hours. According to recent theoretical models (Cox and Cahn 1988, *Ap.J.* 326, 804), a hydrogen-deficient Wolf-Rayet star with an initial mass of $85M_o$ exhibits a first-overtone radial mode of oscillation with a period of 1.23 hours. Additional models by Maeder (private communication) indicate radial oscillations with periods on the order of an hour.

Since the HeII emission line is optically thick, we do not detect the actual pulsation of the star, but, rather, the effects of stellar pulsations on the overlying atmosphere. We believe that radial instabilities in γ^2 Vel give rise to intermittent mass ejection which in turn changes the ionization state of the wind. According to theoretical work by MacFarlane and Cassinelli (1989, *Ap.J.* 347, 1090), as mass propagates outward, it is accelerated to supersonic rates and as it overtakes slower moving gas, both forward and reverse shocks are formed. At the boundary of the two shocks is a sharp density enhancement which modifies the ionization structure of that part of the wind and which in turn, leads to changes in the emission lines.

Preliminary analysis of the higher resolution data, suggests more rapid fluctuations, on the order of 3 - 8 minutes. Although we do not yet understand this result, it may confirm work previously reported by Jeffers, et al. (1973, *Nature Phys. Sci.* 243, 109) of 200 ± 50 seconds.

It is clear that before we can understand the physical processes which are taking place in γ^2 Vel, high resolution continuous photometry spanning several days, at least, are essential.

This work has been supported by NSF grants DPP-8217830, DPP-8414128, DPP-8614550 to the University of Florida, and NASA contract NAS5-2677 to the Space Astronomy Laboratory at the University of Wisconsin.

182

K. A. van der Hucht and B. Hidayat (eds.),
Wolf-Rayet Stars and Interrelations with Other Massive Stars in Galaxies, 182.
© 1991 *IAU. Printed in the Netherlands.*

THE IMPACT OF POLARIMETRY ON WOLF-RAYET STAR MODELS

REGINA E. SCHULTE-LADBECK
Space Astronomy Laboratory, University of Wisconsin - Madison
1150 University Ave., Madison, WI 53706, U.S.A.

Recent observational efforts have presented new results concerning (linear) polarization variations as a function of time as well as wavelength that have considerable impact on WR star models:

Polarimetry throughout orbital phase has provided inclinations of double-line spectroscopic binaries containing WR stars with O star companions, thus improving our understanding of the **masses and evolution of WR binaries** (Schulte-Ladbeck 1989, *A.J.* 97, 1471). New binary-model calculations were presented at the meeting and continuing comparison with observations should prove productive.

Temporal polarization variations in single WR stars show various types of behavior: none, stochastic, along a preferred axis in the Stokes' parameter plane (see Schulte-Ladbeck and van der Hucht 1989, *Ap.J.* 337, 872 and references therein). This is indicative of a variety of atmospheric properties, i.e. occurrence of non-spherical geometries and non-steady dynamics in **WR star winds** (perhaps related to WR subtype?). Polarimetry demonstrates that spherically symmetric, purely radiatively-driven winds do not describe all WR stars. In a growing number of stars, polarimetry implies random "blob" ejections in the winds, while for a few stars a case can be made in favor of rotation. More such observations are clearly needed for the single WR stars.

At the Space Astronomy Laboratory, our interest has focussed on studying the winds of WR stars through spectropolarimetry. Line-polarization features expected from stratification of ionization and excitation in an extended, asymmetric atmosphere have only been detected in three WN5-6 stars but, curiously, in none of the observed WC stars (see Schmidt 1988 in "Polarized Radiation of Circumstellar Origin", p.641). Observations from our Pine Bluff Observatory's 36" telescope covering 3200Å to 7800Å at a resolution of ~30Å confirm the depolarization across emission lines in EZ CMa (WN5). We have obtained a spectropolarimetric data set covering various phases of the star's 3.766-day period which shows that both the continuum and line polarization are variable. The line effect is most pronounced in the strong line of He II at 4686Å, and we find that the polarization changes across the line profile. In the QU plane, the polarization line profile traces a loop. Such behavior has previously been observed in the Balmer lines of Be stars, where it has been modeled through the line absorption in an electron scattering, expanding, rotating and inclined disk. A WR star not previously studied with spectropolarimetry is HD 193793 (WC7+O4-5). We find the polarization spectrum to be smooth and featureless (S/N of about 20 in the polarization); it can be explained as caused primarily by interstellar polarization.

Our optical spectropolarimetric observations of WR stars will soon be extended into the ultraviolet spectral range, thus giving access to different line transitions. Polarization measurements ranging from 1400Å to 3300Å at a resolution of ~6Å are scheduled to be obtained with the Wisconsin Ultraviolet Photo-Polarimeter Experiment (WUPPE). This 20" telescope is part of the Astro-1 space-shuttle mission currently scheduled to fly in mid August, 1990.

It is a pleasure to thank the PBO reducers team. This work is supported by NASA contract no. NAS5-26777.

K. A. van der Hucht and B. Hidayat (eds.),
Wolf-Rayet Stars and Interrelations with Other Massive Stars in Galaxies, 183.
© 1991 *IAU. Printed in the Netherlands.*

GYROSCOPICS AS A MEANS OF DEMONSTRATING ROTATIONAL INSTABILITY IN STARS

M.S.A. SASTROAMIDJOJO
Solar Physics Laboratory, Gadjah Mada University
Yogyakarta, Central Java, Indonesia

ABSTRACT. Light-traces of a light bulb fixed on top of the stem of a spinning top were recorded to simulate rotation. It was found that with dissipation of energy the rotational frequency did not decrease monotonically, but showed 'quasi-periodic' sequences.

1. Introduction

A star is a physical entity of gaseous material hold together by the force of gravity, more or less spherical in shape. A spinning top given initial rotational energy and left by itself will display such features as spin, rotation and precession. It was with this in mind that we tried to simulate stellar rotation in the laboratory, because not only could basic mechanical dynamics be demonstrated, but also phenomena far from equilibrium.

2. Method

The basic set-up needed the following paraphernalia: *(1)* a gyroscope made of a bicycle wheel with a light bulb on the top of the stem and two flashlight batteries in the centre of mass; *(2)* a square grid of chicken wire with a $2 \times 2cm$ mesh, providing a frame of reference; *(3)* a 35-*mm* Pentax camera; *(4)* a ground plate with cup for sustaining the lower end of the gyroscope stem; *(5)* a distance of 70 *cm* from centre of mass to top of stem and a distance of 20 *cm* from centre of mass to lower end.

The procedure was as follows: *(1)* the light bulb was switched on; *(2)* the gyroscope was set spinning; and *(3)* the light-trace was recorded through the wire mesh with 3-second exposures at 5-second intervals, till the gyroscope stopped rotating.

3. Results

The recorded light-traces show a quasi-periodic start with a concentration of light in the center, followed by alternating 'widening' and 'bunching' of the precessions. The same can be said about the rotational frequency, which went through an 'up-and-down' sequence, according to the movement of the gyroscope stem, from equilibrium (standing almost vertical) to far from equilibrium (almost falling down).

It is suggested that the 'new' discipline of 'synergetics', chaos-order-chaos, may perhaps be studied with the gyroscope model.

Acknowledgement

The experiments were done by Indonesian Islamic University (UII, Yogyakarta) first year engineering students, using the facilities of the Gadjah Mada Solar Physics Laboratory.

184

K. A. van der Hucht and B. Hidayat (eds.),
Wolf-Rayet Stars and Interrelations with Other Massive Stars in Galaxies, 184.
© 1991 *IAU. Printed in the Netherlands.*

SESSION IV. BINARIES - *Chair: Anthony F.J. Moffat*

Mrs. Moffat, Tony Moffat, Mrs. Nomoto, Ken Nomoto

WOLF-RAYET BINARIES: OBSERVATIONAL ASPECTS

A.M. Cherepashchuk
Sternberg Astronomical Institute
Moscow University
119899, Universitetsky Prospect
13, Moscow, USSR

ABSTRACT. New data concerning observational aspects of Wolf-Rayet binaries are summarised. WR+O and WR+c binary systems are considered. All the data concerning WR binaries agree well with the modern understanding of WR stars as the helium remnants formed as a result of mass loss by massive O-type stars.

1. Introduction

The problem of Wolf-Rayet (WR) binaries consists of two parts: WR+O systems and WR+c systems. In the latter, relativistic objects (neutron star or black hole) are presumed to exist, mainly on the basis of modern evolutionary considerations.

2. Parameters of WR+O binary systems

The fraction of WR+O binary systems among all WR stars in the Galaxy is of the order of 43% (Moffat *et al.*, 1986). This value is close to the fraction of binary systems among O-stars (Abbott and Conti, 1987). One of the important criteria for duplicity of WR stars is the enhanced X-ray radiation which originates near the front of the shock wave formed near the O-star by the supersonic flow of the WR wind (Cherepashchuk, 1976). At present several new WR+O binaries have been identified using this criterion (Pollock, 1987). In this connection the fraction of WR+O binaries among WR stars is compatible with the above 43%.

A comparison of the galactic distribution of WR+O systems and single WR stars is given by Hidayat *et al.* (1984), Tutukov and Yungelsen (1985), Doom (1987), van der Hucht *et al.* (1988), and Conti and Vacca (1990). In the galactic plane, the increase of the fraction of binary WR stars with increasing galactocentric distance is observed. For $d \leq 4$ kpc $< |z| >$(single WR) = 48 pc, $< |z| >$(SB2+mSB1) = 59 pc, $< |z| >$(lm SB1) = 134 pc (see however, Conti and Vacca, 1990). All these facts agree well with the evolutionary scenario for massive close binary systems (Tutukov and Yungelson, 1973, van den Heuvel, 1976). The system RY Sct may be considered as the progenitor of a WR+O binary (Antokhina and Cherepashchuk, 1988a).

The fraction of binaries among WR stars in other galaxies varies from 40% for the LMC to $\geq$ 70% for the SMC (Moffat, 1988). One should note the absence of SB2 binaries among WN8, 9 stars (Moffat, 1989).

K. A. van der Hucht and B. Hidayat (eds.),
Wolf-Rayet Stars and Interrelations with Other Massive Stars in Galaxies, 187–199.
© 1991 *IAU. Printed in the Netherlands.*

The characteristics of well-studied WR+O binaries are listed, *e.g.*, by Smith and Maeder (1989). Many masses are calculated using values of the inclination of the orbital plane i determined from light curves and from polarisation curves. The basic conclusions concerning the properties of WR+O binaries are the following.

2.1 The mass ratio $q = \frac{m_{WR}}{m_O}$ for WR+O binaries falls in the range $0.17 - 2.67$. The mean value is $q = 0.44 - 0.53$ (here and below), the second value comes from including the system HD92740 with $q \simeq 2.67$). The dependence of q on spectral subtype Sp is presented in Fig. 1. This correlation was pointed out by Moffat (1982) and Moffat *et al.* (1990). In particular, the continuous decrease of mass ratio $q = \frac{m_{WR}}{m_O}$ with hotter WN and especially WC subtypes is observed. For WC subtypes the value of q ranges from ~ 0.5 for WC8 to ~ 0.2 for WC4-WO4. For WN subtypes, the value of q ranges from 2.67 to $0.2 - 0.3$ for WN7 to WN3, respectively. As was pointed out by Moffat et al. (1990), these facts support the idea of evolution of WR stars through the sequence WN-WC8-WC4-WO.

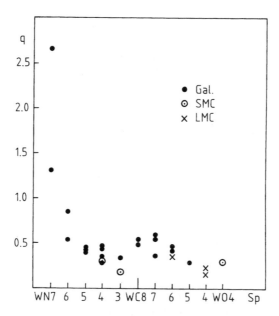

Fig. 1

2.2 The mean mass of WR stars (22 determinations) is $15.6 - 18.4 M_\odot$. The range of masses of WR stars is very wide: from $5\ M_\odot$ to $48\ M_\odot$ and even $77\ M\odot$ (HD92740). The mean mass of WN stars is $17.5 - 22.5 M_\odot$ (12 stars) and ranges from 8 to $48 M_\odot$ and $77 M_\odot$. The mean mass of WC stars (9 stars) is $13.4 M_\odot$ and ranges from $5 M_\odot$ to $27 M_\odot$. The mass of the WO4 star is $14 M_\odot$. Therefore, recent data allow one to suggest that the masses of WC stars on the average are less than those of WN stars. This is in agreement with the modern evolutionary scenario for WR stars. The dependence of the masses of WR stars on their spectral subtypes and the correlation between the masses of WR stars and those of O-stars for WR+O systems are presented in Fig. 2.

2.3 The masses of the O-companions in WR+O binaries range from 14 to 57 $M_\odot$ and on average have the value $32.8 M_\odot$. The mean mass for O-stars in WN+O systems is

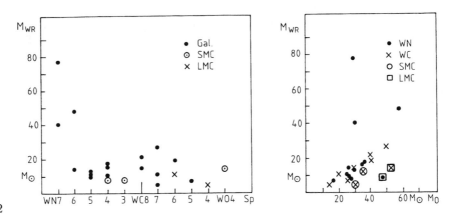

Fig. 2

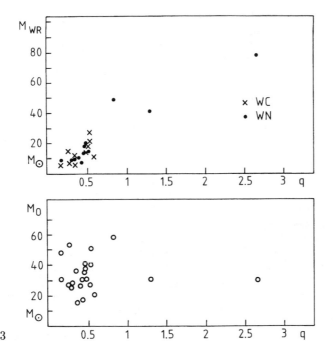

Fig. 3

$31.2 M_\odot$ and for O-stars in WC+O systems is $32.4 M_\odot$. The mass of the O-companion in the WO4+O4V system AB8 is $52 M_\odot$.

2.4 There is a good correlation between the mass ratio q and the mass of WR stars in WR+O systems (Fig. 3): the mass of the WR star increases with q. Between the mass of the O-companion and q there is no clear correlation (Fig. 3).

2.5 The mean value of the total mass of the WR+O systems is $48.5 - 51.1 M_\odot$ and falls in the range $19 - 106 M_\odot$. The mean value of the total mass for the WN+O systems is $50 - 54.7 M_\odot$; for the WC+O systems it is $44.6 M_\odot$; and for the WO+O system it is $66 M_\odot$. These values are almost equal to the mean total mass for O+O systems $\sim 46 M_\odot$ (see, *e.g.*, Svechnikov, 1989). The total mass for WR+O systems seems to increase with q; this is not observed for O+O systems. The total mass for WR+O systems shows a slow decrease with the spectral subtype of the WR component: from $\sim 85 M_\odot$ for WN7+O to $\sim 50 M_\odot$ for (WC4-WO4) + O systems. The O-companions of WR+O binary systems show a large scatter in mass; on average their masses are independent of the spectral subtype of the WR-companion.

2.6 Values of eccentricities of orbits for WR+O systems with periods ≥ 70 days are $e = 0.3 - 0.8$. All the systems with periods $P \leq 14$ days have circular orbits ($e = 0$). In the interval $14^d < P < 70^d$ there are circular and eccentric orbits ($e = 0.17 - 0.5$).

2.7 The bremsstrahlung X-ray luminosity in WR+O binaries (Pollock, 1987) is correlated with orbital period: L_x is greater for systems with shorter periods. In WN+O systems, the mean luminosity $\overline{L}_x = 16 \times 10^{32}$ erg/s for $P < 20^d$ and $\overline{L}_x = 10 \times 10^{32}$ erg/s for $P \geq 20^d$; for WC+O systems $\overline{L}_x = 12 \times 10^{32}$ erg/s for $P < 20^d$ and $\overline{L}_x = 1.5 \times 10^{32}$ erg/s for $P \geq 20^d$. This correlation is consistent with the theory for the generation of X-rays in the stellar winds of WC stars, which are enriched by heavy elements of the CNO group because of the more advanced evolutionary stage of WC stars.

3. X-ray radiation from WR+O binary systems

Cherepashchuk (1967) pointed out that considerable X-ray radiation in WR+O binary systems may be generated in the shocked region formed in front of the O-star by the supersonic wind of the WR companion. This idea was developed by Prilutsky and Usov (1976). Calculations carried out for realistic WR+O systems, taking into account the Compton cooling of hot plasma by optical radiation of O and WR stars (Cherepashchuk, 1976), have shown that the expected X-ray luminosity in the majority of cases is $10^{32} - 10^{34}$ erg/s in the mean energy range kT $= 1 - 10$ keV for mass loss rates from the WR star $\dot{\text{M}}$ $= 10^{-6} - 10^{-5} M_\odot$/yr, respectively. Further development of the theory (Bayramov *et al.*, 1988) has confirmed these results.

Recent results on the analysis of *EINSTEIN* X-ray observations of 48 WR stars (Pollock, 1987) in the range $0.2 - 4$ keV have shown that the X-ray luminosity for WR stars falls in the range $L_x = 10^{31} - 10^{34}$ erg/s. The mean X-ray luminosity of single WR stars is $\sim 5 \times 10^{31}$ erg/s; however, the X-ray luminosity of WR+O binaries is one to two orders of magnitude higher. The orbital phase-dependent variability of the intensity and "hardness" of the X-ray radiation have been observed in several cases of WR+O binaries (Pollock, 1987; Moffat *et al.*, 1982). Using ultraviolet *IUE* observations, the effects of the colliding stellar winds have been discovered recently in the V444 Cyg system by Shore and Brown (1988). Thus, the prediction of shock waves and considerable X-ray radiation in binary WR+O systems (Cherepashchuk, 1967, 1976; Prilutsky and Usov, 1976) is now confirmed by new X-ray and UV observations.

There is some discrepancy between the X-ray observations and the theory: the mean observed X-ray luminosity for WR+O binary systems is more than one order of magnitude lower than the theoretical luminosity, and the mean hardness of the X-ray spectrum is kT ~ 1 keV instead of the theoretically predicted value of kT $\sim$ several keV. It was shown recently (Cherepashchuk, 1990), that the relatively low X-ray luminosity of WR+O binaries may be connected with the ragged, cloudy structure of WR winds (Cherepashchuk *et al.*,

1984, van Genderen *et al.*, 1987, Moffat *et al.*, 1988, Antokhin *et al.*, 1988). Further X-ray observations of WR+O binaries, investigations of the regular orbital variability of their X-ray luminosity and the fluctuations of X-ray flux on a timescale of about 10^3 s would be very important for the determination of the structure and dynamics of stellar WR winds and their chemical composition.

4. Variable linear polarisation of optical radiation from WR+O binary systems

The first discovery of variable linear polarisation in a WR+O system (V444 Cyg) was found in V444 Cyg (Rudy and Kemp, 1978). A major effort to expand polarimetric investigations of WR+O binaries and WR stars in general has been undertaken by Moffat and collaborators (Moffat, 1988; Robert and Moffat, 1989; Robert *et al.*, 1989; Drissen *et al.*, 1987; St.-Louis *et al.*, 1988). Up to now, polarimetric determinations of i and characteristics of the instability of WR winds have been determined for about two dozen WR+O binaries (*e.g.* St.-Louis *et al.*, 1988). For the system V444 Cyg, polarimetric data yield $i = 78.7° \pm 0.5°$ and WR core radius $r_c^{WR} = 3R_\odot$ (Robert *et al.*, 1989; Moffat, 1989) which are in good agreement with light-curve solutions (Cherepashchuk, 1975; Cherepashchuk *et al.*, 1984). In the recent work of Hamann *et al.* (1990), who have made significant progress in modelling Non-LTE extended atmospheres, a new attempt to determine the parameters of V444 Cyg has been carried out. Taking into account the observed spectroscopic estimate of the luminosity of the WN5 star $L_W = 0.167 - 0.254$ (Beals, 1944; Annuk, 1988) we are obliged to choose the light curve solution from the work of Hamann *et al.* (1990) corresponding to $r_c^{WR} \geq 1.4R_\odot$ and $T_c \geq 60000$ K. This result is close to our solution for the light curve of the V444 Cyg system (Cherepashchuk, 1975; Cherepashchuk *et al.*, 1984): core radius and core temperature for the WN5 star: $r_c^{WR} = 2.6 - 2.9R_\odot, T_c = 70000 - 100000$ K.

The parameters of V444 Cyg proposed by Underhill and Fahey (1987) (in particular $r_c^{WR} = 9R_\odot$) strongly disagree with the continuum light curve at $\lambda 4244$ Å. Third light proposed by Underhill and Fahey (1987) for the UV spectral range is negligible in the optical at the $\lambda 4244$ light curve because variations at $\lambda 4244$ between the minima are small ($\sim 0\overset{m}{.}01$).

Up to now, the accuracy of the light curve of the new WR+O eclipsing binary HD5980 is not sufficient for a unique light curve solution. Further observations of HD5980 are needed.

St.-Louis *et al.* (1988) have confirmed on average the increase of $\dot{M}$ with mass M of WR stars from polarimetric data. According to Moffat *et al.* (1988), the WNL stars, with smaller values of wind velocity, exhibit greater amplitudes of random fluctuations in linear polarisation. This may be connected with instabilities in the WR star wind leading to the formation of dense blobs of matter.

5. Ultraviolet IUE observations of WR+O systems; selective atmospheric eclipses

The effect of enhanced opacity in the stellar WR wind has been discovered by Eaton *et al.* (1982, 1985a,b) for the ultraviolet region of the spectrum at $\lambda < 2000$ A. This effect is observed in the eclipsing WR+O binaries V444 Cyg and CV Ser (Eaton *et al.*, 1985a,b; Koenigsberger and Auer, 1985) and in several other spectroscopic binaries (see *e.g.* Hutchings and Massey, 1983). The depth of the atmospheric eclipse strongly increases in the range $\lambda < 1500$ A; this is caused by the absorption of O-star light in the many lines of ions of FeIV, FeV, and FeVI (Koenigsberger, 1988). In the atmospheric eclipse of HD5980,

Koenigsberger *et al.* (1987) did not find the increased absorption at $\lambda\lambda 1300 - 1500$ A; this reflects the relatively poor abundance of heavy elements in the SMC.

A theory of selective atmospheric eclipses in WR+O binary systems has been developed by Khaliullin and Cherepashchuk (1976). For the first time, selective atmospheric eclipses were discovered in the WR+O system CV Ser (Cherepashchuk, 1969, 1971) in the emission band CIII-CIV 4653 Å.

Underhill *et al.* (1988) have criticized our model of selective atmospheric eclipses for WR+O binaries. The considerations of Underhill *et al.* (1988) are fully true for the absolute intensity of an absorption line arising in the spectrum of the O star during atmospheric eclipse in V444 Cyg. But their final conclusion is incorrect because Underhill *et al.* did not take into account the fact that this absorption is veiled by strong, non-eclipsed emission line flux, formed in the WR envelope.

6. New investigations of some eclipsing WR+O binary systems

6.1 THE V444 Cyg SYSTEM

The parameters of this system determined from the light curve solutions (Cherepashchuk, 1975; Cherepashchuk *et al.*, 1984) and from the polarimetric analysis (Robert *et al.*, 1989) seem to be rather reliable: $r_{O6} = 10 R_\odot, i = 78°, L_W \simeq 0.2, r_c^{WR} \leq 2.9 R_\odot, T_c^{WR} > 60000$ K. The structure of the extended atmosphere of the WN5 star from the analysis of multicolor light curves at $\lambda\lambda 2460 Å - 3.5\mu$ has been determined by Cherepashchuk *et al.* (1984). The results concerning V444 Cyg have been used by Pauldrach *et al.* (1985) for the calculation of the model of the WR star wind driven by radiation. Results of the interpretation of the light curve of V444 Cyg have also been used by de Greve and Doom (1988) for a model calculation of the origin and evolution of the V444 Cyg binary. Theoretical analysis of the WN5 companion in the V444 Cyg system has been done by Vanbeveren (1988) and Poe *et al.* (1989). New estimates of the mass loss rate from the WN5 star in V444 Cyg on the basis of orbital period change are: $(1.1 \pm 0.2) \times 10^{-5} M_\odot$/yr (Khaliullin *et al.*, 1984) and $0.6 \times 10^{-5} M_\odot$/yr (Underhill *et al.*, 1990). New spectroscopic elements of the orbit for V444 Cyg have been determined by Underhill *et al.* (1988a) and Acker *et al.* (1989). Rapid oscillations in the intensity of the emission line HeII 4686 in V444 Cyg have been discovered by Zilyaev and Marchenko (1988).

6.2 THE CQ Cep SYSTEM

Analysis of *IUE* ultraviolet spectra (e.g. using new determinations of radial velocities from archive spectra) and UBVJKL photoelectric observations of CQ Cep have been made by Stickland *et al.* (1984). It is shown that CQ Cep belongs to the Cep OB1 association (d $\simeq$ 3.5 kpc, $A_V = 2^m.25$). A photometric solution of the light curves of CQ Cep has been carried out by Leung *et al.* (1983): $q = 1.33, M_{WN7} = 46 M_\odot, M_O = 35 M_\odot, i = 68°, T_c^{WR} = 45000$ K (adopted), $T_O = 41000 - 42000$ K. Polarimetric investigations (Drissen *et al.*, 1986) yield: $i = 78°, M_{WN7} = 42 M_\odot, M_O = 30 M_\odot$. New spectroscopic and photometric observations are also available: Kartashova and Snezhko (1985), Shylaja (1986), Harvig (1987), Underhill *et al.* (1990). The decreasing of the orbital period of CQ Cep is confirmed by Antokhina *et al.* (1987): $\dot{P} = -0.014 \pm 0.004$ s/yr. A new photometric solution of the light curves of CQ Cep, using new information about its distance (Stickland *et al.*, 1984), has been obtained by Antokhina and Cherepashchuk (1988b): $i = 77°, q = 1.3 - 1.6, M_{WN7} = (42 - 66) M_\odot, M_O = (32 - 41) M_\odot, r_O = (9.4 - 10.1) R_\odot, T_O = (40000 - 50000) K, r_c^{WN7} =$

$(10.6 - 12.5)R_\odot, T_c^{WN7} = (60000 - 70000)$ K. If $d < 3.5$ kpc, the values of the masses and temperatures of the components will be smaller. These results are in agreement with those of Lipunova and Cherepashchuk (1982). Other parameters of CQ Cep have been proposed by Underhill et al. (1990); they are in conflict with the shape and amplitude of the light curve.

The disk-like model for WR star envelopes, proposed by Underhill (1984) and Underhill et al. (1988a, 1990), disagrees with observations: up to now there is no correlation between the inclination of the orbit plane and the widths of emission lines and other parameters in WR+O binaries (Aslanov and Cherepashchuk, 1990).

6.3 THE CV Ser SYSTEM

A double envelope nebula structure has been discovered around CV Ser by Gonzalez and Rosado (1984).

Analysis of UV selective atmospheric eclipses in CV Ser using IUE data has been carried out by Eaton et al. (1985). The parameters of CV Ser have been determined from a light curve solution by Lipunova (1982): $i = 67° \pm 3°, r_c^{WC8} = (3.5 \pm 0.5)R_\odot, T_c^{WC8} \geq 55000$ K. It is of great interest to investigate the very peculiar and puzzling variability of CV Ser and the WC9 star HD 164270 on a long time scale. Both stars show considerable ($\sim 0\overset{m}{.}5$) and rather rare (once per several dozens of years) decreases in light.

6.4 THE CX Cep SYSTEM

Polarimetric observations of CX Cep have been carried out by Schulte-Ladbeck and van der Hucht (1989). The orbital inclination is estimated from the polarimetric data: $i = 74° \pm 5°$; corresponding masses are $M_{WN5} = 6M_\odot, M_{O8} = 14M_\odot$. From the light curve solution (adopting $M_{O8} = 22M_\odot$) the values $i = 53°$ and $M_{WN5} = 10M_\odot$ were determined by Lipunova and Cherepashchuk (1982). Our attempt to interpret the light curve of CX Cep with the value $i = 74°$ was not successful.

Intense photometric observations of WR+O binaries have been carried out by Monderen et al. (1987), Moffat and Shara (1986), Lamontagne and Moffat (1987), van Genderen et al. (1987), Balona et al. (1989).

The results of the interpretation of the light curves of WR+O eclipsing binaries are in agreement with modern ideas about the nature of WR stars: WR stars are helium remnants formed during mass exchange in massive close binary systems.

7. WR+c binaries containing possible compact companions

Numerous intense photometric investigations of the microvariability of suspected WR+c stars have been carried out (Balona et al., 1989; Moffat and Shara, 1986; Antokhin and Cherepashchuk, 1989; van Genderen et al., 1987, 1989; Lamontagne and Moffat, 1987; Monderen et al., 1988).

Up to now, only two WR stars may be considered as genuine WR+c binaries: HD50896 and HD197406. These systems are real binaries containing a compact (possibly relativistic) object ($M_c \simeq 1M_\odot$ for HD50896 and $M_c > 4M_\odot$ for HD197406). Evolutionary consider- ations and high values of $|z|$ suggest that the companions in these binary systems are relativistic (neutron star or black hole). However, it is difficult to explain for these systems the absence of strong X-ray radiation (Cherepashchuk, 1981; Stevens and Willis, 1988). One could try to find correlations between basic characteristics of WR+c systems. No correlation is observed between the period P and the semiamplitude of the radial velocity curve K. Also, no correlation exists between the period P and the spectral class (Sp) of the WR star, between the amplitude of the optical variability ΔV and value of K. These facts testify against the intrinsic variability of these stars. Between ΔV and Sp some correlation may be suggested: ΔV reaches a maximum for late WN star types (see Fig. 4). This correlation may be considered as an argument against the binary nature of these stars: at least for some of the 17 suspected WR+c binary systems the microvariability may be connected simply either with the instability of the WR wind (Lamontagne and Moffat, 1987; Drissen et al., 1987; Moffat, 1988b), or with the rotation of star spots. The maximum X-ray luminosity L_X is observed for the minimum value of K. This is not in contradiction with the binary nature of WR+c stars.

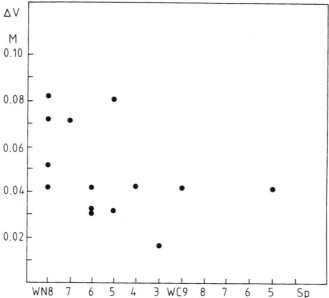

Fig. 4

Theoretical profiles of the emission lines of WR+c stars have been calculated by Antokhin (1986). The results are close to those obtained by Hatchett and McCray (1977) for X-ray binary systems. Very valuable information on the nature of WR+c systems has been obtained from *IUE* spectroscopic data (Koenigsberger and Auer, 1987; St.-Louis et al., 1989; Smith et al., 1986). Detailed investigation of the profiles of the UV lines in the spectrum of HD1922163 (St.-Louis et al., 1989) leads the authors to the conclusion that an accreting neutron star cannot be responsible for the variability of the spectrum of HD192163. A very important effect - the decrease of the colour index of HD164270 with an increase in brightness has been discovered by van Genderen and van der Hucht (1986). This may be connected with the projection of blobs of WR wind material against the hot core of the WR star. A number of publications describing the photometric and spectroscopic variability of suspected WR+c stars as due to nonradial pulsations of single WR stars has appeared (Vreux, 1985, 1987; Gosset and Vreux, 1987). The most probable candidates for binary systems in the evolutionary stage after second mass exchange in massive X-ray binaries are the systems HD197406 and HD50896. As a transition case of evolution between an X-ray binary and a WR+c binary, the object SS433 may be considered (van den Heuvel, 1981; Margon, 1984; Cherepashchuk, 1988).

According to Drissen *et al.* (1986b) HD197406 is supposed to be an X-ray quiet black hole in a WR+c binary, with the mass of the compact object about 12.4 $M_\odot$.

Intense photometric and polarimetric observations of HD50896 have been carried out by Lamontagne *et al.* (1986), Balona *et al.* (1989), and Drissen *et al.* (1989). Radial velocity variations, the linear polarisation variations and variations of X-ray radiation (Moffat *et al.*, 1982) are observed with a period of 3.766 d for HD50896. Long-term variability of HD50896 with a cycle of about 13 years is suspected (van der Hucht *et al.*, 1989). This may be connected with the precession of an accretion disk (Cherepashchuk, 1982). A possible supernova remnant formed after a supernova explosion in the HD50896 system has been discovered by Nichol-Bohlin and Fesen (1986). *IUE* and spectroscopic investigations of HD50896 have been published by Howarth and Phillips (1986) and Willis *et al.* (1986). Physical parameters of the extended atmosphere of the WR star HD50896 have been calculated by Hillier (1987) and Hamann *et al.* (1988). A rather unexpected circumstance has been noted: a genetic connection between WR+c binary systems and WR stars with enhanced oxygen lines (Barlow and Hummer, 1988). In particular, the z-distribution of the WR stars with enhanced oxygen lines seems to be quite similar to the z-distribution for WR+c stars (Rustamov and Cherepashchuk, 1986). It should be also noted that Hogg (1989) was unable to detect radio variability of the WR+c stars HD50896, HD191765 and HD192163 with amplitudes greater than 20% on timescales from 1 to 7 years. The thermal origin of the radio emission for all these WR+c stars is confirmed.

References

Abbott, D.C. and Conti, P.S. 1987, *Ann. Rev. Astron. Astrophys.* **25**, 113.

Annuk, K. 1988, in T. Nugis, I. Pustylnik (eds.), Wolf-Rayet Stars and Related Objects, Acad. of Sci. of Estonia, Tallinn, p. 144.

Acker, A., Prevot, M.-L., and Prevot, L. 1989, *Astron. Astrophys.* **226**, 137.

Antokhin, I.I. 1986, *Astron. Zh.* **63**, 1152 (= *Sov. Astron.* **30**, 680).

Antokhin, I.I. and Cherepashuk, A.M. 1989, *Pis'ma Astron. Zh.* **15**, 701 (= *Sov. Astron. Letters* **15**, 303).

Antokhin, I.I., Kholtygin, A.F., and Cherepashchuk, A.M. 1988, *Astron. Zh.* **65**, 558 (= *Sov. Astron.* **32**, 285).

Antokhina, E.A. and Cherepashchuk, A.M. 1988a, *Pis'ma Astron. Zh.* **14**, 252 (= *Sov. Astron. Letters* **14**, 105).

Antokhina, E.A. and Cherepashchuk, A.M. 1988b, *Astron. Zh.* **65**, 1016 (= *Sov. Astron.* **32**, 531).

Antokhina, E.A., Kreiner, I.M., Tremko, I., and Cherepashchuk, A.M. 1987, *Pis'ma Astron. Zh.* **13**, 417 (= *Sov. Astron. Letters* **13**, 170).

Aslanov, A.A. and Cherepashchuk, A.M. 1990, *Astron. Zh.* **67**, N6 (= *Sov. Astron.* .., ..).

Balona, L.A., Egan, I., and Marang, F. 1989, *Monthly Notices Roy. Astron. Soc.* **240**, 103.

Barlow, M.J. and Hummer, D.G. 1982, in: Wolf-Rayet Stars: Observations, Physics and Evolution, *IAU Symp. No. 99* (Dordrecht: Reidel), p. 387.

Bayramov, Z.T., Pilugin, N.N., and Usov, V.V. 1988, *Astron. Tsirk.* No. 1526, 1.

Beals, C.S. 1944, *Monthly Notices Roy. Astron. Soc.* **104**, 205.

Cherepashchuk, A.M. 1967, *Variable Stars* **16**, 226.

Cherepashchuk, A.M. 1969, *Astron. Tsirk.* No. 509.

Cherepashchuk, A.M. 1971, *Astron. Zh.* **48**, 1201 (= *Sov. Astron.* **15**, 955).

Cherepashchuk, A.M. 1975, *Astron. Zh.* **52**, 81 (= *Sov. Astron.* **19**, 47).

Cherepashchuk, A.M. 1976, *Pis'ma Astron. Zh.* **2**, 356 (= *Sov. Astron. Letters* **2**, 138).

Cherepashchuk, A.M. 1981, *Monthly Notices Roy. Astron. Soc.* **194**, 755.

Cherepashchuk, A.M. 1982, *Astrophys. Space Sci.* **86**, 299.

Cherepashchuk, A.M. 1988, in: R.A. Sunyaev (ed.), *Soviet Sci. Rev. Astrophys. and Space Phys.* **7**, 183.

Cherepashchuk, A.M. 1990, *Astron. Zh.* **67**, .. (= *Sov. Astron.* .., ..).

Cherepashchuk, A.M. and Rustamov, D.N. 1990, *Astrophys. Space Sci.* **167**, 281.

Cherepashchuk, A.M., Eaton, J.A., and Khaliullin, Kh.F. 1984, *Astrophys. J.* **281**, 774.

Conti, P.S. and Vacca, W.D. 1990, *Astron. J.* **100**, 431.

De Greve, J.P. and Doom, C. 1988, *Astron. Astrophys. Suppl.* **74**, 325.

Doom, C. 1987, *Astron. Astrophys.* **182**, L43.

Drissen, L., Moffat, A.F.J., Bastein, P., Lamontagne, R., and Tapia, S. 1986a, *Astrophys. J.* **306**, 215.

Drissen, L., Lamontagne, R., Moffat, A.F.J., Bastein, P., and Seguin, M. 1986b, *Astrophys. J.* **304**, 188.

Drissen, L., Robert, C., Lamontagne, R., Moffat, A.F.J., and St.-Louis, N. 1989, *Astrophys. J.* **343**, 426.

Eaton, J.A., Cherepashchuk, A.M., and Khaliullin, Kh.F. 1982, in: Y. Kondo, J.M. Mead and R.D. Chapman (eds.), Advances in Ultraviolet Astronomy: Four Years of IUE Research, *NASA CP*-2238, p. 542.

Eaton, J.A., Cherepashchuk, A.M., and Khaliullin, Kh.F. 1985a, *Astrophys. J.* **297**, 266.

Eaton, J.A., Cherepashchuk, A.M., and Khaliullin, Kh.F. 1985b, *Astrophys. J.* **296**, 222.

van Genderen, A.M. and van der Hucht, K.A. 1986, *Astron. Astrophys.* **162**, 109.

van Genderen, A.M., van der Hucht, K.A., and Steemers, W.J.G. 1987, *Astron. Astrophys.* **185**, 131.

van Genderen, A.M., van der Hucht, K.A., and Larsen, I. 1989, *Astron. Astrophys.* **229**, 123.

Gonzalez, J. and Rosado, M. 1984, *Astron. Astrophys.* **134**, L21.

Gosset, E. and Vreux, J.M. 1987, *Astron. Astrophys.* **178**, 153.

Hamann, W.R., Schmutz, W., and Wessolowski, U. 1988, *Astron. Astrophys.* **194**, 190.

Hamann, W.R., Wessolowski, U., Schwarz, E., Dunnebeil, G., and Schmutz, W. 1990, in: C.D. Garmany (ed.), Intrinsic Properties of Hot Luminous Stars, Proc. Boulder-Munich Workshop, *A.S.P. Conf. Series* **7**, p. 259.

Harvig, V. 1987, *Tartu Astrofüüs. Obs. Publ.* **52**, 313.

Hatchett, S.P. and McCray, R. 1977, *Astrophys. J.* **211**, 552.

van den Heuvel, E.P.J. 1976, in: P.P. Eggleton *et al.* (eds.), Structure and Evolution of Close Binary Systems (Dordrecht: Reidel), p. 35.

van den Heuvel, E.P.J. 1981, *Vistas in Astronomy* **25**, 95.

Hidayat, B., Admiranto, A.G., and van der Hucht, K.A. 1984, *Astrophys. Space Sci.* **99**, 175.

Hillier, D.J. 1987, *Astrophys. J. Suppl.* **63**, 965.

Hogg, D.E. 1989, *Astron. J.* **98**, 282.

Howarth, I.D. and Phillips, A.P. 1986, *Monthly Notices Roy. Astron. Soc.* **222**, 809.

van der Hucht, K.A. 1990, A Bibliography of Wolf-Rayet Literature 1980 - 1990 (preprint).

van der Hucht, K.A., Hidayat, B., Admiranto, A.G., Supelli, K.R., and Doom, C. 1988, *Astron. Astrophys.* **199**, 217.

van der Hucht, K.A., van Genderen, A.M., and Bakker, P.R. 1990, *Astron. Astrophys.* **228**, 108.

Hutchings, J.B. and Massey, P. 1983, *PASP* **95**, 151.

Khaliullin, Kh.F. and Cherepashchuk, A.M. 1076, *Astron. Zh.* **53**, 327 (− *Sov. Astron.* **20**, 186).

Khaliullin, Kh.F., Khaliullina, A.I. and Cherepashchuk, A.M. 1984, *Pis'ma Astron. Zh.* **10**, 600 (= *Sov. Astron. Letters* **10**, 250).

Kartasheva, T.A. and Snezko, L.I. 1985, *Astron. Zh.* **62**, 751 (= *Sov. Astron.* **29**, 440).

Koenigsberger, G. 1988, *Revista Mexicana Astron. Astrof.* **16**, 75.

Koenigsberger, G. and Auer, L.H. 1987, *PASP* **99**, 1080.

Koenigsberger, G., Moffat, A.F.J., and Auer, L.H. 1987, *Astrophys. J.* **322**, L41.

Lamontagne, R. and Moffat, A.F.J. 1987, *Astron. J.* **94**, 1008.

Lamontagne, R., Moffat, A.F.J., and Lamarre, A. 1986, *Astron. J.* **91**, 925.

Leung, K.C., Moffat, A.F.J., and Seggewiss, W. 1983, *Astrophys. J.* **265**, 961.

Lipunova, N.A. 1982, *Pis'ma Astron. Zh.* **8**, 242 (= *Sov. Astron. Letters* **8**, 128).

Lipunova, N.A. and Cherepashchuk, A.M. 1980, *Astron. Zh.* **57**, 1033 (= *Sov. Astron.* **6**, 193).

Lipunova, N.A. and Cherepashchuk, A.M. 1982, *Astron. Zh.* **59**, 944 (= *Sov. Astron.* **26**, 569).

Marchenko, S.V. 1988, *Kinematics and Physics of Celestial Bodies* **4**, 25.

Margon, B. 1984, *Ann. Rev. Astron. Astrophys.* **22**, 507.

Monderen, P., de Loore, C.W.H., van der Hucht, K.A., and van Genderen, A.M. 1988, *Astron. Astrophys.* **195**, 179.

Moffat, A.F.J. 1982, in: C. de Loore and A.J. Willis (eds.), Wolf-Rayet Stars: Observations, Physics, Evolution (Dordrecht: Reidel), p. 263.

Moffat, A.F.J. 1988a, *Astrophys. J.* **330**, 766.

Moffat, A.F.J. 1988b, in: Polarized Radiation of CS Origin (Vatican), p. 607.

Moffat, A.F.J. 1989, *Astrophys. J.* **347**, 373.

Moffat, A.F.J., Firmani, C. McLean, I.S., and Seggewiss, W. 1982, in: C. de Loore and A.J. Willis (eds.), Wolf-Rayet Stars: Observations, Physics, Evolution (Dordrecht: Reidel), p. 577.

Moffat, A.F.J. and Shara, M.M. 1986, *Astron. J.* **92**, 952.

Moffat, A.F.J., Lamontagne, R., Shara, M.M., and McAlister, H.A. 1986, *Astron. J.* **91**, 1392.

Moffat, A.F.J., Drissen, L., Lamontagne, R., and Robert, C. 1988, *Astrophys. J.* **334**, 1038.

Moffat, A.F.J., Drissen, L., Robert, C., Lamontagne, R., Coziol, R., Mousseau, N., Niemela, V.P., Cerruti, M.A., Seggewiss, W., and van Weeren, N. 1990, *Astrophys. J.* **350**, 767.

Nichols-Bohlin, J. and Fesen, F.A. 1986, *Astron. J.* **91**, 925.

Pauldrach, A., Puls, J., Hummer, D.G., and Kudritzki, R.P. 1985, *Astron. Astrophys.* **148**, L1.

Poe, C.H., Friend, D.B., and Cassinelli, J.P. 1989, *Astrophys. J.* **337**, 888.

Pollock, A.M.T. 1987, *Astrophys. J.* **320**, 283.

Prilutsky, O.F. and Usov, V.V. 1976, *Astron. Zh.* **53**, 6.

Robert, C. and Moffat, A.F.J. 1989, *Astrophys. J.* **343**, 902.

Robert, C., Moffat, A.F.J., Bastein, P., St.-Louis, N., and Drissen, L. 1989, *Astrophys. J.* **347**, 1034.

Rudy, R. and Kemp, K. 1978, *Astrophys. J.* **221**, 200.

Rustamov, D.N. and Cherepashchuk, A.M. 1986, *Pis'ma Astron. Zh.* **12**, 373. (= *Sov. Astron. Letters* **12**, 155).

Schulte-Ladbeck, R.E. and van der Hucht, K.A. 1989, *Astrophys. J.* **337**, 872.

Shore, S.N. and Brown, D.N. 1988, *Astrophys. J.* **334**, 1021.

Shylaja, B.C. 1986, *J. Astrophys. Astron.* **7**, 171.

Smith, L.F., Maeder, A. 1989, *Astron. Astrophys.* **211**, 71.

Smith, L.J., Willis, A.J., Garmany, C.D., and Conti, P.S. 1986, in: E.J. Rolfe (ed.), New-Insight in Astrophysics: 8 Years of Astronomy with *IUE, ESA SP*-263, p. 389.

Stevens, I.R. and Willis, A.J. 1988, *Monthly Notices Roy. Astron. Soc.* **234**, 783.

St.-Louis, N., Moffat, A.F.J., Drissen, L., Bastein, P., Robert, C. 1988, *Astrophys. J.* **330**, 286.

St.-Louis, N., Smith, L.J., Stevens, L.R., Willis, A.J., Garmany, C.D., and Conti, P.S. 1989, *Astron. Astrophys.* **226**, 249.

Stickland, D.J., Bromage, G.E., Budding, E., Burton, W.M., Howarth, I.D., Jameson, R., Sherrington, M.R., and Willis, A.J. 1984, *Astron. Astrophys.* **134**, 45.

Tutukov, A.V. and Yungelson, L.R. 1973, *Nauchn. Inform. Astrosov. Akad. of Sci. USSR* **27**, 58.

Tutukov, A.V. and Yungelson, L.R. 1985, *Astron. Zh.* **62**, 604, (= *Sov. Astron.* **29**, 352).

Underhill, A.B. and Fahey, R.P. 1987, *Astrophys. J.* **313**, 358.

Underhill, A.B., Yang, S., and Hill, G.M. 1988, *PASP* **100**, 1256.

Underhill, A.B. Gilroy, K.K., and Hill, G.M. 1990, *Astrophys. J.* **351**, 651.

Vanbeveren, D. 1988, *Astron. Astrophys.* **189**, 109.

Vreux, J.M. 1985, *PASP* **95**, 274.

Vreux, J.M. 1987, in: H. Lamers & C. de Loore (eds.), Instabilities in Luminous Early Type Stars, Proc. Workshop in Honour of Cornelis de Jager (Dordrecht: Reidel), p. 81.

Willis, A.J. and Garmany, C.D. 1987, in: H. Lamers & C. de Loore (eds.), Instabilities in Luminous Early Type Stars, Proc. Workshop in Honour of Cornelis de Jager (Dordrecht: Reidel), p. 157.

Willis, A.J., Howarth, I.D., Conti, P.S., and Garmany, C.D. 1986, in: A.J. Willis et al. (eds.), Luminous Stars and Associations in Galaxies, *Proc. IAU Symp.* **116**, (Dordrecht: Reidel), p. 259.

Zilayev, B.E. and Marchenko, S.V. 1989, *Tartu Astrofüüs. Obs. Teated* **89**, 139.

DISCUSSION

Vanbeveren: Your conclusion concerning the WR+c systems is very interesting. In 1988 I showed that the fraction of WR+c systems expected from the theory of close binary evolution may be lower than 5% of the total WR population, a result which was confirmed by Meurs and van den Heuvel (1989). As there are about 40 WR stars within 2.5 *kpc* from the sun, one then expects at most 2 WR stars with a compact companion.

Cherepashchuk: Yes, I agree. From our statistical investigation, only a small part of the suspected WR+c binary systems may be real binaries.

Moffat: I would add a comment myself here. Niemela has observed the second best candidate for a compact companion claimed by myself and others (Drissen *et al.*) to be a black hole, and the evidence has been weakened because it is not a run-away object probably as we once thought. This is in press, I think.

Massey: As a follow-up to Vanbeveren's question, let me note that many of the 'WR+c' orbits get an "f" rating in Batten's 8th catalogue (f: suspicious or no orbit). I think that when the semi-amplitude is very small, careful independent observations may be needed to establish whether RV variations are real and periodic.

Cherepashchuk: I agree with you. For the WR+c binaries interpretation of small amplitude radial velocity curves as orbital motions of the components and periodicity of spectral and photometric variability must be checked very carefully.

Underhill: Very often with WR components the semi-amplitude (K) is different for different lines. The question arises which line represents the motion of the center of mass? See the discussion by Underhill *et al.* (1990) for the case of CQ Cep.

Cherepashchuk: For the determination of the masses of WR stars it is necessary to use the emission lines with highest ionisation potential. In this case we can hope that for the binary systems with not very short periods the mass determination for WR components is satisfactory.

Koenigsberger: Is the IR eclipse in V444 Cyg symmetrical? Could it be due to the wind-wind collision region?

Cherepashchuk: The IR eclipse in V444 Cyg is quite symmetrical. But the IR eclipse has much more width and depth than optical ones. It is hard to understand in the model of a continuous WR wind. I suggest that a WR wind consists of many dense small blobs of matter. That allows to understand such a great dimension of the WR envelope in the IR range. The relationship between the shape of the secondary eclipse in the IR range for V444 Cyg and the shock wave around the O6 star needs further investigation.

Shylaja: I also opine that the radial velocity curves cause problems and cannot be generalized as just mentioned by Underhill. Which particular line did you choose in your amplitude *vs.* other parameters graphs?

Cherepashchuk: The autors quoted have used $HeII, NIV, NV, CIII, CIV$ emission lines. The radial velocity curves for some WR+c binaries do not correspond to the true orbital motion. You are right. So for WR+c binary systems the interpretation of radial velocity curves as orbital motion should be very careful. For checking it, we have analyzed the correlations between the values of K and other characteristics of supposed WR+c binaries.

Tony Moffat, Anatol Cherepashchuk

NEW WOLF-RAYET BINARIES

VIRPI S. NIEMELA [1,2,3]
Instituto de Astronomía y Física del Espacio
CC67, Suc 28, 1428 Buenos Aires, Argentina

ABSTRACT. Preliminary radial velocity orbits are presented for three binary systems containing Wolf-Rayet stars, namely Sk-71° 34 in the LMC, and WR8 and WR98 in our galaxy. Sk-71° 34 is found to be a WN3 + O6 double-lined binary with an elliptic orbit of period about 34 days. WR8 and WR98 both have WN type spectra with carbon lines. In WR8 the N and C lines appear to move in antiphase, while in WR98 all emission lines have the same orbital motion.

1. Introduction

Radial velocity studies of stars with Wolf-Rayet spectra are needed for a sufficiently large sample of stars in order to ascertain the binary frequency among these stars. Further studies of binaries will give us fundamental information on the masses and mass limits of Wolf-Rayet stars.

In this paper I present a radial velocity study of three stars with Wolf-Rayet spectra, all of which show velocity variations. Preliminary orbital parameters are also presented.

2. Observations

The observations were carried out at the Cerro Tololo Interamerican Observatory, Chile, with the Cassegrain spectrograph with IT attached to the 1m Yale telescope. All spectrograms have a reciprocal dispersion of 45 Å/mm. Spectra obtained before 1985, December, were recorded on photographic plates; thereafter on a CCD thru 3 images tubes.

The photographic spectra were measured for determination of radial velocities with the Grant oscilloscope engine at IAFE, Buenos Aires. The digital spectra were processed with IRAF at CTIO, La Serena.

---

[1] Senior Visitor, ESO, La Silla Observatory
[2] Member of Carrera del Investigador, CIC, Prov. Buenos Aires, Argentina
[3] Visiting Astronomer, CTIO, NOAO operated by AURA, Inc. under contract with the NSF

K. A. van der Hucht and B. Hidayat (eds.),
Wolf-Rayet Stars and Interrelations with Other Massive Stars in Galaxies, 201–206.
© 1991 *IAU. Printed in the Netherlands.*

202

3. Radial Velocity Orbits

3.1. SK -71° 34:

This hot star in the Large Magellanic Cloud has been classified by Conti and Garmany (1983) as WN3+O or O4f/WN3, depending whether the star may be a binary or not. I have obtained a total of 32 spectrograms, 9 photographic and 23 digital, of this star. The radial velocity of both absorption and emission lines is variable, and they show antiphased variations.

A period search routine applied to the observed radial velocities yields a period of 33.95 days, but other periods close to this cannot be excluded due to the limited data sample. Table 1 lists the orbital parameters of SK-71° 34, and Figure 1 shows the observed radial velocities folded in the period of 33.95 days. The radial velocities of HeII and NV emissions show much scatter, therefore only the amplitude of their variation was considered.

TABLE 1. Preliminary Orbital Parameters of Sk-71° 34

		mean abs.	NV em.
Period	(days)	33.95	
e		0.41	
Vo	(km/s)	189	219
K	(km/s)	100	157
ω	(deg)	41	
Eo	(2,440,000+)	6448.54	
a sin i	($R_\odot$)	60.8	95.5
M sin^3 i	($M_\odot$)	27.6	17.6

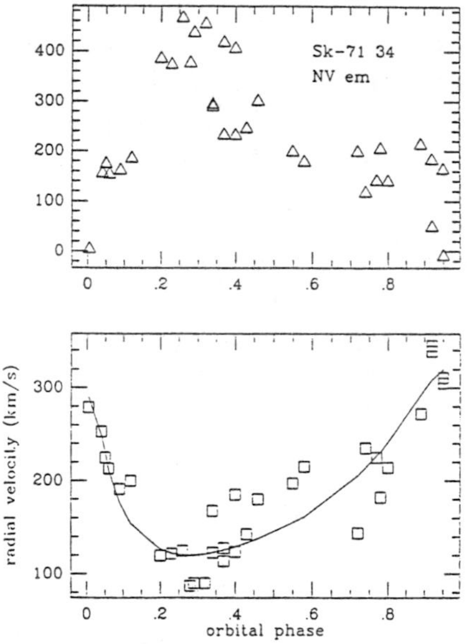

Figure 1. Observed radial velocities of mean absorption and NV emission in the spectrum of SK-71° 34. The curve represents the orbit from Table 1.

3.2. WR 8 = HD 62910:

This WR star has a WN6 type spectrum with carbon lines, it has been classified as WN6+WC4 (cf.van der Hucht et al.1981). A previous radial velocity study (Niemela 1982) showed large amplitude variations of the blue-shifted HeI 3888 absorption, and smaller amplitud velocity variations of CIV 4441 and NIV emissions with much scatter.

Present data, based on 73 photographic spectrograms, confirm the previous variations. However, several periods are still possible, the shortest one being of 38.4 days. Longer periods, of about 55 and 64 days, are equally probable, and they produce elliptic orbits. The blue-shifted absorption of HeI 3888 and CIV 4441 emission seem to move in phase, but the amplitude of the velocity variations of the absorption is much larger than that of the emission.

The NV 4603 and NIV 4057 emissions apparently follow an orbital motion opposite to the HeI 3888 absorption, as shown in Figure 2. Table 2 lists the circular orbital elements for the period of 38.4 days. These elements should be taken only as an exersice, until the true period of HD 62910 will be firmly established.

TABLE 2. Circular orbital element of HD 62910

element		HeI abs. 3888.65	CIV em. 4441.03	NV em. 4603.73
period	(days)	38.4		
Vo	(km/s)	-1238	65	135
K	(km/s)	250	72	23
Eo	(2,440,000+)	4659.6	4659.5	4639.0
a sin i	($R_\odot$)	188.6	54.3	17.3
M $\sin^3$ i	($M_\odot$)		0.8	2.6

3.3. WR 98 = HDE 318016:

The spectrum of this star has been classified as WN7+WC7. No previous radial velocity studies of WR 98 has been published. From 41 photographic spectrograms obtained for the present study, the radial velocity of WR 98 was found to be variable from one epoch to another. A period search routine applied to the observed radial velocities yields as best period 47.8 days. Other periods close to this value are also possible. Table 3 lists the preliminary orbital elements of WR 98, and Figure 3 shows the observed radial velocities together with the orbits calculated with the elements from Table 3. Within the errors, the CIII-IV 4648 and HeII 4685 emissions show the same orbital motion. The NV 4603 and NIV 4057 emission lines also move in phase with the C and He emissions, but their radial velocities show much more scatter.

TABLE 3. Circular orbital elements of WR 98

element		HeIIe 4685	CIII-IVe 4647	NVe 4603	NIVe 4057
period	(days)		48.7		
Vo	(km/s)	0	50	0	-50
K	(km/s)	68	72	107	78
Eo	(2,440,000+)	5257.6	5258.0	5257.7	5258.6
f(M)	($M_\odot$)	1.6	1.8	6.1	2.4

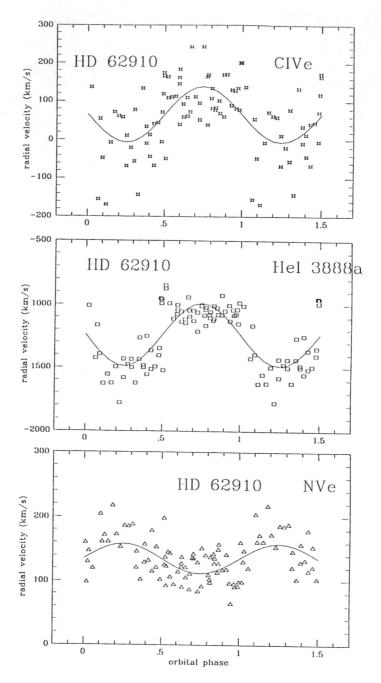

Figure 2. Observed radial velocities of HD 62910. The curves represent the radial velocity orbits from Table 2.

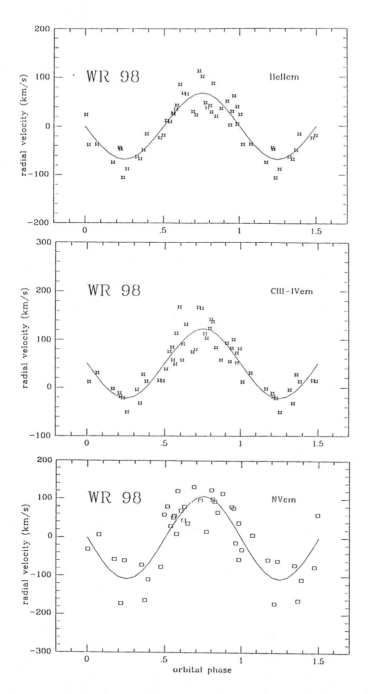

Figure 3. Observed radial velocities of WR 98. The curves represent the radial velocity orbits from Table 3.

4. Conclusions

Two new binary systems with WR components, namely Sk-71° 34 and WR98 have been discovered and another previously suspected binary, WR8, has been confirmed from extensive radial velocity studies. From the preliminary radial velocity orbit of Sk-71° 34, a WN3 + O6 binary in the Large Magellanic Cloud, rather large minimum masses are deduced for both components, about 18 and 28 $M_\odot$ for the WN3 and O6, respectively.

The other two binaries, WR8 and WR98, are galactic stars both with WN+WC type spectra. WR98 is a single-lined binary, while WR8 appears to show C and N spectral features with antiphased motion. WN stars with strong carbon lines in their spectra are few in number. Only 8 are known among all the WR stars in our galaxy and the Magellanic Clouds (cf.Conti and Massey 1989). The binary nature of WR8 and WR98 brings to 5 the number of spectroscopic binaries among these WN/C stars, thus they may all be binaries.

Acknowledgements. This paper was prepared while the author was enjoying the hospitality of ESO at La Silla Observatory in the framework of their Senior Visitor programme. My atendance to the IAU Symposium No.143 was made possible thanks to generous grants from Fundacion Antorchas, Argentina; IAU, and the local organizing committee. I am much indebted to CTIO for the use of their facilities during many observing seasons.

References

Conti, P.S., and Garmany, C.D. 1983, *Publ. Astron. Soc. Pacific*, **95**, 411.
Conti, P.S., and Massey, P. 1989, *Astrophys. J.*, **337**, 251.
van der Hucht, K., Conti,P.S., Lundstrom, I.,and Stenholm, B. 1981, *Space Sci. Rev.* **28**, 227.
Niemela, V.S. 1982, in IAU Symp. No.99, C.de Loore and A.Willis, eds. p.299.

DISCUSSION

Koenigsberger: Do you see the HeI 3888Å absorption actually moving or could the RV variations result from profile variations?
Niemela: I see it moving.

Conti: Concerning HD 69210, is the purported anti-phasing between the CIV and NV statistically significant?
Niemela: Yes.

Schmutz: It is difficult to understand that in a WN6+WC4 binary the P Cygni absorption of $HeI\lambda3888$ belongs to the WC star. "Normal" WC4 stars do not show observable HeI lines in the optical region.
Niemela: Many binaries do show the HeI absorption edges, even if the corresponding emission is not visible. On the other hand, HD 62910 is not a WC4 star.

Massey: For HD 62910, it would be useful to know if the correlation coefficient for NV vs. CIV is statistically significant or not.
Niemela: It seems to be significant at the 90% level.

MASSIVE CONTACT BINARY SYSTEMS

Kam-Ching Leung
Behlen Observatory
University of Nebraska
Lincoln, NE 68588-0111
U.S.A.

ABSTRACT In recent years very massive single stars have been found to be upward of 90 M$\odot$. Massive contact binary systems have been found among the early-type systems, but their masses are far less than those reported for single stars. The most massive component found is about 60 M$\odot$.

It is generally believed that no late-type very massive stars have been detected (Humphreys and Davidson). This may be due to the large amount of mass loss from stellar wind. Recently, several extremely long-period late-type binary systems have been found to be contact systems. Two systems, UU Cnc and 5 Cet, have their primary components with masses exceeding 40 M$\odot$, and K spectra. This result tends to suggest that close or interacting binary stars may be able to preserve the mass loss from stellar wind within the binary systems.

1. INTRODUCTION

Historically, contact binary systems were synonymous with W UMa systems. W UMa contact binaries were normally associated with short-period, late spectral-type (G or later), and low mass (one solar mass or smaller) systems. A decade and a half ago contact systems consisting of massive components were discovered. At present, we have found contact systems in all spectral-types except for M stars. (This exeception could be due to the selection effect being too faint to be discovered effectively. There are only two confirmed eclipsing systems with M spectra and they are detached pairs.) Since we are primarily dealing with Wolf-Rayet stars and massive stars in this Symposium, we will direct our attention to Wolf-Rayet, and massive type systems in this paper.

K. A. van der Hucht and B. Hidayat (eds.),
Wolf-Rayet Stars and Interrelations with Other Massive Stars in Galaxies, 207–212.
© 1991 *IAU. Printed in the Netherlands.*

2. MASSIVE EARLY-TYPE SYSTEMS

If we make an arbitrary definition of massive stars being about 20 M⊙ or more, then there are 10 early-type contact systems in the class: 1 Wolf-Rayet systems, 7 O type systems and 2 B type systems. Some of the spectra of these systems are complicated and the measurement of radial velocities are quite difficult. About half of them have asymmetric light curves. The basic observed and derived quantities of these systems are summarized in Table 1. There is still significant uncertainty concerning the spectral-type and temperature relation among the very early-type stars. Thus, it may be unwise to employ the conventional H-R diagram, effective temperature vs luminosity, since both axes involve temperature calibration. It may be more reliable to use a mass vs radius diagram. A $\log M/M⊙$ vs $\log R/R⊙$ plot of these systems is shown in Figure 1. A straight line links up the two components of the same system for identification. Notice that there is only one point for V348 Car in the diagram since the mass ratio of this system is unity. Six of the systems are found in the vicinity of the ZAMS and TAMS. This suggests that they were evolved contact systems under Case A mass loss. On the other hand, V367 Cyg, RY Sct, UW CMa, and V729 Cyg are found to be located above the TAMS line. It is suggested that these systems are evolved contact systems under Case B mass loss. These systems should be located relatively far away from the main sequence in a regular H-R diagram.

It is believed that very massive stars lose significant portions of their mass through stellar wind. The observations suggest that the very massive systems still evolve to the right of the main sequence, and go through the contact phase of binary evolution similar to the lower mass systems. The system with the latest spectral-type is V367 Cyg (a member of W Ser star) with a B8 spectra. The mass, 19 M⊙ (< 40 M⊙), and the spectral-type, B8, of this system do not violate the Humphreys and Davidson (1979) limit.

3. LATE-TYPE SUPERGIANT CONTACT SYSTEMS

Recently, several late-type (G and K) binary stars (5 Cet, UU Cnc, PW Pup, and possibly HD104901B), with periods of a hundred days or longer, were found to be contact or near contact systems (Leung 1988). The shape OF their light-curves is very similar to those of W UMa and β Lyrae systems, except for their extremely long periodicity. Radial velocity curves are available for two of the systems (UU Cnc and 5 Cet). Both of them are found to be single-line spectroscopic binary systems. Their absolute dimensions are determined by means of photometric mass ratios and the combined photometric and spectroscopic solutions. The dimensions suggests

Table 1. Massive O and B Contact Systems

Name	P (days)	Sp Type	M_H	M_C (M$_\odot$)	R_H	R_C (R$_\odot$)	Percent*	Reference
CQ Cep	1.641245	WN7 + O6	26.4	19.8	9.3	8.1	51	Leung, Moffat & Seggewiss 983
RY Sct	11.12479	O6-7 + O9.5-B0	39:	49:	37:	41.0	41	Milano, Vittone, Ciatti, Mammano & Strazzulla 1981
V729 Cyg	6.5977915	O7fIa + OfIa	59	14	33	17	31	Leung & Schneider 1978b
UW CMa	4.393407	O7 + O7f	46:	19:	34:	22:	24	Leung & Schneider 1978a
V382 Cyg	1.8955143	O7.3 + O7.7	27	19	9	8	6	Bloomer, Burke & Millis 1973
TU Mus	1.3872833	O7.8 + O8.2	23.5	15.8	8.0	6.6	6	Andersen & Grönbech 1975
AO Cas	3.523428	O8.5III + O8.5III	29	25	14	13	3	Schneider & Leung 1978
LY Aur	4.00252	O9.5III + (B0.5III)	31.6	21.0	15.6	12.6	6	Li & Leung 1985
V348 Car	5.562107	(B0.5 III) + BIII	35	35	20.5	20.5	?	Hilditch & Lloyd Evans 1985
V367 Cyg	18.5972	B8III + (A)	19	12	39	31	6	Li & Leung 1987

* Percentage of over contact

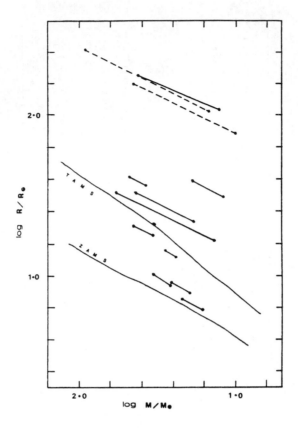

Figure 1. Log M vs Log R diagram of massive contact systems.
Broken lines represent the two solutions of 5 Cet.

Table 2.   Massive Late-Type Contact Systems

Name	P (days)	Sp Type	M_H	M_C	R_H	R_C	Reference
			$M_\odot$		$R_\odot$		
HD104901B	106	FO Ib-II					Leung 1988
PW Pup	156	F2epIab					Leung 1988
UU Cnc	96.7	K2 (4)	44	13	184	109	Lee, Nha & Leung 1991
5 Cet*	96.4	K2	46	10	158	76	Li, Leung & Ding 1988
			94	15	264	108	

* See text

that they are massive supergiant and giant stars. The basic observed and derived quantities are listed in Table 2. Note that there are two entries for 5 Cet since there are two values of photometric mass ratios, 0.21 and 0.16 from the photometric solutions. UU Cnc and 5 Cet are located very far above the TAMS line. This suggests that these systems are evolved contact systems under advanced (very long after the hydrogen exhaustion phase of single star evolution) Case B mass loss. These are very massive systems (both of the primary components are larger than 40 M⊙) and consist of late-type (**K**) supergiants. Generally, single stars (or non-interacting stars) follow the Humphreys and Davidson limit very well. However, for massive stars in an interacting system to be able to evolve to the right of an H-R diagram, the system must be able to preserve mass loss from stellar wind. It would be very interesting to investigate the common envelopes of these systems. It is logical to interpret that the common envelope of such a system must consist of a very deep convective atmosphere. It will be a real challenge to try to make theoretical models for this type of common atmosphere.

The author wishes to acknowledge the partial support of this research through grant INT-8616452 from the NSF.

REFERENCES

Adersen, J. and Gronbech, B. 1975, Astron. Astrophys., 45, 107.

Bloomer, R.H., Burk, E.W., King, C. and Millis, R.L. 1979, Bull. Amer. Astron. Soc., 11, 439.

Hilditch, R.W. and Lloyd Evans, T. 1985, Mon. Not. Roy. Astron. Soc., 213, 75.

Humphreys, R.M. and Davidson K. 1979, Astrophys. J., 232, 409.

Lee, Y.S., Nha, I.S. and Leung, K.C. 1991, Astrophys. J.,

Leung, K.C. 1988, Critical Observations Vs Physical Models For Close Binary Stars ed. by K.C. Leung (New York: Gordon and Breach), p. 93.

Leung, K.C., Moffat, A.F.J. and Seggewis, W. 1983, Astrophys. J., 265, 961.

Leung, K.C. and Schneider, D.P. 1978a, Astrophys. J., 222, 924.

Leung, K.C. and Schneider, D.P. 1978b, Astrophys. J., 224, 565.

Li, Y.F. and Leung, K.C. 1985, Astrophys. J., 298, 345.

Li, Y.F. and Leung, K.C. 1987, Astrophys. J., 313, 801.

Li, Z.Y., Leung, K.C. and Ding, Y.R. 1988, Acta Astron. Sinica, 29, 374.

Milano, L., Vittone, A., Ciatti, F., Mammano, R. and Strazzulla, G. 1981, Astron. Astrophys, 100, 59.

Popper, D.M. 1980, Ann. Rev. Astron. Astropys., 18, 115.

Schneider, D.P. and Leung, K.C. 1978, Astrophys. J., 223, 202.

DISCUSSION

Niemela: Are there other determinations of masses for the red stars that you mentioned, and how do those values compare with yours?

Leung: Both systems are simple line spectroscopic binaries. For 5 Cet, Eaton estimated a mass ratio from line profile and obtained small masses. For UU Cnc, there were estimates from photometric analysis. Unfortunately, they only searched for ratios near unity. That is, they found solutions in localized minima in the $\overline{Z} - g$ plane. The global minimum is around 0.3 instead of near unity!

Vanbeveren: I have a comment on the influence of radiation pressure on the evolution and more specifically on the equipotential surfaces in close binaries. Actually from a evolutionary point of view you are interested in what is going on during the Roche lobe overflow process. Now this process can be considered as two processes: you first have the existence of a critical surface, from where you have very huge mass loss to some kind of Lagrangian point and than you have a second process which has left the star. And how does this behave in such a binary? Now I have shown that if you account for shadow effects then the radiation pressure does not modify at all the equipotential surface. *E.g.*, if you assume the Von Zipal theorem, then you simply come up with the same equipotential surfaces as the usual of Roche lobe. But if you have two massive stars in a binary, once the matter has left the star as a consequence of this critical equipotential surface than you may not use any more, according to me at least, the classical computation done for two stars considered as two point masses, as has been done for low mass binaries in the massive binaries. Because there, in the computation of the trajectories of the particles, may be forming an accretion disk, you have to include radiation pressure, as has been done in the stellar wind theory. And if you do that, then the particles may acquire a velocity which is much larger than the escape velocity of the binary. And this was actually the main concern of the influence of radiation pressure on the Roche lobe. It has to be considered as the influence of radiation pressure on the Roche lobe overflow proces as a whole.

Kam-Ching Leung

WR BINARIES : THEORETICAL ASPECTS.

J.P. De Greve
Astrophysical Institute
Vrije Universiteit Brussel
Pleinlaan 2, B-1050 Brussels
Belgium

Abstract.

The different channels for the formation of WR-stars, as suggested in the literature, are investigated. The presently available tools, in terms of evolutionary recipies, are reviewed and the results investigated of the use of these tools, with respect to the WR-binaries.

Two of the three basic formation channels, mentioned in the literature, may serve as ways to obtain the presently observed group of WR-binaries : stellar wind mass loss in O-stars during the pre-WR phase and mass transfer in a massive close binary system. Discrepancies with observations necessitated the incorporation of stellar wind in the binary scenario.

Theoretical developments in the last years are reviewed and confronted with the observations. The necessity of mass transfer in the formation process is argued for a major fraction of the double-lined WR-binaries. Using specific evolutionary ingredients, the original parameters of the systems are calculated. We also discuss the appearance and disappearance of double-lined WR-systems as well as the existence of eccentric WR-systems after mass transfer. The evolution of a number of specific systems is discussed individually.

1. Introduction.

In what follows I will use the same ingredients for all model estimates, unless otherwise stated (in a very few cases). The ingredients are :

a) The models of Maeder and Meynet (1987) for main sequence stars. These models show an error in the calculation of the stellar wind mass loss, by using the (erroneous) formulae of de Jager and Nieuwenhuyzen (1986) instead of those of de Jager et al. (1988). We stick to them for reasons of consistency with the data in Smith and Maeder (1989), and because we couldn't recalculate the complete series in time at the Institute in Brussels.

b) The theoretical models for Wolf-Rayet stars, published by Langer (1989 a).

c) The formula for mass-dependent mass loss by Langer (1989 b).

d) The characteristics of double-lined WR binaries, as published by Smith and Maeder (1989).

I do not claim that these are the best, or even the right ingredients to model WR-stars, but making a choice has the advantage of avoiding endless discussions on right or wrong, and when largely applicated to observed systems, it can be tested for anomalies anyway. For the mass loss rate of Wolf-Rayet stars,

Bandiera and Turolla (1990) propose a relation $\dot{M} \sim M_{WR}^{2.3}$ using an analytical approach. Other (but

213

K. A. van der Hucht and B. Hidayat (eds.),
Wolf-Rayet Stars and Interrelations with Other Massive Stars in Galaxies, 213–228.
© 1991 *IAU. Printed in the Netherlands.*

often comparable) values for the masses of the components in WR binaries can be found in the papers of van der Hucht et al. (1988), Schulte-Ladbeck (1989) and Moffat et al. (1990), to name but a few.

If we adopt the common idea that WR stars are helium stars, with eventually a small hydrogen envelope, then it is easy to understand why they originate from the more massive stars. If we take $X_{at} \leq 0.3\text{-}0.4$ as a necessary (but not sufficient) condition to have a WR star, then the X_C - q_C relation during core hydrogen burning (with $q_C = M_{CC}/M^*$) for stars with masses 15 to 40 M_O shows that the smaller the mass of the star, the larger the fraction that has to be removed to get a low hydrogen abundance (figure 1).

On the other hand the stellar wind decreases strongly with lower initial stellar mass, so less mass is removed in smaller mass stars (the increase of main sequence lifetime is not able to compensate the decrease in M.

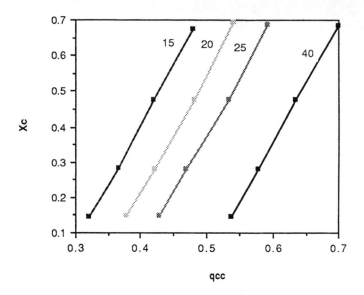

Figure 1. The central hydrogen content by mass (X_C) as a function of the relative convective core mass q_{cc}, for different initial masses of the star (given in M_O at the curves).

However, mass transfer in binaries provides a natural and efficient way (externally provocated) to remove the excess of mass . Moreover, the smaller the mass of the loser, the larger the fraction of mass that is removed (typically 30 % for 40 M_O to 60 % for 15 M_O). Therefore, this external instability can extend the mass range for WR stars downwards.

2. Observations of WR-binaries.

Because this subject is treated in more detail elsewhere, I will confine myself to a brief general description, with emphasis on quantities important for evolutionary models.

Approximately half of the WR stars are found in close binary systems. Although in many WR binaries the radial velocity variations of both components can be measured, the derivation of a spectroscopic orbit remains difficult. The WN4 binary HD90657 is an example of a system where the orbital solution depends on the line measurements.

Doom (1987) reviewed the masses of the WR (SB2) binaries, mostly using values derived by Hellings (1985, see also De Greve et al., 1987). He concluded that the average WN mass in these systems was 16.1 M_O, whereas for the WC stars the mass averages to 13.5 M_O. The respective average mass ratios are 0.52 and 0.42.

The data for the SB2 systems are taken from table 1 of the paper of Smith and Maeder (1989) that lists 27 systems. Within the framework of binary evolution only 16 are of interest, because only for those an estimate of the mass of both components is available. In fact only V444 Cyg and CQ Cep may have reliable mass estimates because the value of their inclination results from core eclipse of the WR star. This table also shows the interaction between models and observations : the inclinations of HD 63099, HD 94546 and HD 190918 (and hence the mass of the WR star) depend on the mass of the O-star. The latter is assigned according to its spectral type.

Let us first recall how good such an approximation is (or how bad).

Using the simple assumption of a main sequence location leads to the following mass ranges for the O-stars of the systems mentioned, if we adopt the models of Maeder and Meynet (other models with different stellar wind mass loss will lead to different mass ranges !) :

WR 9 - HD 63099 :	25 - 90 M_O	$(i = 66° - 37°)$
WR 31- HD 94546 :	20 - 88 M_O	$(i = 43° - 25°)$
WR 133 - HD 190918 :	15 - 85 M_O	$(i = 20° - 11°)$

Can we narrow these ranges using constraints from the luminosity ?

WR 9 (HD 63099) and WR 133 (HD 190918) are both located in a cluster or association. However, the membership of WR 9 (to Pup b or Anon) is doubtful (Smith and Maeder, 1989).

WR 133 is a member of NGC6871 (Lundstrom and Stenholm, 1984), and an $M_V = - 4.3$ is derived for the WR star. The M_V (O9Ib) = - 6.3 leads to M_{bol} = -9.4 (Conti et al., 1983), or $\log L/L_O$ = 5.66. Using the same theoretical models this results in

M (OB) = 40 M_O	M(WR) = 17 M_O	$i = 14°$

The values, determined by Schulte-Ladbeck, purely from observational elements, are M(WR) = 13 M_O and M(O) = 29 M_O, but she consideres it (correctly) as a less well known system.

In the past the WR-star was attributed masses in the range 4 to 9 M_O. Vanbeveren (1989) considers the large period and the late spectral type of the secondary as indications of a relatively late non-conservative case B mass transfer. According to this author the original primary had a mass of _40 M_O. He adopted the same mass value for the WR star, 9.1 M_O, as Massey (1982).

In any case, with the values resulting from the luminosity, the system is no longer among the binaries with small masses of the WR-star.

3. Is there a need...?

Do we need mass transfer to transform O + O type systems into WR + O systems (or WR + B systems) ?

Why is the question interesting in the first place? If mass transfer is the dominant formation process in binaries with moderate masses, we can examine the initial (and perhaps also the final) conditions of the WR stage in a more secure way, because the mass transfer process is fairly well understood (although I can give a 2 pages long list of unsolved or unsufficiently solved problems). And this can throw light on our knowledge of the other formation channels. On the other hand, if mass transfer is not really necessary, we then have to examine the conditions for the mass loss to form the presently known systems.

De Greve, Hellings and van den Heuvel (1988) argue that at least some of the known systems (WR 133, 139, 31, and possibly also systems like WR 21 and WR 9) were formed primarily through mass exchange. The basic argument is the pronounced mass ratio (_ 3) and the rather short period (4-15 days). The idea behind it is that a simultaneous evolution without rejuvenation by accretion leads to a too late spectral type for the secondary, at the time of formation of the WR star.

If the mass ratio is coupled to standard characteristics of stellar evolution of massive stars (main sequence mass loss rate by stellar wind of the order of 10^{-6} M_O/yr, helium burning lifetime equal to 10 % or less of the main sequence lifetime and WR mass loss rate an order of magnitude larger (or more) than the main sequence mass loss rate), then a lower limit can be derived for the main sequence lifetime when only stellar wind is involved. This condition is not fulfilled for a number of systems.

Moffat, Niemela and Marraco (1990) investigated the spectroscopic orbits of WC binaries in the Magellanic Clouds and the implications for WR-evolution. They argue that a mass-luminosity relation for massive stars of the type $L \sim M^3$ allows a greater spread in age and therefore easily explains the early-type main sequence companions in some WC + O binaries, without the necessity of mass transfer to rejuvenate the companion star.

Let us first look at the mass luminosity relation on the main sequence. Because both mass loss and overshooting determine the HRD-position of the star, Figueiredo et al. (1990, preprint) calculated a number of evolutionary series for different values of the stellar wind mass loss rate and the overshooting parameter (masses between 10 and 60 M_O). The central set was calculated with the mass loss rate given by de Jager et al. (1988) and the overshooting parameter a= 0.25. For the central set the relation is

$$\log L/L_O = 1.522 + 2.468 \log M/M_O \qquad (1)$$

implying an exponent of 2.5 in the theoretical models. With such models too late spectral types are encountered if mass transfer is not introduced. Other values for the mass loss rate and the overshooting lead to similar relations with only a small variation of the exponent.

But also if we make a very rough estimate of the masses of the progenitors of the WR stars in WR + O binaries, we derive conditions that unevitably lead to previous interaction between the components of several of the observed WR binaries. To show this, we proceed in the following simple way.

We use the list of masses of Smith and Maeder (1989). If we look at the evolutionary tracks in the mass range 25 to 120 M_O (using the models of Maeder and Meynet, 1986), we find a difference in luminosity between the helium burning stage and the ZAMS. In logarithm this difference varies from

$\Delta \log L/L_O = 0.25$ (25 M_O), over 0.38 (40 M_O), to 0.25 (120 M_O). For the sake of simplicity we take

$$\log L/L_O(ZAMS) = \log L/L_O(\text{helium burning}) - 0.3$$

for the whole mass range.

The luminosities of the observed WR binaries can now readily be transformed into estimated initial masses. The results are given in table 1 for systems with known luminosity of the WR star.

According to this simple exercise the WR stars in O-type binaries originate from stars in the mass range 23 to 85 M_O. The star WR 127 has the smallest progenitor. From a detailed study of V444 Cyg, with theoretical models including mass transfer, De Greve and Doom (1988) found a progenitor mass of 24 M_O, which comes close to the present solution, taking into account their use of models with larger overshooting of the convective core.

If, for each specific mass, we now equal the radius at red point (= right boundary of the main sequence) to the critical Roche radius, we obtain the minimum period for the occurrence of a case B (introduced as P^I by Plavec, 1968). Again for the sake of simplicity, we adopted $q_i = 0.7$ in the calculation. The results are given in table 1, together with the observed periods.

Six out of eleven systems have actually a period smaller than the critical one, implying interactive (and probably nonconservative) processes in their past. But even for the systems CV Ser and HD 186943 a short evolution beyond the main sequence would lead to radii large enough to meet the present critical radius. Hence, unless rightward evolution is prevented (by sudden, enhanced mass loss), most of the observed WR + O systems will have encountered an interactive phase during the pre-WR evolution. However, the mass transfer may have been highly nonconservative, with angular momentum loss determining the outcoming period.

Table 1. Luminosities and corresponding masses of the ZAMS progenitors of WR stars in double-lined binaries, and the corresponding minimum period for a case B of mass transfer. The present periods of the systems are given in the last column for comparison.

WR	System	log L/L_O	M_i/M_O	P^l(d)	P_{obs} (d)
22	HD 92740	6.3	85	7.2	80.35
155	CQ Cep	6.03	59	13.8	1.64
47	HDE 311884	6.03	59	13.8	6.34
79	HD 152270	5.7	41	17.7	8.89
11	γ^2Vel	5.7	41	17.7	78.5
138	HD 193077	5.63	38	17.0	2.32
153	GP Cep	5.59	37	16.5	6.69
113	CV Ser	5.54	35	15.0	29.71
133	HD 190918	5.43	32	12.5	112.8
139	V444 Cyg	5.26	27	8.5	4.21
127	HD 186943	5.1	23	7.0	9.55

A second (though weaker) argument comes from the internal structure and the observed HRD location. WR stars are located at the left side of the HRD (small radii) and are considered to have hydrogen poor surfaces.

In order to obtain a surface hydrogen by mass values of X $\_$ 0.3-0.4, accepted for the onset of the WR stage, enough mass must be removed. If stellar wind is the principal actor also for stars with M_i < 40 M_O, the removal must take place on the main sequence because of the small present periods.

Using figure 1, and the main sequences lifetimes given by Maeder and Meynet, we arrive at the necessary average mass loss rates given in table 2.

We also calculated the average mass loss rate from observations. This value was obtained by averaging the values, published by de Jager et al. (1988), that are closest to the various evolutionary tracks (using at least 5 values). The required mass loss rates are at least a factor 5 larger than the observed ones.

4. Evolution with mass transfer.

The present state of the art is illustrated by a recent computation by Run Huang and Ron Taam, in a paper entitled "On the nonconservative evolution of massive binary systems" (1990). The authors consider the evolution of a massive system of 40 M_O + 25 M_O through case A and case B, both in the conservative and the nonconservative mode. They take into account the enhancements of the stellar wind by tidal effects and irradiation. Only the former turns out to be important, and then only for case A evolution. They also show that the use of radiaton Roche lobes is inappropriate during the mass transfer stage. The importance of the effect of continuum radiation on the shape of the Roche lobe was previously forwarded by several authors (Schuerman, 1972 ; Kondo and Mc Clusky, 1976 ; Vanbeveren, 1976, 1978 ; Zhou and Leung, 1987). However, in practice, when one component fills its Roche lobe, the radiation field is isotropic at large optical depths within the envelope of the star and one cannot include the radiation pressure term together with gravity.

Table 2. Necessary (average) mass loss rates on the main sequence, to expose layers with X = 0.3-0.4 (log M_c, column 2), and averaged observed mass loss rates along the evolutionary tracks of Maeder and Meynet (1988).

M_i/M_O	log M_c	log M_{obs}
15	- 6.1	- 6.6
20	- 5.9	- 6.4
40	- 5.4	-6.3

Also, the optical depth of the material which is ejected through L_1, is high ($\sim$ 100 for a rate of 10^{-4} M_O/yr). The composition is X = 0.602, Z = 0.044, convection is treated using the Schwarzschild criterion, stellar wind mass loss according to the formula of Waldron (1984), slightly adapted to

$$\log M = 1.07 \log L/L_O + 1.77 \log R/R_O - 13 \qquad (2)$$

Effects of the companion's gravity, continuum radiation pressure and centrifugal force are taken into account by assuming that the rate of mass loss is inversely proportional to the average effective gravitational acceleration on the surface.

The most important features of the 'standard' cases are given in Table 3.

Table 3 : Characteristics at the beginning and the end of mass transfer, for the system 40 M_O + 25 M_O (Huang and Taam, 1990, sequences 1 and 4)

			case A (seq. 1)				case B(seq. 4)	
	M_1	M_2	P X_{c2}	X_{c1}	M_1	M_2	P X_{c2}	X_{c1}
begin	37.0	24.5	3.82 0.34	0.18	34.4	24.3	20.45 0.24	0.00
end	15.4	41.9	5.2 0.32	0.00	16.5	42.0	35.17 0.39	0.00

Sybesma (1986) investigated mass transfer in very massive systems, following case A and including overshooting and stellar wind. He found that the extent and effect of the mass transfer depend on the point in the evolution where mass transfer starts.

This aspect was examined in more detail by De Greve and Doom (1989), using the same kind of models, but concentrating on masses between 20 and 40 M_O. They calculated models of the mass-losing component through case AB of mass transfer (overshooting : Doom (1985) ; stellar wind : Lamers (1981) ; conservative mass transfer), using the code of Prantzos et al. (1986), for various masses and periods (in the range 1.5 to 4 days). From the results and similar results in the literature, several relations are derived between initial and final state, for the mass range 20 to 30 M_O. In particular, they give the maximum value of the helium convective core, which can be considered as an estimate of the mass at the onset of the WC phase. From their calculations it follows that for smaller initial periods, smaller remnant masses of the primary are obtained. For a 20 M_O star, the difference can be as large as 5.3 M_O : from M_{if} = 5.8 M_O (P_i = 1.5 d) to M_{if} = 11.1 M_O (P_i = 10 d). Final periods as small as 2.8 d are found.

Schulte-Ladbeck (1989) considered the masses of WR binaries and compared them with results from evolutionary computations. The inclinations of the WR-binaries were derived previously from polarimetric observations (see Moffat, Niemela and Marraco, 1990), hence no model dependence is found in the derivation of the masses of the components. She compared two series of computations, those published by Vanbeveren (1987, classical Schwarzschild convective core, nonconservative mass transfer) and those of

Doom and De Greve (1983, large, parametrised overshooting and conservative mass transfer). She finds a good agreement between observations and nonconservative binary models with classical cores. The models with large convective overshooting fail to describe systems with WR stars with observed masses > 10 M_O.

Although I do not want to enter in a debate on overshooting, a remark is necessary on the foregoing conclusion. It depends on the fact that models with a large amount of overshooting do not exhibit (case B) mass transfer. However, that effect is largely dependent on the formalism adopted for the stellar wind mass loss rate. In the models of Doom and De Greve Lamers (1981) formula was adopted, resulting in an absence of case B for M_i >33 M_O.

This formula gives high mass loss rates for that mass range. If instead we adopt the formalism of de Jager et al. (1988), mass loss rates are obtained a factor 2 smaller (averaged over the main sequence).

With such rates, models with masses >33 M_O will evolve to the right of the HRD, and encounter mass transfer. This will result in much larger secondaries, and remove much, if not all, of the anomaly obtained by Schulte-Ladbeck. Anyway, I fully agree with one of her other conclusions : binary evolution, especially in the large mass range, might eventually help to set a limit on the amount of required overshooting.

Overshooting of the convective core leads to larger luminosities, thus to larger mass loss rates and therefore to smaller masses at the end of the main sequence. Both effects, stellar wind and overshooting, lead to a lower fraction of mass removed from the loser during mass transfer. Therefore, the gainer will be somewhat less massive, the mass ratio less extreme and the period not so large as in the classic case. Additionally, overshooting leads to larger remnants, because the decreasing hydrogen profile is located further outside in the star. The larger remnants in turn bring down the mass of the secondary, the mass ratio and the period.

Summarizing, the additional elements (overshooting, nonconservative mass loss) are necessary to obtain larger mass ratios M(WR)/M(O) and small periods after mass transfer. With them, it is in principle possible to find a satisfying evolutionary solution for each specific WR-system. Of course, all the solutions together must show coherence in the use of the various parameters.

5. Estimating initial parameters through mass transfer.

In this section I determine the initial parameters of the systems with more or less well determined mass estimates, through the application of mass transfer, and taking into account the present status of the WR star. For the latter I considered the following possibilities :

a) For WN-stars :   1. M(WN) = M_f, corresponding to the onset of the WN phase

                           2. M(WN) = 0.5(M_f + M(He)), corresponding to a situation halfway the WN phase

with M_f the mass at the end of the mass transfer, derived from single star models by $M_f = M_{cc}(X_c = 0.3)$. M(He) is the mass of the maximum extent of the helium core. When its outer layers appear at the surface, we may observe the star as a WC star (De Greve and Doom, 1988 a).

b) For the WC stars, we assume that they are at the onset of that stage. However, as Moffat, Niemela and Marraco (1990) argue, a low mass ratio combined with an early WC type may reflect a more advanced WC state.

When severe uncertainties on the mass range are given by Smith and Maeder, we calculate the initial values and the characteristics for the extreme values. The initial mass ratio given is the one assuming conservative mass transfer. The mass loss rates are the values derived from the equation of Langer (1989 b). Where available, we also recall the masses derived earlier on the basis of the simple luminosity assumption of the WR star. The results are given in tables 4 and 5.

Table 4. Characteristics of double-lined WC binaries, and initial parameters assuming case B mass transfer, the mass loss rate is derived from Langer's (1989 b) mass dependent formula ; $M_i(L)$ is the initial mass estimate obtained from the luminosity of the WR star. All masses are in M_O.

System WR	Type	P (d)	M_{WR}	q	M_{i1}	q_i	M_{WC} (10^{-5} Mo/yr)	$M_i(L)$
9	WC 5 + O.7	14.7	10	0.28	30	0.82	3.3	
42	WC 7 + O7V	7.9	8	0.59	25	0.02	1.5	
			14	0.59	38	0.31	7.4	
113	WC 8 + O8-9	29.7	12	0.48	34	0.37	5.6	35
11	WC 8 + O9 I	78.5	17	0.53	44	0.54	13.0	41
			26	0.53	60	0.97	46.0	41
79	WC 7 + O.5-8	8.9	4	0.36	15	0.08	0.4	41
			19	0.36	48	0.92	16.0	41

Table 5. The same as table 4, but for the WN-binaries.

System WR	Type	P (d)	M_{WR}	q	M_{i1}	q_i	M_{WC} (10^{-5} M_O/yr)	$M_i(L)$
139	WN 5 + O6	4.2	10	0.39	24	0.67	3.3	27
					26	0.26	3.2	27
21	WN 4 + O4-6	8.3	10	0.52	24	0.36	3.3	
					26	0.52	3.2	
			14	0.52	31	0.54	7.5	
					34	0.46	7.6	
31	WN 4 + O8V	4.8	7	0.44	18	0.34	1.3	
					20	0.21	1.4	
127	WN 4 + O9Ib	9.6	11	0.5	26	0.45	4.3	23
					29	0.37	4.2	23
			21	0.5	43	0.88	21.0	23
					47	0.83	21.0	23
133	WN 4 + O9Ib	112.8	15	0.43	33	0.83	9.2	32
					36	0.72	9.1	32
47	WN 6 + O	6.3	37	0.9	67	0.61	86.0	59
					72	0.64	92.0	59
			49	0.9	83	0.81	170.0	59
					90	0.91	200.0	59

Nonconservative mass transfer will result in larger q_i values for a given q_{WR}, or, for given q_i, nonconservative mass transfer leads to larger q_{WR} (with q_i and q_{WR} defined by inversed ratios, to obtain always values smaller than 1 : $q_i = \dfrac{M_{2i}}{M_{1i}}$, $q_{WR} = \dfrac{M_{WR}}{M_2}$).

Figure 2 shows the variation of the mass ratio during the WR phase, for an initial system of 30 M_O + 24.6 M_O (used previously for WR9 in the conservative case) for various fractions of accreted mass (expressed in transferred mass). In the conservative case ($\beta = 1$), the mass ratio evolves from 0.4 to 0.28 at the onset of the WC stage (M_{WR} = 9.4 M_O, cfr. WR9), and reaches 0.14 at the end of the helium burning stage (M_{WR} = 4.8 M_O). Only strong mass loss from the system (more than 50 % of the mass transferred) can change this range severely. In the extreme case that no mass is accreted ($\beta = 0$), the mass ratio varies from 0.63 over 0.44 to 0.22 (remark that this extreme case means a mass loss of 25 % of the total mass at the onset of mass transfer, and 22 % of the total initial mass).

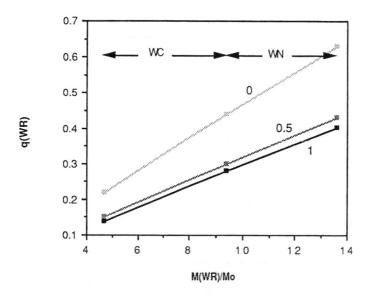

Figure 2. Evolution of the mass ratio during the WR phase, for a system of initially 30 M_O + 24.6 M_O. The WN and WC phase are indicated. The curves are labeled with β, the fraction of transferred mass accreted by the secondary component during mass transfer.

Summarizing, for a system of 30 M_O + 25 M_O, mass transfer (conservative and non-conservative) results in possible mass ratios for WN-binaries in the range 0.63-0.28, and in the range 0.43-0.14 for WC-binaries.

The behaviour of the mass ratio during the WN and WC phase depends strongly on the mass loss rate applied for the WR star (here the mass dependent formula of Langer, 1990) and the definition of the onset of the WC phase. Those two parameters also determine the length of both phases. Figure 3 shows the ratio of the WC lifetime to the helium burning lifetime as a function of the initial WR mass. The smaller this mass is, the smaller the relative duration of the WC phase is. For masses $M_{WRi} \geq 25$ M_O, the star will practically spend the whole helium burning phase as a WC star. Of course, it is possible that the WR phenomenon disappears before the end of the helium burning. This will reduce the WC phase. In figure 3 we assumed the WR phase to stop at 70 % of the helium burning lifetime, resulting in the lower curve.

Finally, we note that the mass loss rates given by the mass dependent equation of Langer, leads to results in agreement with the observations (within a factor 3), eccept for WR 133 (but what is well known for this system ?).

6. What about WR+WR systems and WR+c systems?

The existence of the former systems was predicted by Doom and De Greve (1981), as a result of the evolution of rather massive systems with mass ratios close to 1. Two candidate systems, showing spectral features of a WN and WC star, MR111 and GP Cep, were examined in more detail by Massey and Grove (1989). They found that the lines vary in phase, hence the WN and the WC characteristics are located in one star.

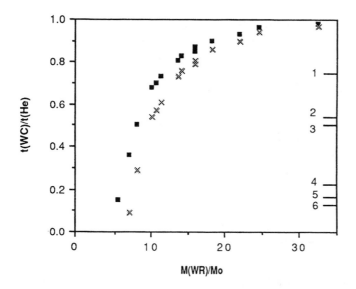

Figure 3. $t(WC)/t(WR)$ as a function of initial WR mass M_{WRi} (upper curve, with $t(WR) = t(He)$). The lower curve is calculated assuming that $t(WR) = 0.70\ t(He)$. For comparison, similar ratios, given by Arnault et al. (1989) for different regions, are shown :

(1) SN 6-7.5 kpc ; (2) SN 7.5-9.5 kpc ; (3) SN 9.5-11 kpc ; (4) LMC ; (5) 30 Dor ; (6) SMC.

For MR111 the unseen companion should have a mass $M > 15\ M_O$ and the WR-star is the less massive one. There is no sign of the companion spectroscopically because of the very large reddening of MR111, which makes it difficult to obtain an adequate signal-to-noise ratio in the blue. GP Cep turns out to be a double pair with $P = 6.7$ days for the O + WR.

The theoretical existence of the WR + WR phase in massive close binary evolution was forwarded by Doom and De Greve (1981). The essential requirements are that the remaining hydrogen burning lifetime of the mass accreting component is shorter than the WR lifetime of the primary, and that the following reversed mass transfer from secondary to primary is highly nonconservative.

We investigated the relevant timescales for $M_{1i} = 40\ M_O$ and $20\ M_O$, $q_i = 0.9$ and 0.95 and $\beta = 1$ (conservative) and 0.5.

We found that no WR + WR phase occurred for $q_i = 0.9$ (both values of β) and for $q_i = 0.95$, $\beta = 0.5$. Only in the conservative case both components become a WR star simultaneously. For the $20\ M_O$ star, this phase lasts 73000 years, during which a WC + WN pair is seen with approximate masses 4 and 11 M_O. For the $40\ M_O$ star, the phase lasts 27000 years and the masses are about $6\ M_O$ and $20\ M_O$. In both cases, the magnitude difference between the two components is about 1. This problem is investigated in more detail elsewhere (De Greve, 1990).

Vanbeveren (1988b) presented a statistical study on the distribution of different groups of massive stars, and outlined the influence of mass transfer on the relative frequency of these groups. According to this study, the number of O- and WR-stars with compact companions is very low (< 5 %). He estimates that 5 to 20 % of the supernovae (from massive stars) results from WR-stars.

7. Conservative or non-conservative ?

Nonconservative mass transfer leads to lower masses for the resulting accretion star. The influence on the period depends strongly on the amount of the angular momentum the leaving gas is taking away.

The amount of mass lost from the system during mass transfer is still a free parameter. For massive stars external mass loss was introduced a long time ago to obtain better agreement between the required initial mass ratios for WR-binaries and the observed mass ratios for massive binaries (Abbott and Conti, 1987).

Recently, Meurs and van den Heuvel (1989) examined the number of evolved early-type close binaries in the Galaxy, under the assumptions of steady state star formation, case B of mass transfer, and stellar wind mass loss.

If for the helium star $M = 6 \pm 1\ M_O$ is adopted ($M_i \geq 20-25\ M_O$) and $\beta \leq 0.5$, then the observed and predicted numbers of WR binaries within 3 kpc from the sun match. In that case the number of WR + compact star is expected to be 0.05 to 0.1 times the number of normal WR binaries. They conclude that a value of $\beta = 0.5$ can readily be adopted, leading to a lower mass limit for the WR stars of $5\ M_O$.

8. Eccentric systems.

Some of the WR binaries have large eccentricities. How does this fit into the binary evolution scheme ?

Tassoul (1987, 1988) discussed the effects of circularisation and synchronisation in early-type, detached close binaries. Synchronisation in early type stars is not only achieved by tidal friction, but also by large-scale meridional flows.

Furthermore, the hydrodynamical spin-down circularizes an eccentric orbit in a much more effective way than Zahn's (1977) radiative damping on the dynamical tide. The equation determining the circularisation takes the form

$$t_{cir}(yr) = 9.4\ 10^{4-N/4}\ \frac{(1 + q)^{2/3}}{r^2\,g}\ \left(\frac{L_O}{L}\right)^{1/4} \left(\frac{M}{M_O}\right)^{23/12} \left(\frac{R_O}{R}\right)^{5} (P_O(day))^{49/12}$$

(3)

The factor $10^{-N/4}$ represents the effect of turbulence on the spin-down mechanism. Physically, it stands for the exchange of angular momentum between the inviscid interior and a thin Ekman-type boundary layer. For early-type binaries N is a small number and we may adopt $N = 0$.

Table 6. Circularisation times t_{circ} calculated with eq. (3), for massive main sequence binaries, compared to the main sequence lifetime t_{ms}. Masses are in M_0, time in years.

M_{1i}	q_i	r^2_g	t_{circ}	t_{ms}
60	0.9	0.085	1.2 E9	3.4 E 6
	0.3		9.5 E8	
40	0.9	0.082	8.0 E 8	4.8 E 6
	0.3		6.2 E 8	
20	0.9	0.074	4.2 E8	8.7 E 6
	0.3		3.3 E 8	

We calculated the circularisation time for ZAMS binaries with primary masses 20 M_0 to 60 M_0 and mass ratios in the range 0.9 to 0.3. The values for the radius of gyration were taken from De Greve (1976), with extrapolation of the 60 M_0 value. The results, given in table 6, show that the proposed mechanism is not able to circularize the orbits of the massive O-binaries during the main sequence. The values were calculated for a period of 10 d. A period of 5 days leads to circularisation times some 40 % of those in table 5.

The circularisation times are always about two orders of magnitude larger than the main sequence lifetime. Therefore, massive binaries with a substantial eccentricity will enter the advanced stages with it. However, as shown by Tassoul $t_{circ} < t_{ms}$ is indeed obtained for smaller masses ($M_i < 10$ M_0).

A study of the origin and evolution of the eccentric, large period system WR 140 ($e = 0.8$, $P = 7.9$ yr) demonstrates that such systems can readily be formed through mass transfer, without circularisation of the orbit, at least not through effects of stellar wind, mass transfer or tidal forces (De Greve, poster, this symposium).

9. Appearance and disappearance of double-lined WR binaries.

For a WR binary to be observable as a double-lined binary, the visual magnitudes of the components must have approximately the same value.

De Greve et al. (1988) describe the characteristics of a system of initially 40 M_0 + 20 M_0, based on models of Doom (1985), at the beginning and end of core helium burning. Using a B.C. of -4.5 (Smith and Maeder, 1989) for the WR star and a B.C. of -3.0 for the 19 M_0 main sequence companion, we can readily arrive at the visual magnitudes :

Begin WR :   M(WR) = 27 M_0, M_{vis} = - 5.5   M(O) = 19 M_0, M_{vis} = - 4.4

End WR :   M(WR) = 13 M_0, M_{vis} = - 4.7   M(O) = 19 M_0, M_{vis} = - 4.5

In this case, the companion star, a O9.5-BO star, will be hardly visible in the spectrum at the onset of the WR phase, but has about the same visual magnitude at the end.

Such a system might start its WR state as a single-lined WN binary ($f(M) = 3.24 \sin^3 i$), and show up as a double-lined WR+O system when the WR star progresses to the WC state. For a larger initial mass ratio, the reverse may take place (cfr. poster of de Loore and De Greve), i.e the system might first evolve as a double lined WN binary, and lateron, during the WC phase, turn into a single-lined O-type binary with a WC component. The same problem was also investigated by De Greve et al. (1988), when investigating the absence of observed WR + B binaries.

More generally, if we assume as before, BC -4.5 for the WR star, and -3.5 for the O type companion (- 3 is more appropriate in case of an early type B companion), we can relate the luminosities to the visual magnitudes M_v through

Table 7. Luminosity difference Δ log L = log L(WR) - log L(O) at the beginning of the WR phase and the WC phase (resp. Δ log L(i) and Δ log L(WC)), and at the end of the WR phase (Δ log L(e)), with t(WR) = t(He). The various solutions for each system correspond to the results in tables 3 and 4.

WR	Δ log L(i)	Δ log L(WC)	Δ log L(e)
9	0.07	- 0.13	- 0.90
42	0.99	0.58	0.12
	0.71	0.51	- 0.37
113	0.54	0.32	- 0.46
11	0.21	0.12	- 1.10
	0.47	0.30	- 0.69
79	0.49	0.13	- 0.17
	0.18	0.04	- 1.03
139	0.10	- 0.31	- 0.75
	0.59	0.20	- 0.30
21	0.40	- 0.01	- 0.45
	0.82	- 0.11	- 0.61
	0.34	0.01	- 0.64
	0.47	0.19	- 0.56
31	0.27	- 0.27	- 0.46
	0.50	0.0	- 0.27
127	0.35	- 0.04	- 0.54
	0.50	0.17	- 0.44
	0.17	0.0	- 0.98
	0.24	0.09	- 0.95
133	0.12	- 0.15	- 0.90
	0.24	0.0	- 0.82
47	0.41	0.41	- 1.04
	0.55	0.52	- 0.84
	0.56	0.51	- 0.80
	0.35	0.35	- 1.13

$$\Delta M_V = M_{V,WR} - M_{V,o} = 2.5(\log L/L_0(O) - \log L/L_0(WR)) + 1 \qquad (5)$$

Applying this to the calculations for the individual systems in section 5, using the mass-luminosity law given in section 3, allows to verify the spectral visibility. We assume that the O-type companion becomes unobservable when the magnitude difference is about 1 magnitude or larger (Hynek, 1961; Massey, 1982). The systems will then be seen as a single-lined WR binary if $\Delta M_V \leq - 1$, or

$$\log L/L_0(WR) \geq \log L/L_0(O) + 0.8. \qquad (6)$$

Using relation (6), we calculated the luminosity difference for the systems in table 4 and 5, at the onset of the WR phase, the onset of the WC phase and at the end of the WR phase (= end of the helium burning phase). The results are given in table 7. for the first two phases, the difference Δ log L = log L(WR) - log L(O) remains positive for the majority of the solution (mostly between 0.5 and 0.2, except for WR 42, with values of 0.99 and 0.71 at the onset), whereas this difference is negative at the end of the WR phase (and exceeding - 0.8 for the systems WR 133, WR 47, and partly for WR 11 and WR 79). One can make the same reasoning in the case that the O-star becomes visually more luminous than the WR star (though the WR star may remain visible for a larger magnitude difference through its emission lines). In that case, the WR star will be unobservable during the last part of the WC phase.

The fraction of time spent as a single-lined O-type binary (with mass function typically $0.1 \sin^3 i$) increases for larger masses of the WR star, larger initial mass ratios, or small mass loss from the system. For further details I refer to the contribution of de Loore and De Greve (this volume).

Acknowledgement.

This work has been supported by the National Fund of Scientific Research (NFWO - Belgium) under grant No. S 2/5 - LV. E96.

References.

Abbott, D.C., Conti, P.S. : 1987, Ann. Rev. Astron. Astrophys. 25, 113

Bandiera, R., Turolla, R.: 1990, Astron. Astrophys. 231, 85

Conti, P.S., Garmany, C.D., de Loore, C., Vanbeveren, D. : 1983, Astrophys. J. 274, 302

De Greve, J.P. : 1976, Ph. D. thesis, V.U.B. Brussels

De Greve, J.P. : 1990, Astron. Astrophys. (submitted A & A)

De Greve, J.P., Doom, C.: 1988 b, Astron. Astrophys. 200, 79

De Greve, J.P., Doom, C.: 1988 a, Astron. Astrophys. Suppl. Ser. 74, 325

De Greve, J.P., Hellings, P., van den Heuvel, E.P.J.: 1988, Astron. Astrophys. 189, 74

de Jager, C., Nieuwenhuijzen, H., van der Hucht, K.A. : 1986, in Luminous Stars and Associations in Galaxies, IAU Symp. 116, eds. C. de Loore, A.J. Willis, P. Laskarides (D. Reidel, Dordrecht), p. 109

de Jager, C., Nieuwenhuijzen, H., van der Hucht, K.A. : 1988, Astron. Astrophys. Suppl. Ser. 72, 259

Doom, C. : 1985, Astron. Astrophys. 142, 143

Doom, C.: 1987, Rep. Progress in Physics 50, 1491

Doom, C., De Greve, J.P. : 1982, IAU Symp. 99 "Wolf-Rayet Stars: observations, physics, evolution", eds. C. de Loore and A. Willis, (Dordrecht: Reidel), p.403

Doom, C., De Greve, J.P.: 1983, Astron. Astrophys. 120, 97

Figueiredo, J., De Greve, J.P., de Loore, C. : 1990, submitted A & A

Hellings, P.: 1985, Ph. D thesis, University of Brussels

Huang, R.Q., Taam, R.E.: 1990, Astron. Astrophys.

Hynek, J.: 1961, Phys. Rev.

Kondo, Y., Mc Cluskey, G.E. : 1976, in Structure and Evolution of Close Binary Systems, IAU Symp. 73, eds. P. Eggleton, S. Mitton and J. Whelan (Dordrecht : Reidel), p. 277

Lamers, H.J.G.L.M. : 1981, Astrophys. J. 245, 593

Langer, N.: 1989a, Astron. Astrophys. 210, 93

Langer, N. : 1989b, Astron. Astrophys. 220, 135

Lundstrom, I., Stenholm, B. : 1984, Astron. Astrophys. Suppl. 58, 163

Massey, P.: 1982, in "Wolf-Rayet Stars : observations, physics and evolution", IAU Symp. 99, C. de Loore, A. Willis (eds.), D. Reidel Publ. Co., Dordrecht, p. 251

Massey, P., Grove, K.: 1989, Astrophys. J. 344, 870

Maeder and Meynet, 1987, Astron. Astrophys. 182, 243

Meurs, E.J.A., van den Heuvel, E.P.J.:1989, Astron. Astrophys. 226, 88

Moffat, A.F., Niemela, V.S., Marraco, H.G.: 1990, Astrophys. J. 348, 232

Paczynski, B.: 1967, Acta Astron. 17,193

Plavec, 1968, Adv. Astron. Astrophys. 6, 201

Prantzos, N., Doom, C., Arnould, M., de Loore, C. : 1986, Astrophys. J. 304, 695

Schuerman, D.W. : 1972, Astrophys. Space Sci. 19, 351

Schulte-Ladbeck, R.E.: 1989, Astron. J. 97, 1471

Smith, L., Maeder, A.: 1989, Astron. Astrophys. 211, 71

Sybesma, C.: 1986, Astron. Astrophys. 168, 147

Tassoul, J.L. : 1987, Astrophys. J. 322, 856

Tassoul, J.L. : 1988, Astrophys. J. 324, L71

Vanbeveren, D. : 1976, Astron. Astrophys. 54, 877

Vanbeveren, D. : 1978, Astrophys. Space Sci 57, 41

Vanbeveren, D.: 1988a, Astron. Astrophys. 189, 109

Vanbeveren, D.: 1988b, Astrophys. Space Sci. 149, 1

Vanbeveren , D.: 1989, Astron. Astrophys. 224, 93

Vanbeveren, D., De Greve, J.P.:1979, Astron. Astrophys. 77, 295

van der Hucht, K.A., Hidayat, B., Admiranto, A.G., Supelli, K.R., Doom, C.: 1988, Astron. Astrophys. 199, 217

Waldron, W.L. : 1984, in The Origin of Non-Radiative Heating/Momentum in Hot Stars, ed. A.B. Underhill, A.G. Michalitsianos (Washington : NASA c.p. 2358), p. 95

Zahn, J. : 1977, Astron. Astrophys. 57, 383

Zhou, H.N., Lemy, K.C. : 1987, Astrophys. Space Sci. 141, 257

DISCUSSION

Conti: It appears that some mass exchange, even in eccentric systems, should lead to matter appearing on the O star companion. For example, γ Vel, a WC star, should have reduced nitrogen rich material on the surface of its O supergiant companion. This is not observed. Do you have an idea as to why we do not see any anomalies?

De Greve: During accretion various mixing processes (due to rotation or a Rayleigh-Taylor instability resulting from the μ-inversion) may lead to almost normal abundances (*cf.* De Greve and Cugier, 1989).

Cherepashchuk: What can you say about the formation of WO+O binary systems?

De Greve: My study did not look into that. However, from the present results WO+O are formed from the most massive systems.

Vanbeveren: I would like to comment on the effect of radiation pressure on the critical Roche lobe. This effect does not affect the Roche lobe of the loser, but it becomes important once the mass transfer sets in.

De Greve: Huang and Taam argue just the contrary. Once the primary fills its Roche volume, the radiation is isotropic to a large optical depth in the envelope, so one cannot include the radiation pressure term together with gravity. On the other hand, the stream surrounding the secondary has a large optical depth due to electron scattering (typical 100 for $\dot{M} \simeq 10^{-4} M_\odot yr^{-1}$), preventing radiation from this star to exert a substantial force on a test particle outside the two components.

Underhill: If you accept the proposal that WR stars are young stars just coming into the ZAMS band, then any combination of O/B absorption line massive star with a less massive WR star can easily occur. Once the WR spectrum star loses its magnetic field then you will have two absorption line stars in a binary system. Because the more massive component evolves faster than the other star it will always be past a possible WR stage before the less massive star.

Maeder: I am always surprised that people in binary evolution consider, apart from mass transfer, that the components in a binary system evolve like single stars. Long before mass transfer occurs, binary interaction could drive internal mixing by tidal interaction and the resulting baroclinic instabilities.

De Greve: I agree that if such effect works efficiently on a timescale comparable to the main sequence, it should be taken into account.

Schulte-Ladbeck: I have a question about the circularization of binaries. EZ CMa has a short period of about 3.8 days, a rather high excentricity of about 0.3, and is also supposed to have a compact companion. Could you comment on its evolution in the framework of your models, please?

De Greve: The short-period eccentric binaries remain a problem, because for short periods (and small eccentricity) tidal circularization should work quite efficiently. But it is still possible, in view of the presence of a compact companion, that the preceding supernova explosion that produced the latter, provoked such a large eccentricity that it is still not reduced to zero.

Smith, Lindsey: To deduce the initial mass, you assume the WC stars have the mass of the He-burning convective core. However, the WC star evolves rapidly by mass loss from a starting value as high as $40M_\odot$ down to 5 or $10M_\odot$ where it becomes a supernova. You will therefore have underestimated the initial masses.

De Greve: I agree with the argument, and I said so in my talk. Adopting an advance WC stage will indeed increase the upper limit to the initial masses.

Moffat: I think the only way to really answer the question of mass transfer is to do hydrodynamics. Benz is starting this, and others are too, in Belgium I believe, with the 3-D SPH code and for V444 Cyg there are very preliminary calculations: density plots with velocity vectors. In the present V444 Cyg system the mass transfer is zero. I find it difficult to believe that in its predecessor, in which the wind strengths will be even more similar so the interface will be more intermediate between the two stars, that there mass transfer actually took place. That is something I do not understand.

DECODING OF THE LIGHT CHANGES IN WOLF-RAYET ECLIPSING BINARIES:
AN APPLICATION TO HD 5980 IN THE SMALL MAGELLANIC CLOUD

J. BREYSACHER
European Southern Observatory
Karl-Schwarzschild-Str. 2, D-8046 Garching bei München, F.R.G.

C. PERRIER
Observatoire de Lyon
Avenue Charles André, F-69561 Saint Genis Laval, France

ABSTRACT. A method of light curve analysis is described which allows the
study of an eccentric partially-eclipsing system containing one component
possessing an extended atmosphere. The effects of transparency as well as
limb-darkening are taken into account. Preliminary results obtained for
the Wolf-Rayet eclipsing binary HD 5980 in the SMC are presented.

1. INTRODUCTION

HD 5980≡SMC/AB 5 (Azzopardi and Breysacher, 1979) is located in NGC 346,
the largest H II region of the Small Magellanic Cloud. The eclipsing
nature of the star was recognized by Hoffmann et al. (1978) but the
correct orbital period, P = 19.266 ± 0.003 days, was found by Breysacher
and Perrier (1980). The obtained light curve revealed a strongly
eccentric orbit: e=0.47 for i=80°, however, the shape of this light curve
was not defined well enough to allow a detailed quantitative analysis.
    The relatively long period together with the large eccentricity making
HD 5980 a potentially interesting object in which to study the structure
of a W-R envelope, numerous new photometric observations in the Strömgren
system were carried out in order to significantly improve the light
curve. What has been achieved in this respect is presented in Figure 1.

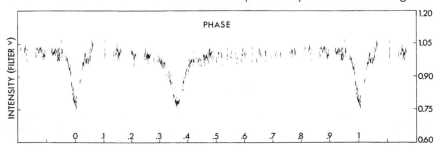

Figure 1.   Light curve in the Strömgren v band of SMC/AB 5 (705
observations) normalized to I=1 around apastron, i.e. for 0.5 ≤ Φ ≤ 0.7 .

K. A. van der Hucht and B. Hidayat (eds.),
Wolf-Rayet Stars and Interrelations with Other Massive Stars in Galaxies, 229–235.
© 1991 IAU. Printed in the Netherlands.

Concerning the interpretation of the data, as none of the existing "tools" turned out to be really suited to our purpose, i.e. the decoding of the light changes of a **partially-eclipsing system** characterized by a **strong eccentric orbit** and containing one component with an **extended atmosphere**, a non-classical approach aiming at the solution of light curves in the frequency-domain (cf. Kopal, 1979; Smith and Theokas, 1980) was attempted.

2. ANALYSIS OF THE LIGHT CHANGES IN THE FREQUENCY-DOMAIN

2.1. The basic equations

Reference being made to Kopal's fundamental work (cf. Kopal, 1979), let us first consider an eclipsing system which consists of two spherical stars revolving around the common centre of gravity in circular orbits, and appearing in projection on the sky as uniformly bright discs. When star 1 of luminosity L_1 and radius r_1 is partly eclipsed by star 2 of luminosity L_2 and radius r_2, then the brightness l of the system (maximum light between minima taken as unit) is given by

$$l(r_1, r_2, \delta, J) = 1 - \int_A J(r) \, d\sigma \qquad (1)$$

where δ is the apparent separation of the centres of the two discs and J represents the distribution of brightness over the apparent disc of the star undergoing eclipse, of surface element $d\sigma$. The assumption that star 1 is uniformly bright gives

$$J(r) = \frac{L_1}{\pi r_1^2} \qquad (2)$$

Combining equations (1) and (2) we obtain for the "loss of light" suffered by the system when an area $A(r_1, r_2, \delta)$ of star 1 is eclipsed

$$1 - l(r_1, r_2, \delta, J) = \frac{L_1}{\pi r_1^2} \int_A d\sigma \qquad (3)$$

As proposed by Kopal, let us focus our attention on the area subtended by the light curve in the $1 - \sin^{2m}\theta$ coordinates (m=1,2,3,...), where θ denotes the phase-angle. The areas A_{2m} between the lines l=1 and the actual light curve l, from the eclipse minimum $\sin^{2m}\theta = 0$ to the first contact $\sin^{2m}\theta_1$, are given by the integrals

$$A_{2m} = \int_0^{\theta_1} (1-l) \, d(\sin^{2m}\theta) \qquad (4)$$

hereafter referred to as **moments of the eclipse**, of index m.

Kopal (1979) has shown that it is possible:

a) to invert this relationship to determine the elements of the system in terms of the moments A_{2m} that can be empirically obtained from the data.

b) to extend this treatment to the case of a partial eclipse of a limb-darkened star by a star surrounded by an extended atmosphere.

A convenient mathematical solution of the problem of atmospheric eclipses is due to Smith and Theokas (1980). To take into account the transparency effects a function F(s) is introduced in equation (1), so that the total amount of light emitted to the observer is now

$$l(r_1, r_2, \delta, J, F) = 1 - \int_A J(r) \, F(s) \, d\sigma \qquad (5)$$

Considering that when a star with an extended atmosphere eclipses an ordinary one it may be advantageous to weigh the data nearer mid-minima, Smith and Theokas (1980) also introduce the moments B_{2m} defined as

$$B_{2m} = - \int_0^{\theta_1} (1-l) \, d(\cos^{2m}\theta) \qquad (6)$$

The expressions for the A_{2m} and B_{2m} moments have been derived by Smith and Theokas (1980) for $m=1\rightarrow4$ and $m=1\rightarrow5$, respectively. We only give here, as an example, the final forms obtained for A_2, A_4 and B_2, B_4

$$A_2 = I_1 R_1 r_2^2 \csc^2 i + \cot^2 iQ , \qquad (7)$$

$$A_4 = I_1 R_2 r_2^4 \csc^4 i - 2I_1 R_1 r_2^2 \csc^2 i \cot^2 i +$$
$$+ I_2 R_1 r_1^2 r_2^2 \csc^4 i - \cot^4 iQ , \qquad (8)$$

$$B_2 = \lambda - \csc^2 i(P - I_1 R_1 r_2^2 - \psi_1) , \qquad (9)$$

$$B_4 = \lambda - \csc^4 i(P - 2I_1 R_1 r_2^2 + I_2 R_1 r_1^2 r_2^2 + I_1 R_2 r_2^4 - \psi_2) , \qquad (10)$$

where i is the orbit inclination. The general expressions for the coefficients $I_m(r_1, J)$, $R_m(r_2, F)$, $Q(r_1, r_2, J, F)$, $P(r_1, r_2, J, F)$, $\psi_m(r_1, r_2, i, J, F)$ and λ can be found in the paper by Smith and Theokas (1980).

2.2. The transparency and limb-darkening functions

While for the transparency function F(s) of the eclipsing W-R star, of radius r_0, Smith and Theokas (1980) simply adopt

$$F(s) = F_y(r_0, \upsilon) = y \left[1 - \upsilon\left(\frac{s}{r_0}\right)^2 \right] \qquad \text{for} \quad s < r_0 \qquad (11)$$

where υ is the coefficient of transparency, we have adopted a law of the form (see Figure 2)

$$F(s) = F_{1-y}(r_3, 0) + F_y(r_2, \upsilon) \qquad (12)$$

where the radius of the opaque core of the W-R star is r_3 and that of the extended eclipsing envelope, r_2.

For the brightness distribution J(r) over the W-R disc, a law very similar to that of the transparency function was taken

$$J(r) = J(0) \left[J_{1-y}(r_3', 0) + J_y(r_2', u) \right] \qquad (13)$$

where J(0) is the central surface brightness and u the coefficient of limb-darkening. J_y is defined as

$$J_y(r_0,u) = y \left[1 - u\left(\frac{r}{r_0}\right)^2\right]^2 \tag{14}$$

When the W-R star is eclipsed the radius of the core, assumed to be of uniform brightness, is now r_3' and that of the limb-darkened envelope r_2'.

Using these laws of transparency and limb-darkening, we have then derived the corresponding expressions for I_m, R_m, Q, P and ψ_m and hence the final equations for the moments A_{2m} and B_{2m}.

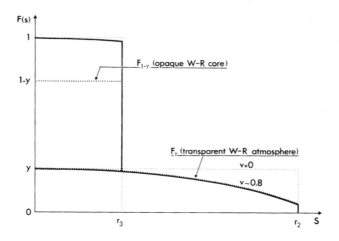

Figure 2. Opacity distribution across the disc of the W-R component.

2.3. The orbital eccentricity

In the case of an elliptical orbit the problem still consists of a determination of the elements of the eclipse from the moments A_{2m} (B_{2m}) derived from the light curve, but taking into account the eccentricity e and the longitude ω of periastron.

In the definition of the A_{2m}'s areas, the phase-angle θ is no longer identical with the mean anomaly M but rather a linear function of the true anomaly v

$$\theta = v + \omega - \frac{\pi}{2} \tag{15}$$

the element $d(\sin^{2m}\theta)$ of integration in the equation defining the moments A_{2m} becomes

$$d(\sin^{2m}\theta) = d\left[\cos^{2m}(v+\omega)\right] \tag{16}$$

As a consequence, the empirical values of A_{2m} cannot be ascertained from the observed data until a proper conversion into the true anomalies has been done. This can be accomplished by a resort to the expansion of elliptic motion (cf. Danjon, 1959)

$$v = \omega + M + (2e - \frac{1}{4}e^3)\sin M + (\frac{5}{4}e^2 - \frac{11}{24}e^4)\sin 2M + \frac{13}{2}e^3 \sin 3M + \ldots (17)$$

The empirical "elliptic" moments of the light curve then furnish the elements of the eclipse exactly as in the "circular" case, care being taken only that the resulting values of the radii have to be reduced to the same unit of length.

For a given value of the inclination i, $\Delta\Phi$ being the phase displacement of the minima, e and ω can be derived by means of the following equations

$$\Delta\Phi = \pi + 2e \left[1 + \csc^2 i - \frac{e}{2} (\frac{8}{3}\cos^2 \omega - 2) \right] \cos\omega \qquad (18)$$

$$e\sin\omega = \frac{d_2 - d_1}{4\sin(\frac{d_1 + d_2}{4})} \{ 1 - \frac{\cot^2 i}{\sin^2(\frac{d_1 + d_2}{4})} \}^{-1} \qquad (19)$$

where d_1 and d_2 are the respective durations of the primary and secondary stellar core eclipses, determined from the light curve.

It is important to note that, although each separate half-eclipse provides a self-reliant solution for the elements, due to the present composite model adopted for the W-R star (Figure 2), with in particular different radii introduced when this component is seen either as an eclipsing or an eclipsed disc, the complete determination of the elements necessarily requires a combination of the solutions furnished by the descending and ascending branches of both minima.

3. APPLICATION TO SMC/AB 5

Using this method we have analysed the light curve of SMC/AB 5 obtained with the Strömgren v filter (Figure 1). The elements for the primary eclipse (O star in front) and the secondary eclipse (W-R star in front) are derived from the moments A_{2m} and B_{2m}, respectively. The empirical values of the "elliptic" moments are determined from smooth curves resulting from a spline fit on the observed points. The solutions are obtained by inversion of Equations (7) to (10) with a Newton-Raphson method for non-linear equations. The ill-defined ascending branch of the primary minimum is, however, not included in the calculation.

No solution is found for i < 86°; the results obtained for i = 86° are summarized in Table 1. All radii are reduced to the semi-major axis of the relative orbit. For the radius r_1 and the luminosity L_1 of the O star as well as for the radius r_3 of the eclipsing W-R core, mean values have been taken as this did not imply any "a priori" assumption concerning these particular elements. SMC/AB 5 being classified WN3+O7: (Breysacher et al., 1982), we have adopted u=0.3 for the limb-darkening coefficient of the O star (cf. Klinglesmith and Sobieski, 1970).

TABLE 1. Elements of the eclipse for i = 86° (e = 0.324, ω = 133°)

	PRIMARY (descending branch)	SECONDARY (descending branch)	SECONDARY (ascending branch)
O star	←--------------- r_1 = 0.163 ± 0.007 ----------------→		
		←---- L_1 = 0.410 ± 0.034 ----→	
WR star	r'_3 = 0.120 ± 0.009	←---- r_3 = 0.112 ± 0.007 ----→	
	r'_2 = 0.227 ± 0.012	r_2 = 0.245 ± 0.018	r_2 = 0.332 ± 0.016
	L_2 = 0.257 ± 0.011		
	y = 0.18 ± 0.06	y = 0.06 ± 0.04	y = 0.16 ± 0.03
	u = 0.5 ± 0.3	υ = 0.37 ± 0.15	υ = 0.28 ± 0.15

The results presented here, which relate to the v filter data only, are obviously too preliminary to allow a thorough discussion of the geometry of the system, nevertheless, some interesting conclusions can already be drawn regarding SMC/AB 5.

1. The size of the W-R core does not change significantly between the primary and secondary eclipses, i.e. when the star is seen as an eclipsed or as an eclipsing disc.
2. The W-R envelope appears highly asymmetrical when occulting the O star. Very different values for r_2 and y are indeed furnished by the descending and ascending branches of the secondary minimum.
3. The fact that L_1+L_2 is far from unity indicates that there exists very likely a third unresolved component in the line of sight, a conclusion which seems to be also supported observationally (Massey et al., 1989).

Absolute radii may be estimated by assuming that the mass-luminosity relation for W-R stars given by Maeder and Meynet (1987) applies to SMC/AB 5. With M_v = -7.3 (Breysacher, 1988) the sum of the masses derived for the O and W-R components, 76.4 $M_\odot$, leads to a semi-major axis of 0.597 AU for this binary. The resulting values for the radii are: 20.9 $R_\odot$ for the O star, about 15 $R_\odot$ for the W-R core and 30 to 40 $R_\odot$ for the W-R envelope.

Although of fundamental importance, an error analysis is beyond the scope of this short communication. The subject will be treated exhaustively in a forthcoming article.

REFERENCES

Azzopardi, M., Breysacher, J.: 1979, Astron. Astrophys. **75**, 120.
Breysacher, J.: 1988, Thèse, Université de Paris VII.
Breysacher, J., Moffat, A.F.J., Niemelä, V.S.: 1982, Astrophys. J.
  257, 116.
Breysacher, J., Perrier, C.: 1980, Astron. Astrophys. **90**, 207.
Danjon, A.: 1959, Astronomie Générale, 2nd edition, Albert Blanchard,
  Paris, 1980.
Hoffmann, M., Stift, M.J., Moffat, A.F.J.: 1978, Publ. Astron. Soc.
  Pacific **90**, 101.
Klinglesmith, D.A., Sobieski, S.: 1970, Astron. J. **75**, 175.
Kopal, Z.: 1979, Language of the Stars, Reidel Publishing Company,
  Dordrecht.
Maeder, A., Meynet, G.: 1987, Astron. Astrophys. **182**, 243.
Massey, P., Parker, J.W., Garmany, C.D.: 1989, Astron. J. **98**, 1305.
Smith, S.A.H., Theokas, A.C.: 1980, Astrophys. Space Sci. **70**, 103.

DISCUSSION

Moffat: Maybe you have seen the poster outside by myself, Niemela and others that the polarization curve, one of the first with an elliptical orbit solution, gives an eccentricity a little bit different than yours, 0.22 ± 0.03-4. It is in the right direction, it is a smaller value.

Jacques Breysacher, Werner Schmutz

IUE AND MASSIVE BINARIES

D. J. Stickland
Space Science Department
Rutherford Appleton Laboratory
Chilton, Didcot
Oxfordshire, OX11 0QX, UK

ABSTRACT. The role of IUE in the determination of orbital elements of massive stars is reviewed, emphasising the complementary value of ultraviolet observations when combined with archival optical data. Results on eight systems are described and the potential of the IUE archive for radial velocity work in general is advocated.

1. Introduction

The factor dominating, above all others, the evolution of a star is its mass; and the only certain way of deriving stellar masses is through the determination and interpretation of binary star orbits. In the case of the populous low-mass stars, statistical analysis of large quantities of such data is feasible, but for the rarer massive stars, it becomes important to study closely the individual cases, in combination with other astrophysical data, in order to estimate as accurately as possible the masses.

Spectroscopic binary work has long been the province of optical astronomers, as inspection of the *Eighth Catalogue* (Batten *et al.* (1989)) will attest. However, the archive of high-resolution spectra secured by the International Ultraviolet Explorer (IUE) now offers an additional, and in many ways complementary, source of data to be exploited. As this archive continues to build up, so it becomes increasingly worthwhile to investigate ways of using it to firm up our data base on the masses of the biggest stars.

2. Why use IUE?

Astronomy is very prone to fashion and I think it is fair to say that the determination of spectroscopic binary star orbits is not one of the trendy pursuits at the present time. So why waste a uniquely valuable facility like IUE getting spectra for something that can surely be done from the ground? Part of the answer, of course, is that virtually all of the binary orbit work done so far has been undertaken using archival data which were secured initially for some far more worthy(?) cause.

However, there are valid reasons for taking ultraviolet spectra for radial velocity purposes. One such is when one needs to avoid spectral contamination by a companion which is causing problems in the optical (*eg*, dragging of the spectrum at phases near the systemic velocity), but which might not, by virtue of a lower temperature, show up in the ultraviolet. Fig. 1 shows optical data on ι Ori where dragging is evident; it seems to be

237

K. A. van der Hucht and B. Hidayat (eds.),
Wolf-Rayet Stars and Interrelations with Other Massive Stars in Galaxies, 237–244.
© 1991 *IAU. Printed in the Netherlands.*

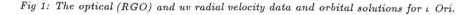

Fig 1: The optical (RGO) and uv radial velocity data and orbital solutions for ι Ori.

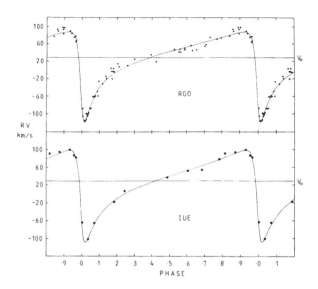

less marked in the uv data. A more extreme case is when one component cannot be seen at all in the optical and one *must* hunt for it in the uv.

Another justification is the richness of the photospheric spectrum in the far ultraviolet in comparison with the optical; furthermore, many of the lines that *are* found in the visible part of the spectrum, and especially those of hydrogen and helium, are prone to disturbance by wind effects. An example of the utility of IUE data here might be the case of V 861 Sco, where optical studies show different lines giving different velocity curves; cross correlation on the rich far-uv spectrum yields a very satisfactory result.

Finally, although perhaps not a sound reason for obtaining new IUE observations, is the modern epoch of this class of data. Much of the pioneering, and indeed definitive work was done in the first half (if not first quarter!) of this century and there are considerable gains to be had in a re-examination and comparison of orbital parameters over a long time base. Periods can be improved, possible changes of period may be detected, and apsidal motion may be measured. The latter two are of considerable significance for massive stars, the first being related to mass loss or exchange and the second to the internal structure of these rapidly evolving stars.

Thus we see that IUE can play a useful role in binary studies in addition to the other tasks of determining the physical conditions that we all know it can do so well. Many investigations can be started right away using the vast archive of over 70,000 images already in the public domain. At least 10,000 are high dispersion spectra of massive stars (mid-B type or earlier) and in numerous individual cases, extensive data runs are available. Although a senior citizen by space experiment standards, it is to be hoped that a few more years of productive service can be anticipated and that the Final Archive of optimally reprocessed IUE data, now being planned by the three agencies involved (NASA, ESA, and SERC), will provide rich pickings for years to come.

Fig 2: All available radial velocity data and the solution for HD 37017; the IUE data are represented by open symbols

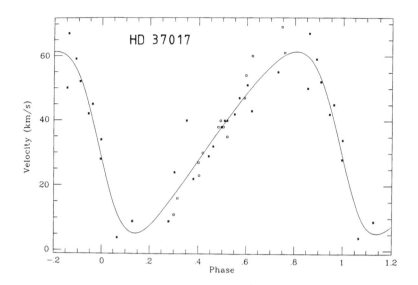

3. Measurements: problems and solutions

In common with most modern detection systems, but unlike the traditional long slit spectrographs, the high-resolution mode of IUE does not admit the impression of both stellar and wavelength calibration spectra on the same image and one does not, therefore, have an immediate reference system against which to measure wavelengths. The wavelength scale is imposed during image processing on the basis of a set of dispersion constants for the echelle spectrograph evaluated from measurements of the positions of lines in the Platinum spectrum secured from time to time as part of the calibration programme. This spectrum is produced by a Pt-Ne hollow cathode lamp mounted behind the primary mirror and whose light is reflected into the spectrographs off the back of the Sun shutter; since it is inadvisable to operate both lamps and mechanisms more often than is absolutely necessary, it is impossible to take calibration spectra along with every programme spectrum. Thus spectrograph (plus image processing) instabilities (ie, zero point drift) cannot be monitored routinely. Furthermore, most spectra these days are secured using a large aperture (10 x 20 arcsec) and although the placement of the (3 arcsec diameter) image within it is generally found to be accurate, it cannot be guaranteed in this slitless system.

How then are we to proceed in order to get radial velocities that can be used in the determination of binary orbits? Probably the best approach for hot stars begins with the recognition of the relatively rich spectrum of interstellar lines to be found in the far ultraviolet and the assumption that their radial velocity is constant in a given star. It may turn out that in many stars such lines are blends of several components, but provided that the same set is used and measured consistently this is not a problem; in this regard, care has to be exercised when producing a set for spectra taken both with the large and small apertures. Laboratory wavelengths (in vacuum) can be applied to these lines (eg. from Morton & Smith (1973)) or an empirical set can be established for use on a particular star.

One merit of using the interstellar lines is that they will share with the stellar spectrum the displacements due to the motion of the Earth and IUE and thus normalization to the interstellar velocity (which can be an arbitary value or that determined by ground-based observers) eliminates the need to evaluate heliocentric corrections.

Proceeding now to the stellar spectra, one is confronted with the problem that ultra-violet astronomy is a fairly youthful science and that approved lists of lines for velocity measurement, such as have been thoughtfully provided in the optical region over many years, just do not exist. Certainly the identifications of some of the stronger features (particularly lines formed in winds) are safe enough, but the same cannot be held true of the bulk of "photospheric" lines. The lazy way to circumvent this hazard is to ignore identifications and to proceed to use one spectrum as a template for the others, taking a selection of 30 or so well defined, consistently-measurable absorption lines, and recording differential velocities with respect to that template. Initially I did this in a rather tedious way, setting on each line with a cursor and recording the wavelength for later computa-tion of the velocity shift. Latterly, however, Ian Howarth has provided me with a nice cross-correlation routine which takes the drudgery out of it and gives essentially identical results; it skips all the wind and interstellar lines but gives much more spectrum than I would have measured manually.

All this is very well but it enables only relative velocities to be obtained. This is fine for many purposes, as will be shown later, but when comparing results from IUE with those from ground-based work, it can be of considerable interest to have a reliable value for the systemic velocity. To this end, work has begun, under the auspices of the UK IUE Project, to establish laboratory-based line lists for use with high-resolution spectra of hot stars. At present, a list has been prepared for SWP spectra of early B-type stars based on ζ Cas (B2IV); this will shortly be joined by one for O9I stars based on HD 209975; this turns out to be a rather sharp-lined star with constant velocity. It is expected that LWR/LWP lists will follow (although the time scale is not clear!). If, then, the uv interstellar line frame can be tied in to optical measures on a given star, absolute stellar velocities can be achieved, probably to within about 5 km/s.

Alternatively, one can pin one's hopes on the IUE calibration and image processing teams and assume that the extracted spectrum has a correct wavelength frame (that of the satellite before 1981 November, heliocentric thereafter) and derive absolute velocities directly, perhaps using the interstellar lines as a check. Work to check out this approach is still underway but it seems probable that a good system will be available in the Final Archive, and that it is actually not too bad at present. The importance of this is in the application it may have for cooler stars where the interstellar spectrum may not be easy (or even possible) to measure.

4. Some case studies

The purpose here is to highlight some of the advantages of using IUE spectra with examples studied thus far. The astrophysical discussion is not exhaustive and readers are referred to the original papers for more detail.

ι Orionis

The first in a series of papers to appear in *The Observatory* making use of IUE data for orbital studies was on ι Ori (Stickland (1987a)); many of the problems discussed above are treated therein. Fig. 1 shows the orbit derived from 19 SWP spectra, together with that produced from optical data at comparable resolution. Although the IUE data set was

Fig 3: The apsidal motion of δ Ori; the most recent point is from IUE data (see Harvey et al. (1987) for a description of the various fits to the data).

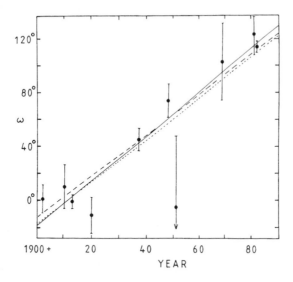

rather limited, particularly around periastron, there are suggestions that the noticeable dragging near the systemic velocity in the optical data is less marked in the SWP spectra. Given the spectral types of O9III and B1III for primary and secondary respectively (Stickland *et al.* (1987)), it is not surprising that the uv spectrum of the primary is less contaminated.

Plaskett's Star

It has been clear for some time that the more massive the star, the stronger the desire seems to be to lose mass. In binary systems, prolific mass loss can result in an increase of period and this can be monitored when a series of data sets are available over a long time scale. In the case of the star with the largest *known* mass (actually a lower limit on the secondary of about 60 $M_\odot$) - Plaskett's Star (O8e) - the period appears to be constant to better than about 0.00004 days on a time scale of decades (Stickland (1987b)); it has to be said, however, that the observed estimates of the mass-loss rate, 5×10^{-6} $M_\odot yr^{-1}$, are just inside this constraint.

HD 37017

Fourteen IUE spectra were used in a hitherto unpublished study of HD 37017 (B1.5V). Earlier work by Blaauw & van Albada (1963), which included recognition of a few archival data from Victoria (Plaskett & Pearce (1931)), gave only a rather approximate period of 18.65 days. Unfortunately the IUE data were all concentrated on the ascending branch of the velocity curve although they do spread far enough along it that the relative IUE velocities may be tied into the system of Blaauw & van Albada fairly well. This enables the period to be refined somewhat and the orbital elements derived from all the available measures are presented below, with the graphical representation given in Fig. 2. A more accurate appraisal will be possible if absolute IUE velocities are derived, as should be

Fig 4: The radial velocity curves of γ^2 Velorum; filled symbols represent the WC8 star and open ones the O9I component

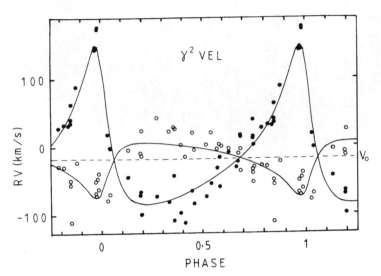

possible employing the ζ Cas line list.

Orbital elements for HD 37017

P = 18.65245 ± 0.00043 days
T_o = JD 2435002.91 ± 0.60
K = 28.9 ± 2.0 km s^{-1}
e = 0.27 ± 0.05
ω = 100 ± 13 degrees
γ = 34.9 ± 1.2 km s^{-1}
r.m.s = 5.9 km s^{-1}

δ Orionis

Among the most satisfying of the IUE binary studies to date (Harvey *et al.* (1987)), δ Ori (O9.5I) yielded results better (in terms of r.m.s. residual) than all of the many previous optical investigations. The value of this modern result was that it enabled a tight constraint to be put on the apsidal motion in this system, 1.67 ± 0.11 degrees per year, as shown in Fig. 3. Taking other system parameters from Koch & Hrivnak (1981), masses of 23 and 9 M$_\odot$ can be deduced with radii of 17 and 10 R$_\odot$, from which apsidal constants can be deduced which agree remarkably well with theoretical predictions.

AO Cassiopeiae

AO Cas (O8.5e + ?) presents numerous problems to the would-be determiner of orbits, as reviewed by Stickland & Lloyd (1988); the stability of the period and the eccentricity have both been questioned while the mass ratio is a vexed matter indeed. The IUE results add weight to the notion of a constant period and are consistent with e = 0;

Fig 5: The radial velocities and solution for V 861 Sco

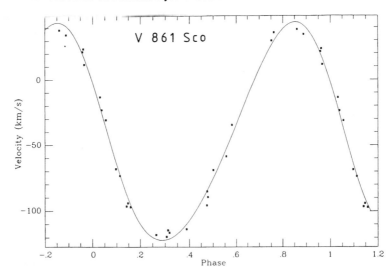

as to the mass ratio, this hinges on whether what has been deemed to be the secondary is really just that. Features were measured on the SWP spectra which appeared to move in antiphase with the primary lines, in the mainstream of optical results, and leading to q = 1.32. The implication is that we are perhaps seeing a system in the full swing of mass exchange, but it feels somehow uncomfortable!

29 Canis Majoris

29 CMa is the prototype Of star (now O8.5If) and was put on the IUE binary programme because of the high rate of mass loss from such objects and the consequent interest in period changes. 28 spectra were available for study and, in combination with results from the past - going back to 1916 - a tight constraint was put on this parameter: 4.393411 ± 0.000003 days. The IUE paper (Stickland 1989), however, does contain an unfortunate error in the time of periastron given for the IUE data: it should have been 2443614.063. At the same time, the uv data perhaps yield the best value of K_1, giving a mass function of 5.5 $M_\odot$.

γ^2 Velorum

Perhaps the major challenge to date and the first double-lined system to be treated, the Wolf-Rayet γ^2 Vel presented a number of problems and gave the least tidy results (Stickland & Lloyd (1990)). Several *absorption* lines in the spectrum of the WC8 star were recorded and a few in the O9I which seemed to move in antiphase (Fig. 4). An important point in favour of the IUE work was the measurement of the N III lines at λ 1750 which could not have arisen anywhere else than the O9I star; this appeared to confirm the lower velocity amplitude of the O star found by Pike *et al.* (1983) rather than that of Niemela & Sahade (1980). That said, the true elements of orbit are still clouded in mystery and dispute!

V 861 Scorpii

V 861 Sco (B0Ia) is the first to benefit from the cross-correlation method. An orbit was obtained, on the basis of 26 spectra, which compares well in most respects with that secured by Wolff & Beichman (1979), although the value of ω is markedly different and more in line with that deduced by Hill *et al.* (1974) several years earlier. However, before wildly speculating about rapid apsidal advance, it has to be noted that the IUE and Wolff & Beichman observations overlap in time. Thus observational error coupled with the small eccentricity is probably the culprit; the r.m.s. residual for the IUE result is again the best. The new elements are set out below and the data displayed in Fig. 5.

Orbital elements for V 861 Sco

P = 7.8484 ± 0.0011 days
T_o = JD 2443734.90 ± 0.23
K = 83.4 ± 2.0 km s^{-1}
e = 0.11 ± 0.02
ω = 64 ± 11 degrees
γ is undefined with relative velocities
r.m.s. = 6.3 km s^{-1}

5. Conclusions

Although IUE has exceeded all expectations several times over, and many of us have come to take it for granted, the reality is that we will not have access to it for ever! So, if enthusiasts in the hot star community can see vital applications, in the area of massive binary stars, which have not already been treated, and suitable archival data do not yet exist, I exhort them to apply for time in the very next round. On the other hand, if your favourite star has been observed, why not see if you can extract radial velocities from the archival data - they may throw up some surprising results.

References

Batten, A. H., Fletcher, J. M. & MacCarthy, D. G. (1989): *P.D.A.O.*, **17**
Blaauw, A. & van Albada, T. S. (1963): *Ap.J.*, **137**, 791
Harvey, A.S., Stickland, D. J., Howarth, I. D. & Zuiderwijk, E. J. (1987): *The Observatory*, **107**, 205
Hill, G., Crawford, D. L. & Barnes, J. V. (1974): *P.A.S.P.*, **86**, 477
Koch, R. H. & Hrivnak, B. J. (1981): *Ap.J.*, **248**, 249
Morton, D. C. & Smith, W. H. (1973): *Ap.J.Suppl.*, **26**, 333
Niemela, V. S. & Sahade, J. (1980): *Ap.J.*, **238**, 244
Pike, C. D., Stickland, D. J. & Willis, A. J. (1983): *The Observatory*, **103**, 154
Plaskett, J. S. & Pearce, J. A. (1931): *P.D.A.O.*, **5**, 1
Stickland, D. J. (1987a): *The Observatory*, **107**, 5
Stickland, D. J. (1987b): *The Observatory*, **107**, 68
Stickland, D. J. (1989): *The Observatory*, **109**, 74
Stickland, D. J. & Lloyd, C. (1988): *The Observatory*, **108**, 174
Stickland, D. J. & Lloyd, C. (1990): *The Observatory*, **110**, 1
Stickland, D. J., Pike, C. D., Lloyd, C. & Howarth, I. D. (1987): *A. & Ap.*, **184**, 185
Wolff, S. C. & Beichman, C. A. (1979): *Ap.J.*, **230**, 519

LONG PERIOD WR BINARIES

K. ANNUK
W.Struve Astrophysical Observatory of Tartu
202444 Tõravere
Estonia, USSR

ABSTRACT. Now it is clear that there exist some long period (periods exceed a year) Wolf–Rayet binaries. This paper presents the results of our radial velocity study for three WR binaries: HD193077 (WN6 + O9), HD193793 (WC7 + O5-6) and HD192641 (WC7 + O9). Our spectroscopic observations confirm that HD193077 and HD193793 are long period systems with the periods of 1538 and 2886 days. Most likely, HD192641 is also a long period WR binary (P $\geq$4400 days).

1. Introduction

Radial velocity studies of HD193077 were previously made by Bracher (1966), Massey (1980) and Lamontagne et al. (1982). Both Bracher and Massey have not found periodic radial velocity variations. However, Lamontagne et al. (1982), combining their own observations with the data by Bracher and Massey suggested that HD193077 is a triple system with periods of 1763 and 2.3238 days. Adding a new observation set spanning over 7 years, Annuk (1990) showed that HD193077 is still a binary system with a period of 1538 days.

Duplicity of HD193793 was first found from the infrared photometry by Williams et al. (1987). They also demonstrated that radial velocities of absorption lines varied with a period of 7.9 years = 2886 days. A more detailed analysis of radial velocities by Moffat et al. (1987) confirmed this period. Recently, Williams et al. (1990) found from additional infrared observations a new period of 2900 $\pm$10 days.

Analogically to HD193793, infrared photometry of HD192641 shows a periodic behaviour with a possible period of 12 years = 4400 days (Williams, van der Hucht and The, 1987). Previous insufficient radial velocity studies (Massey, Conti and Niemela, 1981; Moffat et al., 1986) have not given any periods.

2. Observation and reduction

The spectroscopic observations were carried out at the Tartu Observatory in the years 1980 ÷ 90 using the 1.5 m telescope with Cassegrain spectrographs UAGS (in

245

K. A. van der Hucht and B. Hidayat (eds.),
Wolf-Rayet Stars and Interrelations with Other Massive Stars in Galaxies, 245–250.
© 1991 *IAU. Printed in the Netherlands.*

the years 1980 ÷ 82) or ASP–32 (in the years 1982 ÷ 90) at a reciprocal dispersion of 44Å/mm and 37Å/mm in the blue. The plates IIaO were used.

All plates were traced on the PDS 1010 series microdensitometer at the Tartu Observatory in the photographic density mode with a $10\mu m$ wide slit and $5\mu m$ sampling interval.

Radial velocities of the emission and absorption lines were determined by fitting the Gaussian profiles upon the normalized intensity profiles by the method of least squares. Radial velocities of the strongest emission lines were also computed by the method of bisector. In this case the line profile is sliced up horizontally with a constant intensity step ($\Delta I = 0.05$) previously normalizing the central line intensities to unity. Radial velocities were calculated for every cut.

3. Radial velocities and orbital parameters

3.1. HD193077

We measured the radial velocities of six emission lines: NIV λ4057, HeII λ4686, NV λ4604, HeII λ4541, HeII λ4200 and NIII λ4097. Two stronger and symmetric lines (NIV λ4057 and HeII λ4686) were measured by the method of bisector as well. Radial velocities calculated as a mean of levels $I = 0.10 \div 0.95$ do not differ substantially from those determined by a Gaussian profile fitting routine.

The absorption lines are too broad and shallow. For this reason the measurement of radial velocities of these lines is very inaccurate and here we do not discuss them.

TABLE 1. Orbital elements of HD193077

Element	NIV λ4057	HeII λ4686	NV λ4604	HeII λ4541	NIII λ4097	HeII λ4200
P (day)	1538					
γ (km/s)	-96.0	70.1	59.6	73.2	160.0	79.1
	±1.3	±1.2	±2.3	±2.7	±3.1	±4.8
K (km/s)	30.6	32.1	31.8	34.3	48.5	41.6
	±1.9	±1.9	±4.0	±4.0	±4.0	±7.1
e	0.29	0.14	0.35	0.30	0.30	0.30
	±0.05	±0.06	±0.11			
ω (°)	271	239	207	278	344	271
	±12	±22	±18	±23	±17	
$T_o(2445000+)$	284	235	110	328	540	238
	±39	±93	±59	±86	±63	±28
$a * sini$ (R$_\odot$)	890	967	905	994	1406	1207
	±54	±58	±114	±116	±115	±205
f(m) (M$_\odot$)	4.01	5.13	4.22	5.60	15.82	10.01
	±0.74	±0.92	±1.60	±1.96	±3.90	±5.09
σ (km/s)	12.65	11.74	16.78	21.05	24.73	33.66

In order to determine the period, we used the period search programmes written by Pelt (1980) which are based on the technique of the frequency analysis of temporal series. Combining our velocity measurements with those of Massey (1980), Lamontagne et al. (1982) and Bracher (1966), we analyse the lines NIV λ4057 and HeII λ4686 and their superposition. The period was found to be about 1538 days.

Orbital elements were computed for each set of lines. They are given in Table 1. Using these elements, theoretical radial velocity curves of the NIV λ4057, HeII λ4686 and NV λ4604 lines are compared with the phase data in Figure 1.

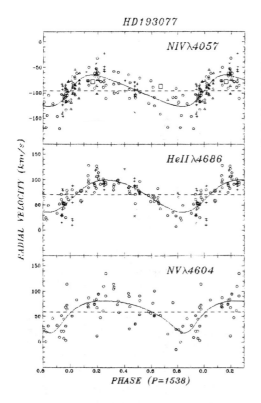

HD193077

NIVλ4057

HeIIλ4686

NVλ4604

RADIAL VELOCITY (km/s)

PHASE (P=1538)

Fig. 1. Velocity curves for HD193077. The symbols have the following meanings: o – from this paper; + – from Lamontagne et al. (1982); $\triangle$ – from Massey (1980); × – individual plates measured by Lamontagne et al. (1982); $\square$ – Bracher's mean velocities referenced by Massey (1980).

Using the derived elements of four emission lines (NIV λ4057, HeII λ4686, NV λ4604 and HeII λ4541), we conclude that K(em) = 32 km/s and e = 0.3. These values yield the mass function f(m) = 4.54 $M_\odot$ and the projected semi–major axis a(WR) $* sini$ = 930 $R_\odot$.

The mass ratio, q = M_{WR}/M_O, of other known WR binaries varies between 0.26 $\div$ 0.83. Assuming the mass ratio for HD193077 to be located in this range, we can estimate the minimum masses of the components: $M_{WR} \approx (1.9 \div 12.6)$ $M_\odot$ and $M_O \approx (7.2 \div 15.2)$ $M_\odot$.

3.2. HD193793

In the blue region of the spectrum of HD193793 the emission line CIV λ4650 is

248

the strongest and best measurable. We measured the radial velocities of this line
by the method of bisector. As CIV λ4650 is blended in red with the emission line
HeII λ4686, we calculated the radial velocities of CIV λ4650 as a mean of levels
$I = 0.55 \div 0.80$, where the blending effect is small.

Radial velocities of absorption are determined as a mean of the Balmer lines (H_β,
H_γ and H_δ) and some helium lines (HeI λ4471, HeI λ4026, HeII λ4541 and HeII
λ4200).

TABLE 2. Orbital elements of HD193793

Element	abs	CIV λ4650
P (day)	2886	
e	0.85±0.02	0.85
γ (km/s)	−5.8±0.6	281.5±2.9
K (km/s)	34.7±2.1	74.0±11.9
ω (°)	34.0±3.5	214=34+180
T_o(2446000+)	120±5	118±20
a * $sini$ (R$_\odot$)	1042±63	2224±357
m * sin^3i (M$_\odot$)	38.3±14.6	18.0±4.3
f(m) (M$_\odot$)	1.83±0.33	17.78±8.57
σ (km/s)	8.55	32.77

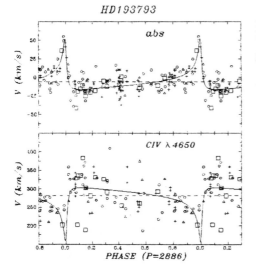

Fig. 2. Velocity curves for HD193793.
The symbols have the following meanings: o – from this paper; + – from
Conti et al. (1984); □ – from Lamontagne et al. (1984); △ – from Moffat et
al. (1987); × – from Galkina (1970).

We adopted the period of 2886 days as it was found by Williams et al. (1987).
The period of 2900 days found recently by Williams et al. (1990) gave a somewhat

worse solution. To combine all available observations with our own observations we made some corrections to radial velocities. In the case of emission lines we added to the data of Lamontagne et al. (1984) +30 km/s; Conti et al. (1984) -25 km/s and Moffat et al. (1987) +35 km/s. In the case of absorption lines we corrected only the data of Conti et al. (1984) with -5 km/s and of Galkina (1970) with -20 km/s. Such corrections are justified by the fact that different authors used different methods in determining the radial velocities.

Orbital elements are given in Table 2 and the velocity curves are shown in Figure 2. Our orbital elements are very close to those found by Williams et al. (1990). A mass ratio q = $M_{WR}/M_O \approx 0.47$ is typical for late–type WC stars.

3.3. HD192641

The emission spectrum of HD192641 is very similar to the spectrum of HD193793. The best line for radial velocity determination is also CIV λ4650. Radial velocities were measured by the method of bisector as a mean of levels $I = 0.45 \div 0.70$.

The absorption lines are weak but not so weak that they cannot be measured. We were able to measure some absorption lines (H_β, H_δ, HeI λ4471 and partially H_γ, HeI λ4026 and HeI λ4143).

Our radial velocities of the emission line CIV λ4650 indicate a clear long–time variation. Although the scatter of radial velocities of absorption is quite large, the antiphase variation with emission line can be detected.

Direct establishment of the period from radial velocities was not available, however, from infrared photometry we may predict the period of about 4400 days or more. Such a period arises also from radial velocities, although so far the observation sets covered a little less than 4400 days.

The preliminary orbital elements are given in Table 3. The velocity curves and an infrared photometry curve are shown in Figure 3. More observations would be needed for this star.

TABLE 3. Orbital elements of HD192641

Element	CIV λ4650	abs
P (day)	4400	
e	0.16±0.04	0.16
γ (km/s)	88.4±0.8	−6.6±3.0
K (km/s)	26.3±1.1	25.0±3.7
ω (^{o})	142.0±14.7	322=142+180
T_o(2447000+)	441±164	441
a * $sini$ ($R_\odot$)	2259±98	2156±314
m * sin^3i ($M_\odot$)	28.9±9.7	30.4±7.0
f(m) ($M_\odot$)	8.01±1.04	6.97±3.04
σ (km/s)	6.56	19.15

250

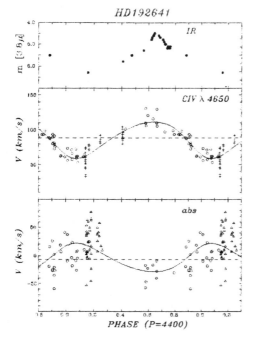

Fig. 3. Velocity curves and an infrared curve for HD192641. The symbols have the following meanings: $*$ – from this paper; $+$ – from Moffat et al. (1986); $\triangle$ – from Massey, Conti and Niemela (1981). Infrared photometry based on the paper of Williams, van der Hucht and The (1987).

References

Annuk, K., 1990, *Acta Astronomica*, (in press).

Bracher, K., 1966, *Thesis Indiana University*.

Conti, P.S., Roussel–Dupre, D., Massey, P. and Rensing, M., 1984, *Astrophys. J.*, **282**, 693.

Galkina, T.S., 1970, *Izvestiya Krymskoi Astrof. Obs.*, **41–42**, 283.

Lamontagne, R., Moffat, A.F.J., Koenigsberger, G. and Seggewiss, W., 1982, *Astrophys. J.*, **253**, 230.

Lamontagne, R., Moffat, A.F.J. and Seggewiss, W., 1984, *Astrophys. J.*, **277**, 258.

Massey, P., 1980, *Astrophys. J.*, **236**, 526.

Massey, P., Conti, P.S. and Niemela, V.S., 1981, *Astrophys. J.*, **246**, 145.

Moffat, A.F.J., Lamontagne, R., Shara, M.M. and McAlister, H.A., 1986, *Astron. J.*, **91**, 1392.

Moffat, A.F.J., Lamontagne, R., Williams, P.M., Horn, J. and Seggewiss, W., 1987, *Astrophys. J.*, **312**, 807.

Pelt, J., 1980, *Frequency analysis of astronomical temporal series*, (Valgus, Tallinn).

Williams, P.M., van der Hucht, K.A. and The, P.S., 1987, *Quarterly J. Roy. Astron. Soc.*, **28**, 248.

Williams, P.M., van der Hucht, K.A., van der Woerd, H., Wamsteker, W.M., Geballe, T.R., Garmany, C.D. and Pollock, A.M.T., 1987, in: *Instabilities in Luminous Early-Type Stars*, eds: H. Lamers and C.de Loore, p.221.

Williams, P.M., van der Hucht, K.A., Pollock, A.M.T., Florkowski, D.R., van der Woerd, H. and Wamsteker, W.M., 1990, *Mon. Not. R. Astr. Soc.*, **243**, 662.

THE ALL-VARIABLE BINARY WR140 (HD 193793, WC7+O4, P=7.94 yr)

K.A. VAN DER HUCHT, *SRON Space Research Utrecht, The Netherlands*
P.M. WILLIAMS, *Royal Observatory Edinburgh, U.K.*
A.M.T. POLLOCK, *Computer & Scientific Co. Ltd., Sheffield, U.K.*
B. HIDAYAT, *Observatorium Bosscha, ITB, Lembang, Indonesia*
C.F. McCAIN, *Jurusan Astronomi, ITB, Bandung, Indonesia*
T.A.Th. SPOELSTRA, *WSRT, NFRA, The Netherlands*
W.M. WAMSTEKER, *ESA-IUE Observatory, Villafranca del Castillo, Spain*

Recently Williams et al. (1990, *M.N.R.A.S.* **243**, 662) presented new observations of the WC7+O4-5 Wolf-Rayet system HD 193793 (WR140) made between 1979 and 1989 at infrared, radio, UV and X-ray wavelengths. Striking variations were evident in all four regimes.

The most complete coverage was in the infrared, where the light curves at JHKLMN defined the 1985 outburst, when the infrared flux increased by up to a factor of ten on a time-scale of weeks before declining more slowly. The infrared observations were interpreted in terms of the formation of $2.8 \times 10^{-8} M_\odot$ carbon in the wind of the WC star at a distance of about $2400 R_*$ between 1985.21 and 1985.54 and the subsequent cooling of the dust as it was carried away in the wind. Combined with earlier data, the infrared light curves led to a new period of 2900 (± 10) days (7.94 yr). Re-analysis of published radial velocities using this infrared photometric period led to new elements for this system: $e = 0.84 \pm 0.04$, $\omega = 32° \pm 8°$ and $T_o(1985) = JD2446160 \pm 29$ (1985.26 $\pm$ 0.08). Periastron passage coincided with the onset of grain formation.

The X-ray luminosity was 1-2 orders of magnitude above that of single O or WR stars, arguing for X-ray generation by interacting winds. Four X-ray observations made with *EXOSAT* showed evidence for greater extinction in the 1 *keV* region in mid-1985 than in 1984, indicating that the X-ray source moved deeper in the Wolf-Rayet star's wind, along with the O4-5 star, during periastron passage. The variation of circumstellar extinction and the dimensions of the orbit were used to determine independent information on the composition of the Wolf-Rayet wind. A *CNO* nuclei content equivalent to a fractional abundance $n_C \approx 0.06$ was derived: intermediate between the solar value and that predicted by contemporary models of evolved massive stars.

The available *IUE* high resolution spectra show 'eclipse' effects, particularly in the *SIV* and *CIV* resonance lines, around phase 0, when the O star is behind the extended atmosphere of the WR star, providing insight in the absorption component of the composite spectrum.

The radio observations show two components to the flux from WR140: the constant free-free emission from the stellar wind and a strong non-thermal source suffering variable extinction as it moves through the WC7 star's wind along the orbit of the O star. The free-free component and assumptions of distance, composition and ionization lead to a mass loss rate of $5.7 \times 10^{-5} M_\odot yr^{-1}$. It is shown that the Wolf-Rayet wind is not isotropic but has a low density cone in the shadow of the O star wherein the attenuation is 1-2% of that of the undisturbed wind. The intrinsic 2-11 *cm* spectral index of the non-thermal source is $\alpha = -0.5$. Recent (March, 1990) *WSRT* 6 *cm* radio observations show the predicted reappearance of the non-thermal radio source, which surprisingly also makes it self know at 21*cm*.

K. A. van der Hucht and B. Hidayat (eds.),
Wolf-Rayet Stars and Interrelations with Other Massive Stars in Galaxies, 251.
© 1991 *IAU. Printed in the Netherlands.*

THE ORIGIN OF WR 140

J.P. DE GREVE
Astrophysical Institute, VUB
Pleinlaan 2, B-1050 Brussels, Belgium.

The formation of the eccentric, long period binary WR 140 (HD 193793, WC7 + O4-5V, e= 0.8, P = 7.9 yr), in terms of an eccentricity modulated late case B of mass transfer, is investigated. Using only stellar wind models, it is impossible to obtain a combination of a spectral type O4-5 together with a WC star.

Assuming that the initial stars have masses 40 and 25 M_O and adopting the present period and radius, the Roche radius of the primary at periastron and apastron is respectively 242 R_O and 2400 R_O. On the other hand, the primary will ascend the giant branch around log T_{eff} = 3.6, log L/L_O = 5.6, implying a radius of 1330 R_O and more. The dynamical timescale of the convective envelope is of the order of periastron passage, i.e. ~ 1 yr. This results in mass transfer, periodically modulated by the large eccentricity.

Fixing the observed mass ratio at 0.22 (Moffat et al., 1987, Williams et al., 1990) and taking the present mass of the O-type star according to its spectral type (40 M_O), we can calculate the initial masses and period, using the models of Maeder and Meynet (1987) and adopting a spherical symmetric stellar wind and conservative mass transfer. The result is a massive system with P_i = 3.2 yr and masses 47 M_O and 39 M_O (the eccentricity was assumed to remain constant).

Several processes might influence the eccentricity : stellar wind (Hadjidimetriou, 1976), mass transfer (Hut and Paczynski, 1984), tidal evolution (Alexander, 1973, Hut, 1981, 1982). We investigated the influence of isotropic mass loss following an Eddington-Jeans law M = - d M^n. Only in the case n > 3 the average eccentricity will decrease. Using various formalisms published in the literature, we found that WR 140 probably had an eccentricity larger than the present one. Mass transfer cannot modify the average value of the eccentricity, though it causes secular changes to occur. Tidal interaction, treated with the weak friction appoximation (Zahn, 1977), affects the eccentricity on a timescale much larger than the evolutionary timescale of the system.

References.

Alexander, M. E. : 1973, Astrophys. Space Sci. 23, 459.

Hut, P. : 1981, Astron. Astrophys. 99, 126.

Hut, P. : 1982, Astron. Astrophys. 110, 37.

Hut, P., Paczynski, B. : 1984, Astrophys. J. 284, 675.

Maeder, M., Meynet, G. : 1987, Astron. Astrophys. 182, 243.

Maeder, M., Meynet, G. : 1988, Astron. Astrophys. Suppl. Ser. 76, 411

Moffat, A.F., Lamontagne, R., Williams, P.M., Horn, J., Seggewiss, W. : 1987, Astrophys. J. 312, 807.

Williams, P.M., van der Hucht, K.A., Pollock, A.M.T., Florkowski, D.R., van der Woerd, H., Wamsteker, W.M. : 1990, M.N.R.A.S. 243, 662.

Zahn, J.P. : 1977, Astron. Astrophys. 57, 383.

K. A. van der Hucht and B. Hidayat (eds.),
Wolf-Rayet Stars and Interrelations with Other Massive Stars in Galaxies, 252.
© 1991 IAU. Printed in the Netherlands.

THE STRUCTURE OF COLLIDING STELLAR WINDS

A. Pollock[1], J. Blondin[2] and I. Stevens[3]
(1) EXOSAT Observatory, ESA/ESTEC, Noordwijk,Holland
(2) Astronomy Dept, University of Virginia, Charlottesville, VA 22903
(3) Code 665, NASA/GSFC, Greenbelt, Maryland, MD 20771

ABSTRACT. We report on progress in the theoretical modelling of colliding winds in WR binary systems, concentrating on two WR+O systems, namely V444 Cygni, and HD 193793.

1. Introduction

A large fraction of WR stars are observed to be in binary systems with either a massive O or WR star companion, or an unseen low mass companion (possibly a compact object). In the first case, both stars have significant stellar winds, and the wind-wind interaction zone will give rise to some very interesting shock related phenomena. We report on progress on the theoretical modelling of these wind-wind interactions in close system such as V444 Cygni (WN5+O6, period=4 days), and very wide systems such as HD 193793 (WC7+O4-5, period=8 years). A full discussion of this work will be presented elsewhere in Pollock *et al.* (in preparation).

2. Results

HD 193793

The (WC7+O4-5) system HD 193793 is a widely separated (semi-major axis$\sim$ 3000 $R_\odot$), and eccentric system ($e = 0.8$), with a period of 2900 days (Williams *et al.* 1990). The WR wind dominates the system ($\dot{M}_{WR}/\dot{M}_O \sim 35$). However, even at periastron the shock front lies away from the surface of the O-star. In HD 193793 the wind interaction is essentially adiabatic throughout the orbit. The cooling parameter χ defined as the ratio of the cooling timescale to the dynamical timescale has the value 2.5 for the wind of the WR star, and a value of ~ 20 for the wind of the O-star.

V444 Cygni

In V444 Cygni the situation is reversed, with χ being small ($\chi = 1.0$ for the O-star wind, and 0.05 for the WR star wind), and cooling is important to the dynamics. Here, the shock structure collapses down almost onto the contact discontinuity between the winds, leading to a thin dense shock structure.

Work is continuing to calculate the X-ray luminosity and observed spectrum (including attenuation, which will be significant at soft X-ray energies) of the systems. With the advent of *ROSAT* observations of such systems, this will allow detailed comparisons of theory and observations for these systems.

References

Williams, P., *et al.* 1990. *M.N.R.A.S.*, **243**, 662.

253

K. A. van der Hucht and B. Hidayat (eds.),
Wolf-Rayet Stars and Interrelations with Other Massive Stars in Galaxies, 253.
© 1991 *IAU. Printed in the Netherlands.*

THE LIGHT CURVE OF V444 CYGNI

W.-R. HAMANN, E. SCHWARZ, U. WESSOLOWSKI
Institut für Theoretische Physik und Sternwarte der Universität,
Olshausenstraße 40, D-2300 Kiel, Federal Republic of Germany

In order to synthesize the eclipse light curve of V444 Cygni, we adopt the following model. The O star revolves round the WR star still within the outer regions of its extended atmosphere. The O star shadows a distinct volume of the WR atmosphere which thus cannot contribute to the total flux seen by the observer. On the other hand, additional radiation emerges from the surface of the O star. Its contribution to the total flux is more or less diminished by absorption when the rays pass through those parts of the WR atmosphere which lie between the O star and the observer. The WR atmosphere is given by our usual models (cf. Hamann and Schmutz, 1987; Wessolowski *et al.*, 1988).

A couple of free parameters describing the WR star (R_*, T_*, $\dot{M}$), the O star (T_{eff}, R), and the orbit (namely, the inclination angle i) can be varied until the synthetic light curve matches the observation. The mass-loss rate has a very characteristic influence, as it determines the width of the first minimum when the O star is occulted by the extended WR atmosphere. We find $\dot{M} = 10^{-4.9}$ $M_\odot$/yr.

Apart from $\dot{M}$ there remain five free parameters in order to fit essentially two properties of the light curve, namely the depths of the two minima. It is not likely that there is a unique solution for that problem. In order to find at least one particular solution, we arbitrarily adopt T_{eff} = 30 kK (O star), T_* = 40 kK (WR star) and i = 78°. Then we vary both stellar radii and achieve a good fit of the observed light curve at all wavelengths (Fig. 1). But we can fit the light curve with different sets of parameters as well. E.g., an increase of the inclination i and a corresponding decrease of both radii leads to the same good agreement with observation. For T_* = 60 kK (WR star) we also find a light-curve solution. Thus the basic result is that we can reproduce the observed light curve of V444 Cygni with our model, but the solution is not unique.

Recall that Cherepashchuk *et al.* (1984) performed a light curve analysis of V444 Cygni and obtained a solution with a "hot" WR star (90 kK effective temperature, related to the radius where τ = 1). We suspect that this is another *particular* solution.

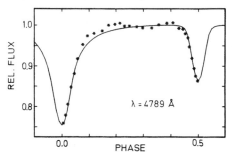

Fig. 1. Light curve of V444 Cygni at 4789 Å. Discrete symbols denote observational data, averaged over several periods, as taken from Cherepashchuk and Khalliullin (1973). The synthetic light curve (continuous line) results from a WR model with R_* = 6.5 $R_\odot$, T_* = 40 kK, $\dot{M}$ = $10^{-4.9}$ $M_\odot$/yr, an O star with T_{eff} = 30 kK, R = 6.7 $R_\odot$, and a circular orbit with $a \sin i$ = 40 $R_\odot$, and inclination i = 78°. Similar fits are obtained for other spectral bands.

References
Cherepashchuk, A.M., Khalliullin, K.F.: 1973, Soviet Astr. - AJ **17**, 330
Cherepashchuk, A.M., Eaton, J.A., Khalliullin, K.F.: 1984, Ap. J. **281**, 774
Hamann, W.-R., Schmutz, W.: 1987, Astr. Ap. **174**, 173
Wessolowski, U., Schmutz, W., Hamann, W.-R.: 1988, Astr. Ap. **194**, 160

K. A. van der Hucht and B. Hidayat (eds.),
Wolf-Rayet Stars and Interrelations with Other Massive Stars in Galaxies, 254.
© 1991 *IAU. Printed in the Netherlands.*

V444 CYGNI AND CQ CEPHEI, KEY WOLF-RAYET BINARY STARS

ANNE B. UNDERHILL
Department of Geophysics and Astronomy
University of British Columbia
Vancouver, B. C. V6T 1W5, Canada

Radial-velocity and photometric observations of eclipsing binaries allow one to estimate the masses of the stars in such systems, their relative brightnesses and sizes. If polarization variations are detected throughout the binary period, one may confirm the inclination of the system. All three types of information are available for V444 Cygni and CQ Cephei, see Underhill, Yang, and Hill (1988a,b), Underhill, Gilroy, and Hill (1990), and the references quoted in those papers.

The properties of the stars are summarized in the following table:

TABLE 1
Orbital Properties of V444 Cygni and CQ Cephei

Quantity	V444 Cygni	CQ Cephei
Period (days)	4.212424	1.6412436
Polarization (%)	≈ 0.5	≈ 5.0
i (degrees)	78 ± 4	74 ± 6
a_{WR} $(R_\odot)$	27.4	9.7
$a_{companion}$ $(R_\odot)$	9.1	8.2
M_{WR} $(M_\odot)$	9.8	13.6^a
$M_{companion}$ $(M_\odot)$	29.6	16.0^a

[a]Assuming $M_{WR}/M_{companion} = 0.85$. This mass ratio is determined by matching the observed brightness ratio in the V band by a theoretical brightness ratio, see Underhill, Gilroy, and Hill (1990). The observed brightness ratio of CQ Cephei in the V band is approximately 0.9.

The effective temperatures, $\log L/L_\odot$, and radii of hydrogen-burning stars having the masses found for the Wolf-Rayet components of V444 Cygni and CQ Cephei can be estimated from the properties of the models by Maeder and Meynet (1988). They are 27,200 K, 3.91, and 4.0 $R_\odot$ for V444 Cygni and 29,000, 4.27, and 5.5 $R_\odot$ for CQ Cephei. These properties, with the estimated properties of the companions, are compatible with the observed light curves. Evolved models of Maeder and Meynet in the stage which they call "Wolf-Rayet" are more than 6 times too luminous with respect to the companion star, and these models will not produce UV and visible light curves like what are observed.

REFERENCES

Maeder, A., and Meynet, G. 1988, *Astr. Ap.*, **76**, 411.
Underhill, A. B., Gilroy, K. K., and Hill, G. M. 1990, *Ap. J.*, **351**, 651.
Underhill, A. B., Yang, S., and Hill, G. M. 1988a, *P. A. S. Pacific*, **100**, 741.
Underhill, A. B., Yang, S., and Hill, G. M. 1988b, *P. A. S. Pacific*, **100**, 1256.

K. A. van der Hucht and B. Hidayat (eds.),
Wolf-Rayet Stars and Interrelations with Other Massive Stars in Galaxies, 255.
© 1991 *IAU. Printed in the Netherlands.*

SPECTROSCOPIC CCD STUDY OF THE TWO-DAY WN5+O8V BINARY CX CEP

DAVID LEWIS, ANTHONY F. J. MOFFAT, and CARMELLE ROBERT
Département de physique, Université de Montréal, Montréal and Observatoire du mont
Mégantic, Canada

After CQ Cep, the system CX Cep has the shortest orbital period known among Galactic
WR+O binaries. However, no definitive spectroscopic study is yet available for CX Cep, probably
because of its relatively faint magnitude (B $\approx$ 13). We have therefore obtained and analyzed some
60 CCD spectra (3700-4900 Å, S/N $\simeq$ 100, 5 Å/2 pixels) in August and October 1987.

Phased radial velocities (RV) show circular orbits with emission and absorption lines generally
moving in antiphase. The best estimate of the RV amplitudes are $K_{\rm WR} = 340 \pm 10$ km s^{-1} (mainly
from NV 4603) and $K_{\rm O} = 240 \pm 15$ km s^{-1} (from Balmer and Pickering absorption lines). With
orbital inclination 74° $\pm$ 5° from polarization observations (Shulte-Ladbeck and van der Hucht 1989,
$Ap.$ $J.$, 337, 872) we find the masses $M_{\rm WR} = 20 \pm 5 M_\odot$ and $M_{\rm O} = 28 \pm 7 M_\odot$ and the orbital
separation $a = 25 \pm 2 R_\odot$. With core radius $R_{\rm WN5} \sim 3 R_\odot$ and $R_{\rm O8V} \sim 9 R_\odot$, the system is therefore
not in contact.

Nevertheless, the relatively close orbit does produce interaction effects as illustrated in the
Figure. Phased equivalent widths of emission lines show that:

(a) NIV 4058 is reduced in strength at phase 0.5 (O star in front) probably by a simple eclipse
effect of the NIV emitting part of the WR wind;

(b) NV 4603/19 is weaker at both phases 0.5 and 0.0 implying that much of the NV emission
must arise between the two stars (via wind collision?);

(c) HeII 4686 (emission) is strongest near phase 0.4, when the O star is still approaching the
observer (bow shock effect?).

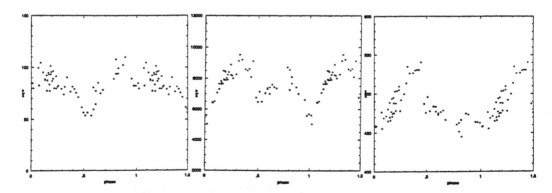

FIGURE. Equivalent width versus phase (WR behind at $\varphi = 0.5$) of emission lines
NIV 4058, NV 4603/19 and HeII 4686 (left to right).

K. A. van der Hucht and B. Hidayat (eds.),
Wolf-Rayet Stars and Interrelations with Other Massive Stars in Galaxies, 256.
© 1991 IAU. Printed in the Netherlands.

POLARIZATION VARIABILITY OF WOLF-RAYET BINARIES: CONSTRAINTS ON WR PARAMETERS

A. F. J. MOFFAT[1], V. S. NIEMELA[2], W. SEGGEWISS[3], A. M. MAGALHAES[4] and M. A. CERRUTI[2]
[1]Département de physique, Université de Montréal, Montréal and Observatoire du mont Mégantic; [2]Instituto de Astronomia y Fisica del Espacio, Buenos Aires; [3]Universitäts-Sterwarte Bonn; [4]Space Astronomy Laboratory, University of Wisconsin and Inst. Astron. e Geofis., Univ. de Sao Paulo

OBSERVATIONS

We have recently used the PISCO and VATPOL polarimeters at the MPI (Chile) and CASLEO (Argentina) 2.2 m telescopes to monitor over several weeks the brightest WR+O systems fainter than 9th magnitude in the Galaxy (V filter) and LMC/SMC (GaAs tube without filter). Each data point is accurate to $\sim 0.02\%$ in polarization P.

All systems observed show double-wave modulation of the polarization with orbital phase. The amplitude (typically several 0.1% in P) and relative phasing of the polarization modulation in Q and U can be used to derive WR mass-loss rates ($\dot{M}_{WR}$; cf. St.-Louis $et\ al.$ 1988) and orbital inclinations (i; cf. Brown $et\ al.$ 1978, 1982; Drissen $et\ al.$ 1986; St.-Louis $et\ al.$ 1987), hence absolute masses M_{WR}, when combined with $M_{WR} \sin^3 i$ from the spectroscopic orbits, which are known for all the systems studied.

RESULTS

For all WR+O systems observed here and published previously (16 Galactic, 4 LMC, 3 SMC), we find the following:

a) $\dot{M}_{WR}$ shows no obvious dependence on metallicity Z ($= 0.02, 0.008, 0.002$ for the Galaxy, LMC, SMC, respectively) as expected, since the main wind opacity elements in WR star winds are internally produced.

b) M_{WR} shows no evident trend with Z for a given subclass, in contrast with the new models of Maeder (1990) for single WR stars. We have no reason however to expect WR stars in all but the very closest binaries to behave differently from single WR stars, in view of the dominating effects of the hot, rapid stellar winds over gravitational perturbations.

c) $\dot{M}_{WR}$ and M_{WR} vary systematically with subtype (hence with luminosity, such that both increase with L_{WR}). We find $\dot{M} \sim M_{WR}$, not $\sim M_{WR}^{2.5}$, as currently proposed (cf. Langer 1989).

REFERENCES

Brown, J. C., McLean, I. S., and Emslie, A. G. 1978, $A.\ Ap.$, **68**, 415.
Brown, J. C., Aspin, C., Simmons, J. F. L., and McLean, I. S. 1982, $M.N.R.A.S.$, **198**, 787.
Drissen, L., Lamontagne, R., Moffat, A. F. J., Bastien, P., and Séguin, M. 1986, $Ap.\ J.$, **304**, 188.
Langer, N. 1989, $A.\ Ap.$, **220**, 135.
Maeder, A. 1990, $A.\ Ap.$, in press.
St.-Louis, N., Drissen, L., Moffat, A. F. J., Bastien, P., and Tapia, S. 1987, $Ap.\ J.$, **322**, 870.
St.-Louis, N., Moffat, A. F. J., Drissen, L., Bastien, P., and Robert, C. 1988, $Ap.\ J.$, **330**, 286.

K. A. van der Hucht and B. Hidayat (eds.),
Wolf-Rayet Stars and Interrelations with Other Massive Stars in Galaxies, 257.
© 1991 IAU. Printed in the Netherlands.

SPECTROSCOPIC CCD SEARCH FOR THE UNSEEN COMPANIONS IN THREE SINGLE-LINE WNL BINARIES

ANDRE GRANDCHAMPS and ANTHONY F. J. MOFFAT
Département de physique, Université de Montréal, Montréal
and Observatoire du mont - Mégantic

We have obtained some 15 CCD spectra at different orbital phases for each of the SB1 systems HD193928 (WN6), HD197406 (WN7), and CQ Cep (WN7), in an attempt to detect the companion, determine its orbit and estimate the masses. The wavelength range of the spectra is $\lambda\lambda 3600 - 4200$Å, with 1.4Å/pixel and $< S/N > \approx 100$/pixel. We compare the best emission-line (N IV λ4058Å) with the best absorption-line (H9 λ3835Å).

For HD193928 we find $K_{WN6} = 128\,\mathrm{km\,s^{-1}}$ and $K_O = 133\,\mathrm{km\,s^{-1}}$. With $i = 71.7°$, based on polarisation data (Grandchamps *et al.*, 1990. in preparation), we obtain $M_{WN6} = 23.8 \pm 8.6\,M_\odot$ and $M_O = 22.9 \pm 3.3\,M_\odot$.

For HD197406 the WN7 star appears to be too luminous to allow a detection of its companion with the present technique.

For CQ Cep the absorption-line follows the same motion as the emission-line and thus arises in the WR star itself.

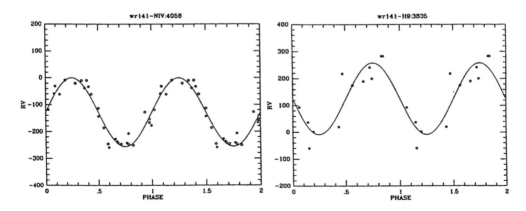

Fig. 1: Velocity sine curve (radial velocities in km s^{-1}) fit to the data of HD193928. The left half refers to N IV λ4058Å emission-line; the right half to H9 λ3835Å absorption-line. The open circles represent the Ganesh and Bappu (1968, *Kodaikanal Observatory Bulletin Ser. A.* <u>185</u>, A104) data; filled circles our data.

K. A. van der Hucht and B. Hidayat (eds.),
Wolf-Rayet Stars and Interrelations with Other Massive Stars in Galaxies, 258.
© 1991 *IAU. Printed in the Netherlands.*

UV OBSERVATIONS OF SELECTIVE WIND ECLIPSES IN THE WOLF-RAYET STAR γ VELORUM

NICOLE ST-LOUIS & ALLAN J. WILLIS
Department of Physics and Astronomy
University College London
Gower Street, London WC1E 6BT
England

1. Introduction

γ Velorum is a well-known WC8+O9I spectroscopic binary system with a period of 78.5002 days. Phase-dependent spectroscopic variations are observed in the ultraviolet and were first attributed to selective eclipsing of the O star light by the dense WC8 wind. Recently, it was suggested that the variations are caused by as a stream of gas moving away from the system at high velocity and arising as a result of the collision between the two stellar winds. In an attempt to shed new light on the problem we have carried out an analysis of *all* IUE high resolution archival spectra of γ Vel, supplemented by 8 *Copernicus* spectra covering the whole orbital period. The study includes all transitions for which variations are found in order to try to obtain a global view of the phenomena taking place.

2. Observations and Results

We have extracted 40 SWP and 30 LWR high resolution, small aperture spectra of γ Vel from the UK RAL archives using the IUEDR STARLINK software package. The 8 *Copernicus* spectra obtained with the U2 phototube have been corrected for charged particle background and for signal pointing errors using the U1 monitoring tube.

Three distinctive types of behaviour are observed. The first, including C IIλ1035,1335, C IIIλ1175,1247,2297, Si IIIλ1206,1295, Si IVλ1125,1722,1727 and S IVλ1062,1072 is qualitatively consistant with atmospheric eclipses of the O star light by the WR wind but the amplitude of the changes suggest the presence of some asymmetric source of emission, visible when the O star is in front and occulted when it is in back. The second type of behaviour is exhibited by the N Vλ1240, Si IVλ1400 and C IVλ1550 resonance doublets. As well as showing a similar behaviour to the first group, these lines develop a high velocity wing on the violet edge of the absorption component at phases when the O star is in front. We interpret this as being caused by a cavity formed in front of the O star as a consequence of the formation of a shock front due to the collision between the two stellar winds. Finally, a broad depression between ~ 1410−1900 Å is associated with a large number of Fe IV transitions between the $3d^44s - 3d^44p$ levels. Instead of being centered on the rest wavelength as expected for atmospheric eclipses, these lines are blue-shifted by 950 km s^{-1}. The C III] λ1909 transition is the only other line showing a similar behaviour, being blue-shifted by ~ 600 km s^{-1}. The behaviour of this third group of lines suggests an asymmetric opacity distribution.

K. A. van der Hucht and B. Hidayat (eds.),
Wolf-Rayet Stars and Interrelations with Other Massive Stars in Galaxies, 259.
© 1991 *IAU. Printed in the Netherlands.*

The Wolf-Rayet + Of star binary AB7: A Warmer in the Small Magellanic Cloud[+)]

Manfred W. Pakull[1)] & Luciana Bianchi[2)]

[1)] Observatoire de Besançon, 25044 Besançon Cedex, France
[2)] Osservatorio Astronomico di Turino, 10025 Pino Torinese, Italy

Summary

Strong nebular HeII λ4686 recombination radiation (λ4686/Hβ = 0.2) has recently been detected in the bright HII region N76 in the Small Magellanic Cloud by Testor & Pakull (1989). The rather symmetric intensity distribution suggests a nebular morphology which consists of a thick outer shell comprising the He^{++} and H$^+$ ionization fronts and a central region of warm, highly ionized gas which is surrounded by a hollow inner shell. The source of the high nebular ionization is identified with the peculiar Wolf-Rayet + Of star spectroscopic binary AB7. Optical and UV spectra (cf. Figure) show in addition to hydrogen and helium absorption lines and the NIII λλ4634-41 complex from the Of star, only broad He II and weak NVemission lines due to the Wolf-Rayet companion. The absence of other diagnostic features qualifies the evolved component as a rare, high-excitation WN star (WN2 or WN1). Subtracting from the optical spectrum of AB7 suitably scaled spectra of SMC O stars up to the point that the absorption lines disappeared, suggests that the WN 1 component contributes about 30 % to the total optical light.

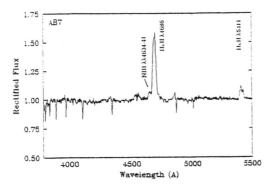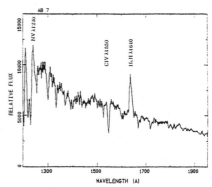

Optical and IUE spectra of the O6 IIIf + WN1 binary AB 7

From the nebular λ4686 flux and the optical brightness of the WN1 star a black body Zanstra temperature of about 80 000 K and a luminosity of 10^6 L$_\odot$ are derived. This result supports evolutionary models which place early WN stars near the helium main sequence of the HR diagram and provides support for the hypothesis of "Warmers", i.e. very hot luminous stars that might be responsible for the high-ionization spectra of certain active galaxies.

Although N76 appears to be predominantly ionization bound we note some leakage of ionizing photons into the surrounding IS medium which might well excite the puzzling λ4686 halo around the nearby oxygen rich SNR 1E0102.2 -7218.

K. A. van der Hucht and B. Hidayat (eds.),
Wolf-Rayet Stars and Interrelations with Other Massive Stars in Galaxies, 260.
© 1991 IAU. Printed in the Netherlands.

Emission line profiles in RY Scuti: a massive binary in a pre-WR stage?*

D. de Martino[1]†, A.A. Vittone[2], C. Rossi[3], F. Giovannelli[4]

[1] IUE Observatory, ESA-Vilspa
[2] Osservatorio Astronomico di Capodimonte, Napoli
[3] Istituto Astronomico dell'Università, Roma
[4] Istituto di Astrofisica Spaziale, C.N.R., Frascati

ABSTRACT. RY Scuti (HD 169515) is a massive $9^{th}mag$ eclipsing binary ($P_{orb} = 11.12^d$) surrounded by a peculiar nebula. High resolution spectroscopic observations in $H\alpha$, He I ($\lambda 5876$), [N II] ($\lambda\lambda 6548, 6584$), [A III] ($\lambda 7136$) and [S III] ($\lambda\lambda 9069, 9532$) are presented. All emission lines show a complex profile with two main structures: a sharp and strong red component and a very broad fainter complex blue one in which at least three sub-components can be detected. We have analysed these profiles in order to derive information on the velocity field in the nebula. Permitted and forbidden lines show the same velocity field indicating a common line forming region. The presence of a multi-structured blue component in all profiles gives evidence that velocity gradients are present within the nebula. Systemic corrected expansion velocities of +30 and -45,-30,-9 Km/s are found for the red and the three blue components respectively. Weaker emission structures, reflecting the same asymmetries of the strong emissions, are observed close to the permitted lines at larger velocities ($V_b = -189 Km/s$ and $V_r = 158 Km/s$). The observed velocity field indicates an asymmetric mass outflow from the system very likely through the second Lagrange point rather than via stellar wind. The system should loose mass during the mass exchange phase. Using the masses ($M_1 = 39M_\odot, M_2 = 49M_\odot$) and radii ($R_1 = 37R_\odot, R_2 = 41R_\odot$) of the two components (Milano et al., 1981), and locating RY Scuti in the [log R, log M] diagram for early type contact binaries from Leung and Schneider (1979), we find that this system has just evolved-off the Terminal Age Main Sequence. Moreover, placing RY Scuti in the [log P, log M_1/M_2] diagram for O+O and WR+OB binaries of Massey (1981) we find that it is in an evolutionary phase just preceeding a WR+OB stage.

REFERENCES.

Leung, K.C. and Schneider, D.P.: 1979, *IAU Symp. 83*, eds. P.S. Conti and C.W.H. de Loore, p.265.

Massey, P.: 1981, *Astrophys. J.*, **246**, 153.

Milano, L., Vittone, A., Ciatti, F., Mammano, A., Margoni, R., Strazzulla, G.: 1981, *Astron. Astrophys.*, **100**, 59.

---

*Based on observations collected at ESO, La Silla, Chile
†On leave from Osservatorio Astronomico di Capodimonte, Naples

K. A. van der Hucht and B. Hidayat (eds.),
Wolf-Rayet Stars and Interrelations with Other Massive Stars in Galaxies, 261.
© 1991 *IAU. Printed in the Netherlands.*

Turolla, Cherepashchuk, de Martino, Matteucci, de Loore, de Groot

SESSION V. MASS LOSS – *Chair: Lindsey F. Smith*

Lindsey Smith

OBSERVATIONS OF WOLF-RAYET MASS LOSS

Allan J Willis
Department of Physics & Astronomy
University College London
Gower Street
London WC1E 6BT, UK

ABSTRACT. Current knowledge of the stellar winds and mass loss rates for WR stars is reviewed. Recent IR spectroscopy and reassessments of UV resonance line P Cygni profiles have led to revisions of terminal velocities, with $v_\infty \simeq 0.75 \times$ previous estimates. Radio and IR ($10_{\mu m}$) free-free emission for well-established thermal sources, coupled with recent considerations of the wind ionisation balance and chemistry, leads to WR mass loss rates lying in the range $10^{-5} - 10^{-4}$ $M_\odot$ yr^{-1}. This scale is confirmed by independent analyses of optical polarisation modulation in WR+O binaries. No significant differences are apparent between the mean mass loss rates of: (a) single and binary WR stars; (b) WN and WC stars, and (c) the WN and WC subclasses. The overall mean WR mass loss rate is $\sim 5 \times 10^{-5}$ $M_\odot$ yr^{-1}. Although WR radiative luminosities are uncertain, there may be a rough scaling of $\dot{M}_{WR}$ with L_*, with a spread of up to an order of magnitude at a given L_*. WR winds have the highest momenta of the hot luminous stars, with values of $\dot{M} v_\infty$ c/L_* in the range 1–30 (WN7,8 and WC9 stars may lie near the lower bound). An additional mechanism to radiation pressure may be required to initiate the high WR mass loss, although thereafter the winds may be radiatively accelerated. Intrinsic variability in optical light, polarisation and emission lines, and in UV P Cygni profiles, indicate significant instability in the WR winds. For extragalactic WR stars in the Local Group, optical line strengths and widths do not suggest substantial differences in wind velocities and mass loss rates of subtypes compared to galactic counterparts.

1. Introduction

It is well established that the predominant emission line spectrum of the WR stars is a direct reflection of their high levels of mass loss (Abbott & Conti 1987). Characterising their stellar winds and mass loss rates as a function of subtype is thus important in developing a quantitative understanding of their physical and chemical properties and evolutionary status from both observational and theoretical standpoints. This paper reviews progress in observational studies of WR mass loss and stellar winds, emphasising recent work over the past five years or so. I shall not dwell much on the theory of WR mass loss – this will be covered in the next review, by Cassinelli (these proceedings). Excellent comprehensive reviews of earlier work in this field have been given by Barlow (1982) and in the exemplary paper by Abbott *et al.* (1986).

265

K. A. van der Hucht and B. Hidayat (eds.),
Wolf-Rayet Stars and Interrelations with Other Massive Stars in Galaxies, 265–280.
© 1991 *IAU. Printed in the Netherlands.*

2. Radio and IR continuum data

Abbott *et al.* (1986, hereafter ABCT) provide the most extensive set of 4.9 GHz radio observations of WR stars from their own data and earlier work by Hogg (1982, 1985), Dickel *et al.* (1980), Florkowski (1982) and Becker & White (1985). In their sample 22 stars have measured 4.9 GHz fluxes, whilst upper limits were obtained for a further 13 stars. ABCT also present radio observations at 14.9 GHz for 6 stars based on their data and that of Hogg (1982). The radio flux is presumed to arise from free-free emission in the outermost regions of the WR winds, where the wind velocity has reached its terminal value, v_∞. In spherical symmetry, the formula developed by Wright & Barlow (1975) may then be used to deduce the rate of mass loss, *viz.*:

$$\dot{M} = 0.095 v_\infty \left(\frac{S_\nu^{0.75} D^{1.5}}{(g\nu)^{0.5}} \right) \left(\frac{\mu}{Z\gamma^{0.5}} \right) \qquad M_\odot \text{yr}^{-1}$$

where S_ν is the radio flux in Jy at frequency ν in Hz; v_∞ in km s^{-1}; D is the distance in kpc; g is the gaunt factor; μ, Z and γ are respectively the mean molecular weight, r.m.s. ionic charge and mean number of electrons per ion. With X_i, M_i and Z_i as the fractional abundance, molecular weight and ionic charge of species i, then:

$$\mu = \frac{\sum X_i M_i}{\sum X_i}; \quad Z = \frac{\left(\sum X_i Z_i^2 \right)^{0.5}}{\sum X_i}; \quad \gamma = \frac{\sum X_i Z_i}{\sum X_i}$$

In the radio region, the free-free spectrum is predicted to have a distribution $S_\nu \propto \nu^\alpha$, with $\alpha = 0.6$, (the departure in α from a value of 2/3 reflects the frequency dependence of the gaunt factor). ABCT considered three methods to test the validity of interpreting the radio data as thermal, free-free emission: (i) interferometric observations of the visibility function compared to predictions of thermal wind models, (ii) consideration of the observed radio or radio–IR spectral index, and (iii) comparison of the observed radio flux with that predicted from the observed optical recombination line emission measure.

Hogg (1985) has used the VLA to resolve the radio emission of γ Velorum at 1.49 GHz, 4.9 GHz and 14.9 GHz (see Fig 1), confirming a thermal origin consistent with spherical outflow. The frequency dependence of the visibility functions requires an electron density decrease more rapid than r^{-2} for wind distances greater than 3×10^{15} cm which is commensurate with helium (the dominant electron source) beginning to recombine. A mass loss rate of $8.6 \pm 1.0 \times 10^{-5}$ M$_\odot$ yr^{-1} and electron temperature of 5600 ± 500 K provided the best fit to the multifrequency radio interferometric data.

Five of the six stars listed by ABCT with 4.9 and 14.9 GHz radio data (but generally too faint to determine visibility functions) show a radio spectral indedx of $\alpha \sim 0.6$, confirming their emission to be thermal (the sixth star, WR 125, has a measured index of -0.5 and is prescribed as definitely non-thermal in nature). For γ Velorum they confirm the result from Purton *et al.* (1982) who deduced $\alpha = 0.58 \pm 0.17$ from data at 5.0, 6.2, 8.1, and 14.5 GHz.

Williams *et al.* (1990) present observations between 1.25μm and 1100 μm of γ Velorum, combined with the radio measurements by Hogg (1982) at 4.9 GHz and by Jones (1985) at 0.843 GHz to examine the spectral distribution of the wind emission (Fig 2). They find $\alpha = 0.69 \pm 0.02$ between the 1100 μm and radio wavelengths, which is consistent with plausible variations of temperature and ionisation in the wind.

Barlow, Smith & Willis (1981, hereafter BSW) present infrared 10μm fluxes for 21 galactic WR stars corrected for the underlying photospheric contribution to estimate the excess wind emission at this wavelength. The 10 μm excess emission data from BSW for 12

Fig 1: Radio emission of γ Vel resolved with the VLA (from Hogg 1985)
Fig 2: The optical–IR–mm–radio emission of γ Vel (from Williams et al. 1990).

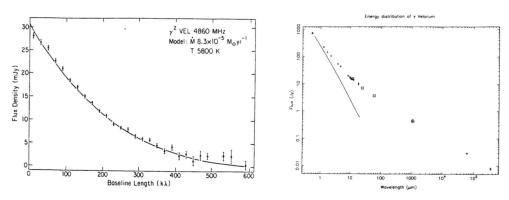

stars with measured 4.9 GHz radio fluxes give IR-radio spectral indices in the tight range of 0.66 – 0.82 (see Table 1). HD 193576 exhibits a 10μm–4.9 GHz index of 0.53 — identical to that expected for a constant velocity isothermal flow at both wavelengths. The mean value (HD 193576 omitted) of α =0.75 ± 0.04 is similar to the mm–radio spectral index determined by Williams *et al.* for γ Velorum (a confirmed thermal source), suggesting that such IR-radio indices also indicate thermal emission (with the difference from 0.53 reflecting changing temperature and/or ionisation conditions in the IR and radio emitting regions).

Table 1 lists the 4.9 GHz fluxes for stars which ABCT have confirmed as thermal emitters using one or more of the above considerations, together with the 10μm data from BSW, including stars with IR but no radio measurements. For such cases, the radio flux can be predicted using the 10μm–radio index of 0.75 (BSW). In § 5, these data are combined with reassessments of wind terminal velocities, chemistry and ionisation balance to update the mass loss rate scale for WR stars.

3. Terminal velocities and v(r)

Until recently (as for analyses of OB mass loss), v_∞ has been taken as the maximum edge velocity of the P Cygni absorption components in the UV resonance lines (e.g. Willis 1982), or deduced from the maximum velocity inferred from the widths of optical emission lines (e.g. for WN stars from Conti Leep & Perry 1983, or for WC stars from Torres, Conti & Massey 1986 who extrapolate the emission line width vs. excitation potential correlations to zero E.P. to predict v_∞). ABCT used these approaches to derive values of v_∞ for the stars in their radio sample noting the good agreement between the optical results and the edge velocities of the UV P Cygni line profiles for stars in common. These approaches in effect estimate the *maximum* velocities which can be associated with any individual WR stellar wind material, v_{max}, which are listed in Table 1. Several recent papers have questioned whether these values of v_{max} are to be identified with the true terminal velocity of the wind associated with the bulk material flow, and suggested lower values of v_∞.

Williams & Eenens (1989) measured the P Cygni absorption velocities in the He I 2.058 μm line in 8 WR stars finding values $\sim$ 0.7 of the terminal velocities deduced from the

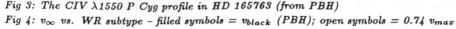

Fig 3: The CIV λ1550 P Cyg profile in HD 165763 (from PBH)

Fig 4: v_∞ vs. WR subtype – filled symbols = v_{black} (PBH); open symbols = 0.74 v_{max}

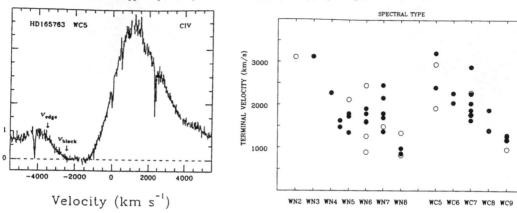

Velocity (km s^{-1})

v_{max} methods discussed above. They argue that this HeI line is formed in the outermost regions of the WR winds and thus provides more suitable measurements of v_∞. Barlow, Roche & Aitken (1988) in their analysis of 12-13 μm spectra of γ Velorum, measured a terminal velocity from the [Ne II] 12.8 μm forbidden line of 1520 km s^{-1}, considerably lower than the value of $\sim$ 2000 km s^{-1} deduced previously from the UV P Cygni profiles. The [Ne II] line exhibits a rectangular profile as expected for an optically thin line formed at large radii and constant velocity, and the typical density for its formation of 5×10^5 cm^{-3} suggesting an emission region radius of $\sim$ 1300R$_*$. The HeI 2.058 μm line in γ Velorum shows a P Cygni central absorption at $\sim$ 1500 km s^{-1}, in accord with the [Ne II] data suggesting that the measurements given by Williams and Eenens (1989) do indeed reflect the true WR wind terminal velocities.

Recently, Prinja, Barlow & Howarth (1990, hereafter PBH) have reassessed inferences on v_∞ from UV P Cygni absorption profile measurements of both OB and WR spectra. They point out (as others have in the past) that the UV resonance line P Cygni saturated profiles generally show extended regions of zero residual intensity (up to some maximum velocity, v_{black}) and additional (non-black) absorption extending to higher velocities up to the maximum value observed, v_{edge} (Fig 3). The observed profiles differ significantly from those expected in Sobolev models of P Cygni line formation with monotonic velocity laws, where zero residual intensity is only reached at v_∞ and the profile should then rise immediately to the continuum. These differences are now generally explained in terms of non-monotonic wind velocity laws and shocks in the winds induced by instabilities in the flows (Lucy 1982, 1983, Owocki, Castor and Rybicki 1988), with $v_{edge} - v_{black}$ a measure of the velocity amplitude of the shocks. PBH argue that v_{black} is a more meaningful measurement of v_∞ and present values for 35 WR stars with available high resolution IUE spectra, finding a mean ratio of $v_\infty/v_{edge} = 0.76 \pm 0.12$. A subset of their data is given in Table 1. For 15 stars in common with the data of PBH and ABCT the mean ratio of $v_{black}/v_{max} = 0.74 \pm 0.08$, and I have adopted this scaling factor for those stars with radio data but no v_{black} measurement to estimate appropriate values of v_∞ in deducing mass loss rates (see below).

Fig 4 plots these revised values of v_∞ vs. spectral type for both galactic WN and WC stars. For the WC stars there is a clear trend of increasing v_∞ with earlier subtype, in keeping with the well known correlation of optical line width with subclass (see Torres et al. 1986). For the WN stars, no clear trend is apparent, although the WN8 stars exhibit low velocity winds, whilst the very earliest types (WN2,3) have the largest values. Most WN subtypes can show a large range of wind velocities.

Little is known directly about the velocity laws, v(r), of WR winds. Cherepaschuk *et al.* (1984) have modelled the IR-UV eclipse data of V444 Cyg to empirically determine the velocity law of the WN5 component wind. Their result is reasonably well fitted with the usual paramaterised law of $v(r) = v_\infty(1-r_*/r)^\beta$ with $\beta \sim 1$. This law has generally been adopted in detailed model atmosphere analyses of WR spectra by Schmutz *et al.* (1988) and Hillier (1987, 1989) who are able to reproduce the observed emission line profiles fairly well. For OB stars, Friend & Abbott (1986) and Pauldrach *et al.* (1986) include the effects of rotation and finite disk of the star and conclude that a value $\beta \sim 0.8$ in the above velocity law provides the best fit to observational data. Thus it may be that the WR winds, like those for OB stars, are accelerated through radiation pressure, although the very high levels of WR mass loss may require an additional mechanism(s) (see below).

4. Ionisation structure

Detailed modelling of the atmospheres of WN and WC stars by Hillier (1987, 1989) characterises the ionisation structure of their winds, showing that even for high core temperature models (e.g. with $T_{eff} = 60,000$ K) the wind electron temperature can be low (e.g. 20000 K at $Ne \sim 10^{11}$ and ~ 9000 K at $N_e \sim 10^9$) giving values of T_e in the radio emitting region comparable to that deduced by Hogg (1985) for γ Velorum. The ionisation stratification in the Hillier models is compatable with observed correlations of optical line width vs. ionisation potential (e.g. Kuhi 1973). Willis (1982) argued that when the UV resonance and low excitation lines were included with optical measurements, line widths correlated better with excitation potential (E.P.) than with I.P., questioning such an ionisation stratification. As discussed by Hillier (1989) this discrepancy may now be understood, since the UV lines, with their increased opacity, may be tapping lower density, higher velocity material associated with shocked gas in the wind, and should not necessarily be considered together with the optical lines.

In the Wright & Barlow formulation the factor 'C' = $(\mu/(Z\gamma^{0.5})$ depends upon the chemistry and ionisation balance presumed for the radio-emitting regions, and recent considerations of this parameter have led to revisions in $\dot{M}$.

ABCT assumed only H and He ions contributed towards the C-factor, and for WN stars estimated the H^+/He^{++} ratio using the Pickering decrement measurements of Conti, Leep & Perry (1983) and the He^+/He^{++} ratio from measured values of the emission measures of optical He I and He II lines. For WC stars the H/He ratio was presumed to be zero, and the He ionisation balance again inferred from the optical He I and He II emission measures. For most stars He^{++} was taken as the dominant ion in the radio region. However, Schmutz, Hamann & Wessolowski (1989) conclude that He^+ is the dominant ion for the great majority of WN and WC subclasses (only three WN2 and WN3 stars in their sample have He^{++} dominant) leading to an upward revision to mass loss rates deduced from the radio data. Van der Hucht, Cassinelli & Williams (1986) for WC stars included the effects of high C-abundances inferred from evolutionary models together with calculated ionisation equilibria in simplified model atmospheres, to propose an upward revision to the mass loss rates of ABCT of about a factor of 2 for WCE stars and by about a factor of 2–3 for WCL stars.

The present view is that He^+ is the dominant helium ion in the radio emitting regions for both WN and WC stars. For WNE stars, with H^+/He^{++} generally $\leq 0.2 - 0.3$ by number (Conti, Leep & Perry 1983), the effect of H on the C-value is negligible, and thus for the radio-emitting regions we can adopt $\mu = 4.0$, $Z = 1.0$ and $\gamma = 1.0$, *viz* $C = 4.0$. For WN7 and WN8 stars the H-contribution is non-negligible in its effect on μ, which can

Star	WR	Spec type	Radio	IR	M_v	D	'C'	V_{max}	V_{black}	$\log\dot{M}$	$\log L_*$	$\dot{M}v_\infty/L_*$	L_w	L_w/L_*
HD 4004	1	WN5	0.5	0.47	-4.6	2.63	4.0	2850	1720	-4.09	5.54	24.29	11.28	0.085
HD 6327	2	WN2	<0.2		-2.3	2.51	1.4	4200		<4.71	4.62	72.02	5.92	0.371
HD 50896	6	WN5	1.0	0.90	-4.8	1.56	4.0	2650	1790	-4.29	5.62	10.44	4.78	0.030
HD 92740	22	WN7+abs		0.48	-6.5	2.63	2.8	2600	2155	-4.21	6.30	2.73	6.22	0.008
HD 93131	24	WN7+abs		0.38	-6.4	2.63	2.8	2800	2455	-4.21	6.26	3.64	9.11	0.013
HD 93162	25	WN7+abs		0.35	-6.2	2.63	2.8	2900	2150	-4.18	6.18	5.34	12.67	0.022
HD 151932	78	WN7	1.4	0.96	-6.7	2.00	2.7	2150	1365	-4.29	6.38	1.44	3.01	0.003
LSS 4064	87	WN7	<0.4		-6.4	2.88	2.6	2000		<4.44	6.26	1.44	2.46	0.003
LSS 4065	89	WN7	0.6		-7.1	2.88	2.9	2000		-4.26	6.54	1.14	3.71	0.003
HD 165688	110	WN6	1.0	0.81	-5.3	1.89	4.0	3300		-4.02	5.82	17.51	17.94	0.071
MR 87	115	WN6	0.4		-5.0	2.19	4.0	1200		-4.66	5.70	1.92	0.54	0.003
BAC 209	124	WN8		0.11	-6.7	5.98	2.9	1800		-4.27	6.38	1.46	2.97	0.003
AS 374	130	WN8	<0.2	0.38	-6.7	4.75	2.4	1100		<4.64	6.38	0.38	0.47	0.001
HD 190918	133	WN4.5+O9Ib	≤0.3	0.25	-4.3	2.09	4.0	1750	1625	≤-4.52	5.42	9.28	2.53	0.025
HD 191765	134	WN6	0.8	0.70	-4.8	2.09	4.0	2300	1905	-4.13	5.82	10.60	8.51	0.034
HD 192163	136	WN6	1.6	1.05	-5.7	1.82	4.0	2000	1605	-4.07	5.98	7.11	6.96	0.019
HD 193077	138	WN6 +abs	0.6	0.29	-4.8	1.82	4.0	1700	1345	-4.46	5.62	5.48	1.96	0.012
HD 193576	139	WN5 + O6	0.3	0.33	-3.9	1.74	4.0	2500	1785	-4.60	5.26	12.30	2.55	0.037
HD 193928	141	WN6	0.6		-5.3	1.82	4.0	1700		-4.49	5.82	2.99	1.58	0.006
HD 68273	11	WC8+O9I	29.0	14.36	-5.0	0.45	4.0	2000	1415	-4.09	5.70	11.38	5.15	0.027
HD 155270	79	WC7+O5-8	0.9	0.86	-5.0	2.00	4.0	3300	2270	-4.04	5.70	20.28	14.73	0.077
MR 66	81	WC9	0.3		-4.8	1.75	4.0	1300		-4.86	5.62	1.56	0.40	0.002
HD 156327	86	WC7 +O7-9	0.5		-4.8	1.95	4.0	2400		-4.36	5.62	9.15	4.30	0.027
HD 156385	90	WC7	<0.3		-4.8	1.97	4.0	3000	2045	<4.46	5.62	8.49	4.62	0.029
HD 157504	93	WC7 +abs	0.9		-5.6	1.74	4.0	3100		-4.13	5.94	9.57	12.14	0.036
MR 74	95	WC9	<0.4		-5.1	1.91	4.0	1300		<4.71	5.74	1.67	0.56	0.003
HD 164270	103	WC9	≤0.2		-4.8	2.54	4.0	1400	1190	≤-4.66	5.62	3.10	0.98	0.006
MR 80	104	WC9	<0.4		-4.5	1.58	4.0	1600		<4.75	5.50	3.31	0.79	0.006
HD 165763	111	WC5	0.3	0.39	-3.8	1.58	4.0	3550	2415	-4.53	5.22	21.36	5.46	0.086
HD 168206	113	WC8 +O8-9IV	≤0.4		-4.6	2.00	4.0	2900	1890	<4.39	5.54	11.06	4.63	0.035
HD 169010	114	WC5	<0.3		-3.6	2.19	4.0	2600		-4.42	5.14	26.31	4.44	0.084
MR 90	121	WC9	0.6	0.35	-4.8	2.06	4.0	1300		-4.76	5.62	1.99	0.51	0.003
HD 192103	135	WC8	0.6		-4.7	2.09	4.0	1900	1405	-4.35	5.58	8.07	2.75	0.019
HD 192641	137	WC7 + OB	0.4		-5.0	1.82	4.0	2700	1885	-4.45	5.70	6.61	3.99	0.021
HD 193793	140	WC7 + O4-5	1.5		-4.8	1.34	4.0	3000	2900	-4.03	5.62	32.02	24.71	0.155
HD 195177	143	WC5	<0.4		-3.7	0.82	4.0	4000		<4.77	5.18	16.14	4.59	0.079
MR 111	145	WN7 + WC4	1.0	0.31	-6.5	1.82	4.0	1500		-4.38	6.30	1.13	1.59	0.002

Table 1: Notes

Radio is the 4.886 Ghz flux in mJy (from Abbott et al. 1986): IR is the 10μm excess emission in Jy from BSW81
M_v's from van der Hucht et al. (1988) except binary values from Smith & Maeder (1989) – D is distance in kpc, from van der Hucht et al. (1988):
V_{max} in km/s from Abbott et al. (1986): V_{black} in km/s from PBH90 – $\log\dot{M}$ in $M_\odot$ yr^{-1}
V_{max} in km/s from Abbott et al. (1986): V_{black} in km/s from PBH90 – $\log\dot{M}$ in $M_\odot$ yr^{-1}
L_w in units of $L_\odot$ assumes B.C. = -4.5 (Smith & Maeder 1989) – $L_w = \frac{1}{2}\dot{M}v_\infty^2$ is in units of 10^{37} erg s^{-1}.

Fig 5: revised $\dot{M}$ vs. subtype – filled and open symbols are single and binary stars respectively

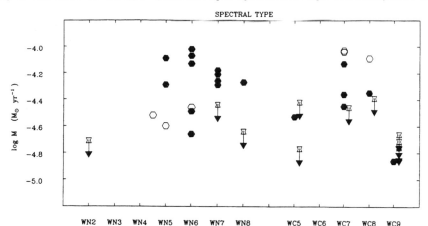

vary from 2.44 – 2.92, Z = γ = 1.0, and thus C lies in the range 2.44 – 2.92, depending on the H-content given by Conti, Leep & Perry (1983). (N.B. if He^{++} were the dominant He ion in the radio region, the corresponding C-values would be: 1.41 (WNE) and 1.29 – 1.39 (WNL)). For WC stars, the most recent estimates of the C/He ratios have been given by Smith & Hummer (1988) from analyses of near-infared recombination lines. They find C/He = 0.3 for WC6–9 stars; = 0.5 for WC5 stars and = 0.7 for WC4 stars. With He^+ as the dominant He ion in the radio region, (Smith & Hummer confirm this in general), then the C-value depends only on the presumed carbon ionisation balance. Adopting C^{2+} as dominant leads to C-values of $\sim$ 4 for all WC subclasses (increases in μ from 5.85 (WC6-9) to 7.30 (WC4) are compensated by corresponding increases in Z and γ). (N.B. for He^{++} dominant, C-values would be $\sim$ 2). The C-values adopted herein, based on the above discussion are also listed in Table 1.

5. Radio mass loss rates

Table 1 presents revised mass loss rates for galactic WR stars, confirmed as thermal sources, taking into account revisions to the C-factors and v_∞ since the work of ABCT, discussed above. Distances are taken from van der Hucht *et al.* (1988). Mass loss rates have been calculated using the radio 4.9 GHz fluxes or upper limits, or (for stars with no radio observations) by predicting 4.9 GHz fluxes by scaling the excess 10μm data using a spectral index of 0.75. The terminal velocity is taken as v_{black} from PBH or as 0.74×v_{max}.

The WR mass loss rates lie in the range 10^{-5} – 10^{-4} $M_\odot$ yr^{-1}, and are plotted as a function of WN and WC subtype in Fig 5. The overall mean value (excluding upper limits) is 5.3 $\pm$ 2.3 x 10^{-5} $M_\odot$ yr^{-1}, with no significance difference between this value and the means for (a) single or binary stars, (b) WN or WC stars. (c) WNE or WNL, (d) WCE or WCL stars. The apparently similarity in $\dot{M}$ between WN and WC stars, and between WCE and WCL stars is open to uncertainty, given the assumptions about the carbon ionisation balance in the radio-emitting regions. ABCT, BSW and van der Hucht *et al.* (1986) have argued that $\dot{M}$ for WC stars may be systematically larger (by $\sim$ × 2) than for WN stars.

6. Mass loss rates – other methods

Other independent methods of deriving mass loss rates for WR stars are generally based on studies of WR+O binary systems. For V444 Cyg, Khaliullin *et al.* (1984) have measured the period change of the system (using data over 1902–1983) to be $\dot{P} = 0.22 \pm 0.04$ sec yr^{-1}, which modelled in terms of a spherically symmetric outflow beyond the system leads to a mass loss rate determination of $\dot{M} = 1.1 \times 10^{-5}$ M$_\odot$ y^{-1}.

St-Louis *et al.* (1988) have studied the phase-dependent variation in optical linear polarisation for 10 WR+O systems, which they interpret in terms of scattering of the companion star light off free-electrons in the WR wind to derive a new, independent method of calculating WR mass loss rates in such systems. The polarisation data also yields values for the individual orbit inclinations and thus stellar masses. Their results, shown in Fig 6, give values of $\dot{M}$ in the range $10^{-4} - 10^{-5}$ M$_\odot$ y^{-1}, comparable in scale to that derived from the radio-IR data.

In the case of both WN+O and WC+O binaries, strong phase-dependent variability is also a common property in the ultraviolet — even for systems which show no optical eclipses. Pronounced line profile changes in the UV resonance and low-excitation lines are seen interpreted in terms of selective line-eclipses of the O-star light as it shines through the WR wind (Willis & Wilson 1976, Koenigsberger & Auer 1985). Such eclipse effects in the CIII] λ1909 line in γ Velorum (Willis *et al.* 1979), and in CV Serpentis (Howarth *et al.* 1982) have been modelled to yield mass loss rates for the WC8 components in these two systems of 9×10^{-5}, and 7×10^{-5} M$_\odot$ y^{-1} respectively.

Additionally, the detailed model atmosphere analyses of the helium line spectra of 30 galactic WR stars by Schmutz *et al.* (1989) provide estimates of mass loss rates for individual stars, yielding values in the range $10^{-5.3} - 10^{-3.9}$ M$_\odot$ yr^{-1}, in excellent agreement with results from the radio data.

7. $\dot{M}$: L relations & the wind momentum problem

It is now well established that for OB stars $\dot{M} \propto L^{1.6}$ (ABCT, Howarth & Prinja 1989), and that this observed relation agrees well with predictions of radiation pressure mass loss theory. Attempts to ascertain a corresponding scaling for WR stars are plagued by uncertainties in the luminosities to be applied, given that the values for T$_{eff}$ and bolometric corrections are very uncertain for this stellar class.

ABCT found for five WR+O double-lined systems a tentative relation between WR mass loss rate and mass of the form $\dot{M} \sim 7\times10^{-8}(M/M_\odot)^{2.3}$ M$_\odot$ yr^{-1}, which, together with the Maeder (1983) WR mass–luminosity relation ($\log(L/L_\odot) = 3.8 + 1.5\log(M/M_\odot)$) implies $\dot{M} \sim 7\times10^{-14}(L/L_\odot)^{1.6}$ (similar in slope to that for OB stars). They further deduced a rough relation between $\dot{M}$ and the quantity M$_v$ – <M$_v$> (the difference for an individual WR star between its measured M$_v$ determined from cluster membership and the average value for its spectral type), providing additional support to the assertion of a WR mass loss rate – luminosity relation.

Smith & Maeder (1989) used new data on galactic WR stellar masses and values of M$_v$, to show (Fig 7) that for both WN and WC stars in WR+O binary systems, a reasonably well-defined relation exists in M$_v$ vs. M_{WR}, which has the same slope as that predicted theoretically for stellar evolution models for He-burning stars with initial masses between 120 and 40 M$_\odot$, from Maeder & Meynet (1987) ($\log(L/L_\odot) = 4.02 + 1.34\log(M/M_\odot)$). This comparison yielded a B.C. of -4.5 mag., which they suggest can be applied to all WR subclasses. Applying this B.C. for stars with known M$_v$ and mass loss

Fig 6: $\dot{M}$ vs. M_{WR} from polarisation studies of WR+O systems (from St-Louis et al. 1988)
Fig 7: M_v vs. M_{WR} for WR binaries from Smith & Maeder (1989) implying a B.C. = -4.5 mag when compared to the evolutionary models of Maeder & Meynet (1987).

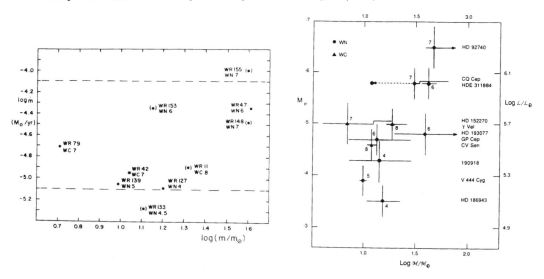

rates from radio and/or IR data from ABCT and BSW, they deduce a rough scaling of $\dot{M} \propto L^{0.7}$, with a spread in $\dot{M}$ for any given L of an order of magnitude (see Fig 8). The slope is considerably lower than the value of 1.6 deduced by ABCT. The revised values of $\dot{M}$ given in Table 1, plotted against L, would not alter their conclusions. The results from St-Louis et al. (1988) (see Fig 6) indicate a rough scaling with $\dot{M} \propto M_{WR}^{0.8-1.3}$, comparable to the Smith & Maeder (1989) conclusions. At present therefore, available data would suggest a rough scaling of WR mass loss with radiative luminosity: $\dot{M}(WR) \propto L^{\sim 1}$, shallower than for OB stars, and with a spread of up to an order of magnitude.

For WR stars characterising the ratio of the momentum in the wind to the single scattering radition pressure limit, viz. $\dot{M} v_\infty c/L_*$, again suffers from uncertainties in L_*. Adopting a single B.C. = −4.5 mag for WR stars from Smith & Maeder (1989) results in the values of L_* given in Table 1, which can probably be taken as upper limits to the radiative luminosities. The resulting momentum ratios, given in Table 1, show values in the range ∼ 1–30, with a trend for lower values for late-type WN and WC9 stars: mean values are: 2.2 (WN7,8), 10.3 (WNE), 13.5 (WC5–8) and 2.3 (WC9). These values are larger than appropriate for OB stars which typically have values $\gtrsim$ 1. Adopting lower B.C's and L_*'s proposed by BSW and Schmutz et al. (1989) will only exacerbate this problem, since then values of the momentum ratio would rise up to 50 or so. The results indicate a mechanism other than radiation pressure may be needed to initiate the WR mass loss, unless multiple scattering is extremely effective. Abbott (1982) concluded that chemical composition effects alone could not enhance the radiative driving for WN stars, and only by $\leq$ × 3 for WC stars, whilst attempts so far to include multiple scattering have led to gains of $\leq$ × 5 (e.g. Friend & Castor 1983). Smith & Maeder (1989) propose that (in addition to radiation pressure effects) an increase in mass loss rate might be promoted by mechanical instability due to the ϵ-mechanism whose effect will be enhanced with changing mass and μ within the star. The expectation is thus that for WN7 and WN8 stars, which generally show appreciable atmospheric hydrogen, radiation pressure

Fig 8: $\dot{M}$ *vs.* L_{WR} *from Smith & Maeder (1989) assuming B.C. = -4.5 mag.*

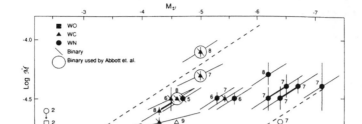

is probably the major driving mechanism. As the last of the hydrogen is removed, the instability mechanism is activated, and the mass loss rate enhanced. Alternatively, Poe, Friend & Cassinelli (1989) have developed an axisymmetric model for WR winds involving a slow-dense equatorial flow and a fast, radiation-driven wind at higher latitudes, with high rotation ($\sim$ 85 percent maximum) and substantial, open magnetic fields ($\sim$ 1500G) needed to accommodate the observed radio data and high wind speeds. The theory of WR winds and mass loss will be addressed further by Cassinelli (these proceedings).

Whatever mechanism(s) is driving the WR mass loss it is clear that these stars are a major source of mechanical energy input (and chemical enrichment) into the interstellar medium. Using the results for WR stars within 3 kpc of the Sun, this topic has been discussed by ABCT and updated by van der Hucht *et al.* (1986) who give the following inputs to the ISM: wind mass flux = 1.6×10^{-4} $\dot{M}$ yr^{-1} kpc^{-2}, wind momentum ($\dot{M} v_\infty$) = 2.3×10^{30} g cm s^{-2} kpc^{-2}: and wind energy input = 3.2×10^{38} erg s^{-1} kpc^{-2}, noting that whilst for WR+OBA stars, the former supply only $\sim$ 8 % of the input radiative luminosity, they supply $\sim$ 70 % of the mechanical mass, momentum and energy input. The effects of these inputs will be discussed elsewhere at this meeting.

8. WR wind instability

In addition to variability that can clearly be ascribed to phase-dependent effects in WR+O binary systems, it is becoming increasingly clear that the stellar winds of WR stars are intrinsically variable. Single WR stars often show optical continuum light variability at the level of $\sigma_\nu \sim$ 0.003–0.03 mag. (Lamontagne & Moffat 1987), as well as continuum polarisation varaibility at a level of $\sigma_p \sim$ 0.01–0.15% (Robert *et al.* 1988). High precision optical spectroscopy in the He II λ5411 line of HD 191765 (WN6) presented by Moffat *et al.* (1988) reveal time-dependent structure (emission bumps) in the emission profile, which appear to accelerate outward along with the ambient wind on a time-scale of hours. Moffat *et al.* (1988) interpret these data in terms of rapid ejections of material condensations ('blobs').

In the ultraviolet, IUE spectroscopy has provided strong evidence for intrinsic WR line profile variability particularly evident in the P Cygni absorption components. Results for HD 50896 (WN5) indicate a variability timescale of $\sim$ 1 day (Willis *et al.* 1989, St-Louis *et al.* 1990), with a similar timescale for HD 192163 (WN6) (St-Louis *et al.* 1989). For HD 96548 (WN8) a longer timescale of UV variability of ~ several days seems present (Smith *et al.* 1986). Both the UV and optical data indicate that the amplitude of this short-timescale variability can alter with epoch – for instance long term photometric monitoring of HD 50896 shows extended periods of relative 'inactivity', with abrupt changes to higher levels of variation (van der Hucht *et al.* 1990). Similarly, IUE monitoring has shown that HD 50896 can undergo periods of stability in its UV line profiles (Willis *et al.* 1989).

Whether or not radiation pressure is *the* mechanism initiating the higher levels of WR mass loss, it is likely that radiation pressure dominates the acceleration in their winds. It is now recognised that the radiation-driven winds of OB stars are highly unstable and develop strong reverse shocks (Owocki, Castor & Rybicki 1989) which may provide a natural explanation for their observed X-ray emission, extended blackness in the absorption components of UV resonance line P Cygni profiles, and the occurence and variability in discrete absorption components seen in these lines (Howarth & Prinja 1989). The WR UV P Cygni profiles also show extended regions of blackness; the X-ray emission from WR stars is (generally) comparable to that of OB stars (Pollock 1987), and therefore the WR winds may also be permeated by strong shocks, with the intrinsic WR wind variability caused by the same mechanism that generate the wind instabilities in the OB stars, with a greater amplitude reflecting enhanced WR wind densities.

9. WR mass loss in other galaxies

OB stars in the LMC and SMC show significantly lower stellar wind terminal velocities, and (possibly) lower mass loss rates from studies of their UV resonance line P Cygni profiles (e.g. Prinja 1987, Garmany & Fitzpatrick 1988), which is broadly in line with expectations of radiation pressure-driven wind models for lower metallicity environments. Whilst no attempts have been made to actually determine mass loss rates for WR stars in the MCs, its appears from the overall nature of their optical and UV spectra that there may be no significant differences in $\dot{M}$ and v_∞ with galactic counterparts. Estimates of v_∞ for LMC stars, both from IUE P Cygni profile data and optical line widths, agree with galactic values of corresponding subtypes (Abbott & Conti 1987). Smith *et al.* (1990) find a tight correlation between the ratio of the widths of $\lambda5808/\lambda4650$ and the $\lambda5808$ line width itself which is the same for both LMC and galactic WC stars.

The strength of the optical emission line spectrum may be taken as indicative of the mass loss rate (*cf* the optical emission measure *vs.* radio flux correlation from ABCT). Conti & Massey (1989) find that the line strengths of leading optical WN and WC emission lines are very similar in the LMC and the Galaxy, whilst Conti, Massey & Garmany (1990) suggest that the optical helium and nitrogen lines in the WNE stars in the SMC may be systematically weaker in strength (but not in velocity width) than galactic counterparts, although here the very small number of stars and binary companion effects may be playing a role. Further afield, Massey *et al.* (1987) find that WN and WC stars in M33 show similar optical line strength *vs.* line width correlations as found for galactic counterparts, whilst the few WR stars in M31 studied show poorer agrement. On balance, available UV and optical data do not point to gross mass loss or wind velocity differences between extragalactic and galactic WR stars *of the same subtypes*. What is striking, of course, is the substantial differences between the total WR populations and subtype occurence in the different galactic environments. If differing global galaxy metallicity is playing a role with regard to WR stars, its effect is possibly more pronounced in the WR precursor phase, than in significantly modifying the mass loss and winds of WR stars once they have formed.

References

Abbott, D.C., 1982, *Proc IAU Symp. No.99*, (eds C. de Loore & A.J. Willis), D. Reidel, p 185

Abbott, D.C., Bieging, J.H., Churchwell, E., & Torres, A.V., 1986, *Astrophys. J.*, **303**, 239

Abbott, D.C., & Conti,P.S., 1987, *Ann. Rev. Ast. Astrophys.*, 25, 113

Barlow, M.J., 1982, *Proc IAU Symp. No.99*, (eds C de Loore & A J Willis), D. Reidel, p 149

Barlow, M.J., Smith, L.J., & Willis, A.J., 1981, *Mon. Not. R. astr. Soc.*, **196**, 101

Barlow, M.J., Roche, P.F., & Aitken, D.K., 1988, *Mon. Not. R. astr. Soc.*, **232**, 821

Becker, R.H., & White, R.L., 1985, *Astrophys. J.*, **297**, 649

Cherepaschuk, A.M., Eaton, J.A., & Khaliullin, Kh.F., 1984, *Astrophys. J.*, **281**, 774

Conti, P.S., Leep, E.M., & Perry, D., 1983, *Astrophys. J.*, **268**, 228

Conti, P.S., & Massey, P., 1989, *Astrophys. J.*, **337**, 251

Conti, P.S., Massey, P, & Garmany, C.D., 1990, *Astrophys. J.*, **341**, 113

Dickel, H.R., Habing, H.J., Isaacman,R., 1980, *Astrophys. J. Lett.*, **238**, L39

Florkowski, D.R., 1982, *IAU Symposium No. 99*, (eds C de Loore & A J Willis), D. Reidel, p 63

Friend, D.B., & Castor, J.I., 1983, *Astrophys. J.*, **272**, 259

Friend, D.B., & Abbott, D.C., 1986, *Astrophys. J.*, **311**, 701

Garmany, C.D., & Fitzpatrick, E.L., 1988, *Astrophys. J.*, **332**, 711

Hamann, W-R., 1981, *Astr. Astrophys.*, **93**, 353

Hamann, W-R., Schmutz, W., & Wessolowski, U., 1988, *Astr. Astrophys.*, **194**, 190

Hillier, D.J., 1987, *Astrophys. J. Suppl.*, **63**, 965

Hillier, D.J., 1989, *Astrophys. J.*, **347**, 392

Hogg, D.E., 1982, *IAU Symp. No. 99*, (eds C de Loore & A J Willis), D. Reidel, p 221

Hogg, D.E., 1985, *in Radio Stars*, (eds R Hjellming & O Gibson), D.Reidel, p 117

Howarth, I.D., Willis, A.J., & Stickland, D.J., 1982, *ESA SP-176*, p 331

Howarth, I.D., & Prinja, R.K., 1989, *Astrophys. J. Suppl.*, **69**, 527

Hummer, D.G., Barlow, M.J., & Storey, P.J., 1982, *IAU Symp. No. 99*, (eds C de Loore & A J Willis), D.Reidel, p 79

Jones, P.A., 1985, *Mon. Not. R. astr. Soc.*, **216**, 613

Khaliullin, Kh.F., Khaliullina, A.I., & Cherepaschuk, A.M., 1984, *Pis'ma Astron. Zh.*, **10**, 600 (= Sov. Astron. Letter., **10**, 250)

Koenigsberger, G., & Auer, L.H., 1985, *Astrophys. J.*, **297**, 255

Kuhi, L.V., 1973, *Proc IAU Symp. No. 49*, (eds M Bappu & J Sahade), D.Reidel, p 205

Lamontagne, R.L., & Moffat, A.F.J., 1987, *Astr. J.*, **94**, 1008

Lucy, L.B., 1982, *Astrophys. J.*, **255**, 278

Lucy, L.B., 1983, *Astrophys. J.*, **274**, 372

Maeder, A., 1983, *Astr. Astrophys.*, **120**, 113

Maeder, A., & Meynet, G., 1987, *Astr. Astrophys.*, **182**, 243

Massey, P., Conti, P.S., & Armandroff, T.E., 1987, *Astr. J.*, **94**, 1538

Moffat, A.F.J., Drissen, L., Lamontagne, R.L., & Robert, C., 1988, *Astrophys. J.*, **334**, 1038

Owocki, S., Castor, J.I., & Rybicki, G.B., 1988, *Astrophys. J.*, **335**, 914

Pauldrach, A., Puls, J., & Kudritzki, R.P., 1986, *Astr. Astrophys.*, **164**, 86

Poe, C.H., Friend, D.B., & Cassinelli, J.P., 1989, *Astrophys. J.*, **337**, 888

Pollock, A.M.T., 1987, *Astrophys. J.*, **320**, 283

Prinja, R.K., 1987, *Mon. Not. R. astr. Soc.*, **228**, 173

Prinja, R.K., Barlow, M.J., & Howarth, I.D., 1990, *Astrophys. J.*, **361**, in press

Purton, C.R., Fieldman, P.A., Marsh, K.A., Allen, D.A., & Wright, A.E., 1982, *Mon. Not. R. astr. Soc.*, **198,** 321

Robert, C., Moffat, A.F.J., Basten, P., Drissen, L., & St-Louis, N., 1989, *Astrophys. J.*, **347,** 1034

Schmutz, W., Hamann, W-R., & Wessolowski, U., 1989, *Astr. Astrophys.*, **210,** 236

Smith, L.F., & Hummer, D.G., 1988, *Mon. Not. R. astr. Soc.*, **230,** 511

Smith, L.F., & Maeder, A., 1989, *Astr. Astrophys.*, **211,** 71

Smith, L.F., Shara, M.M., & Moffat, A.F.J., 1990, *Astrophys. J.*, **348,** 471

Smith, L.J., Willis, A.J., Garmany, C.D., & Conti, P.S., 1986, *ESA SP-263*, p 389

St-Louis, N., Moffat, A.F.J., Drissen, L., Bastien, P., & Robert, C., 1988, *Astrophys. J.*, **330,** 286

St-Louis, N., Smith, L.J., Stevens, I.R., Willis, A.J., Garmany, C.D., & Conti, P.S., 1989, *Astr. Astrophys.*, **226,** 249

St-Louis, N., Smith, L.J., Willis, A.J., Garmany, C.D., & Conti, P.S., 1990, *ESA SP-310*, in press

Torres, A.V., Conti, P.S., & Massey, P., 1986, *Astrophys. J.*, **300,** 379

van der Hucht, K.A., Cassinelli, J.P., & Williams, P.M., 1986, *Astr. Astrophys.*, **168,** 111

van der Hucht, K.A., Hidayat,B., Admironto, A.G., Supelli, K.R., & Doom, C., 1988, *Astr. Astrophys.*, **199,** 217

van der Hucht, K.A., van Genderen, A.M., & Bakker, P.R., 1990, *Astr. Astrophys.*, **228,** 108

Williams, P.M., Eenens, P.R.J., 1989, *Mon. Not. R. astr. Soc.*, **240,** 445

Williams, P.M., van der Hucht, K.A., Sondell, G., & The, P.S., 1990, *Mon. Not. R. astr. Soc.*, **244,** 101

Willis, A.J., 1982, *Mon. Not. R. astr. Soc.*, **198,** 897

Willis, A.J., & Wilson, R., 1976, *Astr. Astrophys.*, **47,** 429

Willis, A.J., et al., 1979, *The 1st Year of IUE (UCL)*, p 394

Willis, A.J., Howarth, I.D., Smith, L.J., Garmany, C.D., & Conti, P.S., 1989, *Astr. Astrophys. Suppl.*, **77,** 269

Wright, A.E., & Barlow, M.J., 1975, *Mon. Not. R. astr. Soc.*, **170,** 41

DISCUSSION

Smith, Lindsey: As co-author of the quoted low power of $\dot{M}$ dependence on M(or L), I should point out that the other author, Maeder, is now using $\dot{M} \propto M^{2.5}$ (suggested by Langer) with spectacular success. Also remembering that Langer suggests a higher rate for WC than WN, it looks like the stars, as they evolve zig-zag across the observed band in the $\dot{M}$-L diagram.

Willis: I am aware that the stellar evolutionists are employing a steeper $\dot{M}$ *vs.* M relation than implied by your work and that from the polarization results. My remit was to concentrate on observational results, which at the present do not necessarily support the theoretical viewpoint directly. I cannot help that, but there may still be considerable uncertainties in both M_{WR} and L rather than in mass loss rate, which is masking the true observational picture.

Conti: Why do we not find a single $\dot{M}$-L relation for WR stars? After all, there is one for O stars. The stellar structure people predict a $\dot{M}$-mass relation. We believe W-R winds are radiatively driven. Could one instead assume there is a unique (displaced from the O star one due to chemistry) $\dot{M}$-L relation for WR stars and use this to infer the WR stars L (and $B.C.$) since the $\dot{M}$ seems well known?

Willis: Again, the recent results in the literature based on observations of mass loss rate, L and $\dot{M}_{WR}$, do not point to a well-defined $\dot{M}$-L relation for WR stars, nor one with the same slope as for OB stars. Whilst radiation pressure is undoubtedly going to be important in WR winds, it may not be the only mechanism driving their mass loss. In that event a simple relation with L may not necessarily be expected. On the other hand, I agree it would be interesting to see the implications of assuming a relation that you propose on the inferred L and $\dot{M}_{WR}$.

Sreenivasan: Questioning observations is supposed to be in bad taste. You say that mass loss rates are uncorrelated with spectral type but Langer says he finds good agreement with observations if he uses a mass loss rate $\propto M^{2.3-2.5}$. This is confusing. Is it still in bad taste to question the observations and/or the interpretations? (From a theoretical point of view, radiatively driven winds give mass loss rates proportional to a power of luminosity and there definitely are other sources/mechanisms of mass loss in WR stars.).

Willis: I think the observations (radio, IR, polarization and UV data) are in pretty good shape, but the interpretation is still subject to some uncertainty. For instance modelling the radio emission does require a knowledge of the ionization balance (and chemistry) in the radio region, which at least for WC stars is not too well known. Factors of two or so uncertainty cannot be ruled out, say between WCE and WCL stars. As for other mechanisms driving WR winds, I can fortunately defer that aspect to Joe Cassinelli in the next review talk!

Pakull: You mentioned bolometric corrections of about -4 to -4.5.

Willis: -4.5.

Pakull: I will tell you tomorrow that there are some WR stars, namely WN, very early ones, which have bolometric corrections of about -7. These are very very hot stars which also ionize large $HeIII$ regions around them and in these stars you also have of course in a radio-emitting region He completely ionized. And this may be a reason why you find for the WN2 stars very low $\dot{M}$ whereas this is only because the He is completely ionized also in the radio-emitting region.

Willis: As far as the bolometric correction is concerned, I was attempting to illustrate the mass loss rate momentum problem by picking a luminosity which I felt was probably an upper limit. I took the Smith-Maeder results, partly because I knew Lindsey Smith was chairing this session. For the bulk of the stars, the Schmutz values alone already raise the momentum question. So, it was an attempt to summarize the lower level of the problem, if a problem it is, and of course we have the bolometric correction problem... If you have -7, which I have not come across before I might add, that is going to ease it considerably. But, -7 will not apply to the bulk of the WR subclasses quite demonstratively, it is not going to affect the momentum ratio problem in general.

Schmutz: First thing we did after analyzing 30 stars, was looking at the mass loss rate and the luminosity and at the moment we consider our models to be good within 1 magnitude for the bolometric corrections, as absolute. But by doing so, comparing stars which have been analyzed by the same methods, you actually have problably the same systematic errors for all. So, your trend would show up much easier, and there simply is no trend of mass loss rate with luminosity. I do not believe there is any.

Moffat: A more recent compilation of polarization data of WR+O binaries in the Galaxy and the Magellanic Clouds (*cf.* poster at this meeting) show that $\dot{M}_{WR}$ and M_{WR} correlate rather well with spectral subclass for WN and WC sequences, with $\dot{M}_{WR}$ and M_{WR}. Possibly the radio $\dot{M}$'s are inflicted with large scatter due to uncertainties in the distances which are not needed for the polarization $\dot{M}$'s.

Willis: This is an interesting and important result that we need to consider. I am impressed with the utilization of polarization variability in constraining mass loss rates and relations with M_{WR} and subtype.

Smith, Lindsey: A thing that should be pointed out is that the success of the models using Langer's mass loss rate $\propto M^{2.5}$ is pretty impressive, as I am sure we will hear more about in the future.

Underhill: Just to remind you again about V444 Cyg. Fourty years of photo-electric observations saw a very distinct rate of period change. Rate of period change is supposed to be connected with a rather simple formula to rate of mass loss from the system, from one star as I should say, actually both stars got winds. The momentum rate of mass loss you can get from your observed rate of period change is $6 \times 10^{-6} M_\odot yr^{-1}$. I think that it is a more fundamental way of estimating rate of mass loss than using the radio flux and saying that it is free-free in a spherical distribution. The fact that you got an index of 0.6 to 0.7 only means that, if it is free-free, you are observing a body of gas with a density gradient in it. It does not tell you that it is a sphere going out. That radio flux could have three sources.

Willis: I might add actually, that the $10\mu - 6cm$ spectral index for V444 Cyg is 0.52, which is almost exactly what you would expect (0.53) for a constant velocity isothermal flow between the IR and radio region. V444 Cyg is the only star in our (BSW, 1981) 10μ sample that does this.

There seems to be some discrepancy between the actual period change, between Cherepashchuk and yourself, that does not worry me too much, actually. Essentially, the radio data agree with the polarization measurements in sofar as both interpretations give a mass loss rate of about 10^{-5}. That would agree with the Cherepashchuk period change. It is a factor or two larger than your measurement of period change of $0.08\ s \cdot yr^{-1}$.

Underhill: (???)

Willis: What I am trying to say is that the mass loss rate inferred from these other analyses is at the lower bound of the range that one gets from the WR stars as a bunch and frankly I never get worked up about factors of two; factors of ten yes, but not factors of two.

Vanbeveren (to Underhill): You contineously repeat your $\dot{P}/P$ argument. However, according to me, you are not consistent with your own WR model. One can derive $\dot{M}$ from

$\dot{P}/P$ if the $\dot{M}$ is spherically symmetric by using an easy formula. However, when you have a large accretion disc around the WR star (as you propose) then this easy relation between $\dot{P}/P$ and $\dot{M}$ disappears.

Willis: I should point again out that Underhill's measurement of $\dot{P}/P$ for V444 Cyg is a factor or two lower than the measurement of Khaliullin *et al.* (1984). The latter would give very good agreement with the radio mass loss rate. The question of who is right about the period change is something for these authors to sort out, but I am not particularly concerned about a factor of two in this game. Moreover, as we know, the mass loss rate for V444 Cyg is amongst the lowest for the WR stars, and the momentum problem is still there for most objects in this class.

Schmutz: I would like to point out that V444 Cyg is not a typical WN5 WR star. "Normal" strong lined WN5 stars have different line profiles. The most important difference is that weak high ionization lines of V444 Cyg are very narrow, *e.g.* $NV\lambda4940$, in contrast to the broad profiles of strong-lined single WN5 stars. This difference is not due to different v_∞ since the v_∞'s are similar. The narrow lines of V444 Cyg indicate therefore that we see down to low wind velocities whereas we do not see slow moving material in "normal" WN5 stars. Therefore V444 Cyg has not a typical velocity law which is in line with its mass loss rate that is at the lower limit of the observed rates.

Willis: I agree with you that the mass loss rate of the WN5 component in V444 Cyg (as deduced from the period change and the radio data) lies at the lower bound for WR stars, $\sim 1 \times 10^{-5} M_\odot yr^{-1}$, and as such we can probably "see" further into its wind down to low velocity parts. The reason for quoting the Cherepashchuk *et al.* (1984) velocity law for V444 Cyg was that, as far as I know, it is the only one which has been observed somewhat directly. On another topic, I concentrated on using the $B.C. = -4.5$ mag from Smith & Maeder (1989), rather than your rather lower values (generally around $B.C. = -3$ mag), partly because I wanted to emphasise effectively the *lower* limits to the momentum problem, and partly because I knew Lyndsey Smith would be chairing this session, and it seemed a wise and prudent thing to do!

Cherepashchuk: Perhaps one could try to take into account the cloud ragged structure of the WR wind. If the WR wind in the infrared and the radio region is cloudy and ragged we can obtain the same radio and IR flux for a lesser value of $\dot{M}$. It may help us to eliminate some discrepancies between $\dot{M}$ values obtained from IR and radio data and those obtained from period change in binary systems.

Schulte-Ladbeck: I just want to remind ourselves that we have heard yesterday and today that there are large-scale inhomogeneities in WR star winds: blobs, shocks, cloudlets, disk, and different effects seem to prevail in different stars. While we all agree that radiation pressure is the first order effect, would you not expect to find a large scatter in the $\dot{M}$-L relation for this very reason?

Willis: I have emphasised the new observations of intrinsic variability in WR stellar winds, both from optical and UV data. It may be that much of this can potentially be related to radiatively-induced instabilities in the winds of the kind that Owocki has talked about. On the other hand, other effects may well be operating and contributing to the $\dot{M}$-L scatter.

INFRARED OBSERVATIONS OF MASS LOSS FROM MASSIVE STARS

M.J. BARLOW
Department of Physics & Astronomy
University College London
Gower St., London WC1E 6BT.

ABSTRACT: The future use of space-borne IR spectroscopy to determine the ionization structure and abundances in the outer winds of WR stars is described. A mass loss rate of 1.7×10^{-5} M$_\odot$ yr^{-1} has been derived from 10 μm photometry of the WO2 star Sanduleak 5 (WR 142). The He/H number ratios in the winds of P Cyg and AG Car have been derived from a recombination line analysis of their 1–4 μm spectra and mass loss rates of 2.2×10^{-5} M$_\odot$ yr^{-1} and 3.7×10^{-5} M$_\odot$ yr^{-1} have been respectively derived for them.

1. Introduction

In this paper I describe a number of recent developments in the field of infrared spectrsocopy and photometry that are relevant to the study of WR and related massive stars. Section 2 describes how spectrophotometry of infrared fine structure lines can be used to derive terminal velocities and ionic abundances in the outer regions of winds. In Section 3, the mass loss rate of a WO star is determined for the first time. In Section 4, the He/H ratios and mass loss rates of P Cyg and AG Car are derived.

2. Wind Terminal Velocities and Abundances from IR Fine Structure Lines

Aitken, Roche & Allen (1982) discovered the presence of the IR fine structure lines of [S IV] 10.5 μm and [Ne II] 12.8 μm in the 8–13 μm spectrum of γ Velorum, while van der Hucht & Olnon (1985) found the [Ne III] 15.5 μm line in the *IRAS* LRS spectrum of the same star. These fine structure lines originate at very large radii in the wind, their emission peaking in the wind regions corresponding to their critical densities of 10^4–10^6 cm^{-3}. At these distances, the stellar wind is moving at terminal velocity and so the line profiles should be flat-topped, with FWZI's equal to twice the wind terminal velocity. Barlow, Roche & Aitken (1988) confirmed that the [Ne II] 12.8 μm line in the spectrum of γ Velorum did have a flat-topped profile and determined a wind terminal velocity of 1520$\pm$200 km s^{-1} for the WC8 star responsible for the emission. The terminal velocity of this star had previously been uncertain due to the presence of the O9.5 I companion spectrum in the ultraviolet. Recently, Prinja *et al.* (1990) have derived a terminal velocity of 1415 km s^{-1} for the wind of the WC8 component of γ Vel, from the black absorption edge velocity of the C IV 1549 Å profile, in good agreement with the IR value. Williams & Eenens (1989) have also recently shown that the velocity of the displaced absorption component of He I 2.058 μm in WR

K. A. van der Hucht and B. Hidayat (eds.),
Wolf-Rayet Stars and Interrelations with Other Massive Stars in Galaxies, 281–287.
© 1991 *IAU. Printed in the Netherlands.*

spectra yields the wind terminal velocity. Spectroscopy of the 2 μm He I line, or of the various IR fine structure lines, can therefore provide a means in the future for determining wind terminal velocities for the numerous WR stars which are too heavily reddened for high resolution UV spectroscopy to be feasible.

The fluxes of the infrared fine structure lines emitted by stars with high mass loss rates can be used, together with the free-free continuum fluxes emitted by the same regions of the wind at radio wavelengths, to derive accurate ionic abundance ratios; relative to hydrogen in the case of OB stars, or relative to helium in the case of WR stars. Barlow et al. (1988) derived $Ne^{2+}/Ne^{+} = 3.8$ for the outer wind of γ Vel, from the fluxes in the [Ne III] 15.5 μm and [Ne II] 12.8 μm lines, and derived a total neon mass fraction that was a factor of two larger than the solar value. However, evolutionary models for WC stars (e.g. Maeder 1984) predict that neon should be enhanced by a factor of twelve relative to its solar abundance, due to the conversion of all the initial C,N and O, first into ^{14}N by the CNO cycle during the H-burning phase, and then into ^{22}Ne by successive α-particle captures onto ^{14}N during the He-burning phase. There is therefore a large discrepancy between observations and current theoretical predictions for the neon abundance in WC stars.

A significant advance in our understanding of abundances and ionization structures in the outer winds of massive stars should occur with the advent of ESA's *Infrared Space Observatory (ISO)*, currently scheduled for launch in mid-1993. *ISO* will carry two spectrometers, one covering the 3-45 μm wavelength region with a resolving power of 1–2×10^{3} and the other covering the 45-200 μm wavelength region with resolving powers of 250 or 10^{4}. A number of fine structure lines of C, N, O and Ne that are not observable from the ground will therefore become accessible to these spectrometers: [C II] 158 μm; [N II] 122 μm and [N III] 57 μm; [O I] 63 & 146 μm, [O III] 52 & 88 μm and [O IV] 26 μm; plus [Ne III] 15.5 & 36 μm and [Ne IV] 14.3 & 24 μm. With projected *ISO* sensitivities, it should be possible to derive ionic and elemental abundances from these lines for WN and WC stars out to distances of at least 3 kpc from the Sun. The same should also be true for high luminosity B supergiants – the line fluxes are proportional to $(\dot{M}/v_{\infty})^{3/2}$ (Barlow et al. 1988), so the lower mass loss rates of these stars, relative to WR stars, are compensated for by their lower terminal velocities.

3. The Mass Loss Rate of the WO2 star WR 142

The WO stars are a rare group of WR stars whose optical spectra show very strong O VI 3811,34 Å emission. Two WO stars have been identified within our own Galaxy but to date no radio detections have been obtained that would allow their mass loss rates to be derived. Given that these objects appear to be the most evolved WR stars so far identified, their mass loss rates are clearly of interest. It is now possible to estimate the mass loss rate for the closest WO star, Sanduleak 5 ($\equiv$WR 142$\equiv$ST 3), using infrared photometry obtained with the 3.8m *UKIRT*.

The mean infrared-radio spectral index (10 μm – 6 cm) found for WR stars is about 0.75 (Bieging et al. 1982), compared to the value of 0.60 predicted for an optically thick stellar wind of constant degree of ionization (Wright & Barlow 1975). It has been realised recently (e.g. Schmutz & Hamann 1986; Hillier 1987) that the main reason for this discrepancy is that the degree of ionization in WR winds decreases outwards, so that the effective charge per emitting (helium) ion is 2 in the IR-emitting region, but only 1 in the radio-emitting region. It is therefore valid to derive WR mass loss rates from 10 μm free-free excess fluxes,

Sand 5 + 80000K black-body normalised to J

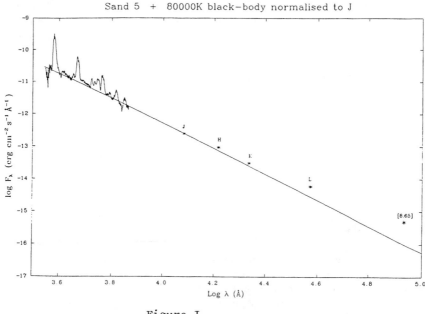

Figure I

P Cyg (dereddened E(B-V)=0.6) + Kurucz 18000K log G=2.05

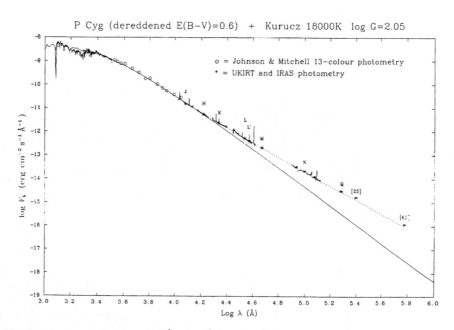

Figure 2

provided the appropriate wind composition and degree of ionization are adopted.

Sanduleak 5 is located in the heavily obscured Galactic cluster Be 87, at a distance of 1 kpc (Turner & Forbes 1982). Figure 1 shows its optical–IR energy distribution after dereddening by $A_V = 6.32$ magnitudes. The optical flux distribution comes from Barlow & Hummer (1982), the JHKL' data comes from Williams (1982) and the 8.65 μm datum comes from *UKIRT*. After subtraction of an 80,000 K blackbody normalised to the optical and J data (Figure 1), an excess flux of 0.10 Jy at 8.65 μm is derived. For a radio–IR spectral index of 0.75, a 6 cm radio free-free flux of 0.13 mJy is predicted from Sanduleak 5, compared to the upper limit of 0.6 mJy set by Abbott *et al.* (1986). To estimate its mass loss rate, a wind terminal velocity of 5500 km s^{-1} is adopted, along with $n(C^{4+})/n(He^{2+}) = 0.20$ and $n(O^{6+})/n(He^{2+}) = 0.03$ (Kingsburgh & Barlow 1990). A temperature of 40,000 K is adopted for the IR-emitting region of the wind (the Gaunt factor is very weakly dependent upon this parameter). A mass loss rate of 1.68×10^{-5} M$_\odot$ yr^{-1} is derived.

This mass loss rate is at the lower end of the range found for WR stars. The average found for those WC stars having well-determined radio fluxes and terminal velocities (Prinja *et al.* 1990) is 6×10^{-5} M$_\odot$ yr^{-1}. However, the mass loss rate found by the latter authors for the WC5 star HD 165763, 3.2×10^{-5} M$_\odot$ yr^{-1}, is intermediate between the WC mean mass loss rate and the rate found above for Sanduleak 5. Since HD 165763 has the earliest spectral type (and the one closest to that of Sand. 5) amongst the WC stars with a radio-derived mass loss rate, one might speculate that this is consistent with WR stars having a mass loss rate which is a strong function of mass (Abbott *et al.* 1986; Langer 1989), since the mass loss rate of a WR star should then decline as it evolves by mass loss stripping from a WC to WO type.

4. The He/H Ratios and Mass Loss Rates of P Cyg and AG Car

The mean 6 cm radio flux for P Cygni is 6.7±0.2 mJy (van den Oord al. 1985), its wind electron temperature in the radio-emitting region is 6000 K (Becker & White 1985) and its wind terminal velocity is 206 km s^{-1} (Lamers *et al.* 1985). For a distance of 1.8 kpc, its mass loss rate is therefore $7.2 \times 10^{-6} \mu$ M$_\odot$ yr^{-1}, where μ is the mean mass per ion in the radio emitting zone. The emitting ions are assumed to be singly ionized. However, Drew (1985) has shown that helium will be neutral beyond about 5 stellar radii in the wind of P Cyg, i.e. in the radio-emitting region, so 7.2×10^{-6} M$_\odot$ yr^{-1} is only the hydrogen mass loss rate of P Cyg, and to determine its total mass loss rate we need to know the He/H abundance ratio, and thus μ, in its wind.

Figure 2 shows the UV–IR energy distribution of P Cyg, from Deacon & Barlow (1990). Shown are the flux-calibrated *IUE* high-resolution data, Johnson & Mitchell 13-colour optical photometry, *UKIRT* and *IRAS* 1.25–60 μm photometry and 1–4μm spectrophotometry. The flux distribution has been dereddened by E(B−V) = 0.60, and the UV–1.25 μm flux distribution has been fitted by an 18000 K, log g = 2.05 LTE model, enabling the exccess f-f and b-f fluxes due to the wind to be derived for wavelengths longwards of 2 μm. The excess flux at 10 μm is 5.4 Jy. A stellar radius of 92.5 R$_\odot$ is implied by the model atmosphere normalisation, corresponding to a luminosity of 8.1×10^5 L$_\odot$.

Cohen & Barlow (1980) published IR photometry out to 10 μm for AG Car, obtained on March 31 1977. Hutsemekers & Kohoutek (1988) have published a long-term light curve for AG Car, along with high-resolution optical spectroscopy acquired on February 22/23 1977, only one month before the IR photometry of Cohen & Barlow; this showed AG Car at the time to have had a spectrum very similar to that of P Cyg. In addition, the flat-topped [Fe II] lines in the optical spectrum of AG Car yielded a wind expansion velocity of 205 km s^{-1}, almost identical to that of P Cyg. After dereddening by E(B-V) = 0.63 (Humphreys *et al.* 1989) and normalising an LTE 17000 K, log g = 1.9 model to the J photometry, an excess flux of 1.35 Jy at 10 μm is derived (Deacon & Barlow 1990). For a distance to AG Car of 6 kpc (Humphreys *et al.* 1989), the model atmosphere normalisation implies a stellar radius of 121.8 R$_\odot$ and a luminosity of 1.12×10^6 L$_\odot$.

Since the spectrum and wind terminal velocity of AG Car at the epoch of the IR photometry were so similar to those of P Cyg, it seems reasonable to assume similar wind velocity laws, in which case the ratio of their mass loss rates is determined by the ratio of their absolute excess fluxes at 10 μm. The 10 μm excess flux for P Cyg is 4.0 times that for AG Car, while the square of the distance to AG Car is 11.1 times larger than that to P Cyg. Therefore, from Wright & Barlow (1975), $(\dot{M}/\mu)^{4/3}$ for AG Car is 2.78 times larger than that for P Cyg. To compare their mass loss rates, we need to know the He/H ratios, and thus the values of μ, in each of their winds.

The He$^+$/H$^+$ ratio in the IR-emitting region of the wind of P Cygni can be derived from an analysis of the He I and H I recombination line fluxes observed in the 1–4 μm spectra shown in Figure 2. These spectra, obtained as service observations with CGS 2 at *UKIRT*, have a resolving power of 300–600, depending on the wavelength region. The recombination coefficients of Hummer & Storey (1987 for H I; in preparation for He I) were used, with $T_e = 15000$ K and $n_e = 10^{11}$ cm^{-3}. In order to avoid optical depth self-absorption effects, only H I recombination lines with $\Delta n > 3$ were used – this left eight H I IR lines suitable for the analysis. Four He I lines between 1.2–2.5 μm were used – they corresponded to angular momentum sub-components of the 4–3, 5–3 and 6–4 terms. The 2s–2p lines of He I at 1.083 μm and 2.058 μm were not used, as these transitions are strongly affected by optical depth and collisional excitation effects. A He$^+$/H$^+$ ratio of 0.5±0.1, by number, was derived for the IR emitting zone of P Cyg. A value of μ(IR) = 2.0 results, while μ(radio) = 3.0, since helium is neutral there. The total mass loss rate for P Cyg given by the radio data and μ = 3.0 is therefore 2.16×10^{-5} M$_\odot$ yr^{-1}.

McGregor *et al.* (1988) obtained JHK spectroscopy of AG Car with a resolving power of 500 during 1984 and 1985. Their March 1984 spectra were taken at an epoch when AG Car had a similar V magnitude to 1977 February/March, according to the light curve of Hutsemekers & Kohoutck (1988). The line fluxes measured by McGregor *et al.* from their March 1984 JHK spectra have therefore been analysed in the same manner as described above for P Cyg. Three H I and two He I lines were suitable for the analysis and yielded He$^+$/H$^+$ = 0.25±0.08, by number, for the IR-emitting region of AG Car. This corresponds to μ(IR) = 1.6.

$\dot{M}/\mu$ was found above to be 2.15 times larger for AG Car than for P Cyg. Since μ(IR) is 1.25 times larger for P Cyg than for AG Car, the mass loss rate of AG Car in 1977 March must therefore have been 1.7 times larger than that of P Cyg, implying $\dot{M} = 3.7 \times 10^{-5}$ M$_\odot$ yr^{-1} for AG Car.

Maeder & Meynet (1987) and Langer (1990) have found that a star of 60 M$_\odot$ initial mass should evolve into a WNL-type star when the surface hydrogen mass fraction, X_H, drops below 0.23–0.30. The analysis above implies that $X_H = 0.5$ for AG Car, while $X_H = 0.33$ for P Cyg. Therefore both stars are well on their way to becoming WNL stars according to this interpretation, consistent with the proposals by Lamers *et al.* (1983) and Humphreys *et al.* (1989) that P Cyg and AG Car are currently evolving towards the WN stage.

286

References

Abbott, D. C., Bieging, J. H., Churchwell, E. & Torres, A. V., 1986. *Astrophys. J.*, **303**, 239.

Aitken, D. K., Roche, P. F. & Allen, D. A., 1982. *Mon. Not. R. astr. Soc.*, **200**, 69P.

Barlow, M. J. & Hummer, D. G., 1982. *IAU Symp. No. 99*, p. 387, eds. de Loore, C. W. H. & Willis, A. J., Reidel, Dordrecht, Holland.

Barlow, M. J., Roche, P. F. & Aitken, D. K., 1988. *Mon. Not. R. astr. Soc.*, **232**, 821.

Becker, R. H. & White, R. L., 1985. *Radio Stars*, p. 139, eds. Hjellming, R. M. & Gibson, D. M., Reidel, Dordrecht, Holland.

Bieging, J. H., Abbott, D. C. & Churchwell, E. B., 1982. *Astrophys. J.*, **263**, 207.

Cohen, M. & Barlow, M. J., 1980. *Astrophys. J.*, **238**, 585.

Deacon, J. R. & Barlow, M. J., 1990. These Proceedings.

Drew, J. E., 1985. *Mon. Not. R. astr. Soc.*, **217**, 867.

Hillier, D. J., 1987. *Astrophys. J. Suppl.*, **63**, 947.

van der Hucht, K. A. & Olnon, F. M., 1985. *Astr. Astrophys.*, **149**, L17.

Hummer, D. G., Barlow, M. J. & Storey, P. J., 1982. *IAU Symp. No. 99*, p. 79, eds. de Loore, C. W. H. & Willis, A. J., Reidel, Dordrecht, Holland.

Hummer, D. G. & Storey, P. J., 1987. *Mon. Not. R. astr. Soc.*, **224**, 801.

Humphreys, R. M., Lamers, H. J. G. L. M., Hoekzema, N. & Cassatella, A., 1989. *Astr. Astrophys.*, **218**, L17.

Hutsemekers, D. & Kohoutek, L., 1988. *Astr. Astrophys.*, **73**, 217.

Kingsburgh, R. L. & Barlow, M. J., 1990. These Proceedings.

Lamers, H. J. G. L. M., de Groot, M. J. H. & Cassatella, A., 1985. *Astr. Astrophys.*, **128**, L8.

Lamers, H. J. G. L. M., Korevaar, P. & Cassatella, A., 1985. *Astr. Astrophys.*, **149**, 29.

Langer, N., 1989. *Astr. Astrophys.*, **220**, 135.

Langer, N., 1990. *Properties of Hot Luminous Stars*, p. 328, ed. Garmany, C. D., Astr. Soc. Pacific Conf. Series, Vol. 7.

Maeder, A., 1984. *Astr. Astrophys.*, **120**, 113.

Maeder, A. & Meynet, G., 1987. *Astr. Astrophys.*, **182**, 243.

McGregor, P. J., Hyland, A. R. & Hillier, D. J., 1988. *Astrophys. J.*, **324**, 1071.

van den Oord, G. H. J., Waters, L. B. F. M., Lamers, H. J. G. L. M., Abbott, D. C., Bieging, J. H. & Churchwell, E., 1985. *Radio Stars*, eds. Hjellming, R. M. & Gibson, D. M., Reidel, Dordrecht, Holland.

Prinja, R. K., Barlow, M. J. & Howarth, I. D., 1990. *Astrophys. J.*, **361**.

Schmutz, W. & Hamann, W.-R., 1986. *Astr. Astrophys.*, **166**, L11.

Turner, D. G. & Forbes, D., 1982. *Publ. astr. Soc. Pacific*, **94**, 789.

Williams, P. M., 1982. *Mon. Not. R. astr. Soc.*, **199**, 93.

Williams, P. M. & Eenens, P. R. J., 1989. *Mon. Not. R. astr. Soc.*, **240**, 445.

Wright, A. E. & Barlow, M. J., 1975. *Mon. Not. R. astr. Soc.*, **170**, 41.

DISCUSSION

Moffat: Another mass estimate of a WO star comes from a study of the orbit of AB8 in the SMC for which the WO4 star has a mass of $\simeq 5M_\odot$. With Sand 5 this confirms that WO stars do have relatively low masses, regardless of the ambient metallicity.

Barlow: Yes. In addition, the poster paper by Kingsburgh and Barlow shows that the O star component of AB8 (= Sk 188 = Sand 1) has a spectral type of O7±0.5. If it is a dwarf, then the implied mass for the WO4 component would be consistent with your new estimate.

Taylor: We have monitored P Cyg (spectropolarimetrically) during its recent outburst. These data show that there is a non-steady mass outflow and the wind is non-spherically symmetric. We expect that there is a circumstellar wind with intermittent "blob" ejections.

Barlow: The high resolution IUE observations of Lamers et al. (1985) showed moving "shell" in the wind of P Cyg, which might be related to the non-steady behaviour that you observed. The high spatial resolution VLA data of Becker and White (1985) indicate that, out in the radio emitting zone, the wind of P Cyg is fairly symmetric. Schmidt-Burgk (1982) has shown that very large asymmetry factors are needed before radio-derived mass loss rates are seriously affected.

Leitherer: AG Car is known to be surrounded by a jet-like structure indicating strong departures from spherical symmetry (binary, magnetic field?). P Cyg in contrast is believed to have a spherically symmetrical wind. This comes from VLA observations of the wind. Did you take the different geometry into account for the $\dot{M}$ you derive?

Barlow: The assumption made was that AG Car had the same distribution of matter in its wind as P Cyg. The Australia Telescope should shortly carry out similar observations on AG Car as the VLA did on P Cyg. AG Car is certainly near to a jet-like structure, but it is also of course surrounded by a remarkably symmmetric ring nebula.

Cassinelli, Taylor, Schulte-Ladbeck, Leitherer, Barlow, Cohen, Felli, Walborn

WOLF-RAYET STELLAR WIND THEORY

J. P. CASSINELLI
Department of Astronomy
University of Wisconsin-Madison
475 N. Charter Street
Madison, WI 53706

ABSTRACT: Two possible solutions to the Wolf-Rayet wind momentum problem are discussed: purely radiation driven wind theory, with multi-line effects, and Luminous Magnetic Rotator theory. Several recently developed radiative processes for enhancing Ṁ or v_∞ are described, and it is concluded that only the winds of rather hot luminous Wolf-Rayet stars could possibly be driven by radiation. These stars should show evidence of acceleration at large radial distances. For the rapid rotators, it is possible to drive a dense equatorial outflow. Limits are discussed regarding the needed surface magnetic fields. With this model, the wind momentum problem is solved in a piece-wise fashion by having the large radio flux of Wolf-Rayet stars come from the equatorial zone and the broad P Cygni lines, arising in the polar wind. The Luminous Magnetic Rotator model can also be tested through observation, primarily through spectropolarimetry.

1. Introduction

This paper will focus on the most perplexing aspect of Wolf-Rayet winds - the wind momentum problem. Cassinelli and Castor (1973) derived an expression for the maximum mass loss rate that could be driven by the transfer of the stellar radiative momentum: $\dot{M}_{max} = L/v_\infty c$. This is called the "single scattering maximum" because it assumes every stellar photon transfers its momentum, $h\nu/c$, just once. The ratio of the actual mass loss rate to $\dot{M}_{max}$ is given by the "efficiency factor for ejecting matter by radiation", η, where

$$\eta = \frac{\dot{M} v_\infty}{L/c} \qquad (1)$$

In the period 1979-1981, it became clear that Wolf-Rayet stars posed a serious problem, for their η values were found to be in the range $\eta = 3$ to 30 (Barlow, Smith and Willis 1981). That is, there is about one order of magnitude more momentum in the wind than in the radiation field that is supposed to have driven it. (Current estimates of η are given by Schmutz et al. 1989 and are in the range, 5 to 74). The initial reaction by most wind theorists was that surely the luminosity or stellar temperature had been underestimated. However, after a general consensus was reached that the momentum problem was real, two approaches were followed in attempting to explain the problem: (1) to modify and

K. A. van der Hucht and B. Hidayat (eds.),
Wolf-Rayet Stars and Interrelations with Other Massive Stars in Galaxies, 289–307.
© 1991 *IAU. Printed in the Netherlands.*

improve radiation driven wind theory to extract more momentum from the radiation field, or (2) to invoke "other" forces or non-spherical effects that could drive a fast wind. In this review I will first discuss the significant improvements that have been made over the past decade in line driven wind theory. Then I will discuss the "Luminous Magnetic Rotator" model that gives rise to a non-spherically symmetric picture for Wolf-Rayet stars.

2. Purely Radiation Driven Winds

It is convenient to focus on the two factors in the wind momentum rate product, $\dot{M}v_\infty$. The Wolf-Rayet stars have "massive" winds and "fast" winds. As a background theme for discussing and classifying the nature of recent developments regarding $\dot{M}$ and v_∞, it is helpful to use the Leer and Holzer (1980) "laws of wind theory":

(I) If additional Momentum or energy is added only to the supersonic (super critical point) region of the outflow, there will only be an increase in the terminal velocity but no increase in the mass loss rate.

(II) If momentum or energy are added to the sub-critical point region, there will be an increase in $\dot{M}$.

Let us illustrate these two laws using simple radiation driven wind theory in which we modify the value of Γ, the ratio of the radiative acceleration to gravity

$$\Gamma = g_{Rad}/g = k_F L/(4\pi c\, GM) \tag{2}$$

and solve the wind momentum equation,

$$v\frac{dv}{dr} + \frac{1}{\rho}\frac{dp}{dr} + \frac{GM}{r^2}(1 - \Gamma) = 0 \tag{3}$$

The results are shown in Figure 1. The top panel shows the nominal model, with no modifications in Γ. The second panel shows the effect of modifying Γ in the supersonic region, note that only $v(r)$ changes. The last panel shows the effect of having some momentum addition occurring in the subsonic region. Note that now both $\dot{M}$ and $v(r)$ change, as does the location of the critical point. The reason the mass loss rate changes when there is additional deposition deep in the wind is that the subsonic region is nearly hydrostatic and the deposition increases the scale height of the density distribution. Therefore the product of $4\pi \rho_s a\, r_s^2 = \dot{M}$, where a, is the sound speed, can increase greatly because of the change in density ρ_s at the critical point. With this basic result of wind theory, let us now explore four radiation related effects that tend to modify either the deposition deep in the wind (and hence increase $\dot{M}$), or the deposition far out in the wind (and hence increase v_∞):

(1) The finite cone angle correction factor, or "Modified CAK" theory as developed by Friend and Castor (1983); Pauldrach, Puls and Kudritzki (1986); Friend and Abbott (1986).

(2) The "multi line" or multiple scattering effects of Panagia et al. (1981); Friend and Castor (1983); Abbott and Lucy (1986); and Puls (1987).

(3) Anisotropic line diffuse radiation field effect as explored by Puls and Hummer (1987).

(4) The increase in continuum optical depth as the star evolves at constant $\dot{M}$, L to smaller radius and the accompanying effects of enhanced line opacity (Conti and Abbott 1987).

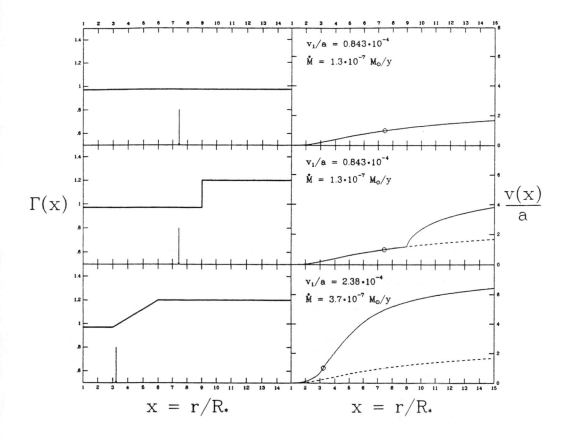

Figure 1. The left panels show three distributions of the radiation acceleration/gravity ratio, Γ, versus radial distance. The arrow indicates the location of the wind's sonic point. The right panels show the corresponding velocity v versus radial distance, with the sonic point marked on the curves. The velocity at the base of the wind, v_1, and the mass loss rate $\dot{M}$, are also shown. The top figure shows the nominal velocity law, i.e. with no change in Γ. The second set of figures shows the effect of adding momentum to the supersonic portion of the flow alone, and it is to be noted that only the velocity law is changed relative to the nominal case, but not the mass loss rate. In the bottom figure, momentum is added to both the subsonic and supersonic part of the flow and now both $v(r)$ and $\dot{M}$ change. (Figure prepared by D. van den Berk).

The relative importance of these four effects in regards to the momentum problem will be discussed at the end of this section. There we will also discuss the limitations on the two laws of Leer and Holzer.

2.1. MODIFIED CAK LINE DRIVEN WIND THEORY

The classic papers on line driven wind theory are those of Lucy and Solomon (1970) and Castor, Abbott, and Klein (1975). Lucy and Solomon showed that one line (C IV 1550Å) located at the peak of the spectral energy distribution of an OB supergiant could drive a wind with $\dot{M} = 10^{-8}$ $M_\odot$/yr to speeds of 3000 km/sec. The velocity is about right but the mass loss rate is too low. Castor, Abbott, and Klein (hereafter CAK) developed a clever approach for treating thousands of lines. They found that a wind with $\dot{M} \geq \dot{M}_{observed}$ and $v_\infty \approx 1.1$ v_{escape} could be driven. Their terminal velocity was too small, as observations have $v_\infty \approx 2 - 4$ x v_{escape}. A straight-forward modification of the CAK theory led to drastic improvements in the comparison between the theoretical predictions and the observations. This modification was the incorporation of the finite cone correction factor. In the original CAK paper and subsequent papers by Abbott (1982) it had been assumed that the radiation field was radial, as if coming from a point source. The radial assumption has two effects on the line acceleration. A smaller stellar luminosity produces a larger flux and hence larger acceleration. The radial approximation leads to an over-estimate of the penetration probability $\beta = (1 - e^{-\tau}/\tau)$ that is used to calculate the radiative acceleration in Sobolev theory. With the non-radial incident radiation field used in the modified CAK theory, the available stellar luminosity produces less acceleration close to the star, and the penetration probability, β, is reduced in regions close to the star. As an immediate consequence, the predicted mass loss rates were reduced by ~ factor of 2-3, bringing them more in line with observations of O and OB stars. Farther out, in the supercritical regions the radiation field is indeed nearly radial as in the CAK models and since a smaller mass is being driven, a faster flow can thereby result. Friend and Abbott (1986) and Pauldrach, Puls and Kudritzki (1986) show that there is now excellent agreement between theory and observation for most O and B stars.

Now, as for Wolf-Rayet stars, Pauldrach et al. (1985) surprised many of us by showing that modified CAK theory could explain the wind of the well studied Wolf-Rayet star V444Cyg, which has a momentum efficiency factor of $\eta \approx 3$. Subsequently, Friend, Poe and Cassinelli (1987) studied this ability of modified CAK theory to produce η factors greater than unity, and their results are shown in Figure 2. This is a plot of $\dot{M}$ vs. v_∞, it shows the locations of three Wolf-Rayet stars, as well as results of modified CAK theory for the noted values of M, R, and Γ_{es}.

Several things should be noted in regards to Figure 2. (a) The $\dot{M}$ and v_∞ data for the three stars can be fitted only by using rather large values for $\Gamma_{es}=\sigma_{es} L/4\pi$ c GM; ranging from about 0.65 for V444 Cyg to values of about 0.88 for γ^2 Vel. These are unacceptable values for Γ_{es} because the value derived from theoretical interior models is only about 0.4. So a new rule has been added to the game: wind theorists should only use values of Γ_{es} allowed by interior considerations. (b) Note that V444 Cyg has a rather mild momentum problem as compared to the other two stars shown, so that an explanation of the wind momentum of this star does not mean that a solution of "the Wolf-Rayet problem" has been found. (c) The figure indicates that all three stars could be fit using some (albeit unacceptable value of Γ_{es}). Why is this? The reason is that a basic assumption in CAK and modified CAK theory is simply not valid. Recall that in CAK theory, (see Abbott, 1988) the force on a "resonance shell" is calculated as if the shell receives the full unattenuated stellar intensity. The models ignore the scattering or absorption by line resonance shells that lie closer to the star, i.e. "line overlap" is not accounted for. Therefore the CAK theories tend to overestimate the momentum deposition and the fit to

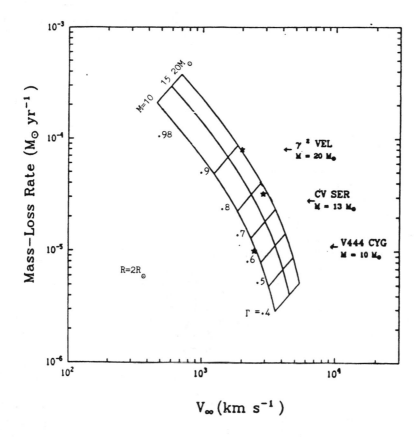

Figure 2. A plot of mass loss rate versus wind terminal velocity. The symbols show the values for Ṁ, and v∞ for three Wolf-Rayet stars. The grid shows results of the "modified-CAK" models of Friend, Poe, and Cassinelli (1987). These models were carried out for a star with radius equal to 2 solar radii, and for the seven values of Γ and three values for the stellar mass as indicated. The calculations used the CAK parameters k=0.3 and α=0.6 (Abbott 1982).

the stellar data shown in Figure 2 is not valid. There is a need to account for "line overlap" considerations or multi-line effects, discussed below.

2.2 MULTILINE EFFECTS

Overlapping spectral lines are those that are separated by less than the Doppler range in frequencies through the wind, i.e. $\Delta v < \Delta v_{wind} = 2\, v_0\, v_\infty/c$. We have discussed how the lines could reduce the momentum deposition at some radius because of scattering deeper in the wind, however, overlapping lines can also <u>increase</u> the net momentum deposition! This can occur in two ways: (a) Multiple scattering which can lead to a multiple deposition of photon momentum hv/c, and (b) Thermal emission of a line photon from a shell closer to the star produces photons that can be scattered by atoms farther out in the wind. Both effects have been studied by Puls (1987), here we will discuss only the effects of multiple scattering. The multiple scattering was explored for a statistical distribution of lines by Friend and Castor (1983) who found that there could be an increase in the total momentum deposition by a factor of 3-5. Abbott and Lucy (1986) used a realistic line distribution versus frequency, assumed only scattering occurs (no absorptions), and followed the momentum deposition using a Monte Carlo method. They assumed a velocity law, and used the computed energy deposition to derive a mass loss rate. Their result for the O4f star ζ Pup is shown in Figure 3. Note that there are a number of very short steps which tend to deposit photon momentum about equally in the forward and backward direction. However, there are also several very long mean free paths. Because of a long mean free path a photon scattered in a inward direction can encounter its next scattering while again having a net outward component. With this occuring, a photon can deposit more than one times hv/c. The deposition history is shown in Figure 4, where we see that for the ζ Pup model the escaping photon has deposited about 4 x hv/c. For a Wolf-Rayet star, it is likely that this multiplying factor could be even larger for reasons discussed in section 2.4. Where is this extra force coming from, and is there any limit? During the scattering processes in an expanding envelope, not only is there a shift in photon direction but also a red shift of the photon, an energy $h\Delta v$ is transferred to the flow by this redshift. Consider then the wind energy flow or wind luminosity

$$L_w \equiv \frac{1}{2}\, \dot{M}\, v_\infty^2 \tag{4}$$

on using equation 1, this becomes:

$$L_w = \frac{1}{2}\, \eta\, L\, \frac{v_\infty}{c} \tag{5}$$

$$\approx \frac{.01}{2}\, L_*\, \eta$$

So for WR stars with $\eta = 3$ to 30, we have $L_w = 1.5$ to 15 percent of the stellar luminosity. Which means the average photon is required to redshift up to as much as 15%. This is certainly possible in principle, but has not yet been demonstrated by a detailed model.

The multi-line effects increase the momentum deposition rather far from the star, 3 R* $< r < 50$ R*, and there is some observational evidence (Koenigsberger, 1990) that WR

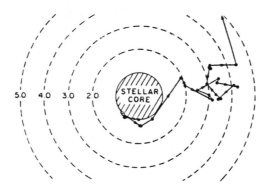

Figure 3. Shows the path of a photon through the wind of ζPup (O4f), experiencing multi-scatterings. The path is shown projected onto a plane. (Abbott and Lucy 1985).

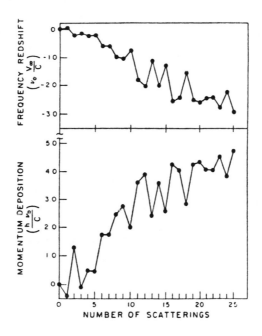

Figure 4. The cumulative frequency redshift (top panel) and momentum transfer (bottom panel), versus the number of the scatterings shown in Figure 3. (Abbott and Lucy 1985).

winds continue to accelerate out to at least 4 R∗. The evidence that there is acceleration out to 50 R∗, found by Hillier (1987), now appears to be unnecessary or at least uncertain (Hillier 1990, these proceedings).

The effect of multi scattering is decreased if photons can escape between large Δv gaps between thick lines. So, for the case of WR stars one wants to consider situations in which the line density vs. frequency is enhanced, as discussed in 2.4.

2.3 THE EFFECTS OF ANISOTROPIC DIFFUSE RADIATION FIELD

In the usual Sobolev approach, e.g. as used in CAK, it is assumed that the continuum radiation field is anisotropic, i.e. coming from the direction of the star in a core-halo picture. It is this anisotropic field that gives rise to the outward force on the resonance shell. It is also assumed that because of multiple scattering within a single resonance shell that the diffuse line radiation is isotropic, and hence this radiation field provides no net outward force. However, for a thick Wolf-Rayet wind this latter assumption is a poor one, as has been shown by Puls and Hummer (1987). They considered the effects of a diffuse radiation field having an angular dependence $\propto \mu \, d \, S_c/dr$, and find that this leads to an extra deposition of momentum close to the star. Figure 5 shows one of their results for the line acceleration. Because the deposition occurs deep in the wind, it gives rise to an increase in the wind mass loss rate, as we would expect from law I.

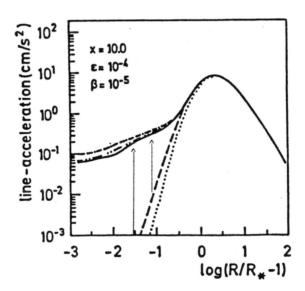

Figure 5. Shows the line acceleration versus logarithm of the height above the photosphere. The lower two curves show results of the Sobolev approximation, without accounting for the force associated with the diffuse radiation field in the line. The upper lines indicate the corrected acceleration, accounting for the diffuse radiation field. The arrows are drawn in to show the significant increase in acceleration that occurs at small heights. (Adapted from Puls and Hummer, 1988).

2.4 SMALL STAR EFFECTS

There are more thick lines at higher wind densities, or equivalently for larger electron scattering Sobolev optical depths $= t = \sigma \rho\, v_o/(dv/dr)$. Recall that a basic feature of CAK theory is that the number of thick lines is proportional to a positive power of t. (N(t) = k t^γ. Also, there are more strong lines in the EUV than in the near UV. Therefore, having Wolf-Rayet stars at a large T_{eff} (> 50,000 K) helps to explain the winds as radiatively driven. Abbott and Conti (1987) plot a boundary on the HR diagram to the left of which one can expect WR stars to be produced by multiple scattering effects. This line corresponds to an iso electron scattering line

$$\tau = \text{Const} = \sigma_{es} \int \rho\, dr \qquad (6)$$

$$= \frac{.2\, \dot{M}}{4\pi\, v_\infty} \frac{1}{R} \ln \left(\frac{v_\infty}{v_o} \right) \qquad (7)$$

for which the Abbott and Lucy (1986) velocity law (β=1) was used. Note that τ is increased as the stellar radius, R_*, decreases. The line for constant τ, is shown in Figure 6a, and 6b shows that for the WR evolutionary tracks of Maeder and Meynet WR stars do, in fact, enter the region where thick winds are expected.

In summary: we have discussed four radiative related effects that have come to light over the past decade. The bottom line is that it is only the multi-line effects that can really lead to an increase in the efficiency factor η to values above unity. The "modified CAK" effects discussed in § 4.1 only redistribute the relative values of $\dot{M}$ and v_∞, but do not change η very much. We have seen that $\eta \gg 1$ can be derived numerically from the modified CAK model, but only by using unacceptable Γ_{es} values, and because line overlap is not accounted for. The anisotropic diffuse field effects, discussed in § 4.3 correct for the core-halo approximation of CAK theories. Since the extra force is deep in the wind, the diffuse field forces tend to increase $\dot{M}$ according to law # II. However, this does not really lend to an increase in the η value because the flow must then be driven to high speeds, which cannot be done unless multi-line effects, or other forces, are accounted for. Thus we see that law # II can be misleading. Let us consider other cases in which subsonic deposition occurs. The most important example is the radiative driving of the winds from Red Giants. Shocks propagating through the outer envelope "bloat" the atmosphere to the extent that the density is sufficiently high at the condensation radius that dust can form (Goldberg, 1987). At that point, radiative acceleration of grains drives the coupled dust/gas mixture out as a slow wind. Nonetheless, as we see in the paper by Jura (these proceedings) the η values for these stars does not exceed unity. In regards to the Wolf-Rayet stars, it is suggested by interior theorists (Maeder, 1990; Langer 1990) that large mass loss rates are driven because the stars are near vibrational instability. The resultant disturbances would presumably affect the subsonic portion of the flow. However, a mechanism for driving the winds to high speed, such as the multi-line effects, would still be needed to explain values of η greater then unity. We have seen in Owocki's paper (these proceedings) that if insufficient momentum is added to the supersonic part of the flow, the material will fall back onto the star.

It is important to note that the observational evidence for multi-line driving of winds with η >1 requires acceleration at large distances from the star. I hope that a major effort is placed on finding evidence for this acceleration in Wolf-Rayet stars.

Tremendous progress is being made by the Munich group in explaining the properties

298

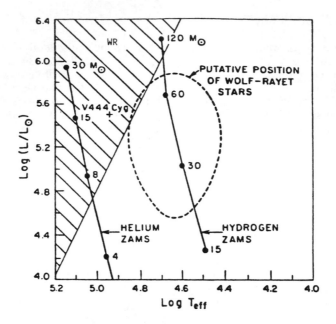

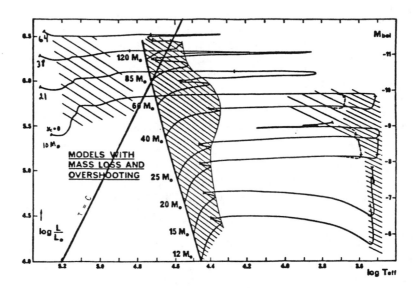

Figure 6. Hertzsprung-Russell diagrams showing properties of luminous stars.
(a) shows the hydrogen and helium zero-age main sequences (ZAMS). The shaded region shows where radiation driven winds are optically thick in the continuum. If W-R stars have the high temperatures and luminosities corresponding to the shaded region, the results of Abbott and Lucy (1985) indicate that their winds might be explained by radiation pressure alone, by way of multi-line effects. (From Abbott and Conti 1987). (b) shows the evolutionary tracks of Maeder and Meynet (1987). The dark line, (marked τ=C) corresponds to the boundary shown in figure (6a). Thus we see that interior models indicate that stars should enter the region where multi-line effects are important.

of radiation driven winds, and I refer you to the recent paper by Pauldrach et al. (1990).

As an introduction to the next section, it is important to realize the following: The multiple scattering should also be decreased if there is a non-spherical density distribution, such as an equatorially enhanced wind. This is because photons can more easily escape in the polar direction. Thus if observations show evidence for rotational distortion, not only is this evidence that another driving mechanism is operating, but the effect of radiative driving through multiple scattering is diminished.

3. Luminous Magnetic Rotators

As a second approach for explaining the Wolf-Rayet momentum problem, let us consider a non-spherical, rotating model. Wolf-Rayet stars are located in the HR diagram rather near to the hypergiant B[e] stars. Zickgraf et al. 1985 proposed that these stars have a two component structure: (1) a fast polar wind detected by the UV resonance lines of Si IV, C IV, N V, and which is presumably driven by the radiation field from the broad polar zone of the star and, (2) a slower, denser flow in the equatorial zone, where the state of ionization is lower and at some distance dust formation can occur. Poe, Friend, and Cassinelli (1989) developed a two component picture for Wolf-Rayet winds incorporating rotational and magnetic forces in the equatorial zone. Cassinelli (1990) has called this type of model a "Luminous Magnetic Rotator", because radiation forces play a major role in driving the flow. Such a picture for Wolf-Rayet stars has several advantages in regards to explaining some WR phenomena: (a) The model can explain the polarization that is seen in some WR stars (Schmidt 1988) and in B[e] stars (Zickgraf and Schulte-Ladbeck 1989). (b) The model provides an evolutionary link to the B[e] and post-luminous blue variable stars. (c) It provides a structure conducive for the formation of dust grains that occurs in some B[e] and late WC stars. (d) The rotational model allows for an explanation of the Wolf-Rayet momentum problem, (e) and perhaps at a stellar "effective temperature" lower than that required for purely radiation driven winds. (f) A new "spin-down problem" is introduced, but this can be overcome with a properly chosen B field as shown by Poe et al. (1989).

Cassinelli, Schulte-Ladbeck, Poe and Abbott (1989) argued that the rapid rotator picture is at least plausible from an evolutionary point of view, and they showed results of a simple model to follow the evolutionary change in the ratio $\alpha = v_\phi/v_{crit}$. The stars must be rotating with an equatorial velocity near the critical speed:

$$v_{crit} = \sqrt{\frac{GM(1-\Gamma)}{R}} \qquad (8)$$

Near the Humphreys Davidson limit, stars lose about 40% of their mass before becoming sufficiently homogeneous to evolve to the left in the HR diagram. The models of Maeder and Meynet (1987) indicate that not only is the mass significantly smaller during the leftward evolution, but also the luminosity is increased. The change in M and L increases Γ (equation 2). Hence v_{crit} can be reduced by a factor of 2 relative to the rightward (pre LBV) phase of evolution. Since v_ϕ varies as $(MR)^{-1}$, and v_{crit} is smaller, α will be significantly closer to unity. In addition ,the paper by Linda Smith (these proceedings) shows that at least for some LBV's the mass loss appears to be bipolar. If the mass loss is in fact primarily from polar regions when stars are near the HD limit, the star can retain a larger angular momentum per unit mass, which makes the rapid rotator model more likely. The rapid rotator model is perhaps not needed for all Wolf-Rayet stars.

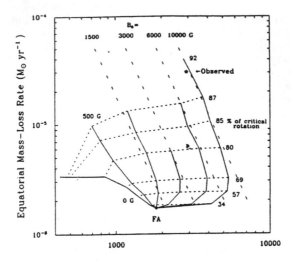

Figure 7. Shows the mass loss rates and terminal velocities that are expected for winds that are driven out of the equator by a combination of radiation pressure, magnetic, and rotational forces. (From Poe, Friend, and Cassinelli 1989). The equatorial mass loss shown is $(4\pi)x$ the mass loss per unit solid angle in the equatorial region. The point FA indicates the $\dot{M}$, v_∞ values derived in the modified CAK theory of Friend and Abbott (1986). The solid lines extending from the FA point are iso-magnetic field lines, ranging in values from 0 to 10,000 Gauss, as indicated. The short dashed, horizontal lines are iso-rotation rate (constant α) lines. The point marked "observed" corresponds to the $\dot{M}$, v_∞ determined for the WR star CV Ser from Radio and UV line profiles. The parameters for the models shown are: M = 13 M_Θ, R_Θ = 8R, Γ = 0.36.

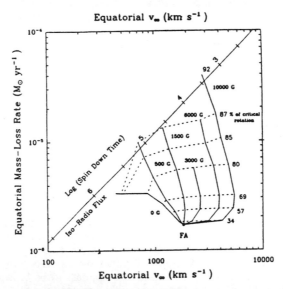

Figure 8: Same as Figure 7, but now contains a line of equal radio flux. This indicates that a wide range of models could potentially explain the observational estimates of $\dot{M}$. Numbers along the iso-radio flux line are estimates of the logarithm of the spin down time of the star, which we see ranging from about 10^3 yrs to over 10^5 yrs. (From Poe et. al. 1989).

Nonetheless, it is relevant for some stars and it is interesting to consider at least this one alternative to the purely radiation driven wind model.

Cassinelli (1990) points out that magnetic rotator wind models are always hybrid wind models, involving at least one additional driving mechanism, which is referred to as the primary mechanism. In the solar case the primary mechanism involves coronal forces, in the case of hot luminous stars it is the radiative driving. The primary mechanism sets the minimal mass loss rate for the star, and the magnetic rotator forces can change the speed of the wind and the mass loss rate. The basic theory for magnetic rotator winds was developed in a classic paper by Weber and Davis (1967). They showed that the wind momentum equation in the equatorial plane could be separated into purely radial and azmuthal (v_ϕ) equations, with associated wind energy and angular momentum constants.

As we have been discussing radiation driven winds in terms of momentum/energy deposition, it is useful here to discuss the magnetic rotator model in terms of the required deposition. The deposition, or transfer of energy from the magnetic field, occurs via the divergence of the Poynting vector, S, (Belcher and Mac Gregor, 1976) where

$$S = \frac{c}{4\pi} (E \times B) \tag{9}$$

where the electric field E is determined by the frozen-in magnetic field constraint. The total energy per gram, ε, is composed of a gas term plus the magnetic term,

$$\varepsilon_{mag} = \frac{S_r}{\rho V_r} \tag{10}$$

where S_r is the radial component of the Poynting vector. Its value at infinity can be expressed in terms of a very convenient quantity, the Michel velocity, V_M;

$$S_{r,\infty} = \rho V_M^3 \tag{11}$$

and V_M is given in terms of the wind mass loss rate, the stellar rotation rate, stellar radius, and surface radial magnetic field as;

$$V_M^3 = \left(\frac{\Omega_*^2 R_*^4 B_{r,*}^2}{\dot{M}} \right) \tag{12}$$

This velocity plays a crucial role in magnetic rotator models. Let V_w be the terminal wind speed associated with the primary wind forces. If $V_M > V_w$ there will be a significant transfer of energy from the field to the wind. This leads to a distinction between two of the three classes of magnetic rotator models. If $V_M < V_w$, the star is a "Slow Magnetic Rotator" (SMR) (as is the current sun). If $V_M > V_w$, the winds radial velocity is affected by the field, and the star is a "Fast Magnetic Rotator" (FMR). The separation occurs at $B_{Min} = (\dot{M} V_w^3)^{1/2} / R_*^2$ which is about 1 Gauss for the sun, 20 Gauss for Be stars and 300 for WR stars. For sufficiently rapid rotation, a third magnetic rotator domain exists, the "Centrifugal Magnetic Rotator" (CMR), a name used by Cassinelli (1990), but discussed as an "extreme FMR" by Hartmann and Mac Gregor (1980). In a CMR, (a) the terminal speed is $V_\infty \approx V_M$, (b) the equatorial subsonic region is in solid body co-rotation with the star, and (c) the sonic point occurs where V_ϕ (solid body) = V(circular) or

$$r_s \Omega_* = \sqrt{\frac{GM}{r_s}} \qquad (13)$$

Thus, we have one of the simplest results for a critical point in all of wind theory (!)

$$(r_s/R_*) = 1/\alpha^{2/3} \qquad (14)$$

In this CMR case, ρ (r) in the subsonic region is determined by gravity and centrifugal forces and is independent of the field. The CMR winds thus provide for a nice clean separation between Ω effects and B field effects. The mass loss rate depends only on $\alpha = \Omega/\Omega_{crit}$, and given this $\dot{M}$ the terminal velocity (= V_m),is determined by $B_{r,*}$. Figure 7 shows a result of Poe et al. for the Wolf-Rayet star CV Ser. Various iso-V_M lines are labeled with the surface field. The point marked FA corresponds to the Friend Abbott (Modified CAK) result for this star. This provides a minimal mass loss rate that is far below the radio estimate labeled "observed". Shown by what I call a flower diagram are the results of luminous magnetic rotator models for various assumed base fields $B_{r,*}$ and rotation rate α ratios. Note that the centrifugal domain, in which the mass loss rate increases, begins for α above about 34% maximum. To achieve the "observed" point would require a field of 10^4 Gauss and $\alpha = 0.90$, as was derived by Hartmann and Cassinelli (1981) [discussed by Cassinelli 1982]. Their solution has several problems. The magnetic field is too large, i.e. would drastically modify the hydrostatic structure of the atmosphere. Secondly, the model suffers from a "spin-down problem", with a rotation e-folding time far below the $\approx 10^5$ year lifetime of Wolf-Rayet stars. Figure 8 from Poe et al. illustrates some of the problems and possible solutions. The diagonal line corresponds to equatorial models that provide the observed radio flux, from which the mass loss rates are derived. This shows that one can explain the radio flux with a range of models, such as a high speed wind with B = 10^4 Gauss and $\dot{M}$ = 3 x 10^{-5} $M_\odot$/yr or with a slower equatorial wind with B $\approx$ 1000 Gauss with $\dot{M}$ = 10^{-5} $M_\odot$/yr. The numbers on the iso-radio flux line give the spin down times. Thus note that the 10^4 gauss model corresponds to a spin down time of only a few thousand years and therefore is unacceptable. Figure 8 also indicates that a wide range of B, $\dot{M}$, v_∞ are possible. It would be useful to narrow the range of say the B field by requiring that the field be consistent with stellar interior constraints. This requirement is analogous to our new rule for the purely radiative driven models, that Γ be consistent with interior values.

Maheswaran and Cassinelli (1988) recently derived constraints on the surface fields in rapidly rotating stars. These are shown in Figure 9, in which is plotted B versus the rotation ratio α. Rapid rotation leads to circulation currents in the interior. If the current speed is larger than the Alfvén speed, the surface magnetic field will be effectively submerged. Hence, the surface field is either approximately zero or it is sufficient to overcome the submersion effect and is above B_{min}, that is shown on the Figure. If the field is very large it dominates the hydrostatic equilibrium equation, and by analogy with radiation driven theory we can say the field can exceed a Magnetic Eddington Limit, also shown on the Figure as B_{max}. These limits, either zero or within the triangular allowed zone, are now plotted on the $\dot{M}$ versus v_∞ plane, and form what I call a "cat" diagram in Figure 10.

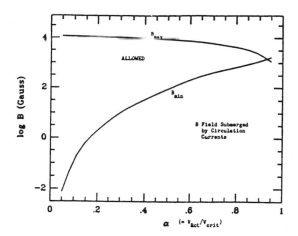

Figure 9: Shows surface magnetic fields versus rotation rate parameter α ($=V_\phi / V_{crit}$). The roughly triangular zone corresponds to the allowed values of the magnetic field for the star discussed in Figures 7 and 8. If the field is above B_{max}, the hydrostatic structure is dominated by the magnetic field/rotation forces. If the field is below B_{min}, the circulation currents have a velocity greater than the Alfvén speed in the outer stellar envelope, and the field will be submerged. The surface field must either be approximately zero, or be in the "allowed region". (From Maheswaran and Cassinelli, 1988).

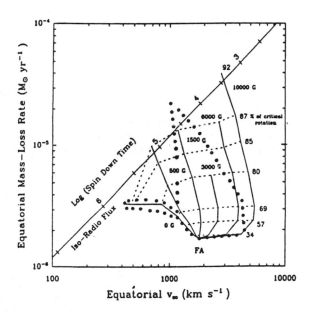

Figure 10. Same as figure 8, but here the dark dots indicated the "Allowed" range of magnetic fields as derived from Figure 9. Note that the Allowed region includes a portion of the iso-radio flux line, with a not-implausible spin down rate.

Thus we have narrowed the allowable range of stellar parameters and it includes a portion of the iso-radio flux line, with a not implausible spin down time. Various parameters associated with the FA point can be adjusted so that the actual length along the iso-radio line could be somewhat larger.

In summary; the rotational model provides an explanation of the momentum problem as follows: the product of $\dot{M}$ v_∞ is made up of two parts which are derived from two quite different parts of the stellar wind. The Wolf-Rayet mass loss rates are derived from radio fluxes and this radiation arises primarily from the denser, slower equatorial regions of a Luminous Magnetic Rotator. The terminal velocity, v_∞, is derived observationally from P Cygni line widths, which are determined by the highest speed along the line of sight in front of the star. The wind outside the narrow equatorial zone is radiatively driven to high speeds, so it is this "polar flow" that determines v_∞. Neither the polar nor equatorial flow violate the momentum maximum $\eta \approx 1$. The momentum problem is thus solved in a piece-wise fashion. This general idea that the high radio flux does not necessarily mean a high mass loss rate should be applicable to other non-spherical pictures such as the cloud model discussed by Cherapascheck and Nugis elsewhere in these proceedings.

4. Summary and Conclusions

In this paper we have discussed two approaches for finding a solution to the Wolf-Rayet momentum problem; purely radiative winds, and luminous magnetic rotators. In regards to the purely radiative wind research, we have discussed four line driven wind effects. The solution of the momentum problem requires the multi-line effects that have been analyzed most thoroughly by Puls (1987). The observational signature of the multi-line radiative driving is the continued acceleration of the flow at large radial distances. Hillier (1987) has described some observational evidence for such acceleration. Also, the observed saturated UV P Cygni lines of Wolf-Rayet stars do not have a sharp violet edge, but rather a gradual rise from zero intensity up to the continuum level (Prinja et al. 1990). Currently this is explained as being caused by absorption in shocks with speeds as high as 1000 km/sec. Perhaps part or most of that high velocity absorption could be caused in the distant acceleration from multi-line effects. It is very important that evidence for the distant acceleration be searched for.

If the multi-line mechanism cannot explain the momentum problem, then clearly we need to look at the alternative models. The best evidence for rotational distortion is the presence of intrinsic polarization and characteristic changes in polarization across strong emission lines (Schmidt, 1988; Schulte-Ladbeck, 1990). Here we have presented some arguments indicating that rapid rotation of Wolf-Rayet stars is a plausible outcome of their evolution. The luminous magnetic rotator model, explains the momentum problem by having the high density and high velocity portions of the wind occuring in different places. Stellar interior theorists favor the vibrational instability mechanisms for driving mass loss. However, from the point of view of stellar wind theory, it is not at all clear how this can lead to the high speed outflows without also utilizing the multi-line acceleration mechanism.

The main conclusions that I would like to stress are that: a) there are at least two ways to explain the Wolf-Rayet momentum problem. Neither theory is, as yet, fully worked out, b) the models have observational consequences that can and should be searched for.

References

Abbott, D.C. (1982), Ap.J., **263**, 723.

Abbott, D.C. (1988), in proceedings of the 6th Solar Wind Conference NCAR TN 306, eds. V.C. Pizzo, T.E. Holzer, D.G. Sime (NCAR), p. 149.

Abbott, D.C. and Conti, P.S. (1987), Ann. Rev. Ast. Astrophys., **25**, 113.

Abbott, D.C. and Lucy, L.B. (1985), Ap. J., **288**, 679.

Barlow, M.J., Smith, L.J. and Willis, A.J. (1981), M.N.R.A.S., **196**, 101.

Belcher, J.W. and MacGregor, K.B. (1976), Ap.J., **268**, 498.

Cassinelli, J.P. (1982), in IAU Symp. 99, Wolf-Rayet Stars: Observations, Physics and Evolution, eds. C.W.H. de Loore and A.J. Willis, (Dordrecht: Reidel), p. 173.

Cassinelli, J.P. (1990), to appear in Angular Momentum and Mass Loss from Hot Stars, eds. L.A. Willson and R. Stalio, (Kluwer: Dordrecht).

Cassinelli, J.P. and Castor, J.I. (1973), Ap.J., **179**, 189.

Cassinelli, J.P., Schulte-Ladbeck, R.E., Poe, C.H., and Abbott , M. (1989), in IAU Colloq. 113, Physics of Luminous Blue Variables, eds. K. Davidson, H.J.G.L.M. Lamers, and A.F.J. Moffatt, (Kluwer: Dordrecht), p. 121.

Castor, J.I., Abbott, D.C., and Klein, R.I. (1975), Ap.J., **195**, 157.

Friend, D.B. and Abbott, D.C. (1986), Ap.J., **311**, 701.

Friend, D.B. and Castor, J.I. (1983), Ap.J., **272**, 259.

Friend, D.B., Poe, C.H., and Cassinelli, J.P. (1987), B. A. A. S., **20**, 1012.

Goldberg, L. (1987), in The M Type Stars, NASA SP-492, eds. H.R. Johnson and F.R. Querci, p. 245.

Hartmann, L. and Cassinelli, J.P. (1981), B. A. A. S., **13**, 795.

Hartmann, L. and MacGregor, K.B. (1982), Ap. J., **259**, 180.

Hillier, D.J. (1987), Ap.J. (Suppl.), **63**, 965.

Koenigsberger, G. (1990), these proceedings.

Langer, N. (1990), these proceedings.

Leer, E. and Holzer, T.E. (1980), J.Geophys. Res., **85**, 4681.

Lucy, L.B. and Soloman, P.M. (1970), Ap.J., **159**, 879.

Maeder, A. and Meynet, G. (1987), Astron. Astrophys., **182**, 243.

Maeder, A. (1990), these proceedings.

Maheswaran, M. and Cassinelli, J.P. (1988), Ap.J., **335**, 931.

Panagia, N. and Macchetto, F. (1981), in IAU Colloq. 59, Effects of Mass Loss in Stellar Evolution, eds. C. Chiosi and R. Stalio (Dordrecht: Reidel), p. 173.

Pauldrach, A., Puls, J., Hummer, D.G., and Kudritzki, R.P. (1985), Astron. Astrophys., **148**, L1.

Pauldrach, A., Puls, J., Kudritzki, R.P. (1986), Astron. Astrophys., **164**, 86.

Pauldrach, A., Kudritzki, R.P. Puls, J., and Butler, K. (1990), Astron. Astrophys., **228**, 125.

Poe, C.H., Friend, D.B., and Cassinelli, J.P. (1989), Ap.J., **337**, 888.

Prinja, R.K., Barlow, M.J., and Howarth, I.D. (1990), Ap. J.,

Puls, J. and Hummer, D.G. (1988), Astron. Astrophys., **191**, 87.

Puls, J. (1987), Astron. Astrophys., **184**, 227.

Schmidt, G.D. (1988), in Polarized Radiation of Circumstellar Origin, eds. G.V. Coyne et al., (Vatican Observatory: Vatican), p. 641

Schmutz, W., Harmann, W.-R., Wessolowski, U. (1989), Astron. Astrophys., **210**, 236.

Schulte-Ladbeck, R.E. (1990), these proceedings.

Weber, E.J., and Davis, L. Jr. (1967), Ap.J., **268**, 228.

Zickgraf, F.J. and Schulte-Ladbeck, R.E. (1989), Astron. Astrophys., **214**, 274.

DISCUSSION

Conti: I will be ready to agree that rotation may play some role in WR stars. But how important is it physically? One star we know the rotation of is HD 50896 (from the periodic polarization changes). Its rotation period is 3.76 days, implying a very slow equatorial velocity. You might see what effect such an equatorial velocity actually does to HD 50896. I would think it small. It cannot by itself create the non-spherical shape of HD 50896.

Cassinelli: It appears from some of the poster papers that we really do not understand the nature of the 3.76 day resonance seen in HD 50896. In fact there is some evidence for a period of 1/3 of that. So I think the possibility of rapid rotation is still open. Secondly, the polarization QU plot of the $HeII$ 4686 line in HD 50896 shows a loop if you go from one side of the line to another. This is good evidence for a disk - *i.e.* material that exists all around the star and not just on one side towards a possible companion.

Sreenivasan: These massive stars have strong winds which spin down the surface layers rapidly to zero well before hydrogen is exhausted in the core. Angular momentum transferred from the core to the surface is also carried away efficiently. Magnetic fields, if present, will also very effectively slow down the star (*e.g.* the Sun). So the Michel velocity would, I am afraid, not help very much. (It does in the case of pulsars). Have you actually checked that this is not a problem? The co-rotation radius (where field lines rotate at the speed) is quite far away and would not be below the sonic point!

Cassinelli: Yes, we have done some modelling of the rotation rate *vs.* time, Cassinelli *et al.* (1989). The B[e] stars could only be explained as occurring after the mass loss at the Humphreys-Davidson limit, at which the stars eject $\sim$ 40 % of their mass and Γ, allowing the post HD evolution to occur with a smaller maximal rotation speed $\sqrt{GM(1-\Gamma)}/R$. It of course helps tremendously if the mass loss at the HD limit occurs primarily at the poles. Now in regards to your second comment, recall that the Maheswaran limits say that the field is either zero or above a certain limit. So one can have a post-main-sequence/pre-Humphreys-Davidson limit occur with zero field, so as to minimize the angular momentum loss. The mass ejection at the HD limit would of course expose the deeper layers of the star and magnetic effects might become important only after this phase. The fact that B[e] stars occur in the pre-WR location in the HR diagram makes this picture plausible. The picture is very complicated both from the point of view of mass loss evolution and surface magnetic field evolution.

Koenigsberger: (1) I have a poster outside on HD 193077, where I claim to see the gravity darkening effect pretty good where you have a rapidly rotating star. Regardless of whether we believe my photospheric absorptions or not, Massey (1980) pretty much concluded that the O star spectrum in the optical region is definitely rotationally broadened to 550 $km \cdot s^{-1}$. This is much larger than any O star measured by Conti and Ebbets. So, definitely there is evidence for rapid rotation in HD 193077. (2) (question to De Greve) When you have close binary systems we assume that they are co-rotating. So, if you start out by two O stars which are co-rotating, and one of them goes to a WR star, we are talking about rotational velocity periods of about three days, if this co-rotation is maintained into the WR phase. This is something no evolutionary model has taken into account.

Cassinelli: (1) It is good to know that there is evidence for rapid rotation of WR stars. It is often thought that WR stars should be slow rotators. (2) The thing about these very luminous stars is that they do not live very long, so a lot of these co-rotation effects might not be important. The other thing is, it happens when a star is undergoing very rapid phases in evolution like shrinking when it is going to the left into the HR diagram, it is collapsing, so it is going to spin up. So, even if it had been locked at some point earlier, it is going to get delocked and spin up.

Montmerle: If the magnetic rotator model works for WR stars, it should also play a role in the O star progenitor, *i.e.*, over the preceding few 10^6 years. Then one can expect a very efficient spin down (note: magnetic fields must be present in O stars, since many of them are non-thermal radio emitters).

Cassinelli: We would have to argue, from the Maheswaran results, that the surface field is small or zero during the O star phase of evolution, *i.e.* the circulation currents submerge the surface field, while the stars are in the pre-HD phase. I will admit that there must be some field in the winds of some O stars. We should consider how that occurs.

Walborn: One property of B[e] stars is that, unlike OB supergiants and LBVs, they show no evidence of CNO anomalies, so I seriously doubt that they are very highly evolved objects or are related to WR stars.

Cassinelli: That is an interesting point. It is the location of the B[e] stars in the HR diagram that we have tried to explain. The stars are very luminous and well separated from the main sequence, and show evidence for a two-component structure, fast polar flow and slower, denser equatorial flow (the Zickgraf model). It is difficult to explain this except as an phenomenon that occurs as post-HD limit stars.

Hamann: If WR winds would be less dense, but faster at the poles, compared to equatorial regions, this would severely affect the formation of the spectral lines. However, mass loss rates we derive from detailed fitting of line profiles by means of spherically symmetric models are in excellent agreement (within ±25%) with the rates derived from the radio emission. In my opinion, this rules out any drastical deviations from spherical symmetry.

Cassinelli: I think that one could also fit line profiles and radio fluxes with a non spherical picture. For example, Rumpl (1980) was able to get excellent fits to the UV line profiles of HD 50896 using a model in which the density in the equatorial zone greatly enhanced. Another thing I would like to stress is that there should be a significant flow out the polar regions, *i.e.* the winds would be as strong as, say, Of stars, out the polar zone. So you should not picture these stars as having mass outflow primarily in the equatorial zone, as is commonly envisioned for the Be stars. Finally, I think that there are enough adjustable parameters that one could fit both the UV line profiles and radio fluxes as well as does your spherical models. But of course there is a need to really do the calculations to verify this statement.

Barlow: If the mass loss rates of most WR stars are strongly influenced by rotation, why is HD 50896 the only "single" WR star which shows a well-defined period plus the polarization effects indicative of an asymmetric mass distribution?

Cassinelli: There are other stars that show intrinsic polarization effects. However, perhaps one reason why more WR stars do not show larger polarization is because the polar flow is not insignificant. The stars are not like Be stars in which there is negligible mass loss in polar regions. These stars have strong polar winds. We assume that the polar zone winds satisfy the radiation momentum limit, but as we know from the OB stars this corresponds to a significant wind. The point is that the polarization can be reduced because of cancellation that occurs because of scattering from electrons in the polar zone. In Be stars that cancellation effect does not occur.

Lozinskaya: I would like to remind you that the X-ray image of the WR star in NGC 6888 looks like this: we see two X-rays spokes inside the bubble, which may be considered as an indication of bi-polar wind.

Joe Cassinelli

DECELERATION ZONES IN THE WINDS OF WR AND P CYGNI TYPE STARS

TIIT NUGIS
W.Struve Astrophysical Observatory of Tartu
202444 Tõravere
Estonia, USSR

ABSTRACT. Specific features both of continua and of line spectra of WR and P Cygni type stars can probably be explained if an extensive deceleration zone is proposed to exist in their stellar winds. The outflowing matter is first accelerated near the stellar surface, then follows the deceleration of the flow and after that the final acceleration of the outflowing matter takes place (i.e. the wind has an ADA–structure: acceleration–deceleration–acceleration). Such a structure of the wind probably arises due to the multiscattering of photons in the envelope having two detached shells, which are optically thick in resonance lines (these shells can form if ionization stratification is present in the envelope).

1. Observational evidence

The stellar wind theory and in most cases also the modelling of line spectra and continua of hot stars allow to conclude that the velocity of the outflowing matter increases with the distance from the stellar surface until an asymptotic regime of expansion with a constant velocity $v = v_\infty$ is achieved.

In the case of some types of hot stars it was concluded that a deceleration zone must be present in the wind.

For P Cygni the presence of a deceleration zone in the wind was proposed already by Kuan and Kuhi (1975) from the analysis of HI line profiles. But the analysis of HeI by Oegerle and Van Blerkom (1976) and also of HI performed by some other astronomers (e.g. Kunasz and Van Blerkom 1978) showed that an agreement with line profiles can be obtained by a monotonically increasing velocity run in the envelope.

The analysis of the continuous spectral energy distribution in IR and radio wavelengths and also the study of HI line profiles which was presented in the papers of Nugis et al. (1979a, b), showed that a deceleration zone must be present in the wind. In these papers the ADA–structure (acceleration–deceleration–acceleration) was proposed for the wind of P Cygni. In the study of Waters and Wasselius (1986) the long–wavelength spectral energy distribution was analyzed too. They concluded that an agreement with observations can be obtained with a monotoni-

K. A. van der Hucht and B. Hidayat (eds.),
Wolf-Rayet Stars and Interrelations with Other Massive Stars in Galaxies, 309–313.
© 1991 *IAU. Printed in the Netherlands.*

cally increasing $v(r)$ only on the assumption that the $60\mu m$ flux is caused mainly by circumstellar dust. The second possibility suggested by them is that in the interval $15R_* < r < 50R_*$ there is a stationary shell with the density exceeding about 5 times the density of the wind. These explanations seem to be too artificial.

In the case of WR stars the presence of a deceleration zone in their winds was found by Nugis (1984, 1990). Here also the wind seem to have the ADA–structure as well.

Using new observational information about P Cygni and WR winds our recent study confirms the previous results (Nugis, 1990).

One of the main observational facts pointing to the deceleration zone in the WR winds is the presence of quite broad absorption components of HeI lines. HeI lines cannot arise close to the stellar surface because helium is doubly ionized there. The velocity change, taking place far from the stellar surface in a non–deceleratingly expanding envelope, cannot cause broad absorption components, because the velocity, corresponding to the long–wavelength edge of the HeI absorption component, is less than the velocity derived from line widths of subordinate lines of high–IP ions, which can arise only quite close to the stellar surface (e.g. $\lambda4945(NV)$ in the case of WN 5–6 stars). The width of HeI absorption components cannot be explained by local broadening mechanisms. This follows from the estimates for known broadening mechanisms and also from the fact that otherwise (when local broadening takes place) narrow absorption components should be present in many weak lines because absorption in the centre of the local line profiles is much stronger than in far wings. The question about the presence of macroturbulence in the envelopes of WR stars remains open, but neither macroturbulence nor microturbulence can explain the run of the spectral energy distribution in the IR spectral region. At first glance one can conclude from qualitative considerations that the spectral energy run in the IR spectral region of WR stars points to the envelope structure with a constant velocity run or, in some cases, even to the acceleration of the matter flow. This can be concluded only when assuming a homogeneous ionization structure. If the actual ionization structure is taken into account, then due to the fact that IR fluxes are effectively formed in that part of the envelope where helium changes from HeIII to HeII, matter density should decrease with radius more slowly than for a $v = const$ regime (that is, the deceleration zone must be present). The ratio of IR and radio fluxes of WR stars indicates that quite far from the stellar surface $(r \geq 50R_*)$ a substantial acceleration of the matter flow takes place.

The presense of the ADA–structure in the P Cygni wind can most easily be proved by the analysis of the spectral energy run in the IR spectral region.

In the case of P Cygni variable narrow displaced absorption components in the profiles of many strong lines have been observed (Kolka 1983; Lamers et al. 1985). Quite strong radio–flux variations have also been detected (van den Oord et al. 1985). These are probably due to the change in the kinematic structure of the envelope, and cannot be explained by variations of the mass loss rate because emission lines and IR–fluxes do not undergo substantial changes. In the studies of Waters and Wesselius (1986) and Felli et al. (1985), it was proposed that strong radio–flux variations are caused by the lowering of ionization state in the radio emission region, i.e. hydrogen was proposed to become neutral from time to time. According to our study this explanation cannot be adopted because hydrogen cannot become neutral in the region where radio–fluxes are effectively formed. Although the optical depth in the principal series continuum of HI becomes quite high ($t_1 \approx 20-25$) already close to the stellar surface, hydrogen is kept ionized by photoionizations from excited states. Hydrogen becomes neutral only at distances $r \geq 3000R_*$, i.e., at much larger distances as compared to the effective radio–flux formation region.

It must be said that in the paper by Lamers et al. (1985) it was alledged that small velocity shifts of absorption components of resonance lines of FeII in the UV spectrum of P Cygni exclude the possibility of the existence of the second acceleration zone. We made some estimates and found that if iron would be singly ionized in the interval $25R_* \leq r \leq 75R_*$, then we must observe high velocities of absorption components of FeII resonance lines. Actually we found that iron is in that region predominatingly doubly ionized and due to low T_e remains chiefly in the ground state. Therefore, the ADA–structure of the P Cygni envelope is not in conflict with low edge velocities of resonance FeII absorptions and of FeIII lines arising from metastable states. High velocities of the absorptions of some resonance lines of low–IP ions found from the IUE spectrum of P Cygni by Luud and Sapar (1980) confirm the presence of the ADA–structure. The maximum edge velocities of resonance absorptions found by Luud and Sapar (1980) tend to be $\approx 500km/s$ just as follows from our models if the mean value of $7mJy$ is adopted for the radio–flux at 6cm.

In our previous study of P Cygni (Nugis et al. 1979 a, b), we tried to explain by high expansion velocity also the presence of broad wings of H_α and H_β lines. However, it was shown by the calculations of Kolka (1980),that these wings are partly due to expansion and partly due to electron scattering, therefore a somewhat lower asymptotic velocity v_∞, as compared to the study of the ADA–models of Nugis et al. (1979 a, b), seems to be quite justified. The higher radio–flux value used in our previous study seems to correspond to the phase of the minimum v_∞.

2. A possible mechanism

Gravitation cannot cause the deceleration of the matter flow in the winds of WR and P Cygni stars.

We propose that the mechanism, which causes the formation of the deceleration zone and of the ADA–structure, is the multiscattering of photons between two different detached regions (shells) of an envelope, which are optically thick in resonance lines (Fig.1). Such detached shells can probably exist, if ionization stratification is present in the wind and resonance lines of non–abundant elements are optically thick at large distances from the star. These conditions are probably satisfied in the case of WR and P Cygni stars.

The multiscattering of resonance photons between two shells causes the deceleration of the inner shell and acceleration of the outer shell, whereas the multiscattering of photons between the opposite sides of the outer shell contributes to an additional acceleration of that shell. The latter process is identical to the mechanism of Panagia and Macchetto, 1982. In their model, instead of the inner shell, there is a star itself and all the photons, which hit the star after being backscattered, would be thermalized and are therefore removed from the process. This strongly suppresses the effectiveness of the multiscattering process (especially near the stellar surface).

In the case of our two–shell model, the photons which hit the inner shell, are not thermalized. They will be backscattered and continue the process of multiscattering until they either acquire such a frequency that no resonance transition can absorb them (these quanta escape from the envelope or from the multiscattering region) or the photons will be destroyed.

Fig. 1. A schematic representation of the mechanism of the multiscattering of photons between two shells, which are optically thick in resonance lines ((a) and (b) – two travelling photons, which are taking part in the multiscattering process).

Of course, the effectiveness of the multiscattering of resonance photons in the winds of hot stars needs further investigation. Abbott and Lucy (1985) and Friend and Castor (1983) concluded that multiscattering takes place in the stellar winds of OB stars, but that it is not so effective in accelerating the wind as proposed by simple estimates of Panagia and Macchetto (1982). A detailed investigation of this mechanism is needed for hot stars. In the case of OB stars probably some multiscattering of resonance photons between different shells may also take place. This may explain the formation of narrow displaced absorption components seen in some strong resonance lines.

References

Abbott, D.C. and Lucy, L.B. 1985. *Astrophys. J.*, **288**, 679.
Felli, M., Stanga, R., Oliva, E. and Panagia, N., 1985. *Astr. Astrophys.*, **151**, 27.
Friend, D.B. and Castor, J.I., 1983. *Astrophys. J.*, **272**, 259.
Kolka, I., 1980. *ENSV TA Preprint* A–4.
Kolka, I., 1983. *ENSV TA Toimetised. Füüs. Matem.* Vol. 32, 51.
Kuan, P. and Kuhi, L.V., 1975. *Astrophys. J.*, **199**, 148.
Kunasz, P. and Van Blerkom, D., 1978. *Astrophys. J.*, **224**, 193.

Lamers, H.J.G.L.M., Korevaar, P. and Cassatella, A., 1985. *Astr. Astrophys.*, **149**, 28.

Luud, L. and Sapar, A., 1980. *Tartu Astrofüüs. Obs. Teated*, No 60, 3.

Nugis, T., 1984. *Tartu Astrofüüs. Obs. Publ.*, **50**, 101.

Nugis, T., 1990. *Astrofiz.*, **32**, 85.

Nugis, T., Kolka. I. and Luud, L., 1979a. *Mass Loss and Evolution of O-Type Stars, IAU Symp.* No 87, p.39, eds Conti, P.S. & de Loore, C.W.H., Reidel, Dordrecht, Holland.

Nugis, T., Kolka, I. and Luud, L., 1979b. *Tartu Astrofüüs. Obs. Publ.*, **47**, 191.

Oegerle, W.R. and Van Blerkom, D., 1978. *Astrophys. J.*, **224**, 193.

Van den Oord, G.H.J., Waters, L.B.F.M., Abbott, D.C., Bieging, J.H. and Church-well, E., 1985. *Radio Stars*, p.111, eds Hjellming, R.M. & Gibson, D.M., Reidel, Dordrecht, Holland.

Panagia, N. and Macchetto, F., 1982. *Astr. Astrophys.*, **106**, 266.

Waters, L.B.F.H. and Wesselius, P.R., 1986. *Astr. Astrophys.*, **155**, 104.

DISCUSSION

Cherepashchuk: In your model some sort of Rayleigh-Taylor instability must be present in the acceleration zone. This instability may form the blobs.

Nugis: Yes, I think that such a structure must be unstable.

Owocki: (1) I do not understand what specific observations require you to invoke deceleration zones. (2) In dynamical wind models I have computed, it is easy to obtain such deceleration zones if parameters are chosen so that $\dot{M} \gtrsim \dot{M}_{CAK}$.

Nugis: (1) The main observational facts which point to the presence of deceleration are the shapes of the long wavelength continua and in the case of some types of WR stars the wide HeI violet shifted absorption components which indicate that the expansion velocity at some place in the zone of formation of HeI lines must be less than the expansion velocity of the region, where the subordinate lines of some high IP lines are formed (the latter are formed closer to the stellar surface). (2) Your information about the presence of deceleration zones in the dynamical wind models in the case when $\dot{M} \gtrsim \dot{M}_{CAK}$ is very important and must be taken into account in further study.

Werner Schmutz

RADIO OBSERVATIONS OF MASSIVE OB STARS

IAN D. HOWARTH
Dept. Physics & Astronomy
University College London
Gower Street
London WC1E 6BT
England

ALEXANDER BROWN
JILA
University of Colorado
Boulder
Colorado 80309-0440
USA

ABSTRACT. The mass-loss rates of O stars and B supergiants are of interest because of their influence on the evolution of these massive stars (among other matters). In principle, the 'safest' (*i.e.* most model-independent) method of determining $\dot{M}$ is to measure the free-free emission from stellar winds at radio wavelengths. This method is complicated, however, by the existence of poorly understood non-thermal emission in some stars, and by the possibility of hydrogen recombination in the winds of B supergiants.

We are in the process of carrying out a VLA survey of OB stars, initially at 3.5cm, to a flux limit of ~0.1mJy. Because *all* our targets should have thermal emission at detectable levels (based on mass-loss rates from Howarth & Prinja 1989 and terminal velocities from Prinja, Barlow & Howarth 1990), the survey is yielding an *unbiassed* estimate of the frequency of non-thermal emission. The improved sensitivity of our survey over earlier work defines the $\log \dot{M} - \log L$ relationship much more precisely than was previously possible, over a large range in luminosities; and allows us to make definitive statements on recombination in B supergiant winds. Our sample includes the first radio detections of an OC star, of a massive X-ray binary, and of thermal emission from a main-sequence star.

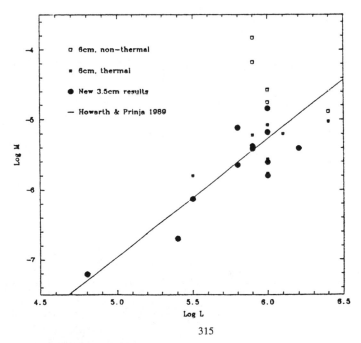

K. A. van der Hucht and B. Hidayat (eds.),
Wolf-Rayet Stars and Interrelations with Other Massive Stars in Galaxies, 315.
© 1991 IAU. Printed in the Netherlands.

PHOTOMETRIC DETERMINATION OF MASS LOSS RATES AND ORBITAL INCLINATIONS FOR WR BINARIES

R.LAMONTAGNE, C.ROBERT, A.GRANDCHAMPS, N.LAPIERRE, A.MOFFAT
Département de Physique
Université de Montréal, Montréal, Canada
L.DRISSEN, M.M.SHARA
Space Telescope Science Institute
Baltimore, U.S.A.

ABSTRACT Most Wolf-Rayet binaries show phase-dependent light variations with a broad dip occurring at phase zero, when the WR star passes closest to the observer. When the orbital period is long, or the inclination is low, this dip is buried in the noise. When eclipses of the stars occur, another dip is seen when the companion passes closest to the observer; in this case, both dips are relatively deep and sharp (e.g. V444 Cygni, HD5980).

In the cases when only one dip at phase zero is seen, a simple model is derived. This involves electron scattering of companion starlight as the orbit makes it systematically traverse different amounts of the WR wind, assumed to be spherically symmetric and to follow a monotonic velocity-radius law. The shape and amplitude of the dip yield estimates of $\dot{M}$ and i (hence M when combined with $M \sin^3 i$ from spectroscopic orbits).

Mass-Loss rates as a function of WR subtypes. Stars are identified by their WR No. (CV Ser = WR113, GP Cep = WR153).

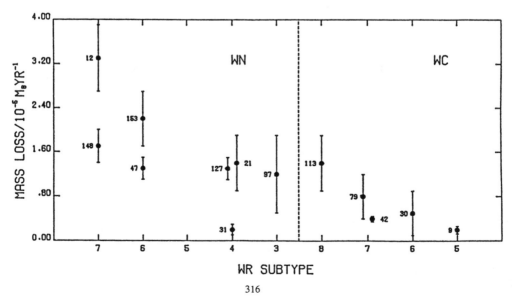

K. A. van der Hucht and B. Hidayat (eds.),
Wolf-Rayet Stars and Interrelations with Other Massive Stars in Galaxies, 316.
© 1991 IAU. Printed in the Netherlands.

TERMINAL VELOCITIES FOR A LARGE SAMPLE OF O STARS, B SUPERGIANTS, AND WOLF-RAYET STARS

R. K. PRINJA, M. J. BARLOW
Department of Physics & Astronomy
University College London
Gower Street, London WC1E 6BT
England

I. D. HOWARTH
JILA
University of Colorado
Boulder, Colorado
U.S.A.

ABSTRACT. We argue that easily measured, reliable estimates of terminal velocities for early-type stars are provided (1) by the central velocity asymptotically approached by narrow absorption features in unsaturated UV P Cygni profiles, and (2) by the violet limit of zero residual intensity in saturated P Cygni profiles. We use these estimators and high resolution IUE data to determine terminal velocities, v_∞, for 181 O stars, 70 early B supergiants, and 35 Wolf-Rayet stars. For OB stars our values are typically 15-20% smaller than the extreme violet edge velocities, v_{edge}, while for WR stars $v_\infty = 0.76 v_{edge}$ on average. We give new mass-loss rates for WR stars which are thermal radio emitters, taking into account our new terminal velocities and recent revisions to estimates of distances and to the mean nuclear mass per electron. We examine the relationships between v_∞, the surface escape velocities, and effective temperatures.

The full paper is due to be published in the 1990 October 1 issue of the *Astrophysical Journal*.

K. A. van der Hucht and B. Hidayat (eds.),
Wolf-Rayet Stars and Interrelations with Other Massive Stars in Galaxies, 317.
© 1991 *IAU. Printed in the Netherlands.*

STELLAR WINDS FROM MASSIVE STARS: THE INFLUENCE OF X-RAYS ON THE DYNAMICS

I. Stevens[1], G. Cooper[2] and S. Owocki[2]
(1) Code 665, NASA/GSFC, Greenbelt, Maryland, MD 20771
(2) Bartol Research Inst., University of Delaware, Newark, DE 19711

ABSTRACT. We report on a theoretical investigation of the X-rays observed from early type stars on the global wind dynamics.

1. Introduction

X-rays are observed from early type stars with $L_x \sim 10^{31} - 10^{34}$ erg s^{-1} , and are believed to be produced from shocks in the wind. The wind model of Castor *et al.* (1975, CAK), and subsequent modifications, remains the most realistic and complete description of line driven winds. Here, we use a modified CAK model which includes the effects of X-ray ionization (XRI) on the force multiplier $M(t)$, and some simple estimates about the distribution of X-rays in the wind to investigate the dynamical impact of the X-rays. The model has been previously used to investigate gas dynamics in MXRB's (Stevens and Kallman, 1990).

2. Results

- For values of $\log_{10} P \geq 21.5$ ($P = L_x v_\infty / \dot{M}$) major changes in wind dynamics can occur.
- $\dot{M}$ is largely unchanged by XRI, while v_∞ tends to be reduced; by $\sim 50\%$ for higher values of P.
- The observed parameters suggest that for a number of early type stars XRI can alter the wind dynamics.
- Stars most likely to be dynamically affected by XRI are those with relatively high values of $L_x / L_{bol} \sim 10^{-6}$ (Chlebowski *et al.* 1989).
- In most WR stars XRI will not significantly alter the wind dynamics.
- However, some WR stars might be affected, particularly those with higher X-ray luminosities such as HD 93162, HD 193793, and HD 104994, though in these stars it is possible other mechanisms are at work (colliding winds, wind-ISM interactions).

References

Castor, J., Abbott, D., and Klein, R., 1975. *Ap. J.*, **195**, 157.
Chlebowski, T., Harnden, F., and Sciortino, S., 1989. *Ap. J.*, **341**, 427.
Stevens, I., and Kallman, T., 1990, *Ap. J.*(submitted).

318

K. A. van der Hucht and B. Hidayat (eds.),
Wolf-Rayet Stars and Interrelations with Other Massive Stars in Galaxies, 318.
© 1991 *IAU. Printed in the Netherlands.*

AN ANALYTICAL APPROACH TO THE PROBLEM OF MASS LOSS FROM WOLF–RAYET STARS

R. TUROLLA
Dept. of Physics, University of Padova
Via Marzolo 8
35131 Padova
Italy

We investigate the dependence of the wind parameters on the properties of the parent star by means of an approximate analytical approach, largely based upon dimensional arguments (Bandiera and Turolla, 1990). The main assumption we use are

a) WR stars can be treated as homogeneous, pure helium structures;

b) nuclear energy generation is due only to 3–α burning;

c) line driving is the mechanism responsible for accelerating the wind;

moreover, all assumptions contained in the original CAK (Castor, Abbott and Klein, 1975) treatment are retained.

Using a simple politropic model for the stellar interior we show that mass loss from WR stars can be characterized just by the stellar mass; it is found that a correlation between mass and mass loss rate of the form $\dot{M} \propto M_*^p$ ($p \sim 2$) exists. Such a relation is similar to those ones obtained by Abbott et al. (1986) from observations of WR stars in binary systems and by Turolla, Nobili and Calvani (1988) from the integration of dynamical stellar models.

References

Abbott, D. C., Bieging, J. H., Churchwell, E. and Torres, A. V. 1986, Astrophys. J., **303**, 239

Bandiera, R. and Turolla, R., 1990, Astron. Astrophys., **231**, 85

Castor, J. I., Abbott, D. C. and Klein, R. I. 1975, Astrophys. J., **195**, 157

Turolla, R., Nobili, L. and Calvani, M. 1988, Astrophys. J., **324**, 899

K. A. van der Hucht and B. Hidayat (eds.),
Wolf-Rayet Stars and Interrelations with Other Massive Stars in Galaxies, 319.
© 1991 *IAU. Printed in the Netherlands.*

Debray, Matteucci, Lewis, Nomoto

SESSION VI. ENRICHMENT – *Chair: Joseph P. Cassinelli*

Anne Underhill, Joe Cassinelli chairing

ENRICHMENT BY WOLF-RAYET STARS AND OTHER MASSIVE STARS IN GAS, DUST AND
ENERGY

MARTIN COHEN
Radio Astronomy Laboratory
University of California
Berkeley
USA

ABSTRACT. I update previous estimates of the separate contributions for
radiative energy, integrated total stellar wind mass and dust mass from
Wolf-Rayet stars and other massive (OBA) stars. In the context of the
intriguing dusty WC9 stars, I: (1) discuss the observability (or otherwise) between 0.4 and 23 μm of the condensation route from hot gas to
carbon-rich grains; (2) urge caution in the use of 10 μm infrared
spectra of these luminous stars to deduce the importance of silicates as
a component of the interstellar medium; and (3) speculate on a possible
new method for discovering new members of this relatively rare subtype
based on IRAS Low Resolution Spectra. I review the observational
evidence for dust condensation around SN 1987A.

1. INTRODUCTION

The first part of my talk will focus on an updated version of a popular
topic, namely the relative importance of the Wolf-Rayet (WR) stars'
influence on the interstellar medium compared with the effects of other
stars in the Galaxy ([1],[2]=[HCW],[3],[4]). What role do the WRs play
as contributors of energy, momentum, ionizing photons, overall
luminosity, and dust through their radiation and powerful stellar winds?
It is customary to compare the attributes of WRs with OBA stars for
their energetics. To assess the WRs' dust contribution requires consideration of the abundant lower mass stars whose red giant phases
provide most of the dust to the interstellar medium.
        My second subject will be the intriguing WC9 stars and the likely
pathway to dust condensation in their carbon-rich winds. This will be
discussed in the context of a combination of recent airborne spectroscopy with IRAS Low Resolution Spectrometer (LRS) data. Detailed LRS
surveys of successively fainter IRAS point sources have resulted in a
comprehensive spectral classification scheme now validated by optical
spectroscopy and IRAS photometric color indices. These surveys have
brought to light an unusual shape of LRS energy distribution which may
correspond uniquely to dusty WCL stars. I would like to present this as

K. A. van der Hucht and B. Hidayat (eds.),
Wolf-Rayet Stars and Interrelations with Other Massive Stars in Galaxies, 323–334.
© 1991 IAU. Printed in the Netherlands.

a potential method for seeking new WC stars with dust.

The final item will cite the evidence for dust condensation in SN 1987A, again through airborne infrared spectra.

2. THE FLUX OF RADIATION, MASS, MOMENTUM, AND ENERGY FROM MASSIVE STARS

2.1 OB and WR Star Counts and Caveats

It has been traditional when undertaking a review of this field to build heavily upon Abbott's [1] presentation for both WR and OBA stars and merely to reflect changes in the WR literature that directly affect estimates of mass loss rates. The literature does not convey the impression that there are any fundamental shortcomings with the OB catalogue of Garmany, Conti, and Chiosi [5] beyond a minor concern about its completeness, particularly for the dwarf component. However, Garmany [6] has expressed a number of serious caveats about "GCC". Among those concerns are: (1) the absence of two-dimensional spectral classifications for a substantial number of stars; (2) potentially specious O-star classifications; (3) the inclusion of stars in "OB associations" which not may represent truly physical groupings. These affect the value of the catalogue as a whole for stellar census work in which distances and estimates of physical parameters, such as mass loss rate and even bolometric luminosity, are important.

Consequently, since I wished to re-evaluate all aspects of the problem that might have undergone change since previous reviews, I have followed [1] but have used a subset of "GCC" kindly supplied to me by Dr. Garmany. This subset consists of 981 stars in the northern hemisphere between galactic longitudes $59°$ and $148°$ that are definitely established to belong to real OB associations. My subsample consists of 406 O,B, or A members of associations whose distances are <3 kpc, and spans all luminosity classes. This quadrant provides perfectly valid estimates for the parameters of interest per kpc^2, the customary format, without the need to study the entire 3 kpc volume centered on the sun and without recourse to the considerable uncertainties in distance and spectral type inherent in the entire version of "GCC". The earlier WR catalogue by van der Hucht et al. [7] has been superseded by an updated version [8](=HHASD) which includes new spectral subtypes and distances for an appreciable number of stars. The WR content of the 3 kpc vicinity centered on the sun has consequently changed significantly since earlier reviews of this field ([2],[3]). I have, therefore, revised the 3 kpc WR census using [8]. There are 61 WRs within this volume: 24 WN, 36 WC, and 1 WO type.

Conti and Vacca ([9]=CV) have very recently studied the distribution of WR stars in the Galaxy based upon new spectra, some obtained after those incorporated into [8]. CV also assign distances on the basis of a different set of intrinsic colors for WRs [10] than HHASD used. This leads to a somewhat different WR census for the 3 kpc zone: 65 stars, 35 WN, 29 WC, and 1 WO type. I, therefore, recalculated all parameters described below, separately for HHASD and CV.

2.2 Luminosity and Ionizing Flux

As part of a substantial study of O-stars, Howarth and Prinja [11]
provide a recalibration of the physical parameters, in particular their
luminosities. I have incorporated this revised calibration into the
assessment of radiative output from the O-stars. I have reverted to
Panagia [12] for the early B types and to Allen [13] for the late B and
early A types. The estimates of ionizing flux for all types still
follow the tabulations of N_L in [12].

Barlow, Smith, and Willis [14] present bolometric luminosities
and ionizing photon fluxes for most WR types. Schmutz, Hamann, and
Wesselowski [15] model WRs and their luminosities agree well (≤ 0.2 dex)
with those in [14] except for the earliest WN and WC types. Smith and
Maeder [16] argue for a constant B.C. (-4.5) for all WRs based on theoretical mass-luminosity relationships. I adopted the scale of [14],
replacing the values for WN3 and 4 (4.5), and incorporating the data for
WN2, from [15]. For WO2 and WC9 I took M_v from HHASD and a B.C. of
-4.5. For types lacking N_L I used the average ratio (from [14]) of
$N_L/L_{bol} = 3.80 \times 10^{43}$.

2.3 Mass Loss Rates

The recent analysis of IUE high-resolution spectra by Howarth and Prinja
[11] contains a re-examination of the issue of terminal velocities in O-
star flows. Turbulence and shocks can lead to maximum observed
velocities exceeding the true far-field velocity (e.g. in rarefactions:
[17]). They argue [18] that the terminal velocity is best estimated
from the central velocity asymptotically approached over time by narrow
absorption components or, statistically equivalent, a scaled down value
for the violet limit (zero residual intensity) of saturated P Cyg profiles. Use of these revalued terminal velocities yields a robust pre-
dictor for the mass loss rate of O stars of all luminosity classes
dependent only on luminosity, namely, $\log(\dot{M}/M_\odot \text{ yr}^{-1}) = 1.69 \log(L_*/L_\odot)$
- 15.4. A very similar slope resulted from previous empirical determin-
ations ([19],[20],[1]) and is consistent with evolutionary models for
high mass stars [21]. I, therefore, used this formula for the OB (and
A0) stars from Garmany's [6] new subcatalogue.

Mass loss estimates for WRs have undergone several recent revisions
stemming from four independent causes, namely: carbon abundances in the
winds; low ionization state of the winds; redetermination of distances
to some stars (e.g. HHASD; CV); and revisions in terminal velocities.
HCW increased the estimates made from VLA radio continuum fluxes [22] to
include the high abundance of carbon in the stellar winds determined by
Prantzos et al. [23]. Support for these high C/He values has come from
other work ([24],[25],[26],[27],[28]). However, Smith and Hummer [26]
suggest that HCW's increases are too large because they used too low a
carbon ionization balance. Both Schmutz and Hamann [29] and Hillier
[30] have shown that the radio-emitting region in WRs shows a strong
gradient in He ionization with radius so that He^+ is more important
than He^{++}. Allowance for the reduced free-free emissivity elevates
the original mass loss rates [22] by a factor ~ 2 for all types [29].

Work in press on terminal velocities by Prinja, Barlow, and Howarth
[31], kindly advanced to me by Dr. Barlow, re-evaluates terminal velocities in WRs by the same methods as those applied to O-stars [11]. The
average change in v_∞ is 0.76 for the WRs, leading to reduced values
of mass loss rate. In the discussion below, I have used the new results
for individual stars to determine mean velocities and mass loss rates
for WR types because these authors have incorporated all the above
changes. For WR types not treated in [31], I have scaled the terminal
velocities used by HCW by 0.76 and have extrapolated the known mass loss
rates to types for which these are not known exactly as HCW did, but
using Prinja et al.'s work [31]. There are ~10 WC9s within 3 kpc of the
sun; therefore, it is important to estimate their mass loss rate carefully rather than merely setting it equal to that for WC8s. $\dot{M}$ depends
on ionization balance through three factors: the Gaunt factor; the mean
number of free electrons per nucleon; and the r.m.s. ionic charge. I
rescaled HCW's mean WC9 $\dot{M}$ according to the change in these three
parameters between their own ionization balance (of He and C) and that
in [26], and by a further factor of 0.76 for the terminal velocity. $\dot{M}$
for type WO2 comes from Barlow's paper at this meeting.

2.4 Momentum and Energy Output

The momentum and energy carried by these hot stellar winds are taken to
be $\dot{M}v_\infty$ and $0.5\dot{M}v_\infty^2$, respectively. Table 1 summarizes the adopted mean
parameters of the WRs and their outputs of mass, momentum and energy per
star per kpc^2 and their numbers within 3 kpc according to both HHASD and
to CV. Only spectral types necessary to the 3 kpc census are listed.

Table 2 compares the physical influence of WR and OBA stars upon
the local medium, separating fractions attributable to WR, O, BA stars
and giving the totals per kpc^2 near the sun. In spite of the obvious
differences inherent in the two WR catalogues (HHASD and CV), the
analyses for Table 2, and even the total quantities in the final column,
differ remarkably little ($\leq 5\%$). Table 2, therefore, presents the
averages derived from both HHASD and CV.

There have clearly been so many changes between different versions
of Table 2 that it is not surprising to find differences. However, the
WRs still dominate the input of both mass and momentum to the local
medium. The principal new change is that their energy output now more
closely resembles Abbott's [1], essentially because the reduction in WR
terminal velocities affects their wind energy by the cube of 0.76, and
the new mass loss scale for the OBA stars [11] elevates their mass loss
rates over those used in [1]. Other differences, principally in the
totals of the quantities, may well reflect the usage of the new subset
[4] of hot stars and the implicit assumption that this quadrant is representative of the whole 3 kpc neighborhood. Garmany's [32] Fig. 4
suggests that this is likely to be a reasonable assumption but one must
view the distribution of hot stars as mapped out by her entire catalogue
with some caution.

2.5 Enrichment of the medium in heavy elements by WRs

Following [1], one can calculate the degree of enrichment of the inter-stellar medium by the nucleosynthetic products carried by WR winds. The abundances of heavy elements used by HCW [23] were converted from a table by number to elemental mass fractions (Table 3). The yield of any element is estimated as the product of the difference in mass fraction between WR wind and local medium and the mean mass loss rate per kpc^2, summed over all WN, WO, and WC stars ([1],[4]). Mean mass loss rates within 3 kpc averaged over HHASD and CV are: WNs, 5.61(-5); WOs, 6.11(-7); and WCs, 5.01(-5) $M_\odot$ yr^{-1} kpc^{-2}. These yields are normalized by the net rate of input of gas to the medium from star formation [4], 4(-3) $M_\odot$ yr^{-1} kpc^{-2}, and are tabulated for He, C, N, and O (Table 3).

Comparing with the expected yields from supernovae alone (Tables 1 and 2 of [33]), WRs provide ∼50% of the enrichment in He, and 33% of that in CNO combined.

3. DUST

3.1 Dust Output in the WRs and other Massive Stars

The dust contribution of WRs to the interstellar medium is that of only the WC7 and later stars, dominantly that of the WC9s ([34],[35],[36]). By contrast, OBA stellar winds are not observed to condense dust.

3.2 Dust Output of Lower Mass Stars

Both Gehrz [37] and Tielens ([38],[39]) have estimated the magnitude and/or composition of stardust for a wide variety of dust-producing objects on a Galactic scale. They do not always agree on dust-to-gas ratios, or average mass loss rates. [38] isolates the carbon stardust. To make estimates more relevant to the solar neighborhood I utilized the detailed Galactic model by Wainscoat et al. [40] to predict the break-down of stellar population close to the sun in terms of its AGB M and C stars, red supergiants, and planetaries. Table 4 combines results from the model predictions with those in [37], [38], and [39]. The model agrees better with Tielens's more conservative estimates than with Gehrz's, and I have therefore used the average of the model's and Tielens's numbers in case of disagreements. For supernovae there is only a lower bound on dust formation (>0.004 Table 4 units) from the small amount observed in SN 1987A (Wooden [41]: section 5 below). Tielens's figures (in []), reproduced here, represent the maximum amounts assuming all C and Si condense into grains in SN. Gehrz [37] estimates ∼5 (Table 4 units) for the dust yield of all supernovae, or one third of the sum of Tielens's maxima for type I and type II SN combined!

Table 1 shows that WC8 and 9 stars contribute 12.6 (Table 4 units) of gas to the 3 kpc vicinity, implying (from Table 3) 4.9 units solely in carbon. For a dust-to-gas ratio of 0.01 I recover Tielens's global estimates of 0.05-0.06 units in carbon dust ([38],[39]).

TABLE 1. The adopted mean parameters of WR stars and star counts within 3 kpc of the sun according to HHASD and CV

Spectrum	V_∞	$\dot{M}$	$\dot{M}$	$\dot{M}V_\infty$	$0.5\dot{M}V_\infty^2$	HHASD	CV
	km s^{-1}	$M_\odot$ yr^{-1}	$M_\odot$ yr^{-1}	10^{29} g cm s^{-1}	10^{37} erg s^{-1}	#	#
WN 2	3200	3.0(-5)	1.06(-6)	0.21	0.34	1	0
WN 3	2300	3.0(-5)	1.06(-6)	0.15	0.18	2	2
WN 4	1950	3.0(-5)	1.06(-6)	0.13	0.13	0	1
WN 4.5	1450	3.0(-5)	1.06(-6)	0.10	0.070	1	2
WN 5	1450	4.0(-5)	1.41(-6)	0.13	0.094	2	6
WN 6	1550	8.5(-5)	3.01(-6)	0.29	0.23	8	8
WN 7	1600	4.9(-5)	1.73(-6)	0.17	0.14	8	8
WN 8	825	4.9(-5)	1.73(-6)	0.090	0.037	1	7
WN 9	1150	4.9(-5)	1.73(-6)	0.12	0.070	0	1
WO 2	5600	1.7(-5)	6.11(-7)	0.22	0.60	1	1
WC 4	3050	3.2(-5)	1.13(-6)	0.22	0.33	2	1
WC 5	2550	3.2(-5)	1.13(-6)	0.18	0.23	7	4
WC 6	2150	3.2(-5)	1.13(-6)	0.15	0.16	6	4
WC 7	2000	7.2(-5)	2.55(-6)	0.32	0.32	8	9
WC 8	1450	6.3(-5)	2.23(-6)	0.20	0.15	3	3
WC 9	1100	1.8(-5)	6.26(-7)	0.044	0.024	11	8

TABLE 2. The influence on the medium of WR and OBA stars

Quantity	% contribution by			Total Rate
	WR	O	BA	(kpc^{-2})
Number of stars	4	23	73	59.65
Radiative luminosity	4	69	27	2.75×10^{40} erg s^{-1}
Ionizing photons	9	88	3	2.33×10^{50} s^{-1}
Mass	81	17	2	1.32×10^{-4} $M_\odot$ yr^{-1}
Momentum	73	25	2	1.54×10^{30} g cm s^{-1}
Kinetic energy	38	61	1	2.82×10^{38} erg s^{-1}

TABLE 3. Element enrichment by WR stellar winds

Element	Relative mass fractions				
	X(WN)	X(WO)	X(WC)	X(ISM)	Yield
He	0.97	0.09	0.32	0.22	1.2×10^{-2}
C	0.00038	0.21	0.39	0.0034	4.9×10^{-3}
N	0.022	0.00	0.00	0.0012	2.8×10^{-4}
O	0.0011	0.66	0.25	0.0082	3.2×10^{-3}
CNO	0.024	0.87	0.64	0.013	8.4×10^{-3}

TABLE 4. A comparison of Galactic producers of "stardust"

| Source | Contribution (10^{-6} $M_{\odot}$ yr^{-1} kpc^{-2}) | | |
	carbon dust	silicates	SiC
M-giants		2.5	
C-giants	3		0.1
Red supergiants		0.2	
Novae	0.3	0.03	
Planetary nebulae	0.03		
WC8 & WC9 stars	0.05		
Type I SN (max.)	{2}	{12}	
Type II SN (max.)	{0.3}	{2}	

4. THE WC9 STARS

4.1 Dust Condensation in WC9s

The WC9s are the most prodigious dust producers of the WRs but precisely
what they condense has been an ongoing problem. The high C/He suggests
that the grains are carbon-rich, probably amorphous. One might imagine
that any condensation route would involve neutral carbon, yet none is
detected in their spectra. Indeed, Williams et al. [36] commented that,
because most of the carbon in WC9s is CII and CIII, there is not sufficient CI for grain formation by the process that operates in C-giant
atmospheres and grain growth by accretion of C ions was precluded by
the positive charges on those ions. Smith and Hummer [26] find that the
dominant C ion in WC9s is CIII, exacerbating this apparent problem.

Not until airborne 5-8 μm spectra were obtained of WC9s were dust
features recognized [42]. When these spectra, of Ve 2-45 and GL 2104,
are combined with IRAS LRS spectra and compared with blackbodies, a
broad feature is seen in both stars between $\sim$7.5 and 9 μm, with peak
near 7.7 μm. No associated 6.2 μm (PAH) band is detected, nor any 3.3 or
11.3 μm feature (CH stretching and bending modes). The 7.7 μm peak
corresponds to vibrations of the aromatic skeleton, but neither 3.3 nor
11.3 μm features is expected in the H-poor environment of a WC9 star.

All these characteristics derive from small ($<$30Å) aromatic
domains, contained within small (perhaps 80Å) disordered grains. The
disorder results from highly supersaturated winds with low temperature
grain formation [42]. Possible pathways to dust formation in C-rich
circumstellar envelopes are: neutral radicals, ions, PAHs, polyacetyl-
enic chains, and fullerenes ("soccer balls"); all act as intermedi-
aries to soot formation in flames. C-giants form dust via acetylene;
C-rich planetaries via PAHs (with significant hydrogenation); but WC
stars represent a situation more akin to laser vaporization of graphite
in a H-poor atmosphere than to flames. In WC stars the absence of H
precludes the acetylene route. Probably dust in the WC9s is built from
polyynes (acetylene-like chains), into C_2, C_3, C_4, etc. [38], then large
flexible structures ($>$20 atoms) that form aromatic rings but with many

unfilled valences at the molecular peripheries (H would normally saturate these dangling edge bonds). Minimization of "surface free energy" drives dust formation in the WC9s via pentagons in the aromatic structures, which lead to warping, curling, and eventually closure in the form of fullerenes (C_{60}, C_{70}: [38],[42]).

Because ion-molecule reactions are faster than neutral-radical ones, the first molecules are thought to form via ions in the late WCs, by radiative association of C with C^+ and C_2^+. C_3^+ may recombine dissociatively [38] leading to C_2 and initiating the sequence from acetylenic chain to large aromatic structures via ion-molecule reactions.

On a global scale the fullerenes are expected to be a very minor form of interstellar dust [38] because C-giants dominate the C-stardust scene (Table 4). Fullerenes may represent a dead-end in the dust formation process in WC stars too; consequently, they may be abundant around WC9s. There is rumored to be a fullerene band near 3860Å. It would be very interesting to search for such features in WC9 stars and for the bands of C_2 and C_3 ions (because one can already set limits on the neutral molecules from the wealth of WC9 spectra).

4.2 Silicate Absorptions and Aromatic Emissions

I simply want to urge caution in the interpretation of mid-infrared spectra of the WCL stars. Without airborne (5-8 μm) spectra it is all too easy to dismiss the overall spectral shape as a featureless hot continuum suffering substantial 9.7 μm interstellar silicate absorption. However, any continuum must fall <u>below</u> the aromatic emission structures near 8 μm, thereby removing an appreciable amount, if not all, of the supposed silicate feature [42]. Consequently, it is inadvisable to determine the visual extinction from τ(9.7 μm) without complete infrared spectra of these stars or, conversely, to assess the ratio A_v/τ(9.7 μm) from WC9s without due caution.

4.3 An IRAS LRS Search for New WC9 Stars

As part of an ongoing study of the 170,000 extracted LRS spectra, Volk and Cohen [43] have provided a complete catalogue of all IRAS point sources having S(12)>40 Jy, to supplement the published LRS Atlas [44]. The spectrum of GL 2104 appears in the Atlas; those of Ve 2-45 (Roberts 80) and GL 2179 in [43]. Their shape is quite unlike that of other LRS spectra although the closest match is to "group P" (PAH emission) spectra. This similarity is strengthened by the airborne spectra described above [42].

Among new, unidentified group P sources are IRAS 17380-3030 and 18405-0448. Fig. 1 is a montage of normalized λF_λ plots for 4 known WC9/10 stars and these 2 new potential WC9s, showing their kinship. It is the combination of broad polcyclic aromatic cluster emission ~7-9 μm and some possible interstellar silicate absorption near 10 μm that create this archetypical spectrum. As yet there is no proof that either of these IRAS sources <u>is</u> a WC9. Van der Hucht, Williams, and The (1989-90) have obtained infrared photometry but optical spectra are still lacking. Both are extremely faint optically but an AAO service request

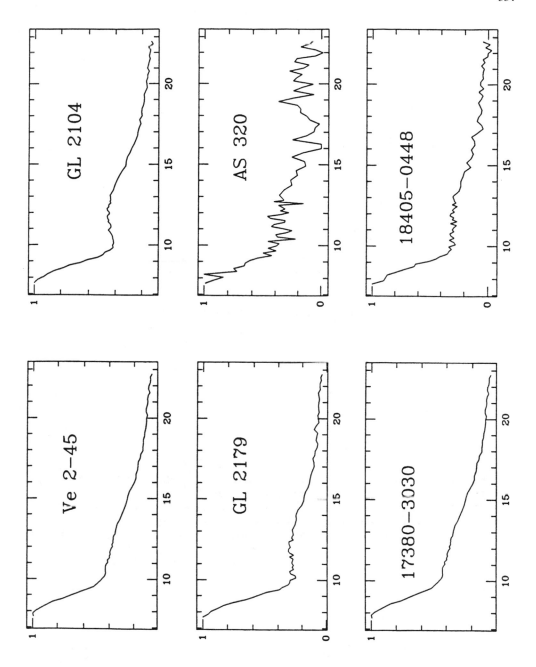

FIG. 1: Montage of LRS spectra for WC9/10 stars and the two possible new
WCL objects, IRAS 17380-3030 and 18405-0448. Abscissa is wavelength in microns; ordinate is normalized λF_λ .

for spectroscopy of 17380-3030 is pending. We have our fingers crossed!

5. DUST IN SN 1987A

Whether SN form dust grains is an interesting topic [45]. Dust observed in the Crab may be swept up, not condensed ([46],[47]). Critical observations by Wooden [41] that show that SN 1987A did produce grains are: 1. The bolometric luminosity of the SN declined faster after 600^d than predicted from Co^{56} decay while the infrared luminosity beyond ~10 μm increased, and the sum of UV+optical+IR energy matched the Co decay curve (nor were any excess γ- or X-rays observed concurrently). 2. Her montage of infrared spectra includes the critical 5-8 μm region and illustrates the decline in luminosity of the IR component consistent with an "IR echo". Wooden calculated the location and temperature of dust during an echo and the mass of emitting dust observed at each epoch. The decline in the IR component halted abruptly near 615^d. The observed dust mass increased dramatically between 400 and 777^d, allied with a strongly increased absorption of visible light. 3. Finally, the profiles of optical and IR forbidden lines appeared blue-shifted after ~500^d, an effect well-shown by the 6.634 μm [NiII] line. Therefore, the red-shifted sides of these emission lines arising in the back of the shell of ejecta were internally extinguished within the line emitting zones after 615^d. Dust, therefore, condensed in SN 1987A.

REFERENCES

1. Abbott, D. C. 1982, Ap.J, 263, 723.
2. van der Hucht, K. A., Cassinelli, J. P., and Williams, P. M. 1986, Astron. Astrophys., 168, 111.
3. van der Hucht, K. A., Williams, P. M., and Thé, P. S. 1987, Quarterly J.R.A.S., 28, 254.
4. Bieging, J. H. 1990, Publ. Astron. Soc. Pacific, Conference Series (Centennial Scientific Meeting, June 1989), in press.
5. Garmany, C. D., Conti, P. S., and Chiosi, C. 1982, Ap.J., 263, 777.
6. Garmany, C. D. 1990, priv. communication.
7. van der Hucht, K. A., Conti, P. S., Lundstrom, I., and Stenholm, B. 1981, Space Sci. Rev., 28, 227.
8. van der Hucht, K. A., Hidayat, B., Admiranto, A. G., Supelli, K. R., and Doom, C. 1988, Astron. Astrophys., 199, 217.
9. Conti, P. S., and Vacca, W. D. 1990, Astron. J., in press (August).
10. Torres-Dodgen, A. V., and Massey, P. 1988, Astron. J., 96, 1076.
11. Howarth, I. D., and Prinja, R. K. 1989, Ap.J. Suppl., 69, 527.
12. Panagia, N. 1973, Astron. J., 78, 929.
13. Allen, C. W. 1973, "Astrophysical Quantities", 3rd. ed. (London: Athlone Press).
14. Barlow, M. J., Smith, L. J., and Willis, A. J. 1981, M.N.R.A.S., 196, 101.
15. Schmutz, W., Hamann, W.-R., and Wesselowski, U. 1989, Astron. Astrophys., 210, 236.
16. Smith, L. F., and Maeder, A. 1989, Astron. Astrophys., 211, 71.
17. Owocki, S., Castor, J. I., and Rybicki, G. B. 1988, Ap.J., 335, 914.

18. Prinja, R. K., and Howarth, I. D. 1986, Ap.J. Suppl., 61, 357.
19. Abbott, D. C., Bieging, J. H., Churchwell, E., and Cassinelli, J. P. 1980, Ap.J., 238, 196.
20. Garmany, C. D., Olson, G. L., Conti, P. S., and Van Steenberg, M. E. 1981, Ap.J., 250, 660.
21. Maeder, A. 1983, Astron. Astrophys., 120, 113.
22. Abbott, D. C., Bieging, J. H., Churchwell, E., and Torres, A. V. 1986, Ap.J., 303, 239.
23. Prantzos, N., Doom, C., Arnould, M., de Loore, C. 1986, Ap.J., 306, 695.
24. Nugis, T. 1982, in Proc. IAU Symp. 99, "Wolf-Rayet Stars: Observations, Physics, Evolution", eds. C. de Loore and A. Willis (Dordrecht: Reidel), p.131.
25. Torres, A. V. 1988, Ap.J., 325, 759.
26. Smith, L. F., and Hummer, D. G. 1988, M.N.R.A.S., 230, 511.
27. de Freitas Pacheco, J. A., and Machado, M. A. 1988, Astron. J., 96, 365.
28. Hillier, D. J. 1989, Ap.J., 342, 392.
29. Schmutz, W., and Hamann, W.-R. 1986, Astron. Astrophys., 166, L11.
30. Hillier, D. J. 1987, Ap.J. Suppl., 63, 947.
31. Prinja, R. K., Barlow, M. J., and Howarth, I. D. 1990, Ap.J., in press (October 1st issue).
32. Garmany, C. D. 1986, in Proc. IAU Symp. 116, "Luminous Stars and Associations in Galaxies", eds. C. de Loore, A. Willis, and P. Laskarides (Dordrecht: Reidel), p.23.
33. Chiosi, C., and Matteucci, F. 1984, in "Stellar Nucleosynthesis", ed. C. Chiosi, A. Renzini (Dordrecht: Reidel), p.359.
34. Cohen, M., Barlow, M. J., and Kuhi, L. V. 1975, Astron. Astrophys., 40, 291.
35. Gehrz, R. D., and Hackwell, J. A. 1974, Ap.J., 194, 619.
36. Williams, P. M., van der Hucht, and The, P. S. 1987, Astron. Astrophys., 182, 91.
37. Gehrz, R. D. 1989, in Proc. IAU Symp. 135, "Interstellar Dust", eds. L. Allamandola and A. Tielens (Dordrecht: Kluwer Academic), p.445.
38. Tielens, A.G.G.M., 1990, in "Carbon in the Galaxy: Studies from Earth and Space", ed. J. Tarter, NASA CP-3061, p.59.
39. Tielens, A.G.G.M., 1990, in "Analysis of Returned Comet Nucleus Samples", eds. S. Chang and D. de Frees, NASA-CP, in press.
40. Wainscoat, R. J., Cohen, M., Volk, K., Walker, H. J., and Schwartz, D. E. 1990, Ap.J., submitted.
41. Wooden, D. H. 1990, Ph.D. dissertation, Univ. of Calif. Santa Cruz.
42. Cohen, M., Tielens, A.G.G.M., and Bregman, J. D. 1989, Ap.J. Letters, 344, L1.
43. Volk, K., and Cohen, M. 1989, Astron. J., 98, 931.
44. "Atlas of Low Resolution Spectra" 1986, Astron. Astrophys. Suppl., 65, 607.
45. Dwek, E., and Werner, M. W. 1981, Ap.J., 248, 178.
46. Marsden, P. L., Gillett, F. C., Jennings, R. E., Emerson, J. P., de Jong, T., and Olnon, F. M. 1984, Ap.J. Letters, 278, L45.
47. Fesen, R. A., and Blair, W. P. 1990, Ap.J. Letters, 351, L45.

DISCUSSION

Shara: In session VIII I will announce the discovery of 13 new WR stars in a 12 degree ($l = 282°$ to $294°$) band centered on $b = 0°$ (where 24 were already known). The present WR census is at least 25% incomplete even locally ($0 - 3 kpc$); and the fraction of WN stars is significantly higher than has been previously claimed.

Cohen: If you can tell me the exact numbers of WR stars of different subclasses that I need to include to make my census complete, that would be very interesting. That means I will have to do all of this again!

Conti: This was a very thorough discussion of hot star contributions to their global environments. The fact that in the mean our and van der Hucht's population totals were not too different suggests such spectroscopic parallax measurements are rather robust. I only would add one caution: using the accurate count of O stars within a single quadrant, compared to all WR stars, might be a little misleading since WR stars are not distributed equally in all four quadrants.

Cohen: I felt it was safer to define the OBA count in this way and assume that the OBA stars were not too differently distributed in the other three quadrants. Since everything is defined per kpc^2 I thought this was a reasonable approach.

Van der Hucht: Conti & Vacca (1990), in their determination of the WR distances in the Galaxy, first used the ones in galactic clusters and associations and in the LMC as M_v calibrators, and subsequently (and contrary to van der Hucht *et al.*, 1988) *recalculated* the distances of WR stars in galactic clusters and associations using their average M_v per subtype values. Since the WR stars in clusters and associations within 3 kpc from the Sun constitute about 50% of all the WR stars in that volume, this inconsistent approach by Conty & Vacca explains part of the discrepancy between the WR galactic distributions determined by Conti & Vacca and those of van der Hucht *et al.* One better leaves WR stars at the distances of their clusters or associations.

Conti: The question of membership of WR stars in galactic associations is by no means settled. A cursory examination of the fields of 40 WR association members indicates a substantial question of a connection for many of them. Most of our (Conti & Vacca) M_v calibration is given by the LMC WR stars. We treated all our distances to galactic stars in a statistical sense. The individual differences to van der Hucht *et al.* can be thought of as indicating the uncertainty in *both* approaches. Association membership for WR stars needs a critical examination, as Garmany is carrying out for galactic O stars.

Sreenivasan: Those of us who have modeled the evolution of the progenitor of SN 1987A have been concerned about whether or not the star goes through a red supergiant phase before the blue supergiant phase because of the detection of dust in a shell. Could you say whether the dust formed in the ejecta is sufficient to decide whether such a red supergiant phase is necessary or not? (You said that IRAS data cannot be used to decide whether there is sufficient swept up dust).

Cohen: I do not feel that the total amount of dust that was observed to condense is significant. The swept up (4-5 light day) shell of pre-existing dust has even less mass. The test for a RSG precursor would surely be through any silicate emission features but none is seen in SN 1987A. Perhaps such grains are present but are either too cool and/or too large to reveal optically thin features. Consequently, the observations of dust in SN 1987A do not address your concern.

Filippenko: We know without a doubt that the progenitor of SN 1987A went through the red supergiant phase before exploding as a blue supergiant. IUE spectra revealed narrow emission lines from a circumstellar shell highly enriched by CNO processing and photoionized by the initial UV flash of the supernova.

DUST FORMATION IN WC-STAR SHELLS

E. SEDLMAYR AND H. GASS
Institut für Astronomie und Astrophysik
Technische Universität Berlin, PN 8-1
Hardenbergstraße 36
D-1000 Berlin 12

ABSTRACT. The conditions of dust formation in the winds of WC-stars are discussed. Simple analytical estimations for the existence of neutral carbon zones where carbon nucleation might occur are given. It is shown that efficient dust formation can take place in the CI region if the effective temperature of the star is below ~ 25000 K and the density at the base of the wind is larger than $\sim 10^{11}$ cm^{-3}.

1. Introduction

The condensation of solid particles requires conditions where the temperature is cool enough and the density is still sufficiently high for allowing clusters to be formed. Typical values for systems of effective dust formation require a kinetic temperature $T \lesssim 1200$ K and a density of the nucleating species $n \geq 10^5$ cm^{-3}. These conditions are usually to be met in the outflows of cool giants and supergiants but also at some stages in the expanding shells of novae and supernovae and in the equatorial planes of B[e]-stars which all are known to be objects of pronounced circumstellar dust formation. Thus, in a first view, it seems to be difficult to understand that the appearance of circumstellar dust formation is not merely confined to these objects where sufficiently cool conditions naturally are expected, but that effective dust formation also occurs in the expanding shells of such high-temperature objects like WR-stars, as is manifested by a pronounced IR excess which has to be attributed to the presence of solid particles formed in their stellar wind.

Still being far from having a full understanding of this phenomenon, we only want to present some basic ideas which might be important for this complex and, possibly, could shed some light on the problem how grains can be formed under such in principle hostile conditions as encountered in the shells of WC-stars.

This concerns in particluar late type WC-stars with effective temperatures below 30000 K where IR-measurements indicate a Planckian emission characteristic which undoubtedly is due to thermal emission of dust grains having an internal temperature around 1000 K. It is the aim of this short contribution to focus on this interesting fact in more detail.

The dust structure of the shells of late WC-stars are determined by
- a hot stellar radiation field: $T_* \simeq 20 \ldots 30 \cdot 10^3$ K
- a high (steady) mass loss rate: $\dot{M} \simeq 10^{-5} \ldots 10^{-3} M_\odot / \mathrm{yr}$

335

K. A. van der Hucht and B. Hidayat (eds.),
Wolf-Rayet Stars and Interrelations with Other Massive Stars in Galaxies, 335–340.
© 1991 *IAU. Printed in the Netherlands.*

- a high expansion veloctiy: $v \simeq 1000$ km/s
- a high carbon (and oxygen) abundance: $\epsilon_C \gtrsim \epsilon_O \gg \epsilon_H$

For more details see Maeder and Meynet (1987), Chiosi and Maeder (1986), and Williams et al. (1987). Assuming $R_* = 10R_\odot$ as a characteristic value for the stellar radius, the above numbers for the mass loss rate and for the expansion velocity provide a typical number density of the expanding material of the order of $n \simeq 10^{11}$ cm^{-3} at $r = R_*$.

Except the chemical composition, these parameters basically are not very different from those characterizing a nova outburst where an expanding shell is expelled from the hot central star. (see e.g. Bode and Evans, 1983):

- effective temperature: $8 \cdot 10^3$K $\leq T_* \lesssim 40 \cdot 10^3$ K
- shell density: $n \simeq 10^{11}$cm^{-3}
- expansion velocity: $= 2000$ km/s
- abundance: $\epsilon_H \gg \epsilon_C > \epsilon_O$

For the expanding shells of novae, we have investigated the process of dust formation. Theoretical model calculations for the evolution of the shells of novae including time dependent chemistry and the equations for dust formation and growth indicate that effective grain formation only can occur at those regions where sufficient neutral atoms or molecules of the dust forming species are present. This concerns in particluar the nucleation phase, i.e. the formation of the critical clusters out of the gas phase which, if formed, are able to grow or at least to survive in the expanding medium. A necessary condition for this region is that carbon is predominantly neutral. For this reason, the time required for carbon ionization is decisive for the dust formation problem.

The time scale τ_{ion} for a complete ionization of the shell is determined by two parameters: The *effective temperature* of the central object and the *density* within the shell. Thus, dust nucleation in a nova shell is controlled by the competition of the characteristic time scale τ_{ion} necessary for the carbon ionization of the shell and the time scale τ_d charactieristic for the formation of macroscopic particles due to gas-kinetic collisions.

If $\tau_{\mathrm{ion}} < \tau_d$, the ionization front overruns the shell and ionizes the dust forming species before effective dust formation can take place. In this case, no dust will form in a nova outburst.

If $\tau_{\mathrm{ion}} > \tau_d$, dust nucleation will take place and the dust formation can even be completed before the shell is ionized.

The question whether in WC-stars the outstreaming matter is ejected in subsequenced shells analoguos to a nova or as a continuos flow like a wind situation is still open. In any case, the above close analogy between a nova shell and the outflow of a WC-star suggests to apply a similar method to investigate dust formation in the outflows of WC-stars.

Both, for novae and WC-stars a spherical symmetric outflow is assumed in our models. Then only two modifications of our novae calculations are necessary for being applied to WC-stars:

i) For a stationary wind of a WC-star, which is the usual assumption in model calculations, one has not to compute the time evolution of the various ionization fronts but only the extension of the corresponding ionization spheres.

ii) The wind material of WC-stars consists mainly of carbon and oxygen (cf. Maeder and Meynet, 1987) and, hence, neutral C atoms will be the dominating dust forming species.

For our further investigations, therefore, we assume a continuos spherical symmetric stationary wind of a WC-star consisting mainly of carbon.

2. The Ionization Structure of a WC-Shell

According to the above arguments, the dust formation region is situated outside the CII ionization sphere, the extension of which is estimated by the following manner:

The degree of ionization of a pure carbon mixture

$$f = \frac{n_{\mathrm{CII}}}{n_{\mathrm{CI}} + n_{\mathrm{CII}}} \tag{1}$$

with n_{CI} and n_{CII} being local number density of carbon atoms and ions, respectively, is given by the local balance of ionization and recombination processes expressed by

$$(1 - f) \left(\frac{R_*}{r}\right)^2 \int_{\nu_0}^{\infty} \frac{\pi F_{*\nu}}{h\nu} \sigma_\nu e^{-\tau_\nu} d\nu - \alpha n f^2 = 0 \quad , \tag{2}$$

where $F_{*\nu}$ is the spectral flux of radiation with frequency ν emerging from the surface of a central object, σ_ν the ionization cross section, ν_0 the threshold frequency for ionization, $\tau_\nu(r)$ the monochromatic optical depth, α the recombination coefficient, r the distance, and $h\nu$ the photon energy. In Eq. (2) is assumed that carbon is only single ionized and, thus, the electron density equals the ion density.

With $n = n_{\mathrm{CI}} + n_{\mathrm{CII}}$ being the total carbon density, $\tau_0 = \tau_{\nu_0}$ the outwards directed optical depth of the shell at ν_0 with $\tau_0(R_*) = 0$ the solution of the quadratic equation (2) is given by

$$f = \frac{a}{2b} - \frac{1}{2c}(a^2 - 4ab)^{1/2} \quad . \tag{3}$$

where the abbreviations

$$a(\tau_0) = \left(\frac{R_*}{r}\right)^2 \int_{\nu_0}^{\infty} \frac{\pi F_{*\nu}}{h\nu} \sigma_\nu e^{-\tau_\nu} d\nu \quad ; \qquad b(\tau_0) = -\alpha n \tag{4}$$

have been introduced.

By a Taylor series expansion of Eq. (3) up to the second order and insertion into the definition equation of the τ_0-scale

$$\frac{dr}{d\tau_0} = (\sigma_0 n(1 - f))^{-1} \quad , \tag{5}$$

with $\sigma_0 = \sigma_{\nu_0}$, one finds

$$\frac{dr}{d\tau_0} = \frac{2a(\tau_0)}{\sigma_0 n \alpha n^2} \quad . \tag{6}$$

Adopting an inverse square law for the dilution of the initial density n_*

$$n = n_* \left(\frac{R_*}{r}\right)^2 \tag{7}$$

and Kramer's rule for the ionization cross section

$$\sigma_\nu = \sigma_0 \left(\frac{\nu_0}{\nu}\right)^3 \quad , \tag{8}$$

integration of Eq. (6) from R_* to a distance R_C where the contribution of the ionizing radiation field vanishes, yields for the radius of the CII sphere R_C:

$$R_C = \left[1 - \frac{2\int_{\nu_0}^{\infty}\frac{\pi F_{*\nu}}{h\nu}d\nu}{\alpha n_*^2 R_*}\right]^{-1} R_* \tag{9}$$

It can be seen from Eq. (9) that for radiation fields with $T_{\text{eff}} \lesssim 25000$ K and $R_* = 10^{12}$ cm the existence of a finite carbon ionization sphere depends very sensitively on the initial particle density n_*. For a Planckian radiation field and $n_* = 5 \times 10^{12}$ cm^{-3}, the ionization sphere is $R_C \cong R_*$ whereas for a particle density $n_* \cong 10^{12}$ cm^{-3}, the ionization sphere is always infinite. For a realistic estimate of the radiation field, the above blackbody radiation field should be modified by a factor $e^{-\tau_*}$ where τ_* accounts for the photoionization of carbon inside the stellar atmosphere. In order to account for this effect, we assume $\tau_* \sim 5$, which seems to be a reliable value for such type of atmospheres. Introducing this factor into eq. (9) yields always a finite value for R_C for effective temperatures ≤ 25000 K and $n_* \geq 10^{11}$ cm^{-3}, which is in good agreement with the observational data as provided by Williams *et al.* (1987) in Table 8.

Outside the CII-sphere, the energy input by photoionization decreases rapidly and the temperature drops abiabatically from the temperature of the CII front, $T_C = 4000$ K, to values where dust formation becomes possible.

3. Dust condensation

3.1 CHEMICAL RESTRICTIONS

Efficient dust formation generally requires the nucleating species to consist of molecules from abundant elements which are not blocked by high bond energies and allow for high temperature condensates. In case of a carbon star, in particular the following molecules contribute to grain formation: C, C_2H, C_2H_2, SiC. For a WC-star, due to a lack of hydrogen, carbon-hydrogen molecules are not present. Hence, the only nucleating species left are C and SiC. As no data of SiC are abvailable to describe the chemistry of SiC-cluster formation, we confine our calculation on the formation of pure C-clusters. Therefore, our results yield only a lower limit for the condensated material expected.

The above assumption of carbon grain formation is based on the fact that all oxygen should be blocked by the CO molecule. This is certainly true in case of C-stars where only expansion velocities of the order ~ 10 km s^{-1} and only moderate UV radiation fields are observed. In case of hot objects like novae and WC-stars, however, this may not always be true because the reaction channels for effective CO formation could be destroyed. As CO formation then has to proceed via rather slow channels, the high expansion velocity of novae and WC-stars may result in an incomplete CO blocking and, thus, in a larger amount of available condensable material.

3.2 DUST FORMATION AND GROWTH

From the arguments and estimations in chapter 2 can be seen that a neutral carbon zone where dust condensation may occur can exist only for stars with effective temperatures $T \lesssim 25000$ K and initial densities $n_* \gtrsim 10^{11}$ cm^{-3}. To investigate dust formation in the neutral carbon zone, we assume an adiabatic temperature T:

$$T = T_C \left(\frac{R_C}{r} \right)^{4/3} \tag{10}$$

and the density structure of eq. (7):

$$n = n_C \left(\frac{R_C}{r} \right)^2 \tag{11}$$

with n_C the carbon density at the inner edge of the CI-region.

With the value for $T_C = 4000$ K given in chapter 2 and $n_C = 10^{11}$ cm^{-3}, for instance, the supersaturation ratio of carbon at $2R_C$ is 10^{10}. At this radius the gas temperature has decreased to 1600 K.

Applying classical homogeneous nucleation theory (Gail, Seldmayr, 1988) one finds that for suitably choosen parameters dust formation is possible within a region $R_C \leq r \leq 1000 R_*$. The results indicate that the nucleation rates J are in the range $10^\circ \leq J \leq 10^9$ cm^{-3} s^{-1}. Thus, the dust formation process can start very rapidly at a well defined radius. This is in agreement with observed energy distributions (Williams et al., 1987). Due to the large consumption of condensating material, the nucleation ceases already after some 10^4 s. The short time for growth allows only the formation of rather small grains. The pertinent results should not be extremely different from those in nova shells, where a mean particle size of 40 monomeres has been found.

Estimating the radii of the inner edge of the dust spheres and the dust densities, the expected results may well fit with the observational data provided by Williams et al. (1987).

4. Conclusions

The above arguments confirm that dust formation quite naturally might occur in the environments of WC-stars with $T_{\text{eff}} \lesssim 25000$ K and densities $n_* \gtrsim 10^{11}$ cm^{-3}. Further quantitative informations concerning the detailed structure of the temperature, the ionization, the dust density and the grain charcteristics in the wind, however, have to be based on more elaborate model calculations.

References

Bode, M.F., Evans, A.: 1983, *Quart. J. Roy. astr. Soc.* **24**, 83

Chiosi, C., Maeder, A.: 1986, *Ann. Rev. Astron. Astrophys.* **24**, 329

Gail, H.-P., Sedlmayr, E.: 1988, *Astron. Astrophys.* **206**, 153

Maeder, A., Meynet, G.: 1987, *Astron. Astrophys.* **182**, 243

Williams, P.M., van der Hucht, K.A., The, P.S.: 1987, *Astron. Astrophys.* **182**, 91

DISCUSSION

Cassinelli: (1) In the model of Zickgraf of B[e] stars, he proposes that the dust formation forms in a disk in the equatorial zone. Is there any reason that you might think this is the case ? (2) Do you think that is the case for WC stars also?

Sedlmayr: This is true, we have done such calculations. If you calculate the radiative transfer, if you take a two component-model with a hot polar wind and a slow equatorial wind with a dense cool equatorial shell, then you can easily arrive at conditions of disk formation. (2) If I take my nova results, they have conditions identical to the WC stars outside the CII regions; we have the same density, we have the same UV field, say, a UV field of 20000K - 25000K, we have the same expansion velocity, and therefore the timescales. Therefore I would guess the same results, of course within an order of magnitude, as we have for novae. I think that dust formation for WC stars is very natural.

Cohen: I think that you can test directly for the presence of dusty disks in WC9 stars by optical polarization. In "the old days", Gary Schmidt and I and Len Kuhi observed some WC9's with spectropolarimetry. We found substantial polarization compared with other WR's, but we felt it was merely interstellar in origin.

Sedlmayr: As I wanted to point out, in our opinion dust formation in winds of late type WC stars resembles very much that of a nova and should occur outside the CII region in a sufficiently dense shell. Disk like dust formation we expect for B[e] stars.

COOL SUPERGIANTS IN THE SOLAR NEIGHBORHOOD

M. JURA
Department of Astronomy
University of California
Los Angeles CA 90024 USA

ABSTRACT. We have identified 21 mass-losing red supergiants (20 M-type, 1 G-type, L $> 10^5$ L$_\odot$) within 2.5 kpc of the Sun. These supergiants are highly evolved descendants of main sequence stars with initial masses larger than about 20 M$_\odot$. The surface density projected onto the plane of the Milky Way is between about 1 and 2 kpc^{-2}. Although with considerable uncertainty, we estimate that the mass return by the M supergiants is somewhere between 1 and 3 10^{-5} M$_\odot$ kpc^{-2} yr^{-1}. In the hemisphere facing the galactic center there is much less mass loss from M supergiants than from W-R stars, but in the anticenter direction, the M supergiants return more mass than do the W-R stars. The duration of the M supergiant phase appears to be between 2 and 4 10^5 years. During this phase a star of initially at least 20 M$_\odot$ returns perhaps 3 to 10 M$_\odot$ into the interstellar medium.

1. INTRODUCTION

For single stars in the mass range between 25 and 60 M$_\odot$, Chiosi and Maeder (1986) propose the following evolutionary scheme: O star $\rightarrow$ Blue Supergiant $\rightarrow$ Yellow/Red Supergiant $\rightarrow$ W-R(N) $\rightarrow$ W-R(C) $\rightarrow$ Supernova. If the mass loss rate is sufficiently low, the W-R phase never occurs, and the star becomes a supernova directly after the red supergiant phase. This scheme ignores mass transfer in binaries which can also lead to the formation of a W-R star (Maeder 1981, Massey 1981, Schulte-Ludbeck 1989). In any case, by comparing the relative numbers of red supergiants and W-R stars, and estimating the mass loss rates from the red supergiants, we can constrain this theoretical picture.

One advantage in studying luminous stars is that they are conspicuous. We are interested in sampling a sufficiently large volume to average over local fluctuations in the surface density of stars yet with sufficient sensitivity to be complete. By analogy with studies of W-R stars (Abbott and Conti 1987), we consider red supergiants within 2.5 kpc of the Sun (Humphreys and McElroy 1984). This is the same volume for which there appear to be reasonably complete catalogs of W-R stars (Conti *et al.* 1983, van der Hucht *et al.* 1988)

Cool stars emit most of their energy at wavelengths longward of 1 μm. In addition, there is less interstellar extinction at infrared than at optical wavelengths, and therefore, infrared

K. A. van der Hucht and B. Hidayat (eds.),
Wolf-Rayet Stars and Interrelations with Other Massive Stars in Galaxies, 341–348.
© 1991 *IAU. Printed in the Netherlands.*

sky surveys are the most comprehensive ways to identify candidate red supergiant stars in the solar neighborhood although optical spectroscopy is still normally required to establish the nature of the star.

One major observational difficulty is to separate the red supergiants that are descended from massive stars from the very luminous red Asymptotic Giant Branch, [AGB], stars (Iben and Renzini 1983) that are descended from intermediate mass main sequence stars. One approach is to restrict the analysis to stars with luminosities in excess of 10^5 L$_\odot$, well in excess of the maximum luminosity on the AGB of $6 \ 10^4$ L$_\odot$. Stars with L $> 10^5$ L$_\odot$ have main sequence progenitors with masses greater than 20 M$_\odot$.

During the past decade, considerable progress has been made in determining the mass loss rates from red stars by using molecular and infrared data (see, for example, Jura 1989, Olofsson 1989). All the identified high luminosity red supergiants – even the G star ρ Cas (Jura and Kleinmann 1990a) – exhibit circumstellar dust and are thought to be losing mass.

2. THE CATALOG

Catalogs of candidate red supergiant stars in the solar neighborhood have been given by Humphreys (1978), Elias, Frogel and Humphreys (1985) and White and Wing (1978). Most of these stars lie in the zone covered by the Two Micron Sky Survey, TMSS, (Neugebauer and Leighton 1969) which was complete in the declination range of -33° $< \delta <$ 81° to m$_K$ = 3.0 mag. If we assume for a cool star that the total luminosity, L, is given by L = 3 νL$_\nu$ for the emission near 2.2 μm [the K-band], and if we assume that m$_K$ = 0.0 mag corresponds to 620 Jy (Beckwith et al. 1976), then stars with L $> 10^5$ L$_\odot$ have M$_K$ < -10. In the absence of interstellar extinction, the TMSS was therefore complete for identifying luminous red supergiants to a distance of 4 kpc from the Sun. With an average extinction of 0.15 K mag/kpc (Jura, Joyce and Kleinmann 1989), we expect the TMSS to have been reasonably complete to a distance of about 3 kpc for stars with M$_K$ < -10. Optical spectroscopy for essentially all the stars in the TMSS has been reported (see Bidelman 1980). We can therefore be reasonably confident that we can identify all the red supergiants in the solar neighborhood; at least those stars in the 75% of the sky covered by the TMSS.

The difficult task is to separate the stars with L $> 10^5$ L$_\odot$ from those red giants with lower luminosities. The fluxes with appropriate corrections for extinction can be well measured; the major uncertainty in determining the luminosity is the distance. In order to include only the high luminosity stars, we restrict our analysis to stars that lie in clusters or otherwise have well-determined distances.

Jura and Kleinmann (1990b) have identified 21 mass-losing red supergiants (20 M-type, one G type, L $> 10^5$ L$_\odot$) within 2.5 kpc of the Sun. The G star is ρ Cas; the M stars are all well known including such famous objects as VX Sgr, NML Cyg, μ Cep, α Ori and VY CMa. The distribution of these stars projected onto the plane of the Milky Way is shown in Figure 1. In marked contrast to the spatial distribution of the 45 W-R stars within 2.5 kpc of the Sun (Conti et al. 1983, van der Hucht et al. 1988, there is no concentration of the red supergiants towards the Galactic Center region.

Because there are 21 stars that lie within a circle of radius 2.5 kpc centered on the Sun, we find that projected onto the galactic plane the surface density of these red supergiant stars is about 1 kpc^{-2}. It seems most unlikely that we have missed more than half of all the relevant stars, and so an upper limit to the surface density of these stars in the solar neighborhood is 2 kpc^{-2}.

Miller and Scalo (1979) estimate that the birthrate of stars with M > 20 M$_\odot$ in the solar neighborhood is about 5 10^{-6} kpc^{-2} yr^{-2}. This implies that the M supergiants spend between about 2 10^5 and 4 10^5 yr in this phase. According to Chiosi and Maeder (1986), stars with initial solar masses of 25 and 40 M$_\odot$ are predicted to have lifetimes in the He-burning phase with effective temperatures less than 6300 K of 4.6 10^5 and 1.2 10^5 yr^{-1}, respectively. Within the uncertainties, there is good agreement between the theoretical and derived lifetimes.

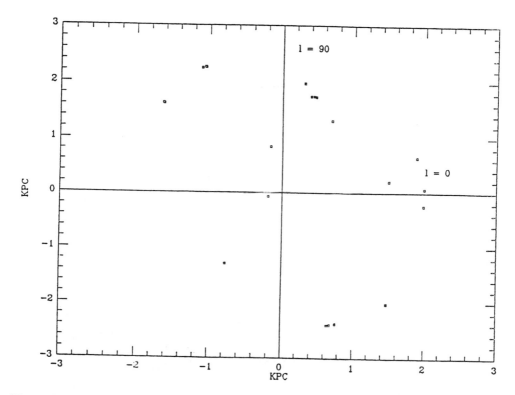

Fig. 1. Plot of the spatial distribution of the 21 red supergiants from Jura and Kleinmann (1990b) projected onto the plane of the Milky Way. The Sun is located at the center of the plot.

3. MASS LOSS RATES

The mass loss rates are inferred from the 60 μm emission measured by IRAS and a standard model for the dust emissivity and dust to gas ratio. If the fraction of dust is very different from the assumed value – as may be the case for μ Cep (Le Borgne and Mauron 1989) and/or α Ori, (Jura and Morris 1981, Glassgold and Huggins 1986) then our inferred mass loss rates may not be correct.

For most stars, the amount of circumstellar dust can be determined relatively accurately from the amount of infrared emission. However, estimates of the amount of gas are more uncertain because it is usually only possible to measure directly the amount of trace species such as CO and then extrapolate to the amount of hydrogen. Consequently, our knowledge of the dust to gas ratios is uncertain. However, there are two stars where, because of special circumstances, the outflowing gas is photoionized, and it is possible to infer the amount of hydrogen that is being lost from its radio free-free emission. One such example is the very luminous star NML Cyg (L = 5 10^5 L$_\odot$, Morris and Jura 1983), where it appears that the dust to gas ratio by mass is close to the value of 0.01 characteristic of the interstellar medium since the inferred mass loss rate derived from interpreting the infrared emission with this assumption is 10 10^{-5} M$_\odot$ yr^{-1} while modelling the radio free-free emission gives 6 10^{-5} M$_\odot$ yr^{-1} (Jura and Kleinmann 1990b).

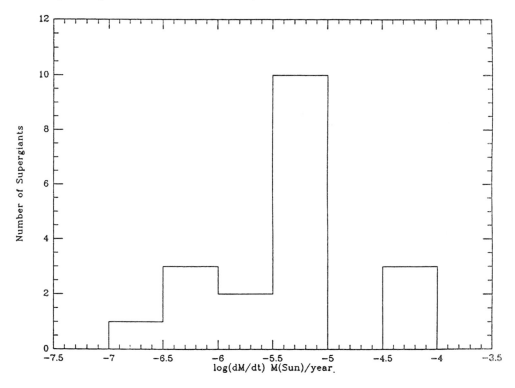

Fig. 2. Histogram of the mass-loss rates of the supergiants in Jura and Kleinmann (1990b).

The other example is the M supergiant α Sco which is not quite luminous enough (L = 5 10^4 L$_\odot$, van der Hucht, Bernat and Kondo 1980) to be listed in our catalog. This star is losing mass, and because it has an early-type companion, much of the circumstellar gas is ionized. An accurate measurement of the gas loss rate is $2 \ 10^{-6} \ M_\odot \ yr^{-1}$ (Hjellming and Newell 1983). Using the known distance, fluxes, luminosity and outflow velocity, and equation (2) of Jura and Kleinmann (1990b), the observed IRAS dust emission gives a total gas loss rate of $1 \ 10^{-7} \ M_\odot \ yr^{-1}$ if we assume a "standard" dust to gas ratio of 0.01 by mass. [We know that the infrared emission from this star is produced by dust even though F_ν varies as ν^2 as from a photosphere because ground-based observations indicate silicate emission near 10 μm from this star (Merrill and Stein 1976).] The way to reconcile the two discrepant results for the mass loss rate is to argue that the dust abundance around α Sco is lower than in the interstellar medium. In any case, with the assumption of a standard gas to dust ratio similar to that in the interstellar medium, a histogram of the mass loss rates for the different stars is shown in Figure 2.

The luminosities of the red supergiant stars vary by a factor of 5: between 1 and 5 10^5 $L_\odot$; in contrast, the range in the mass loss rates is much greater and amounts to a factor of 100. The two stars with the highest inferred mass loss rates are NML Cyg and VY CMa, both of which have values near $10^{-4} \ M_\odot \ yr^{-1}$, but a number of the supergiants have mass loss rates of about $10^{-6} \ yr^{-1}$.

We do not fully understand the physics of the mass loss from the red supergiants. However, in all cases, it appears that radiation pressure on grains can be important since we find that

$$v dM/dt < L/c.$$

The inferred mass loss rates indicate that mass loss is important in the evolution of these stars. If a star loses between 10^{-5} and $3 \ 10^{-5} \ M_\odot \ yr^{-1}$ during the $\sim 3 \ 10^5$ yr that it might spend in this phase of its evolution, the total mass that is lost is in the range of 3 to 10 $M_\odot$. This mass that is lost is not negligible in the evolution of the star if it began with 20 $M_\odot$

The mass-return from the red supergiants is distinctly different from that of the W-R stars. The typical outflow speed from the red supergiants is perhaps 20-30 km s^{-1}; a factor of 100 smaller than for the W-R stars. The red supergiants are all oxygen-rich while some of the W-R stars are carbon-rich. The grains around the red supergiants can be comparable in size to interstellar grains; perhaps 0.1 μm in radius (see Herbig 1972, Jura 1975 for an analysis of the optical scattering properties of the grains around VY CMa). All the W-R stars which have grains are carbon-rich, and it is likely that these grains are quite small [< 0.01 μm] since they are driven supersonically through the wind by radiation pressure, and if they were larger, they would be destroyed by sputtering produced by collisions with the ambient gas (see, for example, Williams, van der Hucht and The' 1987).

The total mass-loss from all 21 stars identified by Jura and Kleinmann (1990b) is $3 \ 10^{-4}$ $M_\odot \ yr^{-1}$. Over half of this total is from the two stars with the highest mass loss rates: NML Cyg and VY CMa. The total mass loss rate projected onto the galactic plane is somewhere between $1 \ 10^{-5}$ and $3 \ 10^{-5} \ M_\odot \ kpc^{-2} \ yr^{-1}$. In contrast W-R stars return about $6 \ 10^{-5} \ M_\odot$ $kpc^{-2} \ yr^{-1}$. Therefore, the W-R stars return more mass to the interstellar medium than do the red supergiants.

As noted above, it seems that the W-R stars are much more concentrated towards the center of the Milky Way than are the red supergiants. Therefore, although the W-R stars dominate the average return of mass from massive stars to the interstellar medium in the solar neighborhood, outside the Solar Circle, the red supergiants may contribute as much matter into the interstellar medium as do the W-R stars.

This work has been partly supported by NASA. I thank Dr. Susan Kleinmann for many useful conversations.

REFERENCES

Abbott, D. C., and Conti, P. S. 1987, *Ann. Rev. Astr. Ap.*, **25**, 113.

Beckwith, S., Evans, N. J., Becklin, E. E., and Neugebauer, G. 1976, *Ap. J.*, **208**, 390.

Bidelman, W. P. 1980, *Pub. Warner and Swasey Obs.*, **2**, No. 6.

Chiosi, C., and Maeder, A. 1986, *Ann. Rev. Astr. Ap.*, **24**, 329.

Conti, P. S., Garmany, C. D., De Loore, C., and Vanbeveren, D. 1983, *Ap. J.*, **274**, 302.

Elias, J. H., Frogel, J. A., and Humphreys, R. M. 1985, *Ap. J. Suppl.*, **57**, 91.

Glassgold, A. E., and Huggins, P. J. 1986, *Ap. J.*, **306**, 605.

Herbig, G. H. 1972, *Ap. J.*, **172**, 375.

Hjellming, R. M., and Newell, R. T. 1983, *Ap. J.*, **275**, 704.

Humphreys, R. M. 1978, *Ap. J. Suppl.*, **38**, 309.

Humphreys, R. M., and McElroy, D. B. 1984, *Ap. J.*, **284**, 565.

Iben, I., and Renzini, A. 1983, *Ann. Rev. Astr. Ap.*, **21**, 271.

Jura, M. 1975, *Astr. J.*, **80**, 3.

Jura, M. 1989, in *Evolution of Peculiar Red Giant Stars, IAU Colloquium 106*, H. R. Johnson and B. Zuckerman, eds., (Cambridge: Cambridge University Press), p. 339.

Jura, M., Joyce, R. R., and Kleinmann, S. G. 1989, *Ap. J.*, **336**, 924.

Jura, M., and Kleinmann, S. G. 1990a, *Ap. J.*, **351**, 583.

Jura, M., and Kleinmann, S. G. 1990b, *Ap. J. Suppl.*, in press.

Jura, M., and Morris, M. 1981, *Ap. J.*, **251**, 181.

Le Borgne, J. F., and Mauron, N. 1989, *Astr. Ap.*, **210**, 198.

Maeder, A. 1981, in *The Most Massive Stars*, S. D.'Odorico, D. Baade, and K. Kjar eds. (Garching: ESO), p. 173.

Massey, P. 1981, *Ap. J.*, **246**, 153.

Merrill, K. M., and Stein, W. A. 1976, *P.A.S.P.*, **88**, 285.

Miller, G. E., and Scalo, J. M. 1979, *Ap. J. Suppl.*, **41**, 513.

Morris, M., and Jura, M. 1983, *Ap. J.*, **267**, 179.

Neugebauer, G., and Leighton, R. B. 1969, *Two Micron Sky Survey* (NASA SP-3047).

Olofsson, H. 1989, in *Evolution of Peculiar Red Giant Stars, IAU Colloquium 106*, H. R. Johnson and B. Zuckerman, eds. (Cambridge: Cambridge University Press), p. 321.

Schulte-Ludbeck, R. E. 1989, *Astr. J.*, **97**, 1471.

van der Hucht, K. A., Bernat, A. P., and Kondo, Y. 1980, *Astr. Ap.*, **82**, 14.

van der Hucht, K. A., Hidayat, B., Adminranto, A. G., Supelli, K. K., and Doom, C. 1988, *Astr. Ap.*, **199**, 217.

Williams, P. M., van der Hucht, K. A., and The', P. S. 1987, *Astr. Ap.*, **182**, 91.

White, N. M., and Wing, R. F. 1978, *Ap. J.*, **222**, 209.

DISCUSSION

Underhill: Can the picture be turned around? Say, WR stars are young objects in not fully dispersed molecular dust clouds. Sedlmayr has noted that having a disk helps dust to form. I would suggest that none of the observations discussed sofar this morning indicate that WR stars are the end-products of massive stars, and relevant to the density of interstellar material. [But Sedlmayer also noted that he needs high densities, *i.e.* high mass loss rates, and high carbon abundances. (Eds).]

Jura: Our results do not demonstrate that red supergiants are pre-WR stars.

Walborn: This is a good investigation of one channel of WR formation, but one must keep in mind that there are at least two others: stars with initial masses greater than $50 M_\odot$ which never become red supergiants, and mass transfer binaries. In addition, it is possible that steady-state red supergiant mass loss is not the dominant mechanism of core stripping: some objects exhibit LBV-like shell episodes, and the entire RSG envelope might be lost in a final, very short timescale event. Finally, lest some be confused by Underhill's remarks, the space distribution of WR stars is entirely inconsistent with a pre-main sequence interpretation: they are always found in evolved young regions, and never in regions of current star formation.

Humphreys: The three high mass loss red supergiants in your data set NML Cyg, VY CMa, and VX Sgr are all OH/IR stars which are the high mass loss phase of the most evolved red supergiants. What was your $\dot{M}$ for IRC+10420? It is higher than the CO $\dot{M}$.

Jura: IRC+10420 was not included in our sample because it is too far away: the inferred mass loss rate is close to $10^{-3} M_\odot yr^{-1}$.

Sreenivasan: You considered only winds and mass loss from radiation pressure active on grains. Not all red supergiants become WR stars and not all stars become red supergiants. Formation of the WR phase is regulated by two time-scales, one for peeling the outer layers of the star and the other for hydrogen to burn to helium. Red supergiants have extensive outer convection zones and have non-thermal winds like the Sun - probably much stronger. All these have to be taken into account before deciding whether or not a star becomes a post-red supergiant WR star. There are also other forms of mass loss beside the two mentioned, *e.g.*, pulsation driven mass loss, *etc.*

Maeder: The nuclear lifetimes for stars in the red supergiant stage depend on the initial masses. The $\dot{M}$ rates for red supergiants are also likely to depend on luminosities. It is the combination of both dependences which determines the mass interval for stars evolving through red supergiant and then WR stages. Thus, please also give us the $\dot{M}$-L relations for red supergiants.

Jura: We have so few stars in our sample that it is difficult to establish firm results. Let me just make one point of clarification. The broad-band infrared colours around the star are very sensitive to the spatial distribution of grains. Because, what we are doing is, we are comparing the amount of cold dust to the amount of warm dust when you are looking at, say, 16μ *vs.* 12 or 25μ, and in fact, generally we find very good agreement, and it is rather exceptional to find disagreement, with R^{-2} distributions of dust grains and $R^{-1.5}$ and $R^{-2.5}$ would produce very different infrared colours than we witness. So, at least for most of the stars we look at, the mass loss rate in the zone where we can probe it, which is not all that big a volume, but it is probably a 1000 light years or so, has been fairly constant.

Leitherer: An important parameter for stellar evolution models is the dependance of mass loss rates of metallicity. In the case of hot stars, theory predicts a dependence of $\dot{M} \sim$

$Z^{0.5-1.0}$, which may (or may not) be supported by observations. (1) Is there a theoretical prediction for red supergiants? (2) What is the observational evidence?

Humphreys: In the LMC and SMC, the red supergiants have smaller 10μ silicate features in the sense they are weaker in the LMC than in galactic red supergiants of comparable luminosity and weaker in the SMC stars than in the LMC. This could be either a metallicity effect or a consequence of lower $\dot{M}$, most likely a combination of both.

Michael Jura

RING NEBULAE AROUND MASSIVE STARS

YOU-HUA CHU
Astronomy Department, University of Illinois
1002 W. Green Street, Urbana, Illinois 61801
U. S. A.

ABSTRACT. Ring nebulae have been found around WR stars, OB and Of stars, and luminous blue variables. Ring nebulae are formed by the interaction between the central stars and their ambient medium via different combinations of stellar winds, ejecta, and radiation. The spectral properties of the nebulae can be used to diagnose the stellar properties, such as luminosity and effective temperature. Correlations between ring nebulae and their central stars may be used to check scenarios of stellar evolution.

1. INTRODUCTION

"Ring nebulae around massive stars" could be the plural form of "ring nebula around massive stars" or "ring nebula around a massive star".

In the former case, the central stars could be one or multiple OB associations or clusters, and the ring nebula would be a superbubble blown by the stellar winds and supernova remnants jointly. Examples are Loop I in our Galaxy (Weaver 1977), N51D and N70 in the Large Magellanic Cloud (LMC). The size of such ring nebulae ranges from a few tens pc to hundreds or even thousands of pc, and their expansion velocities are usually below 50 km/s, with the majority being 10-25 km/s (Georgelin et al. 1983; Rosado 1986). Ring nebulae seen in external galaxies are mostly of this type. These nebulae provide excellent labs for studying the deposition of stellar energies into the interstellar medium and they provide diagnostic information for the integrated stellar properties. However, these applications are not immediately relevant to the theme of this symposium; therefore, I will not cover ring nebulae around OB associations any further, and from now on my "ring nebulae around massive stars" would mean strictly one dominant central massive star in each nebula.

About 25 years ago, a ring-shaped nebula would have to be either a planetary nebula or a supernova remnant. When Johnson and Hogg (1965) found S308, as well as NGC 2359 and NGC 6888, was a ring-shaped nebula centered on a WR star, they suggested that WR stars could interact with their ambient medium to form ring nebulae. Subsequent searches around

349

K. A. van der Hucht and B. Hidayat (eds.),
Wolf-Rayet Stars and Interrelations with Other Massive Stars in Galaxies, 349–363.
© 1991 *IAU. Printed in the Netherlands.*

known WR stars in our Galaxy and the Magellanic Clouds indeed turned up
many more ring nebulae, thus establishing "ring nebulae around WR stars"
a new class of objects. Ring-shaped nebulae are also found around other
types of massive stars, such as Of stars and luminous blue variables.
Originally ring nebulae were recognized morphologically; however, some
nebulae that are known to be shaped by winds or consist of stellar
ejecta have been referred to as "ring nebulae," although their
morphologies are quite irregular.

In this review, I will first describe the different types of ring
nebulae, discuss their formation mechanisms, and use them to diagnose
stellar properties.

2. IDENTIFICATION OF RING NEBULAE

Ring nebulae have been identified in a variety of ways. The classical
method applies the criterion of ring- or arc-shaped morphology to the
optical images. When the IRAS all-sky survey became available, the
infrared images were used efficiently to search for bow shocks around OB
stars and ring nebulae around WR stars. A few HI shells or cavities
were found around WR stars from the HI 21 cm line observations. Some
ring nebulae are well-known stellar ejecta.

2.1. Ring Nebulae around WR Stars

The first three known WR ring nebulae, NGC 2359, NGC 6888, and S308, are
all in the north (Johnson and Hogg 1965). Smith (1967) surveyed the
galactic WR stars and found four ring nebulae in the south - NGC 3199,
RCW58, RCW78, and RCW104. Several incidental discoveries were reported
by, for example, Crampton (1971), Johnson (1975), Lortet, Niemela, and
Tarsia (1980). Two systematic surveys for the 159 stars in the Sixth
Catalog of Galactic WR Stars (van der Hucht et al. 1981) were carried
out independently by Chu (1981) and Heckathorn, Bruhweiler, and Gull
(1982). Since different investigators had different selection criteria,
some reported ring nebulae actually contain several early type stars and
some other ring nebulae are not convincingly associated with their WR
stars. Chu, Treffers, and Kwitter (1983) re-examined all reported WR
rings and concluded, in their Table 2, 15 probable cases and 3 possible
cases. An additional double shell structure was later found around the
binary system CV Ser (WC8 + O8-9V-III) (González and Rosado 1984). The
sizes of these WR rings are mostly in the range of a few pc to 30 pc.

Neutral hydrogen holes or shells have been reported around four WR
stars: HD 113904 (θ Mus), HD 88500, HD 156385, and HD 197406 (Cappa de
Nicolau and Niemela 1984; Cappa de Nicolau, Niemela, and Arnal 1986;
Cappa de Nicolau et al. 1988; Dubner et al. 1990). The three WC stars
have large HI shells with diameters ~100 pc; while the WN7 star, HD
197406, has a much smaller (7.6 pc) HI hole centered on the star. The
expansion velocities are all ≤10 km/s. More HI shells around WR stars
are reported by Niemela and Cappa de Nicolau in this volume.

WR rings often stand out against the background better in the IRAS
60 μm images, because the IR emission from the dust in the nebula is

characterized by a higher color temperature than those in the background due to the heating from the central WR star. This effect is best demonstrated by comparing the appearance of S308 on the Palomar Sky Survey Prints, in the [O III] line (Chu et al. 1982), and in the IRAS 60 μm band (Van Buren and McCray 1988). The IRAS survey is being used to search for IR rings around WR stars by Nichols-Bohlin and Fesen (1990).

WR ring nebulae have also been searched for in the nearby galaxies. Ten WR rings have been identified in the LMC: DEM 39, 45, 137, 165, 174, 208, 231, 240, and 315, and N79W (Chu and Lasker 1980; Chu 1983a; Rosado 1986). A probable new ring nebula in the LMC, N82, that may contain ejecta from a WC9 star is recently reported by Heydari-Malayeri, Melnick, and Van Drom (1990). No WR stars in the Small Magellanic Cloud have ring nebulae. The search for WR rings has been extended to M33 recently; Drissen, Shara, and Moffat (1990) find 11 probable cases and 8 less likely cases in M33. The WR rings in the LMC and M33 are larger than their galactic counterparts. Part of this discrepancy in size is due to an observational effect: the largest extragalactic WR rings (>100 pc) may contain uncataloged OB associations and the smallest WR rings (<5 pc) are yet to be discovered. The median of the smaller WR rings in the LMC or M33 is still larger than the galactic median; this may be due to differences in the interstellar condition among these galaxies.

The smallest ring nebulae are probably the circumstellar shells around the three Ofpe/WN9 stars in the LMC: Sk-67°266, HDE 269858, and HDE 269227 (Walborn 1982). Only the first two nebulae have been resolved spatially (Stahl 1987). The central stars of these nebulae may be LBV's, as described later.

2.2. Ring Nebulae around Of Stars

Lozinskaya (1982) and Lozinskaya, Lar'kina, and Putilina (1984) surveyed 109 Of stars, and found 13 ring nebulae; however, 7 of these rings have more than one O star within. The remaining six are the Bubble nebula (NGC 7635), NGC 6164-5, S22, S119, and two dust shells around λ Cep and HD 153915, respectively. Miranda and Rosado (1987) suggested two additional possible ring nebulae around HD 313864 and HD 172175.

2.3. Bow Shocks around OB Stars

A mass-losing star moving supersonically through the interstellar medium may form a bow shock. Very few bow shocks have been discovered in the optical wavelengths. Two examples are the O9.5V star ζ Oph (Gull and Sofia 1979) and the O5V star BD+50°886 (Deharveng, Israel, and Maucherat 1976; Dyson 1977). Using the IRAS SkyFlux maps, Van Buren and McCray (1988) find 10 additional bow shocks around nearby OB stars.

2.4. Ring Nebulae around Luminous Blue Variables

AG Car was once classified a planetary nebula, but the central star is now known to be a luminous blue variable (LBV). η Car is surrounded by rapidly expanding ejecta, and the star may be an extreme LBV. P Cygni is surrounded by extended (~30" radius) radio emission (Baars and

Wendker 1987), but no optical counterpart has been found (Stahl 1989).

Two of Walborn's (1982) Ofpe/WN9 stars in the LMC are related to LBVs: R127 (≡ HDE 269858) is a confirmed LBV (Stahl et al. 1983), and Sk-67°266 is a spectroscopic twin to R127 in minimum (Walborn 1982). The circumstellar shells of these two stars are resolved (Stahl 1987); their [N II] line profiles are split by 31 and 38 km/s, respectively (Walborn 1982).

2.5. Ring Nebulae around B stars

Six B stars are found to be surrounded by bow shocks or bow waves (Van Buren and McCray 1988). α Sco B (B2.5V) is found to be in a nebula that is rich in [Fe II] line but weak in Hα and absent in [O I], [O II], [N II], and [S II] lines (Swings and Preston 1978)! LSS 3027 (B2V) is surrounded by a parabolic-shaped reflection nebula, which is best seen in the blue continuum and absent in the Hα line (Chu 1983b). The arc-shaped nebula "Herbig 8" around the B star HDE 250550 (Johnson 1982) is brighter in the blue than in the red, hence may be another reflection nebula (Herbig 1960).

3. FORMATION MECHANISMS OF RING NEBULAE

Massive stars can interact with their ambient medium via stellar winds, ejecta, and UV radiation. A ring nebula is presumably formed by a combination of these three modes of interaction.

The first attempt to determine the formation mechanism for a large number of ring nebulae was carried out by Chu and her collaborators, who studied the morphology and internal motion of all known WR rings in the Galaxy and the LMC (Chu 1981, 1982a, b, 1983a; Chu et al. 1982; Chu and Treffers 1981a, b; Chu, Treffers, and Kwitter 1983; Treffers and Chu 1982a, b). The dynamic timescale of a WR ring nebula, approximated as 0.5(radius/expansion velocity), is compared to the lifetime of a WR phase. If the nebular dynamic timescale is larger than the WR lifetime, the nebula is classified R-type, for which the central WR star is only responsible for ionizing but not shaping the nebula. For the nebulae with dynamic timescales smaller than the WR lifetime, their nebular properties are compared to those predicted by wind-blown bubble models (Weaver et al. 1977). A nebula is classified E-type if it consists of mostly stellar ejecta, and W-type if it is shaped by stellar winds.

One major limitation of this kinematics-based classification is its incapability of determining whether a W-type nebula consists of stellar or interstellar material. Recent analysis of elemental abundances and nebular excitation shows that some of the W-type nebulae actually contain shells of stellar ejecta. The classification into E- and W-types is apparently overly simplistic and may be somewhat misleading. I will therefore emphasize below the three modes of interaction, instead of the three types of nebulae, and the discussion is generalized to nebulae around other kind of massive stars. Table 1 gives representative examples for different formation mechanisms; the star names, spectral types, and nebula names are listed.

TABLE 1. Ring Nebulae around Massive Stars

STELLAR EJECTA:	BAC 209	WN8	M1-67
	AG Car	LBV	AG Car
	η Car	LBV	η Car
	α Sco B	B2.5V	anon
BUBBLE/EJECTA:	HD 147419	WN4	RCW104
	HD 192163	WN6	NGC 6888
	HD 96548	WN8	RCW58
	HD 148937	O6.5fp	NGC 6164-5
WIND-BLOWN BUBBLE:	HD 56925	WN4	NGC 2359
	HD 50896	WN5	S308
	HD 89358	WN5	NGC 3199
	HD 92809	WC6	anon(MR26)
	LSS 4368	WO	G2.4+1.4
	BD+60°2522	O6.5III	NGC 7635 (Bubble Nebula)
BOW SHOCKS:	BD+50°886	O5	S206
	λ Cep	O6If	
	λ Ori	OIIIf	
	τ CMa	O9Ib	
	α Cam	O9.5Iae	
	ζ Oph	O9.5V	
	κ Cas	BIae	
	δ Sco	B0.3IV	
	δ Pic	B3III+O9V	
BOW WAVES:	HD 171491	B5	
	δ Per	B5III	
	6 Cep	B3IV	
HII REGIONS:	HD 187282	WN4	anon(MR95)
	MR97	WN8	L69.8+1.74
	HD 117688	WN8	RCW78
	HD 115473	WC5	anon(MR46)
	HD 113904	WC6+O9.5I	anon(θ Mus)
	68 Cyg	O7.5	S119

3.1. UV Radiation

A ring nebula that is ionized but not shaped by the central WR star can be identified by its large dynamic timescale. The classification of type R is still meaningful. The R-type nebulae, consisting mostly of interstellar material, are not expected to show any abundance anomaly. The abundances in the R-type nebula RCW78 are indeed similar to those of its local interstellar environment (Esteban et al. 1990a).

In retrospect, most of the R-type WR ring nebulae would not have been identified, if the selection criteria had been stricter with respect to filamentary morphology, since a large dynamic age is usually a result of slow expansion, which is implicitly guaranteed by the lack of narrow filaments and sharp limb-brightening on a sub-parsec scale.

For the non-WR central stars, an R-type ring nebula can also be diagnosed by the amorphous nebular morphology, but its confirmation needs more than a comparison of timescales as for the WR rings. For example, the ring nebula S119 around the Of star 68 Cyg, appearing quite amorphous on the Palomar Sky Survey Prints, is recently demonstrated to be an R-type nebula instead of a wind-blown bubble with a central cavity around the star, because the IR emission detected in the vicinity of the star implies the existence of dust, and a bow shock around the star is seen in the IR (Wisotzki and Wendker 1989).

The arc-shaped reflection nebulae around B stars may be related to the UV radiation: the radiation pressure pushes the dust out while the ionized gas is relatively unaffected. The B2V star LSS 3027 gives a good example: its reflection nebula is exterior to the round HII region centered on the star (Chu 1983b).

3.2. Stellar Ejecta

"Stellar ejecta" refers to the bulk ejection of stellar surface material, as opposed to the tenuous, continuous, high-velocity stellar winds. Some stellar ejecta are present in the form of a shell, some appear to be bipolar, and some seem to be ballistic clumps. The ejecta nature was originally identified by the kinematic and morphological properties; however, it seems that anomalous abundances may be another good indicator for the ejecta.

Four WR ring nebulae are known to contain stellar ejecta: M1-67, NGC 6888, RCW58, and RCW104. The ejecta nature of M1-67 is concluded from the similarity between the stellar and nebular velocities, which are more than 150 km/s higher than the interstellar velocity expected in the galactic rotation (Treffers and Chu 1982a). The shells of ejecta in RCW58 and NGC 6888 are diagnosed from the nebular morphology and excitation; a [N II]-bright ring enveloped in a [O III]-bright ring can only be explained by a shell of ejecta inside a wind-blown bubble (Chu 1982a; Dufour 1989a; Mitra 1990). The ejecta in RCW104 appear as high-velocity (>100 km/s), [N II]-bright knots near the western rim (Goudis, Meaburn, and Whitehead 1988).

It was once claimed that stellar ejecta were characterized by chaotic internal motion (Chu 1981, 1982a; Treffers and Chu 1982a); however, that claim is apparently an incorrect interpretation of the large aperture Fabry-Perot data of M1-67 and RCW58. Recent long-slit spectra of M1-67, RCW58 and NGC 6888 show smooth bow-shaped line images, indicating regular expansion patterns (Solf and Carsenty 1982; Chu 1988; Marston and Meaburn 1988). The expansion velocities of RCW58 and NGC 6888, 110 km/s and 80 km/s, are the highest two among the WR rings.

All of the above four WR nebulae show He/H and N/O ratios significantly higher than those in the sun and the galactic HII regions (see Table 2). Are enhanced He/H and N/O ratios a sufficient condition for the presence of stellar ejecta? Among the WR ring nebulae in Table 2, S308 is the only other nebula that shows similar abundance anomalies. It has been argued that S308 must consist of stellar material because of its large distance to the galactic plane (Lozinskaya 1983; McCray 1983). Furthermore, even if S308 is a bubble blown out of the interstellar material, the stellar wind is not likely to mix with the interstellar material to cause the observed abundance anomaly (Weaver et al. 1977; Dyson and Smith 1985). Therefore, it is very likely that S308 contains mostly stellar ejecta and the He/H and N/O abundance ratios are indeed an indicator of stellar ejecta.

Among all the ring nebulae around Of stars, NGC 6164-5 is the only one that shows definite stellar ejecta, as the S-shaped nebula has both anomalous abundances (Leitherer and Chavarría-K. 1987; Dufour, Parker, and Henize 1988) and kinematics that cannot be explained by wind-blown bubbles (Pismis 1974).

The ring nebulae around the LBV's all consist of ejected stellar material. The nebula around η Car has many unique properties; its expansion velocity reaches over 2000 km/s (Walborn, Blanco, and Thackeray 1978; Walborn and Blanco 1988; and references therein), and its oxygen abundance is severely depleted while nitrogen abundance is enhanced (Davidson et al. 1986). The nebula around AG Car is relatively moderate with an expansion velocity of ~ 70 km/s (Smith 1990); both oxygen and sulphur abundances are depleted (Mitra and Dufour 1990).

3.3. Stellar Winds

The stellar wind from a massive star can sweep up the ambient medium and form a bubble; the motion of the star relative to the medium can cause the star to be decentered (Weaver et al. 1977), and in the extreme case a bow shock, instead of a closed bubble, will be formed (Gull and Sofia 1979). NGC 3199 is an example that the central WR star is probably moving at about 60 km/s toward the bright shell ridge at the southwest (Dyson and Ghanbari 1989). The bow shocks around OB stars (Van Buren and McCray 1988) are certainly the results of large proper motions and stellar wind interactions.

Ring nebulae that are formed by stellar winds should show shocks. The [O III]-bright shell in RCW58 has been suggested to represent the shock fronts (Chu 1982a). The recent imaging spectral analysis and spectroscopic observations with high spatial resolution show the shocks convincingly for the first time in the Bubble nebula, NGC 2359, and NGC 6888 (Jernigan 1988; Mitra 1990; Dufour 1989a): the [O III]5007/Hβ intensity ratio reaches 15 and 20, and the electron temperature may be as high as 25,000 and 40,000 K for NGC 2359 and NGC 6888, respectively!

Almost all wind-blown bubbles that have been spectroscopically studied have normal abundances relative to their ambient interstellar environments. Examples of the WR rings are NGC 2359, NGC 3199, and the nebula around MR26 (Esteban et al. 1990a; Rosa and Mathis 1990); examples of rings around Of stars are the Bubble nebula and the filamentary bubble in NGC 6164-5, located between the S-shaped ejecta and the outermost dust halo (Jernigan 1988; Leitherer and Chavarría-K.

1987; Dufour, Parker, and Henize 1988). The only exception is S308, which may consist of stellar ejecta, as described in section 3.2.

4. APPLICATIONS OF THE RING NEBULAE

The value of ring nebulae around massive stars goes far beyond their apparent beauty. They provide great opportunities to study the central stars as well as the interstellar environment.

4.1. Nebular Abundances and Stellar Nucleosynthesis

Ring nebulae may carry processed stellar material, and their abundance anomalies will give information about stellar nucleosynthesis. It is long known that some WR rings are enriched in He and N (Parker 1978; Kwitter 1981, 1984). Recent observations have extended to many fainter nebulae; more importantly, great observational efforts have gone into differentiating the ring nebulae and their interstellar environments so that the abundances in the rings can be compared to the true local interstellar abundances (Esteban and Vílchez 1990; Esteban et al. 1990a; Dufour, Parker, and Henize 1988; Mitra 1990; Jernigan 1988). Table 2 summarizes the recent results for a large number of ring nebulae.

It is clear from Table 2 that the rings containing stellar ejecta all have higher He/H than the universal value of 0.1. The N and O abundance anomalies individually cannot be easily discerned from this

TABLE 2.   Abundances of Ring Nebulae

Object	Type	He/H	12+log(O/H)	12+log(N/H)	log(N/O)	Ref.
NGC 6164-5	Of	0.117	8.25	8.13	-0.12	1
η Car	LBV	0.17	≤7.8	~9.0	1.2	2
AG Car	LBV	?	7.2	7.5	0.3	3
RCW 78	WN	0.11	8.92	8.07	-0.85	4
NGC 2359	WN	0.10	8.25	7.27	-0.98	4
NGC 3199	WN	0.095	8.51	7.55	-0.96	5
MR 100	WN	0.115	8.52	7.64	-0.88	5
MR 26	WC	0.09	8.55	7.55	-1	5
S308	WN	0.123	8.2	8.0	-0.2	5
RCW104	WN	0.15	8.54	7.85	-0.69	5
RCW58	WN	0.23	8.72	8.42	-0.3	5
NGC 6888	WN	0.188	8.15	8.45	0.3	6
M1-67	WN	0.22	8.10	8.70	0.60	7
Sun		0.107	8.87	7.96	-0.91	3
MWG HII		0.1	8.7	7.57	-1.13	3
Type I PNe		0.135	8.71	8.88	0.17	3

Ref: 1. Dufour, Parker, and Henize 1988; 2. Davidson et al. 1986;
     3. Mitra and Dufour 1990; 4. Esteban et al. 1990a; 5. Rosa and
     Mathis 1990; 6. Esteban and Vílchez 1990; 7. Esteban et al. 1990b.

table, since the local interstellar abundances, subject to the galactic abundance gradient, are not listed. The N/O ratio, on the other hand, is normally quite constant with log(N/O) ~ -1; any anomaly in N/O ratio can be easily spotted. The nebulae that show enhanced He/H all have N/O ratio greater than the normal value.

A closer examination of the nebulae individually shows that the N/O increase is due to both an enhancement in N abundance and a depletion in O abundance. Dufour (1989b) has noticed that the sum of N and O is nearly constant in ring nebulae and compared them with the Type I planetary nebulae. For the Type I planetary nebulae, Feibelman et al. (1985) suggests that the reaction $O^{16} + 2p \rightarrow N^{14} + \alpha$ is responsible for converting O into N. How about the massive stars? Nucleosynthesis models in massive stars need to explain these abundance anomalies.

The most severe depletion of O occurs in the nebulae around LBV's. In the η Car ejecta, both C and O are depleted, while N is enhanced; the sum of C, N, and O seems to be conserved in η Car ejecta (Dufour 1989a)! Unfortunately, most of the ring nebulae are not detectable in the UV carbon lines, and it is not possible to determine whether (N+O) or (C+N+O) is constant. In either case, the nebular abundances place constraints on the stellar structure and nucleosynthesis models.

4.2. Nebular Spectrophotometry and Stellar Effective Temperature (T_{eff})

The far UV energy distribution of a star cannot be observed directly; nevertheless, this information is carried in the surrounding nebula. By modelling the photoionization structure of a nebula to produce the observed nebular line fluxes and ratios, one can derive information on the T_{eff}. Various diagnostic schemes have been given by, for example, Mathis (1982, 1985), and Evans and Dopita (1985). Note that the value of T_{eff} itself may have large uncertainties, since it depends on the stellar atmospheric models; nevertheless, the ranking of stellar T_{eff} within the same diagnostic scheme should be quite accurate.

Smith and Clegg (1990) find a temperature of 55,000-80,000 K for HD 50896 in S308. Esteban et al. (1990a) find a temperature of 50,000 K for HD 56925 in NGC 2359 and 34,000 K for HD 117688 in RCW78. Rosa and Mathis (1990) use the abundance ratios of S^+/S^{+2} and O^+/O as a diagnostic for T_{eff} (Mathis 1982, 1985); they find S308, NGC 3199, NGC 6888, and NGC 2359 are ionized by hot stars with T_{eff} between 60,000 and 90,000 K, but RCW58, RCW104, MR26, and MR100 have such low excitation that their T_{eff} cannot be reliably determined. In general, the early WN's are hotter than the late WN's, and the WC6 star MR26 seems to be similar to or cooler than the late WN's. This result is quite consistent with the intuitive interpretation of the stellar spectra.

Emission lines from doubly ionized He, often seen in planetary nebulae with high excitation, are not expected in HII regions. However, at least 5 HII regions in the Local Group are known, before this symposium, to emit nebular HeII4686 line: G2.4+1.4 in the Galaxy, N44C and N159F in the LMC, N76 in the SMC, and IC1613#3. G2.4+1.4 and IC1613#3 are each ionized by a WO star, and the extended HeII emission is centered on the WO star (Dopita et al. 1990; Davidson and Kinman 1982). The positive association of nebular HeII emission with the WO stars implies that the effective temperature of WO stars can reach as high as 70,000-80,000 K! This high temperature perhaps could have been expected, since WO stars emit strong OVI lines, and the hottest known planetary nebula nuclei also emit OVI lines. However, not every WO star is hot enough to produce doubly ionized He in the surrounding nebula, as Pakull's (1990, private communication) search for HeII emission around other known WO stars (Barlow and Hummer 1982) yields no positive detection. N159F is ionized by the X-ray binary LMC X-1 (Pakull and Angebault 1986). N44C is apparently centered on an O4-O6 main sequence star (Stasińska, Testor, and Heydari-Malayeri 1986; Pakull and Motch 1989) and N76 centered on a WN3+O binary (Moffat 1988). A high stellar temperature for an O or WN star is quite surprising, and it has been suggested that X-rays and shock excitation may be responsible for the HeII emission (Pakull and Motch 1989; Chu and Mac Low 1990).

In this symposium, three new HeII4686 emitting nebulae are reported around WR stars: Br2 (WN1?) and Br40a (WN3+O6) in the LMC, and AB-5 (WN3+WN6+O) in the SMC (Pakull 1990; Niemela 1990). Perhaps some early type WR stars are indeed hot!

4.3. Ring Nebulae and Bubble Dynamics

The W-type ring nebulae around WR stars (Chu 1981) were thought to be good wind-blown bubbles that can test theoretical models. When the momentum and energy conversion factors were found to deviate from those expected from Weaver et al.'s (1977) bubble models (Treffers and Chu 1982a; Chu 1982a), improved calculations including interstellar cloudlets have been carried out (McKee, Van Buren, and Lazareff 1984) and neutral material in the bubble shells has been proposed to explain the discrepancies (Van Buren 1986).

Now we know that some W-type nebulae actually contain shells of stellar ejecta. Their momentum and energy conversion factors would be quite meaningless because the shell kinetic energy is determined by stellar ejecta and has little to do with the total stellar wind energy output during the evolution of the shell. The discrepancy between the observed and bubble model predicted X-ray luminosities of NGC 6888 (Kähler, Ule, and Wendker 1987; Bochkarev 1988) can be well explained by the shell dynamics being dominated by the stellar ejecta. Even the rings without confirmed stellar ejecta may be contaminated by an unknown amount of stellar ejecta. Stellar winds interacting with the expanding circumstellar material lost by the progenitor cannot be described by the commonly-used model of a bubble in a stationary medium (Dyson and Smith 1985). The significance of these conversion factors may have been overrated; these factors have to be used with great caution.

5. FUTURE WORK

There are various scenarios regarding the evolution for massive stars. Since the abundances and kinematics of a ring nebula reflect the history of the stellar nucleosynthesis and mass loss, detailed studies of nebulae around different central stars may exclude or confirm some evolutionary scenarios. For example, if LBV's evolve into WN's, we must find some nebulae around LBVs that could be evolve into some rings around WN stars. Can a nebula like AG Car evolve into something like RCW58? Do their abundances agree? Are the nebular structures consistent with the evolution of stellar ejecta interacting with stellar winds? The answers to these questions rely on the accurate abundance determination and theoretical calculations of the nebular evolution in the future.

In order to make the aforementioned test, we need a large sample of ring nebulae around different types of massive stars. However, only one Of star is known to have stellar ejecta, a couple Of stars have filamentary bubbles, and 2-3 LBV's have resolved nebulae. Is this small number due to a lack of surveys and observations? Are Of stars less efficient than WR stars in blowing bubbles and ejecting envelopes?

In the past, only cataloged WR stars or Of stars are used to search for ring nebulae. The selection effect may not be as bad, if the ring nebulae in a galaxy is surveyed first, then the central stars are identified and classified. Our Galaxy is hopelessly opaque along the galactic plane, and distant galaxies would have resolution problem. The Magellanic Clouds are the most appropriate for a systematic survey of ring nebulae. A ring with 5 pc diameter will subtend nearly 20" in the LMC, which can be easily resolved from the ground based telescopes. The central stars of ring nebulae can also be identified and classified. An extended survey for ring nebulae and central stars in the Magellanic Clouds is badly needed. Correlations between the nebular properties and the stellar types would be extremely interesting.

Abundance analysis of a ring nebula should try to differentiate the different components from the background. For example, both NGC 6888 and RCW58 have a shell of ejecta inside a bubble, and both nebulae have a stationary interstellar component beside the shell components. Does this stationary component have similar abundances as those in the shell components? Do the bubbles in NGC 6888 and RCW58 have abundances similar to those of the ejecta or the ambient medium?

The IRAS data can be used to search for ring features in the IR, but the interpretation may be tricky. A wind blown bubble is expected to be a soft X-ray source. ROSAT observations of ring nebulae should be made.

The author acknoledges the support of an International Travel Grant from the American Astronomical Society. This work is supported by NSF grant AST-8818192.

REFERENCES

Baars, J. W. M. and Wendker, H. J. 1987, Astr. Ap., 181, 210.
Barlow, M. J. and Hummer. D. G. 1982, in IAU Symposium No. 99, Wolf-Rayet Stars: Observations, Physics, Evolution, eds. C. W. H. de Loore and A. J. Willis (Dordrecht: Reidel), p. 387.
Bochkarev, N. G. 1988, Nature, 332, 518.
Cappa de Nicolau, C. E., and Niemela, V. S. 1984, A. J., 89, 1398.

Cappa de Nicolau, C. E., Niemela, V. S., and Arnal, E. M., 1986, A. J., 92, 1414.

Cappa de Nicolau, C. E., Niemela, V. S., Dubner, G. M., and Arnal, E. M., 1988, A. J., 96, 1671.

Chu, Y.-H. 1981, Ap. J., 249, 195. (Paper I)

Chu, Y.-H. 1982a, Ap. J., 254, 578. (Paper VI)

Chu, Y.-H. 1982b, Ap. J., 255, 79. (LMC paper II)

Chu, Y.-H. 1983a, Ap. J., 269, 202. (LMC paper III)

Chu, Y.-H. 1983b, Pub. A. S. P., 95, 873.

Chu, Y.-H. 1988, Pub. A. S. P., 100, 986.

Chu, Y.-H., Gull, T. R., Treffers, R. R., Kwitter, K. B., and Troland, T. H. 1982, Ap. J., 254, 562. (Paper IV)

Chu, Y.-H. and Lasker, B. M. 1980, Pub. A. S. P., 92, 730. (LMC paper I)

Chu, Y.-H. and Mac Low, M.-M. 1990, submitted to Ap. J.

Chu, Y.-H. and Treffers, R. R. 1981a, Ap. J., 249, 586. (Paper II)

Chu, Y.-H. and Treffers, R. R. 1981b, Ap. J., 250, 615. (Paper III)

Chu, Y.-H., Treffers, R. R., and Kwitter, K. B. 1983, Ap. J. Suppl., 53, 937. (Paper VIII)

Crampton, D. 1971, M. N. R. A. S., 153, 303.

Davidson, K., Dufour, R. J., Walborn, N. R., and Gull, T. R. 1986, Ap. J., 305, 867.

Davidson, K. and Kinman, T. D. 1982, Pub. A. S. P., 94, 634.

Deharveng, L., Israel, F. P., and Maucherat, M. 1976, Astr. Ap., 48, 63.

Dopita, M. A., Lozinskaya, T. A., McGregor, P. J., and Rawlings, S. J. 1990, Ap. J., 351, 563.

Drissen, L., Shara, M. M., and Moffat, A. F. J. 1990, in this symposium.

Dubner, G. M., Niemela, V. S., and Purton, C. R. 1990, A. J., 99, 857.

Dufour, R. J. 1989a, Rev. Mexicana Astron. Astrof., 18, 87.

Dufour, R. J. 1989b, in IAU Symposium No. 131, Planetary Nebulae, ed. S. Torres-Peimbert (Dordrecht: Kluwer), p. 216.

Dufour, R. J., Parker, R.A.R., and Henize, K. G. 1988, Ap. J., 327, 859.

Dyson, J. E. 1977, Ap. and Space Sci., 51, 197.

Dyson, J. E. and Ghanbari, J. 1989, Astr. Ap., 226, 270.

Dyson, J. E. and Smith, L. J. 1985, in Cosmic Gas Dynamics, ed. F. D. Kahn, VNU Science Press, Utrecht, Holland, p. 173.

Esteban, C., Vílchez, J. M., Manchado, A., and Edmunds, M. G. 1990a, Astr. Ap., 227, 515.

Esteban, C., Vílchez, J. M., Smith, L. J., and Manchado, A. 1990b, in preparation.

Evans, I. N. and Dopita, M. A. 1985, Ap. J. Suppl., 58, 125.

Feibelman, W., Aller, L. H., Keyes, C. D., and Czyzak, S. J. 1985, Proc. Natl. Acad. Sci. USA, 82, 2202.

Georgelin, Y. M., Georgelin, Y. P., Laval, A., Monnet, G., and Rosado, M. 1983, Astr. Ap. Suppl., 54, 459.

González, J. and Rosado, M. 1984, Astr. Ap. (Letters), 134, L21.

Goudis, C. D., Meaburn, J., and Whitehead, M. J. 1988, Astr. Ap., 191, 341.

Gull, T. R. and Sofia, S. 1979, Ap. J., 230, 782.

Heckathorn, J. N., Bruhweiler, F. C., and Gull, T. R. 1982, Ap. J., 252, 230.

Herbig, G. H. 1960, Ap. J. Suppl., 4, 337.

Heydari-Malayeri, M., Melnick, J., and Van Drom, E. 1990, Astr. Ap.,
    in press.
Jernigan, T. E. 1988, Ph.D. thesis, Rice University.
Johnson, H. M. 1975, Ap. J., 198, 111.
Johnson, H. M. 1982, Ap. J., 256, 559.
Johnson, H. M. and Hogg, D. E. 1965, Ap. J., 142, 1033.
Kähler, H., Ule, T., and Wendker, H. J. 1987, Ap. Space Sci., 135, 105.
Kwitter, K. B. 1981, Ap. J., 245, 154.
Kwitter, K. B. 1984, Ap. J., 287, 840.
Leitherer, C. and Chavarría-K., C. 1987, Astr. Ap., 175, 208.
Lortet, M. C., Niemela, V. S., and Tarsia, R. 1980, Astr. Ap., 90, 210.
Lozinskaya, T. A. 1982, Ap. and Space Sci., 87, 313.
Lozinskaya, T. A. 1983, Sov. Astron. Lett., 9, 247.
Lozinskaya, T. A., Lar'kina, V. V., and Putilina, E. V. 1984,
    Sov. Astron. Lett., 9, 344.
Marston, A. P. and Meaburn, J. 1988, M. N. R. A. S., 235, 391.
Mathis, J. S. 1982, Ap. J., 261, 195.
Mathis, J. S. 1985, Ap. J., 291, 247.
McCray, R. 1983, Highlights Astr., 6, 565.
McKee, C. F., Van Buren, D., and Lazareff, B. 1984, Ap. J. (Letters),
    278, L115.
Miranda, A.I. and Rosado, M. 1987, Rev. Mexicana Astr. Astrof., 14, 479.
Mitra, P. M. 1990, Ph.D. thesis, Rice University.
Mitra, P. M. and Dufour, R. J. 1990, M. N. R. A. S., 242, 98.
Moffat, A. F. J. 1988, Ap. J.,330, 776.
Nichols-Bohlin, J. and Fesen, R. A. 1990, in preparation.
Niemela, V. S. 1990, in this volume.
Pakull, M. W. 1990, in this volume.
Pakull, M. W. and Angebault, L. P. 1986, Nature, 322, 511.
Pakull, M. W. and Motch, C. 1989, Nature, 337, 337.
Parker, R. A. R. 1978, Ap. J., 224, 873.
Pismis, P. 1974, Rev. Mexicana Astr. Astrof., 1, 45.
Rosa, M. R. and Mathis, J. S. 1990, Properties of Hot Luminous Stars:
    Boulder-Munich Workshop, ed. C. D. Garmany, Pub. A. S. P. Conf.
    Ser., 7, 135.
Rosado, M. 1986, Astr. Ap., 160, 211.
Smith, L. F. 1967, Ph.D. thesis, Australian National University.
Smith, L. J. 1990, in this volume.
Smith, L. J. and Clegg, R. E. S. 1990, Properties of Hot Luminous Stars:
    Boulder-Munich Workshop, ed. C. D. Garmany, Pub. A. S. P. Conf.
    Ser., 7, 132.
Solf, J. and Carsenty, U. 1982, Astr. Ap., 116, 54.
Stahl, O. 1987, Astr. Ap., 182, 229.
Stahl, O. 1989, in Physics of Luminous Blue Variables, eds. K. Davidson,
    A. F. J. Moffat, and H.J.G.L.M. Lamers (Dordrecht: Kluwer), p. 149.
Stahl, O., Wolf, B., Klare, G., Cassatella, A., Krautter, J., Persi.,
    P., and Ferrari-Toniolo, M. 1983, Astr. Ap., 127, 49.
Stasińnska, G., Testor, G., and Heydari-Malayeri, M. 1986, Astr.
    Ap., 170, L4.
Swings, J. P. and Preston, G. W. 1978, Ap. J., 220, 883.
Vilchez, J. M. and Esteban, C. 1990, in this volume.

Treffers, R. R. and Chu, Y.-H. 1982a, Ap. J., <u>254</u>, 569. (Paper V)

Treffers, R. R. and Chu, Y.-H. 1982b, Ap. J., <u>254</u>, 132. (Paper VII)

Van Buren, D. 1986, Ap. J., <u>306</u>, 538.

Van Buren, D. and McCray, R. 1988, Ap. J. (Letters), <u>329</u>, L93.

van der Hucht, K. A., Conti, P. S., Lundstrom, I., and Stenholm, B. 1981, Space Sci. Rev., <u>28</u>, 227.

Walborn, N. R. 1982, Ap. J, <u>256</u>, 452.

Walborn, N. R. and Blanco, B. M. 1988, Pub. A. S. P., <u>100</u>, 797.

Walborn, N. R., Blanco, B. M., and Thackeray, A. D. 1978, Ap. J., <u>219</u>, 498.

Weaver, H. 1977, "The Large-Scale Characteristics of the Galaxy", IAU Symp. No. 84, ed. W. B. Burton et al. (Dordrecht: Reidel), p. 295.

Weaver, R., McCray, R. A., Castor, J., Shapiro, P., and Moore, R. 1977, Ap. J., <u>218</u>, 377.

Wisotzki, L. and Wendker, H. J. 1989, Astr. Ap., <u>221</u>, 311.

DISCUSSION

Moffat: If the WR nebulae that are stellar ejecta show enhanced N/O, the ejecta must come from the WR star itself: (1) Why then are the ejecta clumped or filamentary and, (2) why are they moving so slowly? When the ejecta left the observable part of the WR wind they must have had outward velocities of $v_\infty \approx (1-3)10^3 km \cdot s^{-1}$. (3) If shot off like bullets, what slows them down? (4) Also, can the clumps be related to the blobs we see being ejected in the outer winds of WR stars?

Chu: (1) Stellar ejecta are subject to Rayleigh-Taylor instability, which makes the observed clumpy morphology. The ejecta may or may not be clumping initially. The ejecta in RCW104, the high velocity knots seen in [NII], may be clumping at the ejection. (2) Stellar ejecta and stellar wind result from different physical mechanisms, consequently, the ejection velocity may differ. The ejection velocity of the ejecta is probably of similar order as the escape velocity at the place where ejection occurs. Therefore, the ejection velocity would depend on the radius and mass of the star. (3) If ejecta are shot off like bullets, they will not only slow down gradually by sweeping up the ambient stationary medium, but they can be accelerated by the stellar wind, too.

Dopita: A high $[OIII]/H\beta$ ratio is not necessarily due to shocks. Radiative shocks saturate at a value of the ratio determined by the O abundance. High $[OIII]/H\beta$ ratios can be got by four mechanisms: (1) partially radiative shocks ($T[OIII] \approx 40000K$); (2) fast particle heating ($T[OIII]$ unknown); (3) photoionisation at high specific radiation intensity ($T[OIII] \lesssim 20000K$); (4) X-ray ionization ($T[OIII]$ unknown).

Chu: The $[OIII]/H\beta$ ratio is explained as partially radiative shocks by Dufour, Jerrigan, and Mitra for NGC 2359 and NGC 6888. The X-ray ionisation is unlikely as the X-ray emission comes from the bubble cavity while the $[OIII]$ enhancement is on the outer surface of the shell.

de Groot: On the subject of stellar ejecta, low shocks and proper motions - where you said that you would not know what to do with such information - let me offer a suggestion. An idea of the velocity of the ejecta, when combined with their radii, gives you a timescale since their formation. This can then be combined with the space velocity - proper motion and radial velocity - of the "central" object to find a confirmation for the assumed velocity of ejecta. We did this in the case of P Cyg (1983) where it confirms the ejection of the radio feature some 30,000 years ago, *i.e.* in harmony with ejection at the red supergiant stage. P Cyg also has a much smaller circumstellar shell. In that case the size of the shell fits exactly its ejection some 400 years ago, *i.e.* at P Cyg's outburst in AD 1600 (Leitherer *et al.*, 1987).

Drissen: In our M33 survey fields, which included 50 WR stars, we found 11 good WR ring nebulae. We have to explain why about 75% of WR stars do not blow bubbles.
Chu: The small ones are probably not resolved at the distance of M33, hence have been missed.

Walborn: Davidson *et al.* had two indirect arguments that the C depletion in the η Car knots is not entirely due to grain formation: (1) O is also depleted, but normal Si abundance is observed and (2) as best could be determined, the absolute N abundance is consistent with the normal sum of CNO.

Smith, Linda J.: I would like to make a comment on the ejecta nature of NGC 6888. With WR ring nebulae, it is important to consider the effect of the progenitor wind on the ambient interstellar medium. For NGC 6888 this is particulary important because the central star HD 192163 is a member of the Cyg OB1 association. It is therefore likely that the collective winds and supernovae from the association members will have evacuated the interstellar medium surrounding HD 192163 before the onset of the WR wind. From this then, NGC 6888 has to be composed of almost pure stellar ejecta.

Montmerle: Frequently, I have seen in the literature people deriving ages, *e.g.*, from the radii...
Chu: It should not be called age, it should be called dynamic scale, radius over velocity, it is a convenient scale, it should not really be called age.
Montmerle: I just wanted to add that since in many cases you see that it is more momentum conserving, then energy conserving, then the actual age that you would deduce critically depends on the mechanism that you are assuming. People should watch out for these kinds of things, and even the dynamical age is in most of the cases useless.
Chu: It is a convenience scale, it has not much physical meaning.

Maeder: You say that the ejeca have higher N/O ratio, which is quite consistent with material processed by the CNO cycle and ejected before the WR phase or during the WN phase. But, what about material ejected during the WC phase, because then we would have the supposed strong oxygen and C-enrichment and do you have a signature of that effect?
Chu: No, we do not have an ejecta around WC stars. The ejecta is kind of tricky. Once it is ejected, it breaks up into small chunks. These small chunks will expand, the density goes down very quickly. The density goes down in some fashion, but the emission measure goes down the square of that, so the emission measure goes down quicker than the expansion and very quickly it disappears. It is a detection limit problem!

Smith, Lindsey: Following on the remark of Linda Smith, NGC 3199 and NGC 2359 are much more massive than NGC 6888 and RCW 58. The mechanisms involved are the same; only the density of the surrounding ISM is higher, so the abundance anomalies of the WR wind are diluted by normal abundance of the swept up gas.

You-Hua Chu

RING NEBULAE AROUND Of AND WO STARS

Tatiana A. Lozinskaya
Sternberg State Astronomical Institute, Moscow, USSR

Introduction

To understand the origin of WR-rings it is of interest to consider nebulae associated with the precursors of WR stars and with stars at later stages of evolution. The former are represented by numerous Of stars and the latter by only four WO stars which, according to Barlow and Hummer (1982), are the most extreme WR stars near the end of their lives.

We can briefly summarize the properties and statistics of nebulae connected with Of stars, as compared with WR-rings, following Lozinskaya (1982; 1986):

i) About 80-90% of both WR and Of stars in a distance-complete sample are associated with HII regions, of which 30-40% have a ring-nebula-like morphology. By analogy to WR-rings, the Of-ring nebulae may be subdivided into three types: wind-blown bubbles, ejecta and shell-like HII regions; the characteristics of the WR and Of nebulae defining each type are similar.

ii) The three ways in which stars can influence the ISM — radiation, stellar winds and ejection of slow shells — work together to form multilayer shell-like structures of the ambient gas. The majority of the small-size bubbles and ejecta are located inside extended HII or HII/CO shells or cavities. Many bubbles and ejecta which appear isolated are actually located inside supershells, like the WR ring nebula NCG 6888 inside the Cyg OB1 supershell.

What seems to be unclear regarding the origin of the shells is: why only 30-40% of stars with strong stellar winds are associated with ring nebulae contrary to theoretical predictions (the wind-blown bubbles are even more scarce, about 10%).

One can see at least three explanations:

i)If a strong wind sweeps up, not the interstellar gas, but stellar ejecta of about $1-5M_\odot$, the lifetime of a shell is determined by the emission measure of the expanding ejecta and may be shorter than the duration of the WR-stage. This explanation seems reasonable since the wind of the progenitor star and/or of a young stellar cluster may evacuate the ambient ISM.

ii) The short lifetime of wind-blown bubbles may be caused by a fast dissipation of the hot wind's energy due to hot plasma leakage through the shell, or cloud evaporation and so on.

iii) In a cloudy ISM the cold unshocked wind may leak through tenuous regions between dense clouds like fingers, without forming a shell-like structure.

K. A. van der Hucht and B. Hidayat (eds.),
Wolf-Rayet Stars and Interrelations with Other Massive Stars in Galaxies, 365–369.
© 1991 *IAU. Printed in the Netherlands.*

Oxygen WR star environments

The four Sanduleak WO stars are the most advanced massive stars, probably only several thousand years from a SN explosion. Indeed, our observations of WR 102 (Dopita *et al.*, 1990; Dopita and Lozinskaya, 1990) indicate the absence of He in the star's envelope and point to the absence of it in the interior as well; the star's position on the HR diagram also supports our interpretation of it as the $C - O$ core of a massive star. Similarly, Sand 1 [$M(WR) = 14M_\odot$ and $M(O4) = 50M_\odot$] seems to be the most evolved WR star known in a close binary (Moffat *et al.*, 1985). Thus they provide a unique opportunity to investigate a SN-environment shortly before the explosion.

Since the WO stage is very short (the lifetime of the $C - O$ core of a massive star is only about 1% of the core-He burning duration) we cannot expect WO stars like WR 102 to constitute more than about 1% of the $\sim$ 300 WR stars known; the four objects form quite a representative sample in the Galaxy and MCs.

WO stars are characterized by an extremely fast "superwind": $= v_w = \sim$ 4500 to 7400 $km\cdot s^{-1}$ (Barlow and Hummer, 1982; Torres *et al.*, 1986; Dopita *et al.*, 1990) and the ambient gas is influenced by the previous "normal" outflow of the progenitor WR star. Both facts open new aspects for investigation into stellar wind *vs.* ISM interactions.

The four WO stars appear to be associated with optical and/or IR shell-like structures, although the short WO superwind does not participate in the shell's formation.

The nebula G2.4+1.4 around WR 102 is the only object of the four which was shown to be a classical wind-blown bubble (Dopita *et al.*, 1990; Dopita and Lozinskaya, 1990 and these proceedings). We evaluate the star's mechanical luminosity as $L_w = (0.5 - 2)10^{38} erg\cdot s^{-1}$ which is far in excess of the energy required to blow the shell ($R = 5 - 6$ pc, $v_{exp} = 42$ $km\cdot s^{-1}$, the ambient density ranges from $2 - 4$ to $30 - 60$ cm^{-3}). The shell's dynamical age $t \approx 10^5$ yr is much longer than the WO superwind duration, indicating that the progenitor star created the ring nebula before the onset of the WO stage. IRAS observations of the region reveal an incomplete ring coincident with the edge of the optical nebula, see Figure 1 and Graham (1985).

The second oxygen star in the Galaxy WR 142 is a member of the young open cluster Be 87. The cluster (distance 950 ± 50 pc) is physically associated with an active star-forming region ON2 (Turner and Forbes, 1982; Pitault, 1981) and both are located at the edge of a giant molecular cloud in front of Cyg X-Cyg OB1 (Cong, 1977). There is no prominent optical emission around WR 142 because of heavy absorption. Figure 2 shows the high resolution IRAS image of a small region around the star. It is confused by bright point sources (the two brightest are ON2 and Sh 106) and displays no signs of a shell-like morphology. The large-scale infrared images shown in Figure 3, however, reveal a prominent shell-like structure: a half-ring of about $3°$ seen at both 60 and 100 μm. The reality of the IR shell is supported by a flat depression at 4.8 GHz (see Wendker, 1984) and by a CO arch-like feature (see Turner and Forbes, 1982; Dame *et al.*, 1987).

The radius of the IR shell is about 23 pc; the measured dust colour temperature is $T(60/100) = 29 - 30$ K; for a normal gas/dust ratio of 80, we estimate its mass $M = 10^3(d/1$ $kpc)^2 M_\odot$ (Lozinskaya and Repin, 1990). Assuming it contains the gas evacuated from the interior, the initial density is $n_o = 1$ cm^{-3}. Using $L_w = (0.5 - 2)10^{38} erg\cdot s^{-1}$ as a typical value for WO1-WO2 stars, we evaluate the shell's age $t \approx (4.6 - 7.2)10^4$ yr using Dyson's model (Dyson, 1981), or longer if we take into account a small energy conversion efficiency. Here again the dynamical age seems to exceed the WO phase duration, leading to the conclusion that the short WO superwind does not participate in the shell's formation.

There is another WR star WR 143 inside this IR shell at a similar distance. Moreover, since the IR shell appears to confine the supershell around Cyg OB1 (Lozinskaya and Sitkin, 1988) we cannot exclude a physical connection. If this is the case, then there are 7-8 WR

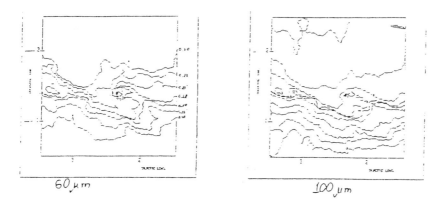

Fig.1. The IRAS 60 and 100 microns images of the region around WR 102

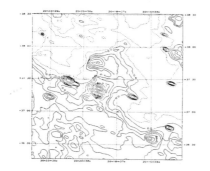

Fig. 2. The high resolution IRAS 60 microns image of a region around WR 142

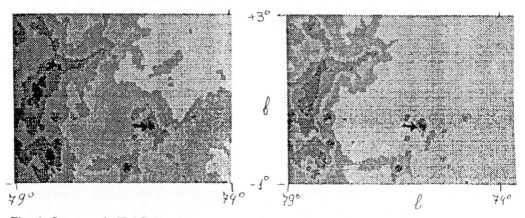

Fig. 3. Large-scale IRAS 60 microns image of a region around WR 142

and Of stars spread over this large region, supplying ionizing and mechanical energy to the common shell. The mechanical luminosity of WR 142 is even sufficient to create the shell over the WR lifetime.

The environment of the WO4+O2V binary Sand 1 in the SMC (the brightest star in NCG 602c) looks similar to WR 142. A weak thin filamentary shell DEM 167 (35′ in diameter) is easily seen on the photographs of the SMC around NCG 602c, while DEM 166 (N90), associated with NGC 602a,b, is a bright compact HII region on its southern edge (Westerlund, 1964; Davies et $al.$, 1976). The high contrast masked print made by Meaburn (1980) demonstrates the complicated structure of his supershell SMC 1 with a smaller filamentary ring (about $30 - 40'$) comprising NCG 602c and Sand 1. The IRAS imaging shows an evident ring of IR emission coinciding with the optical shell (see the IRAS $100\mu m$ image of the SMC in Loiseau et $al.$ (1987)).

Surely, Sand 1 is not the only blue star inside the supershell. However, the difference between the two HII regions - the compact one connected with NGC 602a,b, and the thin ring around NGC602c - may be considered as an indication of the influence of Sand 1. Indeed, NGC 602a,b and NGC 602c form a single association; their stellar contents (except Sand 1) and the associated gas masses: M(DEM166) = $2 \ 10^4 M_\odot$, M(shell) = $3 \ 10^4 M_\odot$ are similar (Westerlund, 1964; Kontizas and Morgan, 1988). The stellar wind of Sand 1 is also powerful enough to create the shell. The HI surface density in the region is of order $N(H) = 1.2 \ 10^{21}$ at $/cm^{-2}$ (McGee and Newton, 1982); the gas is uniformly spread in velocity (Mathewson et $al.$, 1988), implying a uniform distribution along the line of sight. If the depth of the SMC wing is of order $5 - 10$ kpc, the ambient density would be $\sim 0.04 - 0.08$ cm^{-3}. For the WR+O4 wind energy output of $L_w \simeq 2 \ 10^{38} erg \cdot s^{-1}$ ($< \dot{M} > (WR) \simeq 4 \ 10^{-5} M_\odot yr^{-1}$ (Lozinskaya and Sitkin, 1988); $v_w = 4200$ $km \cdot s^{-1}$ (Barlow and Hummer, 1982), the wind duration required to create a bubble of radius $260 - 300$ pc is $t = (2.5 - 5)10^6 yr$, comfortably short for the massive system lifetime.

Sand 2 in the LMC appears to be superimposed on the faint diffuse filamentary HII region DEM 268 of size $65'$ $40'$ (Davies et $al.$, 1976). A smaller filamentary ring comprising the WO star is easily seen of about $20'$ (~ 270 pc) extent. Several OB stars and SGs exist within the large shell DEM 268; excepting Sand 2 (Isserstedt, 1984), their combined UV emission rate and stellar winds are required to produce the shell-like nebula. Deeper photographs and kinematics are required to reveal the nature of the filamentary shell around Sand 2 more fully. The region is quite complicated; the very faint filaments near the 30 Doradus nebula between LMC3 and LMC4 outline a supershell as large as ~ 1 deg (Meaburn, 1980).

Despite the small number of WO stars and the very preliminary character of our examination we may conclude: The four WO stars are located inside optical and/or IR shell-like structures. Only the nebula G2.4+1.4 around WR102 is a classical wind-blown bubble but even here the short "superwind" of the oxygen sequence stage does not seem to be important in the shell's creation. We suppose that the progenitor star forms the ring nebula before the onset of the WO stage and the subsequent superwind produces the characteristic thin-filamentary "scalloped" structure as a result of impacting on the shell, probably accompanied by leakage of the superwind through the holes punched in the shell.

As for the stars inside the supershells, their evolutionary connections are not completely clear. On the one hand the winds of WR stars are sufficient to create the corresponding shells. In the tenous interior of a supershell around a young cluster, the short WO superwind acts as a supplementary energy source re-heating the superbubble.

On the other hand, perhaps a supershell's propagation into a dense molecular cloud will trigger supermassive star formation, a prerequisite to reaching the final oxygen-sequence stage in WR evolution.

References

Barlow, M.J., Hummer, D.G., 1982, in IAU Symp.No.99. Wolf-Rayet Stars: Observations, Physics, Evolution. Eds. C.W.H. de Loore, A.J. Willis (Dordrecht: Reidel), p. 387.

Cong, H.I.L., 1977, Ph.D. Thesis, Columbia University.

Dame, T.M., Ungerechts, H., Cohen, R.S., de Geus, E.J., Gremier, I.A., May, J., Murphy, D.C., Nyman, L.-R., Thaddeus, P., 1987, *Astrophys. J.* **322**, 706.

Davies, R.D., Elliott, K.H., Meaburn, J., 1976, *Mem. R. astr. Soc.* **81**, 89.

Dopita, M.A., Lozinskaya, T.A., McGregor, P.J., Rawlings, S.J., 1990, *Astrophys. J.* **351**, 563.

Dopita, M.A., Lozinskaya, T.A., 1990, *Astrophys. J.* **359**, 419.

Dyson, J.E., 1981, in: Investigating the Universe. Ed. F.D. Kahn (Dordrecht: Reidel), p. 125.

Graham, J.R., 1985, *Observatory* **105**, 7.

Isserstedt, J., 1984, *Astron. Astrophys.* **131**, 347.

Kontizas, E., Morgan, D.H., 1988, *Astron. Astrophys.* **201**, 208.

Loiseau, N., Klein, U., Greibe, A., Wielebinske, R., Haynes, R.F., 1987, *Astron. Astrophys.* **178**, 62.

Lozinskaya, T.A., 1982, *Astroph. Space Sci.* **87**, 313.

Lozinskaya, T.A., 1986, Supernovae and stellar winds: the interaction with the ISM, Moscow, Nauka (AIP, New York, 1991).

Lozinskaya, T.A., Repin, S.V., 1990, *Astron. Zh.* (in press).

Lozinskaya, T.A., Sitnik, T.G., 1988, *Pis'ma Astron. Zh.* **14**, 240 (= *Sov. Astron. Letters* **14**, 100).

Mathewson, D.S., Ford, V.L., Visvanathan, N., 1988, *Astrophys. J.* **333**, 617.

McGee, Newton, 1982, *Proc. Astron. Soc. Austral.* **4**, 195.

Meaburn, J., 1980, *Monthly Notices Roy. Astron. Soc.* **192**, 365.

Moffat, A.F.J., Breysacher, J., Seggewiss, W., 1985, *Astrophys. J.* **292**, 511.

Pitault, A., 1981, *Astron. Astrophys.* **97**, L5.

Torres, A.V., Conti, P.S., Massey, P., 1986, *Astrophys. J.* **300**, 379.

Turner, D.G., Forbes, D., 1982, *Publ. Astron. Soc. Pacific* **94**, 789.

Wendker, H.J., 1984, *Astron. Astrophys. Suppl.* **59**, 291.

Westerlund, B.E., 1964, *Monthly Notices Roy. Astron. Soc.* **127**, 429.

DISCUSSION

Cassinelli: Thank you very much. I have always liked that supermassive star idea... I noticed that some of your nebulae were very massive. From IRAS data you were deriving masses in the order of a 1000 $M_\odot$.

Lozinskaya: The masses of all infrared shells look like something several times $10^3 M_\odot$. You see, Dopita will provide the exact data, but today I am dealing with a large deal of speculation. I think that such a massive star may create a supershell. But on the other hand maybe it is the opposite: maybe a supershell is responsible for the creation of a supermassive star which becomes a WO star.

Tatiana Lozinskaya

G2.4+1.4: AN EXTRAORDINARY MASS-LOSS BUBBLE DRIVEN BY AN EXTREME WO STAR

M.A. DOPITA[1] and T.A. LOZINSKAYA[2]

[1]Mt. Stromlo and Siding Spring Observatories, Australian National University.
[2]Sternberg State Astronomical Institute, Moscow State Astronomical Institute.

ABSTRACT.   The nebula, G2.4+1.4, is shown to be a highly reddened, photoionised, mass-loss bubble of very high excitation powered by WR 102, the most extreme oxygen sequence Wolf-Rayet star known. It lies at a distance of 3±1kpc, and is about 11 pc in diameter. The exciting star, contains neither hydrogen nor helium in its atmosphere, is losing mass at a velocity of 5530 km.s^{-1}, and has the following properties: log (T_{ion}) = 5.20±0.05; log $(R/R_\odot)$ = 0.05±0.20; log $(L/L_\odot)$ = 5.85±0.20. We conclude that the star is the ~20M$_\odot$ core of a supermassive star (M ≤ 60M$_\odot$) seen near the end of its life.

1.   Introduction

The ultra-violet strong object discovered by Blanco *et. al.* (1968), known variously as WR 102, Sk 4 or LSS/LS 4368 is possibly the most extreme example of a Wolf-Rayet star of the Oxygen sequence known.   An extrapolation of the line ratio and line width classification criteria would imply a classification of WC 4.

The nature of the surrounding radio source, the nebula G 2.4 +1.4, has been the subject of considerable debate over the years (Goss and Shaver 1968; Johnson, 1973, 1975; Treffers and Chu, 1982; Chu *et al.*, 1983; Green and Downes, 1987; Caswell and Haynes 1987), and the interpretation of the nebula as a supernova remnant, wind blown bubble around the WO star, or some compound model, have all been considered .

In this paper, which is an abbreviated account of papers published in the *Astrophysical Journal* (Dopita *et al.* 1990; Dopita and Lozinskaya, 1990), we present spectrophotometric and kinematic data on both the star and its surrounding nebula and show that the nebula is in fact a mass-loss bubble powered by the central star. We derive the parameters of the central star and energetic estimates for the nebula.

2.   The Nature of WR 102.

The stellar classification given by Freeman *et al.* (1968): WC 4-5 pec, has subsequently been broadly accepted (see, for example, van der Hucht *et al.* 1981 or Lundstrom and Stenholm, 1979). Sanduleak (1971) listed five WR stars which show very strong OVI -emission, two of them in the Magellanic Clouds. One of the Galactic examples, Sk 3, is the central star of a very old PN (Barlow and Hummer, 1982). According to current understanding the remaining Sanduleak stars should be considered as a separate Population I WO sequence, defined by the relative strengths of O IV, O V and O VI, representing an evolutionary stage following the WC stage (Barlow and Hummer, 1982). The small number of WO stars relative to WN and WC is explained if WO stage corresponds to stars that have reached the end of core He burning and are already burning C in the core. On the basis of its excitation, Sk4 (WR 102) is the most extreme Population I WR star known.

In order to clarify the evolutionary status of WR 102, we obtained spectrophotometry in the range 0.34 - 0.82 μm and infrared spectra in the range 1.0 - 2.5 μm. These spectra are

371

K. A. van der Hucht and B. Hidayat (eds.),
Wolf-Rayet Stars and Interrelations with Other Massive Stars in Galaxies, 371–378.
© 1991 *IAU. Printed in the Netherlands.*

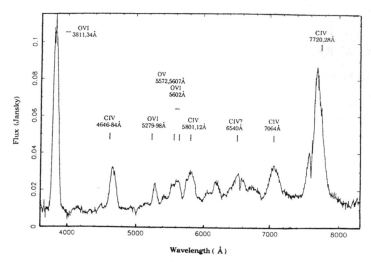

Figure 1. The optical spectrum of WR 102, showing OV, OVI, and CIV lines only.

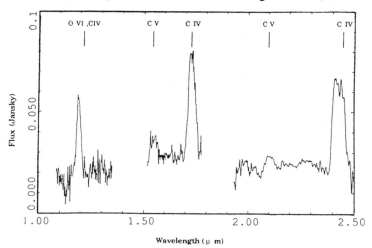

Figure 2. The IR spectrum of WR 102, dominated by CIV and C V emission lines.

shown in Fig 1 and 2. The strongest spectral features are due to resonance lines of O VI and C IV. The near-IR spectrum of WR 102, shown in Figure 2 is dominated by strong broad emission lines of C IV superimposed on a smooth continuum which is essentially flat in F_ν between 1.0 and 2.5μm. The continuum is probably due to optically-thick free-free emission (Barlow and Hummer 1982), and the principal emission lines are those of C IV (Williams 1982). Lines of C V (10-9) 1.55 μm and CIV (11-10) 2.11 μm may also be present.

There is no sign of either hydrogen or helium in this spectrum. In particular, the He I ($2s\ ^1S - 2p\ ^1P$) transition at 2.058 μm and the He I ($3p - 4s$) lines at 2.11 μm are completely absent in WR 102. Thus, we can exclude the possibility of He I lines in the spectrum. The possibility of He II emission is not excluded by the optical data, since the 0.466μm feature may be a blend of CIV with the He II line. However, the IR data show no sign of the He II features at 1.165μm, 1.27 - 1.29μm , 2.037μm , 2.164μm and at 2.31 -

2.37µm (Hillier, Jones and Hyland 1983).These data give the most uneqivocal evidence that the He - rich layers of this star have been completely stripped off, and that here we are seeing the bare He - burnt core of a massive star.

The stellar wind velocity inferred from the FWZM of the unblended or very closely separated lines yields V_w = 5530(±200) km.s^{-1}. (*c.f.* 5500 km.s^{-1} ; Barlow and Hummer 1982 and 5700 km.s^{-1} ; Torres et al. 1986). At this velocity, each solar mass lost carries 3.10^{50} ergs to the interstellar medium (ISM). Thus mass loss from the central star may have already delivered a momentum to the ISM which is greater than that of a supernova, since, for an equivalent energy input, mass loss couples better to the ISM than a point explosion.

The distance to the star is quite difficult to estimate, given the small number of such stars against which it can be compared. The two examples of WO stars in the Magellanic Clouds are both of class WO4 (Barlow and Hummer, 1982). However, the LMC example is a binary, and cannot be used in a comparison of distances. The SMC star, Sand 1, has V = 14.44, (B-V) = 0.08, which implies an absolute magnitude M_V ~ -2.4. If WR 102 has the same absolute magnitude, then with V = 14.64, A_V = 4.3, a distance of 3.5 kpc is derived. Perhaps a better way to estimate the distance is to use the other Galactic example of a WO star, WR 142 (ST 3, Stephenson (1966); Sand 5), of spectral class of WO2, more like that of WR102 (WO1). It lies at a distance of 946±26pc and has <E(B-V)> = 1.7±0.1 (Turner and Forbes, 1982). A direct comparision of the observed magnitudes and reddening V = 13.56, 5.3 < A_V < 6.2 with those of WR 102 (V = 14.64, A_V ~ 4.3) implies a distance in the range 2.5 - 3.6 kpc. On this basis we estimate of 3±1 kpc for the distance to G2.4+1.4.

3. OBSERVATIONS OF G 2.4 +1.4.

3.1. Hα IMAGING

Images of G2.4 +1.4 obtained by Johnson (1975) and Treffers and Chu (1982) show an inner highly symmetrical filamentary shell, 5 arc min. in diameter which is embedded in a larger shell-like structure extending about 8 arc min in the NE-SW direction. The exciting star WR 102 lies some distance from the center of either of these structures.

Following our discovery of extended faint emission, we performed deep Hα narrow-band imaging on the the red arm of the Double-Beam spectrograph (Rodgers *et al.* 1988a) of the

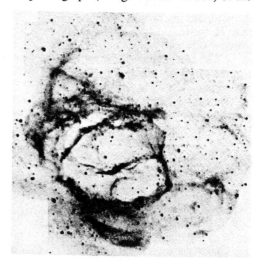

Figure 3.   A deep H-Alpha composite image of G 2.4+1.4. North is at the top, and East to the left.The darkest filaments are those known previously.
Note the scalloped outer shell extending to the North and to the West which this image reveals for the first time.

2.3m telescope at Siding Spring Observatory. The detector was a Photon Counting Array (PCA) with a GaAs image tube front stage, giving a Q.E. close to 20% (Rodgers *et al.* 1988b). The system acted as an f/1.5 camera with a scale at the focal plane of 1.0 arc sec.pixel[-1].

The image resulting from the combination of nine separate pointings is shown in Figure 3 on a logarithmic intensity scale, to bring out the fainter features. The nebula is revealed to be appreciably larger than hitherto supposed, with a complex double shell structure. The outer portions are scalloped in a semi-regular pattern about 3 arc min. in diameter, suggestive of a large-scale instability.

3.2. SPECTROPHOTOMETRY

We have obtained spectrophotometry of the nebula in the range 0.34-0.78µm using the double beam spectrograph of the 2.3m telescope. From four slit positions, the spectra of the ten brightest filaments were extracted, and the data from the brightest of these filaments were coadded to optimise the detection of faint lines. The resultant average spectrum is given in Table 1. The spectrum is like a PN of Excitation Class 7.5, as noted by Johnson (1976).

This spectrophotometry provides a strong constraint on the mode of excitation of the nebula. If shock excited, then models (e.g. Dopita et al. 1984) show that the [S II] and the [N II] lines would be comparable in strength with Hα. Furthermore, these lines should be well-correlated, both in position and intensity, with the Hα emission, since both arise in the recombination zone of the shock. This is inconsistent with our observations, and with the narrow-band imaging observations of Treffers and Chu (1982). The observation of a strong [Ar III] line makes it almost inconceivable that the nebula is shock-excited, since this line is normally emitted in a high-temperature zone in shocks, and as a consequence, the line is always very weak compared with photoionised plasmas.

From these arguments we conclude that the nebula is a radiatively-excited wind-blown bubble. We have constructed isobaric steady-state photoionisation models using the general-purpose modelling code MAPPINGS (Binette, Dopita and Tuohy 1985).

The results of a typical model are given in Table 1 (opposite). The spectrum is not particularly well fitted by the assumption of a Black-Body ionising spectrum, but one would have expected this to be the case, given the extreme nature of WR102.

The effective temperature implied by the He II / Hβ ratio is $T_{ion} > 1.5.10^5$K. The electron density is ~150 cm[-3], and an effective filling factor of 0.06 - 0.14 is estimated. The models also indicate a value for the ionisation parameter of log<Q> = 8.0 (±0.3) cm.s[-1]. The ionised mass of the nebula is therefore in the range 300-1000M$_\odot$.

In order to obtain values of the [O III] λ5007Å/Hβ ratio which approach the observed value, the gas must also have a high metallicity. We estimate a metallicity of at least three times solar.

Table 1: Photoionisation Model for G 2.4+1.4

λ (Å)	Ident.	Flux (Hβ = 100.0) Observed	Model
3728	[O II]	>55:	152
4686	He II	55	40
4861	H β	100	100
4959	[O III]	157	245
5007	[O III]	472	707
5876	He I	5:	11
6300	[O I]	11:	9
6563	H α	272	297
6584	[N II]	42	118
6678	He I	4	3
6717	[S II]	19	20
6731	[S II]	15	17
7165	[Ar III]	23	10
7318,30	[O II]	4:	1

Given the ionisation parameter, stellar temperature, and the nebular density and radius, then the effective radius of the star can be also be estimated. The parameters which we find to best define the central star are:

$$\log (T_{ion.}) = 5.20 \pm 0.05; \quad \log (R/R_\odot) = 0.05 \pm 0.20; \quad \log (L/L_\odot) = 5.85 \pm 0.20$$

Maeder (1983) has shown that WR stars conform to a rather narrow mass / luminosity strip defined by $\log (L/L_\odot) = 3.8 + 1.5 \log (M/M_\odot)$, which implies that WR102 has a mass of order 15 to $30 M_\odot$. However, the initial mass is difficult to estimate from these parameters, since Maeder's evolutionary calculations show that a wide mass range of stars may evolve to endpoints with similar parameters. However the initial mass certainly exceeded $60 M_\odot$, and the position of WR 102 on the H-R Diagram is consistent with a star on the C-O main sequence. Its extreme parameters place it firmly in the régime where the atmosphere is optically thick to electron scattering, which generally results in an effective temperature, T_{eff}, much lower than the ionisation temperatute, T_{ion} (Abbott and Conti, 1987).

3.3. INTERNAL DYNAMICS

We sampled the velocity profiles at some 55 points at a resolution of 8.8 km.s^{-1} using the échelle spectrograph at the Coudé focus of the 1.8m telescope at Mt. Stromlo. Across the whole nebula we find emission at or near $V_{Hel} = 13 \pm 5$ km.s^{-1} ($V_{LSR} = 23 \pm 5$ km.s^{-1}). This feature is broad, and is brighter than the [O III] profile from the nearby sky. Furthermore, the line profiles show substantial variation across the nebula, so this component is certainly a part of the nebula. This feature is narrowest in the diffuse region on the south side of the nebula. We also find a line component with high negative velocity, between -38 km.s^{-1} ≤ V_{Hel} ≤ 5 km.s^{-1} (-28 km.s^{-1} ≤ V_{LSR} ≤ 15 km.s^{-1}) (c.f. Johnson 1975,1976; Treffers and Chu 1982 a). Largest negative velocities are found in the vicinity of WR 102. This further strengthens the case that the central star WR 102 is both the exciting source for the nebula, and is also the source of the kinetic energy of G2.4+1.4.

A simple velocity ellipsoid will not work in the case of G2.4+1.4 because rapid motions characterise only one side of the shell, and because this shell is non-spherical, as seen in projection. We therefore adopted a simple model of a single hemispherical shell with different radii of curvatures in its northwest and its southeast portions, shown in Fig. 4.

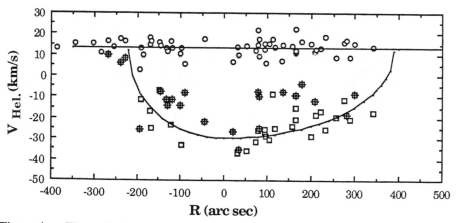

Figure 4.   The velocity ellipsoid for G2.4+1.4. Circles are measured velocities on the farside of the shell which does not show appreciable expansion. The squares are for points on the nearside, and squares with crosses distinguish the brighter filaments. The line is a model of asymmetrical hemispherical expansion.

The velocity of expansion of the nearside of the shell is 42 km.s^{-1}, whereas the far side of the shell shows no systematic expansion. The brighter filaments are characterised by a systematically lower velocity of expansion, of order 20 - 30 km.s^{-1}, a result characteristic of bright filaments in energy-driven bubbles (as well as in old SNRs; Lozinskaya, 1980; Lozinskaya et al. 1988). This is what is expected if the bright filaments result from a cloudy shell which is swept up and accelerated by the wind.

For the case of a bubble evolving into a constant density medium, the relationship beween the radius of the bubble, r, its velocity of expansion ,V_{exp}, the density of the ambient medium, n, and the energy input rate, $\partial E/\partial t$ is given by;

$$r = 1.7(\partial E/\partial t_{36} / n)^{1/5} t_4^{3/5} \text{ pc.}$$
$$V_{exp} = 100 (\partial E/\partial t_{36} / n)^{1/5} t_4^{-2/5} \text{ km.s}^{-1}$$

However, in the case of G 2.4 +1.4, the uniform density assumption is clearly wrong. For an average filament density of 30 - 60 cm^{-3}, and a mean temperature of 8000 K, derived from the photoionisation analysis, we estimate that the pre-shock density on the low density side of the expanding bubble is 2 - 4 cm^{-3}, and on the dense side, about 30 - 60 cm^{-3}. In a medium with a strong density gradient, the shock moving into the denser medium is slowed, and stalls when its velocity drops below the sound speed in the pre-shock medium. Assuming that the bubble expanded in the dense medium to a radius of ~3 pc (about 200 arc sec. on the sky), before breaking out into the lower density medium, and applying the above equations, we estimate an age at breakout of $(1.8\pm0.3) \times 10^5$ years. A mechanical luminosity of $\log(L/L_\odot) = 2.0\pm1.3$ was sufficient to power the bubble. On the other hand, the central star delivers $\log(L/L_\odot) \sim 4.7\pm 0.4$; far greater than what is needed.

Chu (1982) and Treffers and Chu (1982b) derived the kinetic efficiency of a stellar wind bubble , ε, to be the ratio of the kinetic energy in the expanding bubble to the kinetic energy delivered by the central star over its mass losing lifetime. For five wind bubbles associated with W-R stars they found ε is about 0.01. Athough we have derived the ratio of the instantaneous production of mechanical energy to the mean energy production rate, it is clear that our efficiency parameter is also required to be about 0.01, or even less. Van Buren (1986) suggested a mechanism whereby the efficiency can be reduced in the case of evolution in a clumpy medium. Here the clumps inside the bubble are rapidly evaporated, lowering the internal temperaterature of the bubble, and allowing cooling to become important. In this the expansion will become momentum-conserving.If this were the case, then we find an age of $\log(t /yr) = 5.1\pm 0.6$; in good agreement with our estimates of the dynamical age. We therefore conclude that the momentum conserving solution probably applies.

Finally, the characteristic scallops in the outer parts of the bubble to the west, are strongly suggestive of an instability with some characteristic scale length in the ionised shell. The most likely explanation is that here we are seeing a developed Rayleigh-Taylor (R-T) instability. This instability was probably produced at the time of shock breakout, in the layers with a strong density gradient, where continuing acceleration of the nebular shell will occur. This has clearly occurred in G 2.4 +1.4, as we see a dynamical distinction between the bright filaments and the more diffuse material. The bright filaments both lie within the outer shell, and are moving more slowly, both of which would result from the R-T instability in the compressed shell. We estimate the timescale for the development of a RT instability is of order $(\Delta R/g)^{1/2}$, where ΔR is the thickness of the filaments, approximately 5×10^{17} cm, and g is the effective gravity. From both the dynamical data, and the morphology of the nebula, we derive a current age of 2.7×10^5 years for the whole structure (9×10^4 years since breakout). In order to produce the observed scalloping, this requires an acceleration of ~10^{-6} cm.s^{-2} in the ionised layer, which in turn implies a growth time for the RT instability of order 2×10^4 years; comfortably shorter than the evolution time.

4. Conclusions

We conclude that all the optically observed properties of the filamentary nebula G2.4+1.4 are consistent with the effects of violent mass-loss from the extreme WO star, WR102, near its centre. This star is shown to be a stripped C-O core of a Population I star with an initial mass of at least $60M_\odot$. We may also conclude that the surrounding nebula, G2.4+1.4, is a photoionised mass-loss bubble about 10^5 years old, driving into a medium with a strong intrinsic density gradient. This has encouraged the development of Rayleigh-Taylor instabilities which result in the characteristic morphology.

Acknowledgements:

Dr. Lozinskaya acknowledges a grant under the Australian Dept. of Technology, Industry and Commerce under the Australia-USSR Bilateral Science and Technology Agreement, and the receipt of an Australian National University Visiting Fellowship. Without these, this work would not have been possible.

References

Abbott, D.C., and Conti, P.S. 1987, *Ann. Rev. Ast. Astrophys.*, **25**, 113.
Barlow, M.J., and Hummer, D.G. 1982, in *IAU Symp. #99, "Wolf-Rayet Stars:*
    Observations, Physics, Evolution", Eds. C.W.H. de Loore and A.J.Willis, p387.
Binette, L., Dopita, M.A., and Tuohy, I.R. 1985, *Astrophys. J.*, **297**, 476.
Blanco, V., Kunkel, W., and Hiltner ,W.A. 1968a, *Astrophys. J. Lett.* , **152**, L 137.
Caswell, J.L., and Haynes, R.F., 1987, *Ast. Astrophys.*, **171**, 261.
Chu, Y.-H., Treffers, R.R., and Kwitter, K.B., 1983, *Astrophys. J. Suppl. Ser.*, **53**, 937.
Dopita, M.A., Binette, L., D'Odorico, S., and Benvenuti, P. 1984 *Astrophys. J.*,**276**, 653.
Dopita, M.A., Lozinskaya, T.A., McGregor, P.J., and Rawlings, S.J.1990,
    Astrophys. J. , **351**, 563.
Dopita, M.A., and Lozinskaya, T.A. 1990b, *Astrophys. J.* , **359**, 419.
Freeman K.C., Rodgers A.W., and Lynga G., 1968, *Nature*, **219**, 251.
Goss, W.M., and Shaver, P.A., 1968, *Astrophys. J. Lett.*, **154**, L75.
Green, D.A., and Downes, A.J.B., 1987, *M.N.R.A.S.* , **225**, 221.
Hillier, D.J., Jones, T.J., and Hyland, A.R. 1983, *Astrophys. J.*, **271**, 221.
Johnson, H.M. 1973, *Mém. Soc. Roy. Liège, Ser #6*, **5**, 121.
Johnson, H.M. 1975, *Astrophys. J.* , **198**, 111.
Johnson, H.M. 1976, *Astrophys. J.*, **206**, 243.
Lundstrom I., and Stenholm, B.,1979, *Ast. Astrophys. Suppl.*, **35**, 303.
Lozinskaya, T.A. 1980, *Pisma Astron. Zh.*, **6**, 350.
Lozinskaya, T.A., Lomovskij, A.I., Provdikova, B.B., and Surdin, B.G.
    1988 *Pisma Astron. Zh.*, **14**, 909.
Maeder, A. 1983, *Ast. Astrophys.*, **120**, 113.
Rodgers, A.W., Conroy, P., and Bloxham, G. 1988a, *P.A.S.P.*, **100**, 626.
Rodgers, A.W., van Harmelan, J., King, D., Conroy, P., and Harding, P. 1988b,
    P.A.S.P., **100**, 841.
Torres, A.V., Conti P.S., and Massey, P., 1986, *Astrophys. J.*, **300**, 379.
Treffers, R.R., and Chu, Y.-H. 1982a, *Astrophys. J.*, **254**, 132.
    ——————————————— 1982b, *Astrophys. J.*, **254**, 569.
Turner, T.E., Forbes, D. 1982, *P.A.S.P.*, **94**, 789.
Van Buren, D. 1986, *Ap. J.*, **306**, 538.
van der Hucht, K.A., Conti, P.S., Lundstrom, I., and Stenholm B., 1981,
    Space Sci. Rev., **3**, 227.
Williams, P.M., 1982, in *IAU Symp. #99, "Wolf-Rayet Stars: Observations, Physics,*
    Evolution", Eds. C.W.H. de Loore and A.J.Willis, p 73.

DISCUSSION

Niemela: Can you estimate the local standard of rest central velocity of the nebula?
Dopita: Yes, $v_{LSR} = +23km \cdot s^{-1}$. Because of its position, almost towards the galactic centre, this is not very useful for getting an estimate of the distance.

Langer: If you conclude the central star is in a very late evolutionary stage, you should not apply the M-L relation for WR stars to derive its mass, since it is only valid for core He-burning stars. You would over-estimate the mass by that.
Dopita: Agreed, the star is on the "CO" main-sequence rather than the "He" sequence, and so the mass may be less than $20M_\odot$, possibly as low as the 6-$10M_\odot$ as you would like to have.

Vilchez: (1) How does your T_{eff} determination compare with other methods, *i.e.*, is it an "ionization temperature"? (2) How is the systematics of the multiple components in velocities over the nebula?
Dopita: (1) Yes it is an ionization temperature. It is higher than the "Zanstra" temperature (Heydari-Malayeri, this symposium). However, this problem is analogous to the Zanstra discrepancy in PN, and the ionization temperature is probably more reliable. (2) I think this is explained in the text of the paper.

Martin Cohen, Michael Dopita, Tatiana Lozinskaya

ABUNDANCES IN WR NEBULAE

J.M. VILCHEZ and C. ESTEBAN
Instituto de Astrofisica de Canarias
38200-La Laguna, Tenerife, Spain

ABSTRACT. Within an ongoing program of long slit spectroscopy of the known WR nebulae, we show here some results for eight objects of the sample selected from the master list of Chu et al. (1983). We present an analysis of their ionisation structure and determine abundances of oxygen, nitrogen and helium in as many positions as possible. The implications of the abundance results for the chemical enrichment of the Interstellar Medium and stellar nucleosynthesis are briefly reviewed.

1. Introduction and Observations

Of and WR stars are loosing mass at very high rates, exhibiting strong stellar winds which interact with the Interstellar Medium (ISM). According to stellar evolution models, WR stars should be the chemically evolved descendents of massive stars and therefore their stellar winds are expected to be overabundant in helium and heavy elements (e.g. Maeder 1990). Abundances in nebulae sourrounding WR stars can show direct evidence of the chemical processing operating along the evolution of the central star; in particular, WR Ring nebulae offer a unique opportunity to study the chemical and dynamical interaction with the ISM prior to the SN phase.

Most of the abundance studies available for these objects are based on observations of the bright parts of the nebulae or on average spectra without spatial resolution, on the assumption that they are representative of the nebula as a whole (Kwitter 1984; Rosa 1987). However, recent findings suggest that some of these nebulae could be inhomogeneous (Smith et al. 1988; Goudis et al. 1988). Spatially resolved spectroscopy allows sampling zones of the nebulae of differing excitation and (possibly) physical and chemical properties (Rosa and Mathis 1990; Esteban et al. 1990a). Within an ongoing program of long slit spectroscopy of the known WR nebulae, we present here some results for eight objects of the sample selected from the master list of Chu et al. (1983). The observations were intended to fulfil the requirements of wide spectral coverage, spatial resolution and high signal to noise ratios. On the other hand we have also considered necessary for some objects, to combine the preceeding requirements with high spectral resolution in most of the optical range. This method has allowed to isolate different kinematical and/or chemical components, that could be interpreted in terms of central star evolution.

379

K. A. van der Hucht and B. Hidayat (eds.),
Wolf-Rayet Stars and Interrelations with Other Massive Stars in Galaxies, 379–384.
© 1991 *IAU. Printed in the Netherlands.*

The observations were performed using the 2.5m INT and 4.2m WHT at the Observatorio del Roque de los Muchachos (La Palma), and the ESO 3.6m Telescope at La Silla, during different epochs from 1986 to 1990. Intermediate resolution spectra have typical spectral resolution of 3.7 Å and spatial sampling of 1.5 arcsecs. High resolution observations have typical spectral resolution equivalent to 40 km/s FWHM.

2. Results and Discussion

2.1 Ionisation Structure

The study of the ionisation structure in these objects provides a means of checking the role of photoionisation as the fundamental excitation mechanism, at least for the observed areas within the main body of the nebulae. We claim this test has to be done before any attempt to derive the physical conditions, since the reported evidence of shocked-like characteristics among peripheral zones in some objects (Dufour 1989). In addition, it gives us direct information about chemical peculiarities after comparing with the average normal HII region spectra. For our sample of objects we have used the line ratios diagrams : Log [NII]/Hα vs. Log [SII]/Hα , Log [SII] $\lambda\lambda$ 6717/6731 vs. Log Hα/[SII] (Sabbadin et al. 1977) and the series of HII region diagrams from the work of Dopita and Evans (1986). Figure 1 shows two representative diagrams combining [OII],[OIII], and [NII]/Hα; all the observed points match the locus for normal HII regions in the diagram involving oxygen lines, but this is not the case in the other diagram, where the strength of the observed [NII] lines in some objects deviates clearly from the normal HII region sequence, despite their normal [SII]/Hα ratios. This fact is an indication of possible nitrogen overabundances. Overall, the nebulae observed present spectra which are entirely consistent with photoionisation, some of them possibly being nitrogen-rich. Different sets of points located along the sequence in excitation, measured by the [OII]/[OIII] and [OIII]/Hβ ratios, suggest a range in effective temperatures for the central stars. Effective temperatures for the program nebulae have been ranked according to the diagram in Rosa and Mathis (1990) in which, for some objects in common, we find consistent results.

2.2 Chemical Abundances

Ionic and total abundances have been determined for the program nebulae, in as many positions as possible. Direct measurements of the electron temperature have been used in most nebulae though the fainter ones were analysed using the empirical method. Abundances of oxygen, nitrogen and helium mass fraction, Y, are presented in Table 1 for the observed sample and reference objects. For each nebula, spectral type of the central star and estimated galactocentric distance are also quoted. For NGC 2359, S 308, G 2.4+1.4 and NGC 6888 we present average abundances for different zones of the nebulae whose values suggest a non-homogeneous nature. In the case of NGC 6888, the abundances have been taken from the chemo-dynamical study of Esteban and Vilchez (1990), which separates in velocity the spectra of the nebular shell from the ambient component in front of it. Most of the nebulae show Y and N/O values clearly higher than the ones corresponding to their local ISM, being also striking the oxygen-poor nature of a large fraction of them. This result is apparent in Figure 2 where we present Δ Log N/O vs Δ Log O/H (defining Δ Log = Log$_{Nebula}$ − Log$_{ISM}$, i.e. differential abundance with respect to the ISM according to the galactic abundance gradient), for our sample and for some objects from Rosa and Mathis (1990). In this Figure objects with nearly ISM oxygen abundances (-0.2 $\leq$ Δ Log O/H $\leq$

Figure 1.- [OIII]/Hβ vs. [OII]/[OIII] (left) and [OIII]/Hβ vs. [NII]/Hα (right) for the WR nebulae sample. Dotted areas correspond to the HII region sequence after Dopita and Evans (1986).

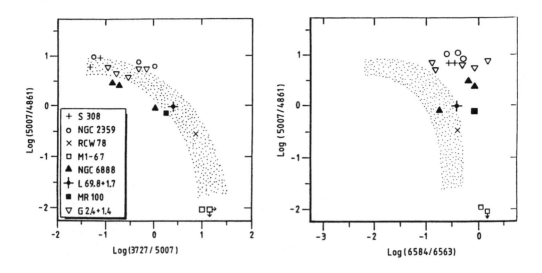

Table 1 .- Average abundances for the observed sample and reference objects

Nebula	Sp. Type	R_G(kpc)[1]	Region	12+LogO/H	LogN/O	Y
NGC 2359	WN5	15	mean	8.24	−0.95	0.30
			"N streamers"	8.19	−0.89	0.33
RCW 78	WN8	8.5		8.92	−0.85	≥0.31
M1–67	WN8	9		8.10	+0.60	≈0.47
NGC 6888	WN6	10	shell	8.15	+0.30	0.43
			ambient	8.75	−0.77	0.29
S 308	WN5	11	A	8.07	+0.24	0.50
			B	8.13	−0.17	0.50
MR 100	WN6	10		8.83	−0.59	0.31
L69.8+1.7	WN8	10.5		8.78	−0.88	≥0.28
G2.4+1.4	WO	7	mean	8.45	−0.60	0.47
			N filament	8.46	−0.25	-
−		15	< HII >[2]	8.35	−1.23	0.26[3]
−		10	< HII >[2]	8.70	−1.12	-
−		8.5	< HII >[2]	8.81	−1.09	0.295[4]
−		7	< HII >[2]	8.90	−1.00	-
ORION		10		8.60[5]	−1.00[5]	0.28[3]
LMC				8.32[5]	−1.44[5]	0.25[3]

[1] Assuming the Sun at R_G = 10 kpc.
[2] Galactic abundance gradient from Shaver et al. (1883).
[3] Peimbert (1986).
[4] Torres–Peimbert et al. (1989).
[5] Rosa and Mathis (1987) and references therein.

0.2) present moderate Δ Log N/O values (within a factor 3 the N/O $_{ISM}$) and low Y fractions (Y $\leq$ 0.34), with a single excursion to Δ Log N/O $\approx$ 0.8 dex for RCW 58 which is a helium rich object (Rosa and Mathis, 1990). On the other hand for oxygen poor objects, the increase in N/O correlates well with oxygen defficiency with a maximun around Δ Log N/O $\approx$ 1.7 for Δ Log O/H $\approx$ -0.7 , all these objects having Y $\geq$ 0.35 . This result implies an overabundance of N/H of at least 0.8 dex for the most oxygen poor nebulae. Within this group are well known nebulae like NGC 6888 and M 1-67 ; for NGC 6888, shell (S) and ambient (A) points are clearly separated in Figure 2; for M 1-67 the almost pure (80%) ejecta nature of the nebula seems well established (Esteban et al. , 1990b). The N/H and O/H shell abundances for these objects behave so that $[X_O + X_N]_{ISM} \approx [X_O + X_N]_{Shell}$ within a factor of two. This behaviour should be investigated for the remaining oxygen poor shells, provided that higher spectral resolution observations are available.

2.3 Discussion

The most straightforward result of this work is that processed material, coming from the central star, is certainly present in many nebulae around WR stars. Spectroscopy of WR nebulae in the optical and far-red has been used to derive abundances that, in particular for oxygen, nitrogen and helium, can show variations with respect to the ISM abundance pattern. Evolutionary models for massive stars (Maeder, 1990) predict evolution of surface abundances from the Zero Age Main Sequence to the end of the central carbon-burning phase. Our results qualitatively agree with the overabundances of nitrogen and helium expected from the models, although a more quantitative comparison needs to account for at least the following aspects:

i) Nebular abundances should, in principle, reflect the history of the nucleosynthesis along the evolution of the central star, rather than the predictions of any particular model, therefore only the integrals of models' *yields* should be compared.

ii) Nebular spectra can be "contaminated" by the ISM; this contamination could act in two ways, a) via real mixing of ejected material and ISM, and/or b) apparent mixture of the nebular spectrum and the spectra of ionised gas in the line of sight (or close) to the nebula, critical for low/intermediate resolution spectroscopy. Both effects, each in a different way, would produce empirical *yields* which are underestimated.

Bearing in mind these warnings, we have brought together in Figure 3 model predictions and observations, with the aim of just making a crude comparison. In this Figure, the ratio of N to O mass fractions, X_N/X_O, is represented vs. the helium mass fraction Y. Predictions from Maeder's (1990) models of 15, 25, 40, and 85 M$\odot$ stars of Z=Z$\odot$ are shown, up to the WN phase, by line-connected symbols; nebular measurements from our work and that by Rosa and Mathis (1990) are also shown. It is apparent from the Figure that objects with Y $\leq$ 0.34 converge, like model predictions, to solar abundances with a spread lower than a factor 3 in X_N/X_O, and therefore they can be interpreted as almost non-mixed wind blown bubbles or alternatively as extremely diluted ejected material; both interpretations mean in practice that their gas should be undistinguishible from pure wind blown ISM. For helium rich objects two kinds of behaviour are present; on the one hand the nebulae (and N-enhanced zones of nebulae) presenting high X_N/X_O ratios (say $\geq$ 1) :

M 1-67, NGC 6888 (Shell) and S 308 A , which can be understood as moderately diluted pre-WN ejected material. On the other hand, those objects showing moderate X_N/X_O values (between 0.2 and 1.) : RCW 104, RCW 58, G2.4+1.4 and S 308 B, which cannot be reproduced by reasonable dilution of ejected material; substantial contamination from ionised ISM in their line of sight could underestimate N/O but this effect can only be operating in crowded environments (perhaps in the direction of G2.4+1.4). For these last objects, a more promising hypothesis can be that of a spatial admixture of N-rich filaments, moving at a different velocity, as has already been suggested for RCW 58 (Smith et al., 1988) and RCW 104 (Goudis et al., 1988).

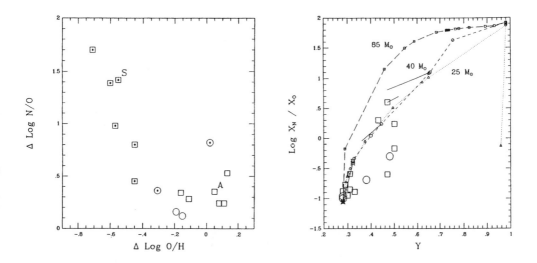

Figure 2 (left).- Differential N/O vs. O/H abundances for the sample nebulae; squares: this work, circles: Rosa and Mathis (1990). Symbols of helium rich objects ($Y \geq 0.34$) are marked by a black dot.

Figure 3 (right).- The ratio of nitrogen to oxygen mass fractions vs. the helium mass fraction. Lines represent model predictions (Maeder, 1990); symbols as in Figure 2. The effect of a 50% dilution with the ISM is showm for the 40 M⊙ model by a continuous line.

References

Chu,Y.-H., Teffers, R.R., Kwitter, K.B. 1983, Ap. J. Suppl., 53, 937.
Dufour, R.J. 1989, Rev. Mexicana Astron. Astrof., 18, 87.
Dopita, M.A., Evans, I.N. 1986, Ap. J., 307, 431.
Esteban, C., Vílchez, J.M. 1990, these proceedings.
Esteban, C., Vílchez, J.M., Manchado, A., Edmunds, M.G. 1990a, Astr. Ap., 227, 515.
Esteban, C., Vílchez, J.M., Smith, L.J., Manchado, A. 1990b, in preparation.
Goudis, C.D., Meaburn, J., Whitehead, M.J. 1988, Astr. Ap., 191, 341.
Kwitter, K.B. 1984, Ap. J., 287, 840.
Maeder, A. 1990, Astr. Ap. Suppl., 84, 139.
Rosa, M.R. 1987, in *Circunstellar Matter*, ed. I. Appenzeller and C. Jordan, Reidel, Dordrecht, p. 457.
Rosa, M.R., Mathis, J.S. 1990, in *Properties of Hot Luminous Stars: Boulder–Münich Workshop*, ed. C. D. Garmany, Pub. A. S. P. Conf. Ser. 7, 135.
Sabbadin, F., Minello, S., Bianchini, A. 1977, Astr. Ap., 60, 147.
Smith, L.J., Pettini, M., Dyson, J.E., Hartquist, T.W. 1988, M. N. R. A. S., 234, 625.

DISCUSSION

Schmutz: (1) What energy distribution did you adopt for the ionizing radiation field? (2) And how sensitive are your abundance determinations to this assumption, especially the helium abundance?

Vilchez: Photoionization model predictions are relatively sensitive to the precise input atmospheres but, more strongly, to the abundances adopted for the gas and to atomic parameters - nominally, in the case of the S^+/S^{++} ratio, to the ionization cross sections taken by the "model makers"-. Our derived abundances are based on observational electron temperature of the gas, so normally we do not apply model nebulae. I should point out however that, in the case of helium, for nebulae with cool stars the total helium abundance could be underestimated - this is not the case of hot ionizing stars -. This problem is general to OB ionizing stars also! (*E.g.*, Vilchez & Pagel, 1988; Vilchez, 1989). It is for cool-ionizing star nebulae that models could help best.

van der Hucht: You suggested that the observed abundances of WR ring nebulae could be used as input for WR atmosphere models. But, do you not think that the abundances in WR ring nebulae reflect pre-WR phases (RSG or LBV)?

Vilchez: I suggest that selfconsistent photoionization models for WR nebulae should adopt gas abundances accordingly with the observations. Whenever possible, input star models have to be chosen to correspond to the metallicity of the zone of galaxy where the nebula is located (*i.e.*, $Z_\odot, 0.5Z_\odot$, etc.). Of course, I agree that the abundances measured could reflect pre-WR phases, even those often result from a mixture of pollution by the star and ISM.

THE DYNAMICS OF THE AG CAR RING NEBULA

LINDA J. SMITH
Department of Physics and Astronomy
University College London
Gower Street, London WC1E 6BT
England

ABSTRACT. High spatial (2″) and spectral (7 km s^{-1}) resolution observations of the ring nebula surrounding the LBV star AG Carinae are presented. The data, covering Hα and [NII], were obtained with the UCL echelle spectrograph on the AAT and cover 4 radial slit positions centred on AG Car. The nebular motions are clearly resolved and show that it is a single shell expanding at 70 km s^{-1} which was ejected from the central star some 7 000 years ago. The data also reveal the presence of a high velocity ($\sim$ 83 km s^{-1}) bipolar mass outflow which has distorted the north-eastern back edge of the shell.

1. Introduction and Observations

AG Carinae is a member of a small group of stars classified as Luminous Blue Variables (LBV's); other members include η Carinae, P Cygni and S Doradus. LBV's are evolved, unstable supergiants which are close to the observed upper luminosity/stability limit in the HR diagram. They are believed to represent a short, unstable phase in the evolution of a massive O star to a WR star. Thackeray (1950) discovered that AG Car is surrounded by a roughly elliptical-shaped (40″ × 30″) ring nebula containing two prominent bright clumps to the NE and SW. More recently, circumstellar shells surrounding other LBV's have been discovered (Walborn 1982; Stahl 1989).

The AG Car ring nebula was first observed spectroscopically by Johnson (1976) who found it to be of low excitation because of the lack of [O III] emission. The dynamics were investigated by Thackeray (1977) who obtained photographic spectra of the Hα, [N II] and [S II] lines for 9 nebular positions. He detected line splitting ($\Delta v \approx 120$ km s^{-1}) at only one slit position although the data overall indicated that complex motions were probably present in the shell. He derived a rough nebular expansion velocity of 50 km s^{-1}. IUE observations by Viotti *et al.* (1988) revealed that the UV nebular spectrum was very similar to the stellar spectrum, thus indicating the presence of circumstellar dust. Far-infrared observations by McGregor *et al.* (1988) in a NE–SW direction showed that the dust is cool and is resolved into two peaks which coincide with the bright optical clumps. Paresce & Nota (1989) imaged the nebula through broad-band optical filters and discovered a prominent bipolar dust structure in the NE–SW direction. In the NE direction, a bright, detached clump of dust is observed whereas in the SW direction, a jet-like feature is seen with a curious helical structure, extending from the star to the edge of the ring nebula. Finally, Mitra & Dufour

385

K. A. van der Hucht and B. Hidayat (eds.),
Wolf-Rayet Stars and Interrelations with Other Massive Stars in Galaxies, 385–390.
© 1991 *IAU. Printed in the Netherlands.*

(1990) present a chemical analysis of the ring nebula; they find that it is oxygen-poor but shows little, if any evidence for a nitrogen enhancement.

The peculiar abundances and the LBV status of AG Car indicate that the ring nebula is composed (at least in part) of stellar ejecta. It is not clear whether the material was ejected as a single shell during a particularly violent mass-loss episode or whether the shell has been formed by the combined action of the stellar wind and multiple, small ejections. For either possibility, it is likely that the strong stellar wind of the progenitor O star evacuated the surrounding interstellar medium. Humphreys *et al.* (1989) have recently revised the distance to AG Car from 2.5 kpc to 6 ± 1 kpc. This has important consequences for the formation and evolution of the ring nebula since the stellar luminosity is now too high for AG Car to have passed through a red supergiant phase before becoming a LBV. Thus the ring nebula cannot have been formed through the present LBV stellar wind sweeping up material expelled by the red supergiant. The new distance leads to a size of $\sim 1.2 \times 0.9$ pc and a mass of ~ 2 $M_{\odot}$ (re-scaling the Hα flux given by Stahl (1987)). These revised values are rather interesting since they are now broadly in line with those expected if LBV ring nebulae are the progenitors of some of the ejecta-type WR ring nebulae.

The best means of understanding the origin of the AG Car ring nebula appears to be through a detailed study of the dynamics. For this purpose, high spatial and spectral resolution observations of the ring nebula were obtained with the UCL echelle spectrograph and a GEC CCD detector at the coudé focus of the 3.9 m Anglo-Australian Telescope (AAT) during 27 April 1989. A slit of dimensions $1''$ by $57''$ was used with an interference filter to isolate a single order covering Hα and [NII]. Exposure times were between 1000 and 2000 s, sufficient to obtain several thousand counts in the brightest portions of the nebula. The data were bias-subtracted, flat-fielded and wavelength calibrated to an accuracy of 0.3 km s^{-1}. They have not been absolutely flux calibrated or sky-subtracted (the observations were taken during dark of moon). The spectral resolution of the data is 7 km s^{-1} (as determined from the FWHM of the arc lines) and each spatial increment represents 1.1 arc sec on the sky (the seeing FWHM was $1.7''$). Hα and [N II] spectra were recorded at four radial slit positions centred on AG Car at position angles $0°$, $30°$, $80°$ and $131°$. Figure 1 shows the exact slit locations superimposed on a [N II] image of the nebula from Stahl (1987) and the corresponding [NII] two-dimensional spectra.

In order to study the nebular kinematics, the [N II] profiles corresponding to each spatial increment were extracted and velocities (relative to the Local Standard of Rest) have been measured for each identifiable nebular component. Velocity maps were then constructed for each slit position and these are shown in Figure 2.

2. Results and Discussion

The spectra displayed in Figure 1, together with the velocity maps given in Figure 2, show obvious nebular line splitting. Overall, they broadly show that the ring nebula is a complete hollow, expanding shell of non-uniform brightness. (The Hα spectra are essentially identical and will not be discussed here because their velocity resolution is lower due to larger thermal broadening.) The most remarkable features observed in the [N II] spectra are two high velocity structures indicative of a bipolar mass outflow. These features are most striking for P.A.$= 0°$ where two oppositely opposed high velocity streams of equal brightness are seen extending from near the star to beyond the edge of the shell. The positive velocity feature is $\sim 19''$ long with $V_{LSR} = +90 \pm 6$ km s^{-1} and the negative velocity counterpart is of similar length with $V_{LSR} = -77 \pm 2$ km s^{-1} (the lower velocity means that it is blended with the approaching side of the shell near the star so is less easily recognisable). The positive velocity counterpart is also seen at P.A.= 30° with $V_{LSR} = +84 \pm 6$ km s^{-1} and a spatial extent of $\sim 14''$. Both high velocity features are also just visible in the data for P.A.= 131° where two faint structures 4″ long can be seen close to the star with slightly higher velocities of -102 ± 2 and $+99 \pm 4$ km s^{-1}.

The angular extent of these features over the nebula seems to be quite small since they are only both fully seen at P.A.= 0° and 180°. The fact that the positive velocity counterpart is only seen at P.A.= 30°, together with the fact that for P.A.= 131°, 311° the two features have a very small spatial extent, leads to an approximate upper limit to the angular size of 30°. There appears to be no obvious corresponding features on the [NII] image in Figure 1. Whether they correspond to the bipolar dust features seen by Paresce & Nota (1989) is less clear. The slit position for P.A.= 30°, 210° crosses the minor axis of the nebula, passing through the detached dust clump (P.A.= 23° − 31°; 9.5″ from the star) and grazing the dust jet (P.A.= 210° − 230°, ending at 15″ from the star). From this, the positive velocity feature extending from 0° − 30° encompasses the dust clump but the negative velocity feature is absent at the position angle of the jet. In addition, the high velocity features cover a much greater spatial extent as they extend beyond the shell.

If we assume that the high velocity features detected at P.A.= 0°, 180° are moving away from the star at the same velocity, the mean velocities given above lead to a projected expansion velocity of 83 km s^{-1} centred on a systemic velocity of +7 km s^{-1}, in remarkable agreement with the stellar radial velocity of +7 km s^{-1} (Wolf & Stahl 1982). We conclude that the high velocity features can be interpreted in terms of a jet-like bipolar mass outflow with a projected velocity of 83 km s^{-1} which is not clearly related (at least at this epoch) to the bipolar dust features of Paresce & Nota (1989).

Turning to the dynamics of the shell itself, the data in Figures 1 and 2 indicate that it is basically expanding away from the star with some distortions from spherical symmetry. These distortions are most evident for the slit at a position angle of 30°. The brightest parts, defining the shell, appear to be symmetrical with respect to the central star, both spatially and kinematically. The north-eastern quadrant, where the material is moving away on the far side of the nebula, is much fainter with higher velocities and a greater spatial extent. This part of the slit corresponds to the faint material which is visible beyond the brightest portion of the ring in the [NII] image shown in Figure 1. Both the dynamical and image information therefore suggest that at this position angle, the shell is expanding more rapidly away from the star. This spatial and velocity discontinuity is also seen, to a lesser extent, at P.A.= 0°. The high positive velocity counterpart of the bipolar mass outflow (discussed above) is also observed at these two position angles, as well as the dust clump at P.A.= 30° (Paresce & Nota 1989). This coincidence suggests that the observed distortion of the shell can be explained if the higher velocity outflow has pushed against the shell and accelerated it.

Images of the ring nebula (*cf.* Figure 1) show that it is slightly elliptical in shape. The shell should therefore to be an ellipsoid, orientated at some angle to the line of sight, rather than a simple sphere. Some ellipticity does appear to be present since a velocity tilt is observed in the data for P.A.= 0° with the upper edge of the shell red-shifted compared to the lower edge. The slit at P.A.= 131° crosses the major axis of the nebula on the sky and shows no obvious velocity tilts. The slit position for P.A.= 80° presents the simplest nebular geometry overall since the back and front of the expanding shell appear to be roughly spherically symmetric. Taken as a whole, the data for the four slit positions

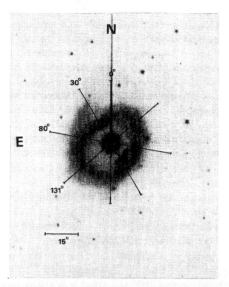

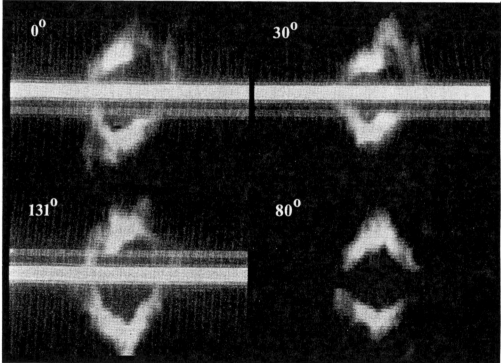

Figure 1: Two-dimensional [N II] spectra covering the four slit positions shown in the [N II] image of the AG Car ring nebula (reproduced with permission from Stahl (1987)). The intensity scale for the spectra is logarithmic; a stellar occulting mask was used for P.A. 80°.

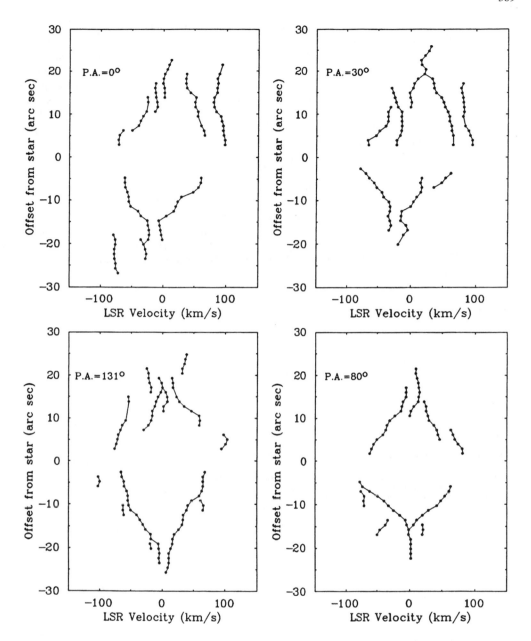

Figure 2: Measured LSR velocities of identified features in the four [N II] images plotted against the offset in arc sec from AG Car. Recognised continuous features in velocity space are depicted by solid lines.

show broadly similar velocity patterns; this overall uniformity suggests that the ring nebula resulted from a single ejection of material from AG Car.

Regarding the expansion velocity of the shell, the velocity maps in Figure 2 (particularly those for P.A.= 80° and 131°) indicate that the shell expansion is roughly symmetric and centred on a velocity of 0 km s^{-1}. The shell expansion velocity is then directly given by the observed approaching and receding velocities at the position of the star; these are measured to be ± 70 km s^{-1}. Additional evidence in support of this expansion velocity is provided by the interstellar Na I absorption lines towards the central star AG Car. Data obtained at the same resolution as the nebular spectra reveal an absorption component at $V_{LSR} = -67$ km s^{-1} which is also seen in Ca II at the same velocity by Wolf & Stahl (1982). The derived expansion velocity is substantially higher than the value of 50 km s^{-1} quoted by Thackeray (1977) and leads to a dynamical age, $t_D \approx R/V_{exp} \approx 7 \times 10^3$ yr for the AG Car ring nebula (assuming that the radius $R = 0.5$ pc). Over this timescale, the stellar wind of AG Car will have contributed at most $\sim 0.5\,M_\odot$ to the nebular mass of $2\,M_\odot$ if the upper limit of $7 \times 10^{-5}\,M_\odot\,\mathrm{yr}^{-1}$ is adopted for the stellar mass loss rate (Lamers 1989). The shell is unlikely to contain any swept-up interstellar material because the progenitor O star will have evacuated the surrounding interstellar medium. The amount of stellar material ejected by AG Car 7×10^3 yr ago is therefore $\sim 1.5\,M_\odot$.

References

Humphreys, R.M., Lamers, H.J.G.L.M., Hoekzema, N. & Cassatella, A., 1989.
    Astr. Astrophys., **218**, L17.
Johnson, H.M., 1976. *Astrophys. J.*, **206**, 469.
Lamers, H.J.G.L.M., 1989. in *Physics of Luminous Blue Variables*, eds. K. Davidson,
    A.F.J. Moffat & H.J.G.L.M. Lamers, Kluwer, Dordrecht, p.135.
McGregor, P., Finlayson, K., Hyland, A.R., Joy, M., Harvey, P.M. & Lester D.F.,
    1988. *Astrophys. J.*, **329**, 874.
Mitra, P.M. & Dufour, R.J., 1990. *Mon. Not. R. astr. Soc.*, **242**, 98.
Paresce, F. & Nota, A., 1989. *Astrophys. J.*, **341**, L83.
Stahl, O., 1987. *Astr. Astrophys.*, **182**, 229.
Stahl, O., 1989. in *Physics of Luminous Blue Variables*, eds. K. Davidson,
    A.F.J. Moffat & H.J.G.L.M. Lamers, Kluwer, Dordrecht, p.149.
Thackeray, A.D., 1950. *Mon. Not. R. astr. Soc.*, **110**, 524.
Thackeray, A.D., 1977. *Mon. Not. R. astr. Soc.*, **180**, 95.
Viotti, R., Cassatella, A., Ponz, D. & Thé, P.S., 1988. *Astr. Astrophys.*, **190**, 333.
Walborn, N.R., 1982. *Astrophys. J.*, **256**, 452.
Wolf, B. & Stahl, O., 1982. *Astr. Astrophys.*, **112**, 111.

DISCUSSION

Walborn: Is there any evidence for chemical composition differences among the various velocity features in your data?
Smith, Linda: I only have $H\alpha$ and $[NII]$ data and have not examined the $[NII]/H\alpha$ ratio in any detail. The appearance of the $H\alpha$ and $[NII]$ two-dimensional spectra are very similar, indicating that the various velocity features show no obvious differences in the $[NII]/H\alpha$ ratio.

Schulte-Ladbeck: I am wondering about the geometry, *i.e.*, the position angle of the bipolar outflow with respect to the Paresce features. Are they perhaps perpendicular?
Smith, Linda: No, they have similar position angles. They are, however, perpendicular to the major axis of the ring nebula, which is probably significant.

Spectrophotometry of ring nebulæ around Wolf-Rayet stars[*])

Manfred W. Pakull
Observatoire de Besançon,
25044 Besançon Cedex, France

ABSTRACT.   Long-slit spectrophotometric observations of HII regions around Wolf-Rayet stars in the Galaxy, the Magellanic Clouds and in IC1613 are employed to measure the ionizing radiation from these stars. The sample includes all known WR stars in the SMC and 14 regions in the LMC comprising most of the known ring nebulæ. Apart from previously known HeII λ4686 emitting nebulæ around a few *WO* stars, HeIII regions were discovered within the nebulæ N76 and N79W which are being photoionized by the strong He^+ Lyman continuum emission from the Magellanic*WNE* stars AB7 and Br2, respectively. The Zanstra blackbody temperatures and luminosities (80 000, 95 000 K and log L/L_O = 5.4, 6.0, respectively) agree remarkably well with more elaborate WN model calculations and suggest a much higher bolometric correction (BC = -5.8 mag) than generally admitted for WN stars. The *nebular* λ4686 fluxes are several times higher than the corresponding broad *stellar* features suggesting that the highly ionized gas would remain detectable at much larger distances (i.e. in HII galaxies) than the underlying hot WNE population.

1. INTRODUCTION

The presence of highly-ionized HII regions in Local Group galaxies and the detection of narrow HeII λ4686 emission lines in X-ray ionized nebulæ (Pakull & Angebault, 1986; Pakull & Motch, 1989a,b) and many HII galaxies (i.e. Campbell, 1988) has motivated us to systematically study the EUV emission of advanced stages of massive stellar evolution. Broad HeII and CIV emission lines in the integrated spectra of many giant HII regions and HII galaxies have revealed large populations of Wolf-Rayet stars. However, the question of whether or not WR stars do also contribute substantially to the ionizing radiation, in particular at photon energies beyond the He^+ Lyman edge (> 4Ryd, hereafter EUV) has remained open since adequate WR model atmospheres have become available only recently. On the other hand, it has been suggested that massive stars in very advanced evolutionary stages (early WC's and WO's) might not only be very luminous, but also extremely hot, reaching "effective" temperatures > 100 000 K (see c.f. Maeder & Meynet, 1987) that are comparable to those of the nuclei of highly ionized planetary nebulæ (PN). Accordingly, such stars might be expected to photoionize large HII regions to PN-like excitation (i.e. presence of He^{++} ions which are not observed in normal HII regions around O stars). This consideration forms also the basis for the Terlevich and Melnick (1985) hypothesis, namely that WO stars ("warmers") might be responsible for the high nebular ionization in at least some types of active galaxies. However, this view has not been widely accepted, not least because sufficently high WO star EUV luminosities had not

---

[*]) *Based on observations collected at the European Southern Observatory, La Silla, Chile, and at the Observatoire de Haute-Provence, France*

K. A. van der Hucht and B. Hidayat (eds.),
Wolf-Rayet Stars and Interrelations with Other Massive Stars in Galaxies, 391–396.
© 1991 *IAU. Printed in the Netherlands.*

been firmly established. Moreover, the *nebular* λ4686 recombination radiation should then be accompanied by even stronger *broad* HeII+CIV λλ4600-4700 emission generated in the stellar winds; this is generally not observed (Pakull & Motch, 1989b).

In this paper I adress the question of ionizing radiation from WR stars and hence, their temperatures and luminosities in a more systematic fashion. The work is based on recent observations of the ionization structure of diffuse or swept-up interstellar gas around WR stars which, in many cases, assumes the morphology of a ring-like structure (ring nebulæ; cf. review by Chu in this volume). Here, I report on the results of long-slit spectroscopic observations of several Magellanic and IC1613 Wolf-Rayet stars and the surrounding interstellar medium that provide crucial informations on the EUV emission of these stars.

2. LONG SLIT SPECTROPHOTOMETRY OF WR STARS

2.1. THE SAMPLE

The sample of Wolf-Rayet stars (c.f. Table 1) was selected on the basis of: (i) presence of ring nebulæ, (ii) associated diffuse nebulosities, and after it turned out that certain WR types are particularly likely to be strong EUV emitters, (iii) WNE and WO subtypes.

TABLE 1: Sample of WR stars / nebulæ[*] observed in the present study

star	type	nebula	λ5007/β	λ4686/β
AB 1	WN 3 + O4:	(RN ?)	5	<0.5
AB 2	WN 4.5/Of	(RN ?)	3 - 5	
AB 3	WN 3 + O4:	-	2	<0.3
AB 4	WN 4.5	-	1	
AB 5	WN 4 + O7I:	N66	6 - 7	
AB 6	WN 3: + O6.5I:	-		
AB 7	WN 1 + O6IIIf	N76 (RN !)	6 - 7	0.3
AB 8	WO 4 + O4V	(-)		
Br 2	WN 1	N79W (RN !)	6-10	0.7
Br 4	WN 1(?)	N83 (?)	3	<0.04
Br 12	WN 3	N16A (RN)	6	<0.03
Br 16	WN 2.5	N105	4 - 5	<0.03
Br 25	WN 3	N44C (RN !)	6	<0.02
Br 26	WN 7	N198 (RN)	3	
Br 29	WN 3/WCE	N138 (RN)	6 - 8	
Br 48	WN 4 + OB	N57C (RN)		
Br 51	WN 3	N62 (RN !)	9 -11	<0.03
Br 52	WN 4	N56 (RN)	6-8	<0.03
Br 65a	WN5 + abs	N59B	1	<0.004
Br 66	WN 3 (+WN3 ?)	N157	4	<0.007
Br 93	WO 4	N157 (RN !)	4 - 7	<0.07
Br 100	WN 4	N74 (RN)		
IC1613	WO	IC1613 #3 (RN)	5	0.3

[*] RN- ring nebulæ; RN! denote Magellanic ring-nebulæ not included in the compilation of Chu & Lasker (1980)

Special emphasis was put on Magellanic Wolf-Rayet stars, not least because the projected spectrographic slit lengths (≈ 3 arc min on the CCD detectors of the 2.2 and 1.5 m ESO telescopes) cover most of the extent of known ring nebulæ in these galaxies. Moreover, the small interstellar extinction towards, and within, the Clouds favors the detection of weak nebular emission lines since intensity is independent of distance as long as the regions are spatially resolved. Table 1 lists the observed WR stars and spectral types together with the associated nebulæ (if present) and diagnostic emission line ratios for most of them. Note that in many cases the "excitation parameter" $I(\lambda5007)/I(H_\beta)$ turns out to be extremely high (> 6) indicating that the ionizing WR stars are probably at least as hot as the earliest O stars.

Three stars in the sample emit sufficiently strong EUV continua to excite nebular HeII $\lambda4686$ recombination radiation in the nebulæ. They are the WO star in IC 1613 #3 (Davidson & Kinman, 1982; Pakull & Motch, 1989b), and the WNE stars AB7 and Br2 in the SMC and LMC, respectively (Pakull et al. 1990; Pakull & Bianchi, this volume; Niemela, this volume). To this list, WR 102 (WO2; Dopita et al. 1990; Dopita & Lozinskaya, this volume) and Br 40a (WNE; Niemela, this volume) must be added. The absence of $\lambda4686$ nebular emission in the ring nebula around Br 93 which was independently discovered also by Heydari-Malayeri & Melnik (this volume) implies that WO stars are not necessarily strong EUV sources; in the case of the SMC WO star AB 8 the situation is less clear since there probably is no nearby interstellar gas which could "reprocess" the EUV radiation into observable line emission.

2.2. THE EARLIEST WN STARS: *WN1*

A continuation of WNE types towards high excitation of the stellar emission lines has recently been proposed by Conti and Massey (1989) who introduced the WN1 classification for the galactic stars WR2 and the LMC member Br4 on the basis of undetectable NV (3p $^2P^\circ$- 3s ^2S) $\lambda\lambda4603,19$ emission, whereas the UV resonance NV (2p $^2P^\circ$- 2s ^2S) $\lambda\lambda1238,40$ doublet remains strong. In AB7 and Br2, the ratios between the NV $\lambda\lambda4609,19$ equivalent width (EW) and the HeII $\lambda4686$ EW, (NV/HeII), are < 0.02 and 0.04, respectively, both values being significantly smaller than in the galactic and LMC stars which Conti and Massey classify as WN2, or later. The (NV/HeII) ratio for the latter stars are all > 0.1, suggesting that we might draw a dividing line between WN1 and the later WN subtypes, according to whether (NV/HeII) is smaller or larger than about 0.07.

It is of some interest to note that whereas the $\lambda\lambda4603,19$ lines become weaker towards earlier WN spectral types, the largely recombination fed NV(7-6) subordinate $\lambda\lambda4933,45$ doublet (Hillier, 1988) remains comparatively strong. The latter emission lines are not part of the classical classification scheme; however, my CCD spectra of the galactic WN1 "standard" WR2 clearly display this emission at about the same strength as in Br2 (EW ≈ 10 Å).

2.3. AB7 AND BR 2

These WN1 stars do completely photoionionize helium in the inner ~20 pc of their ring nebulæ. The electron temperatures, $T_e(OIII) ≈ 16\,000$ K, in both the nebulæ around AB7 (N76) and around Br2 (N79W) indeed strongly suggest that photoionization rather than ionization by shocks is the dominant process that feeds the nebular HeII recombination radiation. These results have several interesting consequences. First of all, the stellar wind layers in which the EUV radiation is formed must be sufficiently hot to strongly emit in the He$^+$-Lyman continuum. Moreover, the outer wind zones have to be optically thin in the EUV, i.e. helium should be completely ionized throughout the wind (see below). Then, using the photon-counting property of the nebular $\lambda4686$ recombination line one can derive the rate of He$^+$-ionizing photons (Q_4; i.e. hν > 4 Ryd) escaping from the Wolf-Rayet wind.

Using the beautifully simple, but nevertheless extremely powerful Zanstra method which compares nebular $\lambda4686$ flux to the stellar flux at observable wavelengths yields blackbody temperatures of 80 000 and 95 000 K, and luminosities, log $L/L_O = 5.3$ and 6.0, for these Magellanic WN1 stars (see Table 2). The analysis is rather straightforeward for the single star Br2; in the case of the WR+O6f binary AB7, the visual magnitude of the WN1 component was estimated by subtracting from the binary spectrum appropriately scaled spectra of a single O6 star until the the H, HeI and HeII *absorption* lines disappeared (Pakull et al. 1990).

One might argue that the extremely high HeII $\lambda4686$/Hβ line ratio (= 0.7 !) in the brightest parts of the ring nebula N79W implies a significantly harder ionizing spectrum than a 95 000 K blackbody. For example, from their analysis of nebular lines ($\lambda4686$/Hβ = 0.55) around the WO star WR102, Dopita et al. (1990, and this volume) derive an ionizing temperature $T_{BB} \approx 150\,000$ K. Moreover, Stasinska & Tylenda (1986) have pointed out that this line ratio levels off at $T_{BB} > 250\,000$ K and never exceeds a value of about 0.6. However, it should be kept in mind that the observed ratio in N79W mainly refers to the He^{++} zone within a larger (ill defined) HII region rather than to the total emission calculated in the model nebulæ. Another effect increasing the $\lambda4686$/Hβ ratio is the possibility that N79W might well be partially transparent to the H-ionizing photons but optically thick in the He$^+$ Lyman continuum (EUV).

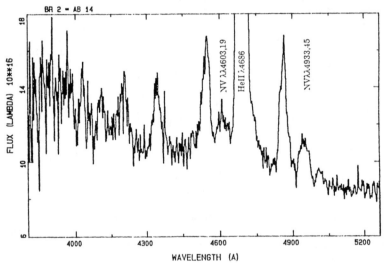

FIG. 1: Blue CCD spectrum of the WN1 star Br2. Note the presence of $\lambda\lambda4933,45$ emission (EW $\approx$ 9 Å) and the very weak NV $\lambda\lambda4603,19$ doublet (EW $\approx$ 6 Å). The equivalent width of the strong HeII $\lambda4686$ line is 160 Å.

3.   COMPARISON WITH WR MODEL SPECTRA

The condition that the (helium) winds of AB7 and Br2 must be completely ionized allows to draw an important lower limit on the rate of He$^+$ ionizing photons, Q_4, emitted by WR "cores". Assuming for simplicity a constant wind velocity law, $v(R) = v_\infty$, and furthermore provided that photoionization from the ground state is the dominant process for the He ionization beyond the radius R_{EUV} where the EUV continuum is formed, this translates to:

$$Q_4 > 1\,10^{49}\ s^{-1}\ [M_{-4}/v_{2500}]^2\ [R_{EUV}/R_O]^{-1},$$

where an electron temperature of 80 000 K has been assumed, M_{-4} is the mass loss rate in units of 10^{-4} M_O/y, and $v_{2500} = v_\infty/(2500$ km/s$)$. This order-of-magnitude estimate suggests that even very hot WR "cores" cannot ionize the wind completely, once the mass loss-rates become sufficiently high. On the other hand, the relatively low M/v_∞ ratios together with the lower limit of the temperature parameter, T_*, which were derived from the optical and UV spectra (see below) already strongly favour the escape of EUV radiation from AB7 and Br2 beyond the stellar wind.

A remarkable result of this simple analysis is that the (pure helium) WR model spectra calculated by the Kiel group (Schmutz, Hamann & Wessolowsky, 1989 and references therein; Schmutz, these proceedings; Hamann, these proceedings) agree rather well with the observed stellar He emission line strenghs and absolute magnitudes of AB7 and Br2, if Zanstra blackbody temperatures are identified with the effective temperatures, T_*, at the "core radius", R_*, which is the inner boundary of the model atmospheres. According to these models, emission lines and continuum emission of WN stars are largely determined by T_* and the radius parameter, $R_* [M_{-4}/v_{2500}]^{-2/3}$.

TABLE 2: WN1 star parameters

Object	M_V	BC	Q_4 $[s^{-1}]$	v_∞ [km/s]	T_{BB} [kK]	L_{BB} $[L_O]$	T_* kK	L_{model} $[L_O]$	M $10^{-4}\ M_O/y$
AB7	-4.4	-5.9	$1.5\,10^{48}$	1600	80	$9\,10^5$	75	$1\,10^6$	0.14
Br2	-2.6	-5.7	$8\,10^{47}$	2500	95	$2\,10^5$	90	$2\,10^5$	0.15

The non-detection of HeI features in WNE stars have so far only allowed to place lower limits on T_*. Now, the observation of *nebular* $\lambda4686$ emission around some of these stars opens the possiblity to substantially narrow down the allowed range of parameters for a given Wolf-Rayet star. After this conference, Werner Schmutz (1990b) has kindly informed me of his extended model calculations concerning the continuum emission down to wavelengths of 100 Å. At least in the parameter range covered by this work the validity of the EUV emitter criterion mentioned earlier is nicely confirmed. In Table 2 the relevant Kiel-model stellar parameters for AB7 and Br2 are summarized together with the results based on naïve blackbody fits. The advocates of a WN "universal" bolometric correction of - 4.5 might in particular note that both BB and more sophisticated modeling of these WN1 stars imply BC < - 5.5.

4. CONCLUSIONS

The main result of the present study concerns the discovery of strong EUV radiation from some WNE stars in the Magellanic Clouds. This constitutes the first *direct* proof of their significantly higher effective temperatures as compared to the O star progenitors. Although the EUV emitters AB7 and Br40a are in binary systems, the lack of a luminous companion of Br2 suggests that also single star evolution might lead to $\approx$100 000 K WN-type remnants.

The discovery of strong HeII λ4686 recombination radiation from the WNE ring nebulæ N76 and N79W also suggests a natural explanation for the previously puzzling presence of this line in a number of giant HII regions and HII galaxies in terms of a population of such WNE stars. In fact, one might call these objects genuine *warmers* since the EUV fed λ4686 emission turned out to be about 8 times stronger than the corresponding broader stellar feature. These *warmers* will not be detectable any more in distant galaxies except for their ionizing radiation.

REFERENCES

Campbell, A. 1988, Astrophys. J. 335, 644
Chu, Y.-H. & Lasker, B.M. 1980, Pub. A. S. P. 92, 730
Conti, P.S. & Massey, P. 1989, Astrophys. J. 337, 251
Davidson, K. & Kinman, T.D. 1982, Pub. A. S. P. 94, 634
Dopita, M.A. & Lozinskaya, T.A. 1990, in this volume
Dopita, M.A., Lozinskaya, T.A., McGregor, P.J. & Rawlings, S.J. 1990, Astrophys. J. 351, 563
Hamann, W.-R. 1990, in this volume
Heydari-Malayeri, M. & Melnick, J. 1990, in this volume
Hillier, D.J. 1988, Astrophys. J. 327, 822
Maeder, A. & Meynet, G. 1987, Astron. Astrophys. 182, 245
Niemela, V.S. 1990, in this volume
Pakull, M.W. & Angebault, L.P. 1986, Nature 322, 511
Pakull, M.W. & Motch, 1989a, Nature 337, 337
Pakull, M.W. & Motch, C. 1989b, ESO workshop on 'Extranuclear Activity in Galaxies', E.J.A. Meurs & R.AE. Fosbury (eds.) p 285
Pakull, M.W., Testor, G. & Bianchi, L. 1990, submitted to Astron. Astrophys.
Pakull, M.W. & Bianchi, L. 1990, in this volume
Schmutz, W., Hamann, W.-R. & Wessolowsky, U. 1989, Astron. Astrophys. 210, 236
Schmutz, W. 1990a, in this volume
Schmutz, W. 1990b, private communication
Stasinka, G. & Tylenda, R. 1986, Astron. Astrophys. 155, 137
Terlevich, R. & Melnick, J. 1985, Mon. Not. R. astr. Soc. 213, 841

DISCUSSION

Cassinelli: Most impressive stars. Radiation pressure would not have any trouble driving their winds. Does this violate the newer picture for evolution?
Langer: You would not really conclude that these positions fit very well to the tracks. However, after the core helium exhaustion, the tracks are nearly predicted to lead to very high effective temperatures and also to somewhat higher luminosities again. So, this picture would really mean that these warmers correspond to the latest evolutionary stage, carbon burning or something like that which would also account for the small number of those objects.
Pakull: These stars are WN's, not WC's.
Langer: But still, you have the possibility of some WN stars being treated as supernova candidates too, so, you must not lead all WR stars to the WC stage.

MASSIVE STARS AND GIANT HII REGIONS: THE HIGH-ENERGY PICTURE

THIERRY MONTMERLE
Service d'Astrophysique,
Centre d'Etudes Nucléaires de Saclay
91191 Gif-sur-Yvette Cedex, France

ABSTRACT. Giant HII regions contain highly energetic objects: luminous, massive stars (including Wolf-Rayet stars) generating powerful winds, as well as, often, supernova remnants. These objects interact with the surrounding gas by creating shock waves. Part of the energy input is radiated away in the form of X-rays; also, protons and electrons may be accelerated in situ and generate γ-rays by collisions with the ionized gas. In addition, the stars themselves (including the accompanying low-mass PMS stars) are sources of X-rays, and W-R stars may emit continuum γ-rays and are associated with nuclear γ-ray lines seen in the interstellar medium. Therefore, both through the stars they contain and through interactions within the gas, giant HII regions are, in addition to their more traditional properties and over nearly 7 decades in energy, important sources of high-energy radiation.

1. Introduction: which "high energies" ?

1.1. WHY ?

"Giant HII regions" are gaseous regions excited by the ionizing flux of luminous, massive stars (O, early B, and Wolf-Rayet stars), which are subject to an intense mass-loss in the form of dense, fast winds (from $10^{-6} \sim 10^{-4}\, M_\odot$, at up to ~ 4000 km.s^{-1}; see Langer 1989, and this volume, for the latest developments). These winds in turn act on the surrounding material and create large cavities, bounded by a shock wave. Additional energy can be input by the explosion of type II supernovae, i.e., the endpoint of evolution of massive stars.

Through a variety of processes, which we will review in this paper, high-energy photons are emitted from the sub-keV range (soft X-rays) all the way to the GeV range (γ-rays). (For a broader perspective, see Montmerle 1987.) Also, high-energy particles are present: non-thermal radio emission reveals the presence of MeV-GeV electrons (depending on the value of the magnetic field), and super-GeV protons may be accelerated in situ and interact with the ionized material. Although the topic will not be discussed here, it is worthy of note that antiprotons and neutrinos may also be produced as secondary particles after high-energy collisions in the ionized gas (see Montmerle 1988 for details).

1.2. HOW ?

Since the physical context of the high-energy phenomena is not discussed elsewhere in this volume, it is perhaps appropriate to briefly summarize here the physical processes relevant to giant HII regions, and their implications.

K. A. van der Hucht and B. Hidayat (eds.),
Wolf-Rayet Stars and Interrelations with Other Massive Stars in Galaxies, 397–407.
© 1991 *IAU. Printed in the Netherlands.*

a) X-rays:
The X-rays are thermal (bremsstrahlung of a $\approx 10^7$ K plasma) when they result from shocks (wind flows, wind bubbles, associated with O and W-R stars), but also from strong flares occurring on low-mass pre-main sequence (PMS) stars. (We shall see below that, because they are very numerous and located in the same molecular clouds as giant HII regions, such stars can make a significant contribution to the overall X-ray budget.) These X-rays are in general optically thin, hence give information throughout the whole emitting region.

But the X-rays may also be non-thermal, in which case the spectrum is a power-law. Compton scattering of the ambient radiation by high-energy electrons (themselves seen by means of their non-thermal radio emission) has been invoked in the case of the winds of W-R stars.

b) Continuum γ-rays:
For instrumental reasons, these γ-rays are known in the Galaxy only above ~ 30 MeV (the threshold for electron pair production in spark chamber detectors). They are produced by collisions of $\gtrsim$ GeV cosmic rays (primarily p and α) with interstellar matter (primarily H and He): protons collisions give neutral pions which decay into 2 γ-rays, as well as charged pions which decay into charged muons and subsequently in electrons and positrons. These secondary e^+ and e^-, as well as primary electrons, also produce γ-rays in the same energy range by bremsstrahlung in the electric field of interstellar atoms. In addition to the standard galactic cosmic rays, protons and electrons can be accelerated locally by shock waves associated with stellar winds and SNRs.

c) γ-ray lines:
In addition to generating high-energy γ-rays as described above, collisions between low-energy cosmic rays with interstellar nuclei, or stellar nuclear reaction products dispersed by SN explosions or stellar winds, may result in the emission of γ-ray lines. This is the case when nuclei in excited states or radioactive isotopes are produced. So far, however, the only positively detected γ-ray line in the Galaxy is the 1.8 MeV line associated with the β^+ decay of ^{26}Al, a product of stellar nucleosynthesis. Since the corresponding half-life is short (~ 10^6 yrs), the presence of this line testifies to the existence of fresh nucleosynthesis in the interstellar medium.

2. X-rays and γ-rays from stars

2.1. MASSIVE STARS

The bulk of what we know today about the X-ray emission from stars has been obtained by the *Einstein* satellite. X-ray emitting stars have been discovered throughout the HR diagram (e.g., Rosner, Golub, and Vaiana 1985), the O stars spanning the luminosity range $L_X \sim 10^{31}$ - 10^{34} erg.s^{-1}. But the O stars have a unique property: the ratio L_X/L_{bol} of their X-ray luminosity to their bolometric luminosity is approximately constant, $\approx 10^{-7}$ (Chlebowski, Harnden, and Scortino 1989), and remarkably independent of main stellar parameters such as mass, radius, rotation velocity, effective temperature, although it may slightly depend on the environment (Chlebowski 1989). The reason for this is not understood. Proposed models for the X-ray emission involve thermal bremsstrahlung from shocks within the wind (see the discussion by S. Owocki, this volume), or non-thermal Compton radiation from the high-energy electrons seen in the radio range.

In the case of W-R stars, the X-ray luminosities span a comparable range, from 10^{32} to 10^{34} erg.s^{-1}. The average X-ray energy (or temperature, depending on whether the X-rays are thermal or non-thermal) is ~ 1 keV, and the column density of interstellar gas is ~ 10^{21} - 10^{22} cm^{-2}. In principle, the shape of the X-ray spectrum should indicate whether we are dealing with thermal (exponential in the case of bremsstrahlung) or non-thermal (power-law) processes.

In practice, however, for lack of spectral resolution, the answer is often ambiguous, and one has to resort to indirect arguments, drawn for instance from radio studies. Non-thermal radio emission is found to be common in the case of O stars (Bieging, Abbott, and Churchwell 1989); until recently, it was thought that, in the case of single W-R stars, the radio emission was thermal (free-free emission from the ionized wind; see, e.g., Hogg 1989, or Williams et al. 1990), but an increasingly large number of non-thermal cases are now being found (see the paper by M. Felli, this volume). This, in turn, reveals the presence of $\lesssim$ GeV electrons (Lorentz factor $\gamma \lesssim 10^3$) in $\gtrsim$ mG magnetic fields, likely accelerated by shocks and/or turbulence in the wind itself. Further, since the wind is an efficient absorber of radio waves because it is ionized, the observed non-thermal radio flux can come only from the outer parts of the wind, i.e., $R > R(\tau_{radio} = 1) \approx$ a few 100 R_*, as shown by Pollock (1987a, b). This has interesting implications: since the electron spectrum is a power law (number density of electrons $\propto \gamma^{-p}$, with p ~ 3), other parts of their spectrum contribute at other wavelengths: keV X-rays are produced by inverse Compton scattering by electrons with γ ~ 10 and leave the star provided that they are emitted at $R > R(\tau_X = 1) \approx 100$ R_*. By the same token, > 100 MeV γ-rays may be produced by bremsstrahlung of electrons with γ ~ 10^3 in the dense layers of the photosphere (where $\tau_\gamma \ll 1$). As a bonus, one may even explain the IR spectrum around 2 μm by synchrotron emission of electrons with γ ~ 10^4 and $B \approx 1$ G also at the photospheric level (where $\tau_{2\mu m} \ll 1$ as well).

2.2. LOW-MASS PMS STARS

At the other end of the mass spectrum (M ~ 0.5-2 $M_\odot$), PMS stars are also X-ray emitters, with luminosities L_X ~ 10^{31-32} erg.s^{-1}. They are thus individually 10^{4-6} times more powerful in X-rays than their main sequence counterparts, and, in particular, the Sun (see, e.g., Montmerle et al. 1983, Feigelson 1987, Montmerle and André 1988). These stars are known under the generic term "T Tauri stars" (which include several sub-classes, see Bertout 1989). They are important in our context because they are much more numerous than massive stars while being also associated with molecular clouds. The X-rays are here produced by thermal bremsstrahlung of powerful flares having kT ~ 1 keV. The observed $N_H \approx 10^{21-22}$ cm^{-2} represents mainly the outers layers of the molecular cloud in which they are embedded. This intense solar-like activity results from the fact that these stars are entirely convective or have deep outer convective zones (the depth depends on the age), which, together with rotation, generates a dynamo effect (e.g., Bouvier 1990).

3. Interaction of massive stars with the surrounding medium

We take here into account the fact that W-R stars generate a lot of mechanical energy ($L_w \approx$ 10^{38} erg.s^{-1}) and that, being the "cousins" of O and B stars grouped in associations, they contribute in a major way to the creation of cavities in the giant HII regions excited by the O and early B stars.

The standard model describing the interaction between massive stars (wind +

ionizing flux) and the surrounding medium is that of the "interstellar bubble" (Weaver et al. 1977). One feature of this model is that it predicts the existence of a *thin* HII shell around the cavity (i.e., $R_{shell} \ll R_{cavity}$); also, the shock is essentially adiabatic, and a large fraction (~ 20%) of the wind energy is imparted to the HII shell. However, there are many famous cases (for instance the Rosette nebula; see also LMC nebulae, e.g., Meaburn et al. 1989) in which the shell is thick ($R_{shell} \gtrsim R_{cavity}$). Also, the kinetic energy imparted to the HII shell is often much smaller than predicted by the standard model (~ 1%, see, e.g., Chu 1983, for the case of W-R nebulae).

The reason for the disagreement lies at least in part in the physical conditions prevailing near or at the shock. Indeed, the basic question is whether or not the shock is adiabatic: in the extreme case in which the shock is isothermal, the ratio of the pressures downstream and upstream of the shock can go to infinity, and all the wind kinetic energy can be radiated away, leaving nothing for the outer motion of the HII shell.

Recent works have suggested possible explanations. Dorland, Montmerle, and Doom (1986), and Dorland and Montmerle (1987) have explored the possibility that electron conduction may play a more important role than previously thought. Indeed, as shown in fusion plasmas ("Tokamaks"), electron conduction is non-linear in the case of steep temperature gradients, i.e. when the characteristic temperature length scale $L_T \lesssim 500 \lambda_0$, where $L_T = T_c / \Delta T_c$, ΔT_c being the temperature gradient obtained with classical conduction, and λ_0 is the electron mean free path, proportional to v^4 (Luciani, Mora, and Pellat 1985; see also Campbell 1984). In that case, electron conduction at a given point becomes "non-local", i.e., depends on the (non-maxwellian) electron distribution function within ~ a few m.f.p. of that point. The result is that the actual temperature gradient is governed by the electrons, giving rise to a softer temperature gradient than ΔT_c. In nebulae like the Rosette nebula, one finds $L_T/\lambda_0 \sim 150$ in the vicinity of the shock, which justifies a nonlinear treatment. Then the plasma downstream of the shock is made up of hot electrons and cold ions: X-ray emission results from bremsstrahlung of the former in the electric field of the latter, hence gives temperatures lower than for a standard downstream plasma: for instance, a 4000 km.s^{-1} wind (typical of W-R stars) impinging on the surrounding HII region gives a temperature in the shock vicinity of $\approx$ 10 keV instead of the $\approx$ 40 keV computed in the standard way (i.e., $kT = \frac{1}{2} \mu m v^2$, where μ is the mean molecular weight). The corresponding ratio $L_X(IPC)/L_w$ of the diffuse X-ray luminosity in the *Einstein* IPC band over the wind power (in a 4π cavity) is ~ 2 x 10^{-4}: for a single W-R star, this amounts to $L_{X,W-R} \approx 4$ x 10^{34} erg.s^{-1} (see Dorland and Montmerle 1987 for details).

In another approach, Breitschwerdt and Kahn (1988, 1989) have shown that acoustic instabilities exist at the wind shock. The growth rate is short enough that turbulence sets in the vicinity of the shock: the characteristic time $t_{wave} \approx t_{sound}(HII$ region) $\ll t_{dyn}(HII$ region). In turn, this leads to a rapid mixing of the hot stellar wind and the much cooler HII gas, such that the cooling rate grows by ~ 1 order of magnitude. Here again, a significant fraction of the wind energy is radiated away near the wind shock, even if, as it turns out, this instability is partly inhibited by magnetic fields. The evolution of the structure of the HII shell and the energy loss have yet to be calculated, but, given the shape of the cooling function (see e.g., Gaetz and Salpeter 1983), it is likely that the energy will be radiated at the peak of this curve, i.e., T ~ 10^6 K, in other words in the sub-keV domain.

4. X-rays from giant HII regions

4.1. STELLAR WIND SHOCKS

Quantitative studies have been done by Dorland and Montmerle (1987) and Montmerle (1987) on the Orion, Rosette, and Carina nebulae. Since the last nebula is the only one of these which contains W-R stars, we will focus our attention on it. The available X-ray data are made up of *Einstein* observations (IPC images for ~ keV X-rays, MPC flux for X-rays of a few keV; Seward and Chlebowski 1982, Chlebowski et al. 1984), and OSO-8 flux for energies up to 20 keV (Becker et al. 1976). More recent data were obtained up to 10 keV by *Tenma* (Koyama et al. 1990). These last three experiments were of scintillating collimator type (i.e., not imaging), but their field of view included the whole nebula. Two components can clearly be distinguished, both fitted by thermal bremsstrahlung spectra: a high-energy one (kT ~ a few keV), and a low-energy one (kT $\lesssim$ 1 keV). Apart from the pointlike emission of massive stars, this last component is diffuse, and covers the whole nebula. As shown by Montmerle and Dorland (1987), this last component may be entirely attributed to the unresolved emission from low-mass PMS stars having the same typical luminosity and surface density as observed in the nearby ρ Oph cloud (Montmerle et al. 1983; see above, § 2.2.). At the distance of the nebula (≈ 2.5 kpc), these stars are too faint to be optically visible individually, but the relevant ones would lie projected on the far side of the HII region, as is possible given the general orientation of the molecular cloud deduced from CO observations, which makes only a small angle to the line of sight. Once the high-energy tail of this component is subtracted (≈ 30% of the luminosity), the remaining X-ray luminosity is consistent with emission from shocks at the boundary of the inner cavity created by the massive stars present (and dominated by η Car and the 3 W-R stars associated with the nebula), according to the model of Dorland and Montmerle (1987). Essentially similar results may be obtained for other nebulae.

Hence, according to these ideas, there is a strong link between the presence of stellar winds inside giant HII regions and emission of X-rays at a temperature of several keV. In turn, this suggests a possible link with the so-called "galactic [X-ray] ridge" (Montmerle 1986). This ridge, discovered by the Japanese satellite *Tenma* (Koyama 1987) and confirmed by EXOSAT (Warwick et al. 1985, 1988), is defined as the diffuse high-energy (several keV) emission seen to be associated with the galactic plane, once known localized sources (supernova remnants, compact binaries, etc.) have been removed. The angular resolution of the resulting composite "image", made up of pointings along different lines of sight, is very poor (a field of view of ~ 1° diameter for the EXOSAT's ME; of ~ 5° for *Tenma*), but it is clear that the emission is thermal (systematic presence of a conspicuous helium-like Fe 6.7 keV line in the *Tenma* spectra) and corresponds to different temperatures along different lines of sight. The total luminosity is $L_{X, tot} \approx 1.2 \times 10^{38}$ erg.s^{-1}.

4.2. W-R STARS AND THE GALACTIC RIDGE

Since many W-R stars lie inside giant HII regions, it is therefore tempting to suggest that, via their powerful winds, these stars make a significant contribution to the galactic ridge. Indeed, taking a typical mechanical luminosity $L_{X, W-R} \approx 4 \times 10^{34}$ erg.s^{-1} (see above), we can explain the observed luminosity with ≈ 3000 such embedded W-R stars. This figure is significantly higher than the estimated total number of W-R stars in the Galaxy (van der Hucht et al. 1988), but since many regions in the galactic plane are heavily obscured by dust, this number is in reality a lower limit: the actual number is unknown but could be compatible with the above figure. At any rate, the possibility of a connection between winds from massive stars (particularly the W-R stars) should be followed up when experiments with a better angular resolution at several keV (in particular the European satellite XMM, planned for the late 1990's) become available.

4.3. THE CONTRIBUTION OF SUPERNOVA REMNANTS

Also efficient at producing X-rays in regions of massive star formation, and in particular in giant HII regions, are supernova remnants. However, if a supernova explosion occurs inside the HII region, the X-rays are efficiently produced only near the edge of the cavity, i.e., when the shock meets a dense medium while retaining a large velocity. Such a situation has been suggested to explain the high X-ray ($\sim$ keV) luminosities observed to be associated with "superbubbles" in the LMC , which reach $\sim 10^{37}$ erg.s^{-1} (Chu and Mac Low 1990 and refs. therein). This interpretation is attractive, since such luminosities are much too high to be explained even by W-R winds.

5. W-R stars and nuclear γ-ray lines

Recent observations by the SMM satellite and by several balloon experiments have revealed the existence of a line emission at 1.8 MeV associated with the galactic plane, with a strong enhancement in the galactic center region. Owing to the very poor angular resolution of the instruments (typically more than $10°$), the geometrical structure of the emission is still controversial. By contrast, the emission process is unambiguous, because the line is the signature of the β^+ decay of ^{26}Al. Since this radioactive isotope has a half-life of $\simeq 10^6$ yrs, the detection of the line means that "fresh" nucleosynthesis is going on; the total mass of ^{26}Al necessary to account for the 1.8 MeV flux is ~ 3 M$_\odot$, but this estimates depends somewhat on the assumed geometry of the emission (Prantzos and Cassé 1986, Prantzos, Cassé, and Arnould 1988).

The origin of the ^{26}Al itself is however unclear. Novae and SN do produce this isotope (see, e.g., Signore and Vedrenne 1988), but would produce too much ^{27}Al if they alone accounted for the observed ^{26}Al. ON stars ("blue stragglers") have also been suggested, but, given the known number of such stars, they could account for only $\simeq 10\%$ of all the ^{26}Al (Walter and Maeder 1989). W-R stars, on the other hand, could explain all the ^{26}Al only provided the galaxy contained some 10^4 of them, which, even taking into account the uncertainties already mentioned above, seems too many (Prantzos, Cassé, and Arnould 1987). More likely, they may contribute to the ^{26}Al only a few times more than the ON stars.

However, in this context, an interesting connection can be made with the "galactic ridge", since the *same* stars could be responsible for at least a significant part of both the galactic ridge (via the interaction of their winds with the surrounding HII regions) and the diffuse nuclear γ-ray line emission (also via their winds, but this time spreading the ^{26}Al synthesized in their cores into the interstellar medium).

6. High-energy γ-ray emission

There may be links between giant HII regions and γ-ray emission even at higher energies. Following the early results by SAS-2, the COS-B satellite has mapped the galactic emission in the ~ 30 MeV-5 GeV range, leading to the discovery of 25 "sources" (Bignami and Hermsen 1983). In reality, because of the poor angular resolution of these experiments (error box of ~ 1° radius), any localized region of diffuse emission within this error box can be proposed as the source counterpart. Molecular clouds, for instance, are good candidates since they represent concentrated gaseous masses immersed in the galactic cosmic ray flux (see above, § 1.2). On the other hand, mainly on the basis of coincidences along the line of sight, Montmerle (1979) had proposed the identification of ~ ⅓ to ½ of the COS-B sources with giant HII regions containing SNRs; others, containing stars with strong stellar winds (especially W-R stars), were soon added to the list (see Montmerle 1985, 1988). The physical mechanism invoked is that in both cases strong shock waves are generated, which can accelerate protons in situ (by the so-called diffusive acceleration mechanism, see, e.g., Drury 1983, Völk 1988). Furthermore, the fact that the acceleration region is in both cases surrounded by large quantities of ionized gas (~ 10^3 to 10^5 $M_\odot$) allows an efficient trapping by Alfvén waves of the accelerated particles, and consequently efficient collisions which produce high-energy γ-rays by pion-muon decay and bremsstrahlung. Using such a mechanism, and modelling giant HII regions like the Carina nebula and their associated molecular cloud in a simplified way, Montmerle and Cesarsky (1981), and Cesarsky and Montmerle (1983), have shown that a high-energy proton density of 10 to 100 times the average density in galactic cosmic rays should be present. Given the large shock energies available, to attain such a density requires an acceleration efficiency of only a few percent, hence should be easily attained. It will be the task of new experiments which a better angular resolution (like GRO, to be launched next fall) to test this model: one of the predictions is that the γ-ray emission of the HII region should be stronger than that of the neighbouring molecular cloud (itself already a strong γ-ray emitter because of the interaction of ambient galactic cosmic rays and its large mass). Giant HII regions containing W-R stars, like the Carina nebula, are therefore prime targets for these experiments.

7. Summary and prospects

Because they contain highly energetic objects which generate shock waves (stars with strong stellar winds, especially W-R stars, and/or supernova remnants), and/or release nucleosynthetic products (W-R stars), giant HII regions are important localized sources of high-energy emission, from ~ keV X-rays to ~ GeV γ-rays. They may be important even on a galactic scale (possible connection with the "galactic [X-ray] ridge" and the 1.8 MeV line diffuse emission). The Table summarizes the various connections discussed in this paper.

   More particularly in the context of W-R stars, it is also apparent from this Table that the study of high-energy emission from giant HII regions can tell us a lot about the external regions of these stars:
-   X-rays tell us about the outer regions of the winds (as does the radio emission);
-   nuclear γ-ray lines tell us about the nucleosynthetic yield of radioactive isotopes;
-   high-energy γ-rays are related to the mass-loss energetics.

   However, in the two last cases, the angular resolution is not sufficient to allow observations of single stars, so that for the time being we mostly have access to quantities integrated over geometrically relatively large regions. In the forthcoming years, several space experiments should help clarify the contribution of individual W-R stars to the high-energy emission of galactic plane, and it is hoped that, at the time of the next IAU Symposium on W-R stars, totally new results will be available in this field.

TABLE.

HIGH-ENERGY EMISSIONS FROM GIANT HII REGIONS

Energy domain	Stars (massive)	Stars (low-mass PMS)	Wind bubble	[SNR]	W. shock region	HII gas
< keV			+		+[1]	
~ keV	+	+		+		
x keV					+[2]	
γ-ray lines (MeV)	+ [3]					
>30 MeV-GeV	+ [4]				+	+
NT radio	+	+		+	? [5]	
CR: e(>GeV)	+ [6]		+	+	+	+[7]
CR: p(>GeV)	? [8]			+	+	+[9]

---

[1] Estimated from Breitschwerdt and Kahn (1988), see text, § 3.

[2] In the model of Dorland and Montmerle (1987).

[3] W-R stars.

[4] W-R stars (Pollock 1987a)

[5] From the wind shock. Presumably buried in the thermal emission and/or absorbed by the ionized gas.

[6] Seen from the non-thermal radio emission.

[7] See Note (5).

[8] Possibly accelerated within the wind flow by shocks (turbulence, instabilites, etc.; see Cesarsky and Montmerle 1983).

[9] Indirectly visible via γ-ray emission.

REFERENCES

Becker R.H., Boldt E.A., Holt S.S., Pravdo S.M., Rothschild R.E., Serlemitsos P.J., Swank J.H. (1976), *Ap.J. (Letters)* **209**, L65.

Bertout C. (1989), *Ann.Rev.Astr.Ap.* **27**, 351.

Bieging J.H., Abbott D.C., Churchwell E.B. (1989), *Ap.J.* **340**, 518.

Bignami G.F., Hermsen W. (1983), *Ann.Rev.Astr.Ap.* **21**, 67.

Bouvier J. (1990), *Astr. Ap.*, in press.

Breitschwerdt D., Kahn F.D. (1988), *M.N.R.A.S.* **235**, 1011.

Breitschwerdt D., Kahn F.D. (1989), *M.N.R.A.S.* **242**, 209.

Campbell P.M. (1984), *Phys. Rev. A* **30**, 365.

Cesarsky C.J., Montmerle T. (1983), *Sp. Sci. Rev.* **36**, 173.

Chlebowski T. (1989), *Ap.J.* **342**, 1091.

Chlebowski T., Harnden F.R., Jr., Sciortino S. (1989), *Ap.J.* **341**, 427.

Chlebowski T., Seward F.D., Swank J., Szymkoviak A. (1984), *Ap.J.* **281**, 665.

Chu Y.-H. (1983), *Ap.J.* **269**, 202.

Chu Y.-H., Mac Low M.-M. (1990), *Ap.J.*, in press.

Dorland H., Montmerle T. (1987), *Astr. Ap.* **177**, 243.

Dorland H., Montmerle T., Doom C. (1986), *Astr. Ap.* **160**, 1.

Drury L. O'C. (1983), *Rep. Progr. Phys.* **46**, 973.

Feigelson E.D. (1987), in *Proc. 5th Cambridge Cool Star Workshop*, eds. J.L. Linsky and R. Stencel (Berlin: Springer), p. 455.

Gaetz T.J., Salpeter E.E. (1983), *Ap.J.Suppl.* **52**, 155.

Hogg D.E. (1989), *Astr.J.* **98**, 282.

van der Hucht K.A., Hidayat B., Admiranto A.G., Supelli K.R., Doom C. (1988), *Astr. Ap.* **199**, 217.

Koyama K. (1987), *Pub.A.S.J.* **41**, 665.

Koyama K., Asaoka I., Ushimaru N., Yamauchi S., Corbet R.H.D. (1990), *Ap.J.*, in press.

Langer N. (1989), *Astr. Ap.* **220**, 135.

Luciani J.F., Mora P., Pellat R. (1985), *Phys. Fluids* **28**, 835.

Meaburn J., Solomos N., Laspias V., Goudis C. (1989), *Astr. Ap.* **225**, 497.

Montmerle T. (1979), *Ap.J.* **231**, 95.

Montmerle T. (1985), Proc. *19th Int. Cosmic Ray Conf.*, San Diego, **1**, 209.

Montmerle T. (1986), Proc. *Advances in Nuclear Astrophysics*, Paris, eds. E. Vangioni-Flam et al. (Gif-sur-Yvette: Editions Frontières), p. 335.

Montmerle T. (1987), Proc. *Starbursts and Galaxy Evolution*, Les Arcs, eds. T.X. Thuan, T. Montmerle, and J. Tran Than Van (Gif-sur-Yvette: Editions Frontières), p. 47.

Montmerle T. (1988), Proc. NATO ASI *Genesis and Propagation of Cosmic Rays*, Erice, eds. M.M. Shapiro and J.P. Wefel (Dordrecht: Reidel), p. 131.

Montmerle T., André Ph. (1988), Proc. NATO ASI *Formation and Evolution of Low-Mass stars*, Viana do Castelo, eds. A.K. Dupree and M.T.V.T. Lago (Dordrecht: Reidel), p. 225.

Montmerle T., Cesarsky C.J. (1981), Proc. Int. School and Workshop *Plasma Astrophysics*, Varenna, ESA SP-161, 319.

Montmerle T., Koch-Miramond L., Falgarone E., Grindlay J. (1983), *Ap.J.* **269**, 182.

Pollock A.M.T. (1987a), *Astr. Ap.* **171**, 135.

Pollock A.M.T. (1987b), *Ap.J.* **320**, 283.

Prantzos N., Cassé M. (1986), *Ap.J.* **307**, 324.

Prantzos N., Cassé M., Arnould M. (1988), Proc. *20th Int. Cosmic Ray Conf.*, Moscow, **1**, 152.

Rosner R., Golub L., Vaiana G.S. (1985), *Ann.Rev.Astr.Ap.* **23**, 413.

Seward F.D., Chlebowski T. (1982), *Ap.J.* **256**, 530.

Signore M., Vedrenne G. (1988), *Astr. Ap.* **201**, 379.

Völk H. (1988), Proc. *20th Int. Cosmic Ray Conf.*, Moscow, **7**, 157.

Walter R., Maeder A. (1989), *Astr. Ap.* **218**, 123.

Warwick R.S., Norton A.J., Turner M.J.L., Watson M.G., Willingale R. (1988), *M.N.R.A.S.* **232**, 551.

Warwick R.S., Turner M.J.L., Watson M.G., Willingale R. (1985), *Nature* **317**, 218.

Weaver R., McCray R., Castor J., Shapiro P., Moore R. (1977), *Ap.J.* **218**, 377.

Williams P.M., van der Hucht K.A., Pollock A.M.T., Florkowski D.R., van der Woerd H., Wamsteker W.M. (1990), *M.N.R.A.S.* **243**, 662.

DISCUSSION

Chu: The Rosette nebula's bubble is in the central cavity; the fat ring represents the ambient imperturbed medium. There does not seem to be disagreement between the nebula morphology and the energy conserving bubble model. Furthermore, the X-ray emission from Rosette is consistent with what the energy conserving bubble model predicts.

Montmerle: What you say can be true only if you forget about the age of the exciting stars. In the energy conserving model, the Rosette nebula should be very young, much younger than the age of the stars you find using mass losing stellar evolutionary models, *i.e.*, several million years, even taking into account more recent work (with different prescriptions for overshooting, mass loss rates, *etc.*). Conversely, if you take the past evolution of the exciting O stars into account, the corresponding energy conserving bubble should now be much larger (radius of 30 *pc*, as compared with the observed $\sim$ 10 *pc*). An extensive discussion on this point is contained in Dorland *et al.* (1986); additional arguments are presented in Dorland and Montmerle (1987).

Walborn: I question the hypothesis that diffuse X-ray emission from giant *HII* regions is due to low-mass pre-MS stars. (1) The existence of such stars with a normal IMF in giant *HII* regions is an open question; (2) in Carina, Herrer and I found a very complex interstellar absorption-line velocity structure corresponding to soft X-ray energies, most likely originating in the massive stellar winds; (3) one of the three similar WN-A stars in Carina is embedded in dense surrounding material, and it only has very high velocity interstellar lines and strong X-ray emission, suggesting interaction between the wind and interstellar medium as the source of both.

Montmerle: First, the X-ray energies corresponding to the thermalization of very fast winds such as those of O stars or WR stars are typically of several *keV* (the exact energy depends on the assumed dissipation mechanism, see §3 of my paper). This is what has been found with OSO-8, the *EINSTEIN* MPC, and *TENMA*. This should not be confused with the sub-*keV* diffuse emission resolved by the "Einstein" IPC and reported by Chlebowski *et al.* As regards this last component, all I am saying is that, if you take the average soft X-ray ($\sim keV$) luminosity per T Tauri star, and assume they have the same surface density (within the outer layers of the clouds, *i.e.*, those still transparent to the X-rays) *as actually observed* in nearby molecular clouds (*e.g.* Montmerle *et al.*, 1983), and integrate over the emissive area, you find a very good agreement with the observed flux in Carina and Orion, after having removed the low-energy tail of the hard X-rays, if any. In this interpretation, the soft X-ray emission would not be diffuse, but simply unresolved. A detailed discussion in the case of the Carina nebula appears in Dorland and Montmerle (1987), and takes into account your work as available at the time. Now the real content in T Tauri stars associated with giant HII regions is admittedly not known (they are too far away), but there is little doubt such stars must be present! And the results on the soft X-rays, in this interpretaion, merely imply that the formation rate of low-mass stars is about the same for all molecular clouds - a really interesting, but not too surprising, conclusion...

Jura: In the case of WR winds, you are looking to a non-standard composition with perhaps a lot of extra C, He. What is the effect of that on their X-rays?

Montmerle: This is more a question for Pollock. To the extent that you are dealing with non-thermal X-rays, of course, then this depends only on the high energy electron content of the envelope. But, of course, if you have a different composition, and if you are, *e.g.*, in a binary system, then the low energy part of the X-rays - if the X-rays are thermal - are probably affected by the presence of lines. But with *EINSTEIN*, there have been a few spectral data that point to this.

Pollock: X-ray observations are a fantastic way of finding out the chemical composition of the winds of hot stars, and soon there are going to be some observations from *ROSAT* which will tell us a lot about those things. I contend a difference between the mean X-ray luminosities of WN and WC stars, which tell you that they are chemically different to each other, and a number of other things as well. But, I also would like to make a couple of comments about what Walborn said about the WR stars in Carina. There are three WR stars there, one is much more luminous than the other two, it is fainter optically, it is hiding behind loads of absorbing material. But, I think that the evidence that big X-ray luminosity has something to do with the interaction with a dense interstellar medium is poor, because there is absolutely no evidence from the high angular resolution HRI observations that that X-ray source is extended. It is unresolved in the HRI and you would expect to see some extension if the X-rays were coming from a bubble of some sort in the interaction with the interstellar medium. And as far as the high energy X-rays go from the Carina Nebula, with particular reference to the OSO-8 measurements, it is true to say that AG Car itself has a very hard X-ray spectrum. In the SSS, which was the good energy resolution instrument aboard *EINSTEIN*, AG Car has an almost flat, in fact slightly rising X-ray spectrum and I think that could account for the sort of OSO-8 picture you were showing there. So, I would just say, careful about that one!

Montmerle: I think this has been taken into account. Knowing the emission from AG Car itself and attributing also into this a kind of wind thermalization mechanism, this can be removed, and I was referring mostly to the X-ray emission without AG Car, even though the picture refers to everything.

Thierry Montmerle

WARMERS

Jorge Melnick and Mohammad Heydari-Malayeri
European Southern Observatory
Casilla 19001, Santiago 19
Chile

ABSTRACT. We present direct observational evidence showing that early type WR stars, and WO stars in particular, are exceedingly hot, and that their ionizing radiation is not fully trapped by their extended atmospheres. We show that extreme WR stars, or *Warmers,* are found in metal poor as well as in metal rich regions in accordance with theory. Images of HeIII regions around Galactic and Magellanic Cloud *Warmers* are shown.

1. INTRODUCTION

The latest stages in the evolution of massive stars are identified with Wolf-Rayet (WR) stars of the carbon sequence. Evolutionary models show that mass loss will remove the outer envelopes of these stars exposing their extremely hot helium burning cores. The models predict that the effective temperatures of these early type WC and WO stars can exceed 100,000K (Maeder and Meynet, 1989). The emergence of these hot stars in young starburst clusters results in a dramatic hardening of integrated ionizing spectra of these clusters, and this effect led Terlevich and Melnick (1985) to introduce the term *Warmers* to refer to massive stars whose effective temperatures are higher than their original ZAMS value. Terlevich and Melnick showed that, not surprisingly, the emission line spectra of model HII regions ionized by clusters with *Warmers* match extremely well the observed spectra of narrow line Seyfert galaxies.

A fundamental uncertainty about these *Warmer* models, however, is whether or not the winds of WR stars are optically thick to ionizing radiation, and in particular, to radiation harder than the HeII limit at 4 Rydbergs.

In this contribution we present observational evidence showing that the hard ionizing radiation from the hottest WR stars is not completely absorbed by their winds. This evidence comes from observations of nebular HeII emission in the integrated spectra of luminous emission-line galaxies, and in nebulae associated with individual WR stars in the Galaxy and the Magellanic Clouds.

2. HII GALAXIES

HII galaxies are gas rich dwarf irregular galaxies whose observable properties are dominated by one or several extremely young giant HII region components (Melnick, 1987). The starburst clusters

K. A. van der Hucht and B. Hidayat (eds.),
Wolf-Rayet Stars and Interrelations with Other Massive Stars in Galaxies, 409–416.
© 1991 *IAU. Printed in the Netherlands.*

in HII galaxies may contain hundreds to thousands of very young massive stars, so HII galaxies are excellent objects to test evolutionary theories of metal poor massive stars. Figure 1 presents the run of the electronic temperature of the nebular gas (Te) as a function of oxygen abundance (O/H). The lines show the predictions of photoionization models using the isochrones of Maeder and Meynet (1989) and Mihalas' (1972) model atmospheres for different effective temperatures. The effect of changing the ionization parameter is illustrated in this Figure by models computed using two extreme values of this parameter which span the range observed in HII galaxies (Melnick and Terlevich, 1987).

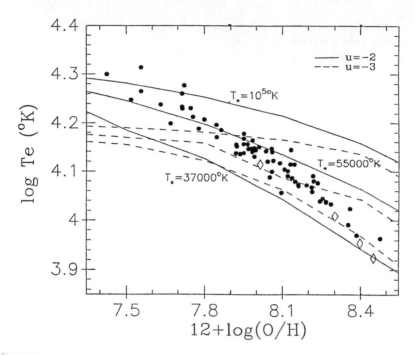

FIGURE 1: Relation between electron temperature and oxygen abundance for the best observed HII galaxies. Typical errors are $\sim 10\%$ in O/H and $\sim 5\%$ in Te. The lines show photoionization models using Mihalas' (1972) NLTE atmospheres for two extreme values of the ionization parameter u. More details about the models are given in Melnick and Terlevich (1987).

Models using other stellar atmospheres are not significantly different from the curves presented in Figure 1 (Melnick and Terlevich,1987). The feature of this diagram which is relevant in the present context is that the electron temperatures of the most metal poor HII galaxies can only be explained if their ionizing clusters contain exceedingly hot stars. Notice that models plotted in the figure are for *single* stars.

The models of Maeder (1990) predict that *Warmers* should appear either in very metal poor, or in very metal rich, systems. This is because the evolutionary time scales decrease with abundance while the mass loss rates increase. Thus, for metal poor stars winds peel off the envelopes at a very advanced stage of evolution. As a consequence, only the most massive stars arrive to the *Warmer* phase and essentially all the WC stars in metal poor clusters should be WOs. For SMC abundances, the minimum mass of stars that reach the WC phase is close to 65 $M_\odot$ (Maeder, 1990).

A further strong evidence for the presence of such stars in metal poor HII galaxies is the observation of *nebular* HeIIλ4686 lines in the spectra of these galaxies. Three examples of HII galaxies with nebular HeII lines are presented in Figure 2.

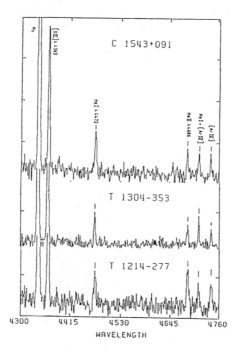

FIGURE 2: Nebular HeII lines in the spectra of three extreme HII galaxies. Notice the strength of the temperature sensitive [OIII]λ4363 lines.

3. INDIVIDUAL WARMERS

At least in principle, a direct determination of the hardness of the ionizing radiation emitted by WO stars could be simply obtained from the analysis of the nebular spectrum of HII regions ionized by these stars. In practice, however, this is made difficult by the fact that the strong winds from WR stars very effectively sweep out the interstellar medium surrounding them and thus prevent the formation of bright HII regions.

As a backup programme for cloudy nights, we have undertaken a search for HeIII regions around hot WR stars in the Galaxy and the Magellanic clouds. Our prime candidates for this search were, of course, the 4 known WO stars with massive progenitors (Sanduleak, 1971). We have obtained images of 3 of these stars: Sand 1=Sk 188 in the SMC, Sand 2=Br93 (Breysacher, 1981), and Sand 4=WR102 (van der Hucht *et al.* , 1981). Sand 1 is a member of the young SMC cluster NGC602c, and shows extremely weak [OIII] emission and no HeII in our images. Br93 is embedded in a bright HII region with a very complex structure in [OIII], but our HeII image is of insufficient quality to reveal any HeII emission.

Sand 4 = WR102 is a galactic WO star associated with the ring nebula G2.4+1.4 (van der Hucht *et al.* , 1981) where Johnson (1975) reported the detection of strong HeII lines. Figure 3 presents a narrow band HeII CCD image of this nebula obtained with the 3.6m telescope on La Silla which shows the presence of a filamentary HeIII region around the star.

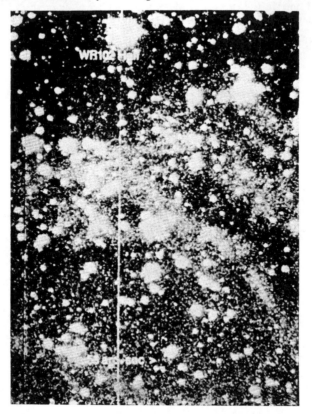

FIGURE 3: Narrow band CCD image of the nebula associated with the WO star WR102 obtained through a 60 Å wide filter centered at the wavelength of the HeIIλ4686 line. The image was exposed for 45 minuted with EFOSC at the 3.6m telescope on La Silla

Johnson (1975) suggested that G 2.4+1.4 was a supernova remnant (SNR) and therefore that the HeII lines could be excited collisionally, but this conclusion was disputed by Green and Downes (1987) who, on the basis of VLA data, conclude that the nebula is most likely a stellar wind blown bubble. This is supported by Dopita *et al.* (1989) and by our own observation that show that G 2.4+1.4 is a wind blown bubble ionised by the WO star.

We have obtained a long slit spectrum of the nebula using EFOSC and the 3.6m telescope on La Silla on a partially cloudy night on June, 16, 1989. Tracings of the spectrum at several positions of the HeIII filaments are presented in figure 4. Table 1 summarizes the relevant line ratios derived from this spectrum.

TABLE 1
WR102: nebular line intensities

Line	Relative intensity
[OII]λ3727	0.5:
[OIII]λ4363	<0.01
HeIIλ4686	0.77
Hβ	1.00
[OIII]λ5007	5.20
C(Hβ) = 2.1	
E(B-V) = 1.5	

From the HeII/Hβ ratio we derive a black-body Zanstra temperature of $\sim 10^5$ °K for the star. The weakness of the temperature sensitive [OIII]λ4363 line implies that the nebula must be metal rich. We estimate an oxygen abundance of $\sim$1.5 times solar on the basis of rough photoionization models. This value is consistent with the galactocentric distance of the nebula derived using the absolute magnitude of WR102 and the reddening derived from the Balmer decrement.

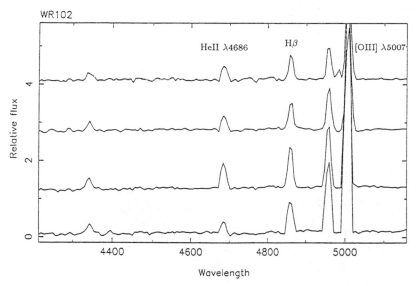

FIGURE 4: Intensity calibrated tracings of the spectrum of the nebula around WR102 for several filaments. The strength of the HeIIλ4686 line relative to Hβ throughout the HeIII regions is clearly seen

Dopita *et al.* (1990) determined line ratios for G 2.4+1.4 which are largely consistent with our values, but they derive a much higher temperature for the star, and also a much higher oxygen abundance. We have not attempted to run very detailed photoionization models because of the uncertainties in the geometry of the nebula, and its abundance. An illustration of these uncertainties is the difficulty encountered both in our models and those of Dopita *et al.* to match the strength of the [OII]λ3727 doublet. We believe, therefore, that the difference between our values for Teff and O/H and those of Dopita *et al.* reflect the range of possible values and not a fundamental difference in the photoionization models.

The observations of WR102 clearly show, however, that WO stars are very hot and that the hard ionizing radiation is not trapped by their extended atmospheres.

Figure 5 presents narrow band HeII images of HeIII regions around WR stars in the Magellanic Clouds. The first panel shows an image of the SMC star AB7 (WN3p+OB) whose HeIII region has been discussed in detail by Pakull and Motch (1989). The second panel shows an image of the HeIII nebula around the LMC star Br2 (WN4) recently discovered by Pakull (1990).

We have also photographed a number of other early type WR stars in the LMC using a narrow band filter centered at the wavelength of HeIIλ4686, without much success. Figure 5 shows a possible candidate around the star Br7 (WC4). The [OIII] image shows that the star appears embedded in a bright HII region. The HeII image suggests the presence of a faint HeIII region near the star, but spectroscopy is required for confirmation.

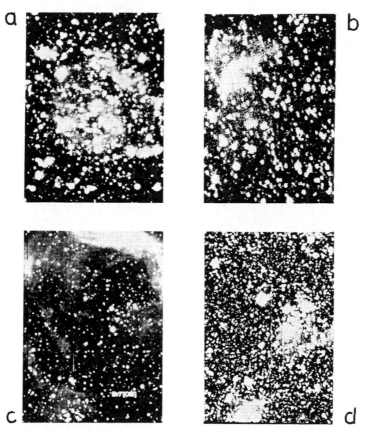

FIGURE 5: Examples of HeIII nebulae around hot stars in the Magellanic Clouds. a) HeII image of AB7 obtained with the Danish 1.5m telescope on La Silla. b) HeII image of Br2 obtained during the commissioning phase of the NTT. c) [OIII] image of Br7, d) HeII image of Br7, both taken with the 3.6m telescope on La Silla.

These observations provide direct evidence for the existence of *Warmers* in low abundance regions while WR102 is an example of a *Warmer* in a metal rich region in the direction to the Galactic centre. Thus, in agreement with the theory, *Warmers* are found both in metal poor as well as in metal rich systems.

Pakull and Motch (1989) comment that, in at least one of their cases, the nebular HeII emission appears significantly weaker than the broad HeII line from the WR star. This leads them to question whether the integrated spectrum of a cluster containing a population of *Warmers* would show nebular HeII lines. Because of the effects of stellar winds, however, the HII regions associated with the individual *Warmers* discussed above are density bounded and therefore the nebular emission lines are weak. In starburst clusters, on the other hand, most of the original gas has not been dispersed by star formation and the HII regions are ionization bounded. Therefore, the stellar lines are diluted by the clusters, while the nebular lines are very strong.

4. CONCLUDING REMARKS

We have showed that the *Warmer* model for nuclear activity in galaxies stands on a solid observational basis in what regards the existence of *Warmers* as hot WR stars that produce copious amounts of hard ionizing radiation.

We also find a substantial agreement between theory and observations regarding the formation of *Warmers* in metal poor systems. As a further test to the theory, it is of crucial importance to determine abundances of the gas in the immediate vicinity of the WR stars.

REFERENCES

Breysacher, J., 1981, *Astr. Ap. Suppl. Ser.*, **43**, 203.

Dopita, M.A., Lozinskaya, T.A., McGregor, P.J., Rawlins, J., 1990, *Ap. J.*, **351**, 563.

Green, D.A., Downes, A.J.B., 1987, *M.N.R.A.S.*, **225**, 221.

Johnson, H.M., 1975, *Ap. J.*, **198**, 111.

Maeder, A., Meynet, G., 1989, *Astr. Ap.*, **210**, 155.

Maeder, A., 1990, *Astr. Ap. Suppl. Ser.in press.*

Melnick, J., 1987, in *Starbursts and Galaxy Evolution*, eds. T.X. Thuan, T. Montmerle, and J. Tran Thanh Van; Editions Frontières, p. 215.

Melnick, J., Terlevich, R., 1987 in *Evolution of Galaxies*: Tenth European Regional Meeting of the IAU, ed. Jan Palous; Prague, August 1987, 111.

Pakull, M.W., Motch, C., 1989 in ESO workshop on *Extranuclear Activity in Galaxies* eds., E.G.A Meurs and R.A.E. Fosbury, p. 285.

Pakull, M.W., 1990, *private communication.*

Sanduleak, N., 1971, *Ap. J. (Letters)*, **164**, L71.

Terlevich, R., Melnick, J., 1985, *M.N.R.A.S.*, **213**, 841.

van der Hucht, K.A., Conti, P.S., Lundström, B., Stenholm, B., 1981, *Space Sci. Rev.* **28**, 227.

DISCUSSION

Langer: If you need many (*i.e.*, more than one or two) warmers to account for some kind of galactic activity you may be in trouble: theory predicts $T_{eff} > 10^5 K$ only for a very brief evolutionary time ($< 10^4$ yr), and observationally very hot WR stars are also very rare ($\sim 1\%$).

Filippenko: A crucial assumption of the Terlevich-Melnick hypothesis is that Warmers form *much* more easily in metal-rich environments than in metal-poor environments. However, from your talk, and especially from Pakull's talk yesterday, I do not see any *observational* evidence for this in studies of individual Warmers.
Heydari-Malayeri: In this work we have concentrated on the Magellanic Clouds, which are metal-poor galaxies. We also show the presence of a Warmer in a metal-rich region of the Galaxy. On the other hand, we do not know at all how the atmospheres of WR stars react to hard UV photons when metallicity is high.

Kunth: Have you attempted to detect $HeIII$ regions around known O-LMC stars?
Heydari-Malayeri: O stars are not hot enough to produce $HeIII$ regions.
Niemela: With Heathcote and Weller we have observed HII regions around O3 stars in LMC, and we did not see any nebular $HeII$ in these HII regions.

Lortet: From Westerlund's BV photometry (1964), there is a red supergiant candidate ($v \approx 14$?) in NGC 602c, the cluster containing the WO4 star of the SMC. Is there any spectrum of this star?
Heydari-Malayeri: We have no spectrum of that star.

Virpi Niemela

RECURRENT DUST FORMATION BY WOLF-RAYET STARS

P.M. WILLIAMS[1], K.A. VAN DER HUCHT[2], P.S. THÉ[3],
P. BOUCHET[4] and G. ROBERTS[5]

[1] *Royal Observatory, Blackford Hill, Edinburgh EH9 3HJ, United Kingdom*
[2] *SRON Laboratory for Space Research,*
   Sorbonnelaan 2, 3584 CA Utrecht, The Netherlands
[3] *Astronomical Institute 'Anton Pannekoek',*
   Roetersstraat 15, 1018 WB Amsterdam, The Netherlands
[4] *European Southern Observatory, Karl-Schwarzschild-Strasse 2,*
   D 8046 Garching bei München, F.R. Germany
[5] *S.A. Astronomical Observatory, P.O. Box 9, Observatory 7935,*
   South Africa

ABSTRACT. A number of Wolf-Rayet stars show variations of up to a factor of ten in their infrared emission on timescales of months to years while their photospheric luminosities remain unchanged. This can be interpreted in terms of variation in the rates at which dust grains form in their stellar winds. Our data show variable circumstellar dust emission from WR 70 (HD 137603), with an episode of enhanced dust formation in early 1989, and fading of emission by dust formed around WR 48a (in 1979) and WR 19 some time before our first observation in 1988. We consider their relation to WR 140 (HD 193793), which forms dust in its wind for a few months at intervals of 7.94 years.

1. WR 48a: a decade of slowly cooling dust

This is a visually faint star discovered in the infrared by Danks *et al.* (1983) as it rose to maximum in 1979. We have been monitoring its steady fading since 1982 and present in Fig 1 a $2.2\,\mu m$ (K) light curve showing fading at a rate of about $0^{m}26$/year. The fading has been slower at longer wavelengths, *e.g.* $0^{m}11$/year at $10\,\mu m$ and $0^{m}22$/year at $3.8\,\mu m$, indicating that the dust is cooling. At shorter wavelengths, *e.g.* $1.25\,\mu m$, the fading was initially faster, as we would expect from cooling dust, but then slowed down and stopped (in 1989) as the dust emission became comparable with that from the stellar photosphere and wind.

K. A. van der Hucht and B. Hidayat (eds.),
Wolf-Rayet Stars and Interrelations with Other Massive Stars in Galaxies, 417–420.
© 1991 *IAU. Printed in the Netherlands.*

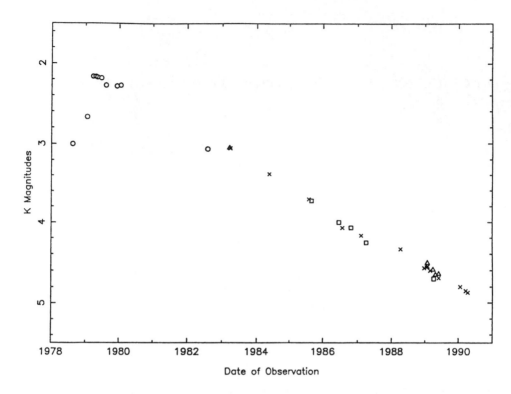

Fig. 1. Infrared light curve of WR 48a based on observations made at ESO (×), the SAAO (△) and the AAT (□) and those published by Danks et al. (○).

In order to model the evolution of the cooling dust shell, as was done for that of WR 140 (Williams *et al.* 1990), we need to know the 'baseline' infrared energy distribution of WR 48a before the dust condensation occurred. Initially, we assume this to be free-free radiation from the stellar wind alone. Modelling the infrared spectra yields the dust temperatures at the times of observation and, from consideration of the thermal equilibrium of the grains in the stellar radiation field, the distances of the dust shell from the central star. We need to know the luminosity of WR 48a to determine the expansion velocity in this way, but the velocities derived ($\approx$ 30 km/s) for reasonable estimates are so much lower than plausible values of the (also unknown) velocity of WR 48a's stellar wind that we have to question our initial assumption about the pre-outburst spectrum of WR 48a. If this showed emission by dust formed in an earlier event which had cooled by expansion, then the dust temperatures derived from the 1979–1990 observations cover a larger range, yielding a higher value for the expansion velocity. This suggests that, although only one has been observed, dust formation episodes from WR 48a may be recurrent – as is the case with WR 140. Besides continued monitoring of the circumstellar dust, we need to learn more about the star in WR 48a, particularly whether or not it is a binary like WR 140, whose dust formation was linked to its orbital motion by Williams *et al.* (1990).

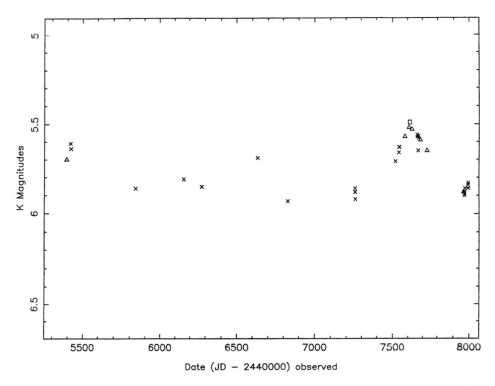

Fig. 2. Infrared light curve of WR 70 (symbols as in Fig 1.).

2. WR 70: a binary with variable dust formation

The WC8+B0 binary system WR 70 (= HD 137603) differs from WR 48a and WR 140 in showing variations on a shorter timescale ($\approx$ 1–2 years) and of significantly lower amplitude ($\approx 0^m\!.5$ in K). Like the other two cases, the amplitude increases with wavelength ($\approx 0^m\!.6$ in M but only $\approx 0^m\!.3$ in H), indicating that it is the amount of circumstellar dust that is varying rather than the underlying star. The (K-L) and (K-[3.8]) colours get progressively redder as the flux rises to and falls from the well-defined maximum around JD 2447650 (1989 March). This observation that the dust is hotter during the rise indicates that the rise is caused by an increase in the amount of dust condensing in the wind at the inner edge of the dust shell for 3–6 months and the fall by the carrying away of this extra dust in the wind after this dust formation ceases. More intensive observations are needed to characterize such events, confirm their recurrence and establish the timescale of the variations.

We recall that, like WR 140, WR 70 is a binary so that the relation of its dust formation episodes to its orbital motion will be of particular interest. The WR 70 system is bright enough in the visible and even UV for detailed study.

3. WR 19: fading dust emission from a WC4 star

Our first observation of WR 19 (LS 3) in 1988 April showed it to have infrared fluxes characteristic of a circumstellar dust shell. Subsequent observations have shown a steady fading of this emission until 1990, when only the free-free emission from the stellar wind was apparent. Like WR 48a, this is a relatively faint star in a difficult field and so will not be easy to study. On the other hand, it is potentially most important, showing that dust formation can occur around stars as early as WC4. As searches for circumstellar dust emission have hitherto been concentrated on stars of WC8 and later, it is likely that dust formation in the winds of Wolf-Rayet stars may be more common than previously thought and that only *continuous* dust formation is restricted to those of the later spectral subtypes.

ACKNOWLEDGEMENTS

We are grateful to the following for contributing observations: David Allen, Brian Carter, Michael Feast, R. de Jong, Dave Laney, A. Leene, Fred Marang, Pieter Mulder, Frank Spaan, R. Vega, Patricia Whitelock, Hartmut Winkler and Dolf de Winter.

REFERENCES

Danks, A.C., Dennefeld, M., Wamsteker, W. and Shaver, P.A., 1983. *Astr. Astrophys.* **118**, 301.
Williams, P.M., van der Hucht, K.A., Pollock, A.M.T., Florkowski, D.R., van der Woerd, H. and Wamsteker, W.M., 1990. *M.N.R.A.S.* **243**, 662.

A TWO-COMPONENT DUST MODEL FOR THE WOLF-RAYET RING NEBULA RCW 58

J.P. CASSINELLI and J.S. MATHIS, *Washburn Observatory, Madison, U.S.A*
K.A. VAN DER HUCHT, *SRON ROU, Utrecht, The Netherlands*
T. PRUSTI and P.R. WESSELIUS, *SRON ROG, Groningen, The Netherlands*

The spectacular ring nebula RCW58 around the Wolf-Rayet star WR40 (HD 96548, WN8) has been observed by the *IRAS*-Survey instrument and by the *IRAS* Chopped Photometric Channel (*CPC*) to study its IR flux and morphology. The survey data were examined with the Groningen Exportable Infra-red High-resolution Analysis system (*GEISHA*) software, and analysed in terms of line and dust contributions.

The *IRAS* flux ratios for RCW 58, and in addition those of two other WR ring nebulae, *i.e.* NGC 6888 and S 308, were compared with predicted flux ratios based on the *MRN* (Mathis, Rumpl & Nordsieck, 1977, *Ap.J.* **425**, 433) grain model, assuming that all grains are in steady thermal equilibrium with the incident radiation. (The standard *MRN* grain mixture has a size distribution $N(a) \propto a^p$ for $a_{min} < a < a_{max}$ with $p = -3.5$, $a_{min} = 0.005\mu$ and $a_{max} = 0.25\mu$.) There is a clear problem, in that all three WR nebulae have anomalously large 25μ fluxes. For RCW58 the $60\mu/100\mu$ flux ratio can be fit if the grains have $T \approx 40K$, while the $25\mu/60\mu$ flux ratio would be too low by an order of magnitude. To increase the $25\mu/60\mu$ flux ratio it is necessary that the grains be hotter. Small grains tend to be hotter than large ones because of their lower emissivities. We have therefore tried a modified *MRN* model for which $a_{min} = 0.002\mu$ and $a_{max} = 0.008\mu$. There is an improved $25\mu/60\mu$ flux ratio, but a fit is not achieved. To fit just the $25\mu/60\mu$ flux ratio of RCW 58 with a *MRN* model requires a temperature of $T \approx 67K$, but for this model the $60\mu/100\mu$ flux ratio would be too large by a factor of 2.5.

Thus (*i*) grains with high temperatures ($T \approx 70K$) are needed to explain the $25\mu/60\mu$ flux ratio; and (*ii*) much cooler grains ($T \approx 40K$) are needed to explain the $60\mu/100\mu$ flux ratio. The need for a hot contribution might possibly be explained by the non-thermal-equilibrium heating (Temperature Fluctuations) of very small grains, whereby the absorption of a photon can briefly increase the temperature of a small grain to temperatures above $100K$.

Incorporating the non-equilibrium temperature distributions of very small grains greatly improves the agreement between model and observations. Very small grains could be produced by sputtering and shattering of larger grains, because of the physical conditions near WR stars.

Our best fit to the *IRAS* data of RCW 58 data uses a two-component model, in which most of the total grain mass is in large *MRN* grains and 10 percent of the grain mass is in the very small non-thermal component. In this model the large grains have a temperature $T \approx 39K$, while the small grains have temperatures ranging between 5 K to 200 K.

From the total IR luminosity of the WR ring nebula RCW 58, it is possible to use the theoretical emissivity to calculate a nebular mass. For our best fit model, we estimate a total dust mass of 0.005 $M_\odot$, which implies a total nebular mass of 0.6 $M_\odot$.

K. A. van der Hucht and B. Hidayat (eds.),
Wolf-Rayet Stars and Interrelations with Other Massive Stars in Galaxies, 421.
© 1991 *IAU. Printed in the Netherlands.*

ON THE CHEMODYNAMICS OF NGC 6888

C. ESTEBAN and J.M. VILCHEZ
Instituto de Astrofísica de Canarias
E–38200, La Laguna, Tenerife, Spain

ABSTRACT. We present preliminary results on an extensive spectroscopical study of the WR Ring nebula NGC 6888. The observations combine high spatial (1.5 arcsec/pixel) and spectral resolution (30 to 50 km s^{-1}) covering most of the optical range – $\lambda\lambda$ 3600 to 6800 Å– at 3 different slit positions along the major axis of the nebula. The spectra of the central parts give an emission system with three different velocities: a) V_{LSR}=-64 km s^{-1}, b) V_{LSR}=+18 km s^{-1}, and c) V_{LSR}=+78 km s^{-1}. Assuming that NGC 6888 is in fact an expanding bubble of gas, as demonstrated by Marston and Meaburn (1988), we can identify components a) and c) as those moving towards us and receding parts of the shell, while b) could be related to the ambient interstellar ionized gas outside the bubble. We have isolated the spectrum for each component in order to analyse the ionization structure and excitation mechanism, as well as to derive their physical conditions and chemical abundances. The use of diagnostic diagrams indicates that the nebula is basically photoionized, without any significant contribution of shock excitation in the zones studied. The ambient interstellar component shows a spectrum typical of an H II region with nearly solar abundances. In the case of the bubble components, their spectra produce [N II]/Hα line ratios which are outside the H II region box, and entering the extended planetary nebulae locus, suggesting a contribution of ejected nitrogen in the bubble. This fact is evident from the abundance analysis we have performed, finding that, with respect to the values quoted for the ambient gas, the O/H is deficient by a factor 4, N/H is 2.5 times higher, and helium is enhanced by a factor 2. These quoted values clearly indicate that a substantial fraction of the gas in the bubble is processed material ejected from the central massive star.

K. A. van der Hucht and B. Hidayat (eds.),
Wolf-Rayet Stars and Interrelations with Other Massive Stars in Galaxies, 422.
© 1991 *IAU. Printed in the Netherlands.*

DETECTION OF STRONG WINDS IN GIANT EXTRAGALACTIC HII REGIONS

H.O. CASTAÑEDA and J.M. VILCHEZ
Instituto de Astrofísica de Canarias
E–38200, La Laguna, Tenerife, Spain

ABSTRACT. The existence of supernova remnants and strong stellar winds in giant extragalactic HII regions should be associated with the existence of gas at very high velocities ($\approx 10^2$ –10^3 km s^{-1}). The detection of large velocity/low intensity gas within the regions could be possible with the use of CCD detectors with long slit intermediate spectroscopy and long integration times. To test this observational approach we have conducted a survey in the brightest giant HII regions of M 101 and M 51. We report on the preliminary results of our programme, and discuss the implications for the energetics of the regions.

K. A. van der Hucht and B. Hidayat (eds.),
Wolf-Rayet Stars and Interrelations with Other Massive Stars in Galaxies, 423.
© 1991 *IAU. Printed in the Netherlands.*

HI BUBBLES SURROUNDING WOLF-RAYET STARS

VIRPI S. NIEMELA
Instituto de Astronomia y Fisica del Espacio
CC 67, Suc 28, 1428 Buenos Aires, Argentina

CRISTINA CAPPA DE NICOLAU
Instituto Argentino de Radioastronomia
CC 5, 1894 Villa Elisa, Argentina

The observations of neutral gas in the neighbourhood of some Wolf-Rayet stars show bubbles blown by the strong winds of these stars. In an attempt to examine how commonly WR stars are surrounded by HI bubbles, we have studied the HI distribution in the vicinity of this type of stars in the section of our galaxy between l= 302 and l= 312 deg. In our study we have used the survey data of Strong et al. and data obtained with the 30 m single-dish antenna of Instituto Argentino de Radioastronomia.

We find that almost all WR stars located at galactic latitudes |b| higher than 2 deg. appear associated with HI bubbles. The parameters of these bubbles are listed in Table 1. The fact that we do not see HI bubbles around WR stars close to the galactic plane, is most probably a selection effect, since due to the greatly increased gas density, these bubbles would be smaller than our spatial resolution. Thus, our results seem to indicate that most WR stars are surrounded by HI bubbles.

The two Wolf-Rayet stars of WN type with associated bubbles, namely WR54 and WR61, are located nearly centrally within their bubbles. In contrast, the WC stars all appear off-center, close to the highest density border of the bubble. This was also the case for the previously studied bubbles around WR17, of spectral type WC5, (AJ 92,1414); and the one surrounding WR90, of spectral type WC7 (AJ 96, 1671). On the other hand, the HI bubbles associated with the WN stars also appear to have larger sizes.

In all cases, the expansion velocities of the HI bubbles appear to be less than 10 km/s. Consequently, the dynamical ages of these bubbles become a few million years, being always larger than the lifetime of the WR stage of the star. Therefore, the progenitors of the now WR stars have greatly contributed to the formation of the bubbles.

TABLE 1. PARAMETERS OF HI BUBBLES AROUND WR STARS

Associated star	WR48	WR52	WR54	WR57	WR61
Spectral type	WC6	WC4	WN4.5	WC8	WN6
LSR Velocity of gas (km/s)	-23	-26	-32	-40	-16
Kinematical distance (kpc)	1.8	2.0	7.6	4.0	10.0
Radius of the Bubble (pc)	45	60	172	35:	95
Swept up mass (1000Mo)	8.5	13	29	3	23
Ambient gas density (cm-3)	0.9	0.4	0.1	0.7:	0.3

424

K. A. van der Hucht and B. Hidayat (eds.),
Wolf-Rayet Stars and Interrelations with Other Massive Stars in Galaxies, 424.
© 1991 IAU. Printed in the Netherlands.

A SEARCH FOR HIGH EXCITATION NEBULAE AROUND WOLF-RAYET STARS IN THE MAGELLANIC CLOUDS

V. S. NIEMELA[1*], S. R. HEATHCOTE[2] and W.G. WELLER[2]
[1]Instituto De Astronomia y Fisica del Espacio, CC67, Suc 28, 1428 Buenos Aires, Argentina
[2]Cerro Tololo Inter-American Observatory, National Optical Astronomy Observatories[**], Casilla 603, La Serena, Chile

We have serendipitously discovered that the WN+O binary Br 40a in the LMC is surrounded by a remarkable high excitation nebula, showing extended, narrow emission lines of HeII. This prompted us to make a systematic search for similar high excitation nebulae around other WR stars in the Magellanic Clouds. This surv ey revealed a second even more extreme example surrounding the WN+O binary AB-7 in the SMC, and one other marginal detection AB-5 (SMC). The detection of nebular HeII emission implies that these WN stars emit a much harder UV spectrum than is traditionally expected. For each of these nebulae we have taken narrow band CCD images at the CTIO 0.9-m telescope, and have obtained spectrophotometry with the 2D-Frutti photon counting detector on the CTIO 1.0-m telescope.

The WN3+O6 star Br 40a (Sk -71° 34) is located in the outskirts of the bright HII region N206 in the LMC. Nebular HeII 4686Å line emission is detected in our two dimensional spectra over a region with an extent of 70 seconds of arc (18 pc). This high excitation zone coincides with a partial ring shaped feature seen in a narrow band [OIII]5007Å image. Figure 1 shows a one dimensional spectrum formed by averaging over the entire extent of the He^{++} region.

The WN3+O6 star AB-7 (AzV 336a) is located near the centre of N76, the second brightest HII region in the SMC. A narrow band HeII 4686Å image shows an approximately circular high excitation zone 144 seconds of arc (37pc) in diameter, centred on the WR star. This He^{++} region fills in a central hole in the annular nebula seen in a narrow band HeI 5876Å image. Figure 2 shows the average spectrum of the high excitation zone.

In our survey we found no evidence for nebular HeII emission in the vicinity of the 27 other WR stars, and 2 hot O3 stars, which are listed in the table. This suggests that either only a small fraction of all WR stars reach such high temperatures, or that the high temperature phase is of extremely short duration, or that most WR stars modify their environment in such a way that a high excitation zone is undetectable.

---

[*]Visiting Astronomer CTIO, NOAO[**]. Senior Visitor ESO. Member of Carrera del Investigador, CIC, Prov. Buenos Aires, Argentina.
[**]CTIO, NOAO, operated by AURA Inc. under a cooperative agreement with NSF.

K. A. van der Hucht and B. Hidayat (eds.),
Wolf-Rayet Stars and Interrelations with Other Massive Stars in Galaxies, 425–426.
© 1991 IAU. Printed in the Netherlands.

Stars included in the survey for which nebular HeII 4686Å emission was not detected

Stars in the SMC		Stars in the LMC					
AB 1	WN3+OB	Br 5	Of/WN6+O7	Br 37	WN3+OB	Br 85	WN4
AB 2	Of/WN4.5	Br 10a	Of/WN6	Br 44	WC4+OB	Br 93a	WN3
AB 3	WN3+O4	Br 16	WN3	Br 46	WN4	Br 95	WN4+OB
AB 4	WN4.5	Br 23	WN3	Br 48	WN4+OB	Br 99	WN4
AB 6	WN3+O7	Br 25	WN3	Br 49	WN3+OB	Br 100	WN3
AB 8	WO4+O4	Br 26	WN7	Br 53	WN4+OB?	Sk-71°51	O3III(f*)
		Br 29	WN3	Br 66	WN3	Sk-67°211	O3III(f*)
		Br 31	WC4+O9	Br 72	B1I+WN3		

AB: numbers from Azzopardi and Breysacher (1979, Astron. Astrophys. **75**, 120)
Br: numbers from Breysacher (1988, Astron. Astrophys. **160**, 185)
Sk: numbers from Sanduleak (1970, *CTIO contributions* N°89)

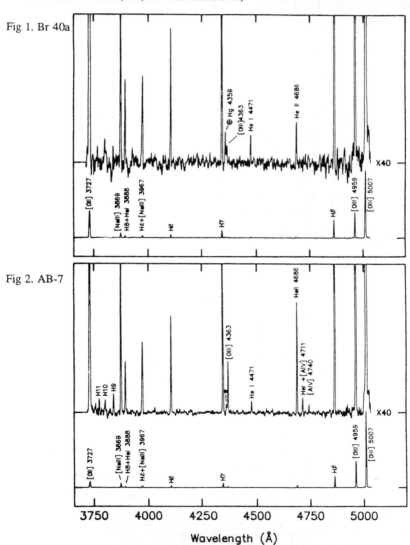

Fig 1. Br 40a

Fig 2. AB-7

Wavelength (Å)

STELLAR CONTENT OF THE GIANT HII REGION NGC 604

Bernard Debray
Astrophysics Division, Space Science Department of ESA
ESTEC, Postbus 299
2200 AG Noordwijk
The Netherlands

BVRI CCD frames of NGC 604, the brightest HII region in the nearby spiral galaxy M33, have been obtained at the Nordic Optical Telescope (Canaries Islands). The seeing was 0.6″ which corresponds to ∼ 2 pc. at the distance of M33. The magnitude of the stars were obtained using the Capella stellar photometry package (Debray et al., 1989).

The $V/B-V$ diagram (Fig. 1) exhibits a narrow and well defined plume of blue supergiants and a much more sparsely populated plume of red stars. The presence of red supergiants confirms the statement by Israel et al. (1990) based on the strong CO absorption seen in their infrared spectra, of the presence of M supergiants in the nebulae together with the hot ionizing stars.

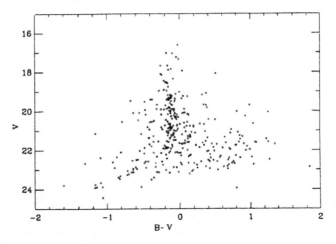

Figure 1. $V/B-V$ diagram of NGC 604.

NGC 604 contains three Wolf-Rayet stars which were once thought to be superluminous star candidates, similarly to R136a, the central object of 30 Doradus. This was first supported by Benvenuti, d'Odorico and Dumontel (1979) and then by Conti and Massey (CM, 1981) and Massey and Hutchings (1983) from IUE observations. The good seeing of the present frames points out clearly that all the bright knots inside the central core of the nebula are groups of at least a few stars (Fig. 2) as already thought by d'Odorico and Rosa (1982).

Taking the values for extinction given by Conti and Massey (1981), and assuming a distance modulus of 24.3 for M33 (Madore et al., 1985), one gets the following M_V magnitudes for the Wolf-Rayet stars : CM11 : -7.9 (instead of -9.1) ; CM12 : -7.7 (instead of -9.5) ; CM13 : -7.4 (instead of -8.7), taking into account the different distance modulus

K. A. van der Hucht and B. Hidayat (eds.),
Wolf-Rayet Stars and Interrelations with Other Massive Stars in Galaxies, 427–428.
© 1991 *IAU. Printed in the Netherlands.*

428

used in CM. These values are now situated in the upper tail of visual magnitudes of WNL stars (Conti, 1986). But the presence of several stellar components for each object within 2 parsecs cannot be excluded.

Similar studies are planned for other giant HII regions in M33 as well as high spatial resolution narrow band imaging to investigate further their WR content.

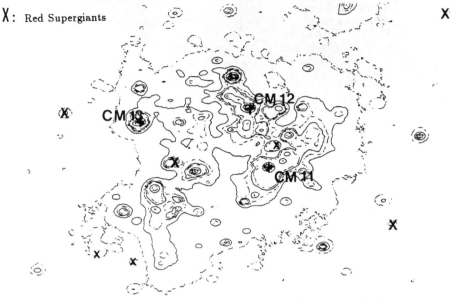

X : Red Supergiants

Figure 2. Isophotes of the central part of NGC 604 in V. One can see for instance clearly that CM13 is made of several components.

References

Benvenuti, P., d'Odorico, S., Dumontel, M. (1979) *Astrophysics and Space Science* **66**, 39

Conti, P.S., Massey, P. (1981) *Astrophys. Journal* **249**, 471

Conti, P.S. (1986) in *IAU Symposium 116, Luminous Stars and Associations in Galaxies*, pp.199-214

Debray, B., Llebaria, A., Dubout-Crillon, R., Petit, M. (1989) *Proceedings of the 1st ESO/ST-ECF Data Analysis Workshop, P. Grosbøl, F. Murtagh and R.H. Warmels. (Eds.)*, 189

D'Odorico, S., Rosa, M. (1981) *Proceedings of the ESO Workshop on "The Most Massive Stars"*, 191

Israel, F.P., Hawarden, T.G., Geballe, T.R., Wade, R. (1990) *Monthly Notices Roy. Astron. Soc.* **242**, 471

Massey, P., Conti, P.S. (1983) *Astrophys. Journal* **273**, 576

Massey, P., Hutchings, J.B. (1983) *Astrophys. Journal* **275**, 578

Madore, B.F., McAlary, C.W., McLaren, R.A., Welch, D.L., Neugebauer, G., Matthews, K. (1985) *Astrophys. Journal* **294**, 560

SESSION VII. EVOLUTION – *Chair: Peter S. Conti*

Conti chairing, Leitherer, Underhill, Niemela, Lozinskaya

Andy Pollock, Lev Yungelson, Arnout van Genderen

STELLAR EVOLUTION IN THE UPPER HRD: THEORY

NORBERT LANGER

Universitäts-Sternwarte Göttingen
Geismarlandstrasse 11, D-3400 Göttingen, F.R.G.

Abstract. Theoretical aspects of the modeling of observable evolutionary phases of massive single stars are reviewed. The SN 1987A progenitor evolution is considered in detail as an example for a star below the WR limit. Formation, structure, evolution, and mass loss of WR stars are discussed, and the impact of supernova research on stellar evolution theory is stressed.

1. Introduction

It is the aim of this review to outline both, recent progress and actual problems of the theory of massive star evolution. Due to space limitations, this work will be far from comprehensive. Instead it will focus on a rather subjective selection of topics with — of course — some emphasis on Wolf-Rayet (WR) stars. This paper concentrates on theoretical aspects of structure and evolution of massive stars. Some effects can be understood easily within a high degree of simplification and abstraction. Therefore, it is sometimes argued in a rather schematic way, e.g. on the basis of schematic diagrams instead of results of actual calculations. It should be noted though, that all conclusions have been verified by detailed model calculations. Finally, the comparison of theoretical stellar evolution results with observations of massive stars deserves more space than it could actually be given here for briefty reasons; the reader is refered to the quoted original literature for that purpose.

The topic of this paper is restricted to observable phases of non-rotating, non-magnetic, massive ($M_{ZAMS} \gtrsim 15\, M_\odot$) single stars (and even for this it cannot be complete). For the non-specialist it may be surprising, that despite such simplifications already the modelling of the earliest evolutionary stages of massive stars are complicated by many problems, as shown in Sect. 2. Relatively much space is devoted to the progenitor evolution of SN 1987A, which is a challenge to stellar evolution theory due to the multitude of tight observational constraints. Sections 3 and 4 deal with the theory of the formation and evolution of WR stars, respectively, while Section 5 briefly outlines the impact of supernova studies on the theory of massive star evolution.

431

K. A. van der Hucht and B. Hidayat (eds.),
Wolf-Rayet Stars and Interrelations with Other Massive Stars in Galaxies, 431–444.
© 1991 *IAU. Printed in the Netherlands.*

2. Early evolutionary phases

Already the modeling of the main sequence phase of massive stars encounters two main problems of stellar evolution theory, namely mass loss and convection. Possibly, massive stars lose a significant fraction of their initial mass on the main sequence, which potentially affects structure and evolution on the main sequence and beyond (cf. Chiosi and Maeder, 1986). Further, massive main sequence stars are very hot, not only at their surface but throughout their interior, as compared to lower mass stars. Radiation pressure is important, which makes convection more likely to occur. Consequently, besides large convective cores, also parts of the envelopes of massive main sequence stars may obtain a superadiabatic temperature stratification, which in connection with mean molecular weigth gradients, leads to the problem of semiconvection (see below). Also the extension of the convective core itself has to be regarded as uncertain due to the unknown efficiency of convective overshooting (cf. Langer, 1986; Renzini, 1987).

a) Main sequence mass loss

In recent years, a quantitative mass loss theory applicable to massive main sequence stars has been developed (Castor et al., 1975; Abbott, 1982; Pauldrach et al., 1986; Owocki et al., 1988), which in principle allows the calculation of the mass loss rate for a given stellar model. However, due to the large numerical effort involved in stellar wind models, completely selfconsistent coupled stellar wind-stellar evolution calculations are not yet available. A step in this direction has recently been done by Langer and El Eid (1990), who performed stellar evolution calculations using the analytical wind solutions of Kudritzki et al. (1989), which approximate hydrodynamic wind models to high precision. The result is — at galactic metallicity — a decrease of the total amount of mass lost during main sequence evolution by a factor 2 − 3 as compared with calculations using empirical mass loss rates. Leitherer and Langer (1990), who calculated selfconsistent ZAMS star-wind models for various metallicities Z and found $\dot{M} \sim Z^{0.6...0.7}$, estimate main sequence mass loss to be negligible (i.e. $\Delta M/M < 5\%$ for main sequence evolution) for stars less massive than $32\,M_\odot$ at solar metallicity, and $50\,M_\odot$ and $80\,M_\odot$ for LMC and SMC, respectively. Thus, according to current theoretical mass loss rates, main sequence mass loss may have been considerably overestimated in recent years.

The reason for the discrepancy of theoretical and empirical mass loss rates on one side but the good agreement of theoretical wind models when compared in detail with observations of individual stars needs further investigation (cf. Pauldrach et al., 1990). For a discussion of evolutionary consequences cf. Chiosi and Maeder (1986), Langer (1990).

b) The role of semiconvection for stars below the Humphreys-Davidson limit

The expression "semiconvection" has different meanings for different astronomers. Here, it is used for the vibrational instability found by Kato (1966) for superadiabatic layers (i.e. $\nabla > \nabla_{ad}$) which contain a positive gradient of the mean molecular weight μ (i.e. $\nabla_\mu := \frac{d\ln\mu}{d\ln P} > 0$). More specifically, this instability occurs for $\nabla_{ad} < \nabla < \nabla_{ad} + \delta\nabla_\mu =: \nabla_L$, where $\nabla_{ad} = \left(\frac{\partial \ln T}{\partial \ln P}\right)_{ad}$, $\nabla = \frac{d\ln T}{d\ln P}$, and $\delta = -\left(\frac{\partial \ln \rho}{\partial \ln T}\right)_P$, i.e. for a situation where the Schwarzschild criterion indicates instability ($\nabla_{ad} < \nabla$) and the Ledoux criterion stability ($\nabla < \nabla_L$). Kato also showed, that the condition for the onset of convection is the Ledoux criterion

(Ledoux, 1941). On the basis of Kato's analysis, Langer et al. (1983) calculated timescale and diffusion coefficient D_{sc} of mixing due to semiconvection as

$$D_{sc} = \frac{\alpha}{6} \frac{\nabla - \nabla_{ad}}{\nabla_L - \nabla} D_{rad}, \qquad (1)$$

with the radiation diffusion coefficient $D_{rad} = \frac{\sigma}{c_P \rho}$, $\sigma = \frac{4ac}{3\kappa\rho} T^3$. α is an efficiency parameter of order 0.1 (Langer et al., 1985).

This understanding of semiconvection is similar to that of Weaver et al. (1978) as discussed in Langer et al. (1990), and recovers what has been called "semiconvective neutrality" (i.e. $\nabla \equiv \nabla_{ad}$, cf. Chiosi and Summa, 1970; Eggleton, 1972; Stothers and Chin, 1976; Iben, 1974) in the case of $\tau_{sc} \ll \tau_{ev}$, i.e. when the evolutionary timescale is large compared to the semiconvective mixing time. Note that $\tau_{sc} \ll \tau_{ev}$ is only valid for semiconvective zones in the envelopes of massive main sequence stars. In this stage, semiconvection is only of little importance (cf. Chiosi and Nasi, 1978). However, semiconvection during H-shell burning (i.e. the contraction phase towards core helium ignition) largely affects the H-profile and thereby the surface temperature evolution (i.e. the HRD track) during core He-burning (Lauterborn et al., 1971ab; Kozlowski, 1971), and semiconvection above the convective He-burning core controls the final mass of the C/O-core and possible blue loops after core He-depletion (Langer et al., 1989; see below). In both cases it is $\tau_{sc} \simeq \tau_{ev}$, which makes a timedependent treatment of semiconvective mixing inevitable.

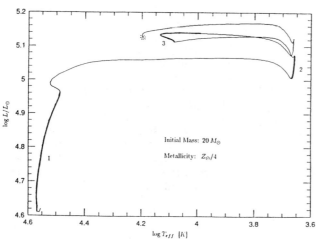

Fig. 1: Theoretical evolutionary track of a 20 $M_\odot$ star of LMC composition computed with semiconvection (see text) as model for the presupernova evolution of the SN 1987A progenitor. Slow evolutionary phases (1=main sequnece, 2=RSG, 3=BSG) are indicated by the thick drawn parts of the track.

An example for the effects of semiconvection, which has recently been studied in Göttingen in great detail, is the progenitor evolution of supernova (SN) 1987A. Stellar evolution sequences for a 20 $M_\odot$ star of LMC composition have been computed using Eq. (1), for different values of the semiconvective efficiency parameter α. All other physical ingredients were up-to-date but standard (cf. Langer et al., 1989, for details). The sequence which can at best account for the observational constraints for the SN 1987A progenitor evolution (cf. Arnett et al., 1989) has been obtained with $\alpha = 0.04$ (which is not too far from our order of magnitude estimate of $\alpha \simeq 0.1$; cf. Langer et al., 1985). The corresponding evolutionary track is displayed in Fig. 1; it is strongly influenced by semiconvection in the following way.

Due to the reduced efficiency of mixing in the presence of μ-gradients as compared to calculations which use the Schwarzschild criterion for convection, the intermediate convection zone which develops during H-shell burning is limited in extension. In particular, the homogenization of the intermediate layers does not extend downwards up to the location of the hydrogen burning shell but only up to a mass fraction q_0. Therefore, the H-burning shell is confined to a region of low hydrogen concentration during early phases of core helium burning, i.e. the shell generates a relatively small amount of luminosity. This is known to lead to a red supergiant (RSG) structure (cf. Stothers and Chin, 1976; Langer et al., 1985) and explains the rapid evolution of our model to the Hayashi line after core H-exhaustion.

During core helium burning the H-burning shell moves outwards and eventually reaches the location q_0, where suddenly the hydrogen concentration becomes high. This leads to an activation of the H-burning shell (the generated luminosity due to H-shell burning almost doubles within a thermal timescale) and consequently the star moves to the blue supergiant (BSG) region in the HR diagram (cf. Lauterborn et al., 1971ab; Fricke and Strittmatter, 1972), where it remains until core He-exhaustion, which leads the star back to the Hayashi-line. This is how the hydrogen profile determines the evolutionary track during core He-burning.

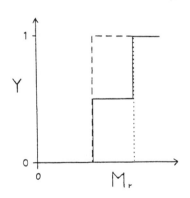

Fig. 2: Schematic He-profiles at time of core He-exhaustion for three different assumptions on convection, i.e. the Ledoux criterion (dashed line), the Schwarzschild criterion (dottet line), and semiconvection (solid line).

Beyond core-He exhaustion a He-burning shell is activated, and consequently the He-profile becomes important. Fig. 2 shows schematically the He-profiles at the time of core He-exhaustion for three assumptions on convection, i.e. according to calculations performed with the Schwarzschild criterion for convection and with the Ledoux criterion for convection, respectively, as well as the case with semiconvection (as defined above). Only the latter case results in a low concentration of helium at the location of the He-burning shell, which is due to the slow semiconvective mixing of helium inside and He-burning products outside the convective core during central helium burning. For the Schwarzschild- or the Ledoux criterion, $Y \simeq 1$ is the result. The He-concentration at the location of the He-burning shell may be particularly important since He-burning due to the 3α-reaction depends on the cube of the He-concentration.

In order to understand the final blue loop of our sequence (cf. Fig. 1) we have to invoke the so called "mirror principle", which is an empirical law stating that an actively burning shell turns a contraction below it into an expansion above it and vice versa (cf. Kippenhahn and Weigert, 1990). Beyond core He-exhaustion the He-burning shell is certainly active and transforms the core contraction towards carbon burning into an expansion of the layers above it. Usually, this expansion is sufficiently strong in order to quickly reduce the temperature at the location of the H-burning shell, which consequently fades away and thus cannot influence the radial motions of the envelope any more. In this case the whole envelope is expanding. In case of semiconvection, however, the activity of the He-burning shell is reduced due to the low concentration of fuel (Fig. 2), which means that the expansion of the overlying layers is less strong as compared with the cases of Schwarzschild or Ledoux criterion. As a result the H-burning shell remains active during the whole contraction phase towards C ignition and changes — according to the mirror principle — the expansion below into an envelope contraction above (see Fig. 3). This is the reason why the star turns into a BSG after core He-exhaustion in the case of semiconvection, but remains in the RSG stage if just the Schwarzschild- or the Ledoux criterion for convection is invoked.

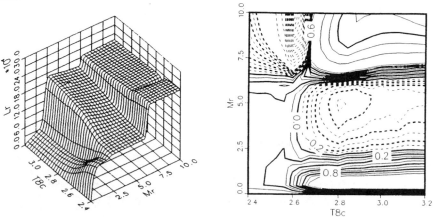

Fig. 3: a) Internal luminosity $L_r/L_\odot$ as function of the mass coordinate $M_r/M_\odot$ (spatial coordinate) and central temperature $T8_c = T_c/10^8 K$ (time coordinate) at and beyond core He-exhaustion. The two steps indicate the locations of the He-burning (at $M_r \simeq 2 M_\odot$) and the H-burning (at $M_r \simeq 7 M_\odot$) shell sources. Note that the H-shell does not fade away when the He-shell source is activated. **b)** Contour plot of the inverse of the local timescale of density variation $d\ln\rho/dt$ for the same $M_r - T8_c$ area as shown in Fig. 3a. Solid lines indicate contraction, dashed lines expansion. The units of $d\ln\rho/dt$ are on an arbitrary linear scale. For $T8_c \gtrsim 2.6$ contracting core, expanding intermediate layers, and contracting envelope can be distinguished.

Note that a low envelope opacity seems also to be required in order to obtain the final envelope contraction, since calculations of a $20\,M_\odot$ star and solar metallicity did not perform this contraction for any value of the semiconvection parameter α. The reason for this metallicity dependence is not yet well understood and deserves future investigation.

Finally, it is important to say that alternative explanations for the (empirically undisputeable) blue-red-blue evolution of the SN 1987A progenitor have been proposed (cf. e.g. Arnett et al., 1989) and cannot be excluded. However, they all include assumptions about rotation or binarity, often in a somewhat arbitrary way. Furthermore, a track like that in Fig. 1 cannot only account for the properties of the SN 1987A progenitor, but also for many general properties of LMC supergiants, which is not discussed here due to space limitations.

3. Formation of WR stars

a) Stars above the Humphreys-Davidson limit

The absence of very luminous RSGs (Humphreys and Davidson, 1979) and the presence of WR stars can be recovered by stellar evolution calculations with standard input physics when a short phase with extremely high mass loss after core hydrogen exhaustion is included (Maeder, 1983). The coincidence, that a class of highly variable luminous stars — the LBVs — appears to have the right properties (i.e. HRD position, mean mass loss rate, number frequency, etc.; cf. Davidson et al., 1989) to fit to this short phase of high mass loss (now called LBV-phase) leads many astronomers to think that the LBV-scenario may in fact be the dominant formation channel for WR single stars. It has been shown that mass loss rate and duration of the high mass loss state needs not to be arbitrarily imposed in a stellar evolution calculation but can be obtained selfconsistently as result of the condition that a certain region of the HR diagram should be avoided by the evolutionary track (Langer, 1990). Typical mass loss rates and timescales obtained in this way are $\dot{M}_{LBV} \simeq 10^{-3} \, M_\odot \, yr^{-1}$ and $\tau_{LBV} \simeq 10^4 \, yr$, but both numbers may vary significantly as function of the main sequence evolution (i.e. as function of the main sequence mass loss rate or assumptions on convection/semiconvection; see above). The product of both numbers, anyway, is constrained by the mass of the hydrogenrich envelope ΔM (i.e. the amount of mass with $X > X_{crit}$; $X_{crit} \simeq 0.25$, cf. Langer and El Eid, 1986; Maeder and Meynet, 1987) left at the time of core H-exhaustion, i.e. $\dot{M}_{LBV} \cdot \tau_{LBV} = \Delta M$.

Note that still not much is known about the physical origin of the high LBV mass loss, which is not surprising in view of the diversity of observational features related with this very inhomogeneous class of stars (see contributions in: Davidson et al., 1989, and cf. Kiriakidis et al., these proceedings, for the case of η Car).

b) Stars below the Humphreys-Davidson limit

The lower ZAMS mass limit for WR formation from single stars M_{WR} is possibly lower than the critical mass limit which corresponds to the upper luminosity boundary for RSGs (i.e. $\sim 10^{5.7} \, L_\odot$, cf. Humphreys and McElroy, 1984; $\Rightarrow M_{ZAMS} \simeq 45 \, M_\odot$) according to Schild and Maeder (1984), Humphreys et al. (1985), and van der Hucht et al. (1988). Consequently, a WR formation channel might exist, which involves a RSG stage. However, this cannot be concluded directly from stellar evolution calculations, basically due to the lack of theoretical predictions or strict observational constraints about RSG mass loss rates (cf. e.g. Jura, this volume). Also, it is not yet known whether an LBV-phase is associated with the RSG-channel of WR formation, either before or after the RSG stage. Note, however, that some LBVs (e.g. R71 or R110 in the LMC; cf. Wolf, 1989; Stahl et al., 1990) have luminosities well below $10^{5.7} \, L_\odot$, and some of their characteristics can be well understood when it is assumed that those objects are post-RSGs (Leitherer and Langer, 1990). Also Lortet (1989) concludes from an investigation of LBV environments that some of them might be post-RSGs (cf. also the contributions of R.M. Humphreys, C. Leitherer, and N. Walborn to this volume).

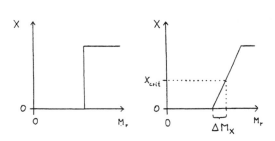

Fig. 4: Schematic H-profiles at time of core He-ignition.

Assuming that the evolutionary connection RSG→(LBV)→WR exists, then whether the subtype of the newly formed WR star is WNL or WNE (or better whether it is a hot or a cool WN star; see bolow) is simply a function of the internal hydrogen profile (cf. Langer, 1987, 1988). Fig. 4 is a schematic sketch of the two principle possibilities, i.e. a "steep" and a "flat" profile. Since all WR stars have small surface hydrogen concentrations (e.g. $X \lesssim 0.25$; cf. Hamann et al., these proceedings) only the part of the profile with $X \lesssim 0.25$ is relevant in this context. In case of a "steep" X-profile, the surface hydrogen mass fraction will change in a short time from a high abundance to zero, which means that the newly formed WR star will be hot (i.e. of type WNE). A "flat" X-profile gives rise to a cool WN phase of duration $\tau_{WN-cool} \simeq \Delta M_X / \dot{M}_{WN-cool}$.

Stellar evolution calculations indicate that the quantity ΔM_X (cf. Fig. 4) strongly increases with increasing ZAMS-mass. Though it is also a strong function of the incorporated physics (esp. again main sequence mass loss rate and convection/semiconvection models), it can be concluded that the WNL-phase (better the cool WN phase) lasts significantly longer for higher initial masses (cf. Langer, 1987; Maeder and Meynet, 1987; Maeder, 1990). Furthermore, when part of core He-burning is spent in the RSG regime before the WR formation phase, the H-burning shell diminuishes ΔM_X due to transformation of H into He. This reduces additionally $\tau_{WN-cool}$ for post-RSG WR stars. Both effects together may account for the scarcity of cool WN stars with relatively low luminosities (Lundström and Stenholm, 1984; Conti, 1986)

4. Structure and evolution of WR stars

a) Effects of hydrogen

The presence of hydrogen in the envelope of a massive star has large consequences for its internal structure and thereby for its visual display and its evolution. Most important, the presence of hydrogen implies the presence of a hydrogen burning shell, due to which — finally as a consequence of the mirror principle; cf. Sect. 3b — the star is much more extended and thus much cooler compared to hydrogenless stars. The larger extension is supported by the larger opacity and the smaller mean molecular weight of H-containing matter compared to pure helium or metalrich mixtures. The radius R_* (i.e. irrespective of the stellar wind) of H-containing WR stars is typically 10-20 times larger than the radius of WR stars of similar mass without hydrogen (cf. e.g. Langer and El Eid, 1986; Maeder and Meynet, 1987; Langer, 1990a). Note that this effect is nicely confirmed by the recent work of Hamann (this volume), who finds a clear correlation of the H-abundance in WN stars with their effective temperature. A further consequence of the presence of hydrogen is a stabilization against vibrational pulsations (Maeder, 1985; Maeder and Schaller, this volume).

Because of the different internal structure and different physical and chemical surface composition of H-containing WR stars compared to WR stars without hydrogen, there is no reason to expect a common mass loss law for both types of WR stars, as discussed in Langer (1989b). Note, e.g., that the momentum problem for WR winds appears to be much smaller for cool WN stars as compared to H-less WR stars (cf. Langer, 1990).

Since the wind of WR stars may have a considerable optical thickness, the consequences of the presence of hydrogen for the apparent radius R_{eff} (i.e. $R_{eff} := r(\tau = 2/3)$) are not easy to predict. However, an order of magnitude estimate can be performed on the basis of Eqs. (8)-(15) of Langer (1989a), where it is shown that the optical thickness of the WR wind $\tau_{wind} = \tau(R_*)$ depends on mean opacity κ, mass loss rate and stellar radius as $\tau_{wind} \sim \kappa \dot{M}/R_*$. Order of magnitudes are $\tau_{wind} \simeq 1$ for cool (H-containing) WR stars, but $\tau_{wind} \simeq 10$ for WR stars without hydrogen, both for $\dot{M} = 3\,10^{-5}\,M_\odot\,yr^{-1}$.

b) Wolf-Rayet stars without hydrogen

WR stars without hydrogen — i.e. hot WNs, WCs, and WOs — have a very simple internal structure compared to H-containing WR stars: they are composed of a large convective He-burning core and a radiative envelope. Due to their high temperature, radiation pressure dominates over gas pressure within almost the whole star except a tiny surface layer. Due to this, and since the main opacity source is electron scattering, the structure of these objects is almost completely independent of their internal chemical composition, but is determined only by their actual mass and — to a lesser extent — by their surface chemical composition (Langer, 1989a).

This finding has several important consequences. First of all, it means that the structure of H-less WR stars is independent of their previous evolution. This allows, e.g., to study those objects theoretically without considering how they have been formed. It explains the existence of a mass-luminosity relation for WR stars (Maeder, 1983) or a relation of the luminosity as a function of mass and surface composition, as well as similar relations for the radii or surface temperatures of WR stars (Langer, 1989a). It implies, e.g., that all H-less (hot) WN stars of the same metallicity should be located on a single line in the HR diagram, at least if rotaion or magnetic fields don't play any role.

A further consequence is — if again rotation and magnetic fields can be ignored; cf., anyway, Cassinelli (this volume) — that the mass loss rates of those objects should depend mainly on their actual mass, whatever its physical origin may be. Langer (1989b) investigated the evolutionary consequences of mass dependent WR mass loss and found a relation of the form

$$\dot{M}_{WR} = (0.6 - 1.0)\,10^{-7} \left(\frac{M_{WR}}{M_\odot}\right)^{2.5} / M_\odot\,yr^{-1} \qquad (2)$$

to yield the best agreement with many observed properties of galactic WR stars. Note that this agreement is rather insensitive to the actual power of the mass loss law, as long as it is $\gtrsim 1$. This result coincides with observational mass loss rates derived by Abbott et al. (1986) and is supported by the work of Smith and Maeder (1989).

One of the most prominent consequences of such mass dependent WR mass loss concerns the initial-final mass relation for massive stars (cf. Fig. 5): a) The final masses for stars which enter the WR phase during their evolution are very small, i.e. well below $10\,M_\odot$, and b) The final masses of stars from a large interval of initial masses (say $\sim 35\,M_\odot - \sim 100\,M_\odot$) turn out to be the same within a very narrow limit $\sim \pm 1\,M_\odot$; cf. Langer, 1989b). Note that for the highest initial masses (i.e. $M_{ZAMS} \gtrsim 100\,M_\odot$), due to an increase of the duration of the H-containing WR phase with initial mass (cf. Sect. 3b), the possibility exists that some hydrogen is kept until the end of evolution. This would imply that those objects would not enter the regime of mass dependent mass loss, and they might remain very massive until the supernova phase (cf. Langer, 1987). The occurrence of this high mass final state depends on the mass loss rates of very luminous WN stars, for which no dependence on stellar parameter could yet be derived.

In a recent grid of stellar evolution sequences for a wide mass and metallicity range, calculated with mass dependent WR mass loss rates, Maeder (1990) confirmed the flat initial-final-mass relation of Fig. 5, and found also good agreement of the metallicity dependence of various WR subtype frequencies with observations (cf. Maeder, this volume).

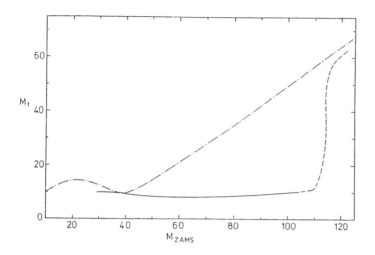

Fig. 5: Schematic initial-final-mass relation for massive stars in our Galaxy in the case where mass-dependent WR mass loss is taken into account (solid line), compared to previous models (dashed-dotted line; from Maeder and Meynet, 1987). The dashed part of the line corresponds to the possible existence of a very massive final state for the highest ZAMS-masses (see text).

Finally a comment on the "WN+WC" spectral type, which has been recently identified as being related to single WR stars rather than to WN+WC binaries (cf. Conti and Massey, 1989; Massey and Grove, 1989; Willis and Stickland, 1990). The implication is a partly mixed region between He-burning convective core and the layers above. In Langer (1990a) it is shown, that stellar evolution calculations which include semiconvection (cf. Sect. 2b) yield a WN+WC phase with a duration of some $10^4 \, yr$, which agrees well with the observed WN+WC frequency. Note that models which only use the Schwarzschild or the Ledoux criterion for convection or models including convective core overshooting do not obtain a WN+WC phase. Therefore, the WN+WC stars present a second argument in favour of semiconvection, which is completely independent of the SN 1987A progenitor evolution.

5. Clues from supernovae for the evolution of massive stars

Due to space limitations, this topic cannot be treated very explicit here. The reader is refered to Langer (1990, 1990b) for more details. Here, just the potential of supernova observations as tool for the analysis of the progenitor evolution is briefly summarized.

- Light curves and spectra of Type II SNe allow estimates of envelope mass and He-core mass of the progenitor, and thus an estimate of the amount of mass lost during the whole hydrostatic evolution (cf. e.g. Woosley, 1988). The inhomogeneity of the Type II SN class (cf. Nomoto, this volume) may reflect the diversity of possible H-containing pre-SN configurations (WN, BSG, RSG; all with different envelope masses).

- WR stars of any subtype may be pre-SN configurations. However, the bulk of galactic WR stars ($40\,M_\odot \lesssim M_{ZAMS} \lesssim 100\,M_\odot$) is supposed to end evolution as low mass WC/WO star (Langer, 1989b, Maeder, 1990).

- A relation of the final low mass WC/WO stage to Type Ib and/or Ic SNe (cf. contributions of A. Fillipenko and K. Nomoto to this volume) cannot be excluded on the basis of the light curve alone (cf. Langer and Woosley, this volume). A confirmation would strongly support the concept of mass dependent WR mass loss.

- The existence of a final high mass WR stage ($\gtrsim 60\,M_\odot$) could in principle be confirmed by observations of the characteristics of an $e^{\pm}$-pair creation supernove (cf. El Eid and Langer, 1986; Herzig et al., 1990).

Clearly, much progress for the understanding of massive star evolution has to be expected from the field of SN studies in the near future.

Acknowledgement. The author is grateful to Prof. B. Hidayat and his colleagues at Bosscha Observatory for their warm hospitality. Travel support by the Deutscher Akademischer Austauschdienst (DAAD) and the Deutsche Forschungsgemeinschaft (DFG) is also gratefully acknowledged. This work has been supported in part by DFG grants La 587/1-2 and Fr 325/28-2.

References

Abbott, D.C.: 1982, Astrophys. J. **259**, 282

Abbott, D.C.,Bieging, J.H., Churchwell, E., Torres, A.V.: 1986, Astrophys. J. **303**, 239

Arnett, W.D., Bahcall, J.N., Kirshner, R.P., Woosley, S.E.: 1989, Ann. Rev. Astron. Astrophys. **27**, 629

Castor, J.I., Abbott, D.C., Klein, R.I.: 1975, Astrophys. J. **195**, 157

Chiosi, C., Summa, C.: 1970, Astrophys. Space Sci. **8**, 478

Chiosi, C., Nasi, E.: 1978, Astrophys. Space Sci. **56**, 431

Chiosi, C., Maeder, A.: 1986, Ann. Rev. Astron. Astrophys. **24**, 329

Conti, P.S.: 1986, IAU-Sympos. **116**, 199

Conti, P.S., Massey, P.: 1989, Astrophys. J. **337**, 251

Davidson, K., Moffat, A.F.J., Lamers, H.J.G.L.M.: 1989, *Physics of Luminous Blue Variables*, Proc. IAU-Coloqu. 113, Kluwer

Eggleton, P.P.: 1972, Mon. Not. Royal Astron. Soc. **156**, 361

El Eid, M.F., Langer, N.: 1986, Astron. Astrophys. **167**, 274

Fricke, K.J., Strittmatter, P.A.: 1972, Mon. Not. Royal Astron. Soc. **156**, 129

Herzig, K., El Eid, M.F., Fricke, K.J., Langer, N.: 1990, Astron. Astrophys. , in press

van der Hucht, K.A., Hidayat, B., Admiranto, A.G., Supelli, K.R., Doom, C.: 1988, Astron. Astrophys. **199**, 217

Humphreys, R.M., Davidson, K.: 1979, Astrophys. J. **232**, 409

Humphreys, R.M., McElroy, D.B.: 1984, Astrophys. J. **284**, 565

Humphreys, R.M., Nichols, M., Massey, P.: 1985, Astron. J. **90**, 101

Iben, I., Jr.: 1974, Ann. Rev. Astron. Astrophys. **12**, 215

Kato, S.: 1966, P.A.S.J. **18**, 374

Kippenhahn, R., Weigert, A.: 1990, *Stellar Structure and Evolution,* Springer

Kozlowski, M.: 1971, Astrophys. Letters **9**, 65

Kudritzki, R.P., Pauldrach, A., Puls, J., Abbott, D.C.: 1989, Astron. Astrophys. **219**, 205

Langer, N.: 1986, Astron. Astrophys. **164**, 65

Langer, N.: 1987, Astron. Astrophys. **171**, L1

Langer, N.: 1988, in: Proc. IAU-Colloq. 113, p. 221

Langer, N.: 1989a, Astron. Astrophys. **210**, 93

Langer, N.: 1989b, Astron. Astrophys. **220**, 135

Langer, N.: 1990, in: *Angular Momentum and Mass Loss for Hot Stars*, proc. 2nd Ames-Trieste Workshop, L.A. Wilson, R. Stalio, eds., in press

Langer, N.: 1990a, in: *Properties of Hot Luminous Stars*, proc. 1st Boulder-Munich Workshop, C. Garmany, ed., A.S.P. Conf. Ser. Vol. 7, p. 328

Langer, N.: 1990b, in: *Supernovae*, proc. 10th Santa-Cruz Summer Workshop, S.E. Woosley, ed., in press

Langer, N., Sugimoto, D., Fricke, K.J.: 1983, Astron. Astrophys. **126**, 207

Langer, N., El Eid, M.F., Fricke, K.J.: 1985, Astron. Astrophys. **145**, 179

Langer, N., El Eid, M.F.: 1986, Astron. Astrophys. **167**, 265

Langer, N., El Eid, M.F., Baraffe, I.: 1989, Astron. Astrophys. **224**, L17

Langer, N., El Eid, M.F., Baraffe, I.: 1990, in: *Supernovae*, proc. 10th Santa-Cruz Summer Workshop, S.E. Woosley, ed., in press

Langer, N., El Eid, M.F.: 1990, in preparation

Lauterborn, D., Refsdal, S., Weigert, A.: 1971a, Astron. Astrophys. **10**, 97

Lauterborn, D., Refsdal, S., Weigert, A.: 1971b, Astron. Astrophys. **13**, 119

Ledoux, P.: 1941, Astrophys. J. **94**, 537

Leitherer, C., Langer, N.: 1990, in: IAU-Symp. **148**, in press

de Loore, C.: 1980, Space Sci. Rev **26**, 113

Lortet, M.C.: 1989, in: Proc. IAU-Colloq. 113, p. 45

Lundström, I., Stenholm, B.: 1984, Astron. Astrophys. Suppl. **58**, 163

Maeder, A.: 1983, Astron. Astrophys. **120**, 113

Maeder, A.: 1985, Astron. Astrophys. **147**, 300

Maeder, A.: 1990, Astron. Astrophys. Suppl. **84**, 139

Maeder, A., Meynet, G.: 1987, Astron. Astrophys. **182**, 243

Massey, P., Grove, K.: 1989, Astrophys. J. **344**, 870

Owocki, S.P., Castor, J.I., Rybicki, G.B.: 1988, Astrophys. J. **335**, 914

Pauldrach, A., Puls, J., Kudritzki, R.P.: 1986, Astron. Astrophys. **164**, 86

Pauldrach, A., Kudritzki, R.P., Puls, J., Butler, K.: 1990, Astron. Astrophys. **228**, 125

Renzini, A.: 1987, Astron. Astrophys. **188**, 49

Schild, H., Maeder, A.: 1984 Astron. Astrophys. **136**, 237

Smith, L.F., Maeder, A.: 1989, Astron. Astrophys. **211**, 71

Stahl, O., Wolf, B., Klare, G., Juettner, A., Cassatella, A.: 1990, Astron. Astrophys. **228**, 379

Stothers, R., Chin, C.-W.: 1976, Astrophys. J. **204**, 472

Weaver, T.A., Zimmerman, G.B., Woosley, S.E.: 1978, Astrophys. J. **225**, 1021

Willis, A.J., Stickland, D.J.: 1990, in: *Properties of Hot Luminous Stars*, proc. 1st Boulder-Munich Workshop, C. Garmany, ed., A.S.P. Conf. Ser. Vol. 7, p. 354

Wolf, B.: 1989, in: Proc. IAU-Colloq. 113, p. 91

Woolsey, S.E.: 1988, Astrophys. J. **330**, 218

DISCUSSION

Niemela: If the progenitor of SN1987A in the LMC had $20M_\odot$, would that preclude smaller mass progenitors for WR stars? In other words: the initial mass for WR stars should be $> 20M_\odot$.

Langer: For single stars, yes. Note, however, that due to the metallicity dependence in stellar evolution you cannot transfer this limit to, *e.g.*, the Milky Way.

Moffat: I agree that $\dot{M} \sim M^\alpha$ with α positive, but observations (*e.g.*, from polarization of WR+O binaries which yield M *and* $\dot{M}$ suggest that $\alpha \approx 1.5$ not 2.5.

Langer: I can live as well with $\alpha \approx 1.5$. My results indicate that the exact value of α is not important. The main point I wanted to make is that it should be clearly $\alpha > 0$, note that in the past $\alpha = 0$ has been used almost exclusively.

Schulte-Ladbeck: I found your illustration of stellar interiors using different criteria very illuminating. I have two questions. The first concerns rotational mixing and goes to you and Maeder: do we now have a number of evolutionary tracks with rotation? The second one is about semi-convection and the smaller cores (*e.g.*, what we need in binaries) and should perhaps be addressed to our binary theorist, De Greve. What is our current standing on binary models with semi-convection (and with conservative/non-conservative Roche-lobe overflow)?

Langer: To your first point: I recently calculated a SN1987A progenitor track including the rotational induced baroclinic instability, which was found to account for the high observed N-enrichment. Also, *e.g.*, Maeder and Sreenivasan calculated massive star track including rotational effects. However, it is my opinion that prescription of the involved rotational physics is still to a large degree arbitrary and/or uncertain. For your second point: I do not know any binary evolution calculation including the effect of semi-convection as I understand it (*e.g.*, as a physical instability).

Vanbeveren: I was very pleased with your model for SN1987A where you were able to explain the progenitor without core overshooting. The OBN binary HD 163181 consists of a $13M_\odot$ (OBN) component and a $22M_\odot$ companion. Since the OBN star is ± 1.5 *mag* brighter than the $22M_\odot$ star, it should be a core helium burning star (after a possible Roch-lobe overflow). But, why then is it not a WR star? A possible solution can be found when non-overshooting models are used and let semi-convection play an important role.

Langer: I would appreciate that binary evolutionists would take semi-convection into account. However, note that in binaries semi-convection is already important during central H-burning for the mass gainer, since its H-burning convective core will tend to grow, leading to a composition discontinuity at its top.

Yungelson: You do predict very low masses of pre-SN. This means that a low amount of mass will be ejected at SN explosions. Would this not lead to overproduction of binary pulsars?

Langer: The low mass of the pre-supernova star does not automatically mean that you create a pulsar. Since the whole star is a metal-rich object and since it "remembers" its high mass origin, you may possibly create a black hole rather than a neutron star.

Barlow: You mentioned that LBV's should become WNE's if there is a sharp composition boundary between hydrogen and helium, while they should become WNL's if there is a smooth composition gradient. The results that I presented on Tuesday on the LBV's P Cyg and AG Car showed that their winds have intermediate compositions with very enhanced He/H ratios. This would seem to argue in favour of there being a smooth composition gradient in the envelopes of these stars.

Langer: Partly this is true. However, they could still transform directly into WNE's, if the hydrogen gradient in the remaining envelope is very steep.

Massey: The designation WNL and WNE has to do with the relative strength of NV and $NIII$, and is not based on H/He ratios. HD 177230, a WN 8 "spectrum standard" (according to Beals!), has no H - even Underhill agreed with that in 1981. One of the three known galactic WN 3 stars *does* show H. I just think the situation is more complicated than implied by using two shoe boxes, and assuming that everything in each box is the same.

Langer: I am sorry if I created confusion by my use of the designations WNL and WNE for hydrogen containing and hydrogen less WN stars. I am aware that the correlation of the hydrogen abundance with WN spectral subtype is not unambiguous, as also demonstrated by Hamann at this symposium.

Nomoto: (1) Regarding the evolution of the SN1987A progenitor, helium enhancement observed in the circumstellar shell would require an additional mixing process in your model. (2) In your model of the SN1987A progenitor, what is the effect of low metallicity for the blueward evolution? (3) How much is the critical value of metallicity that divides the BSG and RSG progenitor?

Langer: (1) If the high helium abundance for the progenitor of SN1987A is confirmed, it poses a problem to our model. (2) Higher Z implies higher opacities which favours extended envelopes. (3) We have done calculations only for $Z = 0.5\%$ (LMC) and $Z = 2\%$ (Milky Way). For the latter case we obtained RSG pre-SN configurations independent of the semi-convection parameters.

Maeder: I think it is fair to say that several explanations of the blue progenitor of SN1987A have been proposed. Among them, the explanation by Nomoto and coworkers looks very attractive and my models also will support it. During the C-burning phase, the external and intermediate convective zones come so close to each other, that any extension of convection would mix the materials from both zones. This results into a blue location in the HRD due to the increased opacity and also this explains the He and N/C observed enhancements. Your criterion for semi-convection explains the blue location of the SN progenitor, but not the He and N/C enhancements. I think this should be properly stated for the clarity of the debate about the SN progenitor.

Langer: Since this is a symposium on WR stars, I tried to concentrate on these objects in my talk and did not take too much time for the discussion of the SN1987A progenitor. Otherwise, I might have been able to convince you that the semi-convection explanation of the pre-SN evolution as well explains many general properties of LMC supergiant stars. Also the high observed N-enrichment in the SN progenitor can be easily accounted for due to the baroclinic instability put forward by yourself a few years ago. A detailed paper about this topic is in preparation. Concerning the SN progenitor model of Nomoto *et al.*, I just am a bit worried, since both the amount of mixing invoked and the time of mixing are not yet physically justified.

Sreenivasan: I think one can produce an acceptable evolutionary pattern of the progenitor of SN 1987A without any *tricks* about the lines outlined in my poster (details in the references). Your method depends upon the way you treat semi-convection! But one cannot self-consistently use the Ledoux gradient without an independent equation to determine the μ-gradient. Once you mix, you alter the μ-gradient according to your mixing scheme and not according to the physics of the interior. That is why using the Ledoux criterion is inconsistent. Finally, the semi-convection you see in the He-burning stage is different from what one sees in the H-burning stage and the kind discussed for horizontal branch stars by Giannone *et al.* (1971). The two arise for quite different reasons and your diffusion coefficient with its α factor is also very curious. It is not very clear what really goes on in your model!

Langer: There is much confusion about what semi-convection means in the literature. The only way I can understand semi-convection is as a physical instability, as discussed by Kato (1966). Especially, the physical origin of the semi-convection in H- and He-burning phases is the same in our models. Furthermore, we do not encounter the numerical problems or inconsistencies you mention. Our method is comprehensively described in Langer *et al.* (1985).

Leitherer: Stars evolve to the red in the HRD, and - depending on how much hydrogen-rich material you remove from the surface - they may evolve back to the blue part of the HRD. What is the actual physical reason for this behaviour?

Langer: There are two possibilities for a star to return to hotter surface temperatures in the HR diagram: (1) so called blue loops and (2) transition to the WR phase, and I presume you refer to the second case. There, I think, the physical reason is twofold. You increase the average mean molecular weight in the envelope (which becomes more He-rich) and reduce the opacity; both tend to decrease the stellar radius. Additionally, the temperature in the H-burning shell is reduced as less and less mass lies above it. Since a shell source increases the total stellar radius appreciably (*cf.* Cox & Giuli), its extinction means that the radius shrinks, *i.e.*, the surface temperature increases.

Norbert Langer

METALLICITY EFFECTS IN MASSIVE STAR EVOLUTION AND NUMBER FREQUENCIES OF WR STARS IN GALAXIES

André Maeder
Geneva Observatory
CH-1290 Sauverny, Switzerland

ABSTRACT. The results of new grids of models of massive stars with metallicities $Z = 0.002, 0.005, 0.020$ and 0.040 and mass loss rates depending on Z are shown. When integrated over the mass spectrum, the models enable us to predict number ratios, such as WR/O, WC/WN, WNE/WR, WNL/WR, WCE/WR, WCL/WR, WO/WR as a function of Z in galaxies.

Comparisons between models and observations in galaxies are made and show, as was suggested by Maeder, Lequeux and Azzopardi (1980), that the effects of metallicity on the mass loss rates are the prime agent responsible for the different distributions of massive stars in galaxies.

1 INTRODUCTION

The changes of the WR populations in galaxies with active star formation were first discovered in the pioneer work by Smith (1968, 1973, 1982). Further observational studies, particularly in the Magellanic Clouds by Breysacher (1981) and Azzopardi and Breysacher (1985) have confirmed the differences in the WR populations. Maeder, Lequeux and Azzopardi (noted MLA, 1980) have proposed an explanation of the observed differences in the number frequency of WR stars by a connection between the local metallicity Z and mass loss. At high Z (inner galactic locations), gas opacities are larger in the outer stellar layers, more momentum is therefore transferred by radiation pressure effects, mass loss by stellar winds in massive O stars is more intense and thus more bare cores are formed.

The proposition by MLA was criticised by Bertelli and Chiosi (1981) who claimed that the galactic gradient in WR stars simply reflects the gradient in O stars. Similar claims against metallicity effects were made by Armandroff and Massey (1985) and by Conti et al. (1983a). However, Meylan and Maeder (1983) showed that the galactic gradient of WR star numbers is steeper than the gradient in O stars. Thus the WR distribution in the Galaxy is not just a reflection of the O–star distribution and the WR gradient cannot be explained only by a change in the IMF.

Up to now, the interpretation of the observed trends have remained essentially qualitative. The aim of this work is to perform a quantitative approach on the basis of new grids of models of massive stars at various metallicities.

445

K. A. van der Hucht and B. Hidayat (eds.),
Wolf-Rayet Stars and Interrelations with Other Massive Stars in Galaxies, 445–451.
© 1991 *IAU. Printed in the Netherlands.*

2 SUMMARY ON MODEL PHYSICS

The models used here to derive the theoretical number frequencies of WR stars are the new grids of evolutionary models of massive stars at various metallicities Z=0.002, 0.005, 0.020, and 0.040 (cf. Maeder, 1990a). These models use up–to–date nuclear cross–sections and new opacity tables made by G. Schaller at Geneva Observatory from the Los Alamos Opacity Programme. Proper account is given in the opacity tables to the fact that the O/Fe and α–nuclei/Fe abundance ratios are larger than solar at low Z. The initial model abundances have also been modified accordingly.

The mass loss rates for Pop. I stars with initial Z=0.020 (except the WR stars) are based on recent data analyses by de Jager et al. (1988). At Z values different from Z=0.020, the mass loss rates $\dot{M}$ have been scaled according to $\dot{M} \sim Z^{\xi}$. An exponent $\xi = 0.5$ has been taken, as indicated by the stellar wind models by Kudritzki et al. (1987).

For WR stars, we no longer use the average observed mass loss rates, as we and others generally did in the past. As shown by Schmutz et al. (1989), this was leading to theoretical WR luminosities that were much higher than those observed. There were several indications in favour of $\dot{M}$–rates for WNE and WC stars depending on the actual masses of WR stars (cf. Abbott et al., 1986; St-Louis et al., 1988; Langer, 1989; Smith and Maeder, 1989). Langer's relation has been used here.

It has repeatedly been suggested (cf. Maeder and Meynet, 1989) that the models should also include the core–overshooting effect. Also, the T_{eff}–values of WR stars have been corrected for the optical thickness of the wind, supposedly due mainly to electron scattering. Due to the T_{eff} correction, the WNE and WC stars follow unique well–defined tracks in the HR diagram, independently of their initial stellar masses. This can be seen, for example, in Fig. 3. These tracks are explained by the fact that as their mass and luminosity decline, the $\dot{M}$–rates and the amplitude of the T_{eff} correction are reduced, thus the WR stars move downwards and bluewards in the HR diagram.

3 MODEL RESULTS CONCERNING WR STARS

Surface abundances play a key role in the identification of models with WR subtypes. As is known, the observed abundances of WNL, WNE, WC and WO stars are consistent with the exposition of nuclearly processed materials. The correspondence applied between WR subtypes and surface abundances is discussed in the detailed paper (Mader, 1990b) on WR distributions in galaxies.

Fig. 1 gives an overview of the lifetimes t_{WR} in the WR stage as a function of mass and Z. The clear trend – easily understandable – is an increase of t_{WR} with initial mass and Z. Also we notice that the minimum mass for WR formation is lower at higher Z. Similar graphs can be established for the lifetimes in the WN–late, WN–early, WC–late and WC–early phases.

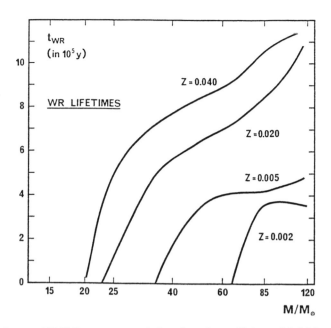

Figure 1 Lifetimes as Wolf–Rayet stars as a function of metallicity and initial stellar masses.

The lifetimes in the various phases can be used to derive relative number frequencies WR/O, WC/WR, WC/WN. Assumptions on the star formation rate (SFR) and the initial mass function (IMF) have of course to be made. For a galaxy or a large galactic ring, the asumption of a constant SFR over the last few 10^7y is a reasonable one. (For a single HII region, this would not be acceptable and aging effects are likely to intervene). For the IMF, two forms $dN/dM = AM^{-(1+x)}$ with x = 1.35 and 1.7 are considered. Table 1 shows the results for the number ratios WR/O, WC/WR and WC/WN at various metallicities. Two cases for the lower T_{eff} limit of O stars are considered, one at log T_{eff} = 4.52 and the other at 4.53.

The results of Table 1 can be compared with the observed number ratios in galaxies and galactic rings in the Milky Way. The main source references for the observations are Smith (1988), Arnault et al. (1989). For the Milky Way, both the data from van der Hucht et al. (1988) and Conti and Vacca (1990, squares in Fig. 2) are considered. As an example, Fig. 2 shows the comparison between the observed and theoretical WC/WN number ratios respectively. On the whole, the agreement is excellent, although there are some sizeable differences between the two sources of galactic data. Fig. 2 gives powerful support to the evolutionary models and to the idea by MLA (1980) that metallicity Z, through its effects on the mass loss rates $\dot{M}$ of O stars and supergiants, is responsible for the enormous differences of the WR populations in galaxies with active star formation.

TABLE 1. Theoretical number ratios for Wolf–Rayet stars as a function of metallicity Z, for 2 slopes x of the IMF.

Z	WR/O log T > 4.53	WR/O log T > 4.52	WC/WR	WC/WN
x=1.35				
0.002	0.0034	0.0032	0.057	0.061
0.005	0.0185	0.0182	0.192	0.237
0.020	0.0790	0.0752	0.640	1.784
0.040	0.1628	0.1557	0.736	2.784
x=1.70				
0.002	0.0021	0.0019	0.058	0.061
0.005	0.0135	0.0133	0.170	0.204
0.020	0.0654	0.0616	0.634	1.732
0.040	0.1438	0.1363	0.744	2.908

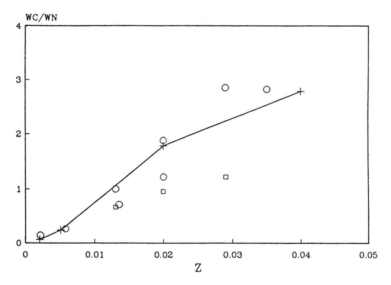

Fig. 2: Comparison of observed (circles, squares) and theoretical (line) number ratios WC/WN of WC stars to WN stars as a function of initial metallicity.

The relative number of WN stars with and without hydrogen, identified with WNL and WNE stars respectively, are also derived (L for late, E for early). The WNL phases generally last much longer for large initial masses, with little dependence on Z. On the other hand, the WNE phases are longer for lower initial masses, also being shorter at Z=0.04 than at Z=0.02. The comparisons with observations of WNL and WNE stars in galaxies show agreement for Z≥0.02. At lower Z, there are more WNE stars than predicted and we conclude that a large fraction of the existing WR stars probably results from binary evolution and that they are mostly WNE stars.

The relative numbers of WC stars with various subtypes are also analysed quantitatively. WC models are classified according to the (C+O)/He number ratios (cf. Smith and Hummer, 1988; Smith and Maeder, 1990):

WC9	(C+O)/He = 0.03 to 0.06	WC5	(C+O)/He = 0.55
WC8	(C+O)/He = 0.1	WC4	(C+O)/He = 0.7 to 1.0
WC7	(C+O)/He = 0.2	WO	(C+O)/He greater than 1.0
WC6	(C+O)/He = 0.3		

The models show that the entry points and lifetimes in the sequence WC9 → WC4, WO are extremely Z– and mass–dependent. The entry into WC9 subtype only occurs for high initial Z and mass (i.e. Z > 0.02 and M ≥ 50 M⊙, cf. Fig. 3). Lower initial M and Z lead to entry at earlier WC subtypes, with a shorter overall WC phase. After integration of the lifetimes over the mass spectrum, we notice that at high initial Z, late and intermediate WC subtypes are favoured, while at lower Z, only early WC and WO stars are found and in much lower frequencies. The theoretical number ratios of late, early WC and WO stars compare quite well with observed number ratios in galaxies of the Local Group. Space and time requirements preclude these results to be shown in detail here, and we therefore refer the reader to the works by Smith and Maeder (1990) and Maeder (1990b).

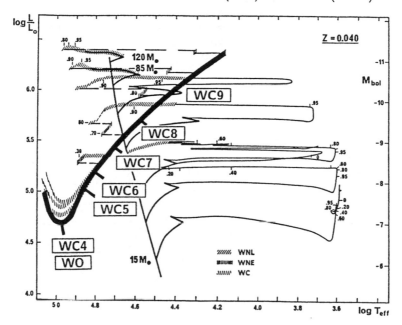

Fig. 3: Evolutionary tracks in the HR diagram for models with Z=0.040. The average intervals occupied by the existing WC subtypes and WO stars are indicated along the thick line, on the basis of the (C+O)/He number ratios.

450

4 REFERENCES

Abbott D., Bieging J.H., Churchwell E., Torres A.V.: 1986, Astrophys. J. **303**, 239

Armandroff T.E., Massey P.: 1985, Astrophys. J. **291**, 685

Arnault P., Kunth D., Schild H.: 1989, Astron. Astrophys. **224**, 73

Azzopardi M., Breysacher J.: 1985, Astron. Astrophys. **149**, 213

Bertelli G., Chiosi C.: 1981, in *The most massive stars*, Ed. S. D'Odorico et al., ESO, Garching, p. 211

Breysacher J.: 1981, Astron. Astrophys. Suppl. Ser. **43**, 203

Conti P.S., Garmany C.D., de Loore C., Vanbeveren D.: 1983a, Astrophys. J. **274**, 302

Conti P.S., Vacca B., 1990: Astrophys. J. in press

van der Hucht K.A., Hidayat B., Admiranto A.G., Supelli K.R., Doom C.: 1988, Astron. Astrophys. **199**, 217

de Jager C., Nieuwenhuijzen H., van der Hucht K.A.: 1988, Astron. Astrophys. Suppl. Ser. **72**, 259

Kudritzki R.P., Pauldrach A., Puls J.: 1987, Astron. Astrophys. **173**, 293

Langer N.: 1989, Astron. Astrophys. **220**, 135

Maeder A.: 1990a, Astron. Astrophys. Suppl. Ser. **84**, 139

Maeder A.: 1990b, Astron. Astrophys. in press

Maeder A., Lequeux J., Azzopardi M.: 1980, Astron. Astrophys. **90**, L17 (MLA)

Maeder A., Meynet G.: 1989, Astron. Astrophys. **210**, 155

Meylan G., Maeder A.: 1983, Astron. Astrophys. **124**, 84

Schmutz W., Hamann W.-R., Wessolowski K.: 1989, Astron. Astrophys. **210**, 236

Smith L.F.: 1968, M.N.R.A.S. **141**, 317

Smith L.F.: 1973, IAU Symp. **49**, 15

Smith L.F.: 1982, IAU Symp. **99**, 597

Smith L.F.: 1988, Astrophys. J. **327**, 128

Smith L.F., Hummer, D.G.: 1988, M.N.R.A.S. **230**, 511

Smith L.F., Maeder A.: 1990, Astron. Astrophys. in press

St. Louis N., Moffat A.F.J., Drissen L., Bastien P., Robert C.: 1988, Astrophys. J. **330**, 286

DISCUSSION

Moffat: This work is very impressive and answers many questions. But one problem is with the masses of WR stars for different ambient Z. Contrary to your predictions, WR star masses in the LMC and SMC are the same as the masses of their galactic counterparts, despite the large range in Z (*cf.* poster on WR masses from polarization data at this symposium). Note that the binary frequency, *e.g.*, of WNL stars is identical in the Galaxy and the LMC (Moffat, 1989) while in the SMC the numbers are too small to say, although certainly not all WR stars in the SMC are binaries. Hence, duplicity cannot play a strong role in WR evolution.

Maeder: I am not sure that the distributions of WR star masses in the Galaxy, the LMC and in the SMC are known sufficiently well to allow us to claim for differences or similarities between them. On the contrary, I do think from available data that at low Z, WR formation mainly occurs from binary evolution.

Lortet: About WC/WN ratios: there are large error bars on observational counts (see section WR inventory). There are huge selection effects, favouring the easy detection of strong lined stars. Even in the LMC, Morgan and Good explain their survey only covers stars with $EW > 30\text{Å}$.

Maeder: You are the observer, and you never put the error bars on your results, thus I look forward to see them in future.

Massey: You compare your model predictions to the number of WR/O stars and the ratio WC/WR stars in a number of local group galaxies. The number of O stars (*e.g.*, massive progenitors) in these galaxies is not known to even astronomical accuracy. I also note the poster paper by Armandroff and Massey: the ratio WC/WN is at most 0.9 in M31, based on the spectroscopically *confirmed* WR stars. In fact, it appears that the WC/WN ratio's in M33 and M31 are quite similar. Metallicity cannot be the only parameter controlling this ratio.

Maeder: Apart from metallicity, the IMF and the star formation rate may also play a role. For example, in a small starburst region the WR/O and WC/WR ratio's may change as the starburst region is aging. But, I doubt that this applies to a galaxy as M31 or M33, where we see the results of star formation rates averaged over very large regions.

Sreenivasan: I have a fundamental concern with your models. It is hard to understand on any physical basis why *all* massive stars have to overshoot by the same amount $d = 0.25$ and why massive stars obey a mass loss rate relationship that depends upon the mass as well as other properties but suddenly switch to obeying a relationship that depends only upon the mass when their surface hydrogen composition drops to 0.2. We must remember that correlations do not imply causal relationships and that statistical fits to observed data may later not be "unfolded" to suit individual cases. This conclusion is evidenced by the fact that your observational input to theoretical modeling does not predict other observed fits!

Maeder: There are no reliable wind models for stars outside the MS band and we have to base the models on the observed mass loss rates. If I remember correctly, are you not also using some kind of parametrisation in your own models?

Walborn: What theoretical parameter determines your linear sequence of WC subtypes in the HR diagram, and is it well established observationally?

Maeder: The parameters determining the sequence of WC subtypes are the $(C + O)/He$ abundance ratios. This is based on the determinations of (C/He) ratios by Smith and Hummer (1988) and on the discussion and calibration established by Smith and Maeder (1990).

Vanbeveren: It must be realised that an uncertainty of a factor 2 in the observed $\dot{M}$ values produces a very large uncertainty in the evolutionary results. We really need $\dot{M}$ values with accuracies better than a factor 2. Maybe for the future it would be better, instead of observing many stars only a few times, if one would observe only a few stars many times.

Maeder: The very high sensitivity of evolution on the mass loss rates is precisely what was emphasised over the last ten years. Thus, as you are mentioning, within the range of uncertainties of the mass loss rates, one may have a certain variety of evolutionary tracks at a given inital mass.

Humphreys: Have you considered the role of other effects such as population statistics on the members of WR stars of different types? *E.g.*, we know that the luminosity of the most luminous star depends on the luminosity of the parent galaxy; the smaller less luminous systems have fewer of the most massive stars. These galaxies also tend to have the lowest metallicity. Is the apparent dependence on metallicity due at least in part to population statistics?

Maeder: R. Schild and myself (1984) made a population synthesis to study this effect.

André Maeder

ESTIMATES OF BOLOMETRIC CORRECTIONS FOR WR STARS AND PROGENITORS OF WN BINARIES.

C. DE LOORE and J.P. DE GREVE
Astrophysical Institute, V.U.B.
Pleinlaan 2, B - 1050 Brussels, Belgium.

Abstract.

Using theoretical models of WR binary evolution, and adopting bolometric corrections for the O-type companions, the B.C. of a sample of WN stars are derived using the luminosity difference and the mass ratio. From the present characteristics of observed WN binaries the initial parameters are estimated assuming conservative case B of mass transfer and stellar wind as mass modifying processes. The influence of nonconservative interaction is discussed.

1. Evolutionary models and observed binaries.

For a number of Wolf-Rayet binaries the masses, period and mass ratio are reasonably well known. We adopted the values given by van der Hucht et al. (1988), other resembling values can be found in Schulte-Ladbeck (1989) and Smith and Maeder (1989). We further calculated the mass ratio evolution of binary systems after a conservative case B of mass transfer, for initial primary mass values between 15 Mo and 80 Mo, and mass ratios $q_i = 0.6$ and 0.9. The luminosity evolution of the WR star was calculated using models of helium stars by Langer (1989a). For the calculation of the mass we adopted his mass dependent mass loss formalism (Langer, 1989b). The mass of the O-type companion was kept constant in view of the short timescale involved, and the much lower mass loss rate.

Table 1. Characteristics of WN + O binaries (masses are in M_O).

HD	WR	Type	P(d)	M(WR)	M(O)	$-M_V$(WR)	$-BC$(WR)
90657	21	WN4 + O46	8.26	12	24	3.5	5.7
94546	31	WN4 + O7	4.83	9	17	3.5	5.7
190918	133	WN4.5 +O9.5	112.8	15	15	5.1	3.5
CX Cep	151	WN5 + O8V	2.13	6	14	4.8	3.6
193576	139	WN5 + O6	4.21	10	26	5.1	3.6
E311884	47	WN6 + O5	6.34	43	51	5.5	5.3
186943	127	WN4 + O9.5V	9.55	16	35	3.7	5.7

K. A. van der Hucht and B. Hidayat (eds.),
Wolf-Rayet Stars and Interrelations with Other Massive Stars in Galaxies, 453–457.
© 1991 IAU. Printed in the Netherlands.

Its luminosity was calculated from the following average main sequence mass-luminosity relation

$$\log L/L_O = 1.522 + 2.468 \log M/M_O$$

derived from our computations for single stars (Figueiredo et al., 1990, preprint).

The results for the mass and the luminosity ratio between the components are shown in Figure 1.

Observed WN + O binaries with rather well known characteristics, are given in table 1. The data are taken from van der Hucht et al. (1988), except for HD 90657 and 94546, where we used the results of Schulte-Ladbeck (1989), but completed with the general BC-values for WN4 stars quoted by van der Hucht et al.

In the following q_i refers to the initial mass ratio and is defined as $q_i = M_{2i}/M_{1i}$, q_f refers to the inverse mass ratio after mass transfer, defined as $q_f = M_1/M_2 \approx M(WR)/M(O)$.

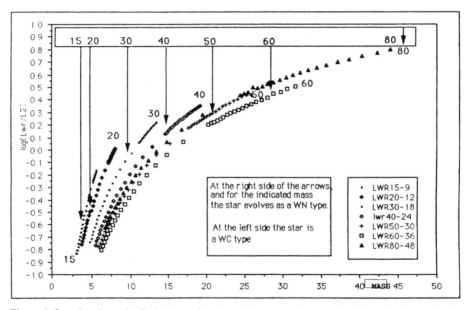

Figure 1. Luminosity ratio (in log) as a function of the mass, for massive binary systems after a case B of mass transfer. Initial masses are in the range 15 to 80 M_O. Vertical arrows indicate the boundary between WN and WC phase.

2. Bolometric corrections for the WN components.

We assume that the mass ratio and the bolometric corrections of the O-stars are accurately known (with respect to the BC of the WR star).

The BC of the WR star can be expressed as a function of BC(O), M_v(WR) and the ratio of the luminosities of the two components, using the well known bolometric magnitude-luminosity conversion. The resulting relation is

$$BC(WR) = M_b(O) - M_v(WR) - 2.5 \log(L(WR)/L(O)).$$

We plotted the log of the ratio of the luminosities as a function of the mass ratios (for two initial values of q_i). Figure 2 shows the result for $q_i=0.6$. The theoretical models occupy a narrow band in the diagram, with $\Delta \log L(WR)/L(O)$ ranging from 0.05 (q = 0.15) to 0.14 (q = 0.45). Using the values of table 1, the observed WN binaries were plotted in the same diagram. We found that they are located well outside the theoretical band. The largest vertical difference 0.8, was found for WR 127, the smallest, 0.12, for WR 139. Next, we assumed that the differences were due to errors in the estimation of the BC of the WR stars (and not in the other parameters involved). We then derived a new estimate of BC(WR) requiring the observed systems to fall on the mean of the theoretical log L(WR)/L(O) - mass ratio relation. The resulting BC values and related quantities, are given in table 2.

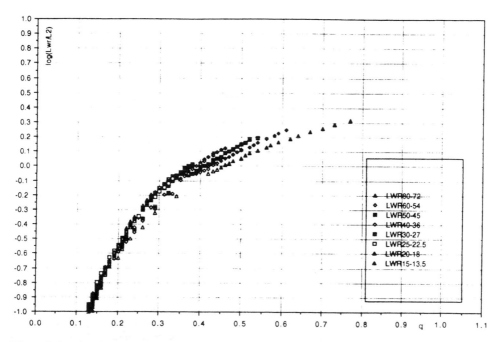

Figure 2. Luminosity ratio (in log) as a function of the mass, for massive binary systems after a case B of mass transfer, for $q_i=0.6$, and initial masses $M_{1i}=15$ to 80 M_0.

Although this method has the advantage of using a narrow relation with mostly well known quantities, it has the disadvantage of being dependent on the initial mass ratio.

However, except for WR47 and WR127, the differences between the two BC values for $q_i = 0.6$ and $q_i = 0.9$ are less than 8 % of the value for $q_i = 0.6$ (see the last column in table 2).

The theoretical luminosity ratio and the mass ratio also depend on the assumption on the mode of mass transfer. Nonconservative mass transfer results in a smaller value for the mass of the O star, hence a larger mass ratio. It also results in a lower luminosity for this component, hence a larger value of log L(WR)/L(O). We calculated the shift of the theoretical luminosity ratio-mass ratio relation, assuming half of the transferred mass is leaving the system. Due to the fact that the points move upward and to the right of the diagram, the overall difference between the new relation and the relation for conservative mass transfer is extremely small (and hence the bolometric corrections). We therefore conclude that, unless the mass transfer is highly non-conservative, or the initial mass transfer very small, the bolometric corrections for WN stars in binaries are fairly well represented by the values derived from models with $q_i = 0.6$ adopting conservative mass transfer.

The average value of BC(WR) from our results is -4.6, which corresponds well with the value quoted by Smith and Maeder (1989). Also the results for the individual systems correspond quite well. The differences with the results of van der Hucht et al. arise from the use of different models and their use of the mass of the WR star, a quantity much less well known than the mass ratio.

Table 2. Bolometric corrections for WN stars, derived from the theoretical log L(WR)/L(O) - mass ratio relation for $q_i = 0.6$ The corresponding bolometric magnitudes and luminosities are also given. The last column shows the values derived using theoretical models with $q_i = 0.9$.

WR	- BC	$M_v(WR)$	$-M_b(WR)$	log L(WR)/L_0	BC(0.9)
21	5.8	3.5	9.3	5.60	5.55
31	5.0	3.5	8.5	5.28	5.25
133	4.5	5.1	9.6	5.72	4.15
139	3.85	5.1	8.95	5.42	3.6
151	3.55	4.8	8.35	5.22	3.3
47	5.30	5.5	10.8	6.20	4.44
127	4.35	3.7	8.05	5.09	3.53

3. Progenitors of WN binaries

Using the foregoing results for the WN stars, we now determine the initial mass and mass ratios of these systems. To obtain these, we compared the mass and luminosity of the WN star with the theoretical models (for $q_i = 0.6$ and 0.9) and looked for the best agreement. The same was done for the mass and mass ratio. The initial parameters leading to the best correspondence (after a conservative case B of mass transfer) between theory and observations in the (M, q, L)-space, were considered as progenitors. The results are given in Table 3.

Under the assumption of conservative case B mass transfer (a combination of 3 assumptions : mass transfer, case B, conservative !) the WN stars have progenitor masses around 30 M_O (except WR 47) and an initial mass ratio larger or equal to 0.7 (except WR 127 and WR 47). Adopting a nonconservative mode of mass transfer results in larger mass ratios M(WR)/M(O) after mass transfer. If we assume that half of the transferred mass is leaving the system, the final mass ratio increases by 0.17 (for $M_{1i} = 15$ M_O) to 0.2 ($M_{1i} = 40$ M_O). This in turn results in larger initial mass ratios for the observed WN binaries. Hence, nonconservative mass transfer brings the initial mass ratio close to 1 for all systems (again with exception of WR47).

Acknowledgement.

This work has been supported by the National Fund of Scientific Research (NFWO - Belgium) under grant No. S 2/5 - LV. E96.

Table 3. Progenitors systems for WN binaries, from the comparison of present mass, mass ratio and luminosity with models after a conservative case B of mass transfer.

WR	M1i/Mo	M2i/Mo	Actual WN state
21	35	28	mid
31	28	22	mid
151	30	21	begin
127	30	15	begin
139	30	27	end
47	60	24	end
133	30	21	end

References

Figueiredo, J., De Greve, J.P., de Loore, C. : 1990, submitted to A & A

Langer, N. : 1989a, Astron. Astrophys. 210, 93.

Langer, N. : 1989b, Astron. Astrophys. 220, 135.

Schulte-Ladbeck, R.E. : 1989, Astron. J. 97, 1471.

Smith, L., Maeder, A. : 1989, Astron. Astrophys. 211, 71.

van der Hucht, K.A., Hidayat, B., Admiranto, A.G., Supelli, K.R., Doom, C. : 1988, Astron. Astrophys. 199, 217.

DISCUSSION

Pakull: I would like to remind you that for some of these stars which have $HeIII$ regions around, we can really determine the luminosity and also the bolometric correction in this case, and for these two stars (whatever you call them, WN1 or WN2) which have $HeII4686$ nebulae around, we derive the bolometric correction, and we find for one of these stars −5.6 and for the other one −7. That is a direct measurement, there is no assumption in it, except that it radiates as a black body, but I was told from the people who do the models that a black body is not a bad approximation for these kinds of stars.

Schmutz: It appears that your B.C.'s are consistent with what we have determined from the spectroscopic analyses.

Bert de Loore

STATISTICS OF WOLF-RAYET BINARIES

L.R. Yungelson and A.V. Tutukov
Astronomical Council of the U.S.S.R.
Academy of Science
48 Pyatnitskaya Str., Moscow, U.S.S.R.

ABSTRACT. A theoretical model of the ensemble of galactic Wolf-Rayet stars is constructed, assuming that all of them are members of either close or wide binaries. The model provides a reasonable explanation of the observed number of WR stars, their distribution over masses, mass ratios of components in binary systems, and spatial velocities. It predicts that up to 10 % of the apparently single WR stars have relativistic companions hidden inside thick stellar winds.

1. The Model

In this contribution we discuss some results pertaining to Wolf-Rayet (WR) stars, obtained by a numerical code, by means of which we attempt to model the binary star population of the Galaxy.

The essence of the method applied is the following. It is assumed that all stars are binaries, either close or wide. From the statistical studies of binary stars one may infer the following equation for the birthrate of binaries (Popova *et al.*, 1982):

$$d^3\nu \approx 0.2 d\log A \ M_1^{-2.5} dM_1 \ f(q)dq$$

where $1 < \log(A/R_\odot) < 6$ is the major semiaxis of the orbit, $0.8M_\odot < M_1 < 100M_\odot$ is the mass of primary component, $q = M_2/M_1$ is the mass ratio of components. The normalization constant corresponds to the formation of about one white dwarf per year in the Galaxy.

Concerning the q-distribution function $f(q)$, one may assume that $f(q) \propto q^\alpha$. There are arguments both for a "flat" ($\alpha = 0$) distribution (Kraicheva *et al.*, 1990) and for a distribution inversely proportional ($\alpha = -1$) to q (Trimble, 1990). We have performed computations both for $\alpha = 0$ and -1.

The evolution of both wide and close binaries is completely determined by their initial M_1, M_2, A. Therefore one may choose a fixed point in the phase space of these parameters to follow all evolutionary transformations experienced by a binary with given M_1, M_2, A. The product of the birthrate of stars with a certain set of initial parameters and of a lifetime in any particular stage of evolution then gives the number of stars in this stage. Simultaneously one obtains all physical characteristics of the binary. Exploring numerically the whole phase space of parameters, one obtains the total numbers of stars of particular types in the Galaxy. This approach in analytical form was successfully applied to cataclysmic

459

K. A. van der Hucht and B. Hidayat (eds.),
Wolf-Rayet Stars and Interrelations with Other Massive Stars in Galaxies, 459–464.
© 1991 *IAU. Printed in the Netherlands.*

binaries, hot subdwarfs, supernovae, etc., in a series of papers by Iben and Tutukov and Tutukov and Yungelson.

The numerical code employing this approach (Tutukov and Yungelson, in preparation) accounts for such processes as mass exchange in close binaries, stellar wind mass loss, angular momentum and mass loss through common envelopes, magnetic braking, gravitational wave radiation, ejection of mass by supernovae and formation of black holes and neutron stars by them. For the masses of remnants after mass exchange or intense stellar wind mass loss stages and the lifetimes of stars we use the analytical approximations of results of evolutionary computations.

The most uncertain phenomenon in close binaries evolution is the mass and angular momentum loss. We treat it as follows. The donor loses ΔM_d of matter in its thermal time scale τ_d, but the companion accretes only a part of this matter ΔM_a in its own thermal time scale τ_a. Thus the efficiency of accretion is about $M_a/M_d \cdot \tau_d/\tau_a$. We split all mass exchange episodes into two phases. In the first one ΔM_a is exchanged and the distance A changes according to the usual conservative formalism. In the second one $(\Delta M_d - \Delta M_a)$ is lost from the system and A changes according to a common envelope formalism (Tutukov and Yungelson, 1979):

$$(M_a + M_d)(\Delta M_a - \Delta M_d)/A_o = \beta(M_a + \Delta M_a)(M_d - \Delta M_d)(1/A_f - 1/A_o)$$

with A_o and A_f being the initial and final separations, and $\beta(\approx 1)$ being a parameter describing the efficiency of orbital energy expenditure on common envelope ejection. Stars with deep convective envelopes were assumed to have only a nonconservative phase of evolution, because of the short time scale of mass exchange episodes.

Stars with initial mass $10 < (M/M_\odot) < 30$ were assumed to produce $1.4M_\odot$ neutron stars, initially more massive stars were assumed to produce $5M_\odot$ black holes.

The initial spatial velocities of stars were estimated under assumption of equipartition of their kinetic energy: $M_1 v_r^2 = 2000 M_\odot (km/s)^2$. Additional components to these velocities are provided by supernovae explosions. The modulus of the resulting velocity is $v = (v_r^2 + v_{SN}^2)^{0.5}$.

Earlier modelling of the ensemble of galactic WR stars assuming that all of them are close binaries has been attempted by Doom and De Greve (1982).

2. The Results

According to current ideas WR stars are core helium burning remnants of initially more massive stars, which have lost their hydrogen envelopes by mass exchange in close binaries (Paczynski, 1967) or by stellar wind (Conti, 1976). If the initial masses of WR stars exceed $10M_\odot$, then their predecessors in close binaries were more massive than $\sim 25M_\odot$. We also assume that components with mass exceeding $50M_\odot$ lose their hydrogen envelopes by stellar wind in the main-sequence stage, both in close and wide pairs. They become WR stars immediately after the main-sequence stage. The separations in this case were changed adiabatically: $(M_1 + M_2) \cdot A = constant$. A strong argument in favour of this picture provides the coincidence of the spatial distribution of WR stars with that of apparently single OB stars with $M/M_\odot > 40 - 50$ (Conti et al., 1983) and of close binaries with $M_{1,2} > 25M_\odot$ (Tutukov and Yungelson, 1985).

Our code generates several scenario's in which bare core helium burning stars more massive than $10M_\odot$ appear. The distribution of numbers of different types of systems containing WR stars is shown in Table 1. The Table displays estimates obtained for two initial distributions over q ($\alpha = 0$ and -1) and for minimum masses of WR stars equal to 8 and $10M_\odot$, as well as some numbers pertaining to observed stars.

Table 1. Statistics of galactic Wolf-Rayet Stars

	TOT.	WR+MS	WR+MS $A \leq 10^3 R_\odot$ $\|\Delta M_b\| \leq 1^m$	WR+NS	WR+DII	WR+WR	WR+SG	WR	TOT. $A \leq 10 R_\odot$	WR+REL $A \leq 10 R_\odot$
$AL=0$	1185	702	197	193	252	20	2	16	134	130
$M \geq 10 M_\odot$		59%	17%	16%	21%	1.7%	0.2%	1.4%	11%	11%
$AL=-1$	977	690	117	101	156	10	1	20	77	74
$M \geq 10 M_\odot$		71%	12%	10%	16%	1%	0.1%	2%	8%	8%
$AL=0$	1544	929	306							
$M \geq 8 M_\odot$		60%	19%							
$OBS.$	157	$v \leq 11^m$	$v \leq 17^m$	13?	3?	3				
		43%	14%	8%	2%	2%				

The total number of galactic WR stars is unknown. Van der Hucht *et al.* (1988) count 46 WR stars inside a 2.5 *kpc* circle around the Sun where the sample is complete. With the 12 *kpc* radius of the Galaxy this results in about 1000 stars in the whole volume. The model estimates for both $f(q)$ are close to this number. Most model WR stars have main-sequence (MS) companions. Among observed relatively bright ($V \leq 11^m$) WR stars the percentage of close binaries is high: $\sim 43\%$. In most observed double-line spectroscopic binaries with WR components the difference of visual magnitudes of components $\|\Delta M_v\| \leq 1^m$. Otherwise the secondary spectrum is not observed. If we pick from the model sample of WR stars only systems with the difference of bolometric magnitudes $\|\Delta M_b\| \leq 1^m$ (this is justified by the proximity of colour temperatures of WR and O stars) we obtain the rate of binarity 12% - 17% which is close to the overall observed value. If one limits the major semiaxes of orbits of model WR+MS systems by $1000 R_\odot$ as in the observed sample, the rate of binarity declines slightly more.

A part of the WR stars has unseen low mass companions, which we had predicted (Tutukov and Yungelson, 1973) to be neutron stars (NS) or black holes (BH). The theoretical relative frequency of them is 26 - 38%, the observed one about 10%. However, it is evident that a lot of low mass companions still have to be discovered.

Only a small portion of the WR stars are really single stars. These are merger products which have attained masses exceeding $50 M_\odot$. The secondaries of systems disrupted by the first SN explosion do not have high enough masses to produce WR stars.

The percentage of WR+WR systems as well as that of WR stars accompanied by supergiants (SG) is very low because such pairs arise only from wide pairs with very close initial masses of components.

About 10% of all WR stars have to have very close ($a \leq 10 R_\odot$, $P \leq 1$ day) companions. Almost all of them are neutron stars or black holes. Both components form a common envelope system inside the thick stellar wind of the WR star. Possibly they manifest themselves by some kind of periodical variability, that can be discovered by special ultraviolet observations.

About 40% of the model WR stars are descendants of stars that have lost their envelopes by stellar wind. This agrees well with the estimate based on the study of the spatial distribution of WR and O stars (Tutukov and Yungelson, 1985).

The mass distribution of all model WR stars is double-peaked (Fig. 1), in obvious contradiction to the mass function of WR stars in double-lined binaries (dots in Fig. 1

462

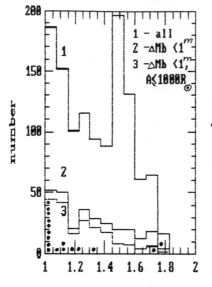

Fig. 1. Distribution of model
WR stars over mass

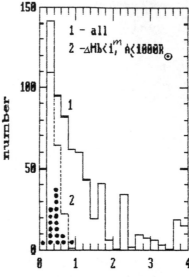

Fig. 2. Distribution of model
WR stars over q

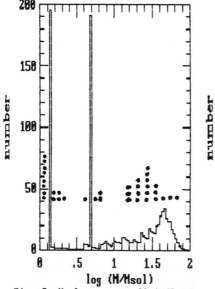

Fig. 3. Number - mass distribution
of companions of WR stars

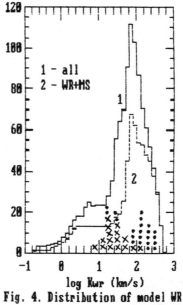

Fig. 4. Distribution of model WR
stars over maximum K_{wr}

as well as in other Figures are related to observed stars (Aslanov *et al.*, 1989)). One may reduce the relative number of massive stars by lowering the minimum mass of WR stars to $\sim 8 M_\odot$ or by increasing to $60 - 70 M_\odot$ the lower mass limit of single stars that become WR stars. However, the discrepancy may be a mere result of observational selection. Limiting the model sample again by $|\Delta M_b| < 1^m$ and $A \leq 1000 R_\odot$, one removes from it most of the massive stars (Fig. 1) and gets agreement with observations. The shape of the initial distribution over q influences the mass spectrum of WR stars only weakly, because most of them descend from initially primary components.

One of the most reliable observed parameters for WR+MS binaries is the mass ratio of the components (Fig. 2). A significant proportion of all model systems have $q > 1$. But by taking into account the same selection effects one gets reasonable agreement with observations.

No definite evidence of low mass companions of single-line observed WR stars being neutron stars or black holes does exist. In Fig. 3 we show the distribution of satellites of model WR stars over mass (for systems with a semi-amplitude of radial velocity $K_{WR} \geq 10 \ km \cdot s^{-1}$ as in the observed sample. Three peaks of it are due to $1.4 M_\odot$ neutron stars, $5 M_\odot$ black holes and $(10 - 100) M_\odot$ OB stars. It is important to note that the overwhelming majority of companions with $M < 6 M_\odot$ are relativistic stars. White dwarfs are absent and only a few companions are normal dwarfs.

Fig. 4 shows the distribution of all WR stars and WR stars with main-sequence companions over the maximum semi-amplitude of radial velocity K_{WR}. The relativistic companions in the systems with $10 \ km \cdot s^{-1} \leq K_{WR} \leq 100 \ km \cdot s^{-1}$ have to be predominantly neutron stars as it is also suggested by observational data (crosses: Aslanov *et al.*, 1989). The systems with $K_{WR} \geq 50 \ km \cdot s^{-1}$ are probably immersed into common envelopes and still wait discovery. The theoretical ratio is $N_{NS}/N_{BH} \approx 0.57$ for $\alpha = o$ and ≈ 0.71 for $\alpha = -1$. The observed ratio is -4. This may indicate that we underestimate the lower mass limit for predecessors of black holes. However, more firm observations are necessary for definite conclusion. Positions of WR stars with MS-companions agree well with the observations (dots).

The first SN explosion in a close binary provides it with a high spatial velocity which it maintains in the second WR stage (Tutukov and Yungelson, 1973). The distribution of observed WR stars over the spatial velocities is unknown. However, the main-sequence lifetime of precursors of second generation WR stars is long enough to increase the scale-height of the whole subsystem of the galactic WR stars. Very roughly, the modulus of z-coordinate relates to spatial velocity as $|z|(pc) - 10v \ (km \cdot s^{-1})$. The ratio of the number of observed WR stars with $|z| > 100 \ pc$ to the number of stars with $|z| < 100 \ pc$ is close to 0.55. The predicted ratio of number of stars with $v > 10 \ km \cdot s^{-1}$ to more slow ones is about the same, in reasonable agreement with observations.

3. Conclusion

The proposed model of evolution of galactic binaries assumes two channels for WR star formation - by mass exchange in close binaries and by stellar wind mass loss in both close and wide very massive binaries. The first channel produces about 60% of all WR stars, the second about 40%. This model, after taking into account some simple observational selection effects, satisfactorily describes the statistical properties of the observed ensemble of galactic WR stars.

Most WR stars with unseen companions have neutron star or black hole satellites, which are descendants of initial primaries.

About 10% of all WR stars may have very close ($A \leq 10R_\odot$) neutron star or black hole companions, immersed into their optically thick stellar winds.

The high spatial velocities of about 40% of the apparently single WR stars are a result of supernova explosions in close binary systems.

The change of the initial distribution of massive binaries over q from a flat one to one inversely proportional to q, does not influence the theoretical sample of WR stars strongly enough to discriminate between them on an observational basis. The main obstacle for this task is the proper account of observational selection.

References

Aslanov, A.A., Kolosov, D.E., Lipunova, N.A., Khruzina, T.S., Cherepashchuk, A.M. 1989, Catalogue of Close Binaries in Late Stages of Stellar Evolution, Moscow Univ. Press.

Conti, P.S. 1976, *Mem. Soc. Roy. Sci. Liege.*, 6-Serie, **9**, 193.

Conti, P.S., Garmany, C.D., de Loore, C., Vanbeveren, D. 1983, *Astrophys. J.* **274**, 302.

Doom, C., De Greve, J.-P. 1982, *Astrophys. Space. Sci.* **87**, 1982.

van der Hucht, K.A., Hidayat, B., Admiranto, A.G., Supelli, K.R., Doom, C. 1988, *Astron. Astrophys.* **199**, 217.

Kraicheva, Z.T., Popova, E.I., Tutukov, A.V., Yungelson, L.R. 1990, *Astrophysics*, in press.

Paczinski, B. 1967, *Acta Astron.* **17**, 355.

Popova, E.I., Tutukov, A.V., Yungelson, L.R. 1982, *Astrophys. Space. Sci.* **88**, 55.

Trimble, V. 1990, *Monthly Notices Roy. Astron. Soc.* **242**, 79.

Tutukov, A.V., Yungelson, L.R. 1973, *Nauchn. Informatsii* **27**, 57.

Tutukov, A.V., Yungelson, L.R. 1979, in P.S. Conti, C.W.H. de Loore (eds.), Mass Loss and Evolution of O-type Stars, *Proc. IAU Symp. No. 83* (Dordrecht: Reidel), p. 401.

Tutukov, A.V., Yungelson, L.R. 1985, *Astron. Zh.* **65**, 604 (= *Sov. Astron.* **29**, 352).

DISCUSSION

Vanbeveren: Your conclusions concerning the WR+c binary frequency very much depends on the adopted model for the mass transfer in close binaries. Could you comment on that?
Yungelson: We had adopted a model that allows to estimate the efficiency of accretion and to estimate the angular momentum loss through common envelopes. It predicts the periods and semi-amplitudes of the radial velocity distribution in a satisfactory agreement with observations. This may be considered as evidence of plausibility of the adopted model.

OBSERVATIONAL ASPECTS OF STELLAR EVOLUTION IN THE UPPER HRD

CLAUS LEITHERER
*Space Telescope Science Institute, 3700 San Martin Drive, Baltimore, MD
21218. Affiliated with the Astrophysics Division of the Space Science Department of the European Space Agency*

1. Introduction: Wolf-Rayet Stars and Their Relation to Massive Stars

Observational evidence strongly favors an evolutionary relation between Wolf-Rayet (WR)
stars and stars with masses above $\sim 10\ M_\odot$. The galactic distribution of WR stars closely
resembles the distribution of O stars both in the z-direction as well as in the spiral-arm
pattern (Hidayat, Admiranto, and van der Hucht 1984, van der Hucht *et al.* 1988, Conti
and Vacca 1990). Lundström and Stenholm (1984) found that the surface density of WR
stars increases with decreasing distance from OB associations as well as from young open
clusters. The fraction of WR stars in clusters turns out to be about the same as that of
O stars. A large number of extragalactic WR stars is situated in giant H II regions, which
also contain high numbers of very massive O stars (see, e.g., Walborn 1990 for 30 Doradus;
Drissen, Moffat, and Shara 1990 for M33). Many WR stars are members of binary systems
with well established parameters so that their masses can be determined directly. In several
cases, WR-star masses of $M \gtrsim 10\ M_\odot$ have been derived (Massey 1981, St.-Louis *et al.* 1987,
Schulte-Ladbeck 1989).

Furthermore, there is overwhelming observational support for the assumption that WR
stars are in a later evolutionary stage than O stars (see Lamers *et al.* 1990). Abundance
analyses of WR stars are consistent with a scenario that we are seeing material processed
by the CNO cycle and by He burning (Hillier 1990). WR stars are evolved objects. The
spectral appearance of some WNL stars closely resembles highly evolved objects, such
as Luminous Blue Variables (Stahl 1986, Walborn 1989). The observed mass-luminosity
relation agrees with the assumption that WR stars are highly evolved objects, rather than
pre-main-sequence stars. Finally, evolved descendants of massive O stars are expected to
be luminous and should easily be detected. If WR stars are **not** descendants of massive
stars, it is hard to understand why such descendants have not yet been discovered (Lamers
et al. 1990).

It should be mentioned that an alternative interpretation has been given by Underhill
(this meeting), who suggests WR stars are predecessors of O stars on their way to the main
sequence. Considering all the observational evidence summarized above, I will assume in
this review that **WR stars are evolved descendants of previously massive stars**.

Our theoretical modeling of massive-star evolution becomes more and more uncertain the
further a star has evolved from the main sequence. Obviously, a thorough understanding

K. A. van der Hucht and B. Hidayat (eds.),
Wolf-Rayet Stars and Interrelations with Other Massive Stars in Galaxies, 465–478.
© 1991 *IAU. Printed in the Netherlands.*

of the pre-WR evolution of a star is required in order to understand the WR phase itself. I will review our current observational database available to study stellar evolution in the upper part of the HRD. The main emphasis will be on aspects which are of direct relevance to WR stars, such as channels which lead to the formation of WR stars, the mass regime of WR predecessors, the location of WR predecessors in the HRD, and consequences of different chemical composition on the relation between WR stars and their progenitors.

2. The Upper HRD — an Overview

The most massive stars are also the most luminous stars — even visually since they follow a mass-luminosity relation of $L \sim M^{\alpha}$, with $\alpha > 1$ (Maeder 1987). Complete samples for the population of massive stars in the solar neighborhood with $M \gtrsim 10\ M_{\odot}$ have been published by a variety of authors, most recently by Blaha and Humphreys (1989). Figure 1 reproduces their observed HRD of all known massive stars which are members of clusters or associations within 3 kpc of the sun. The schematic location of individual WR subgroups taken from Schmutz, Hamann, and Wessolowski (1989) is indicated. WR stars roughly coincide in their location with main sequence O stars — except for WNE-types, which are slightly hotter. I also included the positions of several well-known Luminous Blue Variables (LBV's, Humphreys 1989).

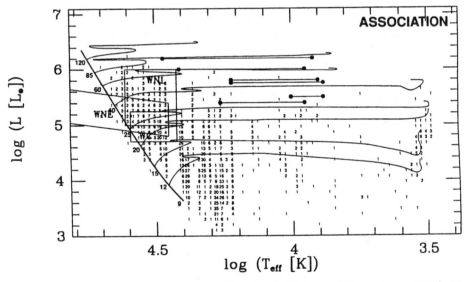

Figure 1. Observed HRD of the solar neighborhood (Blaha and Humphreys 1989). Numbers indicate star densities in the HRD. Solid dots denote the location of LBV's. WR stars are indicated schematically by boxes.

Blaha and Humphreys confirm — using more refined observational data — an important result first detected by Humphreys and Davidson (1979). The most luminous blue stars are about a factor of 6 more luminous (and thus more massive) than the most luminous red stars. This difference in luminosity defines a boundary in the HRD parameterized by Garmany, Conti, and Massey (1987) and Humphreys (1987). The importance of this boundary is obvious: Stars evolve from the main sequence towards cooler temperatures at roughly constant luminosity (e.g. Maeder and Meynet 1988). The observed absence of stars to the right of the boundary may mean that stellar evolution proceeds so fast that the probability of observing red stars with $M \gtrsim 50 \ M_\odot$ is very low. Langer and El Eid (1986) computed evolutionary models for a $100 \ M_\odot$ star evolving into a red supergiant (RSG) and then back into a WR star. They find that the star should spend $10^4 - 10^5 yr$ in the RSG phase. Although this lifetime is rather uncertain (e.g. due to the mass loss in the RSG phase), one would expect that luminous RSG's with $L \approx 10^6 L_\odot$ should be detected if they existed.

Alternatively, the observed boundary in the HRD can be understood if massive stars with $M \gtrsim 50 \ M_\odot$ evolve from the main sequence via the blue supergiant (BSG) phase into WR stars. Maeder (1983) demonstrated that such an evolutionary scenario is possible if a phase of strong mass loss occurs during post-main-sequence evolution in order to remove hydrogen-rich layers from the stellar surface. The star can then evolve into the WR regime without entering the RSG phase. Evidence for such a phase of high mass loss comes from observations of LBV's (Humphreys 1989). Some of these objects are located close to the boundary of the HRD (cf. Figure 1). An estimate of their lifetimes and mass-loss rates suggests that LBV's lose a significant fraction of their total mass in this part of the HRD. Hence, they have subsequently been identified with the phase of strong mass loss *a priori postulated* by stellar evolution models.

The following scenario emerges from the shape of the upper HRD: Stars with initial masses below about 50 $M_\odot$ evolve from the main sequence to the RSG phase and may return back to the blue part to form a WR star, depending on how much mass can be removed from the H-rich envelope via mass loss. Stars above about 50 $M_\odot$ experience mass-loss rates $(\dot{M})$ in the blue part of the HRD high enough that these stars enter the WR phase without passing through the RSG phase. Such a genetic relationship has been proposed by Maeder (1983).

3. Properties of the Observed HRD: Clues for WR Evolution

3.1 THE LACK OF VERY MASSIVE STARS CLOSE TO THE ZAMS

Close inspection of Figure 1 reveals a lack of very massive stars with $M \gtrsim 40 \ M_\odot$ close to the theoretical zero-age main sequence (ZAMS). Such an effect has also been noted by Garmany, Conti, and Chiosi (1982) on the basis of a different data set. From a comparison with theoretical isochrones one finds that the youngest known hydrogen-burning O stars have an age of $1 - 2 \ 10^6 yr$ (see, e.g., Chiosi 1986), so that $\sim 20\%$ of the entire main sequence lifetime is not observed.

Wood and Churchwell (1989) found that $10 - 20\%$ of all O stars in the solar neighborhood (SN) are still embedded in molecular clouds. They are detectable indirectly by radiation from ultracompact H II regions and warm circumstellar dust surrounding the central object. Only about $10^6 yr$ after the onset of H-burning circumstellar gas and dust become transparent and the star becomes visible. This seems to be a plausible explanation for the observed lack of very massive ZAMS stars in the HRD.

The fact that we may underestimate the true number of O stars in the SN by $\sim 20\%$ is of immediate relevance for evolutionary models of WR stars. An important — because easily (!) observable — parameter predicted by models is the number ratio of WR/O stars (see below). The ratio previously derived from observations may have to be revised downward by ∼ 20%.

3.2 THE OBSERVED MAIN-SEQUENCE WIDTH

The observed width of the main sequence provides another important parameter for the evolution on the main sequence and for **post-main-sequence** evolution. Constraints for the amount of overshooting can be derived, which in turn governs the formation of super-giants and WR stars. Bertelli, Bressan, and Chiosi (1984) determined the main sequence width by counting stellar number densities across a synthetic HRD of the SN. Alternatively, Mermilliod and Maeder (1986) fitted theoretical isochrones to the observed color-magnitude diagrams of **individual** clusters. Both studies conclude that the main sequence extends to early- or even late-B stars in the upper HRD so that many OB supergiants are still core-hydrogen burning. Maeder and Meynet (1987) demonstrated that the observed shape of the main sequence can be used to constrain the amount of mass loss and overshooting in evolutionary models.

Bertelli, Bressan, and Chiosi (1984) emphasized the strong sensitivity of the main-sequence width predicted by evolution models on the adopted opacities. Since theoretical opacities are subject to uncertainties, this point should be kept in mind.

Recently, Garmany (1990) reinvestigated the observed main-sequence width in clusters and associations. Preliminary results seem to indicate that earlier studies tend to overesti-mate the main sequence width: if field stars are erroneously assigned cluster membership and/or if stars did not form coevally, the apparent width of the main sequence will be increased.

3.3 THE DISTRIBUTION OF OB SUPERGIANTS IN THE HRD OF THE LMC

With improved spectral classifications becoming available for more and more stars in the LMC, Fitzpatrick and Garmany (1990) are able to construct a complete, homogeneous upper HRD with LMC stars. Figure 2 is taken from their paper. The upper luminosity limit agrees with the results derived for the SN (cf. Figure 1). Comparison of HRD's of several Local Group galaxies covering a wide range of chemical composition reveals no significant variation of the shape of the upper boundary (Blaha and Humphreys 1989). Differences in the luminosities of the most luminous blue stars found in these galaxies can be accounted for by statistical effects (Kennicutt 1983, Schild and Maeder 1983). This does not exclude that there may be differences in stellar evolution and/or the initial mass function from galaxy to galaxy. However, stellar statistics is too poor to answer this question on the basis of the upper limit of the HRD.

Strikingly, there is a pronounced discontinuity in the density distribution among **blue supergiants**. Note that this discontinuity is marginally discernible also in the HRD of the SN but tends to be "washed out" due to uncertainties in the distances of galactic stars. Such a discontinuity can be understood if red supergiants with initial masses of $10\ M_\odot < M < 40\ M_\odot$ undergo a blueward loop shortly before core-helium exhaustion (see Chiosi and Summa 1970, Langer, El Eid, and Baraffe 1990). Note that in these models the stars situated to the left of the discontinuity in Figure 2 are **not** immediate progenitors of WR stars but will return back to the RSG stage until core-helium exhaustion. On the other hand, a recent analysis of the Ofpe/WN star R84 (Schmutz *et al.* 1990) resulted in stellar parameters which place this star right on the density enhancement in the HRD ($log\ L = 5.7$, $log\ T_{eff} = 4.45$). Despite the same location in the HRD as the stars supposedly on the blue loop, R84 shows spectral characteristics of a late WN star, which may indicate a different evolutionary phase for this star.

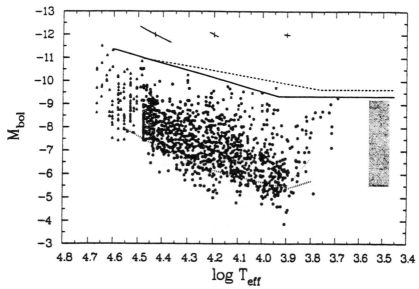

Figure 2. Upper HRD of the LMC published by Fitzpatrick and Garmany (1990). Note the pronounced discontinuity in star density among blue supergiants.

3.4 OBSERVED EVIDENCE FOR CNO- AND HE-ANOMALIES IN EVOLVED STARS

The hypothesis outlined above that many OB stars are actually evolved further than naively deduced from their location in the HRD is further substantiated by observations of CNO processed material in some of these stars. Walborn (1976) first drew attention to the morphological group of OBN/OBC stars, which show nitrogen- and carbon-line strengths different from what is observed in other stars with the same spectral type. It has been suggested (Walborn 1988) that the majority of "normal" B supergiants shows processed material whereas OBC objects are the rare, unevolved objects. Further support for the presence of chemically evolved objects close to the main sequence comes from spectral analyses performed by Kudritzki, Simon, and Hamann (1983) and Bohannan al. (1986) who derived significant He overabundances for several OB stars.

Clearly, the above cases demonstrate a weakness inherent in the HRD as a tool to study stellar evolution. Evolutionary effects are more complex than can be described in a two-parameter space limited to $L(t)$ and $T_{eff}(t)$. The variation of M and the chemical composition on the stellar surface with lifetime may profoundly change the spectral appearance, as in the case of a WR star. On the other hand, these effects may be subtle enough that stars in very different evolutionary phase at the some location of the HRD may be hard to distinguish, as may be the case with certain blue supergiants.

3.5 THE LBV – WR CONNECTION

Luminous Blue Variables are an important link between massive stars still in their core-hydrogen phases and WR stars. They exhibit spectral variations on time-scales of years to decades. At one time they closely resemble other blue supergiants in the same part of the HRD and at the other they are indistinguishable in their spectral morphology from certain WN9 stars (Stahl 1986). The total bolometric luminosity of LBV's remains constant even if T_{eff} varies between $\sim 30,000$ K and $\sim 8,000$ K (cf. Figure 1). Therefore, L can be determined with relatively high accuracy when these objects are in their low-temperature phase — in contrast to WNE- or WC-stars, where high (and uncertain) bolometric corrections must be applied.

Since the wind characteristics of LBV's are not as pronounced as those of WR stars, their photospheres are much better accessible for a detailed analysis (e.g., Kudritzki et al. 1989, Leitherer et al. 1989). The most important results of such studies are a relatively low mass and He enrichment at the stellar surface. Under the assumption that the stellar winds in these objects are driven by radiation pressure, the low wind velocities (as compared with other supergiants at the same position in the HRD) observed in LBV's are an immediate consequence of the low surface escape velocity, and thus of the low mass. It is interesting to note that the WNL stars analyzed by Schmutz, Hamann, and Wessolowski (1989) have the lowest terminal velocities ($900 - 1200$ km s^{-1}) among all WR stars of their sample. This may also indicate the close evolutionary relationship between (some ?) LBV's and WNL stars.

Comparison with evolutionary models suggests that LBV's with ZAMS masses below ~ 40 $M_\odot$ may be post-red supergiants on their way to becoming WR stars. Alternatively, stars in this mass range could already encounter the LBV phase immediately after the BSG phase and never enter the RSG phase. If the latter is true — and if LBV's are in fact progenitors of WN stars — there should be no correspondence between the populations of WN stars and RSG's above a certain threshold mass. It seems more likely, though that LBV's with $M_{ZAMS} \lesssim 40$ $M_\odot$ did pass through the RSG stage because of the presence of a dust shell discovered around R71 (Wolf and Zickgraf 1987). This dust shell may have formed when R71 was still in its RSG phase. With respect to LBV's with higher initial mass, it seems implausible that stars in this mass range ever evolve into the red part of the HRD (cf. Section 2).

4. The Mass Regime of WR Predecessors

The previous section addressed potential connections between massive stars in early and late stages of their evolution. I will now review what mass range of stars on the ZAMS will eventually evolve into WR stars.

Conti et al. (1983) studied the correlation of the O-star population in the SN with the WR population. They found that the WR population correlates best with the population of O stars having initial masses above ~ 40 $M_\odot$. Doom (1987) and van der Hucht et al. (1988) readdressed the issue using a more rigorous statistical analysis and improved observational data. In agreement with Conti et al. they derive a typical progenitor mass of $M_{ZAMS} \gtrsim 35$ $M_\odot$. However, some WR stars may descend from stars down to $M_{ZAMS} \approx 22$ $M_\odot$. On the average, WN stars have lower progenitor masses than have WC stars. Interestingly, by separating those WR stars which are members of binaries from their sample, van der Hucht *et al.* showed that single WR stars preferentially descend from higher mass stars than do WR stars in binaries. This agrees with what would be expected theoretically: stars with initially higher masses experience higher mass loss, which favors the formation of WR stars so that the binary channel to form WR stars becomes less important (see Maeder 1990a).

Schild and Maeder (1984) used the membership of WR stars in clusters and OB associations to determine lower cut-off masses for the formation of WR subtypes. By fitting theoretical isochrones to the observed main sequence turn-off, the mass limits of the WR progenitors can be derived. Schild and Maeder find $M_{ZAMS}(WN) \gtrsim 18\ M_\odot$ and $M_{ZAMS}(WC) \gtrsim 35\ M_\odot$. Contrary to the results derived by van der Hucht *et al.*, no significant difference could be detected for the progenitor masses of binary or single WR stars. Humphreys, Nichols, and Massey (1985) performed a similar type of study and derive $M_{ZAMS}(WR) \gtrsim 30\ M_\odot$, with 80% of all WR stars originating from progenitors above 50 $M_\odot$. WN and WC stars showed no significantly different cut-off mass.

The results of these studies imply that the lower cut-off mass of WR progenitors is around $25 \pm 10\ M_\odot$. WC stars may descend from higher-mass stars than do WN stars but the correlation is not so clear. One should be aware that we are dealing with small-number statistics since the number of O- and WR-stars in individual clusters and associations may be low. Also, the assumption of cluster membership or coeval star formation may not be correct in some cases.

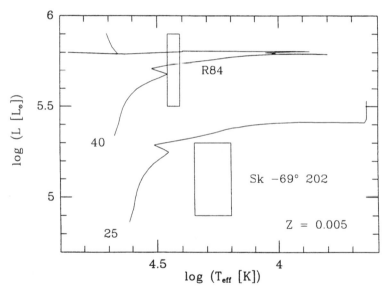

Figure 3. Location of the two LMC stars R84 and Sk −69°202 in the HRD. Evolutionary tracks for LMC abundances are from Maeder (1990b). R84 has been classified as Ofpe/WN9 whereas Sk −69°202 exploded as a supernova without entering the WR phase.

Figure 3 is a HRD giving the location of two well-known LMC stars. One star is Sk −69°202, the progenitor of SN1987A (see Arnett *et al.* 1989), the other is R84, an Ofpe/WN star recently analyzed by Schmutz *et al.* (1990). Sk −69°202 never entered the WR phase whereas R84 exhibits spectral features typical for late WN stars. The progenitor masses of both stars are relatively well-known: $20 - 25\ M_\odot$ for Sk $-69°202$ and $30 - 40\ M_\odot$ for R84. Somewhere between these two mass ranges a critical mass may exist below which the channel for the formation of WR stars is closed. This argument assumes that initial mass (and possibly chemical composition, see below) is the only factor determining this channel. Presently, nothing can be said on the importance of other effects such as, e.g., stellar rotation.

5. The Relative Numbers of WR Stars and Their Progenitors

If WR stars are the evolved descendants of stars having main-sequence masses above $\sim$ $25\ M_\odot$, a relation between the observed numbers of WR stars and their progenitors can be expected. Such a relation, like the ratio $\frac{WR}{O} = \frac{Number\,of\,WR\,Stars}{Number\,of\,O\,Stars}$, reflects the lifetimes of the star spent in each of the two individual phases. The results obtained for the SN are summarized in Figure 4 (taken from Leitherer 1990). The dots in this figure indicate the stellar number counts for O stars (Garmany, Conti, and Chiosi 1982) and WR stars (Conti *et al.* 1983). From this data set one finds $\frac{WR}{O} = 0.12$. Theoretical predictions are indicated by the solid line in Figure 4. These models have been computed with an initial mass function and star formation rate as derived by Garmany, Conti, and Chiosi (1982) and stellar evolution models published by Maeder and Meynet (1988). Observations and model predictions agree for the relative and absolute numbers of O stars. In contrast, the models produce about a factor of 2 less WR stars than are observed.

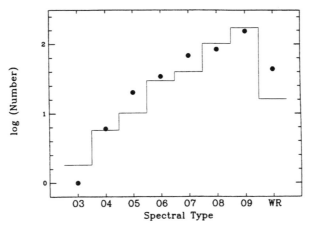

Figure 4.   Observed (dots) and synthesized (histogram) number of O- and WR-stars in the solar neighborhood (from Leitherer 1990).

What are the reasons for this disagreement? On the observational side, I already discussed the fraction of O stars which may still be hidden in primordial gas and dust. These stars are **not** included in Figure 4 and could reduce the discrepancy by $\sim$ 20%. Humphreys and McElroy (1984) argue that the O-star sample in Figure 4 is severely incomplete for the latest types. This would also decrease the observed $\frac{WR}{O}$ ratio. Finally I would like to point out the sensitivity of the results to the adopted T_{eff} versus spectral-type calibration (see Figure

5). At the high-mass end, even slight changes to this calibration result in a significantly different relation between mass and spectral type. At the low-mass end ($M \approx 20\ M_\odot$), minor modifications to the adopted effective temperature of the coolest O star (e.g., O9.5 or O9.7) significantly affects $\frac{WR}{O}$ since the number of late-O stars vastly dominates the total number, given the effect of the inital mass function.

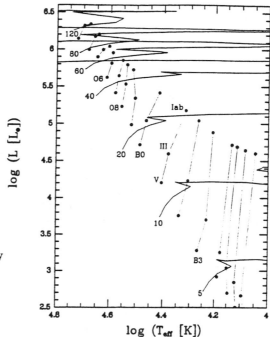

Figure 5. HRD with evolutionary tracks and Schmidt-Kaler's (1982) spectral-type calibration. Dots represent types O3 through B7 in steps of 1 subtype.

On the theoretical side, several uncertainties should be mentioned. Clearly, stellar evolution models themselves may undergo some revision (e.g., uncertainties in the adopted opacities, the amount of overshooting, and the mass-loss rates). The binary channel for the formation of WR stars has been neglected. Maeder (1990a) estimates that binary evolution may slightly increase the theoretical ratio of $\frac{WR}{O}$ but a significant effect should only be expected in a low-metallicity environment, such as the SMC. The models presented in Figure 4 assume a lower cut-off mass for the formation of WR stars of 33 $M_\odot$. Lowering the cut-off mass will significantly increase the theoretical $\frac{WR}{O}$ ratio. As discussed in the previous section, the cut-off mass is still under debate, and values of 20 $M_\odot$ or even lower are not incompatible with observations.

Given the uncertainties — both on the observational and on the theoretical side — I conclude that the observed relative numbers of WR stars and their progenitors in the solar neighborhood are basically in agreement with theoretical models and that they are consistent with the assumption that WR stars are bare cores left over from massive O stars which lost their outer hydrogen-rich layers.

Recent theoretical models (Maeder 1990a) predict a very strong dependence of $\frac{WR}{O}$ on the initial chemical composition of the progenitor star. A lower metal content Z leads to a significantly smaller value of $\frac{WR}{O}$. This is due to several consequences of lower Z: the models predict smaller WR- and longer O-star lifetimes as well as higher cut-off masses for the formation of WR stars. Also, the initial mass of an O star on the ZAMS decreases if Z is lower. All these effects lead to a strong sensitivity of $\frac{WR}{O}$ on Z. In addition, the relative numbers of individual WR-subclasses, such as $\frac{WN}{WC}$ and $\frac{WC}{WR}$ are also expected to vary with Z.

It is difficult to test the variation of $\frac{WR}{O}$ with Z in our Galaxy. The average galactocentric metallicity gradient derived from nebular analyses is $\sim 10^{-0.1}$ kpc^{-1} (Talent and Dufour 1979). A factor-of-2 variation of Z is reached only at a distance of 3 kpc. At such large distances, however, significant incompleteness for O stars sets in (Humphreys and McElroy 1984). It should be noted that Lennon et $al.$ (1990) find no evidence for a galactocentric metallicity gradient within $3 - 4$ kpc of the sun on the basis of quantitative spectroscopy of a sample of B stars. In fact, theory and observations do not agree too well with respect to the dependence of $\frac{WR}{O}$ on Z if this test is performed in our galaxy (see Maeder, this conference).

In principle, a more promising way to study the influence of Z comes from studying stellar statistics in Local Group galaxies which span a wide range of chemical composition. Unfortunately, the stellar census in these galaxies is still rather incomplete, and significant correction factors must be applied to the observations. Maeder, Lequeux, and Azzopardi (1980), Massey and Conti (1983), Freedman (1985), Azzopardi, Lequeux, and Maeder (1988), and Smith (1988) reach different conclusions regarding the dependence of the relative WR numbers to the numbers of their progenitors on Z. At present this is still an open issue but it is to be anticipated that more definitive conclusions can be drawn as soon as more complete statistics are available.

6. Outlook: Massive Stars in Distant Galaxies

Traditionally, constraints on stellar evolution in the upper HRD are almost entirely based on observational data collected in our Galaxy. Progress in telescope- and detector-technology makes it possible to study stellar evolution in galaxies at larger and larger distances. Some of these new data have already been discussed. Some galaxies of the Local Group are ideal laboratories to observe complete samples of massive stars at spatial and spectral resolution still high enough to obtain useful information. The ongoing work of Massey et $al.$ (see, e.g., Massey, Parker, and Garmany 1989) in the Magellanic Clouds and of Moffat et $al.$ (see, e.g., Moffat and Shara 1987) in M31/M33 provides important constraints for the theory of stellar evolution.

At even larger distances, galaxies can no longer be resolved into individual stars. Large populations of WR stars have been detected in the integrated spectra of some galaxies with active star formation (e.g., Kunth and Schild 1986). Although the detail of information is certainly degraded with respect to data from our Galaxy, such observations may well complement results derived locally. The number of stars contributing to the observed, integrated light is very large — one definite advantage over the number statistics in our Galaxy and irregular galaxies of the Local Group, which are mostly rather poor. Arnault, Kunth, and Schild (1989) published a grid of synthetic populations of massive stars with emphasis on the $\frac{WR}{O}$ ratio. If parameters such as, e.g., the initial mass function of massive stars can be constrained independently, observations of the stellar content of such galaxies may serve as an important tool to understand stellar evolution in the upper HRD.

References

Arnault, P., Kunth, D., and Schild, H. 1989, *Astr. Ap.*, **224**, 73.

Arnett, W. D., Bahcall J. N., Kirshner, R. P., and Woosley, S. E. 1989, *Ann. Rev. Astr. Ap.*, **27**, 629.

Azzopardi, M., Lequeux, J., and Maeder, A. 1988, *Astr. Ap.*, **189**, 34.

Bertelli, G., Bressan, A., and Chiosi, C. 1984, *Astr. Ap.*, **130**, 279.

Blaha, C., and Humphreys, R. M. 1989, *A. J.*, **98**, 1598.

Bohannan, B., Abbott, D. C., Voels, S. A., and Hummer, D. G. 1986, *Ap. J.*, **308**, 728.

Chiosi, C. 1986, in *IAU Symposium 116, Luminous Stars and Associations in Galaxies*, ed. C. W. H. de Loore, A. J. Willis, and P. Laskarides (Dordrecht: Reidel), p. 317.

Chiosi, C., and Summa, C. 1970, *Ap. Space Sci.*, **8**, 478.

Conti, P. S., and Vacca, W. D. 1990, *A. J.*, in press.

Conti, P. S., Garmany, C. D., de Loore, C. W. H., and Vanbeveren, D. 1983, *Ap. J.*, **274**, 302.

Doom, C. 1987, *Astr. Ap.*, **182**, L43.

Drissen, L., Moffat, A. F. J., and Shara, M. M. 1990, *Ap. J.*, in press.

Fitzpatrick, E. L., and Garmany, C. D. 1990, *A. J.*, in press.

Freedman, W. L. 1985, *A. J.*, **90**, 2499.

Garmany, C. D. 1990, in *Massive Stars in Starbursts*, ed. T. Heckman, C. Leitherer, C. Norman, and N. Walborn (Cambridge: Cambridge University Press), in press.

Garmany, C. D., Conti, P. S., and Chiosi, C. 1982, *Ap. J.*, **263**, 277.

Garmany, C. D., Conti, P. S., and Massey, P. 1987, *A. J.*, **93**, 1070.

Hidayat, B., Admiranto, A. G., and van der Hucht, K. A. 1984, *Ap. Space Sci.*, **99**, 175.

Hillier, D. J. 1990, in *Properties of Hot Luminous Stars*, ed. C. D. Garmany (Provo: Brigham Young University), p. 340.

van der Hucht, K. A., Hidayat, B., Admiranto, A. G., Supelli, K. R., and Doom, C. 1988, *Astr. Ap.*, **199**, 217.

Humphreys R. M. 1987, in *Instabilities in Luminous Early-Type Stars*, ed. H. J. G. L. M. Lamers, and C. W. H. de Loore (Dordrecht: Reidel), p.3

——. 1989, in *IAU Colloquium 113, Physics of Luminous Blue Variables*, ed. K.Davidson, A. F. J. Moffat, and H. J. G. L. M. Lamers (Dordrecht: Kluwer), p. 3.

Humphreys, R. M., and Davidson, K. 1979, *Ap. J.*, **232**, 409.

Humphreys, R. M., and McElroy, D. B. 1984, *Ap. J.*, **284**, 565.

Humphreys, R. M., Nichols, M., and Massey, P. 1985, *A. J.*, **90**, 101.

Kennicutt, R. C. 1983, *Ap. J.*, **272**, 54.

Kudritzki, R. P., Simon, K. P., and Hamann, W.-R. 1983, *Astr. Ap.*, **118**, 245.

Kudritzki, R. P., Gabler, A., Gabler, R., Groth, H. G., Pauldrach, A. W. A., and Puls, J. 1989, in *IAU Colloquium 113, Physics of Luminous Blue Variables*, ed. K.Davidson, A. F. J. Moffat, and H. J. G. L. M. Lamers (Dordrecht: Kluwer), p. 67.

Kunth, D., and Schild, H. 1986, *Astr. Ap.*, **169**, 71.

Lamers, H. J. G. L. M., Maeder, A., Schmutz, W., and Cassinelli, J. P. 1990, *Ap. J.*, in press.

Langer, N., and El Eid, M. F. 1986, *Astr. Ap.*, **167**, 265.

Langer, N., El Eid, M. F., and Baraffe, I. 1990, *Astr. Ap.*, in press.

Leitherer, C. 1990, *Ap. J. Suppl.*, **73**, 1.

Leitherer, C., Schmutz, W., Abbott, D. C., Hamann, W.-R., and Wessolowski, U. 1989,

Ap. J., **346**, 919.

Lennon, D. J., Kudritzki, R. P., Becker, S. R., Eber, F., Butler, K., and Groth, H. G. 1990, in *Properties of Hot Luminous Stars*, ed. C. D. Garmany (Provo: Brigham Young University), p. 315.

Lundström, I., and Stenholm, B. 1984, *Astr. Ap. Suppl.*, **58**, 163.

Maeder, A. 1983, *Astr. Ap.*, **120**, 113.

———. 1987, *Astr. Ap.*, **173**, 247.

———. 1990a, *Astr. Ap.*, in press.

———. 1990b, *Astr. Ap.*, in press.

Maeder, A., Lequeux, J., and Azzopardi, M. 1980, *Astr. Ap.*, **90**, L17.

Maeder, A., and Meynet, G. 1987, *Astr. Ap.*, **182**, 243.

———. 1988, *Astr. Ap. Suppl.*, **76**, 411.

Massey, P. 1981, *Ap. J.*, **246**, 153.

Massey, P., and Conti, P. S. 1983, *Ap. J.*, **273**, 576.

Massey, P., Parker, J. W., and Garmany, C. D. 1989, *A. J.*, **98**, 1305.

Mermilliod, J.-C., and Maeder, A. 1986, *Astr. Ap.*, **158**, 45.

Moffat, A. F. J., and Shara, M. M. 1987, *Ap. J.*, **320**, 266.

Schild, H., and Maeder, A. 1983, *Astr. Ap.*, **127**, 238.

———. 1984, *Astr. Ap.*, **136**, 237.

Schmidt-Kaler, T. 1982, in *Landolt-Börnstein, New Series, Group VI*, Vol. **2b**, ed. K. Schaifers, and H. H. Voigt (Berlin: Springer), p. 1.

Schmutz, W., Hamann, W.-R., and Wessolowski, U. 1989, *Astr. Ap.*, **210**, 236.

Schmutz, W., Leitherer, C., Hubeny, I., Vogel, M., Hamann, W.-R., and Wessolowski, U. 1990, *Ap. J.*, in press.

Schulte-Ladbeck, R. E. 1989, *A. J.*, **97**, 1471.

Stahl, O. 1986, *Astr. Ap.*, **164**, 321.

Smith, L. F. 1988, *Ap. J.*, **327**, 128.

St.-Louis, N., Drissen, L., Moffat, A. F. J., Bastien P., and Tapia, S. 1987, *Ap. J.*, **322**, 870.

Talent, D. L., and Dufour, R. J. 1979, *Ap. J.*, **233**, 888.

Walborn, N. R. 1976, *Ap. J.*, **205**, 419.

———. 1988, in *IAU Colloquium 108, Atmospheric Diagnostics of Stellar Evolution*, ed. K. Nomoto (Berlin: Springer), p. 70.

———. 1989, in *IAU Colloquium 113, Physics of Luminous Blue Variables*, ed. K. Davidson, A. F. J. Moffat, and H. J. G. L. M. Lamers (Dordrecht: Kluwer), p. 27.

———. 1990, in *IAU Symposium 148, The Magellanic Clouds*, ed. R. F. Haynes, and D. K. Milne (Dordrecht: Kluwer), in press.

Wolf, B., and Zickgraf, F.-J. 1987 in *Instabilities in Luminous Early-Type Stars*, ed. H. J. G. L. M. Lamers, and C. W. H. de Loore (Dordrecht: Reidel), p. 245

Wood, D. O. S., and Churchwell, E. 1989, *Ap. J.*, **340**, 265.

DISCUSSION

De Greve: As a consequence of the interplay of stellar wind and overshooting, the evolutionary tracks show a rightward extension around $M_{bol} = -8$ and a decrease in width above -9. How accurate can observations confirm this behavour in both regions?

Leitherer: One should be aware that this specific behaviour of the evolutionary tracks is a consequence of careful fine-tuning of the models by observations - rather than an independent prediction. Maeder and collaborators used a sample of young clusters to derive the amount of overshooting necessary to reproduce the observations. Therefore, the tracks will very closely resemble the observed main sequence width. However, it should be kept in mind that new results by Garmany *et al.* indicate that severe problems exist in the cluster membership of many OB stars. This will of course affect Maeder's results.

Lortet: A good place to look for ZAMS hot stars is the SMC where the dust content is low and we are able to see stars in less evolved regions. We found a few in Henize N 83 and 84 (the K1 region of the SMC).

Leitherer: It is indeed true that the SMC is an interesting place to study the ZAMS. Due to the low metal content of the SMC, theoretical models predict a significant shift of the ZAMS towards hotter temperatures. Unfortunately, we are still rather incomplete in our census of stars close to the ZAMS in the SMC (cf. the HRD of the LMC published by Fitzpatrick and Garmany). This work is still ongoing (mostly due to Conti and collaborators) and we will probably have to wait for a few more years to obtain statistically significant results.

Humphreys: Comparing the HRD's of the Galaxy and the LMC, you see a gradient in the number of stars stretching all the way from the hottest most luminous to the A-Type supergiants. This is not the same line as drawn by Fitzpatrick and Garmany. It is observed in the Galaxy, LMC and SMC and even in other Local Group galaxies, even though the data sets are much less complete. This is a question for the model-makers: What is the evolutionary origin of this gradient? Is it an artifact of how we are treating the data, due to observational selection or is there a physical explanation?

Sreenivasan: You make two observations, (1) that the Fitzpatrick-Garmany ridge in the LMC represents an ensemble of stars in different evolutionary phases, (2) the evolutionary origin of WR stars is determined entirely by their initial mass. I agree with the former and *one* possible explanation for it can be found in my poster outside but I disagree with the latter. A WR star phase is determined by two simultaneous processes, one that removes the outer layers and the other that transforms the core. The two related time-scales are determined not only by the initial mass but by other phenomena that determine stellar wind mechanisms and core growth.

Leitherer: I emphasized in my talk that the initial mass may not be the only parameter determining the late evolution of very massive stars. Other factors could be stellar rotation or the magnetic field. Cassinelli's models should be mentioned in this respect. I agree with you that it may be too naive to exclude *a priori* parameters other than mass.

Wilson: What is the effect of obscuration by circumstellar dust shells on the observed distribution of luminosities for red supergiants?

Leitherer: Humphreys did several thorough studies to address this problem. We observe many red supergiants with massive dust shells. They are bright in the infrared where we can determine the bolometric luminosity. They are *not* above the Humphreys-Davidson limit. I should mention the case of M 101: If the canonical value for its distance modulus is taken then many red supergiants (and also blue supergiants) are *above* the H-D limit. The distance to M 101, however, is still under debate, and the stars may well be *under* the limit.

Yungelson: (1) Comment: You cannot go with masses of WR progenitors as low as $\sim 15 M_\odot$, because stars of such a low mass do not have massive enough helium cores, predecessors of WR stars, that would produce helium stars hot enough to produce stellar winds able to give rise to WR phenomenon. (2) Question: When you are telling about deficiency of very young O stars relying on isochrones, do you make allowance for unresolved binarity, that can shift the ZAMS upwards?

Leitherer: (1) I agree with your comment. Actually, my mean low cut-off mass of $25 \pm 10 M_\odot$ does not conflict with your value. (2) No allowance for a shift of the ZAMS due to binarity has been made. This may decrease the problem of the observed lack of very young stars somewhat. However, I think we still need an additional explanation - like the ultracompact *HII* regions - to account for the entire effect.

Langer: (1) The diagonal iso-density lines in the observed HR diagram Humphreys referred to in her question, may well reflect the well known fact that blue loops from the Hayashi-line extend to higher surface temperatures for higher masses (luminosities). (2) If this is the case (*cf.* Fitzpatrick and Garmany, 1990), it would mean that almost all BSG's are post-RSG stars and should therefore be *N*-enriched, since blue loops are obtained only for tracks which ignite core helium burning as RSG's. This would be well in line of Walborn's OBN/OBC interpretation.

Claus Leitherer

LMC OB associations containing Wolf-Rayet stars

H. Schild
Department of Physics and Astronomy
University College London
Gower Street, London WC1E 6BT
England

M.C. Lortet and G. Testor
Observatoire de Paris
Section de Meudon/DAEC
F-92195 Meudon Principal Cedex
France

ABSTRACT.
We observed four OB associations in the Large Magellanic Cloud which contain Wolf-Rayet stars. The associations are located in the neighbourhood of 30 Dor. They contain 15 WR stars of WCE and most WN subtypes. The observations include UBV photometry for several thousand objects and spectroscopy in the range from 3500 to 5000 Å for about 250 stars. Two new extreme Of stars and a new WR star as well as many hitherto unknown O stars were identified. Cluster HR diagrams reveal the age structure of the association stars and yield tentative conclusions about evolutionary connections between Wolf-Rayet stars in the LMC.

1. Introduction

The coincidence of Wolf-Rayet stars with OB associations provides an opportunity to determine ages of WR stars. If we assume that WR stars are the end product of standard stellar evolution, we can also in principal assign an initial mass to the WR progenitors. In the case of galactic WR stars, this has been done by Schild and Maeder (1984) and Humphreys et al. (1985).

The Wolf-Rayet stars in the LMC also have a high coincidence rate with OB associations (e.g. Breysacher 1986). The stellar content of these associations is however very poorly known. Up to now, the only association with a WR star examined in detail is LH 39 which contains the WN9-10 star Brey 18 (Schild 1987). We therefore embarked on an observing program with the aim to determine the stellar content of some selected LMC OB associations. In a fist step we focussed on associations near 30 Dor.

2. Observations

During two observing runs in Dec 1988 and Dec 1989/Jan 1990 with the ESO 3.6m telescope and the multi-purpose instrument EFOSC, we obtained spectra of about 250 and UBV colours of several thousand stars in the associations LH 89, LH 90, LH 99 and LH 104.

The CCD field of view was 3.'6 x 5.'6. The photometric observations were reduced with DAOPHOT. It was possible to resolve many tight clusters. We estimate the internal photometric errors to be between 0.02 and 0.05 mag for the bright ($m_V = 12$) and faint ($m_V = 20$) stars respectively. The main error source in the Johnson magnitudes comes from the scarcity of suitable calibration stars in the CCD fields.

The spectroscopic observations covered the range from 3900 to 5100 Å with a resolution of 7Å. The EFOSC instrument was used in its multi-object mode through aperture

479

K. A. van der Hucht and B. Hidayat (eds.),
Wolf-Rayet Stars and Interrelations with Other Massive Stars in Galaxies, 479–484.
© 1991 *IAU. Printed in the Netherlands.*

holes with a diameter of 2.1 arcsec. The data were reduced in a standard way by using STARLINK software available at UCL. Nebular HII emission lines present in many stellar spectra were usually not removed except in cases where spectral classification lines like HeIλ4471 were affected.

3. New Wolf-Rayet and extreme Of stars

Our spectra revealed the presence of a hitherto unknown WR star as well as an extreme Of star in a tight cluster of LH 90 (Fig. 1). Brey 65 and star 1 are Wolf-Rayet stars of WN7 and WN3 type respectively, whereas star 2 is an extreme Of star. All three stars are of about the same $m_V \approx 12.9$ mag (Testor and Schild 1990). The fact that an early and late WN star are only separated by 3 arcsec may explain the different spectral types previously attributed to Brey 65: OB + WN (Smith 1968), WN3-5 (Fehrenbach et al. 1976), O: + WN5-6 (Walborn 1977) and WN4.5 + OB (Breysacher 1981), WN/Of? (Torres-Dodgen and Massey 1988), WN7 + O (Moffat 1989).

Fig. 1: B filter image of the central part of the OB association LH 90 with the WR and extreme Of stars. North is at the top, East to the left.

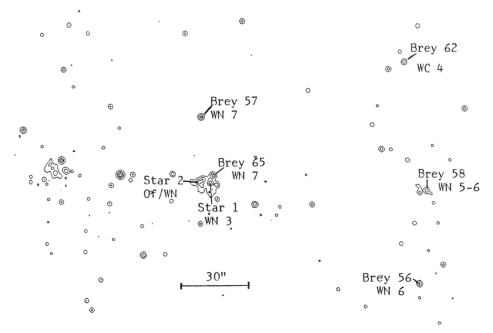

A further extreme Of star was found in a bright nebular region of LH 99 (for identification see Testor and Schild 1990). The star is of visual magnitude 13.6 and has a strong and broad HeIIλ4686 emission line. According to the classification criteria of Walborn (1986) it is of O3If/WN6 spectral type.

4. Stellar content of OB associations

4.1 LH 99

The spectroscopically surveyed area contains the three WR stars Brey 71, MGWR 4 and Brey 73 which are of WN7, WN3-4 and WN7 subtype, respectively. In addition, we find an O3If/WN6 transition object about 30" to the SE of Brey 73 (Testor and Schild 1990). LH 99 certainly is a very young stellar association. We find almost 40 O stars, half a dozen of which are earlier than O5. The HR diagram shows a well populated main sequence going up to $M_{bol} \approx -10$ (Fig. 2). The WR stars and the Of/WN star form the luminous tip in the diagram. We used the bolometric corrections of Smith and Maeder (1989) for the WR stars and Massey et al. (1989) for the OB stars.

We note the coexistence of a WN3, WN7 stars and an Of/WN6 star in an association which apparently is very young. In addition, we find the luminosities of all these stars to be very similar. These factors indicate that there may be an evolutionary link between these stars. Presently we can however not exclude the possibility that one or several WR stars are not physical members of the association. MGWR 4 indeed is strongly reddened (Morgan & Good 1987) and might be a background star. More detailed results are in preparation and will be presented elsewhere (Lortet et al. 1990).

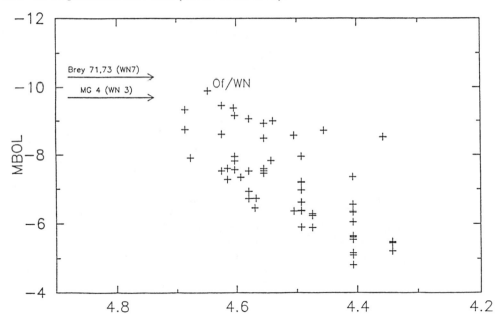

Fig. 2: HR diagram M_{bol} versus $log(T_{eff})$ of the OB association LH 99. The luminosity of the WR stars is indicated by arrows.

4.2 LH 90

The central part of LH 90 is shown in Fig 1 where also the positions of the WR stars are indicated. The earliest stars in LH 90 are of O5 spectral type. All O stars are located in the western and central part of the association. The six WR stars now known in LH

90 are also situated in the same area. The WR population consists of four WN5 to WN7 stars, a WC4 and a WN3 star. Three of the WR stars are located in tight stellar clusters (Fig 1).

The cluster located in the center of the association (labelled β by Lortet and Testor 1984) is particularly interesting because the three visually brightest objects in it are a WN7, a WN3 and an O3If/WN6 star (Testor and Schild 1990). Spectra of these objects are shown in Fig 3. The cluster stars are very likely coeval and physiscal members.

Fig. 3: Spectra of the three visually brightest stars in the central tight cluster in LH 90. For identification see Fig. 1.

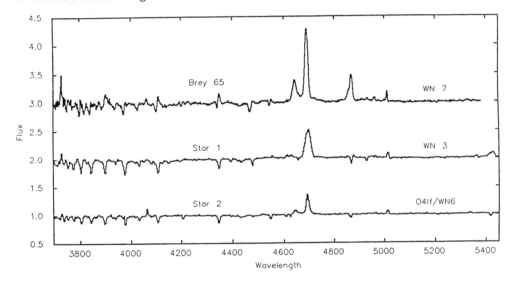

4.3 LH 104

Our photometric observations cover the association almost completely but the spectroscopic observations were centered on an area of 3' x 2.4' which contains the three WR stars Brey 94, 95 and MGWR 6. We thus obtained spectra of 60 stars located in the southern half of the association as defined by Lucke (1972).

The earliest stars in the surveyed part of LH 104 are of O5 and O6.5 spectral type. It therefore seems to be of a similar age to or slightly older than LH 90. A more detailed description of this association and its environment can be found in Lortet and Testor (1988).

4.4 LH 89

We concentrated the spectroscopic observations on the area around the WR stars Brey 60, 61 and 64 which are of WN3, WN4 and WN9-10 spectral type, respectively. We obtained spectra of more than 60 surrounding stars and the earliest stars turned out to be of O9 spectral type. The most luminous stars in this association have $M_{bol} \approx -9$ which matches well the luminosities of the three WR stars (Fig. 4). LH 89 is the oldest of the associations surveyed here. The WR/O ratio is 1.5 which roughly is an order of magnitude larger than in LH 99.

Fig. 4:HR diagram of LH 89. The luminosities of the WR stars are indicated by arrows.

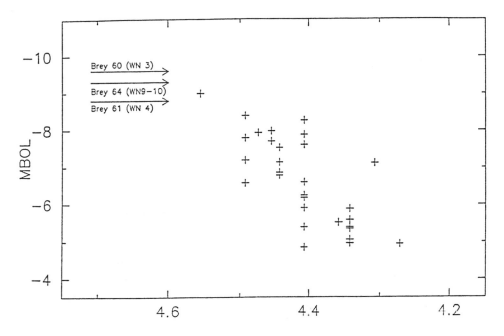

5. Discussion

In Table 1, the surveyed OB associations are ordered according to the earliest stars present. An increasing age sequence from LH 99, to LH 90, LH 104 and LH 89 is thereby indicated. The presence of very hot stars is also reflected by HII emission as seen e.g. on $H\alpha$ images: LH 99 is completely embedded in strong nebular emission; LH 90 and LH 104 show only moderate central nebular emission but are surrounded by strong filaments. The central parts of these associations seem to have already been swept clear by stellar winds and SNe. LH 89 finally is practically free from gaseous emission.

Table 1. Hot star population in the surveyed OB associations. Spectral types of earliest stars, slash stars and WR stars as well as the WR/O number ratios are given.

	earliest stars	/stars	WR stars	WR/O
LH 99	O3V,O3-4V,O3-4III	O3If/WN6	WN7, WN7, WN3-4	0.16
LH 90	O4-5V, O5V, O5I	O4If/WN6	WN7,WN7,WN6,WN5-6 WC4, WN3	0.40
LH 104	O5I, O6.5V, O8V	—	WN4, WC5, WC5	0.50
LH 89	O9V, O9.5V, B0V	—	WN3, WN4, WN9-10	1.5

There is also a marked increase in the WR/O ratio from LH 99 to LH 89. As seen in LH 89, it is possible that under special conditions the number of WR stars can be of the same order or even exceed the number of O stars. This also indicates that WR progenitors can have initial masses at least as low as late O stars.

The subtype distribution of the WR stars also shows some remarkable age preferences with the exception of WN3 and WN4 stars which seem to occur in clusters of all ages. The WN7, WN6 and WN5 stars are confined to the two youngest associations whereas the WC4 and WC5 stars occur in the medium age associations. The WN9-10 star finally occurs in the oldest association. This is in agreement with a previous study on another WN9-10 star (Brey 18) which is located in an even older association (Schild 1987).

The occurence of WR stars in the four associations surveyed suggests the following evolutionary connections between O and WR stars in the LMC:

$$\text{early O} \rightarrow \text{OIf/WN6} \rightarrow \text{WN7} \rightarrow \text{WN3 ?}$$

$$\text{mid O} \rightarrow \text{WN6-4} \rightarrow \text{WC4,5}$$

$$\text{late O} \rightarrow \text{WN3,4 or WN9-10}$$

The hottest stars present in a stellar association only provide a lower age limit for the aggregate. There may however be a considerable age spread in these associations. We are planning to examine this age dispersion with colour-magnitude diagrams in the near future. In relation to the evolutionary status of WR stars, a possible age spread is however not of crucial importance because as it can be seen in Figs 2 and 4, the WR stars usually are among the most luminous stars in an association. This implies that they originate from the most massive and therefore youngest subpopulation present in the cluster.

References

Breysacher, J.: 1981, Astron. Astrophys. Suppl. **43**, 203

Breysacher, J.: 1986, Astron. Astrophys. **160**, 185

Fehrenbach, C., Duflot, M., Acker, A.: 1976, Astron. Astrophys. Suppl. **33**, 115

Humphreys, R.M., Nichols, M., Massey, P.: 1985, Astron. J. **90**, 101

Lortet, M.C., Testor, G.: 1984, Astron. Astrophys. **139**, 330

Lortet, M.C., Testor, G.: 1988, Astron. Astrophys. **194**, 11

Lortet, M.C., Schild, H., Testor, G.: 1990, in preparation

Lucke, P.B.: 1972, Ph.D. Thesis, Univ. Washington

Massey, P., Parker, J.W., Garmany, C.D.: 1989, Astron. J. **98**, 1305

Moffat, A.F.J.: 1989, Astrophys. J. **347**, 373

Morgan, D.H., Good, A.R.: 1987, Mon. Not. R. astr. Soc. **224**, 435

Schild, H.: 1987, Astron. Astrophys. **173**, 405

Schild, H., Maeder, A.: 1984, Astron. Astrophys. **136**, 237

Smith, L.F.: 1968, Mon. Not. R. astr. Soc. **140**, 409

Smith, L.F., Maeder, A.: 1989, Astron. Astrophys. **211**, 71

Testor, G., Schild, H.: 1990, Astron. Astrophys, accepted

Torres-Dodgen, A.V., Massey, P.: 1988, Astron. J. **96**, 1076

Walborn, N.R.: 1977, Astrophys. J. **215**, 53

Walborn, N.R.: 1986, in IAU Symp. **116**, 185

THE WOLF-RAYET CONNECTION - LUMINOUS BLUE VARIABLES AND EVOLVED
SUPERGIANTS

ROBERTA M. HUMPHREYS
University of Minnesota
Department of Astronomy
116 Church Street S.E.
Minneapolis, Minnesota   55455

ABSTRACT.   The physical characteristics and behavior of evolved
massive stars in three different mass ranges are reviewed with
application to whether they may eventually evolve to the WR stage   1.
>40-50 $M_\odot$ as LBV's, 2. $\sim$30-40 $M_\odot$ as cool hypergiant-OH/IR stars and
3. $\sim$10-30 $M_\odot$ as red supergiant-OH/IR stars.   I emphasize the
importance of the relatively short but high mass loss phases as LBV's
and as OH/IR stars in determining the fate of massive stars from 10
to 100 $M_\odot$.

1.   Introduction

About ten years ago it was realized that the observed HR diagram has
an upper luminosity limit.   Humphreys and Davidson (1979) first drew
attention to the empirical boundary and its physical significance for
the evolution of massive stars.   We proposed then that an instability
causing rapid and unsteady mass loss was the basic explanation for
the upper luminosity boundary.   We emphasized the temperature-
dependence of the boundary for the most luminous hot stars, the lack
of cooler counterparts at similar luminosities and the temperature-
independent limit to the luminosities of the cool hypergiants.   All
of the stars that lie on or near this boundary are highly unstable
and the highest mass loss rates are observed along this luminosity
limit (de Jager 1984, 1988).   The temperature-dependent boundary for
the hot stars is marked by the presence of some very luminous
unstable stars including such famous stars as η Car, P Cyg, S Dor and
the Hubble-Sandage variables in M31 and M33, now known collectively
as the Luminous Blue Variables (LBV's).   The temperature dependence
for the hot stars suggests that this boundary defines a critical
location in the HR diagram one that is mass dependent.   This plus the
lack of cooler counterparts, indicates that stars above some critical
mass probably do not evolve to cooler temperatures.   Based on
observed HR diagrams in combination with evolutionary tracks, this
critical mass is probably near 40-50 $M_\odot$.

2. Luminous Blue Variables - The Evolution of Stars >40-50 $M_\odot$

The LBV's are evolved, very luminous, unstable hot supergiants which

485

K. A. van der Hucht and B. Hidayat (eds.),
Wolf-Rayet Stars and Interrelations with Other Massive Stars in Galaxies, 485–497.
© 1991 *IAU. Printed in the Netherlands.*

suffer irregular ejections. The cause of their instability is very
likely radiation pressure, resulting in a greatly enhanced average
mass outflow ($\sim 10^{-4}$ $M_\odot$/yr) which leads to the formation of a pseudo-
photosphere (Leitherer et al. 1985) at visual maximum. At this stage
the slowly expanding (100-200 km/s) envelope is cool (8000-9000 K)
and dense (N $\sim 10^{11}$ cm^{-3}), and the star resembles a very luminous A-
type supergiant. At minimum light, or the quiescent state, the LBV
is at its "normal" high temperature (>15000-20000 K) and the mass-
loss rate is lower. During these variations the bolometric
luminosity remains essentially constant. The visual light variations
are caused by the apparent shift in the star's energy distribution
driven by the instability. The transitions from the minimum to
maximum state at constant bolometric luminosity are shown for the
best studied LBV's on the accompanying HR diagram (Figure 1).

The close connection between LBV's and Of/WN9 stars has been
known for some time and provides strong support for suggestions that
LBV's are predecessors of WR stars. R127 in the LMC is perhaps the
best example. Originally one of the few known Of/WN9 stars first
recognized by Walborn (1977), it recently erupted (Stahl et al.
1984). It is still at maximum light with a spectrum nearly identical
to S Dor. AG Car at minimum light and HDE 269582 = MWC112 also have
spectra that very closely resemble the Of/WN9 stars. Two other
Of/WN9 stars in the LMC, R84 and R99, are also variable but with
smaller amplitudes (Stahl et al. 1984) and are potential LBV's.
Obviously this group of stars should be closely monitored for LBV-
like behavior and we should try to identify their counterparts in our
galaxy.

Most LBV's show some evidence for an excess of IR radiation and
the presence of circumstellar ejecta (Humphreys et al. 1984;
Leitherer et al. 1985; McGregor et al. 1988) produced by the high
mass loss and ejection of shells from the LBV's. Some of this ejecta
is clearly visible as in the famous homunculus of η Car. In η Car
the dust is thick enough to obscure the star (Westphal and Neugebauer
1969). The presence of a ring nebula as in the case of AG Car
(Thackeray 1977; Stahl 1987) is fairly common. Stahl (1987)
described ring nebulae around AG Car, He3-519 a candidate LBV-Of/WN9
star, R127 and S61. Another example is the ring nebula RCW58 around
the known WN8 star HD96548 (Chu 1982; Smith et al. 1988) which very
closely resembles those around the LBV's. With its low expansion
velocity of only 87 km/s, like the slow winds of LBV's, the nebula
may be a "fossil" from the previous LBV stage. Smith et al. (1988)
however concluded that the low velocity, the presence of dust and N
and He enrichment in the nebula were evidence of a red supergiant
progenitor, but they are more consistent with an LBV.

Quantitative analyses of the ejecta from LBV's and the
atmospheres of LBV's and related stars such as the Of/WN9's show that
they are nitrogen and helium-rich (η Car, Davidson et al. 1982, 1986;
η Car, Allen et al. 1985; AG Car, Dufour and Mitra 1987 and reviews
by Walborn 1988, 1989).

The actual amount of mass lost as an LBV is especially relevant
to possible evolution to the WR stage. The duration of the LBV stage
is often estimated at $\leq 10^4$ yrs (see Maeder 1989) by comparing their
numbers with WR stars, but I want to caution that the numbers of
LBV's is incomplete. Because their quiescent phase may be decades
long we may have missed several. R127 and the recent announcement of

R110 (Stahl et al. 1990) are excellent examples. In the LMC there are now 5 confirmed LBV's. I will also include R84 and R99 as candidate LBV's for a total of 7 compared with 115 WR stars. Then N(LBV)/N(WR) is ≈ 0.06. With a total lifetime of $\approx 5 \times 10^5$ yrs (Maeder and Meynet 1987) for WR stars, this gives an LBV lifetime of about ≥ 25000 yrs.

Lamers (1989) gave a thorough review of the various mass loss rates during the LBV cycle and therefore I will only give a brief summary below for typical LBV's.

1. minimum light: Ex. R71 $\sim 6 \times 10^{-7}$ $M_\odot$/yr
2. Maximum light [Mass loss during the maximum light of the moderate variations (~ 2 mag) of the typical LBV is about 10 to 100 times greater than during quiescence.]

AG Car	$2-7 \times 10^{-5}$ $M_\odot$/yr
R71	2×10^{-5}
S Dor	3×10^{-5}
R127	6×10^{-5}

Based on the normal mass loss rates of LBV's between minimum and maximum light and assuming the star spends half the time in each phase, Lamers (1989) determined a time-average normal mass loss rate of $\approx 10^{-5}$ $M_\odot$/yr.

However there is increasing evidence that the LBV's pass through a much more violent stage, that we call eruptions or explosions, during which the mass loss rate is much higher; discussed by both Lamers (1986, 1989) and by Humphreys (1989). Eta Car is of course the most famous example. During its 1840's outburst, η Car probably lost 2-3 $M_\odot$ and its current M is 10^{-4} to 10^{-3} $M_\odot$/yr. More typical examples are P Cyg, AG Car and R127. In its current quiescent stage, P Cyg's M is $\sim 1.5 \times 10^{-5}$ $M_\odot$/yr. If it is like other LBV's then M during its outburst was 10-100 times greater or 10^{-4} to 10^{-3} $M_\odot$/yr. Its extended Hα and NII emission measured by Leitherer and Zickgraf (1987) from previous ejections corresponds to continuous mass loss of 4×10^{-4} $M_\odot$/yr. From direct imaging of circumstellar shells around AG Car, R127 and others Stahl (1987) estimated their kinematic ages and masses and concluded that their average mass-loss rate must be $> 10^{-4}$ $M_\odot$/yr. We (Humphreys et al. 1989) have also shown that a shell of photoionized gas around AE And contains $> 6 \times 10^{-3}$ $M_\odot$. If this is the material ejected during its last maximum, which lasted ~ 20 years, then the M was $> 3 \times 10^{-3}$ $M_\odot$/yr during the ejection event.

The mass loss rates during these more violent eruptions as measured from the circumstellar ejecta are summarized below:

	M	Duration	Age
η Car	10^{-1}	25 years	150 years
P Cyg	4×10^{-4}:	60 years	400 years
AG Car	2×10^{-4}	mean rate over	10^4
R 127	1.7×10^{-4}	mean rate over	1.7×10^4 years
AE And	$> 3 \times 10^{-3}$	20 years	80 years
R71	7×10^{-5}	-	400 years

The total mass lost during the lifetime of an LBV is of course uncertain because we don't know the frequency of these more violent eruptions. There is evidence from the proper motions of older ejected material around η Car that it has undergone more than one eruption at intervals of several hundred years. Lamers gives a

suggested time-averaged mass loss rate of $\sim$2x10^{-4} M$_\odot$/yr over the LBV
lifetime for the violent eruptions. Thus assuming that all LBV's
pass through one or more violent eruptions, they very likely shed >5
M$_\odot$ during their 25,000 years. This is very close to the mass loss of
5-10 M$_\odot$ that a 50-100 M$_\odot$ star must shed after core H-burning to
become a WR star (based on the models of Maeder and Meynet 1987).
The total mass shed as an LBV may be mass dependent, so that a star
like η Car may lose more total mass.
    The characteristics of the LBV's; their instability and
resulting high mass loss rates, circumstellar ejecta and atmospheres
enriched in nitrogen and helium, their very obvious connection to the
Of/WN9 stars, and of course their crucial location in the HR diagram
all lead to the conclusion that LBV's are progenitors of WR stars.
Or, depending on one's perspective, the WR stars are post-LBV's.
Consequently for the most massive stars (>40-50 M$_\odot$) the following
evolutionary scenario is now generally accepted:

$$\text{MS O star} \rightarrow \text{Of} \rightarrow \text{LBV} \leftrightarrow \text{Of/WN} \rightarrow \text{WR} \rightarrow \text{SN (Type Ib)}$$

    Based on the spatial distributions of WR stars and O stars
Conti et al. (1983) suggested that most WR stars derive from star >40
M$_\odot$ and independently Humphreys, Nichols and Massey (1985) looking at
their membership in associations and clusters showed that the lower
limit to the initial masses of WR stars are >30 M$_\odot$ with the majority
>50 M$_\odot$. Van der Hucht et al. (1988) concluded that WN stars derive
from progenitors in the mass range 28-35 M$_\odot$ and WC from 25-60 M$_\odot$
stars. Thus the evolution of evolved stars near the upper luminosity
limit is critical and some of these stars may lose sufficient mass to
become WR stars.

3. THE COOL HYPERGIANTS - THE LUMINOSITY/STABILITY LIMIT AND THE EVOLUTION OF STARS NEAR 40 M$_\odot$

The upper limit to stellar luminosities is usually assumed to be set
by the balance between the acceleration due to gravity and the
radiation pressure gradient a la Eddington. However the observed
luminosity boundary is composed of two components - the temperature-
dependent boundary for hot stars and its turnover at the cool star
upper limit. The classical Eddington limit due to electron
scattering does not show the dependence on temperature for the hot
stars. However as the temperature decreases below 30000 K the
opacity increases due to ions of HI, FeII, et al. A modified or
opacity-dependent Eddington limit which decreases with temperature
has been proposed and discussed by sedveral investigators (Humphreys
and Davidson 1984; Appenzeller 1986; Lamers 1986; Davidson 1987;
Lamers and Fitzpatrick 1988). The opacities reach a maximum and the
Eddington luminosity a minimum at 10000 K. The modified Eddington
limit will then turn up again in the 8000-10000 K temperature range
in agreement with the observed turnover in the luminosity/stability
limit. Stars below the corresponding critical mass could then evolve
to the red supergiant region.
    But the situation may be more complicated. First attempts to
calculate the location of the modified Eddington limit on the HR
diagram have not been entirely successful (Lamers and Fitzpatrick
1988). The F, G, K and M hypergiants in our galaxy and other local

group galaxies define the observed upper luminosity limit for the cooler stars and these stars are all highly unstable. De Jager (1980, 1984), has suggested that the instability in these stars is produced by a turbulent pressure gradient due to the dissipation of mechanical energy. In a series of papers, he and his collaborators have measured supersonic microturbulent motions in the atmospheres of many of these stars.

I think that the observed lumiosity/stability limit is a consequence of 1) <u>radiation pressure</u>, i.e., the <u>modified Eddington limit</u>, which dominates in the hot stars, and 2) the <u>turbulent pressure gradient</u> in the atmospheres of the cool hypergiants which sets an upper boundary to their luminosities independent of radiation pressure in the hot stars.

The instability in the atmospheres of the cool hypergiants is evidenced by their variability in light and in their spectra, by high mass rates (up to a few x 10^{-4} $M_{\odot}$/yr), and the presence of extensive circumstellar dust shells around many. Many of the intermediate-type hypergiants show evidence for shell ejections. ρ Cas (F8Ia) is especially well known for its shell episode (1946-47) in which it decreased 1.5 mag and had the spectrum of an M star. HR8752 (G0-G5Ia+) one of the most luminous hypergiants has shown considerable spectroscopic variation usually attributed to shell ejection. But neither ρ Cas or HR8752 has the large IR excess due to circumstellar dust from high mass loss observed around many very luminous G supergiants (Humphreys et al. 1971). HR5171a (G8Ia+) has one of the largest 10μ silicate features observed in late-type supergiants. This star is especially interesting because it has been getting fainter and redder with time suggesting that the amount of dust and the obscuration are increasing perhaps due to continuous high mass loss or to unstable fluctuations in its outer atmosphere.

Variable A in M33 is perhaps the most enigmatic of all of the cool hypergiants (Humphreys et al. (1988). It was one of the original Hubble-Sandage variables but its behavior is bizarre even for them. In 1950 it was one of the visibly brightest stars in M33, with the spectrum of a very luminous F supergiant. It then rapidly declined in brightness by 3.5 mag becoming faint and red after slowly increasing in brightness during the previous 50 years. It is still faint and red and has the spectrum of an M supergiant not an emission-line hot star! It also has a large infrared excess and is today as bright at 10μ as it was at its visual maximum in 1950. Variable A is a very luminous ($M_{Bol} \approx$ -9.5 mag), highly unstable (2×10^{-4} $M_{\odot}$/yr) star. Its present spectrum is probably produced in an expanded pseudo-photosphere and is shedding its mass in a high-density, low-velocity wind.

The most luminous M supergiants include such well known stars as μ Cep but in addition to the relatively normal red supergiants with circumstellar dust there are also the supergiant OH/IR sources. They are likely the most evolved M supergiants; M supergiants that have lost sufficient mass that their dust shells are now optically thick. The mass loss rates from the supergiant OH/IR sources may be very high 10^{-4} to 10^{-3} $M_{\odot}$/yr. A few with optically thin shells are visibly bright, like VY CMa (M3-5eIa), VX Sgr (M4-8eI), and S Per (M4eIa). Two highly obscured, highly luminous late-type stars in the LMC (Elias, Frogel and Schwering 1986) discovered by IRAS presumably belong to this group; OH emission has just been detected from one

(Wood et al. 1986). What will eventually become of these supergiant OH/IR sources? Could they evolve to Wolf-Rayet stars, analogous to their less massive counterparts which become the central stars of planetary nebulae?

IRC+10420 may be a good candidate for such a star in transition from red supergiant to WR star. IRC+10420 has the spectrum of a very luminous F supergiant (F8Ia+) plus a very large IR excess from a circumstellar dust shell (Humphreys et al. 1973). It is also one of the earliest (warmest) known OH/IR sources (Giguere et al. 1976). Recent OH observations show that the 1665 MHz feature is weakening while the 1612 MHz feature is growing (Lewis et al. 1986). This is what we would expect if the dust shell were dissipating. Interestingly it has also been getting visually brighter (Gottleib and Liller 1978). If this trend continues a very plausible model for IRC+10420 will be a post M supergiant-OH/IR star blowing off its cocoon of dust and gas as it evolves to the left to warmer temperatures on the HR diagram.

I have estimated the lifetime of the supergiant OH/IR stage from the numbers of known M supergiants (122) and supergiant OH/IR sources within 3 kpc of the sun (VX Sgr, VYCMa, NML Cyg and S Per. Also my colleague T.J. Jones (1990) estimates that there should be three OH/IR supergiants within 3 kpc of the sun from the statistics of the OH surveys. This is good agrement, and therefore assuming the red supergiant lifetime is 10^6 yrs the OH/IR star stage should be $\sim$30,000 yrs. Similarly in the LMC there are 95, confirmed red supergiants brighter than the AGB limit ($M_{Bol} \approx$ -7.2 mag) but only one detected OH/IR source. This is because of the limits of sensitivity of the radio telescopes not because they don't exist. There are two other excellent candidates, IRAS 05346-6949 and MG46, both very dusty, high luminosity red supergiants. With three candidate OH/IR sources in the LMC, the estimated duration of this stage is again $\approx$30,000 yrs.

Using these lifetimes for the OH/IR stage combined with mass loss rates for the red supergiant stage from the compilation by de Jager et al. (1988) and those observed for the OH/IR stage plus the lifetimes in various stages from the models by Maeder and Meynet (1988), I have estimated the amount of mass lost as an evolved supergiant for stars of mass 40, 25 and 20 $M_\odot$. The numbers are summarized in Table 1.

By the time the 40 $M_\odot$ star has passed through the region of the cool hypergiants for the second time it will have lost half of its initial mass. From the tables by Maeder and Meynet (1988) the relative surface abundance of hydrogen for a star with $\sim$20 $M_\odot$ remaining will be down to $\sim$.2. It will also be close to the onset of carbon fusion and the star could be well on its way to becoming a WR star. Adopting the evolutionary "funnel" for WR stars presented by Moffat (1989) at the colloquium on LBV's (see Figure 1), we see that the cool hypergiants might enter the WR sequence as late WC's. Could the remnant circumstellar dust shells of the OH/IR phase account for the dust shells around the late WC-type stars? The dust shells associated with the WC9 star are carbon-rich while the red supergiant-OH/IR stars have oxygen-rich circumstellar dust (Cohen et al. 1975). Thus the WC9 dust shells cannot be the fossil remnants of a previous red supergiant stage. Furthermore, the WC dust shells are sufficiently close to the star (Dyck et al. 1984, Ve2-45, 4×10^{10} km)

Table 1. Mass Lost as an Evolved Supergiant (solar abundances)

Initial Mass MS ($M_\odot$)	Initial Mass RSG ($M_\odot$)	RSG			OH/IR			Blue Loop (to Log T_{eff} = 4.0)			Remnant Mass $M_\odot$
		$\dot{M}$ ($M_\odot$/yr)	T (yrs)	M ($M_\odot$)	$\dot{M}$ ($M_\odot$/yr)	T (yrs)	M ($M_\odot$)	$\dot{M}$ ($M_\odot$/yr)	T (yrs)	M ($M_\odot$)	
40	32.2	3×10^{-5}	$3 \times 10^{4*}$	1.0	3×10^{-4}	$2 \times 10^{4*}$	6	$\sim 10^{-4}$	4×10^{4}	4	21
25	22.8	6.3×10^{-6}	7.4×10^{5}	4.7	1.5×10^{-4}	3×10^{4}	4.5	$\leq 3 \times 10^{-6}$	3.6×10^{5}	1	13
20	19	1.6×10^{-6}	1.25×10^{6}	2	1.5×10^{-4}	3×10^{4}	4.5	--------			12.5

*Among the six known M supergiants in our galaxy between $M_{Bol} \sim$ -9 and -9.5 mag, two are OH/IR stars, VX Sgr and VY CMa. The duration of the red supergiant stage in the 40 $M_\odot$ models is very short, only 5×10^{4} yrs. Since the OH/IR stars are one-third of the red supergiants at these luminosities, I divided the time between them.

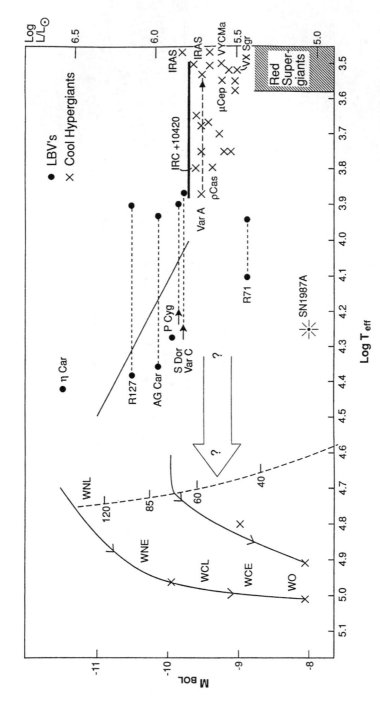

Figure 1 – A schematic HR diagram show the location of the LBV's at maximum and minimum light, the cool hypergiants, red supergiants and SN1987A. The WR "funnel" from Moffat 1989 is also shown.

that the material must have been ejected recently. The WN stars are the more likely antecedents of red supergiant mass loss.

The results for the stars near the luminosity/stability limit are ambiguous but it is possible for some cool hypergiants to lose sufficient mass to become WR stars; although there is no direct connection with a particular subclass of WR star. However, this scenario depends critically on the properties and interpretation of one star, IRC+10420 as an evolved supergiant shedding its remaining hydrogen envelope. Thus for stars in the 30-40 $M_\odot$ range, there are still two possibilities.

$$\text{MS O star} \rightarrow \text{BSG} \rightarrow \text{YSG} \rightarrow \text{RSG} \rightarrow \text{OH/IR} \rightarrow \text{YSG} \rightarrow \text{WR} \rightarrow \text{SN (Type Ib)}$$

$$\text{or RSG} \rightarrow \text{OH/IR} \rightarrow \text{SN (Type II)}$$

4.   EVOLVED SUPERGIANTS IN THE 15 TO 30 $M_\odot$ RANGE

We know that the progenitors of many Type II supernovae must have very extended atmospheres or envelopes that are then illuminated by the supernova explosion. The most likely candidates are the red supergiants and they are generally adopted in the models for the Type II supernovae (Woosley and Weaver 1986). SN1987A and its progenitor Sk-69°202 showed us that some red supergiants in this somewhat lower mass range evolve back to the blue and explode as supernovae as a hot more compact supergiant.

Do some evolved supergiants of lower initial masses also reach the WR stage before becoming supernovae? The numbers in Table 1 show that stars in this mass range do not shed their entire hydrogen envelopes in the RSG-OH/IR stage. Again referring to the tables by Maeder and Meynet, by the time stars of initial masses of 25 and 20 $M_\odot$ are down to 13 and 12 $M_\odot$ respectively, their hydrogen surface abundances would still be $\approx.4$, far from the bare hydrogen-deficient core of the WR star. An additional high mass loss phase would still be required. The two low-luminosity relatively cool LBV's, R71 and R110 might qualify as post RSG's for stars in the 25-30 $M_\odot$ initial mass range, and would be an additional stage where high mass loss could occur. What would be the cause of an instability at this stage in the star's evolution?

We know that Sk-69°202, a B3 supergiant, had previously been a red supergiant from the detection of its circumstellar shell of gas and dust (Frannson et al. 1988; Wampler and Richicci 1988) which was ejected about 10,000 yrs. before. According to Woosley (1988) and others, 6 $M_\odot$ was left in the helium core of Sk-69°202, and its hydrogen envelope still had 5-10 $M_\odot$. Thus at the time of the explosion Sk-69°202 still had 11-16 $M_\odot$ remaining from its initial mass of 18-20 $M_\odot$. The numbers in Table 1 are within this range, although on the low end; however the mass loss rates could be lower for the LMC red supergiants.

The poster at this meeting by Rolf Kudritzki, Hans Groth and myself describes a group of A-type supergiants in the Magellanic Clouds with anomalously strong hydrogen lines and U-B colors that are too red. On the HR diagram, these stars are predominantly between $M_{Bol} = -6$ and -8 mag corresponding to initial masses of 10-20 $M_\odot$. We show that the hydrogen line profiles can be reproduced by an enhanced helium abundance, of the order of .5 by number. This may seem rather high, but although the Maefer and Meynet tracks for 15 and 20 $M_\odot$ do not evolve back to the blue after the red supergiant stage, the 25 and 40 $M_\odot$ tracks show relatively high He abundances at the surface during the blue loop in this temperture range. Spectroscopically normal A-type supergiants are also found in the same luminosity/ temperature region of the HR diagram as these peculiar stars in both galaxies. Thus we are suggesting that these stars are post red supergiants and are excellent candidates for the type of star that eventually exploded as SN1987A.

It is interesting that none of these stars are found brighter than $M_{Bol} \approx -8$ mag. This stage is either very short or perhaps it doesn't exist for stars much more massive than 20 $M_\odot$. It is also important to determine whether these stars exist in our galaxy or whether this blue loop at these masses is a characteristic of lower metallicity systems.

5. FINAL REMARKS

During the past decade most of our interest in massive star evolution and our greatest uncertainties have focussed on the uppermost part of the HR diagram and the evolution of the most massive stars. Although there is still a lot of physics to be learned about the origin of the luminosity/stability limit, I think the greatest uncertainties now concern the final stages in the 10-30 $M_\odot$.

Some possibilities include:

O or B star MS → BSG→ RSG → OH/IR → SN (Type II)

or RSG(OH/IR) → YSG → BSG → SN (Type II)

or RSG → YSG → BSG → RSG → OH/IR → SN (Type II)

In this brief review, I have tried to emphasize the importance of the relatively short but high mass loss phases of the LBV's and OH/IR supergiants in determining the fate of stars from 10 to 100 $M_\odot$.

References

Allen, D.A., Jones, T.J. and Hyland, A.R. (1985), Astrophys. J. 291, 280.

Appenzeller, I. (1986), IAU Symposium #116, Luminous Stars and Associations in Galaxies, pg. 139.

Chu, Y.-H. (1982), Astrophys. J. 254, 578.

Cohen, M., Barlow, M.J., and Kuhi, L.V. (1975), Astron. Astrophys. 40, 291.

Conti, P.S., Garmany, C.D., de Loore, C. and Vanbeveren, D. (1983), Astrophys. J. 274, 302.

Davidson, K. (1987), Astrophys. J. 317, 760.

Davidson, K., Dufour, R.J., Walborn, N.R. and Gull, T.R. (1986), Astrophys. J. 305, 867.

Davidson, K., Walborn, N.R. and Gull, T.R. (1982), Astrophys. J. (Letters) 254, L47.

de Jager, C. (1980), The Brightest Stars (Reidel: Dordrecht).

de Jager, C. (1984), Astron. Astrophys. 138, 246.

de Jager, C., Nieuwenhuijsen, H. and van der Hucht, K. (1988), Astron. Astrophys. 193, 375.

Dufour, R.J. and Mitra, P. (1987), Bull. A.A.S. 19, 1020.

Dyck, H.M., Simon, T. and Wolstencroft, R.D. (1984), Astrophys. J. 277, 675.

Elias, J.H., Frogel, J.A. and Schwering, P.R.W. (1986), Astrophys. J. 302, 675.

Fransson, C., Cassatella, A. Gilmozzi, R., Kushner, R.P., Panagia, N., Sonneborn, G. and Wamsteker, W. (1989), Astrophys. J. 336, 429.

Giguere, P.T., Woolf, N.J. and Webber, J.C. (1976), Astrophys. J. (Letters) 207, L195.

Gottlieb, E.W. and Liller, W. (1978), Astrophys. J. 225, 488.

Humphreys, R.M. (1989), IAU Colloquium #113, Physics of Lumious Blue Variables, p. 3.

Humphreys, R.M., Blaha, C., D'Odorico, S., Gull, T.R. and Benvenuti, P. (1984), Astrophys. J. 278, 124.

Humphreys, R.M. and Davidson, K. (1979), Astrophys. J. 232, 409.

Humphreys, R.M. and Davidson, K. (1984), Science 223, 243.

Humphreys, R.M. and Davidson, K., Stahl, O., Wolf, B. and Zickgraf, F.-J. (1989), IAU Colloquium #113, Physics of Luminous Blue Variables, p. 303.

Humphreys, R.M., Jones, T.J. and Gehrz, R.D. (1988), Astron. J. 94, 315.

Humphreys, R.M., Strecker, D.W., Murdock, T.L. and Low, F.J. (1973), Astrophys. J. (Letters) 179, L49.

Humphreys, R.M., Strecker, D.W. and Ney, E.P. (1971), Astrophys. J. (Letters) 167, L35.

Jones, T.J. (1990), private communication.

Lamers, H.J.G.L.M. (1986), IAU Symposium #116, Luminous Stars and Associations in Galaxies, p. 157.

Lamers, H.J.G.L.M. (1987), in Workshop on Instabilities in Luminous Early-Type Stars, p. 99.

Lamers, H.J.G.L.M. (1989), IAU Colloquium #113, Physics of Luminous Blue Variables, p. 135.

Lamers, H.J.G.L.M. and Fitzpatrick, E. (1988), Astrophys. J. 324, 279.

Leitherer, C., Appenzeller, I., Klare, G., Lamers, H.J.G.L.M., Stahl, O., Waters, L.B.F.M. and Wolf, B. (1985), Astron. Astrophys. 120, 113.

Leitherer, C. and Zickgraf, F.-J. (1987), Astron. Astrophys. 174, 103.

Lewis, B.M., Terzian, Y. and Eder, J. (1986), Astrophys. J. (Letters) 302, L23.

Maeder, A. (1989), IAU Colloquium #113, Physics of Luminous Blue Variables, p. 15.

Maeder, A. and Meynet, G. (1987), Astron. Astrophys. 182, 243.

Maeder, A. and Meynet, G. (1988), Astron. Astrophys. Suppl. 76, 411.

Moffat, A.F.J. (1989), IAU Colloquium #113, Physics of Luminous Blue Variables, p. 229.

McGregor, P.J., Hyland, A.R. and Hillier, D.J. (1988), Astrophys. J. 324, 1071.

Smith, L.J., Pettine, M., Dyson, J.E. and Hartquist, T.W. (1988), Mon. Not. Roy. Astron. Soc. 234, 625.

Stahl, O. (1987), Astron. Astrophys. 182, 229.

Stahl, O., Wolf, B., Leitherer, C., Zickgraf, F.-J., Krautter, J. and de Groot, M. (1984), Astron. Astrophys. 140, 459.

Stahl, O., Wolf, B., Klare, G., Juttner, A. and Cassatella, A., Astron. Astrophys., 228, 379.

Thackeray, A.D. (1977), Mon. Not. Roy. Astron. Soc. 180, 95.

van der Hucht, K.A., Hidayat, B., Admiranto, A.G., Supelli, K.R. and Doom, C. (1988), Astron. Astrophys. 199, 217.

Walborn, N.R. (1977), Astrophys. J. 215, 53.

Walborn, N.R. (1988), IAU Colloquium #108, Atmospheric Diagnostics of Stellar Evolution, p. 70.

Walborn, N.R. (1989), IAU Colloquium #113, Physics of Luminous Blue Variables, p. 27.

Wampler, J. and Richicci, A. (1988), ESO Messenger 52, 14.

Wood, P.R., Bessell, M.S. and Whiteoak, J.B. (1986), Astrophys. J. (Letters) 306, L81.

Woosley, S.E. (1988), Astrophys. J. 330, 218.

Woosley, S.E. and Weaver, T.A. (1986), Ann. Rev. Astron. Astrophys. 24, 205.

DISCUSSION

Leung: Earlier in my paper I brought up two contact systems with combined masses, of over 50, or about 100 $M_\odot$. They suggest that in double stars evolution could preserve some of their, say, wind lost. Presumably, there are large amounts of material surrounding the components. The IUE spectra are extremely complicated and do not look like ordinary K supergiants at all. As a matter of fact, they look very similar to U Gem stars with most of the characteristic strong emission lines. These systems have periods of about 100 days or longer.

Chu: If LBV's turn into WN's, why is oxygen severely depleted in LBV ejecta (η Car and AG Car) but not depleted much at all in ejecta around WN stars?

Humphreys: Oxygen is not overabundant in LBV ejecta. We think their N and He enrichment is the result of CNO process. If WN's are more evolved they may be closer to onset of C-fusion. It would be appropriate for the interior people to consider this problem.

Maeder: Allow me to make a general constructive comment on observations about evolutionary connections. Ideally, one should use (1) location in the HRD, (2) surface composition, (3) pulsations. In that respect, I emphasize that the periods are not expected to be the same for stars on bluewards or redwards tracks. Maybe in some years observations will lead to such multiple comparisons, which would be very constraining.

Humphreys: I did not discuss the periods of variability. I merely mentioned that the cool hypergiants are variable in light as additional evidence of their instability. Most of the cool hypergiants show irregular light variations.

Cherepashchuk: I would like to say something about the progenitors of WR binaries. According to our interpretation (1988), the massive O+O binary system RY Scuti may be considered as the progenitor of the WR+O binary, because the more massive component in this system is surrounded by a thick accretion disk and the less massive companion has weak emission of $HeII4686$ and is an intensively mass loosing star, becoming a WR star. On the other hand, the SS433 object may be considered as the progenitor of the WR+c binary system, because, according to our light curve solution for this eclipsing binary system (Cherepashchuk, 1988), the normal star $(M \approx 20M_\odot)$ is filling its Roche lobe and is loosing mass with a high rate on thermal time scale: $\dot{M} \approx 10^{-4}M_\odot \cdot yr^{-1}$. So our results confirm the suggestion made by van den Heuvel (1981) that SS433 is a massive binary observed at a second mass exchange phase of evolution and that the normal star in SS433 is evolutionary linked with the WR phase.

Humphreys: My remarks about massive stars only address *single star* evolution. The situation for initial masses could be very different in binaries and undoubtedly is.

Langer: RSG's, as any massive star, cannot directly evolve into WC stars, *i.e.*, skip the WN stage, simply because He-burning convective cores of massive stars are always smaller (in mass) than the H-burning convective cores.

Humphreys: I agree that the RSG's do not become WC's. If they become WR stars, they most likely become WN's. As yet, there is no direct connection as in the same sense as for the LBV's.

Massey: If Gallagher were here, he might bring up the issue of where the Hubble-Sandage variables in M33 are found. *Are* the known LBV's in M33 located in the same places as WR stars, or are they preferentially located in the inner-arm regions?

Humphreys: The Hubble-Sandage variables in M33 including Var A are in associations, in the spiral arms and in star forming regions. However, they are not in the large HII regions.

Owocki: In our attempt to understand the physical cause of the LBV phenomenon, I think it helpful to focus on the time scales. Outburst time scale of every few years is far longer than the dynamical time (*e.g.*, sound travel time), and this rules out mechanisms like line-driven instability as well as interior perturbation. On the other hand, this is much shorter than an evolutionary time scale. Thinking of intermediate scales, perhaps we should consider the cooling time scale of the star's outer envelope or deep atmosphere.

Humphreys: In 1979 we suggested that it was interior evolution that brought the star to the stability limits resulting in some large eruption. The subsequent moderate variations (~ 2 *mag*) may be the star's attempt to continue to adjust its atmosphere to remain stable as interior evolution again pushes it to the stability limit.

Cassinelli: Is it true that the LBV's are up in the same general part of the HRD as the B[e] stars?

Humphreys: The B[e] supergiants do not exactly overlap with the LBV's on the HRD. They actually tend to occur seemingly systematically less luminous than the LBV's. They are not variable in light and the evidence of Zickgraf's work is, that they very likely have equatorial disks around. LBV's may also have equatorial disks, but we do not have the strong evidence as he does. He has the dual component model for the emission lines, where some of the emission is coming from the equatorial disk producing the very narrow lines while the broader emission lines that he sees in the same spectrum are coming from a polar region, perhaps. B[e] supergiants are not LBV's. My own suspicion is that the latter will proven to be binary systems.

Roberta Humphreys, Peter Conti, You-Hua Chu

AG CARINAE AND THE WR PHENOMENON IN LUMINOUS BLUE VARIABLES

R. Viotti[1], G.B. Baratta[2], C. Rossi[3], A. di Fazio[3]
[1] Istituto Astrofisica Spaziale (CNR), Via E. Fermi 21,
00044 Frascati RM, Italy
[2] Osservatorio Astronomico, Viale del Parco Mellini 84,
00136 Roma
[3] Istituto Astronomico, Università La Sapienza, Via Lancisi 29,
00161 Roma

ABSTRACT. AG Car is a Luminous Blue Variable which recently evolved from AIe to Ofpe/WN9 in four years at about constant bolometric luminosity, while in the visual the star faded by two magnitudes. This change is probably associated with variable opacity of an unstable massive expanding envelope of a hot star. We discuss the main spectral features of the star and of its ring nebula, and the spectral variations.

1. Introduction

The galactic Luminous Blue Variable (LBV) AG Car is unique for assembling several interesting features, namely: (1) the light history typical of the Hubble-Sandage variables; (2) the peculiar emission line spectrum, and the large spectral variations; (3) the strong IR excess, and especially (4) the small ring nebula closely resembling the typical PNs. The light curve of AG Car shows irregular long term variations, with periods of minimum luminosity with $V > 8$, and rapid brightenings up to $V = 6$ (Mayall, 1969; Sharov, 1975). The star is also variable in the IR (Whitelock et al., 1983). The large IR excess is attributed to f-f emission of a massive wind (Bensammar et al., 1981; Lamers, 1987). A mass loss rate of $3 \cdot 10^{-5} M_\odot \cdot yr^{-1}$ was derived by Wolf and Stahl (1982) from the fit of the $H\beta$ scattering wings. Caputo and Viotti (1970) found that during 1949-1959 the optical spectrum varied between AIeq and B0eq.

In recent years, AG Car underwent large photometric and spectroscopic changes. Its visual magnitude gradually varied from $V = 6$ in 1981 to $V = 8$ in 1985 (Hutsemekers and Kohoutek, 1987), while there was an increase of the line excitation in the optical spectrum from Aeq in 1981 to Beq in 1983, and to Ofpe/WN9 in 1985 (Stahl, 1986; Viotti et al., 1984; Wolf and Stahl, 1982). At maximum and during fading the UV spectrum was dominated by $FeII$ lines which disappeared at minimum (Viotti et al., 1989). The continuum slope became hotter with an increase of the far-UV flux (Viotti et al., 1984). Actually, from the study of the IR-to-UV energy distribution Viotti et al. found that the recent large variations of AG Car occurred at almost constant bolometric magnitude of about -8.3. This 'WR-phase' of AG Car which is lasting since many years, is a useful tool to investigate the relation between LBVs and WR stars.

K. A. van der Hucht and B. Hidayat (eds.),
Wolf-Rayet Stars and Interrelations with Other Massive Stars in Galaxies, 499–504.
© 1991 IAU. Printed in the Netherlands.

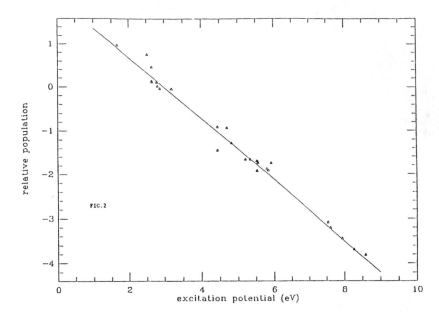

Figure 1a. The excitation of the $FeII$ levels in the optical spectrum of AG Car in December 1981 ($V = 6$). The line corresponds to $T(ex) = 7200 \pm 350K$.

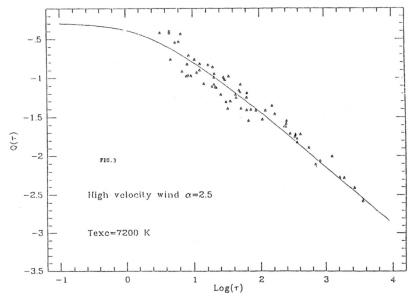

Figure 1b. The Self Absorption Curve for $FeII$ in AG Car. The line points are fitted with a theoretical high velocity accelerated wind model.

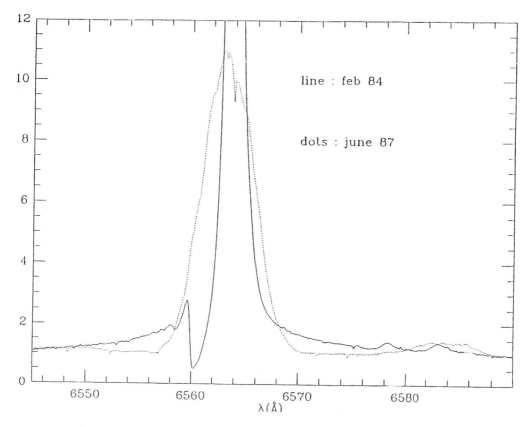

Figure 2. The high resolution profile of H in February 1984 and in June 1987 when AG Car was in the Be and Of/WN9 phases (ESO, CAT-CES).

2. The optical spectrum of AG Carinae

The optical spectrum of AG Car near maximum shows strong emissions of H and $FeII$ with P Cygni profiles. We have analyzed the $FeII$ emissions in the December 1981 spectrum using the Self Absorption Curve method (SAC) described by Friedjung and Muratorio (1987). Figure 1a shows the population of the $FeII$ levels. A remarkable result is the fact that the level population follows a Boltzmann-type law from 1.6 to 8.6 eV, in spite of the fact that the lines are formed outside LTE. The corresponding mean excitation temperature is $7200 \pm 350K$. In Figure 1b the SAC of the $FeII$ lines is fitted with a SAC model of an accelarated wind ($\alpha = 2.5$) with a wind velocity larger than the random volocity. It is also suggested that the $FeII$ region is confined to a limited zone around the star.

During the fading phase of AG Car large profile variation was noted (Bandiera *et al.*, 1989). In February 1984 $H\alpha$ displayed a strong emission peak with a P Cygni absorption at about -150 $km{\cdot}s^{-1}$ extending to -180 $km{\cdot}s^{-1}$ (Figure 2). Broad scattering wings were present. At minimum the line was present with a broad (FWHM = 250 $km{\cdot}s^{-1}$) emission and no wings. Note the weak CII emissions in 1984, and the broad $[NII]$ emission in 1987 (see Stahl, 1986). Broad wings were also present at maximum in the other Balmer lines

502

and in the strongest $FeII$ emissions. These wings can be explained by electron scattering with $T_e = 9500 - 15000K$ and $\tau = 1.0$ for the Balmer lines, and $T_e = 7000K$ and $\tau = 0.5$ for the $FeII$ 5018$\overset{\circ}{A}$ line. This difference could be explained by line formation at different envelope depth. The fit of the $H\beta$ wings in February 1983 suggests $T_e = 19000K$ and $\tau = 1.8$.

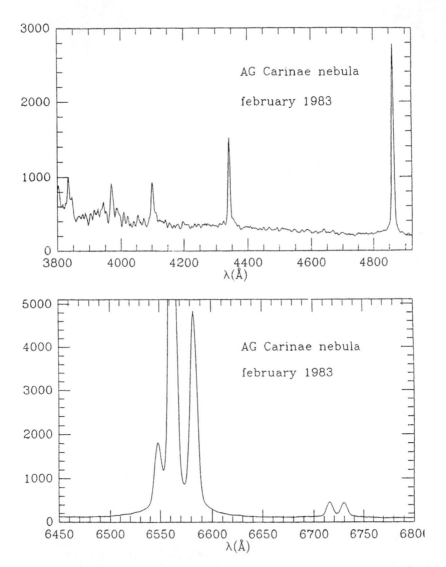

Figure 3. The low resolution optical spectrum of the AG Car nebula in February 1983 (ESO, 1.52 m telescope).

3. The ring nebula

There are many examples of circumstellar nebulae associated with LBVs and WR stars, and AG Car is the most conspicuous one. From the study of its UV brightness Viotti et al. (1988) concluded that scattering by dust particles should be the main source of the UV radiation. The presence of cool dust in the nebula was confirmed by the Kuiper Airborne observations of McGregor et al. (1988). We have observed the optical spectrum of a bright part of the nebula 15″ to NE) in February 1983, and found a rather low excitation spectrum with strong Balmer lines, and weak emissions of HeI, $[NII]$, $[SII]$, and $FeII$ (Figure 3). The Balmer decrement, when compared with case B emission, suggests a colour excess of $E(B - V) = 0.60$. The spectrum is similar to that observed one year later (when the star was 0.6 mag fainter) by Mila Mitra and Dufour (1990). They found N/O and N/S ratios higher than in nearby HII regions. Recent coronographic observations of Paresce and Nota (1989) revealed the presence of bipolar and helical structures in the nebula, possibly associated with asymmetric mass ejection from the central object.

4. Conclusions

Table 1 summarizes the main data on AG Car. AG Car well probably is an evolved massive star with a hot core surrounded by an extended envelope produced by its large mass outflow. At the maximum the expanding envelope is opaque to the UV photons from the core, so that the emerging spectrum is that of a P Cygni Ae or Be star with an effective temperature around $10000K$. After 1981 the envelope was gradually ionized and became more transparent, so that the effective radius decreased by about a factor ten, while the effective temperature increased to more than $30000K$. The star now displays WR features but with a low expansion velocity close to the wind velocity at maximum. The origin of this evolution of AG Car is still uncertain. The unstable structure of the envelope probably has a crucial role, but we cannot anticipate how long the present high temperature phase will last. The asymmetry of the mass outflow and the bipolar structure of the nebula could suggest the presence of a binary system, which is however difficult to support observationally. The dust in the nebula should have been produced some thousand years ago, possibly during an 'η' Car phase of the star. As suggested by Viotti (1987) a consistent increase of the mass loss rate could produce a drastic cooling of the envelope, or of some parts of it, even below dust condensation temperature and give rise to dust formation. In this regard, it would be important to investigate the chemical anomalies of the star and of its nebula which are similar to those of η Car itself.

Table 1. Basic data on AG Car (period 1981 to 1987)

Visual magnitude	6 to 8	Hutsemekers and Kuhoutek, 1987
Spectral type	1981 AOIeq	Wolf and Stahl, 1982
	1983 B Ieq	Viotti et al., 1984
	1985 Ofpe/WN9	Stahl, 1986
$E(B - V)$	0.60 ±0.05	Viotti et al., 1984
distance	2500 pc	Bensammar et al., 1981
$M(bol)$	-8.3 ±0.3	Viotti et al., 1984
mass loss	$3 \cdot 10 - 5 M_\odot y^{-1}$	Wolf and Stahl, 1982

We are grateful to Otmar Stahl and Bernard Wolf for having put at our disposal their optical spectrum of AG Car of December 1981, and to O. Stahl for his CAT CES spectrum of AG Car of June 1987.

References

Bandiera, R., Focardi, P., Altamore, A., Rossi, C., and Stahl, O. 1989, in: *Physics of Luminous Blue Variables*, H. Lamers and C. de Loore (eds.), Kluwer Academic Publishers, p. 279.

Bensammar, S., Gaudenzi, S., Rossi, C., Johnson, H.M., Thé, P.S., Zuiderwijk, E.J., and Viotti, R. 1981, in: *Effects of Mass Loss on Stellar Evolution*, C. Chiosi and R. Stalio (eds.), Reidel, Dordrecht, p. 67.

Caputo, F. and Viotti, R. 1970, *Astron. Astrophys.* **7**, 266.

Friedjung, M. and Muratorio, G. 1987, *Astron. Astrophys.* **188**, 100.

Hutsemekers, D. and Kohoutek, L. 1987, *Astron. Astroph. Suppl.* **73**, 217.

Lamers, H.J.G.L.M. 1987, in: *Instabilities in Luminous Early Type Stars*, H. Lamers and C. de Loore (eds.), Reidel, Dordrecht, p. 99.

Mayall, M.W. 1969, *J. R. Astron. Soc. Canada* **63**, 221.

McGregor, P.J., Finlyason, K., and Hyland, A.R. 1988, *Astrophys. J.* **329**, 874.

Mila Mitra, P. and Dufour, R.J. 1990, *Mon. Not. R. Astr. Soc.* **242**, 98.

Paresce, F. and Nota, A. 1989, *Astrophys. J.* **341**, L83.

Sharov, A.S. 1975, in: *Variable Stars and Stellar Evolution*, W.A. Sherwood and J. Plaut (eds.), Reidel, Dordrecht, p. 275.

Stahl, O. 1986, *Astron. Astrophys.* **164**, 321.

Viotti, R. 1987, in: *Instabilities in Luminous Early Stars*, H. Lamers and C. de Loore (eds.), Reidel, Dordrecht, p. 257.

Viotti, R., Altamore, A., Barylak, M., Cassatella, A., and Rossi, C. 1984, in: *Future of Ultraviolet Astronomy Based on Six Years of IUE Research*, NASA CP-2349, p. 23.

Viotti, R., Cassatella, A., Pontz, D., and Thé, P.S. 1988, *Astron. Astrophys.* **190**, 333.

Viotti, R., Altamore, A., Rossi, C., and Cassatella, A. 1989, in: *Physics of Luminous Blue Variables*, K. Davidson *et al.* (eds.), Kluwer Academic Publishers, Dordrecht, p. 268.

Whitelock, P.A., Carter, B.S., Roberts, G., Whittet, D.G.B., and Baines, B.W.T. 1983, *Mon. Not. R. Astr. Soc.* **205**, 577.

Wolf, B. and Stahl, O. 1982, *Astron. Astrophys.* **112**, 111.

DISCUSSION

Barlow: A comment about the $HeI5876$ profile you showed. You interpreted the edge 250 $km \cdot s^{-1}$ as a terminal velocity. It is now being argued that in the UV the edge velocities are influenced mainly by shocks and turbulences in the wind, and it is a deep black absorption edge that really gives the terminal velocity. By analogy, one might argue that it is the deepest point in that absorption profile which corresponds to terminal velocity, and it is about just over 200 $km \cdot s^{-1}$ which is very similar to what the $[FeII]$ lines give at other epoches.

Niemela: How did you determine that AG Car is member of Car OB2?

Baratta: There are at least two reasons to believe that AG Car belongs to the Key-Hole Nebula complex. First, because this complex includes many peculiar luminous stars (in particular η Car), and AG Car is a nearby peculiar luminous star. In addition, the radial velocity of the interstellar lines is nearly the same. The recent suggestion of a larger distance of 6 kpc (instead of 2.5 kpc), is essentially based on the theoretical (and questionable) need of a higher luminosity for this LBV star.

NEW OBSERVATIONS OF LBV ENVIRONMENTS

NOLAN R. WALBORN[1] and IAN N. EVANS
Space Telescope Science Institute[2]
3700 San Martin Drive
Baltimore, Maryland 21218, USA

EDWARD L. FITZPATRICK[1]
Princeton University Observatory
Peyton Hall
Princeton, New Jersey 08544, USA

MARK M. PHILLIPS
Cerro Tololo Inter-American Observatory[3]
Casilla 603
La Serena, Chile

ABSTRACT. Two observational programs which provide new information about particular LBVs through investigations of their immediate surroundings are described. (1) Digital spectral classification of OB supergiants in compact groups apparently associated with Radcliffe 127 and S Doradus has revealed several interesting objects and indicates which of them are likely to be generically related to the LBVs. (2) Velocity-resolved images of the Eta Carinae shell show qualitatively new features, which will contribute substantially to the interpretation of its complex spatial/kinematical structure.

1. OB Supergiants Associated with Radcliffe 127 and S Doradus

Normal OB stars associated with massive peculiar objects may provide indirectly, valuable information about the physical nature of the latter, such as the initial mass, age, and evolutionary status, although even in the Large Magellanic Cloud uncertainties with regard to coeval formation and chance alignments may arise. Several Luminous Blue Variables (Davidson, Moffat, and Lamers 1989) and Ofpe/WN9 objects (Bohannan and Walborn

---

[1] Visiting Astronomer, Cerro Tololo Inter-American Observatory, [3]National Optical Astronomy Observatories, operated by the Association of Universities for Research in Astronomy, Inc., under contract with the National Science Foundation.

[2] Operated by the Association of Universities for Research in Astronomy, Inc., under contract with the National Aeronautics and Space Administration.

K. A. van der Hucht and B. Hidayat (eds.),
Wolf-Rayet Stars and Interrelations with Other Massive Stars in Galaxies, 505–512.
© 1991 *IAU. Printed in the Netherlands.*

1989) in the LMC are apparently located in compact OB groups or multiple systems (Walborn 1989), only one of which has been investigated in detail to date (Schild 1987). Two further outstanding examples are provided by the currently active LBVs R127 = HDE 269858 = Sanduleak (Sk)−69°220 (NGC 2055, Figure 1) and S Dor = R88 = HD 35343 = Sk−69°94 (Fig. 2 of Leitherer et al. 1985). NRW and ELF have obtained blue-violet digital classification spectrograms of four OB supergiants in the former group and one in the latter, with the two-dimensional photon-counting system at the CTIO 1m telescope during (with one exception) December 1989. These data have a resolution of 1.5 Å and S/N per resolution element of about 80; they were obtained and processed identically to those in the extensive OB digital atlas of Walborn and Fitzpatrick (1990), with respect to which they have been classified.

Four of these spectrograms are reproduced in Figure 2 in the WF Atlas format; the fifth, of R128 = HDE 269859 = Sk−69°221, will be included in the extensive discussion of LMC B-supergiant spectra by Fitzpatrick (1990). These spectra are of considerable interest in their own right: all of them display emission lines, CNO anomalies, and/or variability. Sk−69°217 is a previously unknown LMC Of supergiant, with broad emission underlying the narrow N III $\lambda\lambda$4634, 4640–42 and He II λ4686 emission lines. Sk−69°218 and 226 are new OBN supergiants: note the extreme weakness of C III $\lambda\lambda$4070, 4650. The former spectrum shows weak N III λ4640 emission, while the latter is definitely variable and warrants further investigation: half of the four summed individual observations were obtained on each of two consecutive nights, between which substantial changes in the intensities of several He I, Si IV, N II, and N III lines occurred. In striking contrast with the previous two spectra, that of S Dor No. 1 is nitrogen deficient: note the weakness of N III λ4097 in the blue wing of Hδ, in comparison with the other two. Finally, Sk−69°221 (not shown here), also has a nitrogen-deficient spectrum.

The spectral classifications of these stars are given in Table 1, along with other observed and derived parameters as follows: M_V from the spectral-type calibrations of Walborn (1972, 1973); V and $B − V$ from Ardeberg et al. (1972) or Isserstedt (1975), and $E_{B−V}$ based on the intrinsic colors of Johnson (1966—Fitzpatrick 1988 has derived somewhat redder values for LMC OB supergiants, but the differences are not significant for the present purpose), except for S Dor No. 1 whose values are derived from a photographic B magnitude estimate and the $E_{B−V}$ of S Dor quoted by Leitherer et al. (1985); M_V derived from the photometry with $R = 3$ and a distance modulus of 18.6 for the LMC; M_{bol} corresponding to the latter M_V, and $\log T_{eff}$, both from the calibration of Humphreys and McElroy (1984); and finally, estimates of the initial stellar masses determined by plotting the preceding two parameters on the theoretical HR diagram for $Z = 0.005$ by Maeder (1990), as further discussed below.

The large discrepancies between the spectroscopic and photometric M_V's for three of the late-O/early-B supergiants in Table 1, whose luminosity classes are based on Si/He line ratios relative to galactic standards, but not for the Of object whose luminosity class depends on the He II emission, are identical to the situation found by Walborn (1977), and may well be due to the LMC metal deficiency, similar to but less extreme than the SMC case (Walborn 1983). The definition of primary classification standards in the Magellanic Clouds is an important problem for future work. Curiously, there is no significant discrepancy in the case of S Dor No. 1, which could indicate a range of Si abundances in the LMC, as was suggested in the SMC (Walborn 1983), but note that the photometry of this star is uncertain. The photometric M_V's have been used for the M_{bol} determinations.

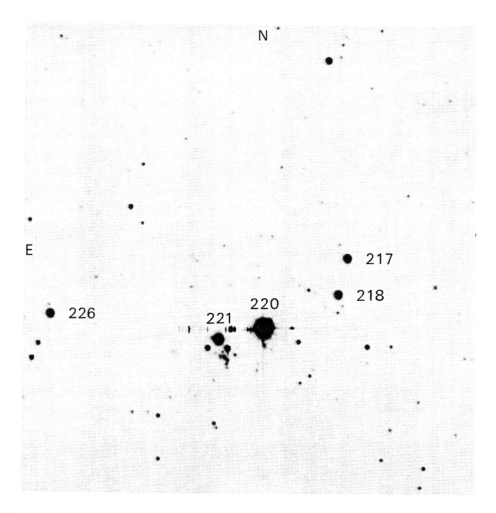

Figure 1. The field of R127 (NGC 2055). Twenty-sec R CCD frame kindly obtained by Dr. N. Suntzeff at the CTIO 0.9m on Oct. 28, 1988 (the overexposed image of R127 has produced some CCD artefacts). Identification numbers from the −69° zone of Sanduleak (1970) are given. Sk−69°217 and 218 are separated by 9″ in declination.

Although the uncertainties and discrepancies among current evolutionary tracks for massive stars are well known, those by Maeder (1990) are the most comprehensive and applicable to the present investigation. While further observational and theoretical work is required, there is growing evidence that the relative CNO abundances in OB supergiant spectra provide essential evolutionary diagnostics (Walborn 1988), specifically the direction of current motion in the HR diagram (redward or blueward) for a given mass range. Accordingly, the mass estimates in Table 1 have been derived with the assumption that

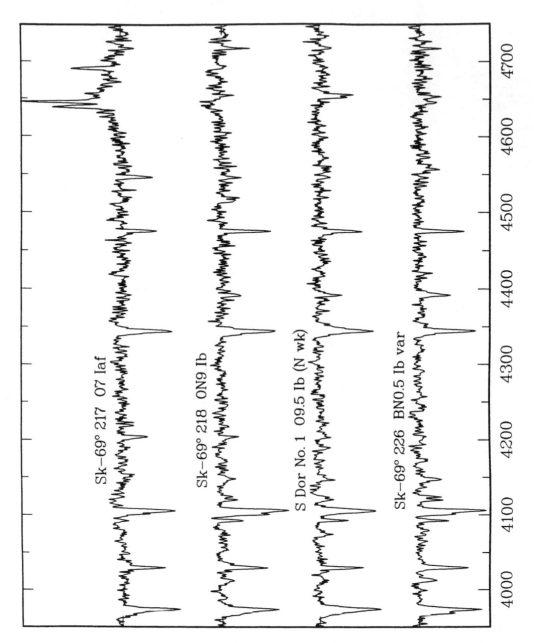

Figure 2. Blue-violet digital spectrograms of LMC OB supergiants near R127 and S Dor. Wavelengths in Å are given and the ordinate ticks are separated by 0.5 linear-intensity, continuum-flux units. See Walborn and Fitzpatrick (1990) for identification of spectral features and the text for discussion.

TABLE 1. Parameters of OB supergiant companions to R127 and S Dor

Sk/Name	Spectral Type	M_V (S.T.)	V	$B - V$	E_{B-V}	M_V (D.M.)	M_{bol}	$\log T_{eff}$	$M_\odot$
$-69°217$	O7 Iaf	-7.0	11.97	-0.11	0.21	-7.3	-10.6	4.54	60
$-69°218$	ON9 Ib	-6.2	12.00	-0.11	0.20	-7.2	-10.3	4.51	55
$-69°221$	B2 Ia (N wk)	-7.0	10.63	-0.06	0.12	-8.3	-9.7	4.26	40
$-69°226$	BN0.5 Ib var	-6.0	11.94	-0.11	0.14	-7.1	-9.1	4.36	30
S Dor 1	O9.5 Ib (N wk)	-6.2	12.35:	-0.25:	0.05:	-6.4:	-9.2:	4.48	35:

the nitrogen-deficient objects are moving redward, and the nitrogen-enhanced ones blue-ward. For comparison, the M_{bol}'s of R127 and S Dor are -10.5 and -9.8, respectively (Humphreys 1989, Wolf 1989), implying initial masses of 60 and 40 $M_\odot$ from Maeder's diagram. Somewhat disappointingly, one concludes that only Sk$-69°217$ and 218 are likely to be generically associated with R127, while Sk$-69°221$ and 226 cannot be, a result also consistent with their relative E_{B-V}'s (see Stahl et al. 1983, although their value of 0.20 for R127 may be subject to some uncertainty). The results for S Dor No. 1 are consistent with a physical association with S Doradus.

2. Velocity-Resolved Images of the Eta Carinae Shell

The morphological, kinematical, and spectroscopic intricacies of Eta Carinae's outer shell have been discussed by Walborn (1976), Walborn, Blanco, and Thackeray (1978), Davidson, Walborn, and Gull (1982), Davidson et al. (1986), and Walborn and Blanco (1988). These studies have revealed a complex system of nebular condensations with high expansion velocities and large N/C, N/O, and He/H ratios, indicative of processed material ejected by an evolved object. Other recent investigations have provided detailed information about structure in the IR (Mitchell et al. 1983; Hackwell, Gehrz, and Grasdalen 1986; Russell et al. 1987) and in X-rays (Chlebowski et al. 1984), IR spectra (Allen, Jones, and Hyland 1985), very high radial velocities and polarization (Meaburn, Wolstencroft, and Walsh 1987; Dufour 1989), UV spectra (Viotti et al. 1989), and photometry from coronagraphic images (Burgarella and Paresce 1990).

As a further contribution toward disentangling the complex kinematical structure of this object, which exhibits large changes on very small angular scales, NRW and MMP obtained velocity-resolved images with the Rutgers Fabry-Perot (Schommer et al. 1988) at the CTIO 4m telescope on the night of Feb. 10, 1988. The Hβ line was selected since it is strong and in a relatively clear spectral region throughout the object. The instrument provided a spectral resolution of 2.5 Å (154 km/sec); the detector was a 512-square, 15 μ-pixel TI CCD. Two series of 29 images each extending from 4847.5 Å to 4882.5 Å (on axis) in steps of 1.25 Å were obtained: one with exposure times of 60 sec behind a central occulting wire for the fainter outer structure, and the other of 1 sec unocculted for the inner structure. Unfortunately, the available Hβ pre-selecting interference filter suffered

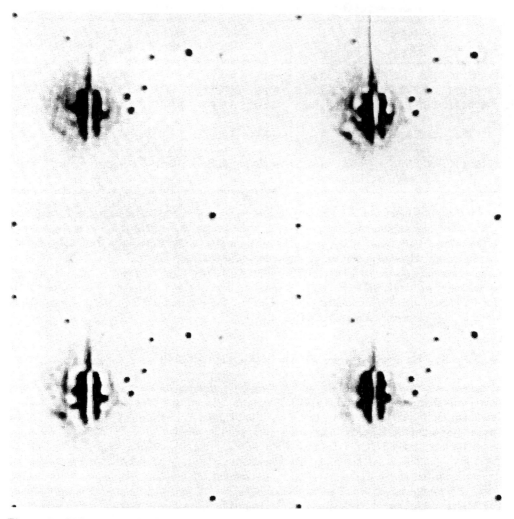

Figure 3. Velocity-resolved images of Eta Car. The heliocentric radial velocities at the object center are −174 km/sec in the upper left panel, +58 km/sec in the upper right, +289 km/sec at lower left, and +521 km/sec at lower right. The vertical spike is caused by the central occulting wire, which is in PA 130°. The two similar stars immediately to the right of the object are separated by 5″.

from internal reflections and produced a triangular stellar PSF in the images. This problem has been addressed for the long-exposure series with maximum-entropy techniques by INE at the STScI (by means of software kindly made available by John Skilling), resulting in a substantially improved, symmetrical PSF with a typical FWHM of 1″2. The structure in the outer homunculus and the outer condensations has likewise been strikingly enhanced.

The systematic velocity behavior of several bright outer condensations shows good agreement with the spectroscopic descriptions of Davidson *et al.* (1982), Meaburn *et al.* (1987), and Dufour (1989), but many other spatial/velocity structural relationships are revealed for the first time. Four of these images are shown in Figure 3; they have not yet been corrected for the FP off-axis wavelength variation, but the velocity range across the object images is only 35 km/sec, which is hardly significant relative to the velocity resolution and displayed intervals. Further analysis of these remarkable images promises to advance qualitatively our knowledge of the Eta Carinae shell.

References

Allen, D. A., Jones, T. J., and Hyland, A. R. (1985) *Ap. J.*, **291**, 280.

Ardeberg, A., Brunet, J.-P., Maurice, E., and Prevot, L. (1972) *Astr. Ap. Suppl.*, **6**, 249.

Bohannan, B. and Walborn, N. R. (1989) *Pub. A.S.P.*, **101**, 520.

Burgarella, D. and Paresce, F. (1990) in preparation.

Chlebowski, T., Seward, F. D., Swank, J., and Szymkowiak, A. (1984) *Ap. J.*, **281**, 665.

Davidson, K., Dufour, R. J., Walborn, N. R., and Gull, T. R. (1986) *Ap. J.*, **305**, 867.

Davidson, K., Moffat, A. F. J., and Lamers, H. J. G. L. M. (eds.) (1989) IAU Colloq. 113, *Physics of Luminous Blue Variables* (Kluwer).

Davidson, K., Walborn, N. R., and Gull, T. R. (1982) *Ap. J. (Letters)*, **254**, L47.

Dufour, R. J. (1989) *Rev. Mexicana Astron. Astrof.*, **18**, 87.

Fitzpatrick, E. L. (1988) *Ap. J.*, **335**, 703.

_____. (1990) in preparation.

Hackwell, J. A., Gehrz, R. D., and Grasdalen, G. L. (1986) *Ap. J.*, **311**, 380.

Humphreys, R. M. (1989) in IAU Colloq. 113, *Physics of Luminous Blue Variables*, ed. K. Davidson, A. F. J. Moffat, and H. J. G. L. M. Lamers (Kluwer), p. 3.

Humphreys, R. M. and McElroy, D. B. (1984) *Ap. J.*, **284**, 565.

Isserstedt, J. (1975) *Astr. Ap. Suppl.*, **19**, 259.

Johnson, H. L. (1966) *Ann. Rev. Astr. Ap.*, **4**, 193.

Leitherer, C., Appenzeller, I., Klare, G., Lamers, H. J. G. L. M., Stahl, O., Waters, L. B. F. M., and Wolf, B. (1985) *Astr. Ap.*, **153**, 168.

Maeder, A. (1990) *Astr. Ap. Suppl.*, in press.

Meaburn, J., Wolstencroft, R. D., and Walsh, J. R. (1987) *Astr. Ap.*, **181**, 333.

Mitchell, R. M., Robinson, G., Hyland, A. R., and Jones, T. J. (1983) *Ap. J.*, **271**, 133.

Russell, R. W., Lynch, D. K., Hackwell, J. A., Rudy, R. J., Rossano, G. S., and Castelaz, M. W. (1987) *Ap. J.*, **321**, 937.

Sanduleak, N. (1970) *Cerro Tololo Inter-Am. Obs. Contr.*, No. 89.

Schild, H. (1987) *Astr. Ap.*, **173**, 405.

Schommer, R. A., Caldwell, N., Wilson, A. S., Baldwin, J. A., Phillips, M. M., Williams, T. B., and Turtle, A. J. (1988) *Ap. J.*, **324**, 154.

Stahl, O., Wolf, B., Klare, G., Cassatella, A., Krautter, J., Persi, P., and Ferrari-Toniolo, M. (1983) *Astr. Ap.*, **127**, 49.

Viotti, R., Rossi, L., Cassatella, A., Altamore, A., and Baratta, G. B. (1989) *Ap. J. Suppl.*, **71**, 983.

Walborn, N. R. (1972) *A. J.*, **77**, 312.

————. (1973) *A. J.*, **78**, 1067.

————. (1976) *Ap. J. (Letters)*, **204**, L17.

————. (1977) *Ap. J.*, **215**, 53.

————. (1983) *Ap. J.*, **265**, 716.

————. (1988) in IAU Colloq. 108, *Atmospheric Diagnostics of Stellar Evolution*, ed. K. Nomoto (Springer-Verlag), p. 70.

————. (1989) in IAU Colloq. 113, *Physics of Luminous Blue Variables*, ed. K. Davidson, A. F. J. Moffat, and H. J. G. L. M. Lamers (Kluwer), p. 27.

Walborn, N. R. and Blanco, B. M. (1988) *Pub. A.S.P.*, **100**, 797.

Walborn, N. R., Blanco, B. M., and Thackeray, A. D. (1978) *Ap. J.*, **219**, 498.

Walborn, N. R. and Fitzpatrick, E. L. (1990) *Pub. A.S.P.*, **102**, 379.

Wolf, B. (1989) in IAU Colloq. 113, *Physics of Luminous Blue Variables*, ed. K. Davidson, A. F. J. Moffat, and H. J. G. L. M. Lamers (Kluwer), p. 91.

DISCUSSION

Montmerle: A question about "coevality". Why do you rule out some stars as part of the R127 group on the basis of their masses (hence lifetimes)? In a given association, stars are not necessarily born at the same time: the differences in *ages* may be larger than the lifetimes of the stars themselves.

Walborn: At least in rich OB clusters one can distinguish Orion Nebula, Carina Nebula, Scorpius OB1, and h/χ Persei phases, in order of increasing ages from $\lesssim 10^6$ to $\lesssim 10^9$ years. Therefore, the spread in formation times within a given object must be less than the age differences among these phases, *i.e.*, a couple of million years. Of course, in large associations there can be sequential formation of different subgroups. My purpose was to find objects coeval with the LBV's.

Smith, Lindsey: I noticed in the spectra of the Large Cloud supergiants that you showed, that the nitrogen emission lines were significantly brighter than $HeII4686$. Is that unusual, is it significant, and do you know what controls these relative brightnesses?

Walborn: There is a range of strengths both in the Galaxy and in the LMC. There must be a range of envelope densities. *e.g.* within the association Sco OB1. In the Galaxy one has HD 151804 and HD 152408 right next to each other.

Underhill: Because the relative populations of the levels between which $CIII4650$ and $NIII4096,4103$ occur are very sensitive to NLTE effects, due chiefly to dielectronic recombination, you should be cautious about interpreting changes in the apparent strength of these lines to abundance differences.

Walborn: Of course, a definitive quantitative analysis is required to derive abundances. However, if I see three spectra of identical temperature/luminosity classes, one with both C and N strong, another with $C > N$, and the other with $N > C$, my intuition suggests abundance anomalies as the most reasonable hyothesis.

NOMENCLATURE PROBLEMS

M.-C. Lortet
D.A.E.C.
Observatoire de Meudon
F-92195 Meudon Cedex
France

I would make a few points on behalf of the IAU Comm.5 (Documentation and Astronomical Data) Working Group on Designations.

1. ABBREVIATIONS

. Always try to be clear. Clearly identified objects can be entered in SIMBAD Data Base immediately.
. Don't use 1 or 2 letters designations for newly discovered objects.
. Use Brey rather than B or Br for Breysacher.
. Use MGWR, not MG, for Morgan and Good new WR stars in the LMC (MG is Mendoza and Gomez, 1973, P.A.S.P. 85, 439, red stars).
. You can always obtain an advice by e-mail LORTET@FRMEU51 or BORDE@FRIAP51 for SIMBAD.

2. SPECTRAL PECULIARITIES OF WR STARS

2.1 WN A and B
This is a distinction related to the <u>width</u> not the strength of the emission lines (B stands for broad)

λ HeII 4686	Hiltner and Schild R.E., 1966, Ap.J. 143, 770			A		B
λNIV 4059	Walborn 1974, Ap.J. 189, 269	A	A(B)	(A)B	B	

2.2 Slash stars (Of/WNE and Of/WNL)

Of/WNE   Sk-67 22   Walborn, 1982 ; see Sect.3.
                Sk-71 34   Conti and Garmany, 1983 ; see Sect.3.
                Five Melnick stars (Melnick 30, 35, 39, 42, 51), Walborn 1986, Proceed. IAU Symp. 116, 185.

Of/WNL   Bohannan and Walborn, 1989, P.A.S.P. 101, 520.
                Ten stars, out of which only three at present have a Breysacher number, namely :
                Brey 18 = R 84 = Sk-69 79 = HD 269227 = BE 543
                Brey 64 = BE 381
                Brey 91 = Sk-69 249C = HD 269927C.

K. A. van der Hucht and B. Hidayat (eds.),
Wolf-Rayet Stars and Interrelations with Other Massive Stars in Galaxies, 513–514.
© 1991 *IAU. Printed in the Netherlands.*

3. NEW WOLF-RAYET STARS IN THE LMC

Since Breysacher's 1981 Catalogue, apart from the slash stars described in Sect.2, 15 new Wolf-Rayet stars have been discovered. They are listed in Table 1, and have been given a Brey number, in agreement with J. Breysacher. The stars are ordered by right ascension, e.g. the star 3a is inserted between stars 3 and 4 of the original catalogue.
Table 1 will be complemented later with the five Melnick stars in 30 Dor (crowded field, accurate coordinates needed) and the 7 remaining Of/WNL, as quoted in Sect.2.

TABLE 1. New Wolf-Rayet stars in the LMC (since Breysacher, 1981)

Brey	Other Names	Sp Type	Ref	Neb.	Assoc.
3a	-	WC9	4	N 82	-
10a	Sk-67 22	O3 If*/WN6-A	9	-	-
16a	MGWR 1	WC5+O	5	N 105	LH 31
19a	MGWR 8, BE 456	WN3	7	-	-
40a	Sk-71 34	O4 f/WN3	2	near N 206	-
44a	AB-18	WN8-9	1	-	-
63a	MGWR 3	WN3	5	-	LH 89
65a	MGWR 2	WN5	3	N 59	LH 88
		WN4	5		
65b	TSWR 1, HD 269828C	WN3+0B	8	N 157C	LH 90
65c	TSWR 2, HD 269828E	O4 If/WN6	8	N 157C	LH 90
70a	MGWR 4	WC:	5	N 157B	near LH 99
		WN3-4	6		
74a	TSWR 3	O3 If/WN6	8	N 157B	LH 99
90a	MGWR 5	WC4	6	N 157	near LH 100
93a	MGWR 7	WN3-4	5	N 160D	LH 103
95a	MGWR 6	WC5+O6	5	N 158	LH 104

References for Table 1
1 Azzopardi M., Breysacher J., 1985, Astron. Astrophys. 149, 213.
2 Conti P.S., Garmany C.D., 1983, P.A.S.P. 95, 411.
3 Cowley A.P., Crampton D., Hutchings J.B., Thompson I.B., 1984, P.A.S.P. 96, 968.
4 Heydari-Malayeri M., Melnick J., 1990, these Proceedings.
5 Morgan D.H., Good A.R., 1985, M.N.R.A.S. 216, 459.
6 Morgan D.H., Good A.R., 1987, M.N.R.A.S. 224, 435.
7 Morgan D.H., Good A.R., 1990, M.N.R.A.S. 243, 459.
8 Testor G., Schild H., 1990, Astron. Astrophys., in press.
9 Walborn N., 1982, Astron. J. 77, 312

DISCUSSION

Conti: I am glad somebody is keeping-up with this, but I have problems with this use of WNA and WNB for narrow and broad lined subtypes. The measured line widths form a continuum of widths. If one is going to use A and B to isolate the extremes, well, alright, but let us not use any "intermediate" A(B) or B(A) or suchlike.

EXPLOSIONS OF HELIUM STARS AND TYPE IB/IC/IIB SUPERNOVAE

K. NOMOTO

Department of Astronomy, Faculty of Science, University of Tokyo, Tokyo 113

ABSTRACT

Theoretical models of supernova explosions of helium stars with various masses are reviewed to examine possible connections between Wolf-Rayet stars and Type Ib/Ic/IIb supernovae. Nucleosynthesis, Rayleigh-Taylor instabilities, and light curves are compared with observations. Maximum brightness and the fast decline of the light curves of typical SNe Ib/Ic can be well accounted for by the helium star models if the helium star mass is as low as 3–5 $M_\odot$. These low mass helium stars can form from stars of 12–18 $M_\odot$ after Roche-lobe overflow in close binary systems. Probably progenitors of typical SNe Ib/Ic are not classified as Wolf-Rayet stars.

1. TYPE IB/IC SUPERNOVAE

Supernova explosions of Wolf-Rayet stars may in principle be observed as Type I supernovae (SNe I) since their maximum light spectra must lack of hydrogen lines. In particular, Wolf-Rayet stars have been suggested to be promising candidates for the progenitors of Type Ib/Ic supernovae (SNe Ib/Ic), since most of SNe Ib are associated with star-forming regions (see, e.g., Harkness and Wheeler 1990 for a review and references).

Unlike the classical Type I supernovae (now designated as SNe Ia), SNe Ib exhibit strong absorption lines of He I at early times, and prominent emission lines of [O I], [Ca II], and Ca II at late times. SNe Ic spectroscopically resemble SNe Ib, except that the He I lines are absent at early times (Harkness et al. 1987). The light curves of SNe Ib/Ic, generally similar to SNe Ia, are very likely to be powered by the decays of ^{56}Ni and ^{56}Co. Well-observed SNe Ib are about 1.5 mag fainter than SNe Ia at maximum (Uomoto and Kirshner 1985; Branch 1986; Panagia 1987).

These spectroscopic and photometric features have let to the Wolf-Rayet star models with a wide range of masses (Wheeler and Levreault 1985; Begelman and Sarazin 1986; Gaskel et al. 1986; Uomoto 1986; Schlegel and Kirshner 1989; Schaeffer, Casse, and Cahen 1987; Ensman and Woosley 1988; Nomoto et al. 1988a,b).

K. A. van der Hucht and B. Hidayat (eds.),
Wolf-Rayet Stars and Interrelations with Other Massive Stars in Galaxies, 515–528.
© 1991 *IAU. Printed in the Netherlands.*

However, the previous Wolf-Rayet star models have some difficulties 1) in reproducing the light curves of typical SNe Ib which decline as fast as SNe Ia (Panagia 1987; Leibundgut 1988), and 2) in producing enough ^{56}Ni to attain the maximum luminosities of SNe Ib in relatively low mass helium star models (Ensman and Woosley 1988). In particular, the light curve of SN Ic 1987M has two important features that challenge the Wolf-Rayet models: (1) maximum brightness is only 0.6 mag dimmer than in SNe Ia, and significantly brighter than in typical SNe Ib; and (2) the decline of the light curve is faster than those of SNe Ia and SNe Ib (Filippenko et al. 1990; also SN Ic 1983I reported by Tsvetkov 1985).

Recently Shigeyama et al. (1990) have shown that the fast declines of the SNe Ib/Ic light curves are well reproduced if mixing of ^{56}Ni occurs in 3–5 $M_\odot$ helium stars. Here we review nucleosynthesis, Rayleigh-Taylor instabilities, and light curves of exploding helium stars, and conclude that the helium stars of 3–5 $M_\odot$ (which form from stars with initial masses $M_i \sim$ 12–18 $M_\odot$ in binary systems) are the most likely progenitors of typical SNe Ib/Ic.

2. NUCLEOSYNTHESIS IN EXPLODING HELIUM STARS AND THE MASS OF ^{56}Ni

In the helium star models for SNe Ib, two scenarios are possible for the presupernova evolution. 1) A fairly massive single star lost its hydrogen-rich envelope in a wind. 2) A star in a close binary system becomes a helium star by Roche-lobe overflow.

Shigeyama et al. (1990) adopted the second scenario and performed hydrodynamical calculations of the explosion of the helium stars of masses $M_\alpha = 3.3$, 4, and 6 $M_\odot$, which are presumed to form from the main-sequence stars of masses $M_i \sim$ 13, 15, and 20 $M_\odot$, respectively. These stars eventually undergo iron core collapse as in Type II supernovae (SNe II). A shock wave is then formed at the mass cut that divides the neutron star and the ejecta, propagating outward to explosively synthesize ^{56}Ni and other heavy elements (e.g., Hashimoto et al. 1989; Thielemann et al. 1990).

Since the mechanism that transforms collapse into explosion is unclear, the mass cut and explosion energy are not known. The adopted presupernova models (Nomoto and Hashimoto 1988) have the following important difference from the previous models, i.e., the iron core masses are as small as 1.18 $M_\odot$ and 1.28 $M_\odot$ for $M_\alpha = 3.3 M_\odot$ and 4 $M_\odot$, respectively, significantly smaller than 1.4 $M_\odot$ in the 6 $M_\odot$ star, due to the larger effect of Coulomb interactions during the progenitor's evolution. Because of steep density gradient at the outer edge of the iron core, it is reasonable to assume that the neutron star mass M_{NS} is equal to the iron core mass (Shigeyama et al. 1990). The final kinetic energy of explosion is assumed to be $E = 1 \times 10^{51}$ erg.

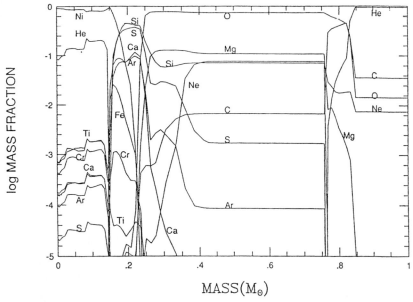

Fig. 1: Explosive nucleosynthesis in the 4 $M_\odot$ helium star (Shigeyama *et al.* 1990). Composition of the innermost 1 $M_\odot$ of the ejecta is shown. (The outermost 1.72 $M_\odot$ helium layer and the 1.28 $M_\odot$ neutron star are not included in the figure.) About 0.15 $M_\odot$ ^{56}Ni and 0.43 $M_\odot$ oxygen are produced.

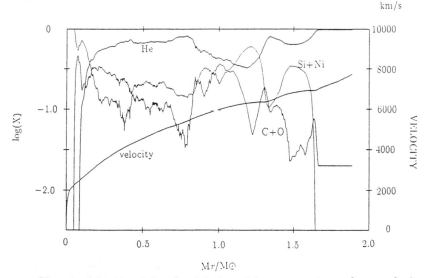

Fig. 3: Chemical composition for $M_\alpha = 3\ M_\odot$ at $t = 180$ s after explosion is plotted against the radial mass coordinate, M_r. Mass fractions of He, C+O, and Si+Ni are shown by the thick, intermediate, and the thin lines, respectively. The mean radial velocity is also shown (thick solid line). ^{56}Ni is mixed to the layer at $M_r = 1.7 M_\odot$, i.e., 0.4 $M_\odot$ beneath the surface (Hachisu *et al.* 1990b).

Nucleosynthesis behind the shock front is determined by the temperature T, which is approximately given by $E = 4\pi r^3/3 \; aT^4$ for the sphere of radius r (e.g., Woosley 1988). For $T > 5 \times 10^9$ K, materials are processed into nuclear statistical equilibrium (NSE) composition, mostly ^{56}Ni. This region corresponds to a sphere of radius $\sim 3700 \; (E/10^{51} \; \mathrm{erg})^{1/3}$ km which contains a mass M_{NSE} (~ 1.44–1.46 $M_\odot$ for $M_\alpha = 3.3$ and 4 $M_\odot$). Therefore the mass of ^{56}Ni is approximately given by $M_{\mathrm{NSE}} - M_{\mathrm{NS}}$. Distribution of the nucleosynthesis products for $M_\alpha = 4 \; M_\odot$ is shown in Fig. 1. The ^{56}Ni masses are 0.26 and 0.15 $M_\odot$ for $M_\alpha = 3.3$ and 4 $M_\odot$, respectively, which are large enough to account for maximum brightness of SNe Ib. The oxygen masses are 0.21 and 0.43 $M_\odot$ for $M_\alpha = 3.3$ and 4 $M_\odot$, respectively, and could be consistent with those inferred from the late time spectra of SNe Ib/Ic because the latter is sensitive to the temperature of the ejecta (e.g., Uomoto 1986).

The above results are in contrast to the previous models (Ensman and Woosley 1988) in which the mass of ^{56}Ni synthesized by helium stars decreases with decreasing progenitor mass. Given the constraint that SN 1987A produced only 0.07 $M_\odot$ of ^{56}Ni (e.g., Shigeyama et al. 1988; Woosley 1988), the helium star progenitor of SNe Ib, which produce $\sim 0.15 \; M_\odot \; {}^{56}$Ni to attain the observed maximum luminosities, must have been more massive than 6 $M_\odot$; this is incompatible with the requirement from the light curve shape (§4).

3. RAYLEIGH-TAYLOR INSTABILITIES AND MIXING

As will be shown in §5, the helium star models are in good agreement with the observed SNe Ib light curves only if the extensive mixing of ^{56}Ni takes place. Mixing and clumpiness in SNe Ib is also inferred from the late time emission line features (Fransson and Chevalier 1989; Filippenko and Sargent 1989). Such a mixing of radioactive elements has been observed in SN 1987A (e.g., Kumagai et al. 1989 and references therein) and successfully accounted for by the Rayleigh-Taylor instability in the explosion (Arnett et al. 1989; Hachisu et al. 1990a; Den et al. 1990; Yamada et al. 1990; Fryxell et al. 1991).

The Rayleigh-Taylor instability in helium stars develop as follows. When the shock wave hits the helium envelope, the expansion of the inner core is largely decelerated, which forms a reverse shock. Then a pressure inversion appears (i.e., the pressure increases outward) in the layer between the forward shock and the reverse shock. The interface between the core and the helium envelope becomes most strongly Rayleigh-Taylor unstable because the density decreases outward steeply and thus $(dP/dr)(d\rho/dr) < 0$. The instability continues to grow until the forward shock reaches the low density surface; then a rarefaction wave propagates inward from the surface to stabilize the interior. Note that in the 20 $M_\odot$ model of SN 1987A by Hachisu et al. (1990a), the most unstable is the hydrogen/helium interface due to the massive hydrogen-rich envelope.

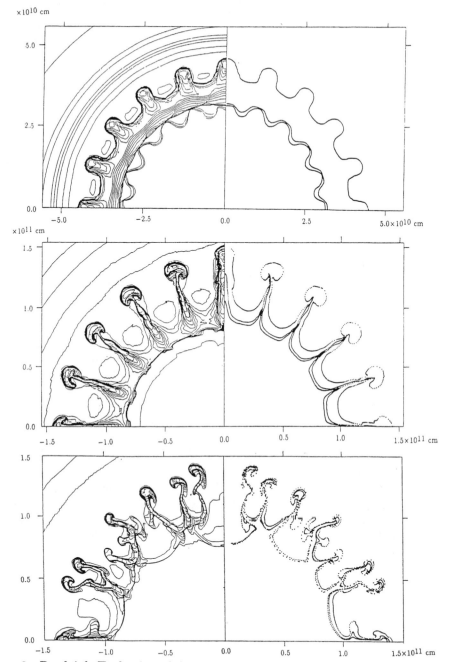

Fig. 2: Rayleigh-Taylor instabilities in the exploding helium stars of $M_\alpha = 6\ M_\odot$ ($t = 84$ s) (a), $4\ M_\odot$ ($t = 200$ s) (b), and $3.3\ M_\odot$ ($t = 180$ s) (c). Shown are the density contour map (*left*) and the marker particles at the composition interfaces, He/C+O, O/Si, and Ni/Si from the outerside (*right*) (Hachisu *et al.* 1990b).

For the helium stars of $M_\alpha = 3.3$, 4, and 6 $M_\odot$, Hachisu *et al.* (1990b) have carried out 2D hydrodynamical calculations to follow the Rayleigh-Taylor instability. As seen in Fig. 2a–c, the instability leads to only a limited mixing and clump formation for $M_\alpha = 6$ $M_\odot$, while it does induce a large scale mixing for $M_\alpha = 3.3$ and 4 $M_\odot$. For $M_\alpha = 3.3$ $M_\odot$, ^{56}Ni is mixed to the layer of 0.4 $M_\odot$ beneath the surface (Fig. 3), which is close to the extent of mixing as required from the light curves.

Such a mass dependence of the Rayleigh-Taylor instability can be understood from the difference in the stellar structure as follows.
1) For smaller M_α the mass ratio between the helium envelope and the core (excluding the neutron star) is larger (i.e., 2.5, 2.7, 1.0, and 0.45 for $M_\alpha = 3.3$, 4, 6, and 8 $M_\odot$, respectively) so that the deceleration of the core and the pressure inversion are larger.
2) Smaller mass stars have steeper density gradient near the composition interface.
3) The stellar radius is larger for smaller M_α, so that it takes longer for the shock wave to reach the stellar surface and the instability grows for a longer time.

We should emphasize the importance of the density structure rather than stellar mass. For example, a single Wolf-Rayet star which reduces its mass down to 4–5 $M_\odot$ by wind could be a SNe Ib/Ic progenitor (Langer 1989; Maeder 1990). However, such a star would not undergo extensive mixing despite the small mass, because its helium envelope would be too small to largely decelerate the core.

4. LIGHT CURVE MODELS FOR TYPE IB SUPERNOVAE

Figures 4 and 5 show the calculated bolometric light curves of the exploding helium stars with $M_\alpha = 3.3$, 4, and 6 $M_\odot$ for two cases of the elemental distribution: 1) the original stratified composition structure with ^{56}Ni being confined in the innermost region (Fig. 4) and 2) the mixed distribution with almost uniform distribution of elements (Fig. 5). In order to clarify the dependence of the light curve shape on M_α and mixing, the amount of ^{56}Ni is assumed to be 0.15 $M_\odot$ for all M_α.

These figures also show three observed light curves: SN Ib 1984L (visual: Wheeler and Levreault 1985), SN Ib 1983N (bolometric: Panagia 1987), and SN Ic 1983I (visual: Tsvetkov 1985). The visual light curve is regarded to be close to the bolometric one. The maximum luminosities and their dates are shifted to those of the corresponding calculated curves.

In the calculated bolometric light curves, an ultraviolet burst occurs at the shock breakout, but it is too brief to be observed because of the small radius of the progenitor. Later the light curve is powered by the radioactive decays of ^{56}Ni and ^{56}Co. Peak luminosity is reached when the time scale of heat diffusion from the radioactive source becomes comparable to the expansion time scale. Maximum brightness is higher if the ^{56}Ni mass is larger and the date of maximum earlier. After the peak,

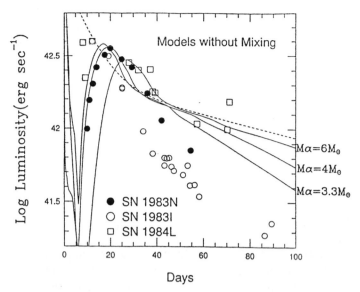

Fig. 4: Calculated light curves of exploding helium stars of $M_\alpha = 3.3$, 4, and 6 $M_\odot$ (Shigeyama *et al.* 1990). All the models assume the production of 0.15 $M_\odot$ ^{56}Ni, kinetic energy of explosion $E = 1 \times 10^{51}$ erg, and no mixing (i.e., stratified composition structure). Filled and open circles are the observed light curves of SN Ib 1984L (visual), 1983N (bolometric), and 1983I (visual). The dotted curve is the energy generation rate of the ^{56}Ni-^{56}Co decays.

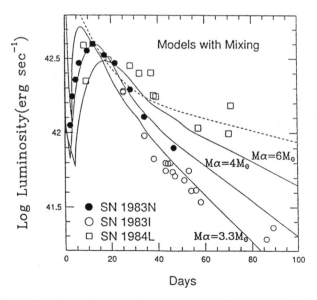

Fig. 5: Same as Fig. 4 but with *mixed* abundance distribution similar to Fig. 3.

the optical light curve declines at a rate that depends on how fast gamma-rays from the radioactive decays escape from the star without being thermalized. Since the escape probability of gamma rays is determined by the column depth to the Ni–Co layer, the optical light curve declines faster if the ejected mass is smaller and if ^{56}Ni is mixed closer to the surface.

For the unmixed cases in Fig. 4, the calculated light curve tail declines faster for smaller M_α. However, these tails (even the fastest decline of $M_\alpha = 3.3\ M_\odot$) are much slower than that of SN 1983N and 83I after $\sim$ day 40.

For the mixed cases in Fig. 5, the light curve shape is significantly different from the unmixed cases:
1) The maximum luminosity is reached $\sim$ 10 days earlier and thus being higher than in the unmixed cases because of the earlier radioactive heating of the surface layers.
2) The decline of the tail is much faster than the unmixed cases.

Figure 5 shows that the bolometric light curve of $M_\alpha = 4\ M_\odot$ is in excellent agreement with SN Ib 1983N from the pre-maximum through day 50. The calculated curve for 50–100 days continues to be close to the bolometric light curve obtained by Leibundgut (1988). Other well-observed SNe Ib generally have early-time visual and bolometric light curves whose shapes are nearly identical to those of SNe Ia (Porter and Filippenko 1987; Leibundgut 1988). On the other hand, the light curve of the mixed 3.3 $M_\odot$ model, rising and declining faster than SN 1983N, is in good agreement with SN Ic 1983I (see the next section).

5. LIGHT CURVE MODELS FOR TYPE IC SUPERNOVAE

As mentioned in §1, the light curve of SN Ic 1987M provided new constraints on the models (Nomoto et al. 1990). In Fig. 6 the filled circles show a quasi-bolometric (3280–9100 Å) light curve constructed from flux-calibrated spectra, where maximum brightness is assumed to be subluminous relative to SNe Ia by 0.6 mag as observed in the B band (Filippenko et al. 1990). Figure 6 also shows the observed bolometric curves of SNe Ia 1972E and 1981B, and of SN Ib 1983N for comparison. In each case, the observed light curve has been shifted along the abscissa to match the corresponding theoretical curve. The peak bolometric luminosities of the two SNe Ia are assumed to match the theoretical predictions of model W7 which corresponds to $H_0 = 60$ km s^{-1} Mpc^{-1} (Arnett et al. 1985; Nomoto 1986).

Figure 6 demonstrates two important features of SN Ic 1987M: 1) its brightness fell somewhat more rapidly than that of SNe Ia and SN Ib 1983N, and 2) SN 1987M was considerably more luminous than SN 1983N, which was $\sim$ 1.5 mag fainter than SNe Ia at maximum (Panagia 1987). These features are very difficult to understand with the previous massive helium star models.

In Fig. 6, illustrated is the light curve of the exploding 3.3 $M_\odot$ helium star with

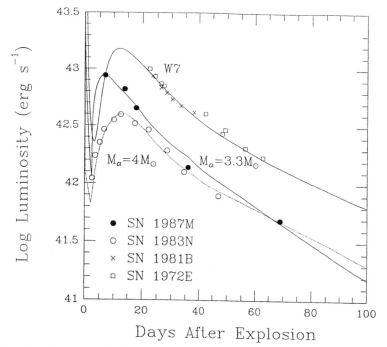

Fig. 6: Quasi-bolometric ($\lambda\lambda 3280$–9100) light curve of SN Ic 1987M, together with the bolometric light curves of SNe Ia 1981B and 1972E (Graham 1987) and of SN Ib 1983N (Panagia 1987). The predicted curves of the 3.3 $M_\odot$ model for SN 1987M, the 4 $M_\odot$ model for SN Ib, and the W7 model for SN Ia are indicated by solid and dotted lines (Nomoto *et al.* 1990).

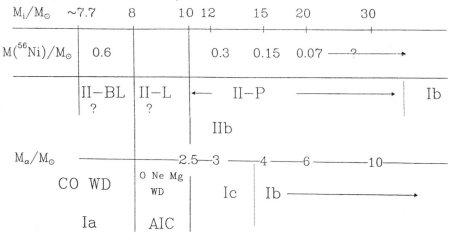

Fig. 7: Hypothetical connection between supernova types and their progenitors for single stars (upper) and close binary stars (lower). $M_{\rm i}$ and M_α are the initial mass and the helium star mass, respectively. AIC stands for accretion-induced collapse of white dwarfs.

0.26 $M_\odot$ of ^{56}Ni being mixed almost uniformly from the center through the layer at 0.2 $M_\odot$ beneath the surface. It is clear that the calculated bolometric light curve of the 3.3 $M_\odot$ model closely matches the quasi-bolometric curve of SN 1987M, both in shape and in peak luminosity. Figure 6 also shows the 4 $M_\odot$ helium star model for SNe Ib (where the ^{56}Ni mass is 0.15 $M_\odot$) and the white dwarf model W7 for SNe Ia (where the ^{56}Ni mass is 0.58 $M_\odot$; Nomoto et al. 1984).

Compared with the 4 $M_\odot$ model for SNe Ib, the 3.3 $M_\odot$ model has several attractive features for SN 1987M. First, maximum brightness is higher by more than a factor of two because of the larger ^{56}Ni mass and the earlier date of maximum brightness. Second, decline in the tail is noticeably faster due to the more extensive mixing and smaller ejected mass. Compared with the W7 model for SNe Ia, the 3.3 $M_\odot$ model gives a lower maximum brightness (by about 0.6 mag) and a faster decline, just as observed in SN 1987M.

6. PROGENITORS OF TYPE IB/IC/IIB SUPERNOVAE

Figure 7 summarizes the currently proposed initial masses M_i of the progenitors for the various types of supernovae. The upper and lower rows respectively show the cases of single stars and helium stars of masses M_α (or white dwarfs) in close binary stars. The produced masses of ^{56}Ni inferred from light curves are also given.

The main conclusion of the previous sections shown in Fig. 7 is that typical SNe Ib and Ic are the explosions of helium stars of 3–5 $M_\odot$ in close binary systems (see also Uomoto 1986). The peak luminosities and the fast decline of the light curves of SNe Ib/Ic can be well modeled because these low mass stars 1) eject smaller mass, 2) undergo more extensive mixing, and 3) synthesize larger amount of ^{56}Ni. In particular, the light curve of SN Ic 1987M supports the theoretical prediction (Shigeyama et al. 1990) that low-mass helium stars in close binary systems produce larger amounts of ^{56}Ni than do high-mass helium stars.

Helium stars with 3–5 $M_\odot$ can form from stars of 12–18 $M_\odot$ after Roche-lobe overflow in close binary systems. The fact that SN Ic 1987M was superposed on a spiral arm of NGC 2715, but not on a very luminous H II region (Filippenko et al. 1990), is consistent with a progenitor having this initial mass. It is unlikely that these supernova progenitors would have been classified as Wolf-Rayet stars with broad emission lines, since their mass-losing wind was probably weak.

Radio observations of some SNe Ib/Ic 1983N, 1984L, and 1990B imply that the progenitors of these supernovae had dense circumstellar shells (Sramek et al. 1984). In the 3–5 $M_\odot$ helium star models, mass exchange during the early binary evolution may make the companion a massive OB star. The wind from such a companion star could interact with the progenitor's wind to form a dense circumstellar shell around the helium star progenitor. Also some gas could have been left after previous episodes of mass exchange.

For larger M_α, helium stars of $M_\alpha \sim 6\ M_\odot$ produces too small amount of ^{56}Ni ($\sim 0.07\ M_\odot$) to be consistent with SNe Ib. More massive helium stars might produce larger amount of ^{56}Ni. Also extensive mass loss from very massive stars could result in the formation of Wolf-Rayet stars with final masses as small as 4–5 $M_\odot$ (Langer 1989; Maeder 1990). However, such a progenitor does not undergo a strong Rayleigh-Taylor instability, since its helium envelope is not sufficiently massive to greatly decelerate the core. Without mixing and the associated formation of clumps, the resulting light curve is too broad to explain SNe Ib/Ic. Furthermore, the birth rate of very massive stars is too small to explain the occurrence frequency of SNe Ib/Ic (Evans et al. 1989).

Figure 7 suggests that SNe Ib/Ic and typical Type II-P supernovae (SNe II-P) originate from the similar stellar mass range, where SNe Ib are from close binaries while SNe II-P are single stars. It is suggestive that the light curve tail of the best observed SN II-P 1969L follows the ^{56}Co decay rate and the inferred ^{56}Ni mass is $\sim$ 0.13 $M_\odot$ if $H_0 \sim 60$ (Kirshner and Kwan 1974); this is close to the ^{56}Ni mass from SNe Ib.

It is interesting to compare SNe Ib/Ic and SN 1987K whose spectral classification changed from Type II to Type Ib/Ic as it aged, thereby being called as SNe IIb (Filippenko 1988). The decline of the SN 1987K light curve is as fast as SNe Ia (and thus SNe Ic) (Turatto et al. 1990). The early time spectra of SN 1987K are very similar to SNe Ic (1983V and 87M), and, very recently, hydrogen feature is identified in the early time spectrum of SN Ic 1987M (Jeffery and Branch 1990). This strongly suggests that the difference in the spectral feature between SNe Ic and Ib is due to the presence of a thin envelope of hydrogen in SNe Ic immediately prior to the explosion. Some hydrogen can be left on the helium star after mass exchange for a certain combination of the initial mass and the close binary separation (Yamaoka and Nomoto 1990).

To examine the above hypothesis, theoretical spectra should take into account non-LTE excitation of hydrogen and helium lines due to mixed radioactive materials (Branch 1988; Wheeler and Harkness 1990). Differences in relative distribution of H, He, and ^{56}Ni and their amounts after mixing might be related to the spectral differences among SNe Ib, Ic, and IIb.

White dwarfs have also occasionally been considered as progenitors of SNe Ic. However, the light curve produced by the off-center single detonation model (Branch and Nomoto 1986; Nomoto 1982; Woosley et al. 1986) declines too fast. The carbon deflagration model with slow flame speed (Woosley 1990) releases smaller kinetic energy than do SNe Ia, thereby forming a light curve which declines more slowly than that of SNe Ia. This appears to be inconsistent with observations of SN 1987M (Fig. 4). Therefore supernova explosions of accreting C+O white dwarfs become mostly SNe Ia and O+Ne+Mg white dwarfs undergo accretion-induced collapse (AIC).

SNe II-L (linear) and II-BL (bright linear, like SN 1979C; Branch, private communication) do not show a plateau, thereby being suggested to have hydrogen-rich envelope of smaller than $\sim 1\ M_\odot$. In Fig. 7, SNe II-BL and II-L are tentatively assumed to be the explosions of AGB stars having degenerate C+O cores (carbon deflagration) and O+Ne+Mg cores (electron capture collapse), respectively (Swartz et al. 1990), since the AGB stars have lost most of their hydrogen-rich envelopes at the explosion. However, the progenitors having small hydrogen might also be related to the Wolf-Rayet stars.

I would like to thank T. Shigeyama, I. Hachisu, T. Matsuda, T. Tsujimoto, M. Hashimoto, F.-K. Thielemann, and A.V. Filippenko, for collaborative work on SNe Ib/Ic. I am also grateful to D. Branch and J.C. Wheeler for stimulating discussion on this subject. This work has been supported in part by the grant-in-Aid for Scientific Research (01540216, 01790169, 02234202, 02302024) of the Ministry of Education, Science, and Culture in Japan, and by the Japan-U.S. Cooperative Science Program (EPAR-071/88-15999) operated by the JSPS and the NSF.

REFERENCES

Arnett, W.D., Branch, D., Wheeler, J.C. 1985, *Nature* **314**, 337

Arnett, W. D., Fryxell, B. A., and Müller, E. 1989, *Ap. J. (Letters)*, **341**, L63.

Begelman, M.C., and Sarazin, C.L. 1986, *Ap. J. (Letters)*, **302**, L59.

Branch, D. 1986, *Ap. J. (Letters)*, **300**, L51.

———. 1988, in *IAU Colloquium 108, Atmospheric Diagnostics of Stellar Evolution*, ed. K. Nomoto, *Lecture Notes in Physics*, **305**, 281.

Branch, D., and Nomoto, K. 1986, *Astr. Ap.*, **164**, L13.

Den, M., Yoshida, T., and Yamada, Y. 1990, preprint (KUNS1004).

Evans, R., van den Bergh, S., McClure, R.D. 1989, *Ap. J.* **345**, 752

Ensman, L., and Woosley, S.E. 1988, *Ap. J.*, **333**, 754.

Filippenko, A.V. 1988, *Astr. J.*, **96**, 1941.

Filippenko, A.V., Porter, A.C., and Sargent, W.L.W. 1990, *Astr. J.*, in press.

Filippenko, A.V., and Sargent, W.L.W. 1989, *Ap. J. (Letters)*, **345**, L43.

Fransson, C., and Chevalier, R.A. 1989, *Ap. J.*, **343**, 323.

Fryxell, B.A., Arnett, W.D., and Müller, E. 1991, *Ap. J.*, in press.

Gaskell, C.M., Cappellaro, E., Dinerstein, H.L., Garnett, D.R., Harkness, R.P., and Wheeler, J.C. 1986, *Ap. J. (Letters)*, **306**, L77.

Graham, J.R. 1987, *Ap. J.*, **315**, 588.

Hachisu, I., Matsuda, T., Nomoto, K., and Shigeyama, T. 1990a, *Ap. J. (Letters)*, **358**, L57

———. 1990b, *Ap. J. (Letters)*, in press.

Harkness, R.P., and Wheeler, J.C. 1990, in *Supernovae*, ed. A. Petschek (Springer-Verlag), p. 1.

527

Harkness, R.P. *et al.* 1987, *Ap. J.*, **317**, 355.

Hashimoto, M., Nomoto, K., and Shigeyama, T. 1988, *Astr. Ap.*, **210**, L5

Jeffery, D., and Branch, D. 1990, private communication.

Kirshner, R.P., and Kwan, J. 1974, *Ap. J.*, **193**, 27.

Kumagai, S., Shigeyama, T., Nomoto, K., Itoh, M., Nishimura, J., and Tsuruta, S. 1989, *Ap. J.*, **345**, 412.

Langer, N. 1989, *Astr. Ap.*, **220**, 135.

Leibundgut, B. 1988, Ph.D. thesis, Universität Basel.

Maeder, A. 1990, *Astr. Ap. Suppl.*, in press.

Nomoto, K. 1982, *Ap. J.*, **257**, 780.

———. 1986, *Ann. New York Acad. Sci.* **470**, 294

Nomoto, K., Filippenko, A.V., and Shigeyama, T. 1990, *Astr. Ap.*, submitted.

Nomoto, K., and Hashimoto, M. 1988, *Physics Reports*, **163**, 13.

Nomoto, K., Shigeyama, T., and Hashimoto, M. 1988a, in *IAU Colloquium 108, Atmospheric Diagnostics of Stellar Evolution*, ed. K. Nomoto, *Lecture Notes in Physics*, **305**, 319.

Nomoto, K., Shigeyama, T., Kumagai, S., and Hashimoto, M. 1988b, in *Proc. Astr. Soc. Australia*, **7**, 490.

Panagia, N. 1987, in *High Energy Phenomena Around Collapsed Stars*, ed. F. Pacini (D. Reidel), p. 33.

Porter, A.C., and Filippenko, A.V. 1987, *Astr. J.* **93**, 1372

Schaeffer, R., Casse, M., and Cahen, S. 1987, *Ap. J.*, **316**, L31.

Schlegel, E.M., and Kirshner, R.P. 1989, *Astr. J.*, **98**, 577.

Shigeyama, T., Nomoto, K., and Hashimoto, M. 1988, *Astr. Ap.*, **196**, 141.

Shigeyama, T., Nomoto, K., Tsujimoto, T., and Hashimoto, M. 1990, *Ap. J. (Letters)*, in press.

Sramek, R.A., Panagia, N., and Weiler, K.W. 1984, *Ap. J. (Letters)*, **285**, L59.

Swartz, D.A., Wheeler, J.C., and Harkness, R.P., 1990, *Ap. J.*, submitted.

Thielemann, F.-K., Hashimoto, M., and Nomoto, K. 1990, *Ap. J.*, **349**, 222.

Tsvetkov, D.Yu. 1985, *Sov. Astr.*, **29**, 211.

Turatto, M., Cappellaro, E., Barbon, R., Della Valle, M., Ortolani, S., and Rosino, L. 1990, *Astr. J.*, in press.

Uomoto, A. 1986, *Ap. J. (Letters)*, **310**, L35.

Uomoto, A., and Kirshner, R.P. 1985, *Astr. Ap.*, **149**, L7.

Wheeler, J.C., and Levreault, R. 1985, *Ap. J. (Letters)*, **294**, L17.

Wheeler, J.C., and Harkness, R. 1990, *Phys. Rep.*, in press.

Woosley, S.E. 1988, *Ap. J.*, **330**, 218.

———. 1990, In *Supernovae*, ed. A. Petschek (Springer-Verlag), p. 182.

Woosley, S.E., Taam, R.E., and Weaver, T.A. 1986, *Ap. J.*, **301**, 601.

Yamada, Y., Nakamura, T., and Oohara, K. 1990, preprint (KUNS1019).

Yamaoka, H., and Nomoto, K. 1990, in preparation.

DISCUSSION

Vanbeveren: A star ($\sim 13 M_\odot$) in a close binary still has some hydrogen in the outer layers after the Roche lobe overflow process. Can you make a rough guess how much H may still be present in order not to see it during the SN?

Nomoto: From the similarity between the early time spectrum of Type Ic SN1987M and that of Type IIb SN1987K, I have speculated that a small amount of hydrogen might exist in Type Ic supernovae. That is why I speculate that SN IIb might also originate from low-mass helium stars. Either hydrogen or mixed heavy elements may be the source of difference between the Ic and Ib spectra. Non-LTE calculation of synthetic spectra with non-thermal radioactive heating is necessary to estimate a limit to the hydrogen mass.

Shara: How large were the asymmetries and/or perturbations you assumed in order to get the plumes you showed?

Nomoto: We applied two types of perturbations, a periodic sinusiodal one and a random one, to the velocity by 5 percent. We are now studying how the results depend on the magnitude and wavelength of perturbation with more accurate methods.

Matteucci: What is the physical reason for having the amount of iron produced by massive stars decreasing with initial stellar mass?

Nomoto: For the evolution of smaller mass stars the effect of Coulomb interaction to reduce pressure is larger, which eventually makes the iron core mass smaller (as small as $1.18\ M_\odot$ for the $13 M_\odot$ star. This results in the formation of a more massive Si-rich layer above the iron core. If the neutron star mass is close to the iron core mass at the explosion, larger mass of ^{56}Ni is produced from the Si-rich layer. This is the case for the 12-16$M_\odot$ stars. However, the mass of ejected iron depends strongly on the mass cut that divides the neutron star and the ejecta, which can be constrained only by the observations until the explosion mechanism becomes clear. The rather small mass of ^{56}Ni in SN1987A and the faintness of the Cas A explosion suggest that the mass cut could be significantly larger than the iron core mass for stars more massive than $20 M_\odot$.

Maeder: I would appreciate a few comments from you regarding the observed abundances in SN1987A and how they compare with your nucleosynthetic yields.

Nomoto: The abundances obtained from explosive nucleosynthesis calculation for the $20 M_\odot$ model of SN1987A (Nomoto *et al.*, 1990) are in good agreement with the abundances determined from the ESO observations (Dantziger, 1990).

ARE WOLF-RAYET STARS THE PROGENITORS OF TYPE Ib/Ic SUPERNOVAE?

ALEXEI V. FILIPPENKO
Department of Astronomy, and
Center for Particle Astrophysics
University of California
Berkeley, CA 94720 U.S.A.

ABSTRACT.   I discuss evidence for and against the hypothesis that Type Ib and Type Ic supernovae (SNe) are produced by core collapse in massive, evolved progenitors. A key object is SN 1987K, whose spectroscopic classification changed from Type II to Type Ib/Ic as it aged. The progenitor of SN 1987K may well have been a massive star which experienced incomplete mass loss, leaving a thin outer envelope of hydrogen. However, several arguments are used to conclude that in most SNe Ib/Ic, the pre-supernova mass loss cannot be caused entirely by strong winds as in Wolf-Rayet stars. Mass transfer in close binary systems is probably important, but in such cases the supernova progenitor is not necessarily a Wolf-Rayet star; instead, it may be a relatively quiescent, hot, low-mass helium star that explodes via core collapse. For example, the rapid decline of the light curve of the Type Ic SN 1987M, and its seemingly low ejected mass, are consistent with this idea. It is also possible that some, but not all, SNe Ib/Ic arise from deflagrations or detonations of white dwarfs.

1. INTRODUCTION

The defining property of Type I supernovae (SNe I) is that their optical spectra do not exhibit lines of hydrogen, unlike spectra of SNe II. Three observationally distinct subtypes of SNe I have been identified. During the first month past maximum, "classical SNe I" (now called SNe Ia) show a deep trough near 6150 Å which is believed to be produced by blueshifted Si II λ6355. The early-time spectra of SNe Ib do not show this feature; instead, strong absorption lines of He I are present (Harkness et al. 1987). SNe Ic lack both the 6150 Å trough and the He I lines (Wheeler and Harkness 1986). Although physical continuity between SNe Ib and Ic has not yet been proved, some investigators refer to SNe Ic as "helium-poor SNe Ib" (Wheeler et al. 1987).

Several months past maximum, the optical spectra of SNe Ia are dominated by blends of hundreds of emission lines, primarily those of [Fe II], [Fe III], and [Co III]. By contrast, late-time spectra of SNe Ib and Ic exhibit strong, relatively unblended emission lines of [O I], [Ca II], and Ca II, with weaker lines of Mg I], Na I, O I, and [C I]; see Figure 2 of Filippenko (1988a), or Figure 3 of Filippenko, Porter, and Sargent (1990; hereafter FPS). No observational differences have yet been found between spectra of SNe Ib and Ic during their "supernebular" phase.

Aside from these defining spectroscopic characteristics, SNe Ib and Ic are known to differ in many ways from SNe Ia. Whereas SNe Ia are generally found in elliptical galaxies and the interarm regions of spiral galaxies, all known SNe Ib and Ic have occurred in late-type spiral galaxies (Sb through Sd), usually in the spiral arms and quite often in luminous

K. A. van der Hucht and B. Hidayat (eds.),
Wolf-Rayet Stars and Interrelations with Other Massive Stars in Galaxies, 529–536.
© 1991 IAU. Printed in the Netherlands.

H II regions. Their blue luminosities are generally lower than those of SNe Ia by 1–2 mag, probably indicating a smaller yield of radioactive Ni^{56}, and their colors near maximum brightness are redder. Their infrared light curves differ significantly from those of SNe Ia, although the optical light curves of most SNe I are similar. Unlike SNe Ia, at least some SNe Ib and Ic exhibit strong radio emission within one year after the explosion. This radiation is produced by the interaction of the ejecta with a circumstellar envelope.

SNe Ia are thought to arise from carbon deflagrations of white dwarfs in binary systems. Most of the observed characteristics of SNe Ib and Ic, on the other hand, suggest that their progenitors are massive, hydrogen-deficient stars, and that the explosion mechanism is core collapse as in SNe II. Specifically, this hypothesis appears to be consistent with (a) the absence of hydrogen, (b) the absence of Si II at early times, (c) the strength of emission lines of intermediate-mass elements at late times, (d) the proximity to spiral arms and H II regions, (e) the low luminosity compared with SNe Ia, and (f) the presence of radio emission. Indeed, many authors (e.g., Begelman and Sarazin 1986; Filippenko and Sargent 1986; Gaskell *et al.* 1986; Schaeffer, Cassé, and Cahen 1987) have concluded that SNe Ib and Ic represent the explosions of Wolf-Rayet (WR) stars. Particularly impressive are the late-time synthetic spectra computed by Fransson and Chevalier (1989) under the assumption that the SN Ib 1985F resulted from core collapse in a 8 $M_\odot$ helium star [25 $M_\odot$ zero-age main sequence (ZAMS) mass].

Here I critically examine the evidence for and against the hypothesis of massive-star progenitors for SNe Ib and Ic. I am primarily concerned with the distinction between the two main types of explosion *mechanisms* — core collapse versus deflagration/detonation — rather than with the question of whether a particular progenitor has sufficiently prominent mass-losing winds to be formally classified as a WR star. White dwarf models of SNe Ib/Ic include those of Branch and Nomoto (1986), Khokhlov and Ergma (1986), Iben *et al.* (1987), and Woosley (1990).

2. SN 1987K: EVIDENCE FOR MASSIVE PROGENITORS

Observationally, one of the most direct links between SNe Ib/Ic and massive stars is SN 1987K in NGC 4651, an object whose spectroscopic classification changed from Type II to Type Ib/Ic as it aged (Filippenko 1988b). Figure 1 shows a series of spectra of SN 1987K. At early times the P Cygni profile of Hα is unmistakable, although its strength is smaller than usual; see Filippenko (1988a) for a comparison with SN 1987A. At late times, there is no evidence for broad Hα emission. Instead, the spectrum is dominated by [O I], [Ca II], and Ca II emission, as are SNe Ib/Ic. (The continuum is stronger than in most SNe Ib/Ic, but contamination by starlight is probably responsible.) This "metamorphosis" is the only one ever seen in a supernova; however, it is possible that significant numbers of SNe II would be observed to undergo similar transformations if they were monitored over sufficiently long time intervals.

The simplest interpretation (Filippenko 1988b) is that the progenitor of SN 1987K was a massive star that lost most, but not all, of its outer layer of hydrogen. At early times, the photospheric spectrum naturally exhibited hydrogen, but emission lines of intermediate-mass elements from the star's interior began to dominate as the density of the ejecta decreased. A progenitor that suffered greater mass loss would have been even more hydrogen-deficient when it exploded via the core collapse mechanism. Consequently, its maximum-light spectrum would have more closely resembled those of normal SNe Ib/Ic.

If the progenitor of SN 1987K was a very massive star (ZAMS mass $\gtrsim$ 30–40 $M_\odot$, perhaps 20–30 $M_\odot$), either single or a member of a wide binary system, the mass loss would have occurred entirely through winds. Immediately prior to the explosion, the progenitor might have been classified as a late-type WN star (e.g., WN8), provided the outer skin of hydrogen was thin enough. Of course, a very massive star in a close binary system might lose part of its mass through transfer to the companion, yet still have the strong winds typical of WR stars.

Suppose, on the other hand, that the progenitor of SN 1987K had relatively *low* ZAMS mass (e.g., 12–20 $M_\odot$). Most likely, it could not have supported a sufficiently high mass-loss rate through winds to be classified as a WR star. If it was a member of a close binary system, however, it may have lost much of its outer envelope by mass transfer onto its companion. Despite not officially being a WR star, it would have been a relatively massive, hydrogen-deficient star that exploded via core collapse, rather than a white dwarf that exploded via deflagration or detonation.

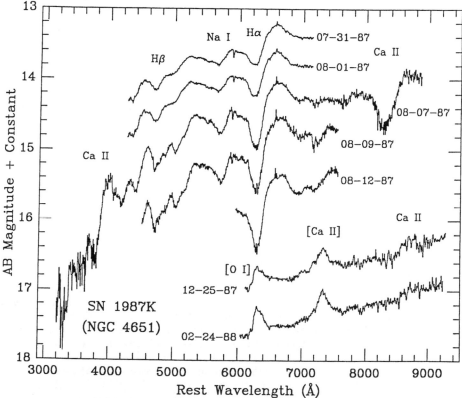

Figure 1: Spectra of SN 1987K, obtained with the 3-m Shane reflector at Lick Observatory. Maximum brightness occurred around 31 July 1987. The redshift of the parent galaxy ($cz = 817$ km s^{-1}) has been removed. AB magnitude $= -2.5 \log f_\nu - 48.6$, where the units of f_ν are ergs s^{-1} cm^{-2} Hz^{-1} (Oke and Gunn 1983). Narrow emission lines, produced by superposed H II regions, have been excised in the late-time spectra for clarity.

3. PROBLEMS WITH THE WOLF-RAYET HYPOTHESIS

There are a number of problems associated with the massive-star hypothesis of SNe Ib/Ic, at least if the progenitors are restricted to *single* stars (or stars in wide binaries) which lose mass through winds alone. These must be examined critically if we are to build confidence in our physical understanding of SNe.

3.1. Progenitor Masses

SNe Ib/Ic are quite *common*; Evans, van den Bergh, and McClure (1989) estimate that collectively, they occur at $\sim 1/4$ the rate of SNe II. (These authors do not distinguish between SNe Ib and SNe Ic.) If all SNe Ib/Ic have very massive progenitors (ZAMS mass $\gtrsim 30$–$40\ M_\odot$) typical of WR stars, there is probably a severe shortage of progenitors. Furthermore, although many SNe Ib/Ic are associated with luminous H II regions, a significant fraction appear to be far from such areas of very active star formation (e.g., SN 1987M in NGC 2715: FPS). Indeed, the poster presented at this symposium by N. Panagia and V. Laidler claims that SNe Ib/Ic and SNe II appear to have approximately the *same* average distance from H II regions. If *all* the progenitors of SNe Ib/Ic were stars with ZAMS mass $> 30\ M_\odot$ (compared with $8 \lesssim M/M_\odot \lesssim 40$ for SNe II), shouldn't SNe Ib/Ic almost *always* be found extremely close to luminous H II regions?

There are three possible solutions to this problem. First, winds in *single* stars of relatively low initial mass ($M \gtrsim 20\ M_\odot$) might, in certain cases, be sufficiently strong to expel a large fraction, if not all, of the outer hydrogen envelope. A more likely possibility, in my opinion, is that a substantial fraction of SNe Ib/Ic arise from relatively low-mass progenitors (e.g., ZAMS mass 10–40 $M_\odot$) in close *binary* systems, with mass transfer being the dominant mass-loss mechanism. As noted for SN 1987K, such a progenitor would not necessarily be a WR star under the usual definition (i.e., evidence for strong mass loss through winds), but it would nevertheless explode via the core collapse mechanism. Finally, *some* SNe Ib/Ic might have white dwarf progenitors.

3.2. Light Curves

Another problem with models involving massive progenitors is that the predicted light curves are usually too broad, compared with observed light curves for helium stars more massive than 4–6 $M_\odot$ (Ensman and Woosley 1988). In conventional models, these helium cores correspond to ZAMS masses of 15–20 $M_\odot$, and mass loss to the WR stage in a single star is probably not feasible. Progenitors having somewhat higher mass might become WR stars, but the remaining helium cores would be too massive according to most calculations.

The poster by N. Langer and S. Woosley at this symposium (see also Langer 1989) may provide a partial solution to the problem. These authors demonstrate that stars having ZAMS mass between 40 $M_\odot$ and 100 $M_\odot$ all end up with helium core masses of 4–6 $M_\odot$, consistent with the requirements Ensman and Woosley (1988) derived from the analysis of light curves. Such stars do, in fact, go through the WR phase. On the other hand, as discussed previously, there are not enough progenitors in the 40–100 $M_\odot$ range to account for the observed frequency of SNe Ib/Ic, and essentially *all* SNe Ib/Ic should be found very close to luminous H II regions. Another problem is that in some cases, such as SN 1987M, the light curve appears to decline even *more* steeply than those of typical SNe I (Fig. 2). These objects seem inconsistent with even the least massive cores discussed by Ensman and Woosley (1988) and in the poster by Langer and Woosley.

In his talk at this symposium, K. Nomoto presented a different solution that may explain many SNe Ib/Ic (see Shigeyama *et al.* 1990). He showed that stars with a ZAMS mass of 12–16 $M_\odot$ in close *binary* systems yield bare helium cores with $M \approx 3$–4 $M_\odot$. These undergo core collapse, explode, and produce light curves with steep slopes, partly because of extensive mixing of Ni^{56} and clumping of the ejecta. Specifically, both the shape of the light curve and the peak luminosity of SN 1987M are well reproduced by such a model (Nomoto, Filippenko, and Shigeyama 1990). Some SNe Ib/Ic could therefore be hot helium stars in close binary systems, as previously discussed in qualitative terms by Uomoto (1986). Of course, it is also possible that white dwarf progenitors may account for those SNe Ib/Ic whose light curves resemble SN Ia light curves.

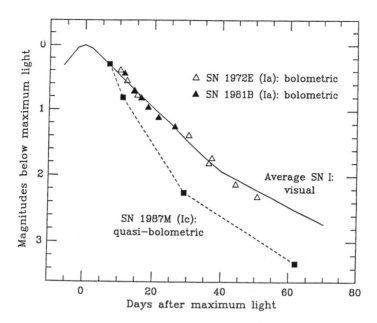

Figure 2: "Quasi-bolometric" ($\lambda\lambda3280$–9100 Å) light curve of SN 1987M (Nomoto, Filippenko, and Shigeyama 1990), compared with the average visual curve of SNe I (Doggett and Branch 1985) and the bolometric measurements of SNe Ia 1972E and 1981B (Graham 1987). The SN 1987M zero points on both axes are uncertain, but the slope of the light curve is unaffected by this. Note that the average early-time *visual* light curve of SNe I (comprised mostly of SNe Ia) provides a good fit to the *bolometric* measurements of SNe Ia. The light curve of SN Ic 1987M, by contrast, is steeper.

3.3. Ejected Mass

The ejected mass of some SNe Ib/Ic appears to be quite small, although there are large uncertainties in the estimates. The mass of ejected oxygen can be calculated from the expression $M_O = 10^8\, f([O\ I])\, D^2 \exp(2.28/T_4)\ M_\odot$, where D is the distance (Mpc) of the supernova, $10^4\,T_4$ is the temperature (K) of the oxygen-emitting gas, and $f([O\ I])$ is the flux (ergs s^{-1} cm^{-2}) of [O I] emission (Uomoto 1986). For some SNe Ib, the mass estimate is quite large, but in the case of SN 1987M it is only $\sim 0.4\ M_\odot$ (FPS). Another method used by FPS gives $M \approx 0.5$–1 $M_\odot$ for the *entire* ejecta. Such small masses seen inconsistent with massive WR progenitors, even if they have pre-supernova masses of only 6 $M_\odot$.

On the other hand, the above method of calculating the oxygen mass depends exponentially on temperature, and the derived number specifically assumed $T = 4700$ K (based on the continuum colors of SN 1987A at a comparable phase in its development). If the temperature were actually 4000 K, we find $M_O \approx 0.94\ M_\odot$, whereas $M_O \approx 2.1\ M_\odot$ if $T = 3500$ K. In principle, the most reliable way of obtaining the temperature is to measure the strength of [O I] $\lambda5577$ relative to [O I] $\lambda\lambda6300, 6364$. However, this is very difficult in practice because [O I] $\lambda5577$ is weak and severely blended with other lines (notably Fe II). The method used by FPS to estimate the total ejected mass is also highly uncertain. Thus, the low ejected masses might not constitute an insurmountable problem, especially in the Shigeyama *et al.* (1990) binary model which invokes the explosion of a 3–4 $M_\odot$ helium core.

If, however, more accurate mass estimates continue to yield very low masses, white dwarf progenitors should be considered seriously.

3.4. Radio Emission

Strong radio emission was detected within one year past maximum in the prototypical SNe Ib 1983N and 1984L, unlike the case in SNe Ia (Panagia, Sramek, and Weiler 1986). If interpreted in the context of the Chevalier (1984) model, the observations of SN 1983N give a value of 5×10^{-7} $(M_\odot \text{ yr}^{-1})/(\text{km s}^{-1})$ for $\dot{M}/v_w$, the mass-loss rate divided by the wind velocity. If the wind velocity is 2000 km s^{-1}, typical of WR stars, the implied mass-loss rate is $\dot{M} \approx 10^{-3}$ $M_\odot$ yr^{-1}, which greatly exceeds the observed rates in WR stars.

Several possible solutions to this dilemma come to mind. WR stars may, for example, experience a very short-lived phase with high $\dot{M}$ immediately prior to exploding. Such a phase would be very difficult to detect by other means, yet it could produce a substantial circumstellar shell consistent with the radio emission. Alternatively, the SN Ib/Ic could have been a relatively low-mass star in a close binary system, as in the model presented by Shigeyama et al. (1990) and by Nomoto, Filippenko, and Shigeyama (1990). If so, the circumstellar shell could be due to a companion with a wind velocity of, say, 10 km s^{-1}. From the observations, the inferred mass-loss rate would then be $\dot{M} \approx 5 \times 10^{-6}$ $M_\odot$ yr^{-1} — quite reasonable for a red supergiant, as discussed by M. Jura at this symposium. The circumstellar gas might also be left over from previous episodes of mass exchange in the close binary system.

It should be noted that, although the radio emission has often been invoked as an argument *against* the white dwarf hypothesis of SNe Ib/Ic, it is actually quite compatible with a white dwarf progenitor in a binary system (Branch 1988). The white dwarf can accrete matter from a red supergiant wind, and the ejecta from the subsequent explosion interact with the wind in the usual manner (Chevalier 1984).

4. CONCLUSIONS

It is clear from the preceding discussion that many SNe Ib/Ic probably do represent the explosions (through core collapse) of hydrogen-deficient, massive stars. However, the question of whether most of these are *Wolf-Rayet stars*, in which mass is lost primarily through winds, is far from settled. Unless single stars of relatively modest mass (ZAMS mass $\approx$ 15–20 $M_\odot$, rather than $M \gtrsim$ 30–40 $M_\odot$) are able to lose their hydrogen envelopes, it seems almost impossible to account for a majority of SNe Ib/Ic with single stars or wide binaries. Many, or even most, SNe Ib/Ic probably come from progenitors in close binary systems, where much of the original mass ($M =$10–40 $M_\odot$) is actually lost through mass transfer. In this case, the pre-supernova winds are not necessarily strong enough to make the progenitors qualify as WR stars.

Despite this conclusion, it is still quite possible that *some* SNe Ib/Ic may be exploding white dwarfs. Hopefully, with more detailed observations of a larger number of SNe, we will find observational distinctions between these explosion mechanisms. It is even possible that SNe Ib and SNe Ic have physically different progenitors and explosion mechanisms, although the simplest current interpretation is that they represent the same type of object, but with different helium abundances or excitation conditions in the outer atmosphere.

I disagree with the suggestion of Panagia and Laidler, expressed in a poster at this symposium, that all SNe Ib/Ic must have ZAMS masses between 6.5 $M_\odot$ and 8 $M_\odot$. This mass range was obtained under the assumption that all stars initially more massive than 8 $M_\odot$ produce SNe II, but this *precludes* the possibility that SNe Ib/Ic have progenitor masses above 8 $M_\odot$. Moreover, according to their analysis, the average distances of SNe Ib/Ic and SNe II from H II regions are comparable; typical SNe Ib/Ic might even be slightly closer to H II regions than are SNe II. This certainly does not imply that SNe Ib/Ic progenitors are significantly *less* massive than the progenitors of SNe II.

Finally, I note that if extreme mass-loss mechanisms such as that of Langer and Woosley *don't* exist, the most massive WR stars may end their lives as black holes not preceded by supernova events. This conclusion is based on the fact that thus far, observers have identified few (if any) SNe which have the expected properties of an exploding, very massive WR star: low optical luminosity, very broad light curve, and absence of hydrogen in the spectrum.

Most of the observations reported here were obtained at Lick Observatory, which receives partial funding from NSF Core Block grant AST–8614510. My research is supported by NSF grant AST–8957063, as well as by the Center for Particle Astrophysics at the University of California, Berkeley, through NSF Cooperative Agreement AST–8809616. Partial financial support for attending this symposium was obtained from the Committee on Research at U.C. Berkeley. I thank N. Langer, K. Nomoto, and L. Yungelson for interesting discussions.

REFERENCES

Begelman, M. C., and Sarazin, C. L. *Astrophys. J. Letters*, **302**, L59 (1986).

Branch, D. In *Atmospheric Diagnostics of Stellar Evolution*, ed. K. Nomoto (Berlin: Springer-Verlag), p. 281 (1988).

Branch, D., and Nomoto, K. *Astron. Astrophys.*, **164**, L13 (1986).

Chevalier, R. A. *Astrophys. J. Letters*, **285**, L63 (1984).

Doggett, J. B., and Branch, D. *Astron. J.*, **90**, 2303 (1985).

Ensman, L. M., and Woosley, S. E. *Astrophys. J.*, **333**, 754 (1988).

Evans, R., van den Bergh, S., and McClure, R. D. *Astrophys. J.*, **345**, 752 (1989).

Filippenko, A. V. *Proc. Astron. Soc. Aust.*, **7**, 540 (1988a).

Filippenko, A. V. *Astron. J.*, **96**, 1941 (1988b).

Filippenko, A. V., Porter, A. C., and Sargent, W. L. W. *Astron. J.*, in press (1990) (FPS).

Filippenko, A. V., and Sargent, W. L. W. *Astron. J.*, **91**, 691 (1986).

Fransson, C., and Chevalier, R. A. *Astrophys. J.*, **343**, 323 (1989).

Gaskell, C. M., Cappellaro, E., Dinerstein, H. L., Garnett, D. R., Harkness, R. P., and Wheeler, J. C. *Astrophys. J. Letters*, **306**, L77 (1986).

Graham, J. R. *Astrophys. J.*, **315**, 588 (1987).

Harkness, R. P., *et al. Astrophys. J.*, **317**, 355 (1987).

Iben, I., Jr., Nomoto, K., Tornambè, A., and Tutukov, A. V. *Astrophys. J.*, **317**, 717 (1987).

Khokhlov, A. M., and Ergma, E. V. *Pis'ma Astron. Zh.*, **12**, 366 (*Soviet Astron. Letters*, **12**, 152) (1986).

Langer, N. *Astron. Astrophys.*, **220**, 135 (1989).

Nomoto, K., Filippenko, A. V., and Shigeyama, T. *Astron. Astrophys.*, submitted (1990).

Oke, J. B., and Gunn, J. E. *Astrophys. J.*, **266**, 713 (1983).

Panagia, N., Sramek, R. A., and Weiler, K. *Astrophys. J. Letters*, **300**, L55 (1986).

Schaeffer, R., Cassé, M., and Cahen, S. *Astrophys. J. Letters*, **316**, L31 (1987).

Shigeyama, T., Nomoto, K., Tsujimoto, T., and Hashimoto, M. *Astrophys. J. Letters*, in press (1990).

Uomoto, A. *Astrophys. J. Letters*, **310**, L35 (1986).

Wheeler, J. C., and Harkness, R. P. In *Galaxy Distances and Deviations from Universal Expansion*, ed. B. F. Madore and R. B. Tully (Dordrecht: Reidel), p. 45 (1986).

Wheeler, J. C., Harkness, R. P., Barker, E. S., Cochran, A. L., and Wills, D. *Astrophys. J. Letters*, **313**, L69 (1987).

Woosley, S. E. In *Supernovae*, ed. A. Petschek (Berlin: Springer-Verlag), p. 182 (1990).

DISCUSSION

Matteucci: Several years ago, Branch and Nomoto proposed off-center detonation of a white dwarf as a model for Type Ib supernovae, in order to explain the early appearance of helium and iron in the spectra. What is the current status of that idea?

Filippenko: The Branch and Nomoto model leads to a light curve that declines far more rapidly than the observed light curves of Type Ib supernovae, because only the outer layer of helium ($0.3M_\odot$) detonates. Thus, it does not explain Type Ib supernovae. The early appearance of strong helium lines is consistent with WR models, since the outer layers of WN stars consist largely of helium. The iron lines are produced by normal iron abundances in the outer atmospheres of these stars, as shown by Wheeler and Harkness.

Langer: Dopita and Lozinskaya told us about an extreme WO star with a wind velocity of about 5000 $km \cdot s^{-1}$. Can this help alleviate the problem of large mass loss rates inferred from the radio observations?

Filippenko: Actually, it makes the problem even worse. The radio observations give us a specific value for mass loss rate divided by wind velocity. Thus, the higher the assumed wind velocity, the greater the inferred mass loss rate. What we actually need are *lower* wind velocities in WR stars in order to deduce more reasonable (*i.e.* smaller) mass loss rates.

Yungelson: The numerical model of the binary population of the Galaxy, I was talking about this morning, predicts for the Galaxy one exploding He-star per ~ 300 years. Most of them have $(2.5 - 10)M_\odot$ masses and most probably would not have strong enough stellar winds to produce the WR phenomenon. Rather, prior to explosions they would be observed as hot subdwarfs. These pre-SN are descendants of binary components with initial masses $\sim (10 - 25)M_\odot$.

Filippenko: Yes, it is true that low-mass helium stars may not have sufficiently strong stellar winds to produce the emission lines characteristic of WR stars. In my *talk* I used a loose definition of WR star, (*i.e.*, stripped helium or carbon-oxygen core), since I only wanted to concentrate on *explosion* mechanisms (core collapse *vs.* deflagration or detonation).

Vanbeveren: New evolutionary computations of core He-burning stars with WR-like stellar wind mass loss rates predict that all stars which are able to evolve into a WR phase will end up as a WO star. Do you not expect then to see oxygen right from the beginning of the SN phenomenon rather than a few weeks after the explosion started?

Filippenko: The [OI] emission lines begin to appear between one and two months past maximum. At even earlier times, the density is so high that the [OI] lines are thermalized; they cannot be distinguished from the continuum. However, the OIλ7774 absorption line *is* very strong right from the start in the spectra of Type Ib supernovae.

X-RAY AND GAMMA-RAY SIGNATURES OF WOLF-RAYET SUPERNOVA EXPLOSIONS

LIH-SIN THE and DONALD D. CLAYTON
Department of Physics and Astronomy
Clemson University, Clemson, SC 29634, U.S.A.

ADAM BURROWS
Departments of Physics and Astronomy
University of Arizona, Tucson, AZ 85721, U.S.A.

ABSTRACT. It is widely speculated that a Type Ib supernova is the explosion of a Wolf-Rayet star. We calculate the X-ray and gamma-ray signatures of models of that type, assuming all hard photons to have originated with Ni decay chains, in hopes of providing diagnostics of the exposed-core models of massive stars, which constitute one model of the Wolf-Rayet stars, calculated by Ensman and Woosley (1988). These provide the characteristic luminosity peak and light curve of Type Ib supernovae for helium-core masses between 4 and 6 $M_\odot$. We compute gamma-ray line shapes and fluxes and the Comptonized X-ray continuum resulting from the decay of the radioactive ^{56}Co and ^{57}Co isotopes that are synthesized by the explosion of the presupernova star (the suggested Wolf-Rayet or post-Wolf-Rayet star) with a Monte Carlo transport code. The expansion velocity, the total mass of the ejecta, the radial mixing of radioactivity in that ejecta, and the ^{56}Ni yield effect both the strength and the evolution of the hard radiation. With the anticipated launch of Gamma Ray Observatory, we can hope to detect Type Ib supernovae to distances of 3 Mpc and utilize the characteristics of the gamma lines and X-ray spectrum to distinguish between differing Type Ib supernova models and to address their suggested relationship to Wolf-Rayet stars.

1 Introduction

Controversy surrounds interpretation of the nature of the Wolf-Rayet stars. Maeder and Meynet (1987) and Langer (1989) argue that cores of massive evolved stars, exposed by large mass loss during the Wolf-Rayet process, provide the characteristics and anomalous abundances of the observed WR stars. Underhill and her colleagues (Bhatia and Underhill 1986, 1988, 1989; Underhill, Gilroy, and Hill 1990) sternly challenge that interpretation, arguing that the abundances are not in fact anomalous and that several other features (e.g. surface brightness) are not in accord with such models. They prefer hydrogen-burning B stars in binaries leading to the observed WR phenomenon. Paralleling this dispute is the

537

K. A. van der Hucht and B. Hidayat (eds.),
Wolf-Rayet Stars and Interrelations with Other Massive Stars in Galaxies, 537–545.
© 1991 *IAU. Printed in the Netherlands.*

emergence of a new class of supernovae, now called Type Ib because they were originally classed Type I owing to absence of hydrogen lines. Wheeler and Levreault (1985) discussed previous analysis of Type Ib structures and advanced the now common paradigm–that Type Ib undergo core collapse and bounce, like Type II, but are small dense objects having light curves dominated by ^{56}Co decay, like Type I. Ensman and Woosley (1988) have constructed detailed evolutionary models of Type Ib, finding that 4-to-6 $M_\odot$ cores of evolved stars of initially greater mass may reproduce satisfactorily the Type Ib light curves, though subject to the same physics uncertainties that surround the actual Type II core bound mechanism. Extensive mass loss during the WR lifetime is supposed to reduce the final mass to 4-to-6 $M_\odot$ in time for the final Type Ib explosion. We will call these progenitors Wolf-Rayet stars, remembering that the identification is contested.

Our contribution to this evolving debate is to explore the usage of gamma-ray lines to elucidate the true structural nature of the Type Ib phenomenon, and thereby its relationship to WR stars. Analysis of supernova structure using gamma-ray line and X-ray continuum has been done by many authors (Clayton, Colgate, and Fishman 1969; Colgate and McKee 1969; Clayton 1974; Ambwani and Sutherland 1988; Woosley and Pinto 1988; Chan and Lingenfelter 1988; Bussard, Burrows, and The 1989: Kumagai et al.1989). However, due to uncertainty of the Type Ib supernovae model, no work has been published predicting the expected gamma-ray line and X-ray fluxes of Type Ib supernovae. In this paper we do this by applying a Monte-Carlo transport code (The, Burrows, and Bussard 1990) to the Type Ib models calculated by Ensman and Woosley (1988). The sources of hard photons are the decay chains ^{56}Ni$\rightarrow$^{56}Co$\rightarrow$^{56}Fe and ^{57}Ni$\rightarrow$^{57}Co$\rightarrow$^{57}Fe (Clayton, Colgate, and Fishman 1969; Clayton 1974; The, Burrows, and Bussard 1990)

2   Models

Ensman and Woosley (1988; hereafter referred to as EW), find that He-cores of 4-to-6 $M_\odot$ give the best fit to the available data of Type Ib peak magnitudes and light curves, although each of their models has its own problems in fitting the data. The Wolf-Rayet supernovae explosion models we use in this paper are those 4-to-6 $M_\odot$ models of EW that produce the best light curves for Type Ib supernovae. Model 4A of EW (hereafter referred to as WR4A) which was initially a 15 $M_\odot$ main sequence star is assumed to have been reduced to a 4 $M_\odot$ Wolf-Rayet star that finally ejects its last 2.68 $M_\odot$ of mass with an explosion energy of 0.5×10^{51} ergs. This model produces 0.06 $M_\odot$ of ^{56}Ni and it is found to be somewhat too faint when compared with SN1983N peak magnitudes (EW). Model 6C (hereafter referred to as WR6C) was a 20 $M_\odot$ main sequence star reduced to a 6.2 $M_\odot$ Wolf-Rayet star that in its explosion of 2.7×10^{51} ergs ejects 4.5 $M_\odot$ and produces 0.16 $M_\odot$ of ^{56}Ni. It produced a good fit to SN1983N light curve (EW).

Because the SN1987A light curve was powered by radiactive decay of ^{56}Co (Catchpole et al.1988; Whitelock et al.1988) and because the early appearance of gamma-ray line showed that ^{56}Co was mixed radially up to its hydrogen envelope (Sunyaev et al.1987; Matz et al.1988; Leising 1988; Sandie et al.1988; Cook et al.1988; Wilson et al.1988; Teegarden et al.1988; Nomoto *et al.*1988; Pinto and Woosley 1988; The, Burrows, and Bussard 1990), in our analysis we also evaluate mixed models; in model WR4Afm and WR6Cfm we mix all material in those entire stars uniformly(fm meaning "fully mixed") and in model WR6Chm, the star inside the He burning shell is mixed uniformly. Such mixing enhances significantly the early X-ray continuum and gamma-ray line fluxes, but it now seems more realistic than an unmixed model, as demonstrated by SN1987A. It is also of interest to seek differences in the X-ray continuum and gamma-ray line fluxes between models having different speeds of expansion; for this we constructed model WR4Ab, which has all the characteristics of WR4A except its speed of expansion is $\sqrt{3}$ times the speed of the original model. We increase the speed of expansion of WR4A by this amount in order to make WR4A more like WR4B of EW, which provides the best overall fit to 1983N. We expect that the parameter space we explore (mixing and speed) covers most of the possibilities for this type of model of Type Ib supernovae. Many of the results we present here (flux ratios and energy fractions) do not depend on the amount of ^{56}Ni and, as EW conclude, this type of model ranges only from 4 to 6 $M_\odot$ helium-core stars, to which it is limited by the widths of bolometric light curves. We find that the characteristics of WR4Ab fall between those of WR4A and WR6C, since the expansion speed of WR4Ab is between that of these models, and also that the behavior of WR6Chm is between that of WR6C and WR6Cfm.

3 Results

The gamma-ray luminosities of these Type Ib supernovae are encouragingly bright. Figure 1 shows that just as for the optical luminosity, Type Ib supernovae 847-gamma-ray-line light curves falling between those of Type Ia and Type II supernovae. For Type Ia model, we show model W7 of Nomoto, Thielemann, and Yokoi (1984), where its gamma-ray and X-ray continuum fluxes have been reported recently by Burrows and The (1990). The Type II model is the model W10hmm of Pinto and Woosley (1988a,b) which is a successful model for almost the entire spectrum emitted by SN1987A. We also note that the peak time of gamma-ray-line light curves of Type Ib supernovae is later than for the Type Ia but earlier than that of the Type II. The 847 keV maximum fluxes for this type of Wolf-Rayet supernova explosion range from $\sim3\times10^{-5}$ to $\sim4\times10^{-4}$ photons/cm^2s for explosions at our standard reference distance of 1 Mpc, which means with OSSE (Oriented Scintillation Spectrometer Experiment, Kurfess *et al.*1983) on Gamma Ray Observatory (with detection limit of 3×10^{-5} photons/cm^2s) we can hope to detect extragalactic Type Ib supernovae gamma-ray lines to $\sim$ 3 Mpc.

In determining the gamma line profiles, we employ the analytic formalism developed by Bussard, Burrows, and The (1989). In Table 1 we show the fluxes at 1 Mpc and properties of the 847 keV gamma line at the time of maximum for those models shown in Figure 1. We find also that the FWHM (full width at half maximum) of gamma line profiles for Wolf-Rayet supernova explosions is between the FWHM's of Type Ia and Type II. One sees from Table 1 that for all supernova types the products T_{max} x FWHM are approximately equal, which is because $T_{max} \sim 1/v$ and FWHM $\propto v$. FWHM of the Ib models studied lie

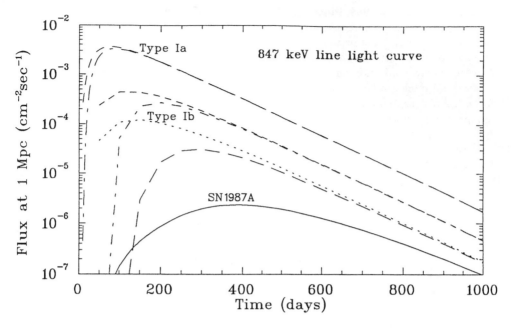

Figure 1. The 847 keV Gamma line light curves of different types of supernovae at distance of 1 Mpc: (Ia) W7; (Ib) WR6C upper and WR4A lower; (SN1987A for Type II) W10hmm. Brighter branches of W7, WR6C, and WR4A are fully mixed counterparts of those models.

between 6 to 23 keV and their shapes depend on the ^{56}Ni distribution. For a photon source distributed in a thin shell, the gamma line profile at maximum has a box shape, while a uniformly distributed source has a parabolic profile (Burrows and The 1990).

Table 1. The 847 keV gamma-ray line maximum fluxes at 1 Mpc, the time at peak (T_{max}), FWHM at T_{max}, and the spectral shape for different types of supernovae.

Type	Model	T_{max}(days)	$F_{max}^{1Mpc}(cm^{-2}s^{-1})$	FWHM(keV)	shape
Ia	W7	90	3.3×10^{-3}	~ 26.9	parabolic
	W7 mixed	80	3.6×10^{-3}	~ 30.0	parabolic
Ib	WR4A	275	3.1×10^{-5}	~ 5.9	box
	WR4A mixed	120	1.2×10^{-4}	~ 13.9	parabolic
	WR6C	180	2.8×10^{-4}	~ 17.4	box
	WR6C mixed	100	4.4×10^{-4}	~ 23.2	parabolic
II	W10hmm	390	2.3×10^{-6}	~ 5.8	parabolic

In Figure 2 we show our calculations of the photon spectrum for four models (WR4A, WR6C, and their fully mixed counterparts) at two different times (100d and 200d). Continuum fluxes greater than $2\times10^{-7}cm^{-2}s^{-1}keV^{-1}$ above E=50 keV have good hopes for detection by OSSE. One sees that the different continua at 100d partially converge at 200d. Line fluxes and ratios between selected X-ray bands are in Table 2.

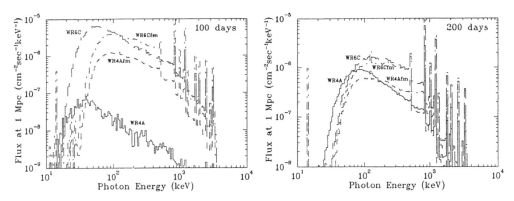

Figure 2. Continuous spectra for model WR4A, WR4Afm, WR6C, and WR6cfm at days 100 and 200. The heights of line-containing bins are plotted as the number of line photons (see Table 2) divided by the width of the energy bin into which they fall. The emergent photons are binned logarithmically from 1 keV to 5 MeV in 188 energy bins.

Table 2. Line fluxes $(10^{-5}cm^{-2}s^{-1})$ for Models at 1 Mpc, the X-ray band ratios, and the ratio of the total X-ray and gamma-ray luminosities.

Model	F_{847}		F_{1238}		$\frac{F(50-100)}{F(550-700)}$		$\frac{F(50-100)}{F(350-400)}$		f_X/f_γ	
	100d	200d	100d	200d	100d	200d	100d	200d	100d	200d
WR4A	4×10^{-4}	1.69	0.002	1.79	3.0	1.8	5.9	3.2	3.7	1.1
WR4Afm	10.9	10.4	8.32	7.65	1.5	1.3	2.0	1.4	0.43	0.28
WR4Ab	0.803	9.04	1.11	7.13	2.8	0.68	4.9	0.95	1.7	0.42
WR6C	5.17	26.7	6.03	20.5	2.7	0.73	5.1	1.0	1.4	0.34
WR6Chm	10.7	25.8	9.84	19.8	2.3	0.73	3.6	0.96	0.50	0.36
WR6Cfm	43.8	35.4	32.8	25.6	1.5	0.85	1.8	0.82	0.36	0.21

The ratio of the total X-ray and gamma-ray luminosities, f_X/f_γ, is also significant because it reveals the time during the early light curve when the X-ray flux is dominant and more easily detected. That same ratio also reveals the extent of mixing (e.g. Leising 1988). We find that f_X/f_γ for fully mixed models is monotonically decreasing after 50 days and $f_X/f_\gamma \leq 0.5$. The WR4Afm example in Figure 3a is such a model. Figure 3a also shows several fractions of total radioactive power going into differing channels as a function of time. For an unmixed model, on the other hand, f_X/f_γ is > 0.5 for times between 50 and 150 days. For any prescription of mixing and kinetic energy, at times after the peak (T_{max}) of the 847 keV line light curve, $f_X/f_\gamma < 0.5$. The effect of mixing dramatically reduces the value of f_X/f_γ for times between 50 and 200 days and this dependence is more pronounced than is that on the speed of expansion. The effect of mixing and expansion on $f_{opt}(= 1 - f_X - f_\gamma)$ is similar to that on f_X/f_γ.

We show in Figure 3b the flux ratios for selected continuum bands as a function of time for fully mixed model WR4Afm. Table 2 shows the effect of mixing and the increased speed of expansion at both 100d and 200d on the $F_{50-100}/F_{550-700}$ and $F_{50-100}/F_{350-400}$ X-ray-

542

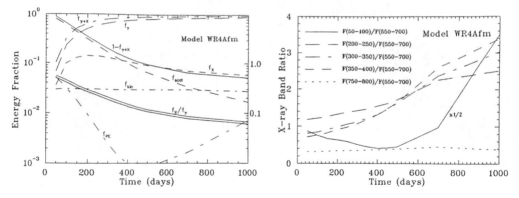

Figure 3a. Fractions of the total radioactive power that emerge as X-ray(f_X) and gamma-lines(f_γ), their total($f_{\gamma+X}$), and the fractions deposited in the plasma by scattering(f_{scat}), photoelectric absorption(f_{PE}), and the ^{56}Co positron(f_{kin}) as functions of time for model WR4Afm. The double curve shows the ratio of f_x/f_γ on the right ordinate. Figure 3b. X-ray band ratios.

band ratios. These band ratios reflect the slope of the X-ray spectrum in Figure 2. Mixing and increased speed cause smaller values of $F_{50-100}/F_{550-700}$ for times when ^{57}Co is not a major contributing source to this band. For a fully mixed model, $F_{50-100}/F_{550-700}$ is < 2 and decreases after 50 days, unless ^{57}Co is present. Its increase in Figure 3b after day 400 is mostly due to the contribution from the 122 and 136 keV ^{57}Co gamma lines, which are beginning to assume a greater role. The other band flux ratios increase with time for any of these models until that time when most of the photons escape directly (only WR6Cfm of our models starts to decrease at 700 days). The crossover of X-ray-band ratios shown in Figure 3b near 550d is due to the increasing contribution of three-photon positronium decays at later times. At early times, most of the photons in the bands below 500 keV come from scattering of the 847 and 1238 keV ^{56}Co gamma lines and from direct escape and scatterings of three photon positronium decays. But at late time most of 847 and 1238 keV gamma lines escape directly, so that their contributions to the X-ray bands become much smaller than direct escape photons from positronium decays.

The X-ray-band ratios and the ratio of X-ray and gamma-ray luminosites of WR4Ab and WR6C are very similar, suggesting that the evolution of these ratios can identify the speed of expansion of the supernova. f_X/f_γ for a higher expansion speed model can be roughly scaled from a lower expansion model by contracting the time scale in proportion to $1/v$. At early time, the X-ray-band ratios for energies larger than 200 keV are similar for mixed and unmixed models and also for slow and fast expansion models, but these ratios become more distinguishable at later times. Differentiating between models is easier in the low 50-100 keV band since its intensity strongly depends on the distribution of elements having high photoelectric cross sections.

We have demonstrated not only that the fluxes, peak time, and FWHM can be used to distinguish different types of supernovae, but also that gamma-line fluxes, shapes, FWHM, X-ray-band ratios, ratio of X-ray and gamma-ray luminosities, and the fraction of energy deposited into the ejecta can reveal the extent of mixing in the ejecta and its speed of expansion. Other characteristics that can be utilized for X-ray diagnosis of a supernovae are line ratios, the X-ray cutoffs, the energy of the peak continuum flux, and turn-on times (Burrows and The 1990). A more complete account of these details is under preparation by us. We can offer hope that good measurements of a Type Ib supernova will reveal its structure, so that if its progenitor was indeed a Wolf-Rayet star, the state of its evolution and prior mass loss may be circumscribed by these measurements.

Acknowledgments: L-ST and DDC research was supported by Naval Research Laboratory Grant N00014-89-J-2034 under the NASA Contract for the OSSE Spectrometer DPRS-10987C on GRO. AB research was supported by the National Science Foundation through grant no. AST-89-14346 and by NASA through the Long Range Space Astrophysics Research Program. We are grateful to Lisa Ensman for providing us with model 4A, 6C, and 8A. L-ST would like to thank TRIUMF, especially C.J. Kost for providing TRIUMF graphics software that we use in preparing the graphs in this paper.

4   References

Arnett, D.W. (1979) 'On the theory of Type I supernovae', *Astrophys. J.*, L37-L40.

Ambwani, K. and Sutherland, P. (1988) 'Gamma-ray spectra and energy deposition for Type Ia supernovae', *Astrophys. J.*, **325**, 820-827.

Bhatia, A.K. and Underhill, A.B. (1986) 'The statistical equilibrium of hydrogen and helium in a radiation fiels, with an application to interpreting Wolf-Rayet spectra', *Astrophys. J. Suppl.*, **60**, 323-356.

Bhatia, A.K. and Underhill, A.B. (1988) 'Carbon and nitrogen lines in the spectra of Wolf-Rayet stars', *Astrophys. J. Suppl.*, **67**, 187-223.

Bussard, R., Burrows, A., and The, L.-S. (1989) 'SN1987A gamma-ray line profiles and fluxes', *Astrophys. J.*, **341**, 401-413.

Burrows, A. and The, L.-S. (1990) 'X-ray and gamma-ray signatures of Type Ia supernovae', to appear in *Astrophys. J.*

Catchpole, R.M., Whitelock, P.A., Feast, M.W., Menzies, J.W., Glass, I.S., Marang, F., Laing, J.D., Spencer Jones, J.H., Roberts, G., Balona, L.A., Carter, B.S., Laney, C.D., Lloyd Evans, T., Sekiguchi, K., Hutchinson, M.G., Maddison, R., Albinson, J., Evans, A., Allen, D.A., Winkler, H., Fairall, A., Corbally, C., Davies, J.K., and Parker, Q. (1988) 'Spectroscopic and photometric observations of SN1987a -II. days 51 to 134', *Monthly Notices Roy. Astron. Soc.* **231**, 75p-89p.

Chan, K.W. and Lingenfelter, R.E. (1988) 'Gamma ray lines from supernovae', in N. Gehrels and G. Share (eds.), *Nuclear Spectroscopy of Astrophysical Sources*, AIP Conf. Proc. 170, New York, pp. 110-115.

Clayton, D.D. (1974) 'Line [57]Co gamma rays: New diagnostic of supernova structure', *Astrophys. J.*, **188**, 155-157.

Clayton, D.D., Colgate, S.A., and Fishman, G. (1969) 'Gamma ray lines from young supernova remnants', *Astrophys. J.*, **155**, 75-82.

Colgate, S.A. and McKee, C. (1969) 'Early supernova Luminosity', *Astrophys. J.*, **157**, 623-643.

Cook, W.R., Palmer, D.M., Prince, T.A., Schindler, S., Starr, C.H., and Stone, E.C. (1988) 'An imaging observation of SN 1987A at gamma-ray energies', *Astrophys. J. Letters*, **334**, L87-L90.

Ensman, L.M. and Woosley, S.E. (1988) 'Explosions in Wolf-Rayet stars and Type Ib supernovae', *Astrophys. J.*, **333**, 754-776.

Kumagai, S., Shigeyama, T., Nomoto, K., Itoh, M., Nishimura, J., Tsuruta, S. (1989) 'Gamma-ray, X-ray, and optical light from the cobalt and the neutron stars in SN 1987A', *Astrophys. J.*, **345**, 412-422.

Kurfess, J.D., Johnson, W.N., Kinzer, R.L., Share, G.H., Strickman, M.S., Ulmer, M.P., Clayton, D.D., Dyer, C.S. (1983) 'The Oriented scintillation spectrometer experiment for the gamma-ray observatory', *Adv. Space Res.*, **3**, 109-112.

Langer, N. (1989) 'Standard models of Wolf-Rayet stars', *Astron. Astrophys.*, **210**, 93-113.

Leising, M.D. (1988) 'Gamma-rays and X-rays from SN1987A', *Nature*, **332**, 516-518.

Maeder, A. and Meynet, G. (1987) 'Grids of evolutionary models of massive stars with mass loss and overshooting. Properties of Wolf-Rayet stars sensitive to overshooting', *Astron. Astrophys.* **182**, 243-263.

Matz, S.M., Share, G.H., Leising, M.D., Chupp, E.L., Vestrand, W.T., Purcell, W.R., Strickman, M.S., and Reppin, C. (1988) 'Gamma-ray line emission from SN1987A', *Nature*, **331**, 416-418.

Nomoto, K., Thielemann, F.-K., and Yokoi, K. (1984) 'Accreting white dwarf models for Type I supernovae. III. carbon deflagration supernovae', *Astrophys. J.*, **286**, 644-658.

Pinto, P.A. and Woosley, S.E. (1988) 'X-ray and gamma-ray emission from supernova 1987A', *Astrophys. J.*, **329**, 820-830.

Sandie, W.G., Nakano, G.H., Chase, Jr, L.F., Fishman, G.J., Meegan, C.A., Wilson, R.B., Paciesas, W.S., and Lasche, G.P. (1988) 'High-resolution observations of gamma-ray line emission from SN 1987A', *Astrophys. J. Letters*, **334**, L91-L94.

Sunyaev, R., Kaniovsky A., Efremov, V., Gilfanov, M., Churazov, E., Grebenev, S., Luznetsov, A., Melioransky, A., Yamburenko, N., Yunin, S., Stepanov, D., Chulkov, I., Pappe, N., Boyarskiy, M., Gavrilova, E., Loznikov, V., Prudkoglyad, A., Rodin, V., Reppin, C., Pietsch, W., Engelhauser, J., Trümper, J., Voges, W., Kendziorra, E., Bezler, M., Staubert, R., Brinkman, A.C., Heise, J., Meis, W.A., Jager, R., Skinner, G.K., Al-Eman, O., Patterson, T.G., Wilmore, A.P. (1988) 'Discovery of hard X-ray emission from supernova 1987A', *Nature*, **330**, 227-229.

Teegarden, B.J., Barthelmy, S.D., Gehrels, N., Tueller, J., Leventhal, M., and MacCallum, C.J. (1989) 'Resolution of the 1,238 keV gamma-ray line from supernova 1987A', *Nature*, **339**, 122-123.

The, L.-S., Burrows, A., and Bussard, R. (1990) 'X-ray and gamma-ray fluxes from SN1987A', *Astrophys. J.*, **352**, 731-740.

Underhill, A.B., Gilroy, K.K., and Hill, G.M. (1990) 'About the eclipsing Wolf-Rayet binary HD214419', *Astrophys. J.*, **351**, 651-665.

Wheeler, J.C. and Levreault, R. (1985) 'The peculiar type I supernova in NGC991', *Astrophys. J. (Letters)*, **294**, L17-L24.

Woosley, S.E. and Pinto, P.A. (1988) 'Gamma-producing radioactivities from supernovae', in N. Gehrels and G.H. Share (eds.), *Nuclear Spectroscopy of Astrophysical Sources*, AIP Conf. Proc. 170, New York, p98-109.

DISCUSSION

Filippenko: Can you distinguish between a 3-4 $M_\odot$ helium core with *no* clumping and mixing, and a 6 $M_\odot$ helium core *with* clumping and mixing?

The: Here we compare model WR4A and WR6Cfm with clumping. Modeling clumps such as done in the paper of Nomoto *et al.* (1989) for SN1987A (where clumping happens inside the *He*-core), at the early time when the ejecta is still thick, most of X-ray and γ-ray radiation come from sources near the surface of ejecta. Therefore, at early time the value of X-ray band ratio and $fx/f\gamma$ for WR6Cfm with clumping do not differ much from WR6Cfm. At later time when sources from deeper layer dominate, the value of X-ray band ratio and $fx/f\gamma$ increases to a larger value than in WR6Cf. $fx/f\gamma$ in WR6Cfm with clumping is larger than $fx/f\gamma$ in WR6Cfm, but cannot be larger than the value in WR4A. However, since clumping effects add significantly to low energy X-rays, the X-ray band ratio $F(50-100)/F(550-700)$ can increase as high as in WR4A.

Montmerle: Have you considered the X-ray emission from longer-lived isotopes like ^{44}Ti? Or do they appear too late to be of importance, given the sensitivity limit of *GRO*?

The: In this calculation we do not include other radioactive sources, only ^{56}Ni and ^{57}Ni decay chains are considered. Other sources are less abundant. It is still important to detect and to know the amount of these radioactive sources in order to have a better understanding of nucleosynthesis theory.

Nomoto: If you take into account clumpiness to the extent as made by Ensman and Woosley, how large is its effect on your results?

The: Clumping increases the X-ray and γ-ray total luminosities. This effect increases significantly the low-energy X-ray, *i.e.*, $F(50-100)/F(550-700)$, but $fx/f\gamma$ increases slightly.

Langer: I would like to point out that a *He*-core of (*e.g.*) $5M_\odot$, which originates from a relatively low mass star (*e.g.* $20M_\odot$), has a much different internal structure compared to a $5M_\odot$ core which originates from a much more massive star (say, $60M_\odot$). For this reason, your models which are based on the Ensman and Woosley calculations, might correspond to the binary scenario discussed by Nomoto rather than to the WR scenario for Type Ib supernova progenitors of Langer and Woosley (*cf.* poster abstract)

The: Chiosi and Maeder have calculated that evolution of the material inside *He*-burning does not depend on the hydrogen envelope. With and without removing the hydrogen, the results should be the same.

Shara: (1) Can *GRO* detect supernovae anywhere in the Milky Way? (2) What would we have learned about SN1987A had *GRO* been aloft there?

The: (1) With *GRO* sensitivity, it will be able to detect all type of supernovae in the Milky Way, unless the explosion happens at the other side of Galactic Center or is blocked by another object along the line of sight. (2) If we had *GRO* on orbit since the first day of SN1987A, we would have got an accurate measurement of γ-line light curve which can be used to uniquely determine the ^{56}Ni distribution and its density structure. Since *GRO* has a narrow FWHM detection limit, measurements of γ-line shapes can map the velocity distribution of radioactive sources. Clumping and mixing can also be determined by X-ray measurements.

Lih-Sin The

IS STELLAR WIND MASS LOSS DURING CORE HYDROGEN BURNING IMPORTANT FOR EVOLUTION.

R. BLOMME[1], D. VANBEVEREN[2], W. VAN RENSBERGEN[2].
(1): Koninklijke Sterrenwacht, Brussels, Belgium,
(2): Dept. of Physics, VUB, Brussels, Belgium.

SUMMARY. The influence of wind mass loss during core hydrogen burning on the evolution of massive stars can be investigated using a relation between $\dot{M}$-L-M-T_{eff} based on

a. observed ($\dot{M}$, L, T_{eff}) values and masses which are determined from evolutionary computations (e.g. Nieuwenhuijzen and de Jager, 1990, *A.&A.* **231**, 134),

b. the theory of radiation driven stellar wind. With the cooking recipe of Kudritzki et al.(1989, *A.&A.* **219,** 205), we determined the predicted $\dot{M}$ of a star at the beginning, in the middle and at the end of core hydrogen burning using the 20 $M_\odot$, 40 $M_\odot$, 60 $M_\odot$ and 85 $M_\odot$ evolutionary computations of Maeder and Meynet (1987, *A.&A.* **182**, 243). The following two formulae then give $\dot{M}$ as a function of L-T_{eff} and M with a sigma value lower than 0.04, i.e.

$$\log(-\dot{M}) = - 13.18 + 1.88 \log L - \log M - 0.4 \log T_{eff} \qquad (1)$$

$$\log(-\dot{M}) = - 10.31 + 1.51 \log L - 0.2 \log M - 0.88 \log T_{eff} \qquad (2)$$

(M, L and M are in solar units)

Formula (1) is determined by assigning equal weights to all points. Formula (2) however gives the relation between $\dot{M}$ and the stellar parameters when the different points are weighted according to the function $M_i^{-3.5}$ (M_i stands for the ZAMS mass of the corresponding evolutionary track); the function describes the observed number density of massive stars corrected for observational selection (Humphreys and McElroy, 1984, *Ap. J.* **284**, 565). With these formulae, evolution with convective core overshooting according to the model of Maeder and Meynet (1987, *A.&A.* **182**, 243) predicts that a 40 $M_\odot$ (60 $M_\odot$, 85 $M_\odot$) star will loose less than 2.7 $M_\odot$ (5.2 $M_\odot$, 7.6 $M_\odot$ respectively) during core hydrogen burning, i.e.

if the theory of radiation driven stellar wind in its present form accurately predicts the mass loss of a massive core hydrogen burning star, then one can conclude that it is a fairly unimportant process as far as evolution is concerned.

The results summarised here will be published in Astron. Astrophys.

K. A. van der Hucht and B. Hidayat (eds.),
Wolf-Rayet Stars and Interrelations with Other Massive Stars in Galaxies, 547.
© 1991 *IAU. Printed in the Netherlands.*

SINGLE-LINED SPECTROSCIC APPEARANCES DURING THE EVOLUTION OF A MASSIVE BINARY SYSTEM.

C. DE LOORE and J.P. DE GREVE

Astrophysical Institute, VUB

Pleinlaan 2, B-1050 Brussels, Belgium

We calculated the evolution of both components of a massive binary system with masses 26 M_O and 23.4 M_O, initial period 27.4 days, through a case B of mass transfer, up to the end of the core helium burning phase of the primary (and further to the end of the main sequence of the secondary). For the WC phase we adopted Langer's (1989) mass dependent mass loss rate $\dot{M} \sim M^{2.5}$). The luminosities were transformed to visual magnitudes, applying bolometric corrections for O stars of Code et al. (1976) and Flower (1977) and BC = - 4.5 for the WR star (Smith and Maeder, 1989). If conservative mass transfer is adopted, the WR star is about 1.2 to 2.0 magnitude less brighter in the visual than the O star during the WN phase (lasting 1.5 10^5 yr). The brightness difference reaches almost 3 magnitudes during the WC + O8 phase (lasting 4.3 10^5 yr), showing the system as a single-lined O-type (!) spectroscopic binary. During the WR phase the mass function of this system varies from 0.67 $\sin^3 i$ to 0.05 $\sin^3 i$! Nonconservative mass transfer and (or) a smaller initial mass ratio may result in the inverse situation, a single-lined WR binary with a large mass function (De Greve et al., 1988). Some of the results are shown in table 1 below.

Table 1 : Characteristics of the WR + O binary at the beginning of resp. the WN and the WC phase, and at the end of the WR phase (Mass loss in 10^{-5} M_O/yr).

	$t/10^6$ yr	M_1	M_2	log L_1/L_O	log L_2/L_O	$-M_{v1}$	$-M_{v2}$	$\dot{M}_{WR}$
WN_i	8.33	11.3	35.2	5.22	5.30	3.9	5.1	4.3
WC_i	8.65	7.3	35.2	4.65	5.31	3.0	5.1	1.6
WR_f	8.88	4.3	35.0	4.46	5.35	2.2	5.2	0.4

References

De Greve, J.P., Hellings, P., van den Heuvel, E.P.J. : 1988, Astron. Astrophys. 189, 74

Code, A.D., Davis, J., Bless, R.C., Hanbury Brown, R. : 1976, Astrophys. J. 203, 417

Flower, P.J. : 1977, Astron. Astrophys. 54, 31

Langer, N. : 1989, Astron. Astrophys. 220, 135

Smith, L., Maeder, A. : 1989, Astron. Astrophys. 211, 71

K. A. van der Hucht and B. Hidayat (eds.),
Wolf-Rayet Stars and Interrelations with Other Massive Stars in Galaxies, 548.
© 1991 IAU. Printed in the Netherlands.

CONVECTIVELY GENERATED TURBULENT PRESSURE: A POSSIBLE CAUSE FOR η CAR - TYPE SHELL EJECTIONS

M. KIRIAKIDIS, N. LANGER and K.J. FRICKE

Universitäts-Sternwarte Göttingen
Geismarlandstrasse 11, D-3400 Göttingen, F.R.G.

A selfconsistent hydrodynamic calculation of a very massive star ($M_{ZAMS} = 200 M_\odot$) including turbulent pressure and energy has been performed. In the contraction phase after core hydrogen exhaustion, the star moves towards cool surface temperatures in the HR diagram (cf. Fig. 1). Consequently, (at $T_{eff} \simeq 8\,000K$) an envelope convection zone developes, and its inner boundery moves inwards with time. First, the envelope remains in hydrostatic equilibrium, with radiation pressure correspondingly decreasing as turbulent pressure increases (gas pressure is small). However, due to the fact, that the gradient of the turbulent pressure is directed inwards at the bottom of the convective zone, this part of the star rapidly contracts. Due to the released contraction energy, the luminosity locally exceeds the Eddington-luminosity. It cannot be transported outwards by convection in the upper part of the convection zone, where convective energy transport is inefficient ($\nabla_c \simeq \nabla_r$). Thus, the local super-Eddington luminosity leads to the ejection of the overlying layers.

In our example we found about $\sim 0.2\,M_\odot$ to be ejected with velocities of about 230 km/s. The stellar surface temperature increases to $T_{eff} \simeq 30\,000K$. Many of our model properties resemble observed properties of η Car or other Luminous Blue Variables. During the further evolution of the remaining star, the process of shell ejection will probably be repeated with a period of some 100 yr.

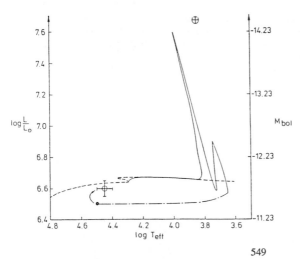

Figure 1: Evolutionary tracks of our 200 $M_\odot$ sequences in the HRD. The solid and dashed lines correspond to computations with and without turbulence, respectively. The encircled crosses show the positions of η Car today and during the great eruption (Davidson, 1986, in: *Instabilities in Luminous Early Type Stars*, H. Lamers, C. de Loore, eds., p. 127; Davidson, 1989, IAU-Colloq. **113**, p. 101). The filled circle corresponds to the last model of our sequence. The arrow indicates the probable further evolution.

K. A. van der Hucht and B. Hidayat (eds.),
Wolf-Rayet Stars and Interrelations with Other Massive Stars in Galaxies, 549.
© 1991 *IAU. Printed in the Netherlands.*

NUCLEOSYNTHETIC ACTIVITY OF WR STARS:
IMPLICATIONS FOR ^{26}Al IN THE GALAXY AND ISOTOPIC
ANOMALIES IN COSMIC RAYS AND THE EARLY SOLAR SYSTEM

N. PRANTZOS

Institut d' Astrophysique de Paris, 98bis, Bd. Arago, F-75014 Paris, France
Service d' Astrophysique, C.E.N. Saclay, F-91191 Gif sur Yvette, France

Abstract: The implications of the nucleosynthetic activity of WR stars are reassessed, in view of recent experimental and observational data. It is confirmed that WR stars may 1) contribute significantly (up to ~20%) to the ~3 $M_\odot$ of ^{26}Al detected in the galactic plane through its 1.8 MeV line, 2) be responsible for the isotopic anomalies of ^{22}Ne and ^{25,26}Mg, detected in galactic cosmic rays (GCR), and 3) be responsible for the inferred presence of ^{26}Al and ^{107}Pd in the early solar system (and, perhaps, some other nuclei as well).

In this work, we calculate nucleosynthesis during H burning (production site of ^{26}Al) and He burning (production site of ^{22}Ne, ^{25,26}Mg, and "light" s-elements, i.e. with mass number 60<A<90) in WR stars with initial mass 50 < $M/M_\odot$ < 100. The injection of the corresponding products in the interstellar medium (ISM) through the WR winds (WN and WC phases) is properly taken into account. We improve upon our previous relative studies ([1],[2],[3]), by using an up-dated set of nuclear data, especially concerning the production and destruction of ^{26}Al (i.e. ^{25}Mg(p,γ)^{26}Al [4], ^{26}Al(p,γ) [5]), the neutron density during He burning (^{22}Ne(n,γ) [6], etc.). The main results are as follows:

1) ^{26}**Al in the Galaxy:** Due to the increase in its production rate [4], the ^{26}Al yield of WR stars is found to be enhanced w.r.t. our previous estimates [3] by ~10% for the 50 $M_\odot$ star, up to ~40% for the 100 $M_\odot$ star. Accordingly, the total quantity of ^{26}Al ejected by galactic WR stars in the ISM during the last million years is estimated to be M_{26} ~ 0.15-0.65 $M_\odot$, depending on the adopted WR distribution (see [3]); these quantities correspond to ~5-20% of the ~3 $M_\odot$ of ^{26}Al in the Galaxy, inferred from observations of its 1.8 MeV line. Thus, it seems that *WR stars may be significant, but not the main, contributors to the galactic content of* ^{26}*Al* (at least, with the current status of stellar and nuclear physics).

2) Isotopic anomalies in GCR: The result of the reduction in the ^{22}Ne(n,γ) cross-section [6] in nucleosynthesis during He-burning is twofold: a) ^{22}Ne is no more a "lethal" neutron poison and is destroyed less than before (in the WC wind and the stellar core), and b) some s-nuclei are produced in larger quantities than previously thought ([1]), due to the higher neutron densities. The conclusions for GCR isotopic anomalies ([1] and more recently [7]) are slightly affected by those new results, but *some new anomalies are predicted, concerning especiallly Se and Sr*.

3) Isotopic anomalies in meteorites: The recent changes in several key nuclear reaction rates affect the ^{26}Al/^{27}Al ratio in the Of-WN winds, which is reduced by a factor of ~2 w.r.t. [2] (the production of ^{27}Al is more enhanced than the one of ^{26}Al), but not significantly the ^{107}Pd/^{108}Pd ratio. The general conclusions as to the (presumed) role of WC stars ([2],[8]) remain valid. Contrary to [8], both in [2] and in our new calculations a substantial production of ^{205}Pb is obtained. A systematic analysis of those effects is in preparation [9].

References

[1] Prantzos N., Arnould M., Arcoragi J.P. 1987, *Ap. J.*, **315**, 209
[2] Arnould M., Prantzos N. 1986, *Nucleosynthesis and its Implications in Nuclear and Particle Physics*, eds. J. Audouze and N. Mathieu (Dordrecht:Reidel), p. 363
[3] Prantzos N., Cassé M., Arnould M. 1987, *Proceedings of 20th ICRC*, Moscow, **2**, p. 152
[4] Iliadis Ch. et al. 1990, *Nuclear Physics*, in press
[5] Vogelaer F. 1989, *Ph. D. Thesis*, unpublished
[6] Winters R., Macklin R. 1988, *Ap. J.*, **329**, 943
[7] Silberberg R., Tsao C. 1990, *Ap. J.*, **352**, L49
[8] Dearborn D., Blake J. 1988, *Ap. J.*, **332**, 305
[9] Arnould M. , Prantzos N. 1990, *in preparation*.

K. A. van der Hucht and B. Hidayat (eds.),
Wolf-Rayet Stars and Interrelations with Other Massive Stars in Galaxies, 550.
© 1991 *IAU. Printed in the Netherlands.*

IS ROTATION A REALLY IMPORTANT PROPERTY FOR THE UNDERSTANDING OF THE EVOLUTION OF MASSIVE STARS?

S.R. Sreenivasan
Department of Physics and Astronomy
The University of Calgary
Calgary, AB, T2N 1N4, Canada

Massive stars have been observed to possess significant rotational speeds. The frequency distribution shows two peaks one centred around 100 km/s and another about 250 km/s. No star has been observed to rotate at or near the critical speed which balances centrifugal force with gravity. This fact has played a large role in the neglect of rotational effects as being unimportant and dismissed as being very uncertain in the literature. On the other hand, standard models do not explain many of the observed features of stellar evolution in this mass-range. It is also customary to employ crude parameterised forms for significant effects such as those of mass-loss, as well as ad hoc variations in the rates of mass-loss which were originally derived by using statistical fits to observed mass-loss rates over a wide range of masses, effective temperatures and luminosities. Further, unrealistic prescriptions for phenomena that are not properly understood such as convective core overshoot are employed to bring about agreement with observed properties of massive stars that are still not completely secure.

It is shown in this work that even relatively low rotational speeds can cause significant effects to the observable properties and distributions of massive stars. Due to mass and angular momentum loss observed rotational speeds of stars do not indicate either the original speed or the history of the rotational speeds as the stars evolve. One can model self-consistently the rotational effects for stars which have relatively low speeds of the order of 1 to 3% in the ratio of centrifugal force to gravity and demonstrate that a number of features of massive stars and their evolutionary characteristics can be understood without the use of ad hoc formulations of the effects. In particular, the evolutionary history of the progenitor of SN 1987A, the characteristics of the evolution of the LMC stars as well as the pulsational characteristics of OB stars can be delineated without much stress. When applied to intermediate mass stars rotational effects can again explain many of the observed features that are not explainable on the basis of standard models. Some examples are: Blue Stragglers, Peculiar Red Giants, High Latitude F Supergiants. We conclude that rotational effects are quite important and cannot be neglected.

K. A. van der Hucht and B. Hidayat (eds.),
Wolf-Rayet Stars and Interrelations with Other Massive Stars in Galaxies, 551.
© 1991 *IAU. Printed in the Netherlands.*

NEW RESULTS IN EVOLUTION IN THE UPPER HRD

E.I. STARITSIN
Astronomical Observatory,
Ural State University,
Sverdlovsk, U.S.S.R.

SUMMARY. A theoretical interpretation of the observed upper luminosity limit is suggested here. Staritsin (1989) considered the core hydrogen and helium burning stages in a 64 $M_\odot$ star. Mixing in a semi-convection zone in a diffusion approximation (Staritsin, 1987) and mass loss by stellar wind (de Jager et al., 1988) were taken into account. During MS evolution the star looses half of its initial envelope. After MS evolution an intermediate convective zone appears. The hydrogen content in the shell source increases. As a result, the star burns helium in its blue supergiant stage. After the hydrogen content in the envelope has decreased to 10% of its mass, the star looses mass with Wolf-Rayet mass loss rates according to de Jager et al. (1988). The star has a WR character during 5% of its full life time.

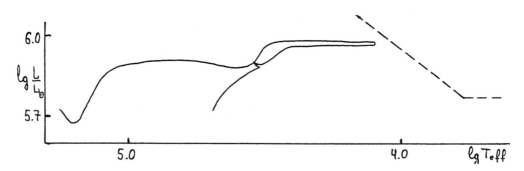

Figure 1. The evolution of a 64 $M_\odot$ star. The dashed line shows the upper luminosity limit according to Humphreys (1987).

References

Humphreys, R.M. 1987, *Publ. Astron. Soc. Pacific* **99**, 5.

de Jager, C., Nieuwenhuijzen, H., van der Hucht, K.A. 1988, *Astron. Astrophys. Suppl.* **72**, 259.

Staritsin, E. 1987, *Nauch. Inf. Astron. Sov.* **63**, 97.

Staritsin, E. 1989, *Astron. Zh.* **66**, 975 (= *Sov. Astron.* **33**, 503).

K. A. van der Hucht and B. Hidayat (eds.),
Wolf-Rayet Stars and Interrelations with Other Massive Stars in Galaxies, 552.
© 1991 *IAU. Printed in the Netherlands.*

THE WR/WR$_{progenitor}$ NUMBER RATIO: THEORY AND OBSERVATIONS.

D. VANBEVEREN
Dept. of Physics, VUB,
Brussels, Belgium.

SUMMARY. A direct comparison between the observed WR/WR$_{progenitor}$ number ratio within 2.5 kpc from the sun and the predicted value (using evolutionary computations of single stars of Maeder and Meynet, 1987, *A.&A.* **182**, 243) reveals a discrepancy of at least a factor of two. In a previous study (Vanbeveren, 1990, *A.&A.* in press) I proposed a solution based on the incompleteness of the observed OB type star sample within 2.5 kpc from the sun. In this summary, I propose a theoretical explanation for the discrepancy. The theoretically predicted WR/WR$_{progenitor}$ number ratio critically depends on the adopted $\dot{M}$ formalism in evolutionary computations during the red supergiant phase (RSG) of a massive star, especially in the mass range 20-40 $M_\odot$. Since any $\dot{M}$ formalism predicts the mass loss rate with an uncertainty of at least a factor of two, I have tried to look for solutions for the WR/WR$_{progenitor}$ problem by using different values of $\dot{M}$ during the RSG (in the mass range 20-40 $M_\odot$); the $\dot{M}$ values and formalism that were adopted were always choosen within the observational uncertainty (i.e. within a factor of two when compared to the formalism used by Maeder and Meynet, 1987).

The results.

When we allow for a ~40 % increase of the RSG $\dot{M}$ values in the mass range 20-40 $M_\odot$ when compared to the values used by Maeder and Meynet (1987), it can be concluded that

a. independent from whether a WR star is a single star or whether it is a member of a close binary, theoretical evolution predicts that WR stars originate from stars with initial ZAMS mass larger than 20 $M_\odot$ although the bulk descends from the M>25-30 $M_\odot$ range corresponding to observations.

b.the theoretically predicted WR/WR$_{progenitor}$ number ratio corresponds to the observed value within 2.5 kpc from the sun.

General conclusion: accounting for the uncertainty of the stellar wind mass loss rate in massive red supergiants and for the possible incompleteness of the OB type star sample within 2.5 kpc from the sun, the theoretically predicted WR/WR$_{progenitor}$ number ratio corresponds to the observed value.

K. A. van der Hucht and B. Hidayat (eds.),
Wolf-Rayet Stars and Interrelations with Other Massive Stars in Galaxies, 553.
© 1991 *IAU. Printed in the Netherlands.*

THE MASS OF WR PROGENITORS.

D. VANBEVEREN
Dept. of Physics, VUB,
Brussels, Belgium.

SUMMARY. A search through literature reveals four methods in order to derive the mass of WR progenitors, i.e.

a. WR stars must be descendant from the most massive stars which share their galactic distribution,
b. the computation of detailed evolutionary models of massive close binaries up to the WR phase, able to explain the observational constraints of these WR binaries,
c. comparing the very narrow mass-luminosity relation of massive core helium burning stars predicted by evolution and estimated bolometric luminosities of WR members of stellar aggregates,
d. the minimum mass of the progenitor of a WR member of a cluster equals the mass of the most luminous star (or the star with the earliest spectral type) in the cluster.

Method d is based on a very uncertain assumption of coeval massive star formation in stellar aggregates. Even then, one may realise that method d is essentially similar to method a (a WR star and the most luminous stars within one cluster have similar galactic coordinates) however method a gives a statistically more significant conclusion. All methods however use different data sets and different evolutionary models and this may lead to different conclusions. I have therefore reapplied methods a, b and c using the WR and OB star catalogues of van der Hucht et al. (1988, *A.&A.* **199**, 217), Smith and Maeder (1989, *A.&A.* **211**, 71), Humphreys and McElroy (1984, *Ap.J.* **284**, 565) and the evolutionary models with and without convective core overshooting of Vanbeveren (1987, *A.&A.* **182**, 207), Maeder and Meynet (1987, *A.&A.* **182**, 243), Vanbeveren (1989, *A.&A.* **224**, 93), Vanbeveren (1990, *A.&A.* in press). The three methods give very similar results, i.e.

method a: on a 95 % significance level the WR stars originate from massive stars with initial mass larger than 28 $M_\odot$ (non overshooting model) and 22 $M_\odot$ (with overshooting),

method b: based on detailed models for 17 well observed WR+OB binaries, it follows that WR components of close binaries originate from OB type stars with initial ZAMS mass larger than 25 $M_\odot$ (no overshooting) and 23 $M_\odot$ (with overshooting),

method c: when the bolometric correction of the majority of WR stars is larger than 4 mag (resp. 3.5 mag), then more than 80 % of the WR stars which are member of clusters or associations originate from stars with initial ZAMS mass larger than 30 $M_\odot$ (resp. 25 $M_\odot$) (no overshooting) and larger than 27 $M_\odot$ (resp. 23 $M_\odot$) (with overshooting).

K. A. van der Hucht and B. Hidayat (eds.),
Wolf-Rayet Stars and Interrelations with Other Massive Stars in Galaxies, 554.
© 1991 *IAU. Printed in the Netherlands.*

THE INFLUENCE OF WR LIKE STELLAR WIND MASS LOSS RATES ON THE EVOLUTION OF MASSIVE CORE HELIUM BURNING STARS.

D. VANBEVEREN
Dept. of Physics, VUB, Brussels, Belgium.

SUMMARY. Evolutionary computations of massive close binaries (MCB) including the effects of stellar wind (SW) and convective core overshooting predict that all massive primaries with ZAMS mass larger than 10 $M_\odot$ start their core helium burning phase (CHeB) as bare helium cores; the hydrogen rich layers are removed on a timescale of the order of 10^4 yrs as a consequence of Roche lobe overflow (RLOF). The CHeB remnant after RLOF resembles closely a zero age CHeB star and its further evolution is entirely independent from its binary nature. Similarly as has been done previously by Vanbeveren and Packet (1979, *A.&A.* **80**, 242), I have performed a phenomenological study on the evolution of massive hydrogen less CHeB stars including the effect of SW mass loss using updated $\dot{M}$ determinations of van der Hucht et al. (1986, *A.&A.* **168**, 111). The SW mass loss rate formalism used in the computations is based on the following requirements:

a. according to the theory of radiation driven winds, I looked for a relation $\dot{M}=aL^b$,

b. there are (only) two WR+OB binaries for which a reasonable good estimate of the mass of the WR component is known and which are also included in the updated $\dot{M}$ list of van der Hucht et al. (1986), i.e. V444 Cyg and γ^2Vel . The WR mass can be transformed into a luminosity using the M-L relation of hydrogen less CHeB stars proposed by Vanbeveren and Packet (1979); this gives us values for a and b (within some uncertainty margin of course),

c. the observed WN/WC number ratio of WR stars with a detected OB type companion ≈ 1.2; varying a and b leads to different predicted WN/WC ratios when the $\dot{M}$ formalism is applied in an evolutionary code.

The resulting relation which reproduces as closely as possible the foregoing requirements is given by

$$\dot{M}=3.2\ 10^{-13}\ L^{1.5} \quad \text{(L in } L_\odot, \dot{M} \text{ in } M_\odot/yr)$$

The evolutionary computations then reveal the following conclusions:

1. all primaries of MCB's with initial mass between 10 M$_\odot$ and 80 M$_\odot$ (possibly up to 100 M$_\odot$) end their life as stars with mass between 1.4 M$_\odot$ and 8 M$_\odot$ respectively,

2. all primaries with initial mass larger than 40 M$_\odot$ end their life as a WO star.

K. A. van der Hucht and B. Hidayat (eds.),
Wolf-Rayet Stars and Interrelations with Other Massive Stars in Galaxies, 555.
© 1991 *IAU. Printed in the Netherlands.*

FROM LBV BINARY TO WR+OB BINARY.

D. VANBEVEREN
Dept. of Physics, VUB,
Brussels, Belgium.

SUMMARY. The LBV phase is generally identified with the hydrogen shell burning phase of a star with initial mass larger than 40-50 M_Θ (see also Humphreys, this volume) and therefore in binaries with ZAMS component masses larger than 40-50 M_Θ and periods larger than 4-7 days the primary may experience a LBV mass loss phase before it reaches its critical equipotential surface (the Roche lobe). I then define the 'LBV scenario' of massive close binaries as follows:

> when a binary component (initial mass larger than 40-50 M_Θ) reaches the LBV phase prior to its Roche lobe overflow phase (RLOF), a stellar wind mass loss phase sets in at rates comparable to the rates encounterd during a RLOF process and which are large enough in order to prohibit the occurence of a RLOF.

Evolutionary models are computed for close binaries with initial primary masses larger than 50 M_Θ and mass ratios ranging between 0.2 and 1. Special attention is given at the predicted spectral type of the secondary component. In order to determine the mass of the primary at the end of its LBV phase, I have used the following general theorem holding for the most massive stars, i.e.

> a hydrogen shell burning mass loser with mass larger than 50 M_Θ in a massive close binary restores thermal equilibrium (and becomes a WR star) when helium starts burning in its core and when its atmospherical hydrogen abundance has dropped to $X_{atm} = 0.2$-0.3 (by weight).

Since LBV mass loss is spherically symmetric, the binary period P increases during the LBV phase (i.e. $P \div (M_1+M_2)^{-2}$, M1 and M2 are the masses of both components) whereas accretion effects can be neglected (i.e. the mass and spectral type of the companion star remain unchanged during the LBV phase). At the end of the LBV phase the binary consists of a stripped helium core (a WR star) and an OB type giant or supergiant; the binary period should be larger than 10-20 days. Possible WR binary candidates which may have evolved through such a LBV phase are HD 92740, HD 68273, HD 137603 and HD 168206.

This summary is part of a paper which has been submitted to Astron. and Astrophys.

K. A. van der Hucht and B. Hidayat (eds.),
Wolf-Rayet Stars and Interrelations with Other Massive Stars in Galaxies, 556.
© 1991 IAU. Printed in the Netherlands.

P CYGNI: WILL IT EVER BECOME A WOLF-RAYET STAR?

MART DE GROOT
Armagh Observatory
College Hill, Armagh
Northern Ireland, BT61 9DG

P Cygni is a BIIa$^+$ hypergiant situated close to the Humphreys-Davidson Limit in the Hertzsprung-Russell diagram. A comparison of its basic physical parameters with computed evolutionary tracks suggests that P Cygni may become a WR star (Lamers et al. 1983, Astron. Astrophys., 123, L8).

However, there are problems with such a scenario.
1. WR stars may not be such evolved objects as commonly believed and their spectral characteristics may be explained by an appropriate treatment of radiative-transfer theory (Bhatia and Underhill, 1988, Ap. J. Suppl., 67, 187).
2. The practice of determining the evolutionary status of a single object like P Cygni from its position in the HRD through a comparison with computed evolutionary tracks is fraught with uncertainties and should be handled with great care.
3. Although WNL stars have much in common with P Cygni, similar luminosities, radii (Doom, 1988, Astron. Astrophys., 192, 170), linewidths, etc., and may even have a two-component stellar wind similar to OB supergiants (see, e.g., Zickgraf et al. 1985, Astron. Astrophys., 143, 421, and references therein; Bhatia and Underhill, 1986, Ap. J. Suppl., 60, 323; 1988, Ap. J. Suppl., 67, 187; Poe et al. 1989, Ap. J., 337, 88), the force driving their winds may be rather different. Whereas in P Cygni the radiation pressure on many lines explains the observed wind phenomena (Lamers, 1986, Astron. Astrophys., 159, 90), the energy present in winds from WR stars is far greater than what can be contributed by radiation pressure alone (Underhill, 1983, Ap. J., 265, 933).

I contend that the conventional picture about the atmospheric parameters and evolutionary status of WR stars has been dictated by the combination of the complexity of their spectra and their, possible, erroneous interpretation in the past, and the plausibility of the arguments brought forward by the modellers of stellar evolution. However, the still existing problems about WR stars, apparent when one considers questions like the one posed in the title of this paper, make it necessary to take a fresh, unbiased look at this whole problem.

557

K. A. van der Hucht and B. Hidayat (eds.),
Wolf-Rayet Stars and Interrelations with Other Massive Stars in Galaxies, 557.
© 1991 IAU. Printed in the Netherlands.

THE INFRARED SPECTRUM OF P CYGNI

J. R. Deacon and M. J. Barlow
Department of Physics & Astronomy
University College London
Gower Street, London WC1E 6BT

ABSTRACT. An energy distribution for P Cyg (B1 Ia$^+$) has been produced using UKIRT photometry and spectra, flux-calibrated IUE high-resolution spectra, Johnson and Mitchell 13-colour photometry and *IRAS* photometry. Infrared excesses due to free-free emission from the wind have then been derived and modelled.

1. The Infrared Excesses

The IR excesses were derived by fitting a Kurucz model atmosphere to the UV and optical data, the best fit being given by a model with T_{eff} =18000K and log g=2.05. Subtracting this model from the observations gave the excess fluxes, which have been modelled assuming a spherically symmetric wind parameterised by either Castor and Lamers (CL) velocity laws of index β or by the linear velocity law used by Waters and Wesselius (1986, *Astr. Astrophys.*, **155**, 104). A full treatment of electron scattering and free-free and bound-free opacity was included, along with the radial dependence of wind electron temperature given by the work of Drew (1989, *Astrophys. J. Suppl.*, **71**, 267).

2. Results

A good fit to the observations was found using the linear velocity law with the following parameters :

$$T_{eff} =18000K \qquad \log g=2.05 \qquad R_*= 92.5\ R_\odot \qquad D=1.8kpc$$
$$v_\infty= 206\ km\ s^{-1} \quad v(R_*)/v_\infty= 0.175 \quad \dot{M} = 2.16\times10^{-5}\ M_\odot y^{-1} \quad n(He)/n(H) = 0.5$$

the value of v_∞ is taken from Lamers *et al.* (1985, *Astr. Astrophys.*, **149**, 29), $\dot{M}$ from radio observations of van den Oord (1985, in *Radio Stars*), $n(He)/n(H)$ was derived from the *JHKL* spectra and R_* by normalising the Kurucz model to the J band magnitude.

3. Discussion

It was found that the linear velocity law gave the best fit to the observations. The CL velocity laws predicted energy distributions which fell off too steeply longward of $\sim$10μm. It appears that longward of $\sim$25μm the adopted model starts to predict smaller excesses than are observed—the possible need to include density fluctuations such as those predicted by Owocki *et al.* (1988, *Astrophys. J.*, **335**, 914) will be investigated.

K. A. van der Hucht and B. Hidayat (eds.),
Wolf-Rayet Stars and Interrelations with Other Massive Stars in Galaxies, 558.
© 1991 *IAU. Printed in the Netherlands.*

A SPECTROSCOPIC STUDY OF η CAR AND THE HOMUNCULUS

D. John Hillier
Joint Institute for Laboratory Astrophysics, University of Colorado
Boulder, Colorado 80309-0440, USA

David A. Allen
Anglo-Australian Observatory
PO Box 296, Epping, NSW 2121, Australia

Spectrophotometric maps of the peculiar massive star η Car and its associated nebulae (the homunculus) were obtained in 1986. The spatial resolution is better than 1.5 arcseconds (0.7 arcseconds per pixel). Spectral coverage extends from 4000Å to 1μm at low (3Å /pixel) resolution, with additional observations of selected regions obtained at moderate resolution (0.7Å /pixel).

We find dramatic differences in line profiles and line strengths across the homunculus. Both lobes show predominantly red shifted emission, with largest velocities occurring in the NW lobe. Our spectra confirm earlier suggestions that the spectrum of the homunculus arises predominantly through outflowing dust scattering light from the central object.

Variations in line ratios between spectra of the central object and homunculus force us to conclude that two distinct regions give rise to the majority of the observed emission lines. Light from the central object is viewed through a dusty disk which also gives rise to the [Fe II] lines. Because the continuum, and also the bulk of the H and He I emission, arises inside the attenuating dust the [Fe II] lines appear strongest (and have the largest equivalent widths) on the central object. In the scattering lobes the continuum starlight is only weakly attenuated; therefore the scattered [Fe II] emission is relatively weak.

Interpretation of the spectra is confused by the possible presence of overlying emission due to material not related to the homunculus — material which may be related to the outlying condensations. We see emission lines of CaII, [Ca II], Na I, [Cr II], [Fe II], and [Ni II] which may be intrinsic to the homunculus. Some of these features show peculiar velocities — we find blue shifted [CaII] emission offset to the NW ($\approx 2\,arcseconds$) from the central object.

A more detailed account of this work is in preparation. A recent introduction to the literature on η Car can be found in the paper of Viotti *et al.* (1990, *Ap. J. Suppl.*, **71**, 983).

559

K. A. van der Hucht and B. Hidayat (eds.),
Wolf-Rayet Stars and Interrelations with Other Massive Stars in Galaxies, 559.
© 1991 *IAU. Printed in the Netherlands.*

THE DISTANCE TO THE S DOR TYPE STAR HR CARINAE

A.M. van Genderen, F.H.A. Robijn and B.P.M. van Esch
Leiden Observatory
Postbus 9513, 2300 RA Leiden
The Netherlands

H.J.G.L.M. Lamers
SRON Space Research Laboratory
Sorbonnelaan 2
3584 CA Utrecht
The Netherlands

ABSTRACT. The distance of HR Carinae is determined with the reddening distance method, resulting into r ~ 5 kpc.

The distance of HR Car is usually taken to be ~ 2.5 kpc (Viotti, 1971). With this distance and a reddening $E(B-V)_J$ ~ 1 (van Genderen et al. 1990) the absolute magnitude turned out to be much too low compared with other S Dor type stars. Applying Wolf's (1989) amplitude luminosity relation, the probable distance turned out to be r ~ 6 kpc (van Genderen et al., 1990).

   Therefore a check by the reddening distance method with the aid of photometry of neighbouring stars was necessary. About 60 stars within an area with a radius of 10' around HR Car (limiting magnitude ~ 14.5) were measured with the VBLUW photometer mounted on the 90-cm Dutch telescope at the ESO, Chile. It appears that HR Car fits satisfactorily in the $E(B-V)_J/r$ diagram with r ~ 5 kpc, which agrees with the new value mentioned above.

   Adopting M_{bol} = -9.5 (for r = 6 kpc) and T_{eff} ~ 14000 K (van Genderen et al. 1990) the position of HR Car in the theoretical HR diagram is now consistent with other S Dor type stars.

References

van Genderen, A.M., Thé, P.S., Heemskerk, M., Heynderickx, D., van Kampen, E., Kraakman, H., Larsen, I., Remijn, L., Wanders, I., van Weeren, N.: 1990, Astron. Astrophys. Suppl. 82, 189
Viotti, R.: 1971, Publ. Astron. Soc. Pac. 83, 170
Wolf, B.: 1989, Astrophys. 217, 87

K. A. van der Hucht and B. Hidayat (eds.),
Wolf-Rayet Stars and Interrelations with Other Massive Stars in Galaxies, 560.
© 1991 IAU. Printed in the Netherlands.

R127 AND ITS SURROUNDING STAR CLUSTER

ANTONELLA NOTA[1,2], CLAUS LEITHERER[1,3], MARK CLAMPIN[4], and
ROBERTO GILMOZZI[1,3]
[1] STScI, 3700 San Martin Drive, Baltimore, MD 21218
[2] On leave from the Observatory of Padova, Italy
[3] Affiliated with the Astrophysics Division, Space Science Department of ESA
[4] Johns Hopkins University, Homewood Campus, Baltimore, MD 21218

The LMC star R127 (=HDE 269858) has originally been classified as a late WN type
(Walborn 1977). Since then, R127 has developed from spectral type B to A (Wolf *et al.*
1988) and is now categorized as a Luminous Blue Variable. R127 is embedded in gaseous
material and a star cluster which are barely resolvable even on the highest-quality images
available. Stahl (1987) studied the circumstellar material using narrow-band CCD images
centered on nebular emission lines. No systematic census of the stars in the immediate
vicinity of R127 has yet been published.

We obtained high-resolution images of R127 with the STScI coronograph mounted on
the ESO/MPI 2.2-m telescope in February 1989 (see Paresce, Burrows, and Horne 1987
for a description of the instrument). R127 is occulted by a long, thin wedge in order to
reduce its brightness ($V \approx 9^m$) relative to the surrounding stars. Figure 1 is a V image
of the region around R127. (Field size: 22"x36"; plate scale: 0.14" per pixel; FWHM of
stellar images: 0.9"). R127 is surrounded by a number fainter stars, most of which were
previously unknown due to the brightness of R127 itself. V photometry of the stars has
been obtained by calibrating the V frame with the standard star Feige 15 (see Table 1). No
color transformations have been performed. The photometric errors are less than 0.2.

R127 has a companion (designated R127B) with $V = 12.9$ at a distance of ~ 3.5"
to the northwest. Another close companion (R127U, $V = 16.1$) is ~ 3.6" to the south.
Although these companions make a negligible contribution to the visual flux of R127, their
influence in the UV at ~ 1500 Å may be crucial if observations with insufficient resolution
are performed. Stahl *et. al* (1983) report that *IUE* spectra of R127 are contaminated
by R127B and another companion (R127U?). Since the UV flux of R127 has decreased
substantially in the meantime, the light contribution in the UV of these companions will
be much more important at present.

All stars in the field of view are most probably LMC stars which are not physically
associated with R127.

Our high-resolution image gives no indication for multiplicity of R127. The point-spread
function is consistent with the assumption that only one star significantly contributes to the
visual light. Our results support the suggestion that R127 is a very luminous ($L \approx 10^6 L_\odot$)
star with an initial mass between 60 $M_\odot$ and 100 $M_\odot$.

561

K. A. van der Hucht and B. Hidayat (eds.),
Wolf-Rayet Stars and Interrelations with Other Massive Stars in Galaxies, 561–562.
© 1991 *IAU. Printed in the Netherlands.*

562

References

Paresce, F., Burrows, C., and Horne, K. 1988, *Ap. J.*, **329**, 318.
Stahl, O. 1987, *Astr. Ap.*, **182**, 229.
Stahl, O., Wolf, B., Klare, G., Cassatella, A., Krautter, J., Persi, P., and
  Ferrari-Toniolo 1983, *Astr. Ap.*, **127**, 49.
Walborn, N. R. 1977, *Ap. J.*, **215**, 53.
Wolf, B., Stahl, O., Smolinski, J., and Cassatella, A. 1988, *Astr. Ap. Suppl.*, 74, 239.

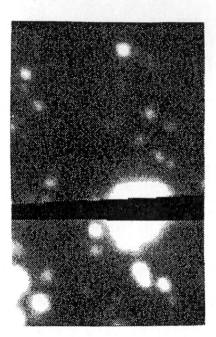
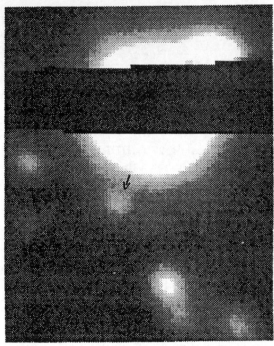

Figure 1. Coronographic V image of the field around R127. The right-hand figure is an enlarged version of the left-hand figure. North is up and east to the left. R127 itself is partly occulted by the wedge. The position of R127U is indicated by an arrow.

Star	x-Pos	y-Pos	V
B	109	117	12.9
C	107	155	18.2
D	131	166	18.3
E	123	141	17.6
F	133	134	18.0
G	81	166	19.0
H	13	208	16.8
I	23	183	17.5
J	95	227	16.0
K	114	176	17.4

Star	x-Pos	y-Pos	V
L	34	119	17.5
M	69	79	15.9
N	69	61	17.2
O_2	107	42	14.8
P	126	34	17.0
Q	134	22	17.9
R	61	28	18.8
S	31	40	16.8
T	25	19	15.0
U	94	67	16.1

Table 1. Photometry of the R127 cluster. The zero point of the coordinate system is in the southeast corner of Figure 1. The northwest corner has $x = 160$, $y = 256$.

PHOTOMETRIC EVIDENCE OF SMALL "S DOR TYPE ERUPTIONS" IN S DOR TYPE STARS IN OR NEAR MINIMUM

A.M. van Genderen
Leiden Observatory
Postbus 9513, 2300 RA Leiden
The Netherlands

ABSTRACT. It is shown that a number of S Dor type stars in or near minimum brightness show small S Dor eruptions: $\Delta V_J < 0^m5$

1. The observations

With the aid of VBLUW photometry van Genderen et al. (1990) have shown that the S Dor type stars AG Car and HR Car (in or close to minimum) show small S Dor type eruptions, causing light variations of $\Delta V_J \sim 0^m3$. They are accompanied by small colour vairations: if the brightness goes up the colour becomes redder and vice versa. This is a typical "S Dor effect". The time scale is of the order of 1/2 yr.

UBV data of the three LMC S Dor type objects: R84, R85 and R99, scattered over a time interval of 25 yr, and collected by Stahl et al. (1984) clearly show this S Dor effect.

De Groot's (1989) light curve in V of P Cyg strongly suggests that small intrinsic micro oscillations ($P \sim 18^d$) are superimposed on two waves of long term variations of $\sim$ 4 months: $\Delta V_J \sim 0^m2$. Probably they are small S Dor eruptions also.

2. Conclusions

Obviously these S Dor type stars in or near minimum, are still subject to mass loss eruptions. It is possible that they are identical with the shell ejections detected in the spectra of P Cyg with a time scale of $\sim$ 1/2 yr. (van Gent and Lamers, 1986; Markova, 1986).

References

van Genderen, A.M., Thé, P.S., Heemskerk, M., Heynderickx, D., van Kampen, E., Kraakman, H., Larsen, I., Remijn, L., Wanders, I., van Weeren, N.: 1990, Astron. Astrophys. Suppl. 82, 189
van Gent, R.H., Lamers, H.J.G.L.M.: 1986, Astron. Astrophys. 158, 335
de Groot, M.: 1989, Armagh Obs. Repr. No. 96
Markova, N.: 1986, Astron. Astrophys. 162, L3
Stahl, O., Wolf, B., Leitherer, C., Zickgraf, F.-J., Krautter, J., de Groot, M.: 1984, Astron. Astrophys. 140, 459

K. A. van der Hucht and B. Hidayat (eds.),
Wolf-Rayet Stars and Interrelations with Other Massive Stars in Galaxies, 563.
© *1991 IAU. Printed in the Netherlands.*

OBSERVATIONAL EVIDENCE FOR EVOLUTIONARY CONNECTIONS BETWEEN WOLF-RAYET STARS AND RED SUPERGIANTS

Leonid N. Georgiev
Department of Astronomy
Sofia University
5 Anton Ivanov Str.
Sofia 1126, Bulgaria.

The main question in our understanding of the nature of WR stars is: which stars are their progenitors and how do they evolve to reach the WR stage? We have some ideas about progenitor's masses and, by mass luminosity relations, about their brightness when they are O stars. But, it is difficult to say anything for their possible Red phase.

We suggest one statistical approach to this problem, based on the recent data for M33.

We have used the data for WR stars from Massey and Conti (1983), Armandroff and Massey (1985) and Massey *et al.* (1987). The stars with coincident coordinates ($3''$ error box) and the same spectral index are excluded. The data for red and blue supergiants are taken from the catalog of Freedman (1984), appended by Ivanov (1990). The combined catalog is complete up to $19^m.5$. Because we are interested in supergiants only, we stop our calculations at this limit. We have accepted as blue supergiants (BSG) the stars with $U - V < -0^m.3$, and for RSG these with $B - V > 1^m.9$ (Humpheys *et al.*, 1985).

The evolutionary theory predicts two main paths for stars with masses greater than $30M_\odot$.

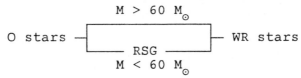

Thence the distribution of O stars (P_{BSG}) must be the same as the distribution of (RSG + WR) stars - P_{WR+RSG}.

Assuming an apparent distance modulus to M33 of about $25^m.0$, we deal with O stars brighter than $M_v = -5^m.5$, or with $M > 30M_\odot$. Also we have a fixed number of WR stars. So, to obtain a coincidence between the distributions mentioned above, we can vary only the magnitude m_1. RSG brighter than m_1 will be added to WR stars and the resulting distribution $P_{WR} + RSG$ is compared with P_{BSG}. The best coincidence occurs for $m_1 = 18^m.5$, and we conclude that the red stars in M33 brighter than $18^m.5$ evolve to WR stars.

Maeder *et al.* (1980) have shown that the lifetime of each phase depends on metal abundance. In favour of this result they suggest an observed gradient in relative numbers of WR stars (N_{WR}) and RSG (N_{RSG}). Using our limit magnitude $18^m.5$ for RSG we confirm the presence of a gradient of N_{WR}/N_{RSG} in M33.

Based on the data of the chemical composition in M33 (Kwitter and Aller, 1981) we have also found that the ratio N_{WR}/N_{RSG} is sensitive to the metallicity Z.

References

Armandroff, T.E., and Massey, P., 1985, *Astrophys. J.* **291**, 685.
Conti, P.S., and Massey, P., 1983, *Astrophys. J.* **273**, 576.
Freedman, W.L., 1984 *Thesis* University of Toronto.
Humphreys, R.M., Nichols, M. and Massey, P., 1985 *Astrophys. J.* **90**, 101.
Ivanov, G., 1990, in preparation.
Kwitter, K.B., and Aller, L.H., 1981 *M.N.R.A.S.* **195**, 939.
Maeder, A., Lequeux, J. and Azzopardi, M., 1981, *Astron. Astrophys. (Lett)* **90**, L17.
Massey, P, Conti, P., Moffat, A., and Shara, M., 1987 *P.A.S.P.* **99**, 816.

K. A. van der Hucht and B. Hidayat (eds.),
Wolf-Rayet Stars and Interrelations with Other Massive Stars in Galaxies, 564.
© 1991 *IAU. Printed in the Netherlands.*

THE ANOMALOUS A-TYPE SUPERGIANTS IN THE MAGELLANIC CLOUDS - EVIDENCE FOR POST-RED SUPERGIANT EVOLUTION

ROBERTA M. HUMPHREYS, University of Minnesota
R. P. KUDRITZKI and H. GROTH, Universitat Munchen

A group of A-type supergiants in the SMC and LMC with spectral types in the range B8-F0 have an anomalously strong Balmer jump and hydrogen lines for their luminosities. The appearance of the Balmer series suggest a luminosity about two magnitudes fainter than derived from their membership in the Clouds. Their colors are also too red in U-B with little or no change in B-V relative to the colors of normal A-type supergiants, consistent with the strong Balmer jump in these star's spectra.

Earlier Kudritzki (1973) had shown that an increased helium abundance will result in stronger hydrogen lines and Balmer jump (and red U-B color). To test this hypothesis, we have obtained high resolution spectra of four of these anomalous supergiants in the SMC and two in the LMC together with spectra of several normal supergiants in both Clouds. The measured equivalent widths of the hydrogen lines confirm their increased strength in these anomalous A-type supergiants while their metallic lines are comparable to those in normal supergiants of similar temperature, luminosity and metallicity.

A grid of NLTE models (8750-14000 K, $\log g = 0.75$-2.00, and $Y = 0.1$ to 1.00) show that an enhanced He abundance up to .5 by number can account for the stronger Balmer series lines. Details of the computations and the fine analysis will appear in a later paper.

Given the location of these anomalous A-type supergiants on the HR diagram plus the strong evidence for enhanced He in their atmospheres, these stars have probably been through a high mass loss stage as red supergiants. They are good candidates for post-red supergiant evolution and the progenitors of the kind of star that became SN1987A.

K. A. van der Hucht and B. Hidayat (eds.),
Wolf-Rayet Stars and Interrelations with Other Massive Stars in Galaxies, 565.
© 1991 IAU. Printed in the Netherlands.

SUPERNOVAE FROM WOLF-RAYET STARS

NORBERT LANGER[1] and STANFORD E. WOOSLEY[2]

1: *Universitäts-Sternwarte Göttingen, F.R.G.*

2: *Lick Observatory, UC Santa Cruz, U.S.A.*

The complete evolution of a star with an initial mass of $60\,M_\odot$ and Z=Z$_\odot$ (i.e. a typical WR progenitor) from the ZAMS through the supernova phase has been investigated. Mass loss in the different evolutionary stages, especially mass dependent WR mass loss, leads to a WO star (surface mass fractions { He,C,O} ={ 0.14, 0.38, 0.48 } ; cf. Fig. 1) of $4.2\,M_\odot$ as pre-SN configuration. The low final mass may be typical for a wide range of initial masses (cf. Langer, 1989, Astr. Ap. 220, 135).

In the ensueing SN explosion ($1.65\,10^{51}\,erg, v_{max} \simeq 1.5\,10^9\,cm\,s^{-1}$) $\sim 0.3\,M_\odot$ radioactive ^{56}Ni are synthesized (Fig. 1). The reason for this large amount of nickel is a shallow density gradient in the pre-SN star, a consequence of its former large initial mass. Composition mixing was found to have only small effects on the peak width of the bolometric light curve. Effects of clumping, simulated by decreasing the e$^-$-opacity, lead to an earlier and brighter maximum and a smaller peak width (Fig. 2).

We conclude that light curves of explosions of low mass WR stars, which originate from massive O-stars, may resemble those of Type Ib supernovae. The large amount of radioactive ^{56}Ni and the spectroscopic signature of the mantel composition may discriminate our model from other proposed Type Ib progenitor scenarios.

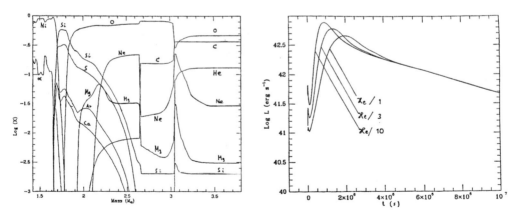

Fig.1 (left): Post-explosion composition of the considered $4.2\,M_\odot$ WR star prior to partial mixing (from: Woosley, Langer, Weaver, in prep.). **Fig.2** (right): Bolometric light curve of the explosion of the partly mixed $4.2\,M_\odot$ WR object for three different values of the e$^-$-opacity (from: Woosley, Ensman, Langer, in prep.).

K. A. van der Hucht and B. Hidayat (eds.),
Wolf-Rayet Stars and Interrelations with Other Massive Stars in Galaxies, 566.
© 1991 *IAU. Printed in the Netherlands.*

PROGENITORS OF TYPE Ib SUPERNOVAE

NINO PANAGIA [1]
Space Telescope Science Institute

VICTORIA G. LAIDLER [2]
Astronomy Programs, Computer Sciences Corporation.

ABSTRACT. We argue that the progenitors of type Ib supernovae are moderately massive stars ($M \sim 7\ M_\odot$) in binary systems whereas the hypothesis that they originate from very massive stars ($M > 20\ M_\odot$) is not consistent with the observational evidence on SNIb.

1. Introduction

The realization that there is a separate subclass of type I supernovae (SNe) to be denoted as type Ib came after the detailed study of the SN 1983N in M83 (Panagia *et al.* 1990; see also Panagia 1985, Wheeler and Levreault 1985, Uomoto and Kirshner 1985). It was immediately clear that SN 1983N is distinctly different from the classical variety of type I SNe (obviously denoted as type Ia SNe) in a number of important aspects. Since then about a dozen SNe have been classified as type Ib SNe, some newly discovered and others found just re-examining old spectra or paying due attention to the comments that the observers gave at the time when the original observations were made (*e.g.* Bertola 1964). In addition to the necessary condition to be called type I SNe, *i.e.* the absence of hydrogen lines from the spectrum, the distinctive characteristics of type Ib SNe can be summarized as follows (cf. Panagia *et al.* 1986, Weiler and Sramek 1988):

- The 6150 Å feature is absent from the spectrum.

- The overall spectral distribution is redder (Δ(B-V)~ 0.5) and fainter (~ 1.5 magnitudes) than for type Ia SNe.

- The optical light curve is essentially "normal", *i.e.* quite similar to that of type Ia SNe.

- The IR light curve is single-peaked, the maximum occurring a few days after the optical maximum.

---

[1] Affiliated to the Astrophysics Division, Space Science Department of ESA; on leave from University of Catania.
[2] Staff member, Space Telescope Science Institute.

K. A. van der Hucht and B. Hidayat (eds.),
Wolf-Rayet Stars and Interrelations with Other Massive Stars in Galaxies, 567–570.
© 1991 *IAU. Printed in the Netherlands.*

- They are strong radio emitters with a steep spectrum and a quick temporal decline.

- They are found only in spiral galaxies.

- They are located in spiral arms and are often projected near an HII region.

In the following we discuss these points in some detail and draw conclusions about the nature of SNIb progenitors.

2. Discussion

2.1. OPTICAL EMISSION

The first four points imply that SNIb originate from rather compact stars (pre-supernova radius less that $few \times 10^{12}$ cm, hence *not* a red supergiant) and have an envelope possibly as small as that of type Ia SNe (because of the similarity of the light curves), but have a chemical composition different from that of type Ia progenitors (absence of the 6150 Å band) and a lower amount of ^{56}Ni synthetized in the explosion (redder and fainter emission).

2.2. RADIO EMISSION

The substantial radio emission of SNIb requires the presence of circumstellar material, that results from pre-supernova mass loss with $\dot{M}/v_{exp} \sim 3 \times 10^{-7}[M_\odot \ yr^{-1}]/[km \ s^{-1}]$ (Weiler *et al.* 1986). Such a flow is rather opaque and may create a *pseudo-photosphere* at the base of the wind itself. For example, assuming a constant wind velocity, the radius at which the optical depth is of the order of unity is given by (Panagia and Felli 1982):

$$R \approx 3 \times 10^{12} \ (\kappa/\sigma_e)\{(\dot{M}/v_{exp})/10^{-7}\} \ [cm] \ > \ 9 \times 10^{12} \ [cm]$$

where κ is the average opacity, σ_e is the electron scattering cross section, $\dot{M}$ is in $M_\odot \ yr^{-1}$ and v_{exp} in $km \ s^{-1}$. A radius like this is about an order of magnitude larger than that of a WR star and is too large for the SNIb progenitor because it would produce a light curve much broader than observed. On the other hand, the value of $\dot{M}/v_{exp} \sim 3 \times 10^{-7} \ [M_\odot \ yr^{-1}]/[km \ s^{-1}]$ is appropriate for a red supergiant. Therefore, the SNIb progenitor has to have a relatively massive companion (*i.e.* several solar masses; Sramek *et al.* 1984) that at the time of the explosion is in the red supergiant phase. Note that the binary system hypothesis to explain the radio emission of SNIb is not applicable to the case of a WR progenitor because the WR wind would completely sweep out the wind of a possible red supergiant companion and, again, the density in the circumstellar material would be far too low to produce the radio phenomenon (*i.e.* to accelerate the relativistic electrons that emit synchrotron radiation and to cause the *f-f* absorption that delays the rise of the flux at long wavelengths).

2.3 LOCATION IN THE GALAXY

In order to satisfy the condition of being located in spiral arms, the lifetime of SNIb progenitors must be shorter than 3×10^8 years, and, therefore their original mass larger than 5.5 $M_\odot$. In fact, *if* type Ib SNe were intrinsically associated with HII regions, their progenitors may be short-lived and, therefore, be appreciably more massive.

Wheeler *et al.* (1987) have claimed that about 50% of type Ib SNe are associated with HII regions. We have made a systematic study of the separations of SNe from HII regions in spiral galaxies by overlaying the best positions of SNe [among those reported in the Barbon *et al.* Catalog of SNe [12] and those astrometrically determined either in the optical or in the radio] on the galaxy images with the use of the GASP[3] software (Panagia and Laidler 1990. For type Ib SNe we find that 6 out of 11 appear to fall within 5" from the center of a knot (*i.e.* presumably an HII region).

Actually, such an "association" with an HII region actually means *close proximity* because the median *absolute distance* of SNIb from HII regions turns out to be about 200 pc: this is much larger than the distance of early O type stars from giant HII regions in our Galaxy. On the other hand, a similar distribution of separations from the nearest HII region is found for SNII in spiral galaxies while for SNIa the distances from HII regions are much larger (the median distance is about 600 pc). This confirms the idea that the progenitors of both SNIb and SNII belong to a young stellar population but is not enough to derive quantitative estimates of lifetimes and/or masses of the progenitors. The main problems here are the limited sample, the poor resolution of ground based observations (we remind that for a galaxy at 10 Mpc a separation of 1" corresponds to a distance of 50 pc!), and insufficient accuracy of recorded positions of SNe (typically affected by errors of 3" or more). These last two effects may be responsible of increasing the dispersion and make the apparent correlation less tight than it may be in reality.

In any case, it is clear that SNIb *cannot* be the result of the explosion of a WR star because the strong radio emission at early epochs cannot be explained with the presence of a WR star. Therefore, the progenitors of type Ib SNe are likely to be only moderately massive stars, *i.e.* in the low part of the mass distribution of massive stars, say, around 10 $M_\odot$.

2.4 SNIb STATISTICS

The range of possible progenitor's masses can be estimated considering that the frequency of type Ib explosions in spiral galaxies is about 1/3 that of type II SNe (Branch 1986, van den Bergh, McClure, and Evans 1987). Therefore, assuming that stars more massive than 8 $M_\odot$ make type II SNe (*e.g.* Maeder 1987) and adopting an initial mass function proportional to $M^{-2.35}$ the possible mass range for SN Ib progenitors turns out to be about 6.5-8 $M_\odot$. This agrees well with the direct estimate of M > 6.5 $M_\odot$ made by Sramek *et al.* (1984) for SN 1983N on the basis of its radio emission. Also, a relatively "modest" mass for the progenitor can naturally explain why the mass ejected in the explosion (as implied by the "normal" optical light curve) is a few solar masses at most (Branch 1988).

3. Conclusions

We conclude that the only viable scenario to account for type Ib events is that of a star with original mass around 7 $M_\odot$, which is member of a binary system in which the companion is slightly less massive (say, $\sim$ 5 $M_\odot$). The primary follows its evolution to the end becoming a rather massive degenerate star, which explodes when the secondary has reached the stage

---

[3]GASP is the Guide Star Astrometric Support Program available at the Space Telescope Science Institute.

of red supergiant. The alternative hypothesis of a very massive progenitor (*i.e.* M > 20-30 $M_\odot$) is ruled out on the basis of the mass loss characteristics required to account for the radio emission and by the low mass envelope implied by the behaviour of the optical and UV light curves.

Finally, let us stress that our point is not that WR stars do not make SN explosions but only that they do not make SNIb events. For example, we wonder if the core collapse of a WR may be "gentle" enough to produce "dim" explosions such as that believed to have given rise to Cas A.

4. References

Barbon, R., Cappellaro, E., Ciatti, F., Turatto, M., Kowal, C.T., 1984, *Astron. Astrophys. Suppl.*, **58**, 735.

Bertola, F., 1964, *Ann. Astrophys.*, **27**, 319.

Branch, D., 1986, *Astrophys. J. (Letters)*, **300**, L51.

Branch, D., 1988, *IAU Colloquium No. 108*, ed. K. Nomoto (Berlin: Springer Verlag), p. 281.

Maeder, A., 1987, *ESO Workshop "SN 1987A"*, ed. I.J. Danziger, p. 251.

Panagia, N., 1985, in *"Supernovae as Distance Indicators"*, ed. N. Bartel (Berlin: Springer), p. 14.

Panagia, N., Felli, M.: in *"Wolf-Rayet Stars: Observations, Physics, Evolution"*, eds. C.W.H. de Loore and A.J. Willis (Dordrecht: Reidel), p. 203 (1982).

Panagia, N., Laidler, V.G., 1990, *Astrophys. J.*, in press.

Panagia, N., Sramek, R.A., Weiler, K.W., 1986, *Astrophys. J. (Letters)*, **300**, L55.

Panagia, N. et al, 1990, in preparation.

Sramek, R.A., Panagia, N., Weiler, K.W., 1984, *Astrophys. J. (Letters)*, **285**, L59.

Uomoto, A., Kirshner, R.P., 1985, *Astron. Astrophys.*, **149**, L7.

van den Bergh, S., McClure, R.D., Evans, R., 1987, *Astrophys. J.*, **323**, 44.

Weiler, K.W., Sramek, R.A., Panagia, N., van der Hulst, J.M., Salvati, M., 1986, *Astrophys. J.*, **301**, 790.

Weiler, K.W., Sramek, R.A., 1988, *Ann. Rev. Astron. Astrophys.*, **26**, 295.

Wheeler, J.C., Levreault, R., 1985, *Astrophys. J. (Letters)*, **294**, 17.

Wheeler, J.C., Harkness, R.P., Cappellaro, E., 1987, *Proc. 13th Texas Symposium on Relativistic Astrophysics*, ed. M.P. Ulmer (Singapore: World Scientific), p. 402.

EVOLUTION OF THE PROGENITOR OF SN 1987A IN THE HR DIAGRAM

H. YAMAOKA[1], H. SAIO[1], K. NOMOTO[1], and M. KATO[2]
[1]Department of Astronomy, University of Tokyo
[2]Department of Astronomy, Keio University

We calculate the blue–red–blue evolution of the progenitor of SN 1987A in the HR diagram by adopting the Schwarzschild criterion for convection and by choosing appropriate parameters for mass loss and mixing [Fig. 1 (left) where metallicity is $Z = 0.005$]. During helium burning, the star moves from the blue to the red due to mass loss from 23 $M_\odot$ to 16 $M_\odot$. Figure 2 (right) shows the lifetime in effective temperature bins normalized to unity (solid) compared with the number of supergiants with $-8 < M_{bol} < -9$ in the LMC, which is also normalized to unity (dashed). The model is qualitatively consistent with the observed histogram.

If the Ledoux criterion is adopted, the star ignites helium in the red supergiant stage, making a loop, and eventually returns to red. The timescale for the star to move from the blue to the red seems to be too fast to reproduce the observed histogram.

The evolution from the red to the blue in Fig. 1 takes place when the surface helium mass fraction is enhanced to $Y = 0.43$. Because of decreased opacity due to low Z and high Y, the surface luminosity exceeds the core luminosity. To compensate this luminosity imbalance, the envelope contracts back to the blue. The deep mixing model can naturally explain the large Y and N/C and N/O ratios observed in the circumstellar materials.

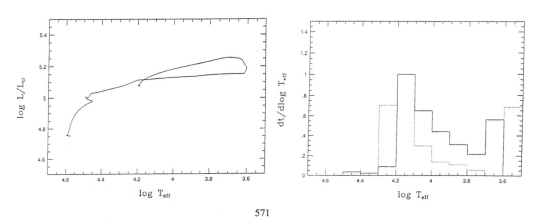

K. A. van der Hucht and B. Hidayat (eds.),
Wolf-Rayet Stars and Interrelations with Other Massive Stars in Galaxies, 571.
© 1991 *IAU. Printed in the Netherlands.*

Koenigsberger, Filippenko, Taylor, Schulte-Ladbeck, Hillier

SESSION VIII. INVENTORY AND DISTRIBUTION – *Chair: JiPi De Greve*

JiPi de Greve chairing

WOLF-RAYET STARS IN NEARBY GALAXIES: FACTS AND FANCIES

Philip Massey and Taft E. Armandroff
Kitt Peak National Observatory
National Optical Astronomy Observatories

ABSTRACT. Surveys for Wolf-Rayet stars in nearby galaxies are briefly reviewed. The completeness and yield of these surveys are discussed in light of recent follow-up spectroscopy. A critical evaluation is made of our current knowledge of the Wolf-Rayet population in nearby galaxies, particularly the WC/WN ratio, the WR/O ratio, the WR surface density, and how these quantities vary within a galaxy and between galaxies, particularly as a function of metallicity. We compare the spectroscopic properties of Galactic and Magellanic Cloud WR stars with those in the more distant systems.

1. INTRODUCTION

The last decade saw a great flurry of investigations of Wolf-Rayet stars in nearby galaxies of the Local Group. Such studies have provided a great deal of new information on the *numbers* of WR stars found in these galaxies, their *types* (WC/WN), their *distributions*, and their *spectroscopic properties*. A partial list of these studies includes Massey and Conti (1983), Moffat and Shara (1983), Armandroff and Massey (1985), Massey, Armandroff, and Conti (1986), Massey, Conti, Moffat, and Shara (1987), Massey, Conti, and Armandroff (1987), Moffat and Shara (1987), and Azzopardi, Lequeux, and Maeder (1988); reviews may be found in Massey (1985, 1986) and Conti (1988).

In preparing today's review talk it became clear to us that not only have these studies resulted in a certain body of "facts", but have also been responsible for a certain number of "fancies". Despite claims of incompleteness and caveats that are mentioned in many of these papers, interpretations of dubious authenticity have become part of our common knowledge. For instance, Massey and Conti (1983) found evidence of a strong galactocentric gradient in the relative number of WC/WN stars in M33, with most WC's found towards the center of the galaxy and most WN's located in the outer regions; this is commonly attributed to a metallicity gradient, although a variety of other explanations (such as a gradient in the IMF) may be responsible. Before we get into arguments as to the cause of this gradient, we might want to consider whether such a gradient exists at all given newer data. Then we might want to examine how well established the metallicity gradient is in M33. Finally we might do well to ask whether the WC/WN ratio at a point in M33 where the metallicity matches that of the LMC is the same as the WC/WN ratio in the LMC. The members of this audience who are regular attendees of these beach symposia may remember Conti's (1982) "An observer's view of stellar theory" diagram from the Cozumel WR meeting, in which a tiny fraction of the rain of observational data falls into the theoretician's black box (held up by the edifice of assumptions and pillar of avoidance) resulting in a steady outflow of predictions leading to muddied waters. It is our self-appointed misson today to critically examine one

575

K. A. van der Hucht and B. Hidayat (eds.),
Wolf-Rayet Stars and Interrelations with Other Massive Stars in Galaxies, 575–586.
© 1991 *IAU. Printed in the Netherlands.*

particular aspect of this process, namely the observational basis for the new mythology. In doing so, it is worth remembering that we are groping in the dark in these studies— no theoretician sat down in 1975, *predicted* that the WC/WN ratio (say) would be a monotonic function of metallicity, and encouraged observers to confirm or refute this. Instead, all such predictions have been *post facto*, and the observations we have made over the years have been at best guided by what a recent Kitt Peak summer student described as "educated, intuitional whims." Today we will examine to what extent these *factos* are in fact facts. Much of the work described was done in collaboration with Peter Conti, who nevertheless should be considered immune from prosecution for any of the material discussed today.

The motivation for studying Wolf-Rayet stars in nearby galaxies is two-fold: (a) what we can learn about the parent galaxies themselves (in which we are basically using WR stars as probes of the massive star content), and (b) what we can learn about massive star evolution (for instance, what evolutionary paths lead to the WC and WN types and how is this affected by such factors as metallicity). Such studies have undergone a revolution in the few years since the Porto Heli meeting on luminous stars in galaxies due to the implementation of multi-object fiber feeds on the world's largest telescopes, coupled to the routine availability of low read-noise, high quantum efficiency CCD's for spectroscopy. Thus in a single night of observations on the Mayall 4-m, it is possible to obtain spectra of >100 faint objects in a nearby galaxy (cf. Barden and Massey 1988). At the time of the Porto Heli meeting, the recent advent of CCD's on large telescopes had allowed sensitive surveys for WR stars to be carried out in NGC6822, IC1613, M33, and M31; the new instrumentation that has come along since that time has allowed the spectroscopy to finally catch up with the imaging surveys. The major punchlines of this spectroscopic work can be characterized as follows: (1) many of the lower significance WR candidates found from the CCD surveys are now spectroscopically confirmed, demonstrating that the older photographic surveys were woefully incomplete, (2) the WC/WN ratio in M31 appears to be no higher than that in M33, suggesting that whatever controls this ratio is not simply metallicity, and (3) the spectra of WR stars in NGC6822, IC1613, and M33 are remarkably *similar* to those in the Milky Way and Magellanic Clouds; the jury is still out on the M31 WR stars.

Below we list the major facts and fancies that have surfaced during the past decade of research; in some instances, there are "strong" and "weak" versions of the same law.
- The WC/WN ratio is a monotonic function of metallicity.
  - The WC/WNE ratio is a monotonic function of metallicity.
  - The WR/O ratio is a monotonic function of metallicity.
  - The (WC+WNE)/O ratio is a monotonic function of metallicity.
- WCL stars are exceedingly rare and are found only in the metal-rich central regions of M31 and the Milky Way.
- M33 WR stars are *not quite like* any others in the universe.
  - FWHM of lines for a given WC subtype is a function of metallicity in M33.
- M31 WR stars are *not quite like* any others in the universe.
- The numbers and types of WR stars in all galaxies of the Local Group are accurately known, or well-established correction factors can be applied.
  - The number of WR stars per unit anything (kpc^2, mass) is a factor of 10 lower in M31 than it is in the Milky Way, conclusively demonstrating that there are few massive stars present in M31.

We will discuss the observational reality of the facts and fancies given above in two general categories. (1) What do we know about the numbers of WR stars in Local Group galaxies (including the WC/WN ratio, the number per unit area or mass)? (2) What do we know about the spectra of WR stars in these galaxies? Are late-type WC stars present and, if so, are they really restricted to the central regions of metal-rich systems (e.g., M31)? More generally let us ask if the line widths and strengths are the same among these extra-galactic systems. This is the same as asking if the stellar wind laws are the same.

2. NUMBERS AND DISTRIBUTION OF WR STARS

Galactic WR stars have been found by a variety of means, mostly incidentally as part of general spectroscopic surveys (e.g., the *Henry Draper Catalogue*). Roughly 160 Galactic WR stars are known (van der Hucht *et al.* 1981). Doubtless there are lightly reddened WR stars of 12th mag and brighter still to be found, and elsewhere in these proceedings Shara describes a dedicated search for WR stars in the Galaxy using optical techniques. Our knowledge of the WR content of the Galaxy is believed to be complete within 3 kpc of the sun or so. In a recent paper, Conti and Vacca (1990) derive distances to Galactic WR stars using spectroscopic parallax and improved estimates of the reddening corrections. Within 2.5 kpc of the sun there are 48 WR stars known, 24 of WN type and 24 of WC type.

The Magellanic Cloud WR stars have been found primarily using objective prism plates with a broad $\lambda4650/4686$ interference filter used to alleviate crowding; details can be found in Breysacher (1981) and Azzopardi and Breysacher (1979). Roughly 100 are known in the LMC and 8 in the SMC. While occasionally a new one shows up in the LMC, these numbers are probably essentially complete. Of the 100-odd WR stars in the LMC, $\approx20\%$ are of WC type. Of the 8 WR stars in the SMC, only 1 is of WC type.

It is worthwhile to pause for a moment here and reflect on what the numbers from these three galaxies alone tell us. First of all, we see that the total number of WR stars in each of these galaxies does not scale in a simple way with mass or area surveyed. The number of WR stars per kpc^2 is 2.3 in the solar neighborhood, and something like 2 in the LMC, but only 1 in the SMC. Of course, we don't know if we have considered equal *volumes* or not, given the uncertainty in the depth extensions of the Clouds. Per unit mass it would appear that the number of WR stars is down about a factor of 3 in the SMC compared to the LMC (Azzopardi and Breysacher 1979), although clearly the masses of these galaxies are rather uncertain. These numbers *may* be telling us that the number of massive star progenitors is down by a factor of 2 or 3 per unit something in the SMC compared to that of the LMC, implying that the star formation rate is lower in the SMC...or they may not.

What about the ratio of WC to WN stars? The progression in this ratio from the Galaxy to the LMC to the SMC is in the same sense as the change in metallicity. The ratio of WC to WN stars is 1.0 in the solar neighborhood, 0.2 in the LMC, and 0.1 in the SMC; the "metallicity" of the Milky Way is the highest, with that of the LMC down by a factor of 0.4 dex and that of the SMC down by another factor of 0.4 dex, using the oxygen abundance as the metallicity indicator. But hold it, there is something remarkably fishy here. In the SMC we are dealing with (very) small number statistics—there is exactly 1 WC star among the 8 WR stars. Given the stochastic probability of finding exactly one of anything in something the size of a galaxy, we must at least take the uncertainty in this quantity to be $\sqrt{N}$: if we looked at some slightly different time, would we see zero or two WC stars? In the case of there being two WC stars, the ratio in the SMC changes to 0.3, *larger* than that of the LMC. Thus the "progression" from the Milky Way to the LMC to the SMC is in fact meaningless! What *is* fair to say is that the relative number of WC to WN stars is far lower in the Magellanic Clouds than it is in the solar neighborhood, but to say any more than that ignores the inherent uncertainties in the small numbers for the SMC. We emphasize that we are not saying that the current numbers of WC and WN stars are poorly known in the SMC—just that there is a certain amount of luck in living at this particular time if we are seeing exactly one star whose lifetime is so short compared to a Hubble time. While we are debunking the "progression" myth of the Milky Way→LMC→SMC, let us note that whatever it is that is primarily responsible for driving the stellar wind, it is unlikely to be oxygen! It is clear from the study by Dufour, Shields, and Talbot (1982) that it is unsafe to assume that all metals scale with O; in particular, they find that carbon is significantly more depleted with respect to hydrogen in the SMC compared to oxygen than it is in the LMC (but see also the abundance analysis of LMC and SMC F supergiants by Russell and Bessell 1989). Does the Fe abundance scale with O or with C? Probably the latter, according to Dufour *et al.* The point here is merely that one has to be a little careful in deciding what "metallicity" to use.

Beyond the Magellanic Clouds, surveys have been made for WRs in NGC6822, IC1613, M31, and M33. Methods have included interference filter photography (Wray and Corso 1972; Corso 1975; Moffat and Shara 1983, 1987; Massey *et al.* 1987) and low-dispersion "grism" or "grens" searches (Westerlund *et al.* 1983; Bohannan, Conti, and Massey 1985; Lequeux, Meyssonnier, and Azzopardi 1987). Both of these methods rely on the detection of the CIII $\lambda4650$/HeII $\lambda4686$ band. These two methods have proven very successful at finding the strongest-lined, brightest WR stars.

However, as is well-discussed by Massey (1985), strong selection effects come into play if one wishes to determine the relative numbers of WC and WN stars within any of these galaxies. The typical equivalent width of HeII $\lambda4686$ in WN stars is nearly a factor of 10 smaller than that of CIII $\lambda4650$ in WC stars (compare, for example, Fig. 4 and Fig. 10 in Massey, Conti, and Armandroff 1987). To be reasonably complete for WN types requires detecting with confidence stars in which the equivalent widths of HeII $\lambda4686$ are 40 Å or so.

In order to accomplish this, an "optimized" interference filter system was designed using spectrophotometry of Galactic and Magellanic Cloud WR stars. These filters were used with CCD's on the KPNO and CTIO 4-m telescopes to survey NGC6822, IC1613, and two fields of M33 by Armandroff and Massey (1985), and to survey eight fields in M31 by Massey, Armandroff, and Conti (1986). Using DAOPHOT to obtain photometry of every star on the frames, the magnitude differences from frame to frame were computed. The error in each star's magnitude was accurately known, and therefore stars that were significantly brighter on one of the on-band exposures could be objectively identified. The advantages of this method are the greater sensitivity (leading to improved completeness), the ability to distinguish between WC and WN stars, and the fact that the method provides a quantitative measure of the significance level of each candidate; the disadvantages are the small field size obtained with the CCD and the fact that the method requires a great deal of work! In addition, the drive for relatively useful completeness implies that a certain number of candidates are bound to be "losers" (non-WR stars) if one is doing a proper job. These surveys have necessitated a great deal of follow-up spectroscopy, made practical primarily thanks to the implementation of multi-object fiber spectroscopy.

It has been our intention that once we understand the fraction of bona fide WR stars among our candidates as a function of significance level in Local Group galaxies (where follow-up spectroscopy *is* possible), these imaging surveys can be extended to more distant galaxies, such as those of the Sculptor and M81 groups. We are not quite there as yet, but almost, and you are invited to view the poster paper by Armandroff and Massey in which we show confirming spectra of many of the lower significance candidates. Despite improvements in the instrumentation, these observations are still hard. For instance, the spectrum shown of AM11 in M33, a WR candidate with a continuum magnitude of 20.5 and a significance level of 3.0σ, is the sum of three 1-hour exposures with the Mayall 4-m telescope. This star is now definitely confirmed as a WR star, but to do so was non-trivial.

The spectroscopic evidence to date is summarized here. For NGC6822, 4 WN stars have been confirmed; there may be 1 or 2 more among the remaining candidates. There are not, at present, any WC stars living in the surveyed area of NGC6822. In IC1613, 1 WC star (previously found by spectroscopy of an HII region by D'Odorico and Rosa 1982) is known, and another star is either a WN or SNR. A few other WN's may be present, although Azzopardi, Lequeux, and Maeder (1988) say not. These small galaxies suffer from the same $\sqrt{N}$ problems discussed earlier for the SMC. The fact that the WC/WN ratio may be 1/0 (i.e., infinite) in IC1613 (a galaxy with a metallicity *lower* than that of the SMC) seems to be conveniently ignored in the same papers that make a big deal of the small WC/WN ratio in the SMC....

In M33 photographic surveys have found $\approx$100 WR stars (e.g., Wray and Corso 1972; Massey *et al.* 1987). Based upon the data then available, Massey and Conti (1983) argued that there is a good correlation of WC to WN ratio with galactocentric distance within the plane of M33.

How good is our knowledge of the WR content of M33? Armandroff and Massey (1985) surveyed two "test" fields in M33 (from Tololo!) in order to test how well these optimized filters performed. The fields were selected only for having lots of previously known WR's. In these two test fields there were 16 previously known WR stars, of which 11 were of WC type and 5 were of WN type; e.g., a WC/WN ratio of 2.2, twice that of the solar neighborhood. Our CCD survey found all of these plus 11 more candidates. Of these, 5 have now been confirmed spectroscopically. All the new ones are of WN type. Thus, suddenly the ratio of WC to WN in these two fields is found to be 1.1, almost identical to that of the solar neighborhood. We may safely conclude that the WC/WN ratio determined in such studies is affected by detection limits and small number statistics.

This of course raises an interesting point: is there in fact a gradient in the relative number of WC/WN stars in M33? The data from the CCD fields suggest that the photographic survey was at least 30% incomplete in these two fields, and of this selectively incomplete for WN stars. This is of no particular surprise, and in fact Massey and Conti (1983) state: "In selecting stars for spectroscopy, we naturally concentrated on stars with the largest differences in our blink comparison, since these were the most certain to be Wolf-Rayet stars. Thus although the overall ratio of WC to WN stars suggested by Table 1 is 1:1, this is likely biased in favor of the WC stars. When making the blink survey we assigned subjective values to how strongly each star blinked; using the distribution of WC and

WN types within these categories, and allowing for the larger incidence of stars found not to be Wolf-Rayet stars in the weaker candidates, we estimate that this ratio could be as low as 1:2. Despite the factor of 2 uncertainty, we believe that the overall WC to WN ratio in M33 more closely resembles that of the solar neighborhood and the LMC than the SMC...." They go on to argue that the change in the relative numbers of WC and WN stars with galactocentric distance probably *is* real. However, one worry in this is that the detection of WR stars is harder near the center of M33 (where the background is higher) and thus disproportionately favors the detection of WC stars in this region. A CCD survey of the center of M33 could answer this, and in fact Drissen, Moffat, and Shara present the results of such a survey as a poster at this conference. They find a number of new WR candidates likely to be WN's, but in fact in much the same ratio as the increase we found in the two fields further out. Thus it appears that the overall gradient with galactocentric distance remains. We will return to the question of what the absolute WC/WN ratio is telling us after we consider M31.

M31 is the Local Group galaxy with more WR mythology than the others combined. "Facts and fancies" include the belief that the WR population is predominately WC's, and that the number of WR stars per unit anything is at least a factor of 10 lower than it is in the Milky Way. Let us consider what is known today about the statistics from the eight fields surveyed in M31 by Massey, Armandroff and Conti (1986).

In these eight fields there are 34 WR candidates of various significance levels. Of these, only two were detected in the photographic survey by Moffat and Shara (1983); another five were rediscovered in their second survey (Moffat and Shara 1987). Of these 34, 19 have now been spectroscopically confirmed. Nine of these are of WC type and 10 of these are of WN type. Thus for the fields in which we feel we have reasonable completeness, the overall WC/WN ratio is 0.9. This must be considered an upper limit since the remaining 15 candidates are expected to be WN's on the basis of our narrow-band photometry. This ratio is already *lower* than that of the two M33 fields, although the oxygen abundance (relative to hydrogen) is 0.6 dex higher in M31 than in M33. So much for the WC/WN ratio being a monotonic function of metallicity.

What about the SFR/IMF massive star deficiency in M31? We find 21 spectroscopically confirmed WR's for the two AM fields in M33. The total area of these two fields is 4.4 kpc^2, leading to a surface density of 5 WR/kpc^2. (The inclination of M33 has been corrected for.) Similarly, in any two "good" fields in M31 there are 8 or 9 spectroscopically confirmed WR stars. The total area (corrected for M31's inclination) is 6 kpc^2, leading to a surface density of 1.5 WR/kpc^2. *Thus regions rich in WR stars in M31 are down by a factor of 3, not 10, compared to rich regions in M33.* For comparison, the number density near the sun is 2 WR stars/kpc^2. Of course, the solar neighborhood is hardly an OB association. For comparison, the numbers (uncorrected for inclination or depth effects) are 2 WR/kpc^2 in the LMC, and 1 WR/kpc^2 in the SMC.

Another part of the M31 massive star myth is that all the star formation takes place in the prominent CO ring. Van den Bergh (1964) shows the de-projected distribution of OB associations in M31 in his Fig. 5; clearly there are young associations that do *not* fall within this ring. Further surveys for WR stars are needed before we can conclude what the galaxy-wide WR content is like; high extinction will always make this a lower limit in any event.

Ultimately we would like to know the relative numbers of WR and O stars in these galaxies. This *can* be determined using existing technology for the Magellanic Clouds, but as recently demonstrated by Massey, Parker, and Garmany (1989), our knowledge of the massive star content of the Clouds is very incomplete. In their study of one association, NGC346, they identified 11 stars of type O6.5 or earlier; this is as many early-type O stars as had previously been known in the rest of the SMC. Given that NGC6822, IC1613, M33, and M31 are 5 to 6 magnitudes further away, we face the fact that the progenitor O stars are a factor of 100 or more fainter. Spectroscopy of the brightest supergiants can be accomplished with existing 4-m telescopes (see, for example, the recent paper by Humphreys, Massey, and Freedman 1990), but such a sample is dominated by lower-mass (10-20 $\mathcal{M}_\odot$) somewhat evolved stars. To make even a dent in this issue requires a significant amount of dedicated time on a 4-m class telescope with multi-object capability, and really an 8-m telescope. The difficulty is that the luminosity functions are dominated by lower-mass evolved supergiants, and that spectroscopy is needed as colors and absolute *visual* magnitudes cannot distinguish between 20 $\mathcal{M}_\odot$ and 100$\mathcal{M}_\odot$ main-sequence stars. These issues are discussed at some length in Massey (1985) and Massey (1990). To illustrate the point here we show in Fig. 1 three luminosity functions. The curve in the

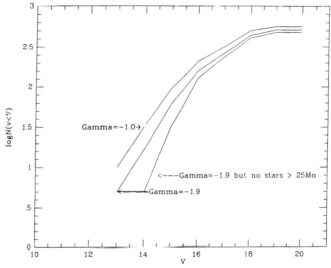

Figure 1: Luminosity functions that result from three different initial mass functions.

middle is the straight observational data taken from the Massey, Parker, and Garmany (1989) study of NGC346 in the SMC. The initial mass function was found to have a slope of $\Gamma = -1.9$. The lower slope is what results from the same data but deleting those stars whose implied masses are greater than 25$\mathcal{M}_\odot$. The upper curve is the luminosity function that results from the same set of stars in NGC346 but with an altered mass function $\Gamma = -1.0$, which would imply a factor of 3 more massive-star progenitors. The slopes of these luminosity functions are virtually indistinguishable. Furthermore, these represent the most optimistic case imaginable: that of an OB association in which all the stars are born at the same time. For a field population (such as what would be found in a galaxy-wide sample) the cooler but visually brighter B supergiants would play a dominant role in the luminosity function, masking any variation at the upper mass end. Therefore, the relative numbers of WR and O stars are significantly more poorly known than the

WC/WN ratio or the surface density of WR's. We do not feel that studies of the WR/O ratio as a function of metallicity will be meaningful until the O star numbers result from spectroscopic surveys.

3. THE SPECTRAL PROPERTIES OF EXTRA-GALACTIC WR STARS

Here we are again guided by "educated intuitional whims" in what we are looking for, since we have no theory driving us in a particular direction. First let us consider the issue of late WC stars. Are there *any* WC stars of type WC8 or WC9 anywhere except in the Milky Way?

- None are present in the Magellanic Clouds.
- None are known in M33.
- None are known for sure in M31.

It is necessary to expound on the last point a little. Moffat and Shara (1983, 1987) classify several of their M31 WC stars as being WC8 or later. However, it is not clear from their spectra that CIII $\lambda5696$ is actually stronger than CIV $\lambda5812$ as it is in Galactic WC stars of this type, and quantitative line ratios are lacking for their data. For the stars in their sample that we have been able to observe, we find no WC stars later than WC7. It is worth noting, however, that the WC8's and WC9's in the Galaxy clearly have the weakest lines of any of the WC's and that there may therefore be selection problems (Massey, Conti, and Armandroff 1987).

Let us now consider the issue of the distribution of WC stars within M31 and M33 a little more carefully. First we will use a somewhat looser definition of what makes a late WC star "late". Massey, Conti, and Armandroff (1987) classified their WC stars using the following criteria:

WC: Neither CIII $\lambda5696$ nor OV $\lambda5592$ could be seen; in all cases CIV $\lambda5812$ was visible, so these stars could not be as late as WC8 or WC9.

WCL: CIII $\lambda5696$ could be seen and was stronger than OV $\lambda5592$.

WCE: OV $\lambda5592$ could be seen and was stronger than CIII $\lambda5696$.

Thus the WCL designation would apply to any WC star of type WC6 or later. Using these simple criteria, do we see a central concentration of the WCL's in M31 or M33? Let us divide each galaxy into an "inner" and "outer" region using de-projected galactocentric distance (e.g., the distance within the galactic disk). We find the following.

TABLE 1
Late WCs in M31 and M33

		#WRs	#WCLs
M31			
	$r < 50'$	15	5
	$r > 50'$	4	0
M33			
	$\rho < 0.6$	34	1
	$\rho \geq 0.6$	29	0

Within the inner regions of M31 we see 15 WR stars of which 5 are WCL's. In the outer regions there are only 4 WR's, and 0 is consistent with the expected number of WCL's given the small number statistics. Similarly the numbers for M33 are completely inconclusive. Thus at present we conclude there is no evidence one way or another for this particular "fact and fancy", but it is one that *could* be answered by simply surveying more regions in M31.

Next let us consider the more general (and interesting) question: are the stellar wind laws in extra-galactic WR stars similar to those of WR stars in our Galaxy? Since stellar winds are driven through resonance lines of highly ionized metals, we might expect that the answer is "no". On the other hand, the compositions of WR stars are mainly "home brewed" and it may be that any differences caused by different "initial metallicity" have long since been lost by the time of the WR phase. To address this question, Peter Conti developed the "inverted hockey stick" plots found in Massey, Conti, and Armandroff (1987). In these plots, the full width at half maximum (fwhm) is plotted against the log of the equivalent width. The former is something like the expansion velocity at the depth where the line is formed, and the latter is something like the wind density. Displayed in such a plot, data from Galactic and LMC stars clearly do not scatter, but fall in a relatively well-defined "inverted hockey stick". Note that the presence of an O-type companion will reduce the log ew but that the fwhm will stay roughly constant; e.g., the presence of a companion will mainly lower a point in this diagram.

Plotting the extra-galactic WR stars in this way using the new data at our disposal reveals the following:
- The Magellanic Cloud WR stars (both WC's and WN's) match the Galactic stars.
- The NGC6822 WN stars pretty well match the Galactic and Magellanic Cloud stars.
- The M33 stars (both WC's and WN's) match the Galactic and Magellanic Cloud stars.
- The M31 stars (particularly the WC's) *don't* match the Galactic and MC stars very well, but the data are still scant.

To this discussion can be added the fact that Hutchings, Massey, and Bianchi (1987) managed to obtain a UV spectrum of the M31 WN star OB69-WR2 with the *International Ultraviolet Explorer* satellite. The spectrum shows a curious absence of the strong emission lines that characterize stars of similar type in the Galaxy and Magellanic Clouds. Not even HeII λ1640 is visible. The optical spectrum of this star shows lines which are somewhat narrow. With the more sensitive UV spectrometer on *Hubble Space Telescope*, it should be possible to extend such studies to many more WR stars in Local Group galaxies.

Although Massey, Conti, and Armandroff (1987) concluded that the M33 WR stars are normal and the M31 stars may be odd, recent spectroscopy by Schild, Smith, and Willis (1990) yields the opposite conclusion—that while the M31 stars in their sample appear "normal", the M33 stars are peculiar. Their reasons for concluding that the M33 stars are odd are two-fold: (1) the FWHM of the M33 WC stars in their sample appears correlated with galactocentric distance, and (2) they find that the FWHM varies within a WC subtype as a function of galactocentric distance.

We have looked for these effects in our considerably larger sample, and our independent data confirms the first of these. Figure 2 shows the distribution of WC stars in the "hockey stick" diagrams for the inner and outer regions of M33. Clearly the WC stars in the inner regions have narrower (and weaker) lines than those in the outer regions.

584

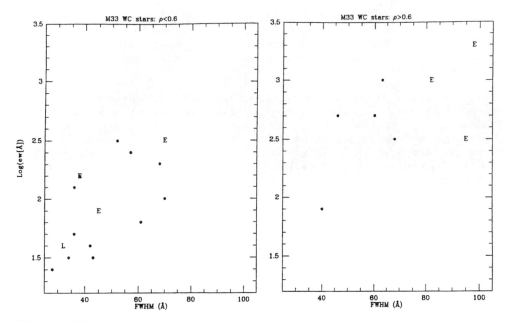

Figure 2: "Hockey stick" diagrams for M33 WC stars divided into two galactocentric distance bins; E denotes WCE's and L denotes WCL's.

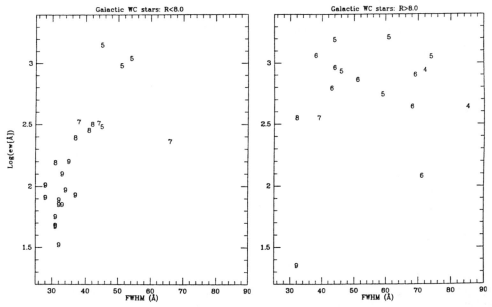

Figure 3: "Hockey stick" diagrams for Galactic WC stars with galactocentric distances less than and and greater than 8 kpc; the numerals denote each star's subtype.

However, examining the data for the Galactic WC stars reveals an identical effect! We have taken values for the equivalent widths and fwhm's from Conti and Massey (1989); the galactocentric distances come from Conti and Vacca (1990). Figure 3 shows the data for the Galaxy broken into "inner" and "outer" regions. Clearly the situation is no different in M33 than in the Milky Way.

Massey, Conti, and Armandroff (1987) did not feel brave enough to assign numerical subtypes to the WC stars in their extra-galactic sample, so their data cannot be used to confirm or refute the second conclusion of the Schild, Smith, and Willis (1990) study that the fwhm within a subtype is dependent upon galactocentric distance within M33. However, Fig. 3 demonstrates that this is certainly the case for the Galactic data: the WC5 stars, for instance, have skinnier fwhm in the inner part of the Galaxy than they do in the outer. So if the M33 stars are similar to those in our Galaxy, we would perhaps not be surprised to see such an effect. Why are these effects present? What is this telling us about stellar wind laws and their dependence upon (say) metallicity? We do not know the answer to this for certain, but whatever it is, it appears to be at work both in M33 and in the Milky Way.

REFERENCES

Armandroff, T. E., and Massey, P. 1985, *Ap. J.*, **291**, 685.
Azzopardi, M., and Breysacher, J. 1979, *Astr. Ap.*, **75**, 120.
Azzopardi, M., Lequeux, J., and Maeder, A. 1988, *Astr. Ap.*, **189**, 34.
Barden, S. C., and Massey, P. 1988, in S. C. Barden (ed.), *Fiber Optics in Astronomy*, (Provo: Astron. Soc. Pacific), p. 140.
Bohannan, B., Conti, P. S., and Massey, P. 1985, *Astron. J.*, **90**, 600.
Breysacher, J. 1981, *Astr. Ap. Suppl.*, **43**, 203.
Conti, P. S. 1982, in C. W. H. de Loore and A. J. Willis (eds.), *Wolf-Rayet Stars: Observations, Physics, Evolution, IAU Symp. No. 99* (Dordrecht: Reidel), p. 3.
Conti, P. S. 1988, in V. M. Blanco and M. M. Phillips (eds.) *Progress and Opportunities in Southern Hemisphere Optical Astronomy*, (Provo: Astron. Soc. Pacific), p. 100.
Conti, P. S., and Massey, P. 1989, *Ap. J.*, **337**, 251.
Conti, P. S., and Vacca, W. D. 1990, *Astron. J.*, **100**, 431.
Corso, G. J. 1975, *Bull. A.A.S.*, **7**, 411.
D'Odorico, S., and Rosa, M. 1982, *Astr. Ap.*, **105**, 410.
Dufour, R. J., Shields, G. A., and Talbot, Jr., R. J. 1982, *Ap. J.*, **252**, 461.
Humphreys, R. M., Massey, P., and Freedman, W. L. 1990, *Astron. J.*, **99**, 84.
Hutchings, J. B., Massey, P., and Bianchi, L. 1987, *Ap. J.*, **322**, L79.
Lequeux, J., Meyssonnier, N., and Azzopardi, M. 1987, *Astr. Ap. Suppl.*, **67**, 169.
Massey, P. 1985, *Pub. A.S.P.*, **97**, 5.
Massey, P. 1986, in C. W. H. de Loore, A. J. Willis, and P. Laskarides (eds.) *Luminous Stars and Associations in Galaxies, IAU Symp. No. 116*, (Dordrecht: Reidel), p. 215.
Massey, P. 1990, in C. D. Garmany (ed.) *Properties of Hot Luminous Stars* (Provo: Astron. Soc. Pacific), p. 30.
Massey, P., Armandroff, T. E., and Conti, P. S. 1986, *Astron. J.*, **92**, 1303.
Massey, P., and Conti, P. S. 1983, *Ap. J.*, **273**, 576.
Massey, P., Conti, P. S., Armandroff, T. E. 1987, *Astron. J.*, **94**, 1538.
Massey, P., Conti, P. S., Moffat, A. F. J., and Shara, M. M. 1987, *Pub. A.S.P.*, **99**, 816.
Massey, P., Paker, J. W., and Garmany, C. D. 1989, *Astron. J.*, **98**, 1305.

Moffat, A. F. J., and Shara, M. M. 1983, *Ap. J.*, **273**, 544.

Moffat, A. F. J., and Shara, M. M. 1987, *Ap. J.*, **320**, 266.

Russell, S. C., and Bessell, M. S. 1989, *Ap. J. Suppl.*, **70**, 865.

Schild, H., Smith, L. J., and Willis, A. J. 1990, *Astr. Ap.*, in press.

van den Bergh, S. 1964, *Ap. J. Suppl.*, **9**, 65.

van der Hucht, K. A., Conti, P. S., Lundstrom, I., and Stenholm, B. 1981, *Space Sci. Rev.*, **28**, 227.

Westerlund, B. E., Azzopardi, M., Breysacher, J., and Lequeux, J. 1983, *Astr. Ap.*, **123**, 159.

Wray, J. D., and Corso, G. J. 1972, *Ap. J.*, **172**, 577.

DISCUSSION

Sreenivasan: You said that WR stars in M33 and M31 are different from those anywhere else. Does this refer to the surface number densities or does it cover other aspects? (The densities you quoted later on appeared similar!).

Massey: Actually, I did not say that. I tried to diffuse those claims. Pundits in this room have claimed that the spectra M33 and M31 WR stars are unlike those in the Galaxy, but quantitative measurements have yet to demonstrate this.

Moffat: (1) In going from regions of extreme star formation (*e.g.*, GHR's) to the field, evidence shows that the WC/WN ratio changes dramatically (*e.g.*, the ratio in 30 Dor is ~ 2 to ~ 0.1, *cf.* the rest of the LMC: $\sim 20/80 \approx 0.25$). (2) While colours of O stars may be degenerate when looking in galaxies where crowding and differential reddening prevail, this is not so for nearby young clusters with accurately observed colours, *e.g.* $\sigma(U - B) \approx 0.01$, since $(U - B)_o = -1.2$ for O3V to -1.1 for O9V.

Massey: (1) Lindsey Smith maintains in her talk that WC4's dominate the integrated WR spectrum of NGC 604, the biggest HII region in the Local Group outside 30 Dor. (2) Much of the galactic calibration of $(U - B)_o$ with spectral type comes from heavily reddened galactic OB associations - as you know, there are no lightly reddened O3V stars around to provide an empirical calibration. Models also give only a tiny difference in $(U - B)_o$ as a function of T_{eff} from $30000 K$ to $50000 K$. For a recent discussion of this, see Massey, Parker and Garmany (1989), and Massey, Garmany, Silkey and Degioza-Eastwood (1989, Astron.J. **97**, 107).

de Groot: If you were born in Northern-Ireland, you would know that there is a different kind of hockey stick according to whether you are born as a boy or as a girl. It would be a hurly or a commogy stick, exactly the shape that you drew.

Massey: Actually, it was Peter Conti that came up with the phrase "inverted hockey sticks". Maybe we need something to distinguish the late WN's from the early WN's. Perhaps we can do it on the basis of sex, rather than the $NV/NIII$ ratios.

GALACTIC WOLF-RAYET DISTRIBUTION AND IMF

BAMBANG HIDAYAT
Observatorium Bosscha, ITB
Lembang, West Java, Indonesia

ABSTRACT: Wolf-Rayet star to OB star number ratios in some selected regions in the Galaxy are presented. The variations of the number ratio as a function of location is found to be marginal.

1. Introduction

The $WR/OB_{progenitor}$ number ratio in the Galaxy is the important parameter for the determination of the initial mass of WR progenitors. Observationally, the number ratio is determined from the observed distributions of WR and OB stars in the Galaxy. The observed values are then compared with predicted values derived from theories of stellar evolution. Discrepancies between the two values have been found (*e.g.*, Conti *et al.*, 1983) and thus complicate the exact determination of the WR initial mass. Introduction into the theory of effects of metallicity and convective core overshooting can largely remove the discrepancy.

However, the initial mass function depends critically on the luminosity function, which, in turn, determines the SFR. While in principle the luminosity function can be determined from a volume-limited sample, it must be assured the the co-spatial sample is complete. The purpose of this note is to indicate the sensitivity of the SFR, in view of the limited sample distribution.

2. The local galactic structure

Smith (1968) was the first to note a radial variation in the galactic WR subtype distribution. Using larger samples of WR stars with newly calibrated parameters, van der Hucht *et al.* (1988) and Conti & Vacca (1990) have found similar effects and confirm that, in general, WR stars delineate the local galactic spiral arms. These studies claim that the WR sample is complete within 2.5 *kpc* – perhaps 3.0 *kpc* – from the Sun.

Within the same volume element OB stars have also been surveyed (*e.g.* by Garmany, 1982). The distribution of the most massive OB stars does show the same pattern of the commonly accepted local spiral structure, while the less massive stars are distributed in a more uniform way.

Therefore the number densities of the WR and OB stars can readily be calculated, and intercompared.There are, however, some subtle differences that one has to take into account, in order to access the meaning of intercomparison.

587

K. A. van der Hucht and B. Hidayat (eds.),
Wolf-Rayet Stars and Interrelations with Other Massive Stars in Galaxies, 587–589.
© 1991 *IAU. Printed in the Netherlands.*

The lack of a 2-dimensional classification for all OB stars excludes the exact determination of the spatial distribution of this type of stars in the Galaxy. Therefore one has to be cautious for a direct comparison of number densities. An other fact is that there is a region which seems to be devoid of WR stars, but still harboring some OB stars. This refers to the direction of $l \approx 245°$, in which FitzGerald and Moffat (1976) have indicated the existence of a 15-kpc arm.

On the other hand, the galactic longitudes between $l \approx 100°$ and $l \approx 140°$-$180°$ seems to be deprived of Population I objects (Raharto, 1990). The questions which may arise from these observed phenomena relate to the problem of the metallicity gradient and/or the IMF in the Galaxy. Maeder (1990, this symposium) has clearly indicated the influential factor of metallicity for the formation of WR stars from massive OB stars.

The complication that may arise by direct comparison of the number densities in the solar neighbourhood lies in the inherent nature of the local galactic structure. There are two, perhaps three branches of the local arm within 3 kpc from the Sun. The smearing process involved in the determination of number densities may become serious if one compares number densities of elements of galactic volumes of different characteristics. At a particular galactic longitude, the line of sight may largely find interarm regions, while in other directions the line of sight is looking alongside the spiral arm. In different galactic spiral arms, processes which lead to the formation of WR stars may be different because of different IMF and metallicity. The metallicity differences can be divided in two components: arm-to-arm differences as well as galactic radial differences.

We suggest to investigate WR and OB number densities in selected galactic longitude zones in order to discriminate the arm and inter-arm regions. This reduces the statistics however. Within 2.5 kpc from the Sun, 50% of the WR stars are found in clusters and associations, while of all known galactic WR stars, 44 (*i.e.* 30%) belong to clusters and associations.

3. The IMF

This parameter cannot be determined independently. It is based on the observed luminosity function, with evidence of complex variations on spatial, temporal and, lightly, luminosity scales. The commonly accepted law for the luminosity function is that it may vary in a simple parameterized way. Van den Bergh (1986) shows that a fixed IMF can produce different luminosity functions, depending on abundances.

Miller and Scalo (1979) indicated that the combined effect of stellar wind mass loss and binary mass exchange changes the exponential law, which causes the variation of massive stars observed in different regions of the Galaxy.

Because differences in IMF have strong effects on the formation of WR stars, as shown by Arnault *et al.* (1989) in the case of both star formation bursts and constant formation rates, the use of an observed luminosity function should be taken with care.

4. Some results

Assuming the recent finding of Blitz and Spergel (1989) would not pose serious changes to the picture within 2.5 kpc from the Sun, the number ratios of WR and OB stars in some clusters and associations are studied. The galactic regions studied have been chosen on the basis of Bok's (1981) local structure.

The lower limits of the initial O star masses have been chosen as 15 $M_\odot$ for the WN stars and 35 $M_\odot$ for the WC stars, following van der Hucht *et al.* (1988). Table 1 gives the WN/OB number ratios in four regions.

Table 1. WR/OB star number ratios.

M_{OB} minimum	Carina	Sag-Sco	Orion	Perseus
15 $M_\odot$	0.20	0.16	0.14	small
35 $M_\odot$	0.30	0.2	0.2	small

4. Conclusion

If compared to the average WR/OB number ratio in the solar neighbourhood of 0.14 (van der Hucht *et al.*, 1988), the results given above show only marginal differences. Whether those reflect real spatial differences in the luminosity function of massive stars requires deeper study.

References

Arnault, Ph., Kunth, D., Schild, H. 1989, *Astron. Astrophys.* **224**, 73.

van den Bergh, S. 1986, *Astrophys. Space Sci.* **118**, 435.

Blitz, L., Spergel, D.N. 1989, *Bull. American Astron. Soc.* **21**, 1189.

Bok, B. 1981, H.N. Russell Lecture, *Publ. Steward Obs.* No. 435.

Conti, P.S., Garmany, C.D., de Loore, C., Vanbeveren, D. 1983, *Astrophys. J.* **274**, 302.

Conti, P.S., Vacca, W.D. 1990, *Astron. J.* **100**, 431.

FitzGerald, M.P., Moffat, A.F.J. 1976, *Astron. Astrophys.* **50**, 149.

Garmany, C.D., Conti, P.S., Chiosi, C. 1982, *Astrophys. J.* **263**, 777.

van der Hucht, K.A., Hidayat, B., Admiranto, A.G., Supelli, K.R., Doom, C. 1988, *Astron. Astrophys.* **199**, 217.

Miller, G.E., Scalo, J.M. 1979, *Astrophys. J. Suppl.* **41**, 533.

Raharto, M. 1990, in: *Proc. 5th Asian-Pacific Regional IAU Meeting*, Sydney, Australia, in press.

Smith, L.F. 1968, *Monthly Notices Roy. Astron. Soc.* **141**, 317.

DISCUSSION

Montmerle: How many of these stars belong to OB associations with giant *HII* regions?
Hidayat: There are five OB associations included in these simple statistics.
Conti: Among known galactic WR stars there is only one *giant HII* region, NGC 3603, with a WR star. Carina, which has three WR stars, is not quite a giant *HII* region by all definitions.

Ian Howarth, Linda Smith, John Hillier, Wolf-Reiner Hamann

A Deep Survey for Faint Galactic Wolf-Rayet Stars

M. M. Shara[1], M. Potter[1], A. F. J. Moffat[2], L. F. Smith[3]
[1] *Space Telescope Science Institute, 3700 San Martin Drive, Baltimore, MD 21218*
[2] *University of Montreal, Dept. of Physics, C.P. 6128, Succ. A., Montreal PQ H3C 3J7, Canada*
[3] *Mount Stromlo and Siding Springs Observatory, Private Bag, P.O. Woden ACT 2606, Australia*

ABSTRACT: Surveys of the Galaxy for Wolf-Rayet (WR) stars are mostly based on objective prism searches, and are generally complete to only about 13th visual magnitude. We are using direct narrowband and broadband Schmidt plates to survey large areas of the southern Milky Way for WR stars to 17-18th magnitude. We expect to find more than 50 new WR stars. The newly detected stars should be among the most distant and/or reddened known in the Galaxy. The survey is also designed to test the completeness of previous bright WR star surveys, and thus to help settle debates over the Initial Mass Function of the most massive stars. We have now located 13 new WR stars in a 40 square degree region in Carina where 24 WR stars were already known. A 25% incompleteness in detection of WR stars as close as 2–3 kpc is suggested.

1 Wolf-Rayet Stars Surveys

Narrowband images taken with CCDs and 4 meter telescopes can reveal 24th magnitude Wolf-Rayet stars at the distance of the M81 group of galaxies; however, our own Galaxy has been completely surveyed only to $m \simeq 13 - 14$ for WR stars (based on the latest Galactic WR catalogue; van der Hucht *et al.* 1981, updated by van der Hucht *et al.* 1988). Only 12 of 157 known Galactic WR stars are fainter than V=15. This is indicative of the crowding problems which limit objective prism surveys in the Galactic plane.

2 Motivation for the Present WR Star Survey

Benefits of the survey are to search a much larger part of the Galaxy for massive star formation (WC and WN stars) than has hitherto been accomplished; to verify and extend the result that the ratio of WC to WN stars varies with Galactocentric distance; and to locate very distant WR stars to act as probes of the interstellar medium, in the UV, visible and IR, in spiral arms far from the Sun.

3 Narrowband-Broadband Imaging and Spectroscopic Followup

By virtue of their generally strong, broad emission lines, WR stars can be easily detected at faint magnitudes in extended regions by comparing direct images taken with narrow on-versus broad on-line interference filters. WR stars appear considerably brighter (0.2–0.7

591

K. A. van der Hucht and B. Hidayat (eds.),
Wolf-Rayet Stars and Interrelations with Other Massive Stars in Galaxies, 591–594.
© 1991 IAU. Printed in the Netherlands.

magnitudes) in the narrowband images than non-WR stars. All plates are digitized, and all HeII-bright candidates are detected in objective fashion via computer analysis of the images.

With this technique (unlike objective prism plates) confusion is not a problem even in the Galactic plane at 18-19th magnitude. These filters successfully detect weak-lined WN stars in very crowded fields (Moffat, Seggewiss and Shara 1985). We note that two of the new objects are members of clusters and a third is only 3" from a previously known WR star. Thus our search technique is particularly useful at finding WR stars in crowded fields missed by previous, confusion-limited objective prism surveys.

Spectroscopic checking of every candidate which passes all the photometric tests must be carried out. This is to ensure that the candidates really are WR stars, and to obtain their subtypes.

4 Results

4.1 Galactic Distribution

The previously noted (Conti and Vacca 1990) tendency for WR stars to be found below the Galactic equator is strongly reinforced. All 13 new WR stars lie at $b < 0°$, which may be indicative of a warp in the Galactic disk.

The 13 new WR stars are significantly fainter and redder than the previously known Carina stars. This increased reddening is, of course, most readily interpreted as being due to increased distance from the Sun. The extension of the Carina arm is more striking than ever. Three of the thirteen new stars are closer than 3 kpc., in comparison with 9 known stars within the same distance. This sugguests an incompleteness in the present WR census of ~25% even for the closeby stars.

While individual stars' Galactocentric distances R are somewhat uncertain, the overall trend is not. The average R of the 13 new stars is 11 or 11.7 kpc (from the van der Hucht *et al.* 1988 and Conti and Vacca 1990 calibrations, respectively). This is further out from the Galactic center than all but eight of the known Galactic WR stars. *Thus our new stars are an extremely useful sample for testing whether the predicted and observed trends among WR stars in galaxies are confirmed in the outer Milky Way.* Probably the most important of these is the decreasing WC/WN number ratio at increasing galactocentric distances.

4.2 Subtypes

One of the most striking results of this survey so far is the remarkable ratio of WC to WN stars found: 2 to 11 (0.18), respectively. Conti and Vacca (1990) noted that the WC/WN ratio is 0.95 inwards of the solar circle (8.5 kpc) and 0.74 outside. Smith's (1988) study demonstrates the WC/WNE dependence on log [O/H] (and hence Galactocentric distance). The dramatic decrease in WC/WN number ratio continues the trend seen in her Figure 2.

The low WC/WN ratio seen in our sample is also supportive of the work of Maeder (1990). His models predict that initial mass and metallicity together determine the lifetimes of massive stars in the WR phase. Thus at increasing R the decreasing metallicity and smaller numbers of very massive stars conspire to strongly decrease the WC/WN ratio, as we now observe.

Future papers in this series will extend our search over much of the Southern Milky Way, and will concentrate on the global distribution of WR stars in the Galaxy.

Spectral classes, photometry, and derived parameters of the confirmed new Galactic WR stars

van der Hucht et al. 1988/Conti and Vacca 1990

Star	Sp	Sou.	α(2000)	δ(2000)	l	b	v	b−v	(b−v)₀	E(b−v)	Aᵥ	vₒ	vₒ−Mᵥ	d(kpc)	z(pc)	R(kpc)	cluster
SMSP1 = WR19a	WN7	4	10:18:53.4	−58°07'53"	283.90	−1.01	17.45	1.71	−.27/−.20	1.98/1.91	8.12/8.10	9.33/9.35	15.9/15.3	15.1/11.2	−270/−200	15.4/12.3	
SMSP2 = WR20a	WN7	4	10:23:58.0	−57°45'49"	284.27	−0.34	14.14	1.38	−.27/−.20	1.65/1.58	6.76/6.70	7.4/7.44	14.0/13.3	6.3/3.7	−40/−25	9.2/8.4	Wester-lund 2
SMSP3 = WR20b	WN7	4	10:24:18.4	−57°48'30"	284.33	−0.35	15.40	1.69	−.27/−.20	1.96/1.89	8.04/8.01	7.36/7.39	14.0/13.3	6.3/4.5	−40/−30	9.2/8.6	
SMSP4 = WR31a	WC7	2	10:57:43.0	−60°34'01"	289.40	−0.72	16.37	0.96	−.31/−.25	1.27/1.19	5.21/5.05	11.2/11.32	16.0/15.5	15.8/12.7	−200/−160	15.3/12.7	
SMSP5 = WR35a	WN6	2	11:00:24.5	−59°59'36"	289.46	−0.06	13.92	0.86	−.28/−.20	1.14/1.06	4.67/4.49	9.3/9.43	14.6/15.3	8.3/11.6	−10/−15	9.7/11.9	
SMSP6 = WR35b	WN4	2	11:01:02.3	−60°14'01"	289.63	−0.24	14.49	1.27	−.27/−.20	1.54/1.47	6.31/6.23	8.2/8.26	11.7/12.5	2.2/3.1	−10/−15	7.2/8.0	
SMSP7 = WR38a	WN6	1,2	11:05:49.0	−61°13'41"	290.57	−0.92	16.21	0.83	−.28/−.20	1.11/1.03	4.55/4.37	11.7/11.84	17.0/17.7	25.1/35.3	−400/−560	23.5/33.3	Sher 1
SMSP8 = WR38b	WC6	3	11:06:18.7	−61°14'13"	290.63	−0.90	16.21	1.50	−.27/−.25	1.77/1.75	7.26/7.42	9.0/8.79	12.6/13.0	3.3/4.0	−50/−60	8.0/8.0	
SMSP9 = WR42a	WN4.5	4	11:12:15.9	−61°05'04"	291.23	−0.49	17.61	1.46	−.26/−.20	1.72/1.66	7.05/7.04	10.6/10.57	15.2/14.8	11.0/9.0	−90/−70	11.2/9.9	
SMSP10 = WR42b	WN3:	5	11:13:03.8	−62°14'18"	291.75	−1.52	16.96	2.43	−.27/−.20	2.70/2.63	11.07/11.15	5.9/5.81	8.7/10.0	0.5/1.0	−10/−20	8.3/8.2	
SMSP11 = WR42c	WN6	2	11:14:01.6	−61°03'47"	291.42	−0.39	16.56	1.36	−.28/−.20	1.64/1.56	6.72/6.61	9.8/9.95	15.1/15.9	10.5/14.8	−70/−100	10.8/14.1	
SMSP12 = WR42d	WN4	2,4	11:14:38.8	−61°11'16"	291.54	−0.48	15.28	1.33	−.27/−.20	1.60/1.53	6.56/6.49	8.7/8.79	12.2/13.0	2.8/4.0	−20/−30	7.9/8.0	
SMSP13 = WR44a	WN3:	5	11:18:43.5	−61°26'37"	292.09	−0.54	16.20	1.55	−.27/−.20	1.82/1.55	7.46/6.57	8.7/9.63	11.5/13.8	3.3/5.8	−30/−50	7.9/8.3	

Sou. (source of spectra):

1: CTIO 4m, CCD, January 1988 (Lasker)
2: CTIO 1m, 2D-Frutti, March 1988
3: CTIO 4m, CCD, June 1988
4: SAAO 1.9m, IDS, April 1989
5: CTIO 4m, CCD + Argus, February 1990

Notes: SMSP4 = WR31a spectrum is contaminated by light of nearby A-type star; magnitudes here refer to the least contaminated (single) spectrum

5 References

Conti, P.S., Vacca, W.D. 1990, *Astron. J.* **100**, 431.

Maeder, A. 1990, *Astron. Astrophys.*, in press.

Moffat, A.F.J., Seggewiss, W., Shara, M.M. 1985, *Astrophys. J.* **295**, 109.

Smith, L.F. 1988, *Astrophys. J.* **327**, 128.

van der Hucht, K.A., Conti, P.S., Lunström, I., Stenholm, B. 1981, *Space Science Reviews* **28**, 227.

van der Hucht, K.A., Hidayat, B., Admiranto, A.G., Supelli, K.R., Doom, C. 1988, *Astron. Astrophys.* **199**, 217.

DISCUSSION

Montmerle: In view of the large incompleteness your work shows to exist in previous census, would you exclude the possibility that the total number of WR stars in our galaxy could be higher by factors 2-3, say, than the $\sim$ 2500 quoted by van der Hucht *et al.* (1988)? This would be important for estimating the high-energy diffuse flux (galactic X-ray ridge, nuclear γ-ray line emission) that has been associated with these stars (Montmerle, this volume), particularly in the inner galaxy.

Shara: If you are talking about the entire galaxy there are numbers that are bandied about of a 1000 of even higher, and I do not think I would find that terribly shocking. There are certainly stars, very far away, that we have found here, but we may be in an unfair advantage, in a sense, by looking along the Carina arm. We may have less reddening, *e.g.*, than in the other regions. I do not think that our results tell us how many WR stars there are in the galaxy. They are only telling us that the present survey is moderately complete down to the fourteenth magnitude. It is really a magnitude limit that is sampled rather than a volume limit. But even at fourteenth magnitude, there are stars waiting to be discovered. Thus, it is certainly possible, but we must expand the survey significantly to make an educated statement.

Massey: "Gradients" often imply two points, rather than one - do you not need to look in another direction with your sensitive technique before knowing if you have found a WC/WN gradient in the Galaxy?

Shara: Absolutely! I was simply taking the WC/WN number ratio close to the Sun ($\sim$ 1) as deduced from the Sixth Catalogue, and comparing with the 2/11=1.18 ratio we find in our survey. Perhaps we see a strong gradient, as our new WR stars are at an average galactocentric distance of 11.4 *kpc*. But, perhaps we are just more complete than all previous surveys for WN stars, and we may find a small WC/WN ratio even when we look inwards in the Galaxy.

Conti: Speaking from a purely personal point of view, I would not be surprised if the total estimate of the number of WR stars galaxy-wide is at least a factor two too low. A factor ten is not impossible. Discovery of additional WR stars is completely dominated by the extinction problem.

van der Hucht: I must compliment you on this virtual monk's work, and I do hope to see your Seventh galactic WR Catalogue in my lifetime. I can understand that it is not always easy to keep up the good spirit during all this hard work. Therefore, I can understand that you once, in a sad mood, confessed to Tony Moffat: "Roses are red, carnations are pink. If I had your talents, I didn't have to blink." But please: carry on!

Shara: I promise to continue *if* you promise not to publish any more poetry.

Hidayat: Would you undertake surveys in the direction of low-absorption galactic fields (such as Puppis, at $l \sim 240°$)?

Shara: We will first do $l = 282°$ to $l = 332°$. If we succeed in completing this, then we will consider increased coverage. We must measure magnitudes and positions of 4×10^7 objects to complete the present survey.

WOLF-RAYET STARS IN LOCAL GROUP GIANT HII REGIONS

LAURENT DRISSEN

Space Telescope Science Institute
3700 San Martin Drive
Baltimore, MD 21218

ABSTRACT. We present the results of our search for WR stars in three of M33's most massive giant HII regions, namely NGC 604, NGC 595 and NGC 592. A few of the putative "superluminous" WR stars are shown to be crowded multiple systems. We also compare the WR populations of the most massive GHRs in the Local Group. WN stars are 5 times more numerous on the average than WC stars in these large complexes. The variations among GHR WR populations are discussed in the context of metallicity and age differences.

1. Introduction

Giant HII regions (GHRs) appear to be the most intense sites of star formation known in normal galaxies, and they offer the unique opportunity to study massive stars at their birthplace. Apart from 30 Dor in the LMC, the most massive GHRs in the Local Group, namely NGC 604 and NGC 595, are located in M33. The presence of WR stars in these regions is well known, but some uncertainties remain. We present our observations in section 2 and 3, and we compare the WR population of the best studied Local Group GHRs in section 4.

2. Observations

Narrow-band images of NGC 604, NGC 595 and NGC 592 were obtained with an RCA CCD detector attached to the Mont Mégantic 1.6m telescope. A set of three filters was used to detect WR stars:
4780/65 (λ_c = 4780 Å ; FWHM = 65 Å) ; reference continuum
4686/35 ; detect He II 4686 in WNs
4650/29 ; detect CIII 4650 in WCs, NIII 4641 in WNs.
    The reference continuum image was subtracted from the on-line image, after scaling in such a way that bright stars outside the nebula disappear completely. Details of the observations and reductions are presented elsewhere (Drissen, Moffat and Shara 1990, hereafter DMS).

3. Results

3.1 NGC 604.

With about 10^6 M$_\odot$ of HII, NGC 604 is an order of magnitude more massive and more luminous than any known HII region in our Galaxy. D'Odorico and Rosa (1981) were

K. A. van der Hucht and B. Hidayat (eds.),
Wolf-Rayet Stars and Interrelations with Other Massive Stars in Galaxies, 595–600.
© 1991 *IAU. Printed in the Netherlands.*

596

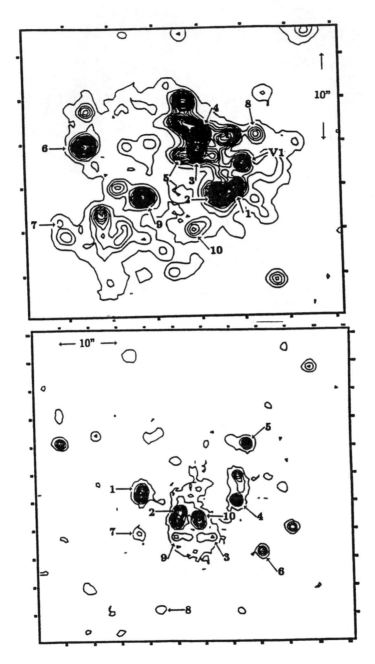

Figure 1. Isophotes of the on-line 4686 Å image of NGC 604 (top) and NGC 595 (bottom), showing the positions of WR candidates. Star V1 in NGC 604 is variable (see DMS). North is at the top, east to the left.

line flux in each object is consistent with that of a single Galactic WR star (DMS), we assume that only one WR star is present at each WR position.

Table 1 presents statistics on the WR population among the best studied and most massive GHRs in the Local Group. The total mass in OB stars has been derived by Kennicutt (1984) from the Hα flux. Although the statistics are rather poor, a few points should be noted. First, the number ratio of WR to O stars, WR/O (obtained by direct counting in resolved GHRs or deduced from the global light output in M33), is much lower than unity. A burst of star formation thus does not necessarily imply a large WR/O ratio, as it has sometimes been claimed (D'Odorico and Rosa 1981). However, this ratio does not seem to be constant from one region to the next. This is particularly obvious when one compares NGC 604 with NGC 595: both regions contain the same number of WR stars, despite the factor of two difference in mass.

TABLE 1. Local Group Giant HII Regions

Region	Galaxy	M_{OB} ($M_\odot$)	N_{WR}	N_{WR}/M_{OB} ($\times 10^{-4}$)	$N(WC)$	$N(WNL)$	[O/H]
30 Dor	LMC	50 000	25	5	5	17	8,41[1]
NGC 604	M33	15 000	7-10	5-7	2	>4	8,51[2]
NGC 595	M33	7000	8-10	11-14	1	7	8,44[2]
NGC 3603	MW	5000	2-3	4-6	0	2	8,39[3]
Carina	MW	2000	3	15	0	3	8,62[1]
NGC 346	SMC	2000	1	5	0	0	8.22[1]
NGC 592	M33	1000	2-3	20-30	0	—	—

Notes: For 30 Dor, we included all WR stars within 150 pc from R136, and considered that 4 WNL stars compose R136 (Moffat *et al.* 1987). The total mass in OB stars is taken from Kennicutt (1984). References for [O/H]: (1) Shaver *et al.* 1983; (2) Vilchez *et al.* 1988; (3) Melnick *et al.* 1989.

Three factors can in principle affect the WR/O ratio in GHRs: metallicity, the slope of the IMF and the age of the burst. Both theoretically and observationally, metallicity has been shown to be the major factor controlling WR/O and WC/WN number ratios (Smith 1988, Maeder 1990; see however Massey 1990). However, [O/H] is very similar in NGC 604 and NGC 595. A flatter IMF could also increase the WR/O ratio, without affecting the WC/WN ratio as much (Maeder 1990). Although this hypothesis cannot be rejected, there is little, if any, evidence for changes in the slope of the IMF as a function of galactocentric distance in M33 (Freedman 1985), or between the field and starburst regions (Blaha and Humphreys 1989, Massey, Parker and Garmany 1989). Another parameter that could affect the WR/O ratio is the age of the burst. Time dependent synthetic WR/O ratios computed by Arnault, Kunth and Schild (1989) show that the WR/O ratio increases very rapidly a few Myrs after the burst, stays constant for a while (the actual duration depending on the metallicity) and decreases rapidly after that. Thus, two young starburst regions the first to discover WR stars in that region: spectroscopy of the brightest knots revealed the presence of WR characteristics in most of them. This led the authors to conclude that an unusually large number of WR stars ($\sim$ 50) were present in that nebula. Conti and Massey (1981, hereafter CM) have also detected WR emission lines in three objects. Assuming that these objects were single stars, they concluded that they were, like R136 in the core of 30 Dor, superluminous.

An isophote map of the on-line 4686 Å image of NGC 604 is displayed in Figure 1, indicating the position of our WR candidates. Stars nos. 1, 2, 3, 4, 6 and 9 are spectroscopically confirmed WR stars (all but no. 1 being WNs; see CM, D'Odorico and Rosa 1981, DMS). Stars 1 and 2 were considered until now as a single WR star (CM11), but they are clearly resolved in our images. Moreover, star no. 2 is elongated and is very likely to be a composite object (see also Debray 1990). Another CM object, no. 12 in their list, is also resolved here into two components: nos. 3 (WN) and 5 (likely to be a WC). Detection levels of candidates 7, 8 and 10 are quite low and spectroscopy is an absolute requirement before considering them as genuine WRs. The absolute visual magnitude (M_V) of the brightest WR stars in NGC 604 is close to -7.7 (Debray 1990), about a magnitude fainter than what was previously thought.

3.2 NGC 595.

This region has received much less attention than NGC 604, although the presence of WR stars has been known since CM obtained spectra of 3 luminous stars in its core. Massey and Conti (1983) and Massey, Conti and Armandroff (1987) have spectroscopically confirmed three more WR stars, and Armandroff and Massey (1985) have detected three more candidates. An isophote map of the on-line 4686 Å image of NGC 595 is presented in Figure 1. Two of the previous "superluminous" WR stars, nos. 1 (CM6) and 2 (CM5) are clearly resolved from adjacent bright stars. Their new estimated magnitudes are still quite high (-6.9 for no. 1 and -7.9 for no. 2), but no more than some WNL stars, e.g. as seen in 30 Dor (Moffat 1989). Lequeux, Meyssonier and Azzopardi (1987) have mentionned objects 4 and 5 in their objective prism survey. Star no. 3 is the only WC star in that region. Armandroff and Massey (1990) have recently obtained spectra of stars 4, 5 and 7 and classify them as WNL.

3.3 NGC 592.

This region is marginal compared to NGC 604 (about 10 times less massive), but nevertheless contains a few WR stars. We have resolved another putative superluminous WR star (CM3) into two components, the fainter one being the WN star. Its previously estimated M_V (-9) is now reduced to -7. Another confirmed WN star (no. 2 in Massey, Conti and Armandroff 1987) is 20″ to the west of CM3. We have also detected a new WR candidate, probably of WN type.

4. THE WOLF-RAYET POPULATION OF GIANT HII REGIONS

It is a difficult task to deduce the exact number of WR stars in each knot from our data alone, because a large number of abnormally weak emission-line Of/WR stars could mimic the appearance of a single, "normal" WR star, at least in the brightest knots. However, there is no need to postulate such an anomaly and since the emission-of slightly different ages can have very different WR/O ratios.

The other important fact to notice from Table 1 is that all these GHRs are completely dominated by WN stars, most of them being of late (6-7) types. The global WC/WR ratio in these regions is ~ 0.17. The lack of WC stars in starburst "WR galaxies" has also been noticed (Vacca and Conti 1990). Metallicity certainly plays an important role in this trend. With the exception of the Carina nebula, all these regions have a low metal abundance (unfortunately, there is no GHR comparable to NGC 604 or 30 Dor in M31, where the metallicity is much higher). However, this does not explain the fact that 30 Dor contains mainly WNL stars while the rest of the LMC is dominated by WNE (Moffat *et al.* 1987). Models (Langer 1987; Maeder 1990) indicate that the duration of the WNL phase increases with the mass of the progenitor. In this context, the observations could be explained if the average mass of the WR progenitors in GHRs is higher than in the field, or that these regions are so young that only the most massive stars have evolved away from the main sequence.

Many points remain unsettled and there is an absolute need for high resolution imagery and high signal-to-noise spectroscopy of the most massive stars of the Local Group GHRs, as well as in larger starburst regions, such as the "hypergiant" HII regions in M101 where WR spectral features have been detected (D'Odorico, Rosa and Wampler 1983).

References

Armandroff, T. E. and Massey, P. 1985, *Ap. J.*, **291**, 685.

Armandroff, T. E. and Massey, P. 1990, this Symposium.

Arnault, P., Kunth, D. and Schild, H. 1989, *Astron. Ap.*, **224**, 73.

Blaha, C. and Humphreys, R. M. 1990, *A. J.*, **98**, 1598.

Conti, P. S. and Massey, P. 1981, *Ap. J.*, **249**, 471 (CM).

Debray, B. 1990, this Symposium.

D'Odorico, S. and Rosa, M. 1981, *Ap. J.*, **248**, 1015.

D'Odorico, S., Rosa, M. and Wampler, E. J. 1983, *Astron. Ap. Sup.*, **53**, 97.

Drissen, L., Moffat, A. F. J. and Shara, M. M. 1990, *Ap. J.*, in press.

Freedman, W. L. 1985, *Ap. J.*, **299**, 74.

Kennicutt, R. C. 1984, *Ap. J.*, **287**, 116.

Langer, N. 1987, *Astron. Ap.*, 171, L1.

Lequeux, J., Meyssonnier, N. and Azzopardi, M. 1987, *Astron. Ap. Sup.*, **67**, 169.

Maeder, A. 1990, *Astron. Ap.*, in press.

Massey, P. 1990, this Symposium.

Massey, P. and Conti, P. S. 1983, *Ap. J.*, **273**, 576 (MC).

Massey, P., Conti, P. S. and Armandroff, T. E. 1987, *A. J.*, **94**, 1538.

Massey, P., Parker, J. W. and Garmany, C. D. 1989, *A. J.*, **98**, 1305.

Melnick, J., Tapia, M. and Terlevich, R. 1989, *Astron. Ap.*, **213**, 89.

Moffat, A. F. J. 1989, *Ap. J.*, **347**, 373.

Moffat, A. F. J. , Niemela, V. S., Phillips, M. M., Chu, Y.-H. and Seggewiss, W. 1987, *Ap. J.*, **312**, 612.

Shaver, P. A., Mc Gee, R. X., Newton, L. M., Danks, A. C. and Pottasch, S. R. 1983, *M. N. R. A. S.*, **204**, 53.

Smith, L. F. 1988, *Ap. J.*, **327**, 128.

Vacca, W. D. and Conti, P. S. 1990, preprint.

DISCUSSION

Vilchez: How would sequential star formation affect the WR/WC ratio?

Drissen: According to Arnault, Kunth and Schild (1989), the shorter the burst is, the higher the WR/O ratio would be. Sequential star formation would then decrease the WR/O ratio.

Maeder: The aging in a starburst region can lead to large WR/O ratios as you have shown. However, this is likely to apply only to regions limited in size. For large regions or galaxies an average star formation rate over the last million years is more likely.

Drissen: I agree that one should distinguish between the field stars and starburst regions. The WR/O and WC/WN ratios may be different in the two cases. However, a burst of star formation does not necessarily mean very high WR/O ratios (close to 1), as we can see in Local Group GHR.

Laurent Drissen

WOLF-RAYET POPULATIONS IN DWARF GALAXIES

LINDSEY F. SMITH
Mount Stromlo and Siding Spring Observatories
Australian National University
GPO Box 4, Canberra ACT 2601
Australia

ABSTRACT. The Wolf-Rayet (WR) feature at 4650 A is observed in about 10% of the dwarf galaxies with high surface brightness knots. The intensity of the feature implies the presence of tens to thousands of WR stars. H$beta$ fluxes imply correspondingly large numbers of O stars. The easily observed intensity ratio WR$bump$/H$beta$ is a measure of the WR/O star numbers.

The metallicity of dwarf galaxies ranges from $Z = Zo/30$ to $Zo/2$, or O/H" = $\log$(O/H)+12 = 7.4 to 8.6. WR$bump$/H$beta$ correlates with O/H" and O/H" > 7.9 appears to be a necessary condition for the presence of the WR feature. Giant HII regions in ordinary galaxies extend to higher than solar metallicities and, in extreme cases, WR/O ≈ 1 are implied.

The subtypes present in giant HII regions in nearby galaxies appear to be exclusively late type WN and, occasionally, early type WC. Spectra of most BCD galaxies are compatible with a similar population. However, some high metallicity giant HII regions in large galaxies appear to have stronger NIII4640 relative to HeII4686 than occurs in WN subtypes in the Galaxy and the Magellanic Clouds.

The data needed for more detailed analysis of dwarf galaxy observations is collected.

1. The Observations

Dwarf galaxies are low mass, low luminosity systems; the dividing line from ordinary luminous galaxies is at about $M_B = -19$. Among the dwarfs, there is a subset with high surface brightness knots, having the appearance of giant HII regions. They were, in fact, first called extragalactic HII regions (Sargent and Searle 1970), but were subsequently found to be embedded in otherwise low surface brightness galaxies. They now go by the name of Blue Compact Dwarf (BCD) galaxies.

The explanation of these galaxies, which appears to be generally accepted, was given by Searle et.al. (1973) and by Gerola et.al. (1980); the BCD galaxies represent a phase of intense star formation in dwarf irregular (Im) galaxies. The regions of star formation appear to involve 1E3 to 1E5 ionising stars; i.e. from numbers a little larger to several orders of magnitude larger than for 30 Doradus. Furthermore, similar compact objects appear to exist up to $M_B \approx -23$ involving 1E7 ionising stars (Hazard 1985). Such objects are very numerous, about 1 per square degree (Hazard 1985). Noting that 30 Dor is larger than any giant HII region in our own Galaxy, the rate of star formation in these regions is prodigious.

Giant and supergiant HII regions in the larger galaxies present observational similarities to those in BCD galaxies - they have unresolved cores containing large numbers of massive stars plus, of course,

601

K. A. van der Hucht and B. Hidayat (eds.),
Wolf-Rayet Stars and Interrelations with Other Massive Stars in Galaxies, 601–612.
© 1991 *IAU. Printed in the Netherlands.*

surrounding nebulosity. For this reason, they are often included in the same studies and will, accordingly, be included in the present review.

Dwarf galaxies are generally of low metallicity; ranging from Zo/30 (Zw 18) to Zo/2; they show no large scale structure and no abundance gradients (Kunth 1985). The regions are small (or the order of 100 pc) so it is fair to assume that they are, or were, chemically homogeneous before the present burst of star formation commenced. From this derives one of the principle reasons for interest in the BCD galaxies - they present an opportunity to study star formation at well defined and low Z, with a very large number of stars being formed in the one burst. By the same token, giant HII regions in ordinary galaxies represent the same phenomenon at higher metallicities - another reason for their inclusion in the papers reviewed here.

Allen et.al. (1976) made the first observation of WR features in a dwarf galaxy, He2-10. The most prominent emission feature, characteristic of all subtypes of WR star, is the broad emission in the vicinity of 4650 A. The flux contributed by one WR star to this feature is of the order of 1E36 ergs/sec (Kunth and Sargent 1981; Arnault et.al. 1989, henceforth AKS). Figure 5 of Kunth and Schild (1986) shows the fluxes observed in the spectra of BCD galaxies and giant HII regions; they range from 1E38 to 1E41 ergs/sec, i.e. 100 to 100,000 WR stars ! It is plotted against O/H" (used as a shorthand here for (logO/H+12) and includes some high Z giant HII regions. The meaning of the (fairly good) correlation is ambiguous. The authors consider that a selection effect may be operating.

Metallicity, however, does appear to play a significant role, as shown by AKS. Their Figure 8, shows the ratio of the WR feature (called WR*bump*) to H*beta* versus O/H". The H*beta* flux is a measure of the number of ionising photons and hence the total number of O (and WR) stars. The *upper limit* of WR*bump*/H*beta* increases with O/H", implying that high O/H" is a necessary condition for high WR/O star numbers. It is also notable that the WR feature is only observed for O/H" > 7.9. To first approximation, WR/O (numbers) $\approx$ WR*bump*/H*beta* (intensity). Note that the values of the latter imply WR/O $\approx$ 1 for some of the high O/H" regions in ordinary galaxies (eg. NGC5128 #13, O/H"$\approx$9, see Rosa and d'Odorico 1986 for population simulation; Mrk309 and NGC6764, see Osterbrock and Cohen 1982).

High O/H" is not, however, a sufficient condition for a large WR presence; low values of WR*bump*/H*beta* also occur. An observational effect contributes - the nebulosity is more extended than the ionising cluster and (depending on resolution and slit size) can get excluded from the observation. However, the scatter is undoubtedly real and is expected (see below) since the WR lifetime is short compared to the O star lifetimes allowing the presence of O stars without any WR stars.

In giant HII regions in nearby galaxies, the subtypes which occur appear to be predominately WN6-7 and, less commonly, WN8 and WC4 (see D'Odorico et.al. 1983; Rosa and D'Odorico 1986; for 30 Dor see Walborn 1986; Melnick 1985a, Moffat et.al. 1987). A conspicuous contribution from WC spectra is unusual in dwarf galaxies (see eg. Kunth and Schild 1986). The latter authors stress that [FeIII] 4658 occurs frequently and, at low resolution, can masquerade as CIV 4650 creating a false impression of a WC contribution. In lower metallicity galaxies, the [FeIII] line is not conspicuous in published spectra - as would be expected. He2-10 is the clearest case of a dwarf galaxy where WC stars appear to contribute (but see the discussion in Section 2.4 below). However, in giant HII regions in "ordinary" galaxies there are many convincing examples of WC contribution (see eg. D'Odorico et.al. 1983). Rosa and D'Odorico (1986) note a strong correlation between regions where there has been a type II supernova and the presence of WC emission.

Subtypes of WR stars in the dwarf galaxies are mostly compatible with the presence of these subclasses. However, it will be shown below that spectra of some giant HII regions defy simulation with a simple combination of WN6-8 and WC4.

2. The expectations.

I now propose to compare these observations with what we expect to see, based on stellar evolution calculations and observations of individual WR stars in nearby galaxies. Specifically:
* the number ratios of WR/OB and WC/WN as functions of metallicity (Section 2.1);
* the subtype distributions (Section 2.2);
* the calibration of the flux in the WR feature at 4650 A (Section 2.3);
* separation of WN and WC contributions using the CIV 5808 feature (Section 2.4);
* the question of whether WR stars are the same in all galaxies (Section 2.5).

2.1. NUMBERS OF WOLF-RAYET STARS

The numbers of WR stars relative to O stars (from which they evolve) and the numbers of WC stars relative to the WN stars vary dramatically from one galaxy to another and within large galaxies as a function of radius. The cause of the variation has been attributed to metallicity (Smith 1968b, Maeder et.al. 1980) and to variations of the IMF (Conti et.al. 1983). The theoretical connection to metallicity is that higher mass loss rates for the progenitor O star are predicted when the metallicity is high (Abbott 1982). This causes the star to reach the WR stage earlier in its evolution and thus increases the overall WR lifetime and the WC fraction of that lifetime (Maeder et.al. 1980; Maeder 1981).

The theoretical connection to the IMF is that only massive stars become (Population I) WR stars and so an increase in the relative numbers of massive stars will increase the WR/OB ratio. Both factors will have an effect. However, the evidence favouring metallicity as the *dominant* factor is now overwhelming. In short:
* The correlation of both WR/OB and WC/WN with metallicity is excellent (Smith 1988; AKS; Maeder 1990b). The criticism of the correlation (eg. Massey 1986; Massey and Conti 1983; Armandroff and Massey 1985) is that regions with similar O/H" to the LMC and SMC have different populations. The answer is (Smith and Maeder 1990) that, in the large spirals, metallicity at a given galactocentric radius covers a wide range (eg. Talent and Dufour 1979). Because higher Z regions have *more* WR stars as well as more WC stars, the average WR population will be weighted towards values characteristic of a higher Z value.
* The agreement between predicted and observed number ratios at different metallicity is excellent (Maeder 1990b and this volume);
* The detailed explanation of the WC subtype distribution in terms of metallicity (Smith and Maeder 1990, and see below) clinches the argument.
* I would further point out that regions of low Z are always low in WR numbers and the effect of Z on the IMF is generally suggested to favour high mass stars (eg. Scalo 1986; Terlevich 1985).

Thus the correlation found for dwarf galaxies and giant HII regions between the upper limit of WR*bump*/H*beta* and O/H" is appropriately interpreted by the AKS as due to more WR stars at higher metallicity. The O/H" lower limit of 7.9 corresponds to $Z \approx .002$ at which the models (Maeder 1990a,b) predict that only stars more massive that 85 Mo attain WR status.

The large range of WR*bump*/H*beta* values at a given O/H" is qualitatively explained by the evolution of a star burst region. AKS calculate WR/O and WR*bump*/H*beta* as a function of time. For continuous star formation, the numbers are, of course, approximately constant. However, for the burst scenario, there is a WR phase which commences at an age of ~3E6 yrs, when the most massive stars become WR stars, and ends with the demise of the lowest mass stars which are capable of loosing their H envelopes in their lifetime; both times are Z dependent, the end point more than the starting point.

At low Z, the cluster's WR phase is very short, ~6E5 yrs at Zo/10, because only the most massive stars attain WR status. The lifetime of the lowest mass O stars is ~6E6 yrs, 10 times longer; this satisfactorily accounts for the observation that only ~1/10 of the BCD galaxies have a detected WR feature.

For solar Z, the WR phase lasts ~3E6 yrs; during that time the ratio of WR/O star numbers and WR*bump*/H*beta* increase, as the number of WR stars decreases more slowly than the number of remaining O stars. The WR population decreases more slowly because, as lower mass stars enter the WR phase, their shorter WR lifetime is offset by their larger numbers.

30 Dor is noteworthy in this regard. Numbers assembled by AKS indicate that the 30 Dor region has a WR/O ratio approximately equal to the average for the whole LMC. However, the age is estimated at only 2E6 yrs (Melnick 1985b), less than expected for the onset of the WR phase, even for a 120 Mo star. The WR population in 30 Dor may increase dramatically in the next million years.

AKS's population simulations do not generate WR*bump*/H*beta* values as high as observed. This is remedied, in part, by higher mass loss rates used in more recent evolutionary models (Maeder 1990a,b). AKS demonstrate that the "burst scenario" generates WR/O values an order of magnitude higher than continuous formation. It remains to be seen whether burst simulations with the new models will generate the highest observed values of WR*bump*/H*beta* $\approx$ 1.

2.2. SUBTYPE DISTRIBUTION

Variation of subtype distribution with radius in the galaxy and between the Galaxy and the Magellanic Clouds was first noted by Smith (1968) and has been extensively studied since (Hidayat et.al. 1982; van der Hucht et.al. 1988). The most notable feature is that the later type WC stars are more strongly concentrated to the Galactic centre and are missing from the Magellanic Clouds. The suggestion that this is due to the different metallicity of the progenitors was made by Smith (1968) on circumstantial evidence. The reason is now elucidated by the models of Maeder (1990a, Smith and Maeder 1990). When the metallicity is low, the WR stage is achieved later in the evolution when the core is more "cooked" and has higher C/He and O/C ratios. (Smith and Hummer 1988) observe a strong increase of C/He abundance to early subtypes. Smith and Maeder (1990) hypothesise that the surface composition is the *primary* parameter determining subtype. The models then predict (correctly) that WC9 stars are only possible when the metallicity is high; in low Z regions such as the LMC, only WC4 stars are possible; in even lower Z regions, such as the SMC, only WO stars are possible

It follows that in dwarf galaxies, with predominantly low metallicity, only WC4 and WO stars are expected. In ordinary galaxies, giant HII regions with higher Z should be able to produce later types; however, with only the strongest lines showing over the O star continuum, WC4-6 cannot be differentiated. To produce WC8&9 stars which are conspicuously different (see Section 2.4), requires regions with Z > Zo; such regions deserve careful observation.

The factors controlling WN subtype are not yet so clear. Identifying WNL/WNE with stars with/without hydrogen (a generalisation with many exceptions), Maeder's (1990b) models show that WNL evolve predominantly from the more massive stars. (This has been suggested often before, eg. Langer 1987, on the basis of the higher luminosity of the WNL.) However, at Z < Zo, where models predict largely WNL, the observations (LMC and SMC) show large numbers of WNE stars. The discrepancy may be due to the increasing importance of binaries.

Thus, our uncertain expectation for dwarf galaxies is that the dominant WN subtype with be WNL. With the signal/noise available in spectra of dwarf galaxies plus the uncertainty of the line ratios, it is hard to tell the difference (see Section 2.4). As a curiosity, note that R136, the central cluster of 30 Dor and HD 97950, the central cluster of NGC 3603 have broader HeII4686 than is characteristic of WNL stars (35 A versus 20-30 A). Line width is not, however, a reliable subtype indicator.

2.3. CALIBRATION OF LINE FLUX IN THE WR*bump* AT 4650.

The estimate of 1E36 ergs/sec for the flux from a single WR star in the 4650 feature originates from Kunth and Sargent (1981) and is based on average EW's and absolute magnitudes. AKS use 1E36 erg/sec for a WN star and 4E36 for a WC star.

With the advent of CCD's, direct measurements of flux have become easy and a significant number of stars have been observed.
For WC stars:
* Smith et.al. (1990a, henceforth SSMa) find that, for LMC WC4-6 stars, the flux in the 4650-4686 feature is the same for all stars: log F(erg/sec) = 36.7∓0.15 (assumes M-m = 18.5 mag, corresponding to log F/f = 47.5.) However, Smith et.al. (1990b, henceforth SSMb) find that the value for Galactic WC5-7 stars appears to be lower by about 0.5 dex.
* Line fluxes for WO star are probably lower than for WC4. For Br 93 in the LMC , the 5808 flux may be 0.0 to 0.5 dex fainter than for the WC4 stars and the 4650 flux even lower. For the WO star in the SMC, the 4650 flux is about 0.6 dex fainter than the value above (for M-m=18.8, EW from Conti 1989, colours from Smith 1968a and equations from SSMa). Care is needed; the WO stars are a very diverse class (see Smith and Maeder 1990; Torres-Dodgen and Massey 1987).

For WN stars:
* Conti and Morris (preprint) find that, for LMC WN stars, the HeII4686 flux covers a wide range. Averaging over the five WN6&7 stars in the 30 Doradus region, yields log F(erg/sec) = 36.5∓0.6 . WNL in other parts of the LMC are few and have are not yet measured for fluxes; WNE stars in Conti and Morris' sample yield log F(erg/sec) = 35.9∓0.3.
* Kunth and Schild (1986) specify that they include only HeII4686 in the flux measurements. AKS's WR*bump* measure includes NIII4640 and HeII4686. Hence the ratio of these two lines is relevant.
* The ratio of NIII4640/HeII4686 is uncertain. Data of Conti and Massey (1989) and unpublished data which I hold in conjunction with Moffat and Shara indicate that NIII4640 is weaker relative to HeII4686 for LMC stars than for Galactic ones. The correction for Galactic WN7 stars is about 0.16 dex and for LMC WN7 stars is about 0.06 dex which we can neglect compared to the scatter of 0.6 dex in the flux of HeII4686.

For both sequences:
* Maeder's (1990a,b) models indicate that all WR stages will be more massive and therefore more luminous in total flux in regions of lower metallicity. Whether this will reflect directly in the line fluxes is not clear, but the difference found by SSMb indicates that it does.

Conclusion: The flux in the WR*bump* is probably metallicity dependent. The average values for LMC stars (assuming M-m=18.5), log F(ergs/sec) = 36.5 for HeII4686 in WN6-7 stars, and 36.7 for CIV4650 + HeII4686 in WC4 stars are probably good average values for single WR stars in regions of moderately low metallicity (O/H" = 8.2 to 8.5). However, for both higher and lower metallicity regions, the WC value is probably too high and the WN value is untested. For Galactic and LMC WN7 stars, the ratios of NIII4640/HeII4686 are approximately 0.5 and 0.15, respectively, representing a small correction on a fairly uncertain flux value for HeII4686.

2.4. OBSERVATIONAL INDICATORS OF SUBCLASS, CONTRIBUTION OF OTHER LINES TO THE SPECTRA.

Two questions appear relevant:
1. What other features may be visible apart from the WR*bump* at 4650 A, which may be useful as subtype indicators?
2. How can we separately estimate the numbers of WN and WC stars contributing to the WR*bump* ?

CCD spectra of WC stars in the Galaxy and the LMC have been published by SSMa&b. In collaboration with Shara and Moffat, I hold a collection of CCD spectra of WN stars in the Galaxy and the LMC; from these, I have extracted the generalisations below regarding strengths of emission lines in the wavelength range 4000 - 6000 A.

In WN spectra, HeII4686 is the strongest feature rivalled only by NIII4640 in WN8 stars. HeII4686 > NIII4640 always. Next strongest is NV4605-20 (WN3-4) or NIV4057 (WN5-7) or NIII4640 (WN6-8), a factor 2-5 down (for Galactic stars), followed by HeII5411, a factor ~5 down compared to HeII4686. HeI 5875 is significant for WN7-8. (Relative strengths here are in peak/continuum flux which is more conspicuous and better defined in blended features.)

WC spectra have two strong features, HeII+CIII/IV4650 and CIV5808. In the 4650 region, 4650 > 4686 always. In WCE spectra, these two lines are by far the strongest; in WCL spectra, CIII5696 becomes equal to CIV5808 in WC8 and exceeds it in WC9. (WC8 stars have been found in M31 but no WC9 stars have yet been seen in an external galaxy.)

These differences should allow simple discrimination between WN and WC dominance and permit estimates of how many of each sequence is present.

Figure 1 shows a combination of a WN7 spectrum (Br90 in the LMC) with a WC4 star (Br10). Br10 is twice as bright, in line and continuum, as any other WC4 star in the LMC, so I count it as two "average" WC4 stars. Notice:
* The flux scale is -2.5 log f(lambda). The top two spectra are at their real magnitudes. The "double" WC4 star is about 2 mag fainter in the continuum, but the flux in the 4650 feature is about 4 times as great.
* The lower three spectra are shifted downwards to obtain good separation and represent a combination of WN7+WC4 in ratios: 1:2, 3:1, 20:1 .
* The shorter wavelength peak is always at 4640 A; the WC feature forms a broad pedestal for the narrower features and becomes inconspicuous in the blend.
* Peak 4686 >> 4640 *always* .
* Even at 20:1 dilution, the CIV5808 is predominantly from the WC star; the WN star contributes about one third of the flux.

Clearly, separation of WC and WN star contributions using 4650 *alone* is not easy. However, the 5808 feature can assist.
* For WC stars, the ratio of 5808/4650 is a slowly varying function of subtype (SSMb). The *flux* ratio for both LMC and Galactic WC4 stars is -0.22∓0.1 dex. (i.e. 4650 is 1.7 times stronger). (The EW ratio is ~+0.2 dex, i.e. 5808 is stronger.) Thus, if WC stars are dominant, the 5808 line will be comparable to 4650; the strength of 5808 can be used directly to estimate the contribution of the WC stars to the blend at 4650. Note that we expect (in these low metallicity galaxies) that most of the WC stars will be WC4; should they be WC6, the flux ratio is ~-0.3 dex (EW ratio ~0.1) only 0.1 dex different from WC4&5 stars. However, if they are WO stars, the EW ratio may be as high as 0.8 dex (Smith and Maeder 1990).

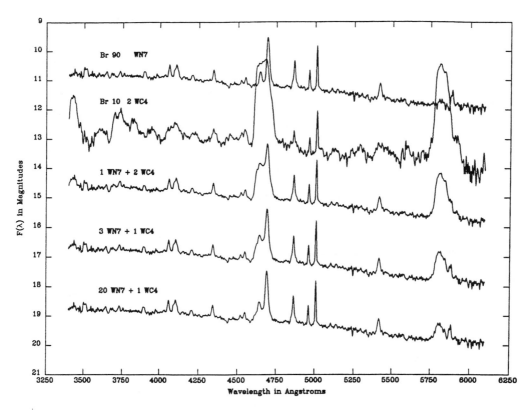

Figure 1. A combination of Br90 (WN7) and Br10 (two WC4 stars). The top two spectra are at their real magnitudes. The lower three spectra are shifted downwards to obtain good separation and represent a combination of WN7+WC4 in ratios: 1:2, 3:1, 20:1 .

TABLE 1. Average log F(erg/sec) for LMC Stars

	CIII/IV4650+HeII4686	CIV5808
WC4	36.7 (a)	36.5 (a)
WN7	36.5 (b)	35.1 (c)

(a) SSMa.
(b) Mean of 30 Dor stars in the sample of Conti and Massey (1989).
(c) Smith, Shara and Moffat (in preparation).

CIV5808 in WN stars is relatively weak but needs to be allowed for since the WN/WC ratio is usually large; the correction is uncertain because the CIV5808 strength in WN spectra is highly variable and probably Z dependent. Table 1 gives the mean fluxes for LMC stars in the two lines. On the assumption that the LMC stars are characteristic of low Z regions such as found in the BCD's, these values allow the quantitative separation of the WN and WC.

Comparing Figure 1 with published spectra of BCD galaxies, we find that most metal poor (O/H" < 8.6) BCD's appear to be pure WN, showing HeII4686 and only a hint of NIII4640 (see eg. Kunth and Sargent 1981).

WC dominated spectra have not been observed in BCD's. The average spectrum of NGC604 (d'Odorico and Rosa 1981) is a nice example from a giant HII region in a nearby galaxy.

Intermediate types are found in a few BCD's and some giant HII regions. However, these spectra all share an important difference from the ones in Figure 1: namely, the peak flux at 4640 > 4686 ! This is not possible with WN6-8+WC4 of the LMC variety. In order of decreasing 4640/4686, we have: Mrk 710 (Kunth and Schild 1986), He2-10 (see below); NGC 6764 and Mrk 309 (Osterbrock and Cohen 1982); NGC 5430 (Keel 1982); NGC 5128#13 (Mollenhoff 1981).

The case of He2-10 is ambiguous. The spectrum given by d'Odorico et.al. (1983) appears to have broad CIV4650 > HeII4686 as expected. The spectrum given by Allen et.al. (1976) (only of the blue region) shows broad NIII4640 > HeII4686. It is probable that the slit positions were different and include different WR stars (see Rosa and d'Odorico 1986, who emphasis this point strongly). What makes the thing odd, is that the shape of the feature is identical in the two observations; only the wavelength scale is different.

The clue to these unexpected spectra appears to be that, with the exception of NGC5430, they all have O/H" > 8.8 (cf. Orion 8.7, Solar 8.9), so that the generalisation "WN6-7+WC5" may not apply. When red spectra are available (Mrk309, NGC5430?, NGC5128, He2-10) , they all appear to have evidence of a WC contribution. If a Galactic WN7 star had been used in Figure 1, the NIII4640 line would be a factor ~3 stronger. However, to get the extreme spectra of Mrk710, a ratio of 4640/4686 >>1 appears to be needed, implying WN stars of low ionisation and/or high N/He abundance ratio.

2.5. DIFFERENCES BETWEEN GALAXIES

The generalisation current in the literature is that stars of the same subtype in different galaxies appear to be the same. However, evidence to the contrary is becoming convincing.
* SSMb find that the flux in the CIV5808 line is 0.5 dex lower in WC5-7 stars in the Galaxy than for WC4-6 stars in the LMC.
*SSMa find that the ratio of C/O lines is greater in the WC4 stars in the Galaxy than in the LMC.
* Smith and WIllis (1983) find that the EW's of HeII4686 are larger in the LMC stars than in the Galaxy.
* New data (as yet unpublished) indicates that HeII4686 is stronger relative to NIII4640, NIV4057, CIV5808 and HeI 5875 for WNL stars in the LMC compared to the Galaxy. This appears to be the same effect as noted above by Smith and Willis. Data of Conti and Massey (1989) indicate the same effect for HeII4686/NIII4640 although they do not comment on it. Note that if the reverse is true and the N and C lines in WN spectra are stronger in high Z regions, it may help to explain 4640 > 4686 in the galaxies noted above.

Conclusions: Evidence for differences in luminosity (total and line) and in line ratios for WR stars in different galaxies is becoming convincing. Care needs to be taken in calibrating these quantities.

3. Overall conclusions

Interpretation of the broad emission in spectra of knots in BCD galaxies as due to the presence of WR stars appears to fit most expectations based on nearby galaxies and stellar evolution models.
* WR/OB star numbers increase with metallicity.
* WC stars are rare at the metallicity characteristic of dwarfs galaxies.
* Flux in the WR*bump* for LMC WC4 stars is log F(ergs/sec) = 36.7 and for WN7 stars is 36.5. These numbers *may* be representative for stars in regions of moderately low Z (O/H" = 8.2 to 8.5) but caution is needed at lower and higher Z.
* Numbers of WN and WC stars may be separated using the CIV5808 feature as well as the WR*bump* at 4650 if the assumption is made that the LMC stars are representative.
* At high Z, the spectra indicate too strong a line at 4640 A for simulation by LMC stars.
* Subtypes are difficult to differentiate due to low signal/noise and evident differences between galaxies.

Acknowledgement. It is a pleasure to thank Mike Potter of STScI for much of the data reduction of our unpublished spectra of WN stars and for preparing Figure 1.

References.

Abbott, D.C. (1982) Ap.J. 259, 282.
Armandroff,T.E. and Massey,P. (1985) Ap.J. 291, 685.
Allen,D.A., Wright,A.E. and Goss,W.M. (1976) M.N.R.A.S. 177, 91.
Arnault,Ph., Kunth,D. and Schild,H. (1989) A&A 224, 73.
Conti,P.S. (1989) Ap.J. 341, 113.
Conti,P.S. and Massey,P. (1981) Ap.J. 249, 471.
Conti,P.S. and Massey,P. (1989) Ap.J. 337, 251.
Conti,P.S., Garmany,C.D., de Loore,C.W.H. and Vanbeveren,D. (1983) Ap.J. 274, 302.
D'Odorico,S. and Rosa,M. (1981) Ap.J. 248, 1015.
D'Odorico,S., Rosa,M. and Wampler,J.E. (1983) A&A Suppl. 53, 97.
Gerola,H., Seiden,P.E, and Schulman,L.S. (1980) Ap.J. 242, 517.
Hazard,C. (1985) in D.Kunth, T.X. Thuan, J. Tran Thanh Van (eds.), Star Forming Dwarf Galaxies, Kim Hup Lee Printing: Singapore, p 9.
Hidayat,B., Supelli,K. and van der Hucht,K. (1982) in C.W.H. de Loore and A.J.Willis (eds.) IAU Symposium 99, Wolf-Rayet Stars: Observations, Physics, Evolution, Dordrecht: Reidel, p 27.
Hutsemekers,D., and Surdej,J. (1984) A&A 133, 209.
Kunth,D. (1985) in D.Kunth, T.X. Thuan, J. Tran Thanh Van (eds.), Star Forming Dwarf Galaxies, Kim Hup Lee Printing: Singapore, p 183.
Keel,W.C. (1982) Publ.A.S.P. 94, 765.
Kunth,D. (1989) in J.E.Beckman, B.E.J.Pagel (eds.), Evolutionary Phenomena in Galaxies, Cambridge University Press p 22.
Kunth,D., and Sargent,W.L.W. (1981) A&A 101, L5.
Kunth,D. and Schild,H. (1986) A&A 169, 71.
Kunth,D. and Sargent,W.L.W. (1981) A&A 101, L5.
Langer,N. (1987) A&A 171, L1.
Massey,P. (1986) in C.W.H.de Loore, A.J.Willis, P.Laskarides (eds.), IAU Symposium No. 116, Luminous Stars and Associations in Galaxies, Dordrecht: Reidel, p 215.
Massey,P. and Conti,P.S. (1983) Ap.J. 273, 576.
Maeder,A. (1981) A&A 102, 401.
Maeder,A. (1990a) A&A Suppl. in press.

610

Maeder,A. (1990b) A&A in press.

Maeder,A., Lequeux,J. and Azzopardi,M. (1980) A&A 90, L17.

Melnick,J. (1985a) A&A 153, 235.

Melnick,J. (1985b) in D.Kunth, T.X. Thuan, J. Tran Thanh Van (eds.), Star Forming Dwarf Galaxies, Kim Hup Lee Printing: Singapore, p 171.

Moffat,A.F.J., Niemela,V.S., Phillips,M.M., Chu,Y. and Seggewiss,W.M., (1987) Ap.J. 312, 612.

Mollenhoff,C. (1981) A&A 99, 341.

Osterbrock,D.E. and Cohen,R.D. (1982) Ap.J. 261, 64.

Rosa,M. and D'Odorico,S. (1986) in C.W.H.de Loore, A.J.Willis, P.Laskarides (eds.), IAU Symposium No. 116, Luminous Stars and Associations in Galaxies, Dordrecht: Reidel, p 355.

Sargent,W.L.W., and Searle,L. (1970) Ap.J. Letters, 162, L155.

Searle,L., Sargent,W.L.W., and Bagnuolo,W.G. (1973) Ap.J. 179, 427.

Scalo,J.M. (1986) in C.W.H.de Loore, A.J.Willis, P.Laskarides (eds.), IAU Symposium No. 116, Luminous Stars and Associations in Galaxies, Dordrecht: Reidel, p 451.

Smith,L.F. (1968a) M.N.R.A.S. 140, 409.

Smith,L.F. (1968b) M.N.R.A.S. 141, 317.

Smith,L.F. (1988) Ap.J. 327, 128.

Smith,L.F. and Hummer,D.G. (1988) M.N.R.A.S. 230, 511.

Smith,L.F. and Maeder,A. (1990) A&A, in press.

Smith,L.F., Shara,M.M. and Moffat,A.F.J. (1990a) Ap.J. 348, 471.

Smith,L.F., Shara,M.M. and Moffat,A.F.J. (1990b) Ap.J. in press.

Smith,L.J. and Willis,A.J. (1983) A&A Suppl. 54, 229.

Talent,D.L. and Dufour,R.J. (1979) Ap.J. 233, 888.

Terlevich,R. (1985) in D.Kunth, T.X. Thuan, J. Tran Thanh Van (eds.), Star Forming Dwarf Galaxies, Kim Hup Lee Printing: Singapore, p 393.

Torres-Dodgen,A.V. and Massey,P. (1987) Ap.J.Suppl. 65, 459.

van der Hucht,K.A., Hidayat,B., Admiranto,A.G., Supelli,K.R. and Doom,C. (1988) A&A 199, 217.

Walborn,N. (1986) in C.W.H.de Loore, A.J.Willis, P.Laskarides (eds.), IAU Symposium No. 116, Luminous Stars and Associations in Galaxies, Dordrecht: Reidel, p 185.

DISCUSSION

Moffat: When I was a young student back in 1968, I remember being strongly influenced by a colloquium you gave at the Bonner Sternwarte on WR stars. Today, I had the same sentiment! One question I still have concerns the predicted masses for WO stars at low Z vs. WO masses at high Z. Observations of masses in WR+O binaries do not agree with the predictions, *e.g.*, the WO4 star in the SMC binary AB8 has a mass of $5M_\odot$ (*cf.* poster at this symposium) compared to $\sim 30M_\odot$ predicted at $Z = 0.002$. (I am not convinced that WR properties, including masses, in all but the *very* closest binaries are any different from single WR star properties).

Smith, Lindsey: I emphasize that the models are for *single* stars and indeed predict much higher minimum masses at low metallicity. However, the phenomenon is due to slower removal of the hydrogen envelope. If you have a binary close enough to cause mass exchange soon after the star leaves the MS, its evolution to a WC star is equally rapid at any Z.

Niemela: Both $CIII$-$IV4650$ and $CIV5808$ are seen as nebular lines in some very hot HII regions. How much would their contribution be to the WR "bumps"?

Smith, Lindsey: Nebular carbon at $4650\mathring{A}$ is not considered by the observers. What does appear to be present is forbidden iron at $4658\mathring{A}$. At low resolution, care is needed to not confuse this with WR carbon.

Vanbeveren: The WR binary frequency in the Galaxy is about 30-40%; in the SMC it is about 20%. So, it looks like the WR binary frequency increases with decreasing Z. Do you not think then that most of the WR stars in these dwarf galaxies are binaries and that a comparison with the evolutionary computations of single stars is inappropriate here?

Smith, Lindsey: Your point is valid. However, the agreement in subtypes predicted by single star evolution and those observed is impressive. It is hard to tell from such a great distance what is going on.

Spurzem: In discussions of the mass loss rate of WR stars is an artificial distinction between single and binary stars. Especially in dense clusters of massive stars, as they may form in starburst regions, frequent close encounters between single stars will enhance the mass loss rate due to tidal interactions.

Smith, Lindsey: It strikes me as an interesting thought, which I have not even discussed with the theoriticians. But if you have a binary, interaction may also affect the internal mixing in the stars, and would therefore also affect the subclass frequency. But I am not a theoritician, that is out of line. [See also Maeder, p. 228. (Eds.)]

Conti: I would like to comment on requiring a high binary frequency in metal-poor environments to provide WR stars. The solar vicinity has a well determined binary fraction of 40%. The SMC has 5 out of 8 WR stars known as established binaries. By binaries, I mean those with an established orbit. Thus the SMC, with low metallicity, could have 100% binaries, considering small number statistics. But what of the LMC with intermediate metallicity? Does it also have relatively many WR binaries? (Again, orbits, not just line of sight companions).

Moffat: The binary frequency among WN6/7 stars (for which the statistics are fairly reliable) is identical in the LMC and the Galaxy ($\sim 57\%$), (Moffat, 1989). The numbers are binary/total WN6/7 = 5/9 in LMC, and 11/19 in the Galaxy, based on confirmed orbits or constant radial velocity. There is no obvious correlation of binary frequency with metallicity.

Underhill: $NIII4634$-40 as strong as or stronger than $HeII4686$ is the signature of Of stars. Thus, when such ratios are observed in galaxies there may be a large population of Of stars present.

Smith, Lindsey: Yes, I noticed that in the spectra shown yesterday. The problem is that the emission line strength in Of stars is much less than in WR stars. The number of WR stars needed is already enormous; the number of Of stars would be even greater.

Walborn: The WR population of giant *HII* regions will be dominated by the descendants of the most massive stars, which are luminous WNL's. However, 30 Dor is embedded in a 10 times larger, less extreme young region containing numerous WC's. At great distances, the angular scales will be very small, and as Lindsey has shown, the early WC can match 20 WNL's spectroscopically. Therefore, the interpretation of stellar populations in distant starbursts is a very difficult problem.

Smith, Lindsey: It is noted in the written manuscript that the 30 Dor cluster is even younger than the predicted onset of the main WR phase. The members of stars in this region may increase dramatically in the next million years.

Maeder: Your superb presentation demonstrates us how blue compact galaxies bring a lot of very useful indications on massive star evolution at low and moderate metallicities. Regarding some comments made on binary evolution, I agree that at low Z evolutionary models predict less WR stars than observed and this might suggest that at low Z the rare WR stars may mainly result from binary evolution.

Shara: Under what circumstances, if any, would you think it prudent to use the constancy of flux in emission lines in WC stars as a distance indicator?

Smith, Lindsey: My *guess* is that the dominant parameter is the metallicity. So, if you have a region with the same metallicity as the LMC, then the LMC star values apply. If you go to higher or lower metallicity, the fluxes appear to decrease.

Mike Shara, Lindsey Smith,
Tony Moffat

LARGE NUMBER OF WOLF-RAYET STARS IN EMISSION-LINE GALAXIES

Jose Miguel MAS-HESSE[1], Daniel KUNTH[2]
[1] Departamento de Astrofisica, Universidad Complutense de Madrid, Spain
[2] Institut d'Astrophysique de Paris, France

ABSTRACT. We present synthetic populations of massive stars that constrain the properties of the star formation episodes, predict the occurence, subtype distribution and evolution of WR star populations at solar and SMC metallicities. Instantaneous burst and continuous star formation rates have been considered. Our predictions have been tested on a sample of 17 galaxies. The evolution of the WR stars is qualitatively well reproduced. The IMF slope is flatter on average than in the solar vicinity and is not correlated with the metallicity. The extinction law is in general similar to that of the SMC, independently of metallicity.

INTRODUCTION

Evolutionary population synthesis models have been built to study the properties of the star formation episodes taking place in galaxies with starbursts. Such starbursts that occur over short periods are excellent test grounds to study the formation and evolution of massive stars. In this contribution we will show how the study of the evolution with time of a massive star cluster can be used to reproduce the output radiation of emission line galaxies at essentially all observable wavelengths from the near UV where starlight dominates down to the far infrared and the radio range. Our models will be explained in broad terms and examples will be given to illustrate the best way to reproduce the observations. A set of 17 galaxies in which starbursts are taking place have been analyzed giving some interesting clues concerning the occurence of large WR numbers, the IMF and the interstellar extinction law.

POPULATION SYNTHESIS EVOLUTIONARY MODELS

Our evolutionary population synthesis models have been explained at length in Arnault, Kunth and Schild (1989) and Mas-Hesse and Kunth (1990). We have used stellar evolutionary tracks from Maeder and Meynet (1987) including mass-loss and overshooting. For a given IMF, we can derive the number of stars per luminosity class (I, III and V) and spectral subtype (O3 to M5). Our models are improved over previous ones by including $Z_\odot$ and $Z_\odot/10$ metallicity stellar evolutionary tracks (tracks of $Z_\odot/10$ have been provided to us by Maeder and Meynet prior to publication) and allowing for very short time steps (≥ 0.05 Myr) to account for very rapid stellar phases such as WR. Two star formation rates (SFR) describe extreme possible situations : an instantaneous burst (IB) and a continuous regime (CSFR) lasting over 20 Myrs. The output of these models are analyzed using fundamental mean properties of standard stars. This step is performed by combining published data together with a library of theoretical model atmospheres (Kurucz, 1979 and Mihalas, 1972). We have also computed the emission of the interstellar gas ionized by massive stars and re-emitting H_β emission as well as the far infrared radiation (FIR) from dust particles and the radio emission from relativistic electrons and thermal gas. The mean features of the models are summarized in Table 1.

K. A. van der Hucht and B. Hidayat (eds.),
Wolf-Rayet Stars and Interrelations with Other Massive Stars in Galaxies, 613–618.
© 1991 IAU. Printed in the Netherlands.

Table 1.

Mup	120, 60 $M_\odot$
Mlow	2 $M_\odot$
α	1, 2, 3
stellar tracks	$Z_\odot$, $Z_\odot/10$
star formation rates (SFR)	instantaneous burst (IB)
	continuous star formation rate (CSFR)
time step	≥ 0.05 Myr
mass loss	$\dot{M} \propto Z^{0.4}$
overshooting	included in $Z_\odot$ models

The IMF is parameterized as IMF = dN/dm proportional to $m^{-\alpha}$. Mup and Mlow are the upper and lower mass limits of the IMF.

Among the parameters we have synthesized the following deserve a particular description:

i) UV absorption lines

In the UV range the best observed absorption lines are those of SiIV and CIV. These lines are signatures of hot stars with strong driven winds. They are useful spectral type indicators and we find that the ratio of their equivalent widths W(SiIV)/W(CIV) is nearly metallicity independant. This ratio is therefore a good tracer of the massive stars that prevail in the cluster, regardless of the metallicity and can be used to estimate quite accurately the age of an instantaneous burst. Moreover the slope of the IMF is constrained by this ratio in the first million years after the onset of the burst. The ratio has also the observational advantage of beeing reddening free and independant of normalisation.

ii) The Wolf-Rayet bump at 4650 Å

Wolf-Rayet stars are detected in star-forming clusters by their characteristic bump at 4650 Å (see L. Smith, this volume). This bump appears as a blend of NIII and HeII lines from WN stars and CIII, CIV and HeII lines from WC ones. Our models predict the WR population of both types as a function of time. Progenitors at $Z_\odot$ have masses $> 35M_\odot$ while at $Z_\odot/10$ they have masses $> 85M_\odot$. We have estimated that the luminosity of the bump due to WN stars is $2\ 10^{36}$ ergs/sec and $6\ 10^{36}$ erg/sec for WC galactic stars. On the other hand their ionizing power compares to that of O6V-B0I type stars, i.e. around 10^{49} photons/sec. Theoretical lifetimes for WR are 0.5 Myr for galactic stars and only 0.14 Myr for WR of $Z_\odot/10$ abundances. Such short lifetimes explain why the WR bump can be seen only over a short period thus giving a strong constraint on the age of a given star formation episode. At low metallicity the detection of the feature is further constrained to a much shorter period explaining the low detection rate of WR in metal deficient galaxies. We have plotted in fig. 1 the evolution of the L(WRbump)/L(H$_\beta$) ratio for different star formation scenarios.

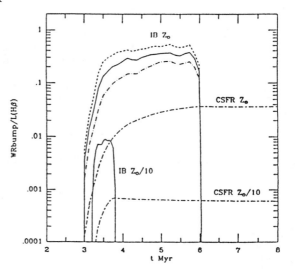

Figure 1: Predicted WR bump to H$_\beta$ luminosity ratio. For $Z_\odot$ metallicity, predictions have been plotted for 3 slopes of the IMF and the IB case, $\alpha = 1$:—, $\alpha = 2$:—, $\alpha = 3$, -.-.-, and for $\alpha = 2$ in the CSFR case. Mup = 60 $M_\odot$ in all the models.

iii) The stellar continuum

The stellar energy distribution (SED) emitted by the stelllar cluster has been computed in the range 912 Å – 3.2μ using the stellar spectral energy distribution of each individual stars. Details of the procedure are given in Mas-Hesse and Kunth (1990). The ionizing flux has been obtained in a similar fashion and we also calculate the effective temperature T_{eff} of the cluster that can be easily compared with the observational effective temperature deduced from the method of Stasinska (1980) or Vilchez and Pagel (1988).

COMPARISON WITH THE OBSERVATIONS

We have analyzed a sample of 17 galaxies in which starbursts are taking place with UV, optical, FIR and radio data. All the results will be published in a forthcoming paper. These galaxies comprise 7 blue compact galaxies, 1 spiral starburst, 1 merger and 3 giant HII regions in M33 and M101. We summarize in Table 2 the most interesting observational data. The strategy adopted for the analysis of the data is the following:

- We first estimate the age, the SFR and the IMF slope using 3 diagnostic plots: $W(H_\beta)$ versus $W(SiIV)/W(CIV)$ at $Z_\odot$ and $Z_\odot/10$ and Log T_{eff} versus $W(SiIV)/W(CIV)$ at $Z_\odot/10$. The first two plots are shown in figs. 2a and 2b.

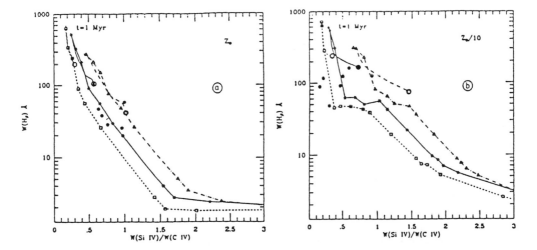

Figure 2a: Predicted $W(H_\beta)$ versus $W(SiIV)/W(CIV)$ ratio at $Z_\odot$. Filled circles represent values taken from Table 2. Symbols along the tracks are time intervals of 1 Myr.

Figure 2b: Same as fig. 2a but for $Z_\odot/10$.

- The extinction law is derived and the E(B-V) is estimated by fitting the predicted SED to the observed one. We show in fig. 3 the synthesized spectrum of IZw 18 superposed on the observed one.
- Further constrains to the models are set using the L(WRbump)/L(H_β) ratio (see fig. 1a in Arnault et al., 1989) and the full set of synthesized parameters. We list in Table 2 model parameters which better reproduce the observational data.

Table 2. Observed and derived parameters of the sample.

Name	Type	O/H	WSi/WC	WH$_\beta$ Å	T$_{eff}$	WR/H$_\beta$	LFIR erg/s	E(B-V)	k$_\lambda$	α	SFR	age Myr
NGC 588	GH	8.3	0.14	97	37500	–	≤3.7E39	0.07	M	1	IB	3.4
NGC 595	GH	8.4	0.54	–	35000	–	≤7.9E40	0.26	G	1	IB	5.5
NGC 2363	IG	7.9	–	300	47500	–	2.7E41	0.06	M	1-2	CR	4.5
IZW 18	BG	7.2	–	135	40000	–	≤6.5E41	0.08	G	1-2	IB	3.4
MRK 710	SG	8.9	0.71	35	37500	0.24	7.4E42	0.16	M	2	IB	5.5
Tol 3	BG	8.2	0.45	98	37500	0.09	5.4E42	0.09	M	2	IB	3.7
NGC 3256	ME	–	1.0	20	–	–	1.9E45	0.22	M	2	IB	7.0
Haro 2	BG	8.4	0.98	30	38500	0.15	1.5E43	0.12	M	2-3	IB	5.6
MRK 36	BG	7.8	0.20	110	45000	0.0	≤3.5E41	0.05	M	1	IB	3.0
NGC 4214	IG	8.3	0.67	49	37500	0.25	5.3E42	0.09	M	2	IB	5.0
IZW 36	BG	7.8	–	300	45000	–	≤1.2E41	0.0		1-2	CR	4.5
NGC 4670	BG	8.4	0.82	27	37500	0.1	3.1E42	0.10	M	2	IB	6.0
MRK 59	IG	8.0	0.91	135	42500	0.08	2.0E42	0.09	G	3	CR	8.5
NGC 5253	IG	8.2	0.59	160	38500	–	4.0E42	0.10	M	2-3	IB	3.3
NGC 5471	GH	8.0	0.50	129	40000	0.02	5.9E41	0.08	G	2	IB	3.5
IIZW 70	BG	8.0	0.30	49	38000	0.0	2.1E42	0.05	M	1	IB	3.8
IC 4662	IG	8.3	1.00	65	38500	–	5.6E41	0.06	M	3	CR	14

Notes:

Type:	*BG: Blue compact dwarf galaxy, IG: Irregular galaxy, SG: Starburst galaxy, ME: Merger, GH: Giant HII region.*
E(B-V):	*Derived from the fit of the predicted and the observed continua.*
k$_\lambda$:	*Extinction law- M: SMC, G: galactic.*
α:	*IMF slope.*
SFR:	*Star formation rate - IB: instantaneous burst, CR: constant rate.*

The table is explained in greater detail in Mas-Hesse (1990).

The main results we have obtained are the following:

i) The WR stars

Our models satisfactorily explain the observed fact that the ratio WR/O (Kunth and Schild, 1986; Azzopardi et al., 1988) is a decreasing function of the oxygen abundance just because the WR progenitors have larger masses at smaller metallicity and a shorter lifetime. Moreover the scatter of the L(WRbump)/L(H$_\beta$) for a given O/H also noted in Kunth and Schild is largely accounted for by the evolution of the starburst region with time. Our models however fail to generate L(WRbump)/L(H$_\beta$) values as high as observed especially at low metallicity. Maeder's new models using higher mass loss rates (1990a,b) seem to reach better agreement but this point will be further checked in our models. The subtype distribution, i.e. the WN to WC ratio is also well reproduced. Indeed most metal poor blue compact

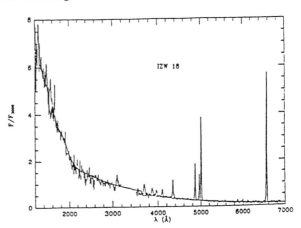

Figure 3: Observed spectrum of the BCD IZw18 from 1200 to 7000 Å. The continuous line is the synthesized spectrum normalized at 3000 Å.

dwarf galaxies with published spectra have only WN stars because WC phases only occur when the metallicity is high (see L. Smith in this volume for a full discussion and references). The occurrence of WC stars in the Ir galaxy NGC4214 at 5808 Å has been predicted from our models and observationally confirmed afterwards with a correct WN/WC ratio as indicated by the ratio of the bumps at 4650 and 5808 Å.

ii) The interstellar extinction

The continuum (SED) of all the objects has been well reproduced in the 1200-7000 Å range using either the SMC or the Galactic extinction laws. Our method is sensitive providing the color excess is larger than 0.05. The most striking result is that the SMC extinction law generally well describe the sample independently of the metallicity. On the other hand, the color excess tends to increase with the oxygen abundance.

The color excess as derived from the continuum fitting is systematically lower than the value derived from the Balmer decrement:

a) The discrepancy can be removed by correcting from underlying stellar absorption using mean equivalent widths around 5 to 6 Å. A justification for that is that a wide slit (10") has been used giving a resolution so low as to fill in the whole stellar absorption. Equivalent widths in the range 8-13 Å have been predicted by A. Diaz (1988) for similar clusters.

b) A widely accepted possibility is that the main contribution to the UV continuum comes from windows where the extinction is relatively low, while at optical wavelengths the contribution from more obscured regions is not negligible (Fanelli et al., 1988). Nevertheless, this hypothesis can not explain why the color excess derived from the continuum is essentially the same in the whole 1200-7000 Å range.

iii) The star formation processes

Only very young bursts adequately reproduce our data. Massive stellar formation is quasi instantaneous with upper mass limits of at least 120 $M_\odot$. In few cases the star formation episodes could have lasted for about 10 Myrs.

The mean age for the star forming epoch is around 4.5 Myrs and generally in the range 3-6 Myrs, probably as a result of a selection effect since the observed emission lines that typify the sample fade drastically 6 Myrs after the onset of the burst. NGC3256 on the other hand has been selected from its strong FIR luminosity and is the most evolved object, hosting a star-forming episode having taken place around 7 Myr ago. The IMF is not universal in the sense that its slope is in the range 1 to 3 with a mean value of 2, i.e. richer in massive stars than the solar neighbourhood. Our results do not support the correlation found by Terlevich and Melnick (1981) between the IMF slope and the oxygen abundance. Such a correlation is postulated by the authors to account for the increase of T_{eff} as O/H decreases and the observational evidence that $W(H_\beta)$ appear to be systematically lower as O/H increases. In fact we find that T_{eff} is only affected by the IMF slope at the very beginning of the burst (2 Myrs). On the other hand, the slower evolution at low metallicity implies a longer lifetime of massive stars and therefore a higher effective temperature in the cluster. Furthermore, the evolution of red supergiants is strongly metal dependent, and their contribution to the luminosity at 5000 Å changes by up to a factor 2 from $Z_\odot$ to $Z_\odot/10$, originating lower $W(H_\beta)$ at solar metallicity.

REFERENCES

Arnault, Ph., Kunth, D. and Schild, H.: 1989, Astron. Astrophys. **224**, 73
Azzopardi, M., Breysacher, J. and Lequeux, J.: 1985, Astron. Astrophys. **143**, 213
Diaz, A.: 1988, Mon. Not. Roy. Astr. Soc. **231**, 57
Fanelli, M.N., O'Connell, R.W. and Thuan, T.X.: 1988, Astrophys. J. **334**, 665
Kunth, D. and Schild, H.: 1986, Astron. Astrophys. **169**, 71
Kurucz, R.L.: 1979, Astrophys. J. Suppl. Ser. **40**, 1
Maeder, A.: 1990a, Astron. Astrophys. Supp. **84**, 139

618

Maeder, A.: 1990b, *Astron. Astrophys.*, in press
Maeder, A. and Meynet, G.: 1987, *Astron. Astrophys.* **182**, 243
Mas-Hesse, J.M.: 1990, Ph. D., Universidad Complutense. Madrid. Spain
Mas-Hesse, J.M., Kunth, D.: 1990, *Astron. Astrophys.*, submitted
Mihalas, D.: 1972, "Non-LTE Model Atmospheres for B and O stars", NCAR-TN/STR-76
Stasinska, G.: 1980, *Astron. Astrophys.* **84**, 320
Terlevich, R. and Melnick, J.: *Mon. Not. Roy. Astr. Soc.* **195**, 839
Vilchez, J.M. and Pagel, B.E.J.: 1988, *Mon. Not. Roy. Astr. Soc.* **231**, 257

Acknowledgements: J.M. Mas-Hesse wishes to thank the hospitality of the IAP, where the major part of this work has been done.

DISCUSSION

Walborn: I want to qualify the remark I made after Lindsey's paper, I think may be relevant to this last point you were making, I think I momentarily forgot my own review on 30 Dor at the Starburst Meetings in Baltimore and Sydney. The point is that if you are isolating the population of the giant *HII* region, then it will be dominated by the evolutionary products of the most massive stars, I think that is why you see predominately WN. But, 30 Dor has emerged in a ten times larger young region which contains many WC stars and undoubtedly mass transfer binaries. At these very large distances, you will be integrating over those different populations, and as Lindsey very nicely showed, one early type WC can equal 20 late type WN's. So, I think the point is that the interpretation of the stellar population and the relationship to the ionized nebulosity becomes extremely complicated and difficult the further you go in distance.

ON THE DETECTION OF WR STARS IN NEARBY GALAXIES

J. LEQUEUX
Observatoire de Meudon
91195 Meudon CEDEX, France
and Ecole Normale Supérieure

M. AZZOPARDI
Observatoire de Marseille
13248 Marseille CEDEX, France

ABSTRACT. The different methods for detecting WR stars in nearby galaxies are reviewed and compared. Each one has its own advantages and drawbacks, and they are complementary rather than competing. Both monochromatic filter imaging and field- or single-object spectroscopy are required for an absolutely safe detection of objects near the limit. The results obtained to-date are shortly reviewed and perspectives are discussed.

1. Introduction

Wolf-Rayet stars are very sensitive tracers of metallicity and age of young stellar populations. At the scale of a large galaxy where recent star formation may be considered as having occured more or less uniformly in time, considerations on the late evolution of massive stars predict that the ratio of the number of WR to O or supergiant stars should be a very sensitive function of metallicity (Maeder et al. 1980). In a recent study, Maeder (1990 and this Symposium) shows that the WO/WN/WC ratios and the distribution between WR subtypes also depends on metallicity. In the case of a starburst the number of WRs with respect to that of other massive stars depends also strongly on time since even the most massive stars take a few million years before turning into WR (Arnault et al. 1989; Mas-Hesse and Kunth 1990). Fortunately, those ratios do not seem to depend much on the initial mass function of massive stars.

It is thus of high importance for understanding the formation and evolution of WR stars and for using them as tracers of metallicity and age of young stellar populations to obtain complete, unbiased samples of WR stars in various environments, e.g. in galaxies or portions of galaxies of different types and metallicities, to secure a spectral classification and photometry for all objects in the sample, and to study their binarity. The present review discusses some of the means for achieving this goal. Another problem, which will not be discussed here, is to obtain for comparison estimates of the number of O stars and of supergiants of various types. I will describe shortly the advantages and inconvenients of the various methods used for detecting WR stars in nearby galaxies: systematic spectroscopy of bright stars, filter monochromatic imaging and field spectroscopy, then discuss the results and their interpretation. A list of WR surveys done with the two last techniques in nearby galaxies is given in Table 1, with the exception of old Magellanic Cloud surveys and of new surveys reported elsewhere in this Symposium.

619

K. A. van der Hucht and B. Hidayat (eds.),
Wolf-Rayet Stars and Interrelations with Other Massive Stars in Galaxies, 619–624.
© 1991 *IAU. Printed in the Netherlands.*

TABLE 1: Surveys of WR stars in nearby galaxies

Gal.	Method	Teles. (m)	Detec.	Filter λ(Δλ) nm	Disp. A/mm	Limit mag.	New(tot.) WR >5σ	Ref.
LMC	Field Sp.	0.40	Phot	465(12)	15	17.5	13 (78)	79AZb
"	"	"	"	"	"	"	4(101)	80AZ
"	"	"	"	"	"	"	2(10 5)	85AZ
"	"	1.2	"	Various	600	19	4	85MOR
"	Mon.Im.	"	"	469(7,150)	–	"	2	"
"	Field Sp.	"	"	498(85)	800	–	0(2)	90MOR
30Dor	Mon. Im.	4	CCD	465,469(4) 470(35)	–	–	1(17)	85MOFb
SMC	Field Sp.	0.40	Pho	465(12)	150	17.5	4(8)	79AZa
M31	Mon.Im.	3.6	Phot	467(9) 450(100)	–	21.5	17	83MOd
"	"	"	CCD	4 filt. (6)	–	21.5	20	86MASa
"	"	"		Im.tube467,445(90)	–	21.5	20	87MOFb
"	Field Sp.	"	Phot	483(95)	2000	20	15(50)	90MEY
M33	Mon.Im.	2.1	Im.tube	445,467(7)	–	21:	25	72WRA
"	"	3.6	Im.tube	467,445(10)	–	21	41	83MASb
"	Field Sp.	4	Phot	420(200)	1500	21:	0(4)	85BOH
"	Mon.Im.	3.6	Im.tube	467,445(9)	–	21	51(115)	87MASb
"	Field Sp.	3.6	Phot.	483(95)	2000	20	0(52)	87LEQ
"	Mon.Im.	2.2	CCD	469(3,5)etc.	–	–	14(59)	90DRI
HII reg	Mon.Im.	3.6	CCD	469,465(3) 478(7)	–	–	10(26)	"
NGC	Field Sp.	3.6	Phot	483(95)	2200	20	1	83WE
6822	Mon.Im.	2.2,3.6	Im.tube	467(9)etc.	–	21.5	0	83MOd
"	"	4	CCD	various	–	21.5	3(4)	85AR
IC	Mon.Im.	4	CCD	various	–	21.5	1	85AR
1613	Field Sp.	3.6	Phot	483(95)	2000	20	1	87LEQ

2. Systematic bright star spectroscopy

This method is obviously very time-consuming although completely safe. It is not surprising that it has only been used in small parts of galaxies, generally in active regions of star formation. See e.g. for giant regions in nearby galaxies, especially M33, d'Odorico et al. (1983) and references herein, and Conti and Massey (1981). The statistics are rarely complete, as more WR stars are generally discovered using other techniques (Drissen 1990). Most of the objects detected in these complexes look overluminous and are probably multiple systems similar to R136 in 30 Dor.

3. Monochromatic filter imaging

WR stars are characterised by strong emission lines which are not all the same for the different WR types. It is thus tempting, in order to search systematically for WRs and to separate the WNs from the WCs, to image the galaxy through narrow-band filters on and off emission lines, and then to compare the images either by blinking them or by doing photometry. This technique has been pionneered by Wray and Corso (1972) on M33 using an image-tube with a photographic plate as the detector, then widely used with photographic plates, intensified or not, and with CCDs. Monochromatic imaging has definite advantages: high sensitivity, minimum confusion by nearby stars in crowded fields, possibility of distinguishing WCs and WNs by using different filters. The choice of the filter width is rather critical: the best results are obtained with filters whose width matches the width of the lines, i.e. 60-100 A, the reference continuum filters being of similar width (Armandroff and Massey,1985).

This method is unfortunately sensitive to seeing and not completely foolproof. Very red stars can be wrong candidates if the central wavelengths of the filters used in the comparison are rather different, as noted by Moffat and Shara (1987), but this can be avoided. At low levels of detection all sorts of objects can be detected as candidates as shown e.g. by Azzopardi et al. (1988). We agree with Drissen (1990) that only those detections announced with a probability larger than 5σ can be considered as reliable. Note also that it is more difficult to detect WNs than WCs as the equivalent widths of their emission lines are generally smaller. Unfortunately we will see later that the luminosity function of WNs extends also to fainter magnitudes than that of WCs, so that the surveys of WCs are always more complete than those of WNs.

Data processing is very important in such work. Usually photometry is secured independently in each color using an appropriate software such as DAOPHOT and the results are compared. If the field is very rich and the background variable as in HII regions, it is probably better to substract directly images obtained with different filters. If the seeing was different there are problems. Drissen (1990) solve them with much success using the CLEAN algorithm applied to each frame. This algorithm has the advantage of giving final images with an arbitrary but identical point-spread function: they can be subtracted with minimal residuals. Presumably this method should be implemented in all future studies.

4. Field low-resolution spectroscopy

This method has been used even earlier than the previous one, at least for the Magellanic Clouds. The idea here is to identify the WRs via their emission lines on objective-prism or objective-grating plates. It is very advantageous to use a filter to restrict the spectral range to what is really needed, usually centering in the 4486 A blend. This limits the length of the spectra, hence image crowding in dense regions, and also the sky background. This was done in the late 70's by Azzopardi and Breysacher (see Table 1) for their Magellanic Cloud survey, then for most subsequent surveys. It is commonplace to work on fields as large as 1 square degree using GRISMs or GRENSes at the focus of large telescopes with photographic plates. This is why this method has been so popular until recently. However our own photographic surveys of the extended local group galaxies, especially M33 and M 31 (see Table 1) will probably be the last of the kind.

Field spectroscopy is a relatively safe way of discovering WRs and has the advantage that other types of objects with strong spectral features in the range of the filter can also be discovered, e.g. planetary nebulae or compact HII regions through Hβ and [OIII] 4959 and 5007 A emission, or carbon stars through C2 absorption bands with heads at 4735 and 5165 A: for a discussion see Azzopardi (1989). As with the previous method, it is easier to detect WCs than WNs. Working in the blue-green spectral region, sky background is not a severe problem. Field crowding is not a major problem either if the dispersion is very low, and masking techniques can ease the photographic detection in crowded fields. However the various spectral orders that show up if GRISMs or GRENSes are used make the examination of the images uneasy. The use of a simple prism, preferably of Fehrenbach type (no deviation at some central wavelength) is certainly better although it has not yet been implemented on a large telescope. A study made for us at ESO by H. Dekker shows that its use is certainly feasible without difficulty.

The use of a photographic plate in most previous surveys, while it allows large fields of view, has the drawback of a lack of sensitivity. It has also the usual limitations of limited dynamic range: saturation for bright objects and detection threshold for the continuum of faint objects which appear as points not specific of WRs of course. There is no reason why CCDs cannot be used. We have made recently some experiments with EFOSC at the ESO 3.6m telescope (not on WRs unfortunately) which have been very successful. There is no doubt that the sensitivity of the method can be made as good as that of filter imaging provided that the dispersion is chosen such that the size of the WR emission-line spots is limited by seeing, not by dispersion. It will keep the advantage of being more foolproof, while it will probably keep the drawback that building an automatic detection algorithm is not going to be easy. Also, one should look at the possibility of separating the WNs from the WCs, which has not been done up to now.

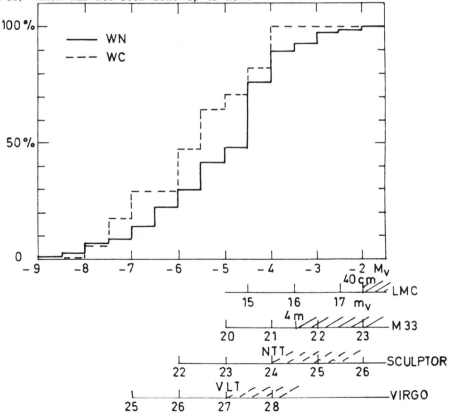

Figure 1. Cumulative distribution of V absolute magnitudes of WRs of the LMC. Scales at the bottom correspond to galaxies at various distances, taking into account their respective interstellar extinctions. Built from Breysacher (1986).

5. Completeness of the surveys, present results

Let us first consider the situation with the Magellanic Clouds. After the completion of the photographic field spectroscopy of Azzopardi and Breysacher there were 104 WRs detected in the LMC and 8 in the SMC. Breysacher (1986) gives visual magnitudes for 94 of the LMC WRs; Fig. 1 displays the cumulative distribution of absolute V magnitudes for these stars. More powerful subsequent surveys have not detected any new WR in the SMC and only a few more in the LMC: see Table 1; also, 4 new WRs have been detected recently by Testor and Schild (1990) and by Heydari-Malayeri et al. (1990). With the exception of 2, these new WRs are in the range of the ones detected with the 40-cm telescope survey. This illustrates the completeness of this survey, and also shows that the distribution of absolute magnitudes given in Fig. 1 can be considered as representative. The limit of V = 17.5 announced for the survey is also realistic and roughly suffices for completeness. This translates into V = 23 for M33, taking into account the average extinction for this galaxy. The published surveys claim a limit of about 21.5. Fig. 1 shows that with this limit only 20 % of the WNs and few WC are missed. It is clear that the photographic survey of Lequeux et al. (1987) which stops at V = 20 should be incomplete by about 50% for the WNs and 20% for the WCs, in agreement with the figures given by the authors (55% and 25% respectively with respect to surveys limited at V = 21.5). The situation should be the same for IC1613 and NGC6822 which are roughly at the same distance as M33, although extinction is larger in NGC6822: thus the (small) figures given in Table 1 for these galaxies should reflect approximate completeness and there is little chance to detect other WRs there. The surveys for M31 are certainly much more incomplete although this galaxy is also at the same distance, because of the large internal extinction which is difficult to quantify. However there is no doubt that this galaxy does not contain a large number of WRs, due to its well-known low star-formation activity.

The results of the WR surveys with respect to the WR/O and to the WC/WN ratios have been discussed most recntly by Azzopardi et al. (1988) and in several papers at this Symposium. The surveys of WR stars in giant extragalactic HII regions (Drissen, 1990) show predominently WN features, and give a WR/O star number ratio apparently similar to that in the field of spiral galaxies, while this ratio is expected to be much higher a few 10^6 years after a burst of star formation (the number of O stars can be evaluated as in Lequeux et al. 1981). All this can be explained by clustering of WR stars. If several grouped WR stars are taken as a single star, the spectrum will often be dominated by WN features from late WNs which are the most luminous (see Breysacher, 1986) and the apparent WR/O number ratio will be smaller than the actual ratio.

6. Perspectives

Whatever the technique, it is necessary for further studies to secure spectroscopy of the candidates. The most economical way to do so, if one is only interested in the total number of WRs in a field and in the WC/WN ratio, is to use both techniques simultaneously. This is possible in a single observing run with imaging and field spectroscopy devices as EFOSC or EMMI on the ESO telescopes. It should be possible with such an equipment to reach securely a detection limit of about V = 24. This would allow an almost complete survey of WRs in galaxies of the Sculptor group (see Fig. 1). With exquisite seeing, e.g. provided by active and adaptative optics, and with the future giant telescopes it might be possible to reach V = 26 or 27 and to detect a large fraction of the WRs in the field of galaxies of the Virgo cluster. This is a very exciting perspective indeed!

REFERENCES

Armandroff, T.E., Massey, P., 1985, Astrophys. J. 291, 685 (85AR)
Azzopardi, M., 1989, in Recent developments of MC research, K.S. de Boer and G. Stasinska, eds. Observatoire de Paris, p. 57
Azzopardi, M., Breysacher, J., 1980, Astron. Astrophys. Suppl. 39, 19 (80AZ)
Azzopardi, M., Breysacher, J., 1979, Astron. Astrophys. 75, 120 (79AZa)
Azzopardi, M., Breysacher, J., 1979, Astron. Astrophys. 75, 243 (79AZb)

Azzopardi, M., Breysacher, J., 1985, Astron. Astrophys. 149, 213 (85AZ)
Azzopardi, M., Lequeux, J., Maeder, A., 1988, Astron. Astrophys. 189, 34
Bohannan, B., Conti, P.S., Massey, P.,1985, Astron. J., 90, 600 (85BOH)
Breysacher, J., 1986,
Conti, P.S., Massey, P., 1981, Astrophys. J. 249, 471
d'Odorico, S., Rosa, M., Wampler, E.J., 1983, Astron. Astrophys. Suppl. 53, 97
Drissen, L., 1990, Thèse de Doctorat, Université de Montréal (90DRI)
Heydari-Malayeri, M., Melnick, J., van Drom, E., 1990, Astron. Astrophys. in press
Lequeux, J., Meyssonnier, N., Azzopardi, M., 1987, Astron. Astrophys. Suppl. 67, 169 (87LEQ)
Lequeux, J., Maucherat-Joubert, M., Deharveng, J.M., Kunth, D., 1981, Astron. Astrophys. 103, 305
Maeder, A., 1990, Astron. Astrophys. in press
Maeder, A., Lequeux, J. , Azzopardi, M., 1980, Astron. Astrophys. 90, L17-20
Massey, P., Armandroff, T.E., Conti, P.S., 1986, Astron. J., 92, 1303 (86MASa)
Massey, P., Conti, P.S., 1983, Astrophys. J. 273, 576 (83MASb)
Massey, P., Conti, P.S., Moffat, A.F.J., Shara, M.M., 1987, Pub. Astron. Soc. Pac. 99, 816 (87MASb)
Mas-Hesse, J.M., Kunth, D., 1990, in preparation
Meyssonnier, N., Lequeux, J., Azzopardi, M., 1990, in preparation
Moffat, A.F.J., Seggewiss, W., Shara, M.M., 1985, Astrophys. J. 295, 109 (85MOFb)
Moffat, A.F.J., Shara, M.M., 1983, Astrophys. J., 273, 544 (83MOd)
Moffat, A.F.J., Shara, M.M., 1987, Astrophys. J., 320, 266 (87MOFb)
Morgan, D.H., Good, A.R., 1985, Month. Not. R. Astron. Soc. 216, 459 (85MOR)
Morgan, D.H., Good, A.R., 1990, Month. Not. R. Astron. Soc. 243, 459 (90MOR)
Testor, G., Schild, H., 1990, Astron. Astrophys. in press
Westerlund, B.E., Azzopardi, M., Breysacher, J., Lequeux, J., 1983, Astron. Astrophys. 123, 159 (83WE)
Wray, J.D., Corso, G.J., 1972, Astrophys. J. 172, 577 (72WRA)

DISCUSSION

Montmerle: There has been some reference to *HST* observations of WR stars. I would like to know, perhaps in more detail, what are the current programs for observing with *HST*?
Shara: There is a mixture of GTO and regular astronomer time in the first year, and indeed there are some programs to look into *HII* regions in M33. So, yes, there is allocated time even in the first year.
Lequeux: The problem with *HST* is, it is only a 2.4 meter telescope, so, I think we would do better in the future with a 8 to 10 meter telescope, and the best seeing we can get.

Walborn: The 30 Dor ionizing cluster subtends $3'$, which translates into $0''.5$ at the Virgo cluster distance. This makes detailed studies of WR stars in Virgo *HII* regions impossible.

Massive stars and galactic evolution

FRANCESCA MATTEUCCI
Istituto di Astrofisica Spaziale, Frascati, Italy

Abstract

The evolution and nucleosynthesis in massive stars are briefly reviewed, and compared with the information derived from SN1987A in LMC . Most of the theoretical models agree with the measured abundances and they can be used in models of galactic evolution.

Models of chemical evolution of galaxies are presented and the role of massive stars in their evolution is discussed.

Finally, the role of Wolf-Rayet stars in galactic evolution is studied, particularly from the point of view of their final fate. It is shown that, if Wolf-Rayet were the progenitors of type Ib supernovae, the Galactic chemical evolution would not change substantially with respect to the case of white dwarfs being the progenitors of type Ib supernovae. However, the predicted frequency of type Ib supernovae in the Wolf- Rayet case would be far too low in comparison with observational estimates.

I.Introduction

Massive stars play a major role in Astrophysics. They are responsible for the nucleosynthesis of the majority of heavy elements and, due to their short lifetimes (several 10^6 years), they dominate the chemical enrichment in the early phases of galactic evolution.

In order to understand the history of chemical enrichment in galaxies one has to know how stars evolve through the nucleosynthetic processes occurring in their interiors and how they restore the newly created and unprocessed material into the interstellar medium (ISM).

As is well known, the life of a star can be described as a sequence of nuclear burnings (H, He, C, Ne, O and Si) occurring in its central core. In stars massive enough ($M \geq 10-12 M_\odot$) all six nuclear burnings take place. This sequence stops with the formation of ^{56}Fe nuclei, after which the collapse of the Fe-core follows.

There are still many uncertainties and open problems in understanding the supernova explosion mechanism in massive stars. Two are the suggested mechanisms: a) prompt and b) delayed explosion. They are related to the shock originating from core bounce or neutrino heating, respectively. In case a) there are problems in causing a succesful explosion, whereas in case b) in reproducing the correct supernova energy (see Woosley, 1986; Hillebrandt 1987 for exhaustive reviews on this subject). In any case, the result of this explosion should be a supernova (SN) of type II.

When the shock wave passes through the stellar mantle, during SN explosion, explosive burnings occur (Si, O, Ne, C, He and H) and a considerable modification can be experienced by Si-Ca elements. All the ejected Fe-group nuclei in type II SNe are practically produced during explosive Si-burning. Recent hydrostatic and explosive nucleosynthesis calculations for massive stars are due to Woosley and Weaver (1986) (hereafter WW), Woosley et al. (1988) (hereafter WPW), Hashimoto et al. (1989), Thielemann et al.(1990) (hereafter THN), Arnett (1990). Most of these works were triggered by the occurrence of SN1987A in the LMC, which represents an optimal observational counterpart for nucleosynthetic models of a $20 M_\odot$ star (e.g. the estimated mass for SK-69202, the progenitor of SN1987A).

K. A. van der Hucht and B. Hidayat (eds.),
Wolf-Rayet Stars and Interrelations with Other Massive Stars in Galaxies, 625–637.

© 1991 IAU. Printed in the Netherlands.

Most of the nuclearly processed material in massive stars is ejected into the ISM during SNII explosion, although these stars loose mass also in a quiescent way (stellar winds) during H- and He-burning phases. The relative importance of the yields of He and metals from type II SNe and stellar winds has been studied in detail by Maeder (1981, 1983, 1985).

As illustrated in Fig. 1, the wind contribution to He becomes dominating with respect to the contribution to He from SNe only for stars with initial masses larger than $\simeq 60M_\odot$, whereas the wind contribution to metals is always negligible with respect to the contribution of SNe, although stars with mass larger than $40 - 50M_\odot$ can loose heavy elements in the wind. The observational counterpart of stars providing such a contribution is likely to be Wolf-Rayet (WR) stars of type WC.

On the other hand, heavy mass loss by stellar winds can affect the amount of metals ejected through a SN explosion. The main effect of a heavy mass loss is, in fact, to reduce the amount of metals which are ejected in SN explosion. In this case, the fact that much He is lost through the wind and not tranformed into heavier elements, contributes to a large reduction of the metal yields (see Maeder, 1990). Usually, in models of chemical evolution of galaxies the occurrence of mass loss in massive stars is taken into account by changing the relation between initial stellar mass, M, and the mass of the He-core, M_α, with respect to stellar models with no mass loss. This is possible when nucleosynthesis calculations are performed on bare He-cores (for example Arnett 1978,1990; THN). However, the net effect of mass loss by stellar wind on M_α is still controversial. Originally, Chiosi et al.(1978, 1979) found a lower M_α as a consequence of mass loss with respect to conservative models. Later Maeder (1981,1983,1985) did not find any sensitive difference with respect to conservative models (within 10%) (but see Maeder, 1990).

Another important effect on the chemical yields from massive stars is represented by assuming core-overshooting in the stellar models (for extensive reviews on this subject see Chiosi and Maeder, 1986; Chiosi, 1986).

Classically, the extension of a stellar convective core is set at the layer where the acceleration of the fluid elements is zero. However, the zero acceleration point does not coincide with the zero velocity point. Consideration of this effect leads to the so-called "overshooting". The most recent work on the subject is from Maeder and Meynet (1989) (hereafter MM89). Given the uncertainties present in the theory of convection in stellar interiors, the amount of overshooting has been parametrized. During recent years various studies of stellar evolution have caused the overshooting parameter to fluctuate. What is clear is that the main effect of overshooting in massive stars is to enlarge the mass of the He-core with respect to classical stellar models. This leads to a net increase in the production of He and metals.

In spite of the large number of theoretical calculations none of those published until now can explain the observational finding (Peimbert, 1986) that the ratio between He and metal production during galactic lifetime, $\frac{\Delta Y}{\Delta Z}$, is of the order of 3 to 5. All the models published insofar predict a $\frac{\Delta Y}{\Delta Z}$ not larger than 1, unless one makes the "ad hoc" assumption that some massive stars end their life by imploding to black holes instead of exploding like SNe (Schild and Maeder, 1984).

Another solution could be represented by adopting a mass loss rate in massive stars much bigger than those adopted up to now. Maeder (1990) in a very recent paper has recomputed stellar tracks with new opacity tables accounting for non solar ratios at low metallicities of α-elements and Na and Al.The deviations from solar proportions are taken from Lambert (1987). A new rate of mass loss is adopted which depends on metallicity and mass of WR stars, resulting in a larger mass loss rate than previously used. Moderate overshooting is also included. In this case, as a result of the huge mass loss rate, the value of $\frac{\Delta Y}{\Delta Z}$ is predicted to be of the order of 2.

In this paper we will review the nucleosynthesis in massive stars compared with the SN 1987A in LMC (section II).

Then the effect of the evolution and nucleosynthesis in massive stars on the evolution of galaxies will be discussed. In particular, the effect of massive stars on the evolution of

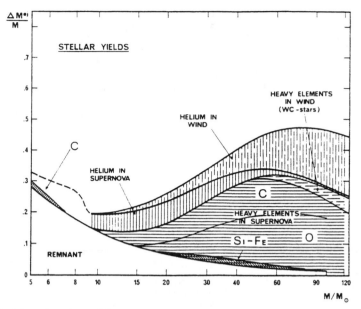

Fig. 1- Mass fraction of He and heavy elements, ejected through stellar winds and supernovae, as a function of the initial stellar mass. The yields from type I SNe are not taken into account. The figure is taken from Chiosi and Maeder (1986).

abundances and abundance ratios in galaxies (section III).

Finally, attention will be devoted to the final fate of massive stars and in particular to the identification of the mass range for progenitors of type II and Ib SNe (section IV). In fact, while there is a general consensus on the fact that massive stars should be the progenitors of type II SNe, although we do not know the upper mass limit for stars becoming such SNe, the progenitors of type Ib SNe are still controversial. Many authors identifie the progenitors of type Ib SNe with WR stars. Alternative models have been proposed involving white dwarfs in binary systems and at the present status of knowledge it is difficult to choose among different scenarios.

II. Massive star nucleosynthesis and SN1987A

The yields of chemical elements, namely the fractions of stellar material in the form of newly created elements, ejected into the ISM through stellar winds and SN explosions, with respect to the material locked up in low mass stars and remnants, depend crucially on stellar evolution and nucleosynthesis. In the following we will focus only on massive stars, which are known to be the major source of heavy elements.

In recent years, several reviews on massive star evolution appeared (see for example Chiosi and Maeder, 1986).

Before discussing the most recent nucleosynthesis results on massive stars we would like to briefly summarize here the most important aspects of the evolution and nucleosynthesis of massive stars:

- $9 \leq M/M_\odot \leq 12$. These stars ignite carbon non-degenerately and those which have cores between 2.2 and 2.5 ignite oxygen in a degenerate Ne-O core. For those with He-cores from 2.5 to 3 $M_\odot$ all six burning stages are ignited non-degenerately and a Fe-core in hydrostatic equilibrium is eventually formed. It is worth noting that the initial mass range $9 - 12M_\odot$ derives from classical stellar models. If overshooting is taken into account this mass range becomes $6.6 - 10M_\odot$ (MM89). These stars end their lives as type II SNe and contribute mostly to the enrichment in He and very little in heavy elements (some C and N, see Hillebrandt 1985). In fact, these stars are only enriched in heavy elements in their collapsing cores and since they leave a neutron star as a remnant it is obvious that they eject material of essentially unprocessed composition.

- $12 < M/M_\odot \leq M_{uSNII}$. These stars are responsible for producing the bulk of heavy elements such as O, Ne, Mg, Si, S, Ar, Ca etc. and possibly r-process elements. Some ^{56}Fe, $\simeq 1/3$ of the total Galactic iron, is likely to be produced in these stars as a result of explosive nucleosynthesis occurring in the Si-Ca layers, whereas the bulk of iron should be produced in type I SNe (Matteucci and Greggio, 1986). The amount of iron measured in SN1987A by Danziger et al. (1990), $\simeq 0.08M_\odot$, confirms that iron production should take place in massive stars. From the theoretical point of view, the amount of iron synthesized in massive stars is quite uncertain, depending on details of the explosion mechanism which is not yet well understood.

M_{uSNII} is the upper mass limit for a star to explode like a type II SN and is practically unknown. One could assume that $M_{uSNII} = M_{lWR}$ where M_{lWR} is the minimum Main Sequence mass necessary for a star to become WR. This mass is also quite uncertain depending on the amount of mass loss and overshooting adopted in stellar models. MM89 suggested $M_{lWR} \simeq 40M_\odot$. Therefore, if only stars in the mass range $9 - 40M_\odot$ become type II SNe, which is the fate of WR stars? They will eventually explode, but lacking the hydrogen envelope their light curve and spectra would be different from type II SNe. It has been suggested either that they could be the progenitors of SNe such as Cas A (see for example Chevalier and Kirshner, 1978) or that they could be the progenitors of subluminous type I SNe, known as type Ib SNe (see for example Gaskell et al., 1986).

Wolf-Rayet stars contribute to He, N , ^{22}Ne, ^{26}Al and ^{12}C enrichment (Maeder, 1981,1983,1985; Dearborn and Blake, 1984; Prantzos et al. 1985). While their contribution to He, N, ^{26}Al and ^{22}Ne is important (probably they are responsible for the whole galactic ^{22}Ne), their contribution to C is negligible when compared to the SN contribution and to the global C production, either because the bulk of this element is likely to come from intermediate mass stars or because of the paucity of WR stars of type WC in the IMF.

Type II SNe are thought to leave compact remnants which can be neutron stars or black holes. The limiting mass for the formation of black holes is very uncertain as well as if stars leaving black holes do explode or implode. The neutron star mass is also an uncertain quantity, due to the uncertainties in the explosion mechanism, although some constraints are given now by the amount of Fe measured in SN1987A, which allows one to fix the mass cut between neutron star and ejecta in a star with initial mass of $20M_\odot$.

- $M > 100M_\odot$. These stars should explode during O burning due to "pair instability". Recent models by MM89 suggest that the range for pair creation SNe should be $100-200M_\odot$, in agreement with prevoius studies. From the point of view of galactic enrichment, these objects would mostly contribute to oxygen (Ober et al.1983).

- Supermassive objects. Are those which either collapse directly to black holes or suffer total disruption due to explosive H-burning. Masses larger than $7.5\,10^5$ should end up as black holes (Appenzeller and Fricke, 1972), whereas masses in the range $4.1\,10^2 - 7.5\,10^5$ should suffer total disruption during H-burning. The results are very sensitive to the initial stellar metal content and explosion seems possible only for solar metal content. These objects produce mostly He and traces of ^{15}N and ^{7}Li (Woosley et al., 1984).

The most recent nucleosynthesis results on massive stars are those that appeared after SN1987A in LMC. In particular, WPW calculated the detailed isotopic composition of the ejecta of stars with masses of 18 and 20 $M_\odot$, respectively. The model for the $18 M_\odot$ star was computed by assuming two different precriptions for the rate of $^{12}C(\alpha, \gamma)^{16}O$ reaction, which is one of the major free parameters in stellar evolution : i)the value suggested by Caughlan et al. (1985) and ii) the very recent one suggested by Caughlan and Fowler (1988) which is a factor of three lower than the previous one. The model with the lower rate predicts an amount of ^{56}Ni of $\simeq 0.07 M_\odot$, in very good agreement with what is observed. However, it should be taken into account the fact that some ^{56}Ni can fall back onto the neutron star. Moreover, the model with the small rate is overly rich in neon and deficient in oxygen, so that the authors concluded that an intermediate value of the rate, between i) and ii) should be preferred.

THN calculated the evolution of a He-star of $6 M_\odot$, corresponding to the He-core of a $20 M_\odot$ star and the composition of its ejecta after explosion. They adopted the rate of Caughlan et al.(1985) for the $^{12}C(\alpha, \gamma)^{16}O$ reaction. The amount of ^{56}Ni in this case was fixed by the observed Fe mass in SN1987A.

Detailed nucleosynthesis for He-core masses between 2.7 and $32 M_\odot$, corresponding to initial masses between 10 and $85 M_\odot$, including, of course, a $20 M_\odot$ star, was very recently presented by Arnett (1990). Also in this case a mass cut was imposed for the $20 M_\odot$ model by the observed iron in SN1987A.

All these calculations differ in the various assumptions on the input physics, and details can be found in the quoted papers. We show in Table I the observed (at day 410 after explosion) amounts of several species in SN1987A from Danziger et al. (1990), compared with different nucleosynthetic models for a $20 M_\odot$ star. It should be noted that the observed values for Fe and Ni seem to be reasonably secure, whereas the estimated mass of oxygen is still very uncertain, going from 0.2 up to $3 M_\odot$.

In spite of the uncertainties both in theory and observations, an inspection of Table I shows that the agreement between predictions (column 3,4 and 5) and data is reasonably good, except for the amount of Ni predicted by THN, which is larger by more than a factor of ten with respect to the observed value. Therefore observations (assuming that all the

Table I:

Observed and predicted abundances for SN 1987A

Species	Observed[1]	WPW(88)[2]	THN(90)[3]	Arnett(90)
C	0.072	0.18	0.114	0.288
O	0.2÷3.0	1.6	1.48	0.774
Si	0.102	0.11	0.085	—
Ar	>0.0008	0.011	0.00377	—
Ca	<0.0105	0.0096	0.00326	—
Fe	0.0825	0.14	0.076	0.07
Ni	0.0022	0.0044	0.0197	—

1) Danziger et al. (1990)
2) Woosley et al. (1988)
3) Thielemann et al. (1990)

Ni mass is observed), can put serious constraints on stellar models. THN discussed the possibility of removing this discrepancy either by i) changing the mass cut or ii) altering the stellar model. Comparison of the theoretical predictions shows differences inside a factor of 2-3, except for the predicted Ni which differ by a factor of 5 between THN and WPW.

III. Massive stars and galactic evolution

Models of chemical evolution for the solar neigbourhood and the whole disk including detailed nucleosynthesis from type Ia, Ib and II SNe, have been recently computed by Matteucci and François (1989) and Matteucci (1990). These models assumed nucleosynthesis prescriptions from WW in the domain of massive stars, whereas for low and intermediate mass stars (between 0.8 and $8M_\odot$) the results of Renzini and Voli (1981) were adopted. For the nucleosynthesis in type Ia SNe (C-O white dwarfs exploding by C-deflagration in binary systems), prescriptions from Nomoto et al. (1984, their model W7) were taken into account. These SNe produce $\simeq 0.6M_\odot$ of iron plus traces of C-Si elements. The progenitor model assumed for type Ib SNe was a C-O white dwarf merging after gravitational wave radiation with an He-non-degenerate star and exploding by He-off center detonation. (Iben et al., 1987; Tornambè and Matteucci, 1987) In this case, a maximum amount of iron of $\simeq 0.3M_\odot$ is produced.

The nucleosynthesis prescriptions of WW refer only to the presupernova configuration and some assumptions on the explosive nucleosynthesis had to be made. In particular, it was assumed (Woosley, private communication) that the ejected iron is produced by explosive nucleosynthesis on Si-Ca elements and that stars with $M < 12M_\odot$ do not produce iron at all, whereas stars with masses between 12 and $20M_\odot$ produce an increasing amount of iron with a maximum of $0.1M_\odot$ for a $20M_\odot$ star, which is a very good approximation when compared to the amount of iron measured in SN1987A. Finally, for stars with masses larger than $20M_\odot$ it was assumed that they produce a maximum of $0.4M_\odot$ of iron (which is probably an overestimate).

It was also assumed that part of the Si-Ca elements falls back into the collapsing core before explosion starts. In particular, the parameter which describes this fraction of material has been constrained to reproduce the more recent nucleosynthesis results of WPW for a $20M_\odot$ star (see Table I).

For elements such as He and N, which are lost mainly through stellar winds in massive stars, prescriptions from Maeder (1981;1983) were taken into account.

In this paper, we show new results obtained with recent yields calculated by Arnett (1990), and we compare them with the previous results and with the observations. In Fig. 2 the predicted $|O/Fe|$ vs.$|Fe/H|$ for the solar neighbourhood region is shown. The two curves refer to the predictions obtained with the two different yields from massive stars, as described above, and they are practically identical. The same is true for α-elements (Ne, and Mg) In Fig. 3 is reported, as an example, the predictions for $|Mg/Fe|$ vs. $|Fe/H|$. No comparison is made for Si and S because in Arnett's calculation Si-Ca elements are not distinguishable.

In Fig. 4 are shown the predicted $|C/Fe|$ vs. $|Fe/H|$ obtained with the two different prescriptions. In this case, the predictions differ in a non negligible way. The reason for this can be attributed either to different treatments of convection or to different rates dopted for the major reaction rates.

From the observational point of view, while the behaviour of O and α-elements seems to be fairly well established (Wheeler et al., 1989), many uncertainties are still present in the data relative to C and N. The carbon/iron ratio, in particular, appears to be roughly solar for stars with $|Fe/H| > -2.0$. However, some of the data indicate that for $|Fe/H| \leq -2.5$, C could be overabundant with respect to iron.

In any case, a comparison between theory and observations is difficult in the case of C, due to the uncertainties present in the nucleosynthesis results for low and intermediate

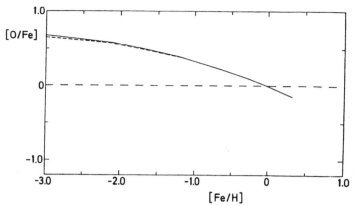

Fig. 2- Predicted $|O/Fe|$ vs. $|Fe/H|$ relations. The continuous line refers to a model adopting Arnett's (1990) yields from massive stars whereas the dotted line refers to a model adopting WW's results.

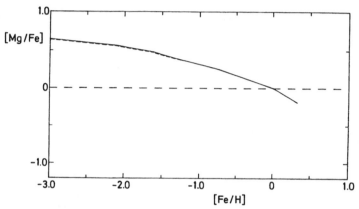

Fig. 3- Predicted $|Mg/Fe|$ vs. $|Fe/H|$ relations. Continuous and dotted lines refer to the same cases as in Fig. 2.

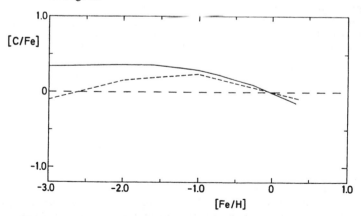

Fig. 4- The same as in Figs. 2 and 3 for $|C/Fe|$ vs. $|Fe/H|$ relations.

mass stars.

From the behaviour of O and α-elements with respect to iron one can easily understand the influence of massive stars on galactic evolution. In fact, it is clear that at low metallicities ($|Fe/H| < -1.0$) massive stars dominate the chemical evolution of the Galaxy, and that the almost constant or slightly declining observed $|O/Fe|$ and $|\alpha/Fe|$ ratios with metallicity reflect the nucleosynthesis in massive stars. The abrupt change in the slope of these relations for $|Fe/H| > -1.0$ is the consequence of the appearence of type I SNe restoring the bulk of iron.

Therefore, the measured $|el/Fe|$ ratios in metal poor stars can impose constraints on nucleosynthesis in massive stars. For example, Matteucci and Greggio (1986) concluded that, although type I SNe should produce the bulk of iron in the Galaxy, type II SNe should also contribute to iron ($\simeq 1/3$ of the total), otherwise the predicted $|O, \alpha/Fe|$ ratios should be continuously declining with the same slope over the whole metallicity range. The measured amount of Fe in SN1987A has confirmed this suggestion. On the other hand, if only type II SNe would produce iron, the $|O, \alpha/Fe|$ ratios would be greater than solar over the whole range of metallicity.

Another and better example of the predominance of massive stars on galactic evolution refers to the Galactic bulge and elliptical galaxies. In a recent paper, Matteucci and Brocato (1990) predicted the $|O, \alpha/Fe|$ vs. $|Fe/H|$ relation for the bulge of our Galaxy, (shown in Fig. 5) and the same conclusions apply for elliptical galaxies. In particular, due to the faster evolution of these systems relative to the other Galactic regions the predicted O and α-element ratios with respect to iron are expected to be greater than solar in the majority of bulge stars, and of the same order as those observed in halo stars, which, in fact, have the same age as the bulge stars. Very recently, Barbuy and Grenon (1990) have measured the O/Fe ratio in some bulge stars and the agreement with the predictions seems quite good. In fact, they they have found an average $|O/Fe| = +0.2$ in stars with an average $|Fe/H| = +0.5$. On the other hand, in systems like the Magellanic Clouds or the external regions of the galactic disk, the predominance of massive stars in the chemical enrichment is restricted to a narrower metallicity range, due to the slower evolution of these systems, as shown in Fig. 5. As a consequence, one should expect to find, in the Clouds, oxygen to be underabundant with respect to iron as compared to the solar neighbourhood of the Galaxy. Observational evidence for this has been found by Russel et al. (1988).

Finally, in Table II are shown the predicted solar abundances obtained by using the two different prescriptions for nucleosynthesis in massive stars and compared with Cameron (1982). As one can see, Arnett's prescriptions give an higher abundance for C (by a factor of 1.25) with respect to the abundance obtained with the WW results, and in better agreement with the observed one. On the other hand, the O abundance is lower by a factor of 1.6 relative to the other theoretical result and to the observed value.

Ne and Mg abundances are higher than the corresponding ones in the case of WW's nucleosynthesis by a factor of 2 and 1.7, respectively, and in better agreement with Cameron (1982). The Si-Ca elements(taken as a whole) in Arnett's(1990) case are slightly lower than in WW's case and again in better agreement with the observed values.

The abundance of iron is practically the same in the two models and in good agreement with the observed one. This is due to the fact that massive stars have a little influence on iron abundance which mostly depends on the assumptions made on nucleosynthesis in type I SNe, and those are the same in both models.

IV. Supernova rates and Wolf–Rayet as progenitors of type Ib SNe

In this section we will discuss the effect on galactic evolution of assuming that WR stars are the progenitors of type Ib SNe.

Type Ib SNe were discovered already 20 years ago by Bertola (1964): they are subluminous with respect to type Ia SNe. In particular, the luminosity of the maximum is

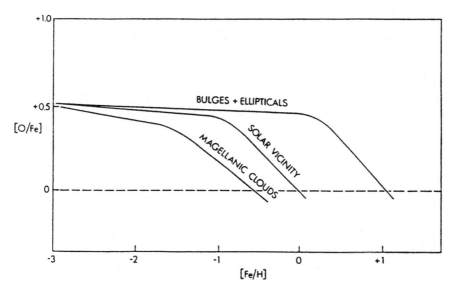

Fig. 5- A rough sketch of the predicted $|O/Fe|$ vs. $|Fe/H|$ relation in different systems, showing their different evolutionary histories. Figure taken from Matteucci and Brocato (1990).

Table II: Predicted and Observed Solar Abundances

	WW(86)	Arnett(90)	Cameron(82)
X_H	0.742	0.74	0.772
X_{12_C}	2.75(−3)	3.45(−3)	3.87(−3)
X_O	8.8(−3)	5.25(−3)	8.55(−3)
X_N	9.2(−4)	7.2(−4)	9.34(−4)
X_{13_C}	3.3(−5)	2.82(−5)	4.64(−5)
X_{Ne}	7.0(−4)	1.4(−3)	1.34(−3)
X_{Mg}	3.5(−4)	6.00(−4)	5.81(−4)
X_{Si}	8.13(−4)	$\left\{ \begin{array}{l} \\ 1.2(-3) \\ \\ \end{array} \right.$	7.49(−4)
X_S	4.57(−4)		4.41(−4)
X_{Ca}	1.0(−4)		7.04(−5)
X_{Fe}	1.44(−3)	1.2(−3)	1.33(−3)

1.5–2.0 magnitudes fainter than that of type Ia SNe. As a consequence, their light curves can be explained by an amount of Fe of $0.1 - 0.2 M_\odot$, as opposed to the $\simeq 0.6 M_\odot$ of Fe necessary to power the light curve of type Ia SNe.

Type Ibs seem to occur in spiral arms or close to HII regions (Porter and Filippenko, 1987), although this finding has been questioned by Panagia (1990). Moreover, in contrast with type Ia, type Ib seem to be strong radio emitters. This has been interpreted in favour of WR stars as progenitors (but see again Panagia, 1990). Frequency estimates suggest that a fraction of 30 to 60% of all type I SNe are of type Ib (Panagia, 1987; van den Bergh, 1990). Alternative models to WR have been proposed and they involve either white dwarfs in binary systems (Branch and Nomoto, 1986; Iben et al., 1986; Tornambè and Matteucci, 1987), or He-stars of $3 - 4 M_\odot$ (with Main Sequence progenitors in the mass range $12 - 16 M_\odot$) in binary systems (Shigeyama et al., 1990).

WR models have problems in reproducing the light curves of typical type I b SNe, in producing the right amount of ^{56}Ni and the observed frequency of these SNe, given the paucity of WR stars in the IMF. In fact, if only stars with masses greater than $40 M_\odot$ become WR stars (MM89) then the expected type Ib SN frequency would be too low, even adopting a Salpeter (1955) IMF, which favours massive stars with respect to other IMFs more suitable for the solar neighbourhood (Scalo, 1986). One possible solution could be to assume that all stars with mass greater than $\simeq 16 M_\odot$ become WR, but this is probably an unrealistic assumption.

On the other hand, by assuming that type Ib SNe originate from white dwarfs in binary systems one does not have any difficulty in reproducing the observed frequencies of all SN types. In a recent review, van den Bergh (1990) concludes that the best current estimate for the total Galactic SN rate is $\simeq 2$ per century. Of these supernovae $\simeq 18\%$ are expected to be of type Ia, $\simeq 17\%$ of type Ib and $\simeq 65\%$ of type II.

Matteucci and François (1989), by adopting Scalo's IMF, predicted the following rates for our Galaxy: $0.4 SNe 100 yr^{-1}$ for type Ia, $0.4 SNe 100 yr^{-1}$ for type Ib and $1.1 SNe 100 yr^{-1}$ for type II SNe, in very good agreement with van den Bergh's estimates. These rates were obtained by assuming that type II SNe come from stars in the mass range $9 - 100 M_\odot$, whereas type I SNe come from C-O white dwarfs in binary system, therefore from stars with initial masses $\leq 8 M_\odot$.

If WR stars, instead of C-O white dwarfs, are assumed as progenitors of type Ibs, the same model predicts: $\simeq 0.4 SNe 100 yr^{-1}$ for type Ia, $\simeq 0.06 SNe 100 yr^{-1}$ for type Ib and $\simeq 1.045 SNe 100 yr^{-1}$ for type II. Therefore, while the total type II rate in not affected by this assumption, the type Ib rate results too low by a factor of $\simeq 6.5$ with respect to the case with white dwarfs as progenitors and to the observed one. Only if one assumes that WR stars originate from all stars greater than $\simeq 15 - 16 M_\odot$, the type Ib rate raises up to $0.35 SNe 100 yr^{-1}$, while the type II rate is still acceptable. Therefore, from the point of view of the SN frequencies the white dwarf model for progenitors of type Ib should be favoured.

From the point of view of the predicted solar abundances, the model with the WR stars as progenitors of type Ibs predict a slightly smaller iron abundance with respect to the other case. In particular, in a model where the Arnett (1990) yields are adopted, predicts a solar iron abundance of 8.410^{-4}, which is still acceptable, while the other abundances are left unchanged. This lower solar iron abundance leads also to slightly lower overabundances of O and α-elements with respect to iron at low metallicities, but always inside the observational uncertainties. In conclusion, the differences between the abundance results obtained for the solar neighbourhood under the two different assumptions on the type Ib progenitors, do not allow us to distinguish between the two scenarios.

References

Appenzeller, I., Fricke, K.J.: 1972, *Astron. Astrophys.* **18**, 10
Arnett, W.D.: 1978, *Astrophys. J.* **219**, 1008
Arnett, W.D.: 1990, in *Chemical and dynamical evolution of galaxies*, ed. F. Ferrini et al., (Giardini:Pisa), in press
Barbuy, B., Grenon, M.: 1990, in *Bulges of Galaxies*, ESO/CTIO workshop, in press
Bertola, F.: 1964, *Ann. Astrophys.* **27**, 319
Branch, D., Nomoto, K.: 1986, *Astron. Astrophys.* **164**, L13
Cameron, A.G.W.: 1982, in *Essays in nuclear astrophysics*, ed. C. Barnes et al., (Cambridge University Press), p.377
Caughlan, G.R., Fowler, W.A., Harris, M.J., Zimmerman, B.A.: 1985, *Atomic Data and Nuclear Data Tables* **32**, 197
Caughlan, G.R., Fowler, W.A.: 1988, *Atomic Data and Nuclear Data Tables* **40**, 283
Chevalier, R.A., Kirshner, R.P.: 1978, *Astrophys. J.* **219**, 931
Chiosi, C.: 1986, in *Nucleosynthesis and chemical evolution* ed. B. Hauck et al. (Geneva Observatory Publ.), p.199
Chiosi, C., Maeder, A.: 1986, *Ann. Rev. Astron. Astrophys.* **24**, 329
Chiosi, C., Nasi, E., Bertelli, G.: 1979, *Astron. Astrophys.* **74**, 62
Chiosi, C., Nasi, E., Sreenivasan, S.R.: 1978, *Astron. Astrophys.* **63**, 103
Danziger, I.J., Lucy, L.B., Bouchet, P., Gouiffes, C.: 1990, in *Supernovae*, ed. S.E. Woosley (Springer Verlag: New York), in press
Dearborn, D.S., Blake, J.B.: 1984, *Astrophys. J.* **277**, 783
Gaskell, C.M., Cappellaro, E., Dinerstein, H.L., Garnett, D.R., Harkness, R.P., Wheeler, J.C.: 1986, *Astrophys. J. (Lett.)* **306**, L77
Hashimoto, M., Nomoto, K., Shigeyama, T.: 1989, *Astron. Astrophys.* **210**, L5
Hillebrandt, W.: 1985, in *Production and distribution of CNO elements*, ed. I.J. Danziger et al. (E.S.O. Publ.) p.325
Hillebrandt, W.: 1987, in *High energy phenomena around collapsed stars*, ed. F. Pacini (Reidel: Dordrecht) p.73
Iben, I. jr., Nomoto, K., Tornambè, A., Tutukov, A.: 1987, *Astrophys. J.* **317**, 717
Lambert, D.L.: 1987, *J. Astrophys. Astr.* **8**, 103
Maeder, A.: 1981, *Astron. Astrophys.* **102**, 299
Maeder, A.: 1983, *Astron. Astrophys.* **120**, 113
Maeder, A.: 1985, in *Nucleosynthesis and its implications on nuclear and particle physics*, ed. J. Audouze and N. Mathieu (Reidel: Dordrecht), p.207
Maeder, A.: 1990, preprint
Maeder, A., Meynet, G.: 1989, *Astron. Astrophys.* **210**, 155 (MM89)
Matteucci, F.: 1990, in *Frontiers of stellar evolution*, ed. D.L. Lambert, P.A.S.P., in press
Matteucci, F., Brocato, E.: 1990, *Astrophys. J.*, in press
Matteucci, F., François, P.: 1989, *M.N.R.A.S.* **239**, 885
Matteucci, F., Greggio, L.: 1986, *Astron. Astrophys.* **154**, 279
Nomoto, K., Thielemann, F.K., Yokoi, Y.: 1984, *Astrophys. J.* **286**, 644
Ober, W., El Eid, M., Fricke, K.J.: 1983, *Astron. Astrophys.* **119**, 61
Panagia, N.: 1987, in *High energy phenomena around collapsed stars* ed. F. Pacini (Reidel: Dordrecht) p.33
Peimbert, M.: 1986, *P.A.S.P.* **98**, 1057
Porter, A., Filippenko, A.: 1987, *Astron. J.* **93**, 1372
Prantzos, N., Arcoragi, J.P., Arnould, M.: 1985, in *Nucleosynthesis and its implications on nuclear and particle physics*, ed. J.Audouze and N. Mathieu (Reidel: Dordrecht), p.293
Renzini, A., Voli, M.: 1981, *Astron. Astrophys.* **94**, 175
Russel, S.C., Bessel, M.S., Dopita, M.A.: 1988, in *The impact of high S/N spectroscopy on stellar physics*, ed. G. Cayrel de Strobel and M. Spite (Reidel: Dordrecht), p.545
Salpeter, E.E.: 1955, *Astrophys. J.* **121**, 161
Scalo, J.H.: 1986, *Fund. Cosmic Phys.* **II**, p.1

Schild, H., Maeder, A.: 1984, *Astron. Astrophys.* **143**, L7
Shigeyama, T., Nomoto, K., Tsujimoto, T.: 1990, *Astrophys. J. (Lett.)* submitted
Thielemann, F.K., Hashimoto, M., Nomoto, K.: 1990, *Astrophys. J.* **349**, 222 (THN)
Tornambè, A., Matteucci, F.: 1987, *Astrophys. J. (Lett.)* **318**, L25
van den Bergh, S.: 1990, in *Supernovae*, ed. S.E. Woosley (Springer Verlag: New York) in press
Wheeler, J.C., Sneden, C., Truran, J.W.: 1989, *Ann. Rev. Astron. Astrophys.* **27**, 279
Woosley, S.E.: 1986, in *Nucleosynthesis and chemical evolution*, ed. B. Hauck et al. (Geneva Observatory Publ.), p.1
Woosley, S.E., Axelrod, T.S., Weaver, T.A.: 1984, in *Stellar nucleosynthesis*, ed. C. Chiosi and A. Renzini (Reidel: Dordrecht), p.263
Woosley, S.E., Pinto, P.A., Weaver, T.A.: 1988, *Proc. ASA* **7**(4), 355 (WPW)
Woosley, S.E., Weaver, T.A.: 1986, in *I.A.U. Coll. n. 89* ed. D. Mihalas and K.A. Winkler, p.91 (WW)

DISCUSSION

Vanbeveren: All stars with initial mass larger than $10M_\odot$ which are member of a close binary end their life as hydrogen deficient stars (as a consequence of Rocke lobe overflow). I remind you that this has nothing to do with the WR phenomenon. If then one assumes $8M_\odot$ as minimum mass for single stars to explode (most of them as Type II), $10M_\odot$ as minimum mass for close binaries to explode, taking a 30% close binary frequency, and assuming some IMF, one ends up with a $\sim 20\%$ massive $(M > 10M_\odot)$ star frequency which will explode hardly showing any hydrogen at all. So, the 17% you give nicely coincides with the $\sim 20\%$ I give here. I therefore conclude that as far as frequencies are concerned Type Ib agrees with the number of massive close binaries.
Matteucci: Yes, I agree.

Maeder: Your recent results on the different behaviours of the O/Fe ratios in elliptical galaxies and other galaxies, the solar vicinity and the Magellanic cloud, are a magnificent piece of work. Do you already have comparisons of your model available with abundances determined in highly redshifted galaxies observed in the Lyman forest of QSO? Because, in these highly redshifted galaxies you should also have some massive stars which contribute to the nucleosynthesis.
Matteucci: Unfortunately, I do not have, but that is a very good point in fact, to compare evolution with iron red shift.

Nomoto: (1) Helium detonation in some accreting white dwarfs is a natural outcome of evolution theory. Its explosion would be observed not as SNIb but SNI-F, *i.e.*, fast decline-type of SNI (like S And) which would easily be missed from observations. So chemical evolution models should not neglect SNI-F; (2) In your model, how is the abundance gradient in the galactic disk formed?
Matteucci: (2) The abundance gradient in the galactic disk, form, in my model, as a consequence of assuming different time scales for disk formation at different galactocentric distances. In particular, the time scale for disk formation increases with galactocentric distance. For more details, see Matteucci and Francois (1989).

Heydari-Malayeri: I saw in one of your viewgraphs that you have considered extremely high mass stars. There is no reliable observational evidence for the existence of stars more massive than $100 M_\odot$. There is a poster by us outside on display. I think that this is an important point to keep in mind.

Matteucci: I only mentioned very massive stars, but in fact I included in my models only stars with masses smaller than $80 - 100 M_\odot$. Even if more massive stars would exist, they would not make much difference, because there would be only a few.

Vilchez: I have not seen in your picture of the O/Fe vs. Fe relationship, the data of Abia and Rabolo. What do you think about that?

Matteucci: I did not show them on purpose, because there is a bit of a discussion on them, and I have the data here, it is important anyway to show it. This is the O/Fe vs. Fe relationship. The black dot is Abia and Rabolo's, the other is Barbieu and others. The difference between the two is that the one refers to giants and the other to dwarfs, and probably there are some problems in both of these kinds of analysis. The giants give always overabundances which are much lower than the dwarfs. But, another difference is that they seem to find that this part is not really constant but increases. Now I have been to many meetings where also Rafael was, and others, and people tend to say: "I am not an observer", so, I have to trust them, but probably the truth is in the middle, because these stars here, the dwarfs, may have the problems of non-LTE and I do not know about the giants. So, at the moment, untill we have a confirmation of this, I would expect that, if this trend exists for oxygen, it should exist for magnesium, *e.g.*, whereas it does not. But, if it was true, what does it mean? If the overabundance of oxygen is so high, it means that the iron in very massive stars must be very little or none. And so, the yield that I use would not give this high value, because at maximum I find 0.6, and I use the yield of Woosley and Waever or Arnett which predicts the mass of iron to increase with the initial mass of the star. Let us use the suggestion of Ken.

Filippenko: I want to clarify one issue regarding the absolute quantity of iron (in the Galaxy) produced by different types of supernovae. If SNeIa, SNeIb, and SNeII produce $0.6 M_\odot, 0.15 M_\odot$, and $0.10 M_\odot$ of Fe, respectively, and the relative frequencies of SNeIa, SNeIb, and SNeII are 18%, 17%, and 65%, then the total production rate of iron in SNeIb and SNeII is *comparable* to that of SNeIa. Taking into account the fact that SNeIb and SNeII were more numerous in the past (where SNeIa were not), the total amount of Fe produced by SNeIb and SNeII *exceeds* that produced by SNeIa. Thus, the *absolute* quantity of Fe is *not* dominated by SNeIa in our galaxy.

Matteucci: Yes, you are right. The different contributions to total galactic iron from the various SN types from my model are: 23% Type II, 25% Type Ib and 50% Type Ia. So, the total contribution of Type II and Type Ib SNe is comparable to the contribution of Type Ia. When I said that Type Ia SNe dominate the galactic production of iron, I meant in comparison with the contribution of Type II and Type Ib SNe, separately.

Francesca Matteucci

THE DISTANCE TO HD 50896 (EZ CMA)

WERNER SCHMUTZ[1] and IAN D. HOWARTH[1,2]
[1] Joint Institute for Laboratory Astrophysics
University of Colorado and National Institute of Standards and Technology
Boulder, CO 80309-0440, USA
[2] Department of Physics and Astronomy, University College London
London, WC1E, 6BT, England

ABSTRACT. The WN5 star HD 50896 lies in the same sightline as the open cluster Cr 121. In order to investigate the possibility of cluster membership, we observed the interstellar Na D lines in the spectra of HD 50896 (=EZ CMa, =WR6) and stars in its neighbourhood (on the plane of the sky) at high spectral resolution (R=80,000 and 105,000). The observations were obtained with the 1.4-m coudé auxiliary telescope and the coudé echelle spectrograph of the European Southern Observatory.

From the strengths and velocity structure of the interstellar features it is immediately clear that HD 50896 is *not* a member of Cr 121; rather, it is a background object. A comparison with spectra of other background stars shows that the line of sight towards HD 50896 is very similar to those of HD 51854 (B1 V) and HD 50562 (B3 III). From intermediate-band and Hβ photometry the distances to these two B stars are 1.75 kpc and 1.95 kpc, respectively. The most red-shifted absorption component in the spectrum of HD 50896 is displaced only slightly more than that of HD 51854, and a little less than that of HD 50562. We conclude that the the most probable distance to HD 50896 is 1.8 kpc.

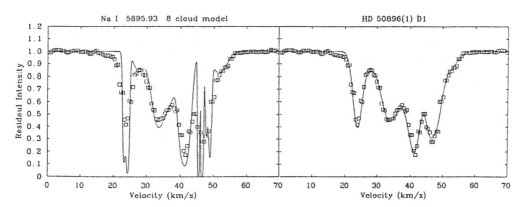

Fig. 1.   Model fit to observation of HD 50896, taking into account both of the D lines, and including the hyperfine structure. The continuous line in the left-hand panel shows the synthesized D1 spectrum; the line in the right-hand panel represents the model spectra after a convolution with the instrumental profile (2.9 km s^{-1} full width). The observed profile is shown as squares.

K. A. van der Hucht and B. Hidayat (eds.),
Wolf-Rayet Stars and Interrelations with Other Massive Stars in Galaxies, 639.
© 1991 IAU. Printed in the Netherlands.

DETERMINATION OF THE INTERSTELLAR EXTINCTION TOWARD WN STARS FROM OBSERVED ubv COLOR INDICES

WILLIAM D. VACCA and WERNER SCHMUTZ
Joint Institute for Laboratory Astrophysics
University of Colorado, Boulder, CO 80309-0440, USA

Recent observations of Wolf-Rayet stars by Massey (1984) and Torres-Dodgen and Massey (1988) have yielded high quality, absolutely calibrated spectra of nearly all known Wolf-Rayet objects. These observations also indicate that discontinuities, or "jumps", are present in the continuum spectra of some Wolf-Rayet stars. Such continuum jumps are predicted by the current theoretical models of Wolf-Rayet atmospheres. In general, these models provide good fits to the observed spectra of Wolf-Rayet stars. The models also indicate that, between jumps, the intrinsic continuum can be closely approximated by a power law in wavelength. In this case, we have the following relation between the intrinsic colors:

$$(u - b)_0 = 0.83(b - v)_0 + D_{3645} \ ,$$

where

$$\frac{\log \lambda_u - \log \lambda_b}{\log \lambda_b - \log \lambda_v} = 0.83 \ ,$$

and D_{3645} is the strength of the He II ($n = 4$) jump at 3645 Å in mags. This relation holds because the central wavelength of the u filter ($\lambda_u = 3650$ Å) is nearly coincident with that of the He II jump. In addition, models of WN stars with helium-dominated atmospheres predict a correlation between D_{3645} and $(b - v)_0$:

$$(b - v)_0 = -0.98 D_{3645} - 0.28 \ .$$

The intrinsic uncertainty in this relation is ±0.02 mags in $(b - v)_0$. Using these theoretical relations, we can derive a formula for calculating the E_{b-v} color excesses for WN stars from the observed narrow-band $u - b$ and $b - v$ color indices:

$$E_{b-v} = [(u - b) + 0.20(b - v) + 0.28]/[k + 0.20] \ ,$$

where $k \equiv E_{u-b}/E_{b-v}$. We adopt the value $k = 0.74$ (which is the average obtained from several extinction laws) and determine color excesses for WN stars in the Galaxy and the Large Magellanic Cloud. We find excellent agreement between our values and those determined by Lundström and Stenholm (1984) for WN stars which are probable or definite members of Galactic associations or open clusters.

640

K. A. van der Hucht and B. Hidayat (eds.),
Wolf-Rayet Stars and Interrelations with Other Massive Stars in Galaxies, 640.
© 1991 *IAU. Printed in the Netherlands.*

THE DISTRIBUTION OF M SUPERGIANTS WITH LARGE INFRARED EXCESS AND WR STARS IN THE GALAXY

M. Raharto
Bosscha Observatory and Department of Astronomy
Bandung Institute of Technology
Indonesia

Summary. From the $m_{12} - m_{25}$ *vs.* spectral type diagram (see for example: Habing, 1987, Figure 5), it can be seen that there are no early M giant stars with $m_{12} - m_{25} > 0.7$. On the other hand all early M stars with $m_{12} - m_{25} > 0.7$ are M supergiants. These early type M supergiants have typically infrared excesses at 12 μm in the range of 20 to 170 times higher than their photosperic fluxes.

The color distinction between early M giants and M supergiants was applied to identify early M giant stars from the Catalogue of Spectral Classifications for Stars of the Catalogue of the Caltech-Two-Micron Survey by Bidelman (1980). The search was confined to $l = 345° - 300°$, through the galactic centre direction. Only good data were used, *i.e.* for objects with flux quality 3 in the scale of IRAS (1984). There are 182 stars classified as M supergiants according to the above mentioned criteria, 58 of which are known spectroscopically as M supergiants. The surface distribution of the M supergiants found is compared to the surface distribution of Wolf Rayet (WR) stars listed in van der Hucht *et al.* (1988). The distribution shows four clumpings of M supergiants, at $(0° < l < 30°, -2° < b < +2°)$, $(60° < l < 90°, -1° < b < +3°)$, $(100° < l < 120°, -2° < b < +2°)$ and $(120° < l < 150°, -5° < b < +0°)$. The first three are found in the direction of high concentrations of WR stars.

In order to know whether there is a spatial coincidence between these two types of stars, the distance estimates of M supergiants were calculated. The distance estimation was facilitated by the calibration of the absolute magnitude of M supergiants at 12 μm which was assumed to be a liniar function of $m_{12} - m_{25}$ (Raharto, 1990) and the interstellar extinction at 12 μm was negligible.

Then the distributions of the two types of stars within 3 kpc from the Sun were compared. Two apparent clumpings of M supergiants coincide with the WR space distribution given by van der Hucht *et al.* (1988). The clumpings of M supergiants at $(100° < l < 120°, -2° < b < +2°)$ and $(120° < l < 150°, -5° < b < +0°)$, however, do not coincide with any of the WR concentrations. It has been known that this direction is devoid of WR stars (Roberts, 1962). The clumping of M supergiants in this direction may indicate that the IMF in different spiral arms of the Galaxy may not the same.

Acknowledgements

The author would like to thank the LOC of the IAU Symposium no. 143 for providing financial support to attend the meeting as well as their hospitality during the meeting. The

K. A. van der Hucht and B. Hidayat (eds.),
Wolf-Rayet Stars and Interrelations with Other Massive Stars in Galaxies, 641–642.
© 1991 *IAU. Printed in the Netherlands.*

author is also indebted to Prof. B. Hidayat for his critical comments and suggestions on
the draft of the manuscript. The author would also like to express thanks to Prof. Habing,
Dr. Piet Schwering and their colleagues in Leiden Sterrewacht for their hospitality and
various help during a two year stay (1984-1986) in Leiden where this work was initiated. A
part of this works was financially supported by Institute for Research - Bandung Institute
of Technology.

References

Bidelman, W.P., 1980, Publications of the Warner and Swasey Observatory, Vol. 2.

Habing, H.J., 1987, Proc. IAU Symposium No. 122, Circumstellar Matter, I. Appenzeller
and C. Jordan (eds.), D. Reidel Publ. Co., p. 197.

van der Hucht, K.A., Hidayat, B., Admiranto, G, Supelli, K.R., Doom, C., 1988 *Astron.
Astrophys.* **199**, 217.

IRAS, 1984, The Explanatory Suplement, eds. C.A. Beichman, G. Neugebauer, H.J.
Habing, P.E. Clegg, T.J. Chester (Goverment Printing Office).

Raharto, M., 1990, in Proc. 5th IAU Asian Pacific Regional Meeting Sydney-Australia,
1990, in press.

Roberts, M., 1962, *Astron. J.* **67**, 79.

ON THE LUMINOSITY FUNCTION OF OPEN CLUSTERS CONTAINING WOLF-RAYET STARS

A.G. ADMIRANTO[1] and B. HIDAYAT[2]
[1] *National Institute of Aeronautics and Space (LAPAN),*
Bandung, West-Java, Indonesia
[2] *Observatorium Bosscha, Institute of Technology Bandung,*
Lembang, West-Java, Indonesia

SUMMARY. It is generally accepted that mass loss and overshooting are the two main processes that play important roles in the formation of Wolf-Rayet stars (Chiosi & Maeder, 1986). Such mechanisms modify the interior of a star. We cannot, however, disregard the possibility that there are external factors, which can influence the formation and evolution of Wolf-Rayet stars.

In order to see how strong external factors (if any) influence the formation and evolution of WR stars, we would like to investigate stellar conglomerates, in which external factors can readily be discerned. In open clusters stars may be close enough to exert some influence on each other, thus producing the kind of external factors we are looking for.

We have investigated the luminosity function of young open clusters with log t < 7, notably ten open clusters with WR 'members' or 'probable members' according to Lundström and Stenholm (1984). The clusters are Be 68, Be 87, Cr 228, NGC 3603, NGC 6231, NGC 6871, Ma 50, Ru 44, Tr 16 and Tr 27. Plots of absolute magnitude *vs.* number of stars in each cluster were generated. Subsequently, the luminosity functions of these clusters containing WR stars was compared whith those of clusters in the same age range but without WR members (Burki, 1977).

We find marked differences between the two samples. The form of the luminosity function of the clusters without WR stars can be approximated by a power law, while the luminosity functions of the clusters with WR members are too irregular to be approximated by power laws. We propose that the differences may be attributed to differences in the initial mass functions. This project will be continued, also with a view to clusters containing CP stars.

A.G.A. acknowledges the support from the ICTP to attend this Symposium. B.H. thanks the ITB Research Foundation for their support in this investigation.

References

Lundström, I., Stenholm, B. 1984, *Astron. Astrophys. Suppl.* **58**, 163.
Burki, G. 1977, *Astron. Astrophys.* **57**, 135.
Chiosi, C., Maeder, A. 1986, *Ann. Rev. Astron. Astrophys.* **24**, 329.

K. A. van der Hucht and B. Hidayat (eds.),
Wolf-Rayet Stars and Interrelations with Other Massive Stars in Galaxies, 643.
© 1991 *IAU. Printed in the Netherlands.*

THE STELLAR ASSOCIATION LH 99

M.C. LORTET and G. TESTOR
D.A.E.C.
Observatoire de Meudon
F-92195 Meudon Cedex
France

H. SCHILD
University College London
Department of Physics and Astronomy
United Kingdom

The stellar content of this stellar association, related to the supernova remnant N 157B = SNR 0538-691, was up to now unknown, except for three Wolf-Rayet stars and a few red supergiant candidates. A thorough study based on UBV photometry, spectra of 95 stars and nebular spectra will be described elsewhere (see also Schild et al. in these Proceedings).The outstanding properties of this region are :

A. the presence of a shell-shaped supernova, which dominates the thermal radio continuum. It is relatively faint in [OIII], strong in [SII], [Fe 4658] and He II 4686. The expansion velocity reported by Chu and Kennicutt (1988) is 180 km/s. In addition, many extensions and filaments are found all over a wide area, and possibly there has been more than one supernova responsible.

B. the existence of about 40 newly discovered O stars including three O3-4 unevolved stars .

C. the WR stars are of type WN7 (Brey 71, Brey 73) and WN3-4 with strong C IV 5812 emission (Brey 70a = MGWR 4, Morgan and Good, 1987). An extreme Of star has just been discovered near Brey 73 (Testor and Schild, 1990).

D. the physical boundaries of the association are quite uncertain. A star concentration 3 or 4' in maximum size (NS) is clearly seen near the SNR, it is surrounded by an obscure corona, itself limited by a conspicuous SE-NW ionization front.

Brey 70a is outstanding by its high luminosity (M_{bol} about -10.0), high reddening, strong and wide lines, and a strong CIV 5812 line, only found in WN/WC stars, such as Brey 29 and Brey 72 in the LMC.

We are exploring the idea that the apparent star concentration containing the brightest parts of the SNR, Brey 73, TSWR 3 and several hot stars is real and distinct from the surrounding areas where hot O, WC, WNE and red supergiant candidates are scattered in a large volume, possibly at different depths along the line of sight.These may span a large range in ages, exactly as is found slightly to the North in the 30 Dor Nebula.

REFERENCES

- Morgan, D.H., Good, A.R., 1987, M.N.R.A.S. 224, 435
- Testor, G., Schild, H., 1990, Astron. Astrophys., in press

K. A. van der Hucht and B. Hidayat (eds.),
Wolf-Rayet Stars and Interrelations with Other Massive Stars in Galaxies, 644.
© 1991 IAU. Printed in the Netherlands.

Multiplicity of Very Massive Stars

M. Heydari-Malayeri

European Southern Observatory
La Silla, Chile

Abstract. Are there very massive stars (VMSs) of mass greater than 100 $M_\odot$? This question constitutes one of the fundamental problems of astrophysics. We present observational evidence against the existence of such stars in the Magellanic Clouds. The multiplicity of VMSs has several important consequences for astrophysics. If VMSs do not exist we need to revise our ideas about the formation and evolution of stars.

The question of the upper mass limit is very important for star formation theories and its related subjects, for example the initial mass function (IMF). The multiplicity of massive star alters the IMF in two ways. It eliminates the stars more massive than about 100 $M_\odot$ and at the same time brings about a large number of stars in the mass interval 30–60 $M_\odot$. This has important implications for several astrophysical problems, e.g. the evolution of galaxies and the choice of extragalactic distance indicators.

Today in spite of recent theoretical progress made in understanding the formation and evolution of massive stars the question of exact upper limit to stellar masses remains open. In this regard the problem with R136, once believed to have a mass of 3000 $M_\odot$ is not fully resolved, since its main component may be about 250 $M_\odot$ (Walborn, 1984). But can stars of this mass exist? Observational results are of vital importance in answering this vital question.

We summarize in Table 1 some results on the previously supposed supermassive stars in the Magellanic Clouds. The observations were carried out using the ESO 2.2m and NTT telescopes. The observations and image processing methods are explained in Heydari-Malayeri et al. (1988, 1989) and Heydari-Malayeri and Hutsemékers (1990).

The global spectral type and visual magnitudes of the stars are given in columns 2 and 3 respectively. The mass previously attributed to each star is listed in column 4. The number of resolved components is presented in column 5. The V magnitude of the brightest component is given in column 6 and the corresponding mass in column 7. The masses were estimated using the new evolutionary models of massive stars for the Magellanic Clouds metallicities (Maeder,

K. A. van der Hucht and B. Hidayat (eds.),
Wolf-Rayet Stars and Interrelations with Other Massive Stars in Galaxies, 645–646.
© 1991 *IAU. Printed in the Netherlands.*

1990).

Table 1. New mass estimates

Star	Sp.	V_p	$(ZAMS)_p$	Components	V_n	$(ZAMS)_n$	Note
LMC							
Sk–66°41	O5 V	11.72	> 120	6	12.2	85	(a)
Sk–69°253	O9.5: I	11.23	120	14	12.0	60	
	or B 0.7–1 I						
SMC							
Sk–157	O9.5: III	12.17	60	11	13.2	40	
NGC 346 #1	O4	12.43	130	3	12.3	85	(b)
NGC 346 #2	O3	13.43	90	1	–	..	(b)

(a) Main component probably multiple
(b) See Heydari-Malayeri & Hutsemékers (1990)

New results on several other so-called VMSs will be presented in a forthcoming paper.

References

Heydari-Malayeri, M., Magain, P., Remy, M.: 1988, *Astron. Astrophys.* **201**, L41

Heydari-Malayeri, M., Magain, P., Remy, M.: 1989, *Astron. Astrophys.* **222**, 41

Heydari-Malayeri, M., Hutsemékers, D.: 1990, in preparation

Maeder, A.: 1990, *Astron. Astrophys. Suppl. Ser.*, in press

Walborn, N.R.: 1984, in *Structure and Evolution of the Magellanic Clouds*, IAU Sympos. **108**, eds. S. van den Bergh, K.S. de Boer, p. 243

A WC9 Star in the LMC

M. Heydari-Malayeri, J. Melnick

European Southern Observatory
La Silla, Chile

Abstract. We have discovered the only WC9 star in the Large Magellanic Cloud on the basis of observations made at several ESO telescopes, especially NTT. The latest Wolf-Rayet stars of the carbon sequence so far observed in the LMC are WC5-6. The lack of late type WC stars is understood in terms of an abundance effect, according to the models of Maeder. Our observations therefore pose a new observational challenge to the theory of the evolution of massive stars.

In the course of a systematic investigation of the HII blobs in the Large Magellanic Cloud, we serendipitously observed a peculiar low excitation, compact HII region which turned out to have several remarkable characteristics. Although this object is catalogued as N82 (Henize, 1956), virtually nothing about it could be found in the literature.

N82 is a relatively very compact HII blob of average radius $1''.3$ corresponding to 0.4 pc. No internal structure is revealed in the Hα image taken with an excellent seeing of $0''.68$ FWHM, apart from the fact that the object seems somewhat elongated in SE-NE direction. No star is detected in N82.

N82 has a peculiar spectrum. The [NII]$\lambda\lambda6584,48$ lines are extraordinarily strong (as large as Hα) and the [O III] $\lambda\lambda4959, 5007$ lines very weak ([O III]/Hβ=1.8). There are also the auroral lines [O III] $\lambda4363$ and [N II] $\lambda5755$. These are due to important abundance anomalies in N82. Nitrogen is enriched by a factor of 5 with respect to the average LMC value and oxygen depleted by a factor 2.5.

The only plausible explanation for these abundance anomalies is that the HII region is ejected by a central W-R star. This is compatible with the fact that the derived electron densities for N82 are unusually large. During the latest stages of massive stars several effects contribute to the change of surface composition. For example, mass loss by stellar winds can progressively reveal nuclearly processed elements, and dredge-up by external convective zones can dilute newly synthesized elements in stellar envelopes. According to models by Maeder (1987),

K. A. van der Hucht and B. Hidayat (eds.),
Wolf-Rayet Stars and Interrelations with Other Massive Stars in Galaxies, 647–648.
© 1991 *IAU. Printed in the Netherlands.*

a massive star of type O evolves into red supergiant, blue supergiant, nitrogen-rich W-R and carbon-rich W-R. At this stage the outer envelopes of the star are thrown out into the interstellar medium. In the case of N82 the central W-R star should have ejected more than 7 $M_\odot$.

It is interesting that the derived oxygen and nitrogen abundances correspond very closely to the values predicted by the models of Maeder (1987) for the material shed by massive stars approaching the WC phase. This is an important support for the theory.

The presence of the W-R star embedded in the compact HII region N82 was confirmed by two deep NTT spectra in the blue and red. The presence of several C III lines, especially the $\lambda4650$ blend and $\lambda5696$ indicate that we are dealing with a carbon-rich W-R. The absence of C IV lines on the other hand indicate that the star is of a very late type, probably WC9.

Our observations also indicate that the WC9 star is likely to be a member of a binary system, with an early O-type companion. This may resolve the apparent discrepancy with the models which are only valid for single star evolution. According to these models (Maeder, 1990), for low metallicities, only the most massive stars manage to remove their outer layers. In other words, for low Z the surface composition departs from low (C+O)/He ratios and enters the WC phase only very late in the helium-burning phase.

In conclusion, N82 is a high density blob of stellar wind material shed by a massive star, probably in a binary system. The W-R component is the only late type WC star known in the LMC. The present results suggest that the massive stars may play a more important role than has been previously acknowledged in the nitrogen enrichment of the interstellar medium.

For more details see Heydari-Malayeri et al. (1990)

References

Heydari-Malayeri, M., Melnick, J., Van Drom, E.: 1990, *Astron. Astrophys. Letters*, in press
Henize,K.G.: 1956, *Astrophys. J. suppl.* **2**, 315
Maeder, A.: 1987, *Astron. Astrophys.* **173**, 247
Maeder, A.: 1990, private communication

MASSIVE STARS IN M31 AND M33 OB ASSOCIATIONS

K. CANANZI and M. AZZOPARDI
Observatoire de Marseille
2 Place Le Verrier
F-13248 Marseille Cedex 4, France

SUMMARY. We observed several OB associations of M31 (OB 33, 42, 59 and 69) and of M33 (OB 4 and 137) in $UBVI$ and through interference filters centered on lines of $HeII$ and $CIII$, as well as a narrow continuum comparison filter at 4749 Å (Armandroff & Massey, 1985). The observations were made at the CFH 3.6m telescope with a RCA-CCD. The stellar photometry was obtained using DAOPHOT.

The present results concern the M31 associations. We selected the blue stars as those with $B - V < 0.2$. For those stars we find an average slope of the luminosity function of 0.51 between $V = 17$ and 22. This slope is in agreement with the average slope for the whole galaxy (Berkhuijsen and Humphreys, 1989). However, they emphasize the fact that the best OB stars selection criterion was $B - V \leq 0.1$ and $U - V \leq -0.9$. They then have found a slope of 0.74! We attempted to reproduce this slope from an Initial Mass Function (IMF) defined as $dn(M)/dlogM = M^x$ (where n is the number of stars formed per unit of time and $x < 0$) in two simple models of star formation: an instantaneous burst and an uniform rate of star formation. The calculated slope of the luminosity function varies between 0.50 and 0.25 for a slope x of the IMF varying between -3 and -2 respectively, in the Burst case, and from 0.88 and 0.62 in the case of an uniform star formation rate. The observed slope of the luminosity function thus logically favours the Constant Star Formation Rate hypothesis.

We detected Wolf-Rayet star candidates in our fields by comparing the stellar magnitudes in the three narrow filters, which also allow a separation between WN and WC stars. Retaining only the very probable candidates, we obtain the following ratios between the numbers of these stars: $N_{WC}/N_{WR} = 0.57$ and $N_{WC}/N_{WN} = 1.33$. Then, according to Maeder (1990), who has plotted the two ratios *vs.* metallicity for WR stars in galaxies with active star formation, this leads to a $Z \approx Z_\odot$.

References

Armandroff, T., Massey, P. 1989, *Astrophys. J.*, **291**, 685.
Berkhuijsen, E.M., Humphreys, R.M. 1989, *Astron. Astrophys.* **214**, 6.
Maeder, A. 1990, *Astron. Astrophys.* (in press).

K. A. van der Hucht and B. Hidayat (eds.),
Wolf-Rayet Stars and Interrelations with Other Massive Stars in Galaxies, 649.
© 1991 *IAU. Printed in the Netherlands.*

OPTICAL SPECTROSCOPY OF WR STARS IN M33 AND M31

H. SCHILD, L.J. SMITH & A.J. WILLIS
Department of Physics and Astronomy
University College London
Gower Street, London WC1E 6BT
England

ABSTRACT. We present new optical spectroscopy obtained with the Faint Object Spectrograph on the William Herschel Telescope of 6 WR stars in the local group galaxy M33 and 6 WR stars in M31. These spectra cover the wavelength range $\lambda\lambda 3500$–9750 Å. We confirm the previous WR classifications of the M31 stars. In the M33 sample, 4 stars are classified as WC4–5, one is confirmed as WN–WCE and a further WN–WCE star has been identified. The widths of the CIV $\lambda 5800$ emission differ by a factor of three amongst the M33 WC4–5 sample, spanning values exhibited by galactic WC8–WC4 stars and confirming the existence of narrow-lined WC4-5 stars in M33. We find that the M33 WCE line widths appear to be correlated with galactocentric distance and loosely connected to ambient metallicity. The narrowest lined WCE star MC 53 is in an unusually high (above solar) metallicity region. Further observations of a larger sample of M33 stars are required to confirm these preliminary results.

The full version of this paper is to be published in Astronomy and Astrophysics.

K. A. van der Hucht and B. Hidayat (eds.),
Wolf-Rayet Stars and Interrelations with Other Massive Stars in Galaxies, 650.
© 1991 *IAU. Printed in the Netherlands.*

Hα EMISSION-LINE STARS IN M33

C. NEESE, T. E. ARMANDROFF and P. MASSEY
National Optical Astronomy Observatories
Kitt Peak National Observatory
P.O. Box 26732, Tucson, AZ 85726-6732, U.S.A.

We have used the KPNO Schmidt+CCD with interference filters to survey the nearby Sc galaxy M33 for sources of Hα emission. The Schmidt+CCD combination provided large area combined with high quantum efficiency and linear response; we were able to survey most of M33 to a hitherto unprecedented depth in Hα. The Hα emitting sources revealed by this survey include about 500 stellar sources with red continuum magnitudes ranging from 17th to 21st. Such Hα emission-line stars in our own galaxy and in the Magellanic Clouds include many extremely interesting objects of high luminosity and high mass-loss rates, such as Of and Oe stars, extreme Be stars, Wolf-Rayet stars, Hubble-Sandage variables, and SS433. In fact, four known Hubble-Sandage variables which are within the area of the survey were easily detected.

Using the KPNO 4-meter telescope with the long-slit cryogenic camera and with the Nessie multi-object fiber-feed, we have obtained low-resolution spectra for over 200 of the stellar Hα emitting sources detected in this survey. Unsurprisingly, about half of these objects turned out to be compact HII regions with no detectable continuum. Twenty percent of the objects observed are red stars. While some contamination of our sample by red stars is expected given the placement of our off-band filter, it is interesting that half of the red stars detected also show Hα emission. These may be Mira variables.

No analog of SS433 was found, although spectra were obtained for all of the strongest Hα sources. We believe that if any SS433-like objects exist in M33, none are oriented in such a way as to be detectable. One bizarre object was found which exhibits extremely broad Hα emission (≈ 50 Å FWHM) but no emission at Hβ, HeII $\lambda4686$, or CIV $\lambda5812$. Higher dispersion spectroscopy for this extremely interesting object is planned. Nine stars appeared to be extreme Be or Oe stars. Three new Wolf-Rayet stars were found, although their detection was due to associated nebulosity, not the HeII Pickering line coincident with Hα. One known supernova remnant was observed, and its spectrum displays evidence of shock heating. We 'rediscovered' four of the five known Hubble-Sandage variables, plus we found an additional five objects which are spectroscopically indistinguishable from the known Hubble-Sandage variables with their forest of FeII and [FeII] emission. These objects are currently being investigated for photometric variability.

651

K. A. van der Hucht and B. Hidayat (eds.),
Wolf-Rayet Stars and Interrelations with Other Massive Stars in Galaxies, 651.
© 1991 *IAU. Printed in the Netherlands.*

CCD SURVEY FOR WR STARS IN THE CENTRAL PART OF M33

L. DRISSEN[1], A. F. J. MOFFAT[2] and M. M. SHARA[1]

[1] Space Telescope Science Institute, Baltimore, MD (USA)
[2] Département de physique, Université de Montréal, Montréal (CANADA)

There have been numerous surveys for Wolf-Rayet stars in M33 (among others: Wray and Corso 1972, Ap. J., **172**, 577; Massey and Conti 1983, Ap. J., **273**, 576; and Massey et al. 1987, P.A.S.P., **99**, 816). About 100 WR stars have been discovered and spectroscopically confirmed so far. All these surveys have used photographic plates and a relatively large ($\sim$ 100 Å) filter centered on the 4640-4686 Å emission-line region. This choice of filter does not favor the detection of the narrow-line WNL stars. We have thus obtained CCD images of the central ($8' \times 20'$) part of M33 with a different set of filters (broad B and narrow 4686/35 Å fwhm) with the 2.2m telescope at KPNO and the 1.6m telescope at Mont Mégantic in an attempt to detect new WR candidates. R and Hα images were also obtained.

Twenty-nine (29) new candidates have been found, half of them being of high quality. In the area surveyed, 43 spectroscopically confirmed WR stars were previously known. We have detected all but two of them (MC34 and MC45). This result shows that even if only our best candidates (detection level $\geq 7\sigma$) are genuine WR stars, the incompleteness of previous surveys may be as high as 30 %. Since most of them are likely to be WNs, WC/WN ratios (and hence previous conclusions about the effects of metallicity and IMF gradients in this galaxy) are likely to be revised. Obviously, the WR content of M33 deserves further study.

Our Hα images allowed the detection of 15 ring nebulae around confirmed WR stars (details will be published elsewhere). Ring nebulae were also detected around 4 of our new WR candidates. WR ring nebulae in M33 are larger, on average, than their Galactic counterparts: while 80 % of the Galactic rings are smaller than 20 pc, most of M33's WR rings are larger than 20 pc. This may be because the interstellar medium in M33 is more favorable to the expansion of such bubbles; but it could also be, at least in part, an observational effect, since we have probably missed all nebulae smaller than 5 pc.

652

K. A. van der Hucht and B. Hidayat (eds.),
Wolf-Rayet Stars and Interrelations with Other Massive Stars in Galaxies, 652.
© 1991 IAU. Printed in the Netherlands.

A WOLF-RAYET CLUSTER IN IC 4662

O.-G. Richter[1,*] and M.R. Rosa[2,*]

[1] Space Telescope Science Institute, Baltimore, MD 21218, U.S.A.
[2] The Space Telescope European Coordinating Facility,
     European Southern Observatory, Garching, F.R.G.
[*] Affiliated to the Astrophysics Division of the
     Space Science Department of E.S.A.

ABSTRACT   An imaging and spectroscopic survey for WR stars in IC 4662 revealed the presence of a cluster consisting of a mix of about 10 WN 7-8 and WC 5 stars plus the equivalent of some 5 OV stars at a projected distance of about 100 pc from the cluster core of the bright central H II complex.

The magellanic irregular galaxy IC 4662 had been included unsuccessfully in a series of spectroscopic searches for WR stars (see Rosa and D'Odorico 1986) which concentrated on the blue associations scattered around the galaxy's body and on the main H II complex. This complex is very similar to the 30 Dor region in the LMC. Judged from the high degree of resolution IC 4662 is possibly an outlying member of the Local Group of Galaxies. Given its small distance (adopted as 2 Mpc) we conducted a comprehensive narrow-band imaging survey through filters off- and on-line the typical WR features.

CCD images were obtained at the ESO 2.2m telescope in April 1984. For details of the instrumental setup and reduction procedures see Rosa and Richter (1988). Comparison between images taken with the different narrow band filters immediately revealed an almost point-like strong WR candidate in the outskirts of the bright H II complex, indicated in Fig. 1. Subsequently, IDS spectra with a resolution of 10 Å were taken with the ESO 3.6m telescope at the candidate position and in the central H II complex (see Fig. 2).

The candidate spectrum shows the characteristic WR features at 4650 Å and 5810 Å, very similar in appearance to those of e.g., NGC 300 # 2 or NGC 604 B in D'Odorico et al. (1983). The main contributors to these features are WN 7-8 and WC 5 stars. This is in accord with previous results (Rosa and D'Odorico, 1986) which show that the WR population in more than 30 giant H II regions in nearby galaxies is dominated by late WN and WC 5 stars. The absolute flux in the WR bands ($\approx 3.0 \times 10^{-14}$ erg/s/cm^2/Å) can be used to estimate the number of WR stars present. The main uncertainties arise from the assumed distance to IC 4662, and the luminosities in the emission bands of the "standard" WR stars used for comparison. Assuming an absolute magnitude of -7^m in the continuum near the 4650 Å band for a typical "mixed" WN 7-8, WC 5 star, and an equivalent width of 50 Å for the 4650 Å band in such a star, about 10 such stars would be present. The

K. A. van der Hucht and B. Hidayat (eds.),
Wolf-Rayet Stars and Interrelations with Other Massive Stars in Galaxies, 653–654.
© 1991 IAU. Printed in the Netherlands.

observed H_β flux of 1.2×10^{-13} erg/s/cm^2 together with an $A_V = 1\overset{m}{.}2$ corresponds to the Lyman continuum luminosity of about 5 OV stars. Another estimate of the number of luminous stars is obtained from the WR cluster's absolute visual magnitude [$m_V = 16\overset{m}{.}6$] of $M_V \approx -11^m$. If most of the blue and visual light is contributed by stars with M_V of order -6^m roughly 80 such stars are present in addition to the 10 WR stars at $M_V = -7^m$.

The relative number of massive stars (WR/O) derived above is indicative of an evolved OB association at an age ≥ 4 Myrs (see Maeder 1990) and a metallicity of [O/H] = 8.3. However, based on the close resemblance of the Im galaxy IC 4662 with the LMC, particularly the 30 Dor complex ([O/H] = 8.25), one might have expected at least a few WR stars in the main H II region. The marked absence of WR features there might be due to a smaller evolutionary age of ≤ 2 Myr. Although the limiting magnitude of our images is at least 2^m fainter than the discovered WR cluster no other WR candidates could be identified. This apparent segregation of massive star clusters in age *and* space in IC 4662, if confirmed, is exceptional.

REFERENCES

D'Odorico, S., Rosa, M., Wampler, J.E.: 1983, *Astron. Astrophys. Suppl. Ser.*, **53**, 97.

Maeder, A.: 1990, in *"Angular Momentum and Mass Loss of Hot Stars"*, L.A. Willson and R. Stalio (eds.), AMES-Trieste NATO Workshop, Kluwer Publ.

Rosa, M., D'Odorico, S.: 1986, in *"Luminous Stars and Associations in Galaxies"*, IAU Symp. No. 116, Reidel, Dordrecht, pp. 355.

Rosa, M., Richter, O.-G.: 1988, *Astron. Astrophys.*, **192**, 57.

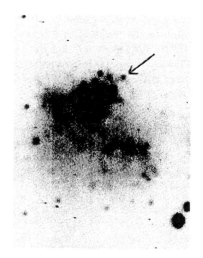

Fig. 1: IC 4662 in the light of the WR 4650 Å feature. The arrow indicates the bright candidate west of the H II complex.

Fig. 2: Spectra for the WR candidate (bold, upper) and the neighboring outskirts of the H II complex (thin, lower). The typical WR features at 4650 Å and 5810 Å are evident.

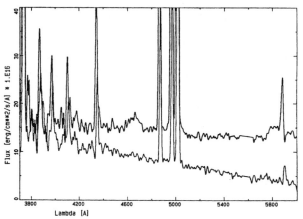

DISCOVERY OF WOLF-RAYET STARS IN THE SBmIII GALAXY NGC 4214

ALEXEI V. FILIPPENKO[1] and WALLACE L. W. SARGENT[2]

[1]Department of Astronomy, and
Center for Particle Astrophysics
University of California
Berkeley, CA 94720 U.S.A.

[2]Palomar Observatory
105–24 Caltech
Pasadena, CA 91125
U.S.A.

NGC 4214 is a nearby ($d \approx 5.4$ Mpc), gas rich, Magellanic irregular galaxy (SBmIII). Sandage and Bedke (1985, *Astron. J.*, **90**, 1992) show an excellent photograph of the galaxy in their Figure 1, Panel 21. As part of an extensive spectroscopic survey of nearby galaxies, we obtained long-slit (position angle 107°) spectra of NGC 4214 around the Hα and Hβ regions, with the Double Spectrograph on the 5-m Hale reflector at Palomar Observatory.

Prominent emission lines attributable to Wolf-Rayet stars are visible in the spectra. The Wolf-Rayet features exist in the nucleus, as well as in several off-nuclear H II regions. The shape and extent of the broad blend at 4660 Å indicates the presence of WC and WN stars. The detection of WC stars is very rare in galaxies outside the Local Group. Our discovery of WC stars is confirmed in spectra obtained with the 3-m Shane reflector at Lick Observatory; the C IV λ5808 line is clearly visible. This is consistent with the relatively high abundance of heavy elements in NGC 4214, compared with extragalactic H II regions. Broad Hα emission, due to Wolf-Rayet stars, is also visible in the nucleus of NGC 4214.

Very steep spatial gradients in the intensities of the Wolf-Rayet lines are observed in NGC 4214. Indeed, it appears that the lines are prominent in some regions, yet almost nonexistent in adjacent areas only 3″ away. Most surprising is the fact that, in general, the strongest Wolf-Rayet lines appear in H II regions having the brightest stellar continuum (e.g., Fig. 1). Weak, if any, Wolf-Rayet features are present in H II regions having strong, narrow emission lines but only a faint continuum. If the intensity of the broad features were entirely a function of the age of a starburst, with Wolf-Rayet stars beginning to appear after a few million years, we would not expect these results; the optical continuum of a starburst does not change appreciably over such short time scales. Similarly, metallicity gradients in NGC 4214 are too small to explain the observations. Instead, this correlation may be indicative of variations in the initial mass function at different locations of NGC 4214.

Details can be found in Sargent and Filippenko (1990, *Ap. J. Letters*, submitted).

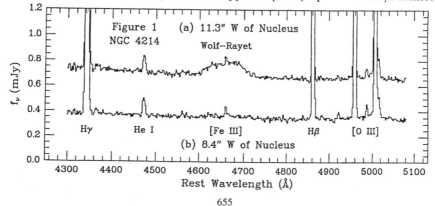

Figure 1 (a) 11.3″ W of Nucleus
NGC 4214 Wolf–Rayet
(b) 8.4″ W of Nucleus

655

K. A. van der Hucht and B. Hidayat (eds.),
Wolf-Rayet Stars and Interrelations with Other Massive Stars in Galaxies, 655.
© 1991 *IAU. Printed in the Netherlands.*

WOLF-RAYET GALAXIES

WILLIAM D. VACCA and PETER S. CONTI
Joint Institute for Laboratory Astrophysics
University of Colorado, Boulder, CO 80309-0440, USA

Wolf-Rayet (W-R) Galaxies are a subset of starburst galaxies (usually, blue compact dwarfs) in whose integrated spectra a broad He II λ 4686 emission feature has been detected (e.g., Osterbrock and Cohen 1982; Kunth and Joubert 1985). This line is a prominent emission feature in the spectra of Galactic WN stars. The presence of 10^2 to 10^5 W-R stars in these galaxies has been inferred from a comparison of the line fluxes in the integrated galaxy spectra with those in the spectra of Galactic WN stars (e.g., Kunth and Schild 1986; Armus, Heckman, and Miley 1988).

In an effort to perform a systematic study of the properties of W-R galaxies, we have obtained moderate resolution ($\sim$ 3.5 Å) spectra of 10 known W-R galaxies and 4 comparison starburst galaxies using the 2D-Frutti detector on the 4-m telescope at CTIO. The spectra cover the wavelength range $\sim$ 3120 to $\sim$ 7000 Å. We report on these observations and a preliminary analysis of the spectra of these galaxies.

In all 10 W-R galaxies, the He II λ 4686 feature is resolved, with a typical FWHM of $\sim$ 14 Å. In the spectra of 3 W-R galaxies a broad (FWHM $\sim$ 20 Å) N III λ4638 emission feature is also detected. Both of these lines are characteristic features of the spectra of late WN or Of/WN stars. The absence of C III and C IV emission lines in the spectra suggest that the WC/WN number ratio in these galaxies is very small, similar to that found in the low metal abundance environments such as the LMC and SMC. The blue continua and strong nebular emission lines (e.g., the Balmer series, [O II], [O III], and [Fe III]) indicate that the W-R galaxies have higher excitation, and therefore, many more hot stars, than the comparison starburst galaxies. The relatively weak or absent Balmer absorption lines and Balmer jumps in the spectra of W-R galaxies indicate that the hot stars must be the dominant contributors to the overall galactic spectra. These galaxies may be greatly scaled-up examples of less energetic extragalactic "giant" II II regions. We interpret W-R galaxies as examples of the youngest phase of the starburst phenomenon, in which massive stars are still being born or star formation has only recently ended (less than a few $\times 10^6$ years ago).

REFERENCES

Armus, L., Heckman, T. M., and Miley, G. K. 1988, *Ap. J. (Letters)*, **326**, L45.
Kunth, D., and Joubert, M. 1985, *Astr. Ap.*, **142**, 411.
Kunth, D., and Schild, H. 1986, *Astr. Ap.*, **169**, 71.
Osterbrock, D. E., and Cohen, R. D. 1982, *Ap. J.*, **261**, 64.

656

K. A. van der Hucht and B. Hidayat (eds.),
Wolf-Rayet Stars and Interrelations with Other Massive Stars in Galaxies, 656.
© 1991 *IAU. Printed in the Netherlands.*

COMPUTER SIMULATION OF THE EVOLUTION OF AGN: SINGLE SUPERMASSIVE STAR VS. A CLUSTER OF MASSIVE WR STARS

R. SPURZEM[1], K. HERZIG[2]
[1] Institut für Theoretische Astrophysik, Olshausenstr. 40,
D-2300 Kiel, F.R.G.
[2] Universitäts-Sternwarte, Geismarlandstr. 11,
D-3400 Göttingen, F.R.G.

ABSTRACT. The evolution of gas and stars in galactic nuclei is reconsidered and numerical models of early evolutionary phases of galactic nuclei are discussed, in particular the evolution of a supermassive gaseous object as a hypothetical precursor of a supermassive black hole.

1. Introduction

Since the discovery of quasars and of other types of active galactic nuclei (AGN) various theoretical models have been proposed to explain the characteristics of these exceptional astronomical objects. Among their extreme properties are the high luminosities ($\approx 10^{47} \mathrm{erg\,s^{-1}}$) produced on very small scales ($\approx$ 1pc), the non-thermal emission spectra, polarization, the variability on short time scales, and the occurrence of jets. As the primary energy source the liberation of gravitational binding energy by the accretion of matter onto a supermassive central black hole (SMS) having a mass in the range of $10^7 \mathrm{M_\odot}$ to $10^9 \mathrm{M_\odot}$ has been suggested (Lynden-Bell, 1969) and is now commonly accepted (cf. e.g. Rees, 1984).

Tidal disruption of stars and accretion of the debris in the vicinity of the black hole cannot achieve such large masses within a Hubble time from a small seed black hole (Marchant and Shapiro 1980). Clues to the history of the supermassive black hole are processes like collisional disruption and/or coalescence of the stellar population of the nucleus and newly formed stars consisting of material liberated in stellar collisions or accumulated in the galactic centre from stellar mass loss in the outer regions.

Two basic scenarios are distinguished; first the Spitzer-Saslaw-Stone picture (Spitzer and Saslaw 1966; Spitzer and Stone 1967; see also Begelman and Rees, 1978; Illarionov and Romanova 1988): gas is liberated from disruptive stellar collisions if the central core velocity dispersion exceeds the escape velocity of individual stars. The gas settles as a thin disk of radius $R_d = \epsilon^{1/2} R$, where R is the initial star cluster's radius and ϵ its initial eccentricity. Predominantly small stars ($0.1 - 0.5 \mathrm{M_\odot}$, Illarionov and Romanova 1988) form out of the disk. Elastic encounters with disk and core stars and the Ostriker-Peebles instability enhance the velocity dispersion of such a stellar subsystem which again will suffer from liberation of gas and star formation. If the total luminosity of the resultant star-gas mixture reaches the Eddington limit star formation should cease and a supermassive star (SMS) containing a population of low-mass stars begins its subsequent evolution. We model the further development of such configuration by a computer simulation; the results are discussed below.

The other basic scenario is due to Colgate (1967) and Sanders (1970). They argue that coalescence of stars in the central region will lead to an agglomeration of several thousand massive stars ($M > 100 \mathrm{M_\odot}$); generally they are believed to produce an enhanced super-

657

K. A. van der Hucht and B. Hidayat (eds.),
Wolf-Rayet Stars and Interrelations with Other Massive Stars in Galaxies, 657–658.
© 1991 IAU. Printed in the Netherlands.

nova rate and a cluster of compact stellar evolution remnants which ultimately collapses relativistically to a supermassive black hole (Zeldovich and Podurets, 1965; Shapiro and Teukolsky 1985). However, if coalescence timescales are short enough to proceed to a coherent supermassive object, one again ends, as in the Spitzer-Saslaw-Stone scenario, at the formation of a dense gas star system.

A more detailed numerical study of this evolutionary phase by means of a multicomponent model taking into account a mass spectrum of stars, coalescence and the dynamics of the liberated gas is subject of present work. For cloud systems in galaxies such a model was already utilized (Yorke et al. 1989).

2. Conclusions

Supermassive stars of more than about $5 \cdot 10^5 M_\odot$ are subject to the post-Newtonian dynamical instability during their H-burning phase; they do not explode and thus form a supermassive black hole, provided their metallicity is below a critical value (Fuller et al., 1986). Usually such stellar evolution calculations deal with the evolution of an isolated supermassive star; Langbein et al. (1990) reported a first attempt to model numerically the evolution of a supermassive star containing a stellar system consisting of normal one solar mass single stars. Their calculations show that the energy generation by dissipation of stellar kinetic energy in the SMS exceeds the nuclear energy generation in certain evolutionary phases. Therefore we stress that the stellar evolution of such an SMS – provided it exists inside a galactic nucleus – may not be calculated without taking into account the stellar dynamical energy generation. The final fate of the SMS cannot yet be determined from the results of the simulations. The total amount of dissipated stellar kinetic energy alone, however, is smaller than the initial binding energy of the SMS. Future work in progress has to account for nuclear energy generation in combination with the stellar dynamical effects.

3. Acknowledgements

This work was supported in part by DFG grants Fr 325/23 and Yo 5/5; R. Sp. thanks the Institute of Astronomy and Astrophysics Würzburg and the Universitätssternwarte Göttingen for their kind hospitality.

4. References

Begelman, M.C., Rees, M.: 1978, *Monthly Notices Roy. Astron. Soc.* **185**, 847
Colgate, S.A.: 1967, *Astrophys. J.* **150**, 163
Fuller, G.M., Woosley, S.E., Weaver, T.A.: 1986, *Astrophys. J.* **307**, 675
Illarionov, A.F., Romanova, M.M.: 1988, *Astrophys. J.* **32**, 148
Langbein, T., Spurzem, R., Fricke, K.J., Yorke, H.W. : 1990, *Astron. Astrophys.* **227**, 333
Lynden-Bell, D.: 1969, *Nature* **223**, 690
Marchant, A.B., Shapiro, S.L.: 1980, *Astrophys. J.* **239**, 685
Rees, M.: 1984, *Ann. Rev. Astron. Astrophys.* **22**, 471
Sanders, R.H.: 1970, *Astrophys. J.* **162**, 791
Shapiro, S.L., Teukolsky, S.A.: 1985c, *Astrophys. J.* **292**, L41
Spitzer, L., Saslaw, W.C.: 1966, *Astrophys. J.* **143**, 400
Spitzer, L., Stone, M.C.: 1967, *Astrophys. J.* **147**, 519
Yorke, H.W., Kunze, R., Spurzem, R.: 1989, in *Structure and Dynamics of the Interstellar Medium, Proc. IAU Coll. No. 120*, eds. G. Tenorio-Tagle, M. Moles, J. Melnick, Springer, Berlin, p. 186
Zel'dovich, Ya. B., Podurets, M.A.: 1965, *Astron. Zh.* , **42**, 963 (engl. translation in *Sov. Astron.*, **9**, 742).

SESSION IX. SUMMARY – *Chair: Karel A. van der Hucht*

Peter Conti

SUMMARY OF SYMPOSIUM

PETER S. CONTI
Joint Institute for Laboratory Astrophysics
University of Colorado, Boulder, CO 80309-0440, USA

"It is a significant fact that the emission and absorption lines of Wolf-Rayet spectra cannot be regarded as being formed in one body of gas at one representative temperature and pressure" (Underhill 1968)

"I think it fair to say that the demonstration of ... diversity has brought the Wolf-Rayet stars from being a curiosity to being an important part of the 'main stream' evolution of all massive stars" (Smith 1982)

It is not easy to put together, from the past four and one half days and into the space of about a half an hour, a fully concise summary of what has been presented here at our meeting. What follows must necessarily be a highly personalized view of these proceedings. My report is colored not only by my pre-conceptions but also by my own active participation in research on Wolf-Rayet stars and my singular interpretation of what is important, new, and newsworthy. I hope you will bear with me if not every contribution was singled out, and you will accept my own rendering of the various research results presented here. With a view to save time (and space) and to avoid potential hard feelings, I will not quote the authors of the contributions which may be readily found in any case in these proceedings.

The quotations that lead off this Summary come from the only two astronomers at this meeting who have also attended both previous IAU Symposia and the Boulder mini-symposium on the topic of Wolf-Rayet (W-R) stars. Their well known "antipodal" attitudes represent dramatically different interpretations of the nature of the spectra, and ultimately the understanding of, W-R stars. The Boulder meeting in 1968, coming some hundred years after the discovery of such stars, may be said to have begun the modern era of work in this field. You have already heard at the after dinner talk earlier in the week a summary of the initial "historical" period by Len Kuhi. In this paper, I shall briefly review the major themes of the previous meetings, so as to put into context the main results of this one. It must be remembered that considerable research and publication has gone on in between these Symposia but in briefly looking back at them we may obtain a flavor of the work that has ensued.

The Boulder mini-symposium (Gebbie and Thomas 1968) contains the modern definition of the W-R phenomena, in terms of a number of spectroscopic (i.e. observed) properties of the class, both those

661

K. A. van der Hucht and B. Hidayat (eds.),
Wolf-Rayet Stars and Interrelations with Other Massive Stars in Galaxies, 661–668.
© 1991 *IAU. Printed in the Netherlands.*

of Population I and certain central stars of planetary nebulae. There was much discussion of the newly emerging quantitative empirical classification of the WN and WC and their numerical subtypes. The overall composition, still very poorly understood, was argued about incessantly. The first indications of the anomalous galactic distribution of W-R subtypes was presented and tentative interpretations offered. Only primitive atmospheric models existed which were little more than "handwaving" in complexity. Considerable "yelling" about such models must have occurred at this meeting as may be inferred from the tone and nature of the proceedings. I was sorry to have missed this conference as my own interests in hot luminous stars were just beginning to stir at about this time. The publication time for the proceedings was under six months, a goal often promised but rarely honored.

In Buenos Aires in 1971 (Bappu and Sahade 1973) a major thrust of Symposium #49 was devoted to binary evolution. At the time, nearly all of us thought that ALL W-R stars were double and their anomalous spectroscopic behavior would be able to be understood in terms of "mass exchange" and "stripping" of the initial primary in close binary systems. The first clear demonstration that the ionization decreased outwards (at least in WC stars) was shown. The first UV data (OAO) and the first radio data were presented. "Single point" atmospheric models were introduced: they gave some confidence to our perception of the anomalous composition of W-R stars. A comparison with optical emission line strengths was in its infancy.

A decade later, in Cozumel in 1981 at IAU Symposium #99 (De Loore and Willis 1982), a major thrust was the discussion of "winds" in W-R stars. This term had been used before only sparingly; now it became a way of conceptualizing the spectra of W-R stars. Massive star evolution scenarios were calculated for single stars; the realization of the importance of mass loss and mixing had occurred in the recent past. The notion was that with sufficient mass loss and mixing, the products of previous H-burning or current He-burning in the core would be exposed on the surface. Several different pathways, or "channels" by which a massive star could evolve to show W-R characteristics were proposed. Most participants, if polled, would have agreed with Lindsey Smith's remark, quoted above, from her summary talk. The concept of WO types as a separate class was introduced. Extensive IUE data and high resolution spectra were presented and analyzed. The first X-ray data were given; near infrared spectrophotometry was provided. The atmospheric models, while still "single point" approximations, were able to predict with some accuracy a few of the lines observed from the UV to the near IR. Considerable progress in binary star orbit analysis was given, with particular attention to the masses of W-R stars and their companions. The concept of and evidence for compact companions in certain W-R systems was argued about at length. The interaction of W-R stars with their environments, in the form of "rings", was offered and some of the first evidence of W-R stars beyond the Magellanic Clouds was shown.

As our Chairman has remarked in his opening address, the number of published papers on W-R stars has increased steadily in the past two decades. At our meeting here, the topics seem to have fallen into eight main categories as follows: i) Standard Atmospheric Models, Composition, Mass loss [20]; ii) Variability, Instabilities [10]; iii) Binaries [16]; iv) Environments [11]; v) Evolution, Luminous Blue Variables [16]; vi) W-R Descendants, SNIb [6]; vii) W-R in Other Galaxies, Initial Composition [11]; viii) W-R in Giant HII regions, W-R Galaxies [7]; unclassified [5]. The numbers following each topic in brackets are "weighted" percentages of the number of scheduled papers: the weighting is 3 times the number of oral papers plus the number of poster papers as listed in the preliminary program. These numbers can be used to give one a feeling for the emphasis in each area at this conference but should not be taken more seriously than that.

I will now consider each of these areas in turn and give my personal opinion of the highlights of the work as reported here, both by the oral reviews and contributions and by the poster papers.

i) Standard Atmospheric Models, Composition, Mass Loss

Finally, and after much work by several groups, it has become possible to follow the sage advice of Ann Underhill in 1968 (as quoted in the beginning of this talk) and derive STANDARD ATMOSPHERIC MODELS which solve <u>self-consistently</u> the radiative transfer and statistical equilibrium equations throughout a moving extended atmosphere. These models assume a pure helium composition for the opacity, spherical symmetry, a monotonic velocity law, time independence and homogeneity. I shall return to these assumptions below but I cannot stress strongly enough that these models, constructed independently by several groups, are able to reproduce the observed <u>continuum</u> over several decades of wavelength and the strength and shape of <u>most emission lines</u> to better than a factor 2. In my view this is the major theoretical advance in the study of W-R stars at this meeting.

Not yet included in the models are line blanketing, particularly the FeV and FeVI emission line blends below 1500Å for which there is now direct evidence. Furthermore, the velocity law adopted was arbitrary (but similar for all models) and has not been derived from first principles. To do so is an important next step that will involve hydrodynamical calculations which may be amenable to treatment in the next decade with the advent of very powerful computers.

In addition to fitting the observed spectrum, the standard atmospheric models give predictions which can be compared with other evidence. For WN stars, the <u>mass loss rates</u> from the line spectra are found to be in the range 10^{-4} to 10^{-5} solar masses/yr, similar to that obtained from the radio free-free estimates. The <u>composition</u> is clearly anomalous with respect to solar values and is in quantitative accord with that predicted by STANDARD EVOLUTION MODELS for massive stars which include mass loss and mixing. The <u>Teff</u> inferred from the models seems to be a little cool for the WNL stars with hydrogen, and some WNE stars, compared to evolution predictions; for the hottest WNE stars the Teff appear to be consistent with such predictions. The <u>luminosities</u>, L, inferred from the standard atmospheric models are nearly all too faint compared to the predictions of the standard evolution models. This has led several investigators to propose that the mass loss rates of W-R stars are proportional to some power of the mass (greater than unity) which is also indicated by the empirical relationship between these parameters. Similarly, the observed mass of many WN stars seems small compared to that expected from evolution of massive stars, unless a mass dependent mass loss rate for W-R stars is adopted.

It appears that in WN stars nitrogen blanketing will not play a major role in the structure of the atmosphere. In WC stars, the standard atmospheric models will need to include the effects of carbon as an opacity source before any meaningful comparisons to luminosity and Teff are made. The mass loss rates and compositions are consistent, as for the WN sequence, with the radio free-free emission, and the predictions of the evolution models, respectively.

Could W-R stars have significant departures from spherical symmetry? This theme came up at several points in our proceedings and I would like to personally address this issue. In the one case (HD 50896) where we are fairly certain of the rotational origin of the variable polarization, the period is 3.7 days and the equatorial velocity relatively low and far from a "critical value". The flat topped profiles of 5696 CIII (and other lines) in WC stars sets an upper limit on departures from sphericity as such a line can only be produced in a spherical shell. I have not seen any quantitative discussion of this straightforward and very model independent observation. Many W-R stars show <u>no</u> changes in polarization with time. There is no evidence of non-sphericity in eclipsing binary systems. All of these points together suggest to me that spherical symmetry is a pretty good approximation for most, if not all, W-R stars.

Could W-R stars have <u>substantial</u> non-homogeneities and/or a non-monotonic velocity law? Here the evidence pro or con is sketchy and clearly needs more work. As I will discuss presently, there is clear evidence for <u>some</u> inhomogeneities in the winds of W-R stars. The extent to which this controls, or is dominated by, the wind is currently unsettled.

A new concept which has recently emerged in the study of winds of W-R (and OB type) stars is that of a distinction between the "edge" velocity of the P Cygni profile, and the "terminal" velocity. The latter is a fundamental parameter of the wind and is needed to connect the radio emission measures to the mass loss rates. It appears that this velocity is best estimated from the violet most part of the "zero continuum (or black)" absorption profile; the edge velocity, typically about 25% larger, is the maximum violet displacement of any portion of the absorption profile. The difference may be due to low density higher velocity material coming from "reverse shocks" in the stellar winds of W-R (and OB type) stars. The realization of this velocity effect lowered previously determined radio mass loss rates by 25%; on the other hand, the recognition that the ionization balance in the outer part of the wind was lower raised previous radio mass loss rates by a factor 2. In any case, the mass loss rate predictions of the standard atmospheric models, and the observed free-free measurements are now in good agreement.

Are W-R winds driven entirely by radiation pressure? While there are those who advocate this, and I count myself among them, there is a problem for a few W-R stars with required multiple scatterings for a given photon of 10 to 40 or more. A consensus among some theorists is that multiple scatterings of up to 10 can reasonably be expected in a radiatively driven wind but larger numbers are not possible. It could be that in those few stars with larger inferred multiple scattering values the L has been underestimated due to an inadequate bolometric correction. This will have to be sorted out in the future.

ii) Variability, Instabilities

There has been considerable progress in observations of the variability of W-R stars involving measurements in the continuum, the emission lines and in polarimetry. <u>Extrinsic</u> variability is produced in close binary systems, due mostly to geometric distortion effects which are reasonably well understood. Of particular value are polarimetric measurements which enable one to solve for the orbital inclination and thus better estimate the stellar masses in well studied orbital systems. Many W-R stars have <u>intrinsic</u> variability in the emission lines at the few percent level; the WN8 and WC9 stars have somewhat larger amplitude changes which has been interpreted as being due to their larger radii and more sluggish winds. A few W-R stars appear to be constant over time in their spectral features.

The cause of such variability would seem to be instabilities in the radiatively driven winds, which are predicted to be present. The theory suggests the presence of "reverse shocks" which lead to low densities at high velocities and high densities at low velocities in the winds. This is exactly what is needed to account for the distinction between the "edge" and "terminal" velocities observed in the P Cygni profiles of certain ions. I would expect considerable quantitative advances in such dynamical wind problems in the next few years.

Radial pulsation has been predicted by some investigators but has not yet been demonstrated to be present from observations. Non-radial pulsations, predicted by other investigators, may have been found in a few cases but such an interpretation is currently controversial and this issue is not settled.

iii) Binaries

The study of binary systems with O type companions continues. Eclipse effects can be used to probe the wind structure of the W-R component: e.g., the establishment of the FeV, FeVI pseudo-continua. A summary of the properties of the known W-R binaries has provided valuable information on the individual masses and the mass ratios. While there may be one established W-R plus W-R binary, the evidence is not compelling and now, in contrast to a decade ago, the double star scenarios suggest none should be found. A detailed analysis of an eccentric orbit eclipsing binary system in the SMC and useful estimation of its parameters needs to be noted.

There now seems to be little evidence for W-R systems with compact companions: neutron stars or black holes, although one or two objects remain as possible candidates.

iv) Environments

A major new result was the findings of nebular 4686 HeII surrounding a few early type W-R stars. A straightforward interpretation of the presence of this line requires equivalent Teff of 70000 or more. A large b.c. would be inferred for such stars, thus possibly easing the multiple scattering limit problem. W-R "rings" can be used as probes of the properties of the exciting W-R stars and they give Teff and compositions consistent with the predictions of the standard atmospheric models.

There was evidence presented for bi-polar outflow in those few cases where there is an observed asymmetric nebulosity. How non-spherical is the ejecting W-R star in such a case? I was disappointed not to have seen this discussed. Massive stars, along with their local and giant HII region environments, produce copious amounts of high energy photons. Surface magnetic fields of few gauss fields might help understand the X-rays observed in W-R winds. There will be a major observational impact on the high energy astrophysics of W-R stars in the next decade with the launch of several new satellites.

v) Evolution, Luminous Blue Variables

It appears from the standard evolution models that mass loss and mixing in massive stars can produce helium rich objects which have masses, luminosities and compositions much like the observed W-R stars. In the current theory, they have a singular dependence on the "metal abundance", z. It would be expected from these models that as z goes down, the lower mass limit above which a single W-R star can be formed will go up. This would affect the W-R/O star ratio. Furthermore, the number ratio of WN/WC subtypes should also be a simple function of z, getting larger as z decreases. Both effects come about because of the dependence of mass loss rate on z and the resultant length of time a massive star will spend in each stripped, or "bare core" stage (H-burning products:WN, or He-burning products:WC).

The observed LBV may also play an important role in massive star evolution. Their major mass loss episodes and an estimated frequency of such occurrences, along with a steady outflow of material in between such events, make them prime candidates for being an important link between Of and W-R phases. An important and unresolved question is whether or not all W-R stars go through an LBV phase; some would appear to do so. Currently, the origin of the LBV phase is thought to be related to the Eddington limit which a massive star may approach at certain times in its evolution. LBV phases have not been included specifically in the standard evolution models but their effect would be similar to raising the effective mass loss rates during core H-burning.

Massive stars that enter a Wolf-Rayet phase will all end up near 10 solar masses, irrespective of their initial masses, given a mass loss rate depending on a power of the mass as proposed by several investigators. This channeling down to a more or less uniform mass is an important prediction of the standard evolution models. It does not appear to be at odds with our current knowledge of W-R star masses which, however, all come from well studied binary systems in which the evolution may have proceeded differently.

There seems to be general assent among those modeling W-R star evolution that the "metal abundance - z" must play an important role. It is clear that this parameter will influence the mass loss rates in the stellar winds; its function in internal mixing is less settled. There is fairly strong evidence, at least from the solar vicinity and the Magellanic Clouds, that the value of "z" may have had a substantial influence on the numbers of WN and WC star produced. Recently, new values of the composition of the Magellanic Clouds has been reported by Dopita (private communication at this meeting; see also the papers in the literature). The new numbers are as follows (they have not been incorporated in the papers presented at this conference):

	O/H	Fe/H	WN/WC
Solar	1	1	1
LMC	1/3	1/2	4
SMC	1/6	1/4	7:

The values of the O/H and Fe/H ratios in the LMC and SMC come from detailed analyses of supergiants, HII regions, planetary nebulae and SN remnants and are consistent with one another. What bothers me about this table is the extreme sensitivity of the WN/WC ratio to the metal abundance: the ratio changes from the solar vicinity to the LMC by a factor 4, twice the difference in the Fe/H ratio (which should control the stellar wind mass loss rate) and larger than that of the O/H ratio (which controls the stellar structure equations). If we take such an interpretation at face value, we could turn the problem around and by simply obtaining the WN/WC ratio in other galaxies, estimate their "z" in a very straightforward and unambiguous manner. This seems just a little too simple for me (see also below).

vi) W-R Descendants

This topic appears for the first time at this meeting. There is a newly recognized class of SN called type Ib, (and Ic) in which the lines of hydrogen are absent, yet they appear, from their galaxy type origin and locations to be associated with massive progenitor stars (rather than low mass stars as for the SN1a). A number of investigators have suggested that such events might be derived from W-R stars which are well known for having lost their hydrogen. However, the SNIb frequency seems to be a little too high to be due to stars with initial masses as large as those proposed for W-R objects. It could be that some W-R binaries might end up as SNIb, but not all SNIB could be derived from W-R progenitors with the current statistics.

From a theoretical point of view, it is possible to carry a single 60 solar mass star through its various mass loss stages to arrive at catastrophic collapse phase which has properties much like those observed for SNIb. Furthermore, if single W-R stars do not become, SNIb, what is their fate? The issue of the W-R descendants is clearly not a settled one and much new observational and theoretical work is to be expected in the future.

vii) W-R in Other Galaxies; Initial Composition

In a deep search in a direction down the Carina Arm some 13 additional W-R stars have been found. This suggests an incompleteness of about 25% in the counts of W-R stars near the solar vicinity. The impact on the total numbers estimated for the Galaxy is potentially important.

There were new discoveries of W-R stars in M33 and M31. There now appears to be little or no gradient in the WN/WC ratio with galactocentric distance in M33 and the value is near unity in M31 (which has a higher than solar value of z). The statistics are seemingly secure given the numbers of stars involved. Thus the inferred WN/WC ratios did not appear to follow the simple prescription of dependence on z as suggested by the relations for the LMC and SMC. What other factors might influence the WN/WC ratio? This will need to be sorted out by additional observations of other galaxies and detailed examinations of completeness of the stellar samples so far considered.

viii) W-R in Giant HII regions, W-R Galaxies

Additional W-R stars have been identified in several of the Giant HII regions of M33. In these recent star forming regions, the WN/WC ratio is large! Clearly these do not follow the simple z dependence predicted by the standard evolution models. Could it be that the relatively bright WN stars in such HII regions are in their H-burning stages, thus lasting a relatively long time? Newly discovered W-R stars in several other nearby galaxies were reported and the search procedures were reviewed.

Wolf-Rayet galaxies are those in which a broadened 4686 HeII emission line has been detected in the integrated spectrum. Some few dozen are known, typically as blue compact galaxies, but with relatively heterogeneous properties of luminosity and composition. In some cases, the W-R stars seem to appear in bright "knots", or supergiant HII regions, within star forming galaxies. An unanswered question is why W-R stars appear in some, but certainly not all, galaxies with other evidence of recent star forming episodes. Population synthesis models of 17 W-R galaxies suggested ages of $3-6 \times 10^6$ years, implying a relatively recent "starburst". The evidence seems to favor a relatively "flat" IMF, compared to the solar vicinity, with no dependence of the slope on z for the studied systems.

This concludes my summary of the papers presented at the meeting. I would like to close by making a list of several issues which have engaged much of our attention in the past but now should be considered as "settled", and then addressing where I think the future decade of research in W-R stars will be leading.

Settled Issues

W-R stars represent a normal ending point of massive star evolution. WN, WC, and WO subtypes have anomalous abundances which represent the <u>expected</u> nuclear burning products in stars whose outer hydrogen rich material has been stripped away.

Massive star evolution models which include mass loss, mixing, and binary interactions <u>can produce</u>, in various combinations of importance, objects with properties similar to W-R stars.

Observed W-R continua and spectra can be modeled by a hot core - dense stellar wind: the standard atmospheric model.

668

Future decade - major themes

Second generation models will need to be constructed which will combine the standard evolution model with a dynamical treatment of the stellar wind. These will need to account for the observed instabilities of W-R stars.

As extreme Population I objects, W-R stars should be excellent indicators of galactic structure. Where are the spiral arms of our Galaxy? Does our Galaxy have 2, or 3, major features? Is it an Sb or Sc type, or something in between?

Extensive population studies of individual W-R stars in more and more distant galaxies will be undertaken. The total numbers compared to the parent O star population and the WN/WC ratios will be used to examine the predicted simple dependence on z. It may be possible to investigate the environments of these galaxies from their massive star populations. Certainly the recent massive star formation rate can be evaluated.

W-R stars will be used to investigate "starburst" phenomena in even more distant galaxies where only the integrated spectrum is available. Is the IMF different in such environments?

"Primeval" blue galaxies may now have been detected at redshifts of about 1. One would expect that in such galaxies the initial massive stars that produced much of our current "metals" would be present. Are W-R stars also evident? Have they played a major role in the initial composition changes of our universe?

In closing, I see many opportunities for research on W-R stars, both those that are relatively nearby that can be studied in detail, and those that are at very distant locations that can only be observed in ensemble. In particular, now that they are better understood as evolved massive stars, they will be guideposts to our comprehension of star formation and galaxy evolution itself.

REFERENCES

Bappu, M.K.V. and Sahade, J. 1973 Wolf-Rayet and High-Temperature Stars, IAU Symposium #49 (D. Reidel, Dordrecht)

De Loore, C.W.H. and Willis, A.J. 1982 Wolf-Rayet Stars: Observations, Physics, Evolution, IAU Symposium #99 (D. Reidel, Dordrecht)

Gebbie, K.B. and Thomas, R.N. 1968 Wolf-Rayet Stars Proceedings of a Symposium, NBS Special Publication 307 (U.S. Government Printing Office, Washington D.C.)

Smith, L.F. 1982 in Wolf-Rayet Stars: Observations, Physics, Evolution, eds. C.W.H. De Loore and A.J. Willis (D. Reidel, Dordrecht), p. 599

Underhill, A.B. 1968 in Wolf- Rayet Stars, eds. K.B. Gebbie and R.N. Thomas (U.S. Government Printing Office, Washington D.C.), p. 185

CLOSING REMARKS BY THE RECTOR OF UNIVERSITAS UDAYANA, DENPASAR

Yang terhormat Bapak Gubernur Kepala Daerah Propinsi,
Bapak Muspida Tingkat I Bali,
Bapak Ketua DPRD Tingkat I Bali,
Distinguished Participants,

On behalf of the Director General for Higher Education of the Ministry of Education and Culture of the Republic Indonesia, I would like to convey my appreciation and thanks to all of you, the participants from overseas as well as from Indonesia, who have been working hard to share ideas and experiences during the course of the discussion on major issues and problems rising up as a consequence of the multi-dimensional view of the topic of the symposium with special emphasis on current aspects of problems of Wolf-Rayet Stars' origin and evolution.

As we understand, this symposium could have been carried out successfully, thanks to the efforts of the Local Organizing Committee, which has laid out the symposium design neatly and has enabled the execution of the symposium very smoothly in cooperation with the Directorate General of Higher Education of the Ministry of Education and Culture, the Indonesian National Committee for UNESCO, and the Ministry of Tourism, Post and Telecommunication of the Republic Indonesia, sponsored by the International Astronomical Union.

We are proud to see that the symposium has been successfully carried out.

I notice that this symposium has been organized with good understanding and has been intellectually delibrated.

Let the outcome of the symposium be a strengthening means for the conduct of cooperation through joint endeavour of scientists from all over the world.

Last but not least may the cultural atmosphere of Bali contribute to the success of the symposium.

We do hope that all of you will extend your stay in Bali to know more about Bali with her thousand smiles, thousand temples, and the intermingled life and culture of Bali.

Finally, may I declare the symposium of International Astronomical Union No. 143 on Wolf-Rayet Stars and the Interrelations with other Massive Stars in Galaxies, on behalf of the Director General for Higher Education of the Ministry of Education and Culture of the Republic Indonesia, officially closed.

Thank you.

Prof.Dr. I.G.N.P. Adnyana

K. A. van der Hucht and B. Hidayat (eds.),
Wolf-Rayet Stars and Interrelations with Other Massive Stars in Galaxies, 669.
© 1991 *IAU. Printed in the Netherlands.*

OBJECT INDEX

WR 30 (= HD 94305) – 316.

WR 31 (= HD 94546) – 215, 220, 225, 316, 453-454, 456-457.

WR 31a (= SMSP 4) – 593.

WR 35a (= SMSP 5) – 593.

WR 35b (= SMSP 6) – 593

WR 38a (= SMSP 7) – 593.

WR 38b (= SMSP 8) – 593.

WR 40 (= HD 96548) – 27, 83-84, 100, 111-113, 115, 119-120, 127, 131, 147, 150-151, 153, 275, 353, 421, 486.

WR 42 (= HD 97152) – 220, 225, 273, 316.

WR 42a (= SMSP 9) – 593.

WR 42b (= SMSP 10) – 593.

WR 42c (= SMSP 11) – 593.

WR 42d (= SMSP 12) – 593.

WR 43 (= HD 97950) – 605.

WR 44a (= SMSP 13) – 593.

WR 46 (= HD 104994) – 27, 40, 83-84, 111, 129, 131-137, 143, 145, 318.

WR 47 (= HDE 311884) – 217, 220, 225, 273, 316, 453, 455-457.

WR 48 (= θ Mus = HD 113904) – 29, 350, 353, 424.

WR 48a (= Danks 1) – 417-420.

WR 50 (= TH17-84) – 27, 111, 129, 131, 137-138, 143.

WR 52 (= HD 115473) – 111, 131, 353, 424.

WR 53 (= HD 117297) – 131.

WR 54 (= TH17-89) – 424.

WR 55 (= HD 117688) – 27, 111, 129, 131, 138-140, 143, 353, 357.

WR 57 (= HD 119078) – 78, 111, 424.

WR 61 (= He3-969) – 424.

WR 66 (= HD 134877) – 111.

WR 69 (= HD 136488) – 111.

WR 70 (= HD 137603) – 116, 417, 419, 556.

WR 71 (= HD 143414) – 111.

WR 75 (= HD 147419) – 83, 119, 353.

WR 78 (= HD 151932) – 78-79, 83, 111, 270.

WR 79 (= HD 152270) – 105, 126-127, 217, 220, 225, 270, 273, 316.

WR 81 (= He3-1316) – 270.

WR 82 (= LS 11) – 111.

WR 86 (= HD 156327) – 99, 111, 129, 131, 140-143, 270.

WR 87 (= He3-1370 = LSS 4064) – 270.

WR 88 (= Thé 1) – 99.

WR 89 (= AS 223 = LSS 4065) – 270.

WR 90 (= HD 156385) – 79, 113, 131, 270, 350, 424.

WR 93 (= HD 157451) – 131, 270.

WR 93a (= Th3-28 = PK359 +3.1) – 21.

WR 95 (= He3-1434) – 270.

WR 97 (= HDE 320102) – 131, 316.

WR 98 (= HDE 318016) – 201, 203, 205-206.

WR 98a (= IRAS 17380-3030) – 330-332.

WR 102 (= Sand 4 = LSS 4368) – 101, 353, 366-368, **371-377**, 393-394, 411-415.

WR 103 (= HD 164270) – 78-79, 111, 131, 147, 149, 151-152, 193, 194, 270.

WR 104 (= Ve2-45) – 270, 329-331, 490.

WR 105 (= AS 268) – 83.

WR 108 (= HDE 313846) – 83-84, 131.

WR 110 (= HD 165688) – 83-84, 131, 270.

WR 111 (= HD 165763) – 25, 41, 61, 63-64, 67-68, 85, 99, 131, 147-148, 151, 153, 268, 270, 284.

WR 112 (= GL 2104) – 329-331.

WR 113 (= HD 168206 = CV Ser) – 111, 126, 191-193, 216-217, 220, 225, 270, 272-273, 293, 300, 302, 316, 350, 556.

WR 114 (= HD 169010) – 270.

WR 115 (= IC14-19) – 270.

WR 118 (= GL 2179) – 126, 330-331.

WR 121 (= AS 320) – 131, 270, 331.

WR 123 (= HD 177230) – 27, 67, 95, 110-111, 129, 131, 142-143, 443.

WR 124 (= BAC 209) – 111, 270, 353.

WR 125 – 111, 266.

WR 127 (= HD 186943) – 216-217, 220, 225, 273, 316, 453, 455-457.

WR 128 (= HD 187282) – 79, 83-84, 100, 111, 353.

WR 130 (= LS 16 = AS 374) – 270.

WR 131 (= IC14-52) – 353.

WR 133 (= HD 190918) – 215, 217, 220, 222, 225, 270, 273, 453, 456-457.

WR 134 (= HD 191765) – 12, 19, 55-56, 83-84, 110-111, 115-117, 119, 122, 147, 151, 195, 270, 274, 356-357, 381.

WR 135 (= HD 192103) – 12, 19, 78-79, 99, 110-111, 126, 270.

QM LIBRARY
(MILE END)